D1704199

Coffee:
Growing, Processing,
Sustainable Production

Edited by
Jean Nicolas Wintgens

Coffee is natural and bliss,
Coffee is fascinating

Meike Hühne

Sponsored by:

Nestlé

Related Titles

Sinha, N.

Handbook of Food Products Manufacturing

2 Volume Set

approx. 2308 pages in 2 volumes
2007
Hardcover
ISBN: 978-0-470-04964-8

Ribéreau-Gayon, P. (ed.)

Handbook of Enology

2 Volume Set

928 pages in 2 volumes
2006
Hardcover
ISBN: 978-0-470-01157-7

Brennan, J. G. (ed.)

Food Processing Handbook

607 pages with 189 figures and 41 tables
2006
Hardcover
ISBN: 978-3-527-30719-7

Flament, I.

Coffee Flavor Chemistry

424 pages
2002
Hardcover
ISBN: 978-0-471-72038-6

Janick, J., Moore, J. N. (eds.)

Fruit Breeding

3 Volume Set

1388 pages
1996
Hardcover
ISBN: 978-0-471-12675-1

Coffee:
Growing, Processing,
Sustainable Production

A Guidebook for Growers, Processors, Traders, and Researchers

Edited by
Jean Nicolas Wintgens

Second, updated Edition

WILEY-VCH Verlag GmbH & Co. KGaA

Editor

Jean Nicolas Wintgens
Chemin des Cornalles 27
1802 Corseaux
Switzerland
e-mail: jean-nicolas.wintgens@vtxnet.ch
Phones: 0041 21 921 2608 (Switzerland)
         0052 333 122 9543 (phone & fax Mexico)
Fax: 0041 21 923 6138

cover   cover picture created by
         Schulz Grafik-Design, Fußgönheim

■ All books published by Wiley-VCH are carefully produced. Nevertheless, authors, editor, and publisher do
not warrant the information contained in these books,
including this book, to be free of errors. Readers are
advised to keep in mind that statements, data, illustrations, procedural details or other items may inadvertently
be inaccurate.

Library of Congress Card No.: applied for

British Library Cataloguing-in-Publication Data.
A catalogue record for this book is available from
the British Library.

**Bibliographic information published by the
Deutsche Nationalbibliothek**
The Deutsche Nationalbibliothek lists this
publication in the Deutsche Nationalbibliografie;
detailed bibliographic data are available on the
Internet at <http://dnb.d-nb.de>

© 2012 WILEY-VCH Verlag GmbH & Co. KGaA,
Boschstr. 12, 69469 Weinheim, Germany

Printed in Singapore.
Printed on acid-free paper.

Cover Design aktvComm, Weinheim
Composition   Hagedorn Kommunikation GmbH,
                 Viernheim
Printing
and Binding   Markono Print Media Pte Ltd,
                 Singapore

ISBN   978-3-527-33253-3

Dedication

To my Parents for all the sacrifices they made for their children.
May the Lord they believed in give them peace and serenity.

To my wife Erna for her unflagging support and encouragement as well
as discerning advice, during the years it took me to complete this book.

To the coffee growers world-wide, many of whom now labour
for so little reward.

To my unfullfilled dreams of becoming a coffee planter myself.

Jean Nicolas Wintgens

Foreword

During the last two decades major changes have been taking place in worldwide coffee production. The most significant of these changes has been the massive increase in the quantities produced. ICO statistics show consistently increasing production levels particularly over the last 15 years reaching a record level of 120 million 60 kg bags in 2002. Over the same period there has been a significant decrease in the average price paid for coffee on world markets which fell to USD cents 47 per pound in 2002 compared to 120 cents in the 1980s.

The basic cause of the decline in coffee prices has been one of oversupply with the increase in coffee production in recent years outstripping demand by several percentage points. There has been a steady accumulation of large stocks of coffee, particularly in importing countries, which have put additional downward pressure on coffee prices. Thus in recent times many coffee producing countries have seen a catastrophic decline in export earnings despite increasing production, and the incomes of many coffee farmers have been reduced to the point at which they are struggling to survive. The impact on the finances of millions of small farmers in lesser developed countries has reached crisis proportions.

This book has been produced against a general background of over production of coffee in relation to prevailing demand and the consequent very low prices presently being paid for coffee on world markets. The over production of coffee may be largely attributed to the development of new land for coffee plantations, especially in countries like Brazil and Vietnam, and to the rapid progress in increasing coffee yields which has occurred in most coffee producing countries in recent decades, as a result of the application of new technologies and the introduction of new coffee varieties.

This book provides a good understanding of the technologies which have been responsible for increasing coffee yields and has obviously been written for a wide audience including coffee specialists, students of tropical and sub-tropical agriculture and practical farmers.

However it has been noted that, in many cases, the adoption of modern intensive husbandry practices (no shade, high fertiliser use, heavy use of agro-chemicals for pest and disease control, etc.) produces high yields for a few years but these high yields are not sustainable in the long term. Consequently the importance of adopting cultivation practices that will support and sustain coffee production into the foreseeable future is a theme that is recurrent throughout the book.

The diversification of cropping within the coffee plantations, the use of locally produced organic matter and the establishment of shade are particularly appropriate for small coffee growers when the prices received for green coffee will no longer support costly inputs of synthetic fertilisers and other agro-chemicals.

In addition, higher quality of coffee, which commands a higher price, can be obtained with more extensive low cost cultivation techniques. Lower costs and higher prices for quality beans may increase crop profitability. In this context the potential for organic coffee production has been amply described.

An excellent account of the coffee production cycle together with a section on modern harvesting techniques and on the latest technology in field processing is also included, as well as aspects of commercialisation, storage and shipment of green coffee.

Finally, a rapid overview of the present economics of coffee production, followed by a glimpse into potential new fields of action for the future, complete the book.

The many authors who have contributed to this book have gained their specialist knowledge of the coffee crop through ex-periences in many different coffee producing countries across the globe. The varied backgrounds of the contributors to this book (in terms of history, language, environmental conditions, economic circumstances, research facilities and experience, etc) greatly add to its value as a reference work for the future. Also the many excellent and easily-understandable illustrations contained in the book greatly enhance its value as a teaching aid.

This book is particularly important in that it signposts the future for coffee as a crop. In the future it is desirable that coffee production should be sustainable, preferably grown by small farmers under extensive systems of cultivation that ensure the protection of the soil and maintain satisfactory yields of good quality green coffee.

Néstor Osorio
Executive Director
International Coffee Organisation
2004

Preface

The origins of coffee and historical development trends

The history of coffee begins in Ethiopia where local people have been drinking coffee for many centuries. The advantages of the spontaneous growth of coffee trees of the species *Coffea arabica* were recognized and developed at a very early stage. At the beginning of the 18th century, visiting Arab traders also found the drink agreeable and, from then on, coffee was shipped to the Arabian peninsular. A demand for coffee quickly developed amongst the peoples of Arabia and coffee plantations were established in Yemen. Later, during the 18th and 19th centuries, there was a steady expansion of coffee drinking throughout the world and arabica coffee was exported to many countries, with the source planting material originating from either Ethiopia or Yemen.

Coffea canephora var. *robusta* was identified later on from more diverse origins in the forests of tropical central Africa and was later transferred to many lowland areas of Africa as well as to tropical countries of Asia and the Americas. Nevertheless, despite the wide distribution of the coffee crop, production remained at a relatively low level through the first half of the 20th century, chiefly because it was still regarded as a luxury item in the developed countries

of the northern hemisphere. Coffee was grown in mixed plantations under shade in conditions which matched the forest conditions where naturally occurring coffee was initially discovered. Efforts to increase coffee yields were largely dependent on the development and improvement of coffee varieties through the selection of superior individual trees and their multiplication, either through seed as in the case of *Coffea arabica* or through vegetative propagation in the case of *Coffea canephora* var. *robusta*.

After the end of the Second World War and during the second half of the 20th century, major changes took place in the lifestyle of the populations in developed countries. As income levels rose, millions of people were able to enjoy luxuries, including coffee, that had previously been the domain of a rich minority. There was a huge increase in the demand for coffee and coffee producing countries rose to the challenge. The major coffee producing countries were in the forefront of the research and development effort that was launched to expand coffee production to meet the rapidly increasing demand. National research efforts have been ably supported by a number of international research bodies and private companies with an interest in coffee.

Before the mid-20th century, efforts to improve coffee yields had concentrated on traditional methods of varietal improvement using simple selection techniques. Following the increasing demand for coffee and associated rising prices paid for the crop, the incentive to exploit every avenue for raising coffee yields was extensively developed. More sophisticated breeding techniques were introduced to raise the inherent productive potential of the varieties being planted and to overcome disease and pest problems. New intensive production methods were introduced on commercial plantations to increase yield levels. Traditional methods of coffee cultivation under shade trees were discarded in favour of monoculture plantations. The use of inorganic fertilisers replaced organic manures and mulching. Clean weeding became normal practice with the introduction of mechanised cultivation and the use of highly efficient selective herbicides, which replaced occasional weeding, carried out by manual labour using simple hand tools and, in some cases, the use of cover crops as a weed control measure. Similarly insect pest problems have been controlled by the application of a wide range of pesticides and new techniques have been developed to deal with recently identified problems of nematodes and diseases.

Current situation in the coffee trade world-wide

Total world production in the coffee year 2003/04 (October-September) will reach 101 million bags (60 kg bags) or 6.06 million tons, a fall of 15 percent from production of 119 million bags, or 7.14 million tons, recorded during the preceding year. In 2003, world consumption is estimated to be just over 112 million bags, or 6.72 million tons, compared to 109 million bags, or 6.54 tons, in 2002. Domestic consumption in exporting countries in 2003/04 is estimated to be 27 million bags, up

slightly from 2002. Consumption in importing countries is estimated at 85 million bags, up significantly from 82 million bags in 2002. Opening stocks in exporting countries were pegged at 20 million bags in 2003/04 compared with 19 million bags in the preceding year, while stocks in consuming countries, including free ports, were estimated at 20 million bags, a similar amount to the preceding year. At the end of the coffee year 2003/04, world stocks of coffee are forecast to fall to 29 million bags, compared to 40 million bags of coffee, or 2.4 million tons, in 2002/03. This current situation marks the reversal of the trend of the previous decade that showed an average production increase of 3.6% per annum over preceding years but a rise of only 1.5% per annum in demand levels.

The new trend towards significant under production is the predictable consequence of over production of coffee that has been accompanied by a drop in coffee prices on world markets. The fall began in 1999 and reached historic lows in 2002. In the 1980's coffee prices on world markets averaged around 120 US cents per lb but 2002 prices fell to 42 US cents per lb, compared to 179 US cents per lb in 1998.

The fall in prices has resulted in a dramatic reduction in the export earnings of coffee producing countries. In the early 1990's the total value of coffee exports on world markets (exports f.o.b.) was US$ 10 – 12 billion but the value shrunk to little more than US$ 5.0 billion in 2002.

Over 125 million people are largely dependent on the income generated by their coffee crop for their livelihood. Moreover, a number of countries rely on the sale of coffee for a major part of their national income and, more importantly, their foreign exchange earnings. In both Uganda and Burundi, for instance, 90% of the GNP is based on coffee exports. At the other end of the chain, millions of consumers drink an estimated 15 billion cups of coffee daily, in spite of the current economic depression. This still represents a turnover of some 10 billion US dollars and places the coffee trade in the top bracket of the overall world market, third after crude oil and wheat. Coffee is the leader of the world's agricultural commodity markets. It ranks higher than staple foods like sugar, corn or meat and also higher than the market for timber, cotton or wool.

The shape of things to come

There has been a general acceptance of the fact that, with the long-term application of intensive cultivation practices, high levels of coffee production may become unsustainable. High coffee yields are increasingly associated with a reduced longevity of plantations and the resulting increase in costs for renewal or replacement. At the same time, the economics of coffee production have changed in recent years, with declining price levels on international markets and increasing costs of inputs. The pendulum has swung in favour of more extensive production methods. Consequently, coffee agronomic research programmes have been increasingly directed towards the development of cultivation techniques that optimise returns to the farmer and which are more environmentally friendly and produce a better quality of coffee.

The advantages of plantations with shade trees and enhanced soil protection have been re-assessed. Applications of lower levels of agro-chemical inputs have been investigated. Techniques have been developed to replace the use of agro-chemicals with alternative strategies such as the development of better adapted varieties, the on-site production of organic matter, the application of biological methods for pest and disease control, etc. In some areas organic coffee production is being developed using methods which rely entirely on these alternative strategies.

As a result of the catastrophic decline in coffee prices on world markets and the drastic reduction in farm incomes, many smallholder coffee growers are likely to make changes in their farming systems to ensure their survival. So far the majority of farmers, in the absence of attractive alternative cash crops, have chosen to retain most of their coffee in the hope that there will be a favourable upward price movement on world coffee markets in the near future. However, if coffee prices remain depressed in the long term, small coffee growers are likely to turn increasingly to crop diversification as a means of reducing their dependence on coffee. Such diversification is likely to include multi-cropping systems, for example, the intercropping of coffee with annual food crops to provide basic subsistence for the farmer and his family and/or the establishment of coffee in mixed stands with other tree crops, which might include a range of fruit trees and multi-purpose forest trees (producing mulching material, livestock fodder, timber, etc.) The trees planted in mixed stands with the coffee will not only generate additional income but may also serve as shade to moderate the micro-climate within the coffee, as wind protection belts and as an aid to erosion control.

Choices

In the future, the small coffee farmer will need to pay more attention to profitability, given the present low coffee prices and the ever increasing cost of agro-chemical inputs. In a situation of overproduction a higher quality of coffee is likely to command premium prices. Thus farmers will have to pay more attention to quality. Low cost production systems are commonly associated with higher quality coffee but yields are lower.

The coffee farmer will have to choose between maximising yields of lower quality coffee, which may prove unsustainable in the long term, ultimately causing damage to the environment, and accepting lower yields using environmentally friendly cultivation practices which involve lower input levels. This choice will be based largely on estimates of crop net returns, which will be influenced by the price premiums being paid for quality, the cost of inputs, including labour, and on the plantation replacement costs (based on the longevity of the plantation).

Farmers should be more interested in "quality" production methods, particularly when corporate and/or government policies are formulated to actively and financially encourage sustainable, environmentally friendly cultivation practices. In the final analysis only quality is truly sustainable.

Ted R. Lingle
Executive Director
Specialty Coffee Association of America
2004

Introduction

This book, that I am pleased to present to coffee professionals world-wide, has been compiled with a view to enhancing the knowledge of sustainable cultivation practices in coffee growing areas. The book describes methods of coffee cultivation and coffee preparation techniques adapted to the evolving modern view of environmental respect which combines sustainability with profitability.

The changed objectives which are needed to move towards more sustainable, environmentally-friendly production systems and which are covered in this book, are listed below:

- the optimisation of yields as opposed to maximum yield levels
- the emphasis on quality of coffee produced as opposed to quantity
- the respect for essential environmental issues through soil conservation and water conservation measures and energy saving
- the use of locally generated sources of mainly organic fertilisers recycling plant nutrients and reducing the need for (imported) inorganic fertilizers
- the selection of planting material adapted to local agro-climatic conditions and meeting the requirements of local planters

- the emphasis on multiplication by vegetative propagation, particularly in the case of allogamous species, rather than seed
- the adoption of locally proven cultivation and processing practices which ensure longevity and sustainability

A few long-standing, conflicting schools of thought are discussed in the light of recent recorded experiences and research findings, particularly those concerning:

- the comparative benefits of pruning on single and/or multiple stems
- the advantages and disadvantages of the use of shade trees and cover plants
- the effects of post-harvest handling on the final coffee quality, particularly coffee fermentation methods and sun-drying practices

New technologies, that are being increasingly applied in modern coffee production systems, are explained and developed and include the following:

- biotechnology applied to coffee-growing
- integrated pest management in coffee
- scientific methods of coffee yield estimation
- mechanized systems of coffee harvesting

- environmentally-friendly processing practices
- bulk handling of coffee beans, storage and container-shipping
- production of organic coffee

In addition, special indexes help the reader's understanding of the book and include acronyms for information sources, the meanings of specialised vocabulary, conversion tables and other technical explanations.

A carefully selected team of international specialists have contributed to the book and each co-author was chosen with respect to his or her special field of competence and expertise. The aim of all the contributors has been to provide specialist information in an easily understandable form and to present useful practical guidelines for many of the people working in the coffee industry. We sincerely hope this goal has been achieved.

Jean Nicolas Wintgens
Corseaux, 2004

Acknowledgement

I would like to thank Nestlé S.A., who gave the initial incentive to this book and provided vital support during the early stages of preparation. It was Nestlé's commitment to the book that convinced WILEY (VCH) in Germany to accept it for publishing.

My heartfelt thanks also go to the 44 co-authors who contributed the chapters. Each one collated and presented his individual knowledge and experience in a specific field. Many of the authors are still professionally active, which has ensured that much of the information provided is the most up to date available at the time of publication.

The translations, the editing and the final revisions were carried out by Stella Bonnet-Evans whose considerable skills and dedicated work have played an essential role in the production of the book. I am greatly indebted to Stella for her contribution.

By the same token, I am specially grateful to my friend Heinz Waldburger who, from the very beginning of the project, encouraged and supported me in many different ways.

It was he who introduced me to Wiley-VCH for the publication of my book and his technical know-how and advice was invaluable.

My appreciation goes also to Aline Sigrist whose experience in graphic design and page-setting greatly enhanced the presentation of this book.

I also want to express my special gratitude to Dr. C. Brando for his judicious advice as well as his continuous search of political and scientific support for the happy conclusion of this book.

My thanks to my friends Robin Goodyear, Graham Quinn and Jacques Snoeck and to my son Jean-Marc Wintgens, all of whom participated willingly and enthusiastically in constructive reviews of a number of chapters.

Finally I want to thank other participants in the production of the book (translators, word processors and graphic designers) for their precious collaboration, namely Erna Wintgens, Colette Fardeau, Bettina Weber, Arturo Guzmán, Michel Glineur, Leticia Laguna G. and Patricia Tejeda M.

J. N. Wintgens

Contributors

Editor

WINTGENS Jean Nicolas

- Agronomist in tropical crops and animal husbandry – Belgium
- Management of large coffee, rubber and cocoa estates in Central Africa, including land survey and research (Cotonco)
- Head of Agricultural Services for Nestlé, Panama
- Establishment and management of extended mechanized farming projects (Nestlé, Panama and Nigeria)
- Worldwide advisor for coffee and cocoa production and supply (Nestlé Headquarters, Switzerland)
- Several publications on coffee and cocoa
- Chairman of the Cocoa Research Policy Committee within the IOCCC
- Representative to the IOCCC for Chocosuisse
- Currently semi-retired in Switzerland/Mexico with part-time consultancy work, mostly on coffee

List of Contributors

ARANDA Delgado Eduardo

- Biologist
- Asociación Mexicana de Lombricultores, AC Xalapa, Ver. Mexico
- Manager of a vermicomposting enterprise (Terranova Lombricultores)

AVELINO Jacques

- Plant pathologist (PhD)
- CIRAD scientist in charge of epidemiological studies on coffee diseases mainly on Coffee Leaf Rust and American Leaf Spot in Mexico and Central America with the PROMECAFE project

BERRY Dominique

- Doctor's degree in biology and plant physiology – University of Clermont-Ferrand, France
- DEA in biology and vegetal physiology – University of Clermont-Ferrand, France
- Research on coffee and cocoa for IRCC/CIRAD in France and Africa
- Head of the coffee programme for CIRAD-CP, France, 1998–2003
- Currently Director of the Tree Crop Department of CIRAD, France

BIEYSSE Daniel

- DEA in agronomy
- Phytopathologist
- Research on coffee plant diseases – Coffee Leaf Rust, Coffee Berry Disease and Wilt diseases
- Currently researcher for CIRAD, France

BLANK Eduard

- Director of Global Transportation, NESTEC Ltd, Vevey, Switzerland

BONNET-EVANS Stella

- Degree in Educational Sciences – University of Geneva, Switzerland
- Postgraduate degree in conference interpretation and translation (double A, English-French) – University of Central London, England
- Teacher of English and Economics at business school level
- Translator for Quality Management and the certification of medicaments for a Swiss firm in Vevey
- Currently retired

BOUHARMONT Pierre

- Ing. agronome group tropical regions (AIALv) – Louvain Belgium
- INEAC – Research and surveys on coffee Arabica in the former Belgian Congo, Rwanda and Burundi
- IRA, IRCC and CIRAD – studies on coffee Arabica and Robusta in Cameroon and France
- Currently retired

BRANDO Carlos Henrique Jorge

- Civil engineer – University of São Paulo (USP), Brazil
- Graduate work at doctoral level – Massachusetts Institute of Technology (MIT), Cambridge, USA
- Manager and Director of Pinhalense coffee machinery manufacturer (Brazil)
- Director and Partner of the coffee consulting, marketing and trading company P & A Marketing International – ES Pinhal, Brazil

CASTILLO PONCE Gladis

- BSc biology
- Master in agricultural sciences – Mexico, Montecillo, Mexico
- Postgraduate in nematology – Maracay, Venezuela
- Currently Head of the national program for industrial crops at INIFAP, Mexico

CLEVES SERRNO Rodrigo

- Ing. Agrónomo – University of Costa Rica
- Specialist in grain technology
- Postgraduate Agric. Engineer College of Costa Rica
- Worked in grain processing and storage in El Salvador
- Director of CICAFE (ICAFE), Costa Rica, 15 years
- Research and publications on coffee processing technology

CHAGAS Cesar Martins

- BSc Agriculture – UFLA, Brazil
- Doctor's degree at São Paulo University, Brazil
- Postgraduate works in France, Germany and Italy
- Research on plant virology, ultrastructural research and diagnosis
- Currently retired

CHARRIER André

- Doctorat d'Etat in Biological sciences (Plant Genetics) – University of Paris XI, Orsay, France
- Emeritus Professor in Genetics and Plant breeding – Montpellier SUPAGRO, France
- Deputy Scientific Secretary in Agronomy and Biotechnology at ASIC Board
- Scientific expertise in agrobiodiversity and plant genetic resources for tropical crops
- Co-author of more than 70 papers in international journals and book chapters on coffee

CHIN Chiew-Lan

- BSc and MSc in agriculture – University of Agriculture, Malaysia, majoring in plant pathology
- Former advisor for Nestlé projects in Asia, Oceania and Africa zone (AOA)

CILAS Christian André-Jean

- DEA of quantitative and applied genetics and certificate in mathematical methods applied to biology – University of Orsay, France
- Works on genetic improvement and biometry of coffee and cocoa – IRCC Togo and CIRAD/CP Montpellier, France

CROWE Terence James

- Read Natural Sciences at Magdalene College – Cambridge, England
- Postgraduate course in Applied Entomology – Imperial College London, England
- MA (Cantab) and DTA Imperial College of Tropical Agriculture, Trinidad
- Research and publications on entomology in coffee and cotton (Kenya)
- Project Manager as entomologist with FAO in Ethiopia and Burma
- Currently semi-retired in Kenya with part-time consultancy work

DESCROIX Frédéric

- Ing. agronome specialized in tropical regions, degree of Etudes supérieures des techniques d'Outre-mer, agro-économie – DESTOM, France
- Research on coffee and cocoa for IFCC/IRCC/CIRAD in France and Africa
- Currently leader of the project for the rehabilitation of the production of coffee gourmet "Bourbon pointu" in Reunion Island

DÍAZ CÁRDENAS Salvador

- Ing. agrónomo esp. en industias Agricolas – Autonomous University of Chapingo, Mexico
- Master of regional and rural development sciences – Autonomous University of Chapingo, Mexico
- Professor and investigator for the CRUO, Autonomous University of Chapingo, Huatusco, Mexico
- Director of fortalecimiento de la oferta de servicios profesionales para el desarollo rural en la SAGARPA

DURÁN O. Leonardo

- Ing. agrónomo specialized in tropical zones – Autonomous University of Chapingo, Mexico
- Expert in vermicomposting
- Technologist for the Regional Agricultural Cooperative of Tosepan Titataniske de Cuetzalam, Puebla, Mexico
- Inspector for organic procedures and certification, CERTIMEX, Mexico

ESCAMILLA PRADO Esteban

- Ing. agrónomo of phytotechnology
- Master of tropical agrosystem sciences
- Doctor's degree in Tropical Agrosystem Sciences
- Currently professor and researcher for CRUO, Autonomous University of Chapingo, Huatusco, Mexico
- Mayor of the Municipality of Chocamán, Veracruz, Mexico

ESKES Albertus B.

- Degree in plant breeding with PhD thesis on durable resistance to Coffee Leaf Rust
- Work with the FAO in Argentina and Brazil, and with the Instituto Agronomico de Campinas in Brazil
- Coffee survey for IBPGR (IPGRI) in Yemen
- Coordinator for coffee and cocoa breeding at IRCC/CIRAD, France
- Currently coordinator of the CFC/ICCO/IPGRI project on "Cocoa germplasm utilization and conservation"
- Chairman of the International Group for the Genetic Improvement of Cocoa (INGENIC)

FARDEAU Colette

- Degree in Educational Sciences – Ecole Normale Supérieure de Liège, Belgium
- High school teacher for Germanic languages
- Multi-language secretary

GOODYEAR Robin

- BSc agriculture – University of Leeds, UK
- Diploma in Tropical Agriculture, University of West Indies, Trinidad
- Worked as Tropical Agricultural Consultant
- Now retired

KITAJIMA W. Elliott

- Agronomist – ESALQ/USP, Brazil
- PhD in plant pathology – ESALQ/USP, Brazil
- Postgraduate training in the USA and The Netherlands
- Research on morphology and cytopathology (plant viruses)
- Currently coordinator of the Electron Microscope Unit at ESALQ/USP, Brazil

LAMBOT Charles

- Ing. agronome of tropical agriculture FSA.Gx. – Gembloux, Belgium
- Scientist in charge of coffee research program at ISABU, Burundi
- Project leader on African raw materials for Nestlé R & D, Abidjan, Ivory Coast
- Currently project leader on coffee at Nestlé Research Center, Tours, France

LASHERMES Philippe

- PhD in Plant Genetics from the University of Clermont-Ferrand (1987)
- Director of Research in the Living Resources department of the Institut de Recherche pour le Développement (IRD), France
- Research Scientist at ICARDA (International Center for Agricultural Research in Dry Areas), Syria (1988–92)

- Research Scientist at the IRD since 1992, in Man (Côte d'Ivoire) then Montpellier, France
- Scientific expertise includes population genetics, molecular evolution and plant breeding
- Co-author of more than 50 papers in international journals and book chapters on coffee
- Co-ordinator of numerous research projects including 3 international projects funded by the European Community
- Focussing on the improved use of coffee genetic resources

LEROY Thierry

- Doctor's degree in crop improvement – ENSA, Rennes, France
- Currently team leader in CIRAD-CP, France

MARTINEZ C. Felix

- Zootechnic Veterinary
- Currently Corporate Affairs Director, Nestlé SA de CV Mexico

MULLER Raoul Amédée

- Ing. agronome – Ecole Nationale Sup. Agr., Nancy, France
- Licencié ès Sciences – Univ. Nancy
- Ing. agronome of tropical agronomy – ESSAT, Paris
- Diploma in phytopathology – ORSTOM, France
- Doctor's degree – University Pierre and Marie Curie, Paris, France
- Phytopathologist for the Government Services Research Center, Cameroon
- Head of the IRCC Research Center, Cameroon
- Head of Phytopathology Services of the IRCC, France and concurrently Scientific Director
- Advisor to PROMECAFE, IICA and IRCC representative for Latin America
- Recently deceased

MUSCHLER Reinhold G.

- Graduate training in Geoecology, University of Bayreuth, Germany, and PhD in Agroforestry and Farming Systems Research, University of Florida, USA
- Former associate professor of agroecology and agroforestry. Head of the CATIE-GTZ Agroforestry Project and of the department of Ecological Agriculture at the Center for Research and Education in Tropical Agriculture (CATIE), Costa Rica
- Past representative of CATIE to PROMECAFE
- Spearheaded organic coffee project with Organic Growers' Association of Turrialba (APOT) with support from CATIE, the UK and Germany
- Technical and training publications on coffee agroforestry systems, coffee quality, sustainability, diversification and organic production
- Currently international consultant on agroecology, agroforestry and organic production in the tropics with the Institute for Development and Education towards Self-sufficiency (IDEAS), Costa Rica

PAES DE CAMARGO Angelo

- Agronomist
- PhD in Agriculture Climatology
- Expert in agrometeorology of the coffee crop
- Scientific Researcher at the Instituto Agronomico de Campinas (IAC), Brazil
- Now retired

PAES DE CAMARGO Marcelo Bento

- Agronomist
- MS/PhD in Agriculture Meteorology – Lincoln/Nebraska, USA
- Expert in coffee crop modelling
- Scientific Researcher at the Instituto Agronomico de Campinas (IAC), Brazil
- Currently team leader of the Agrometerology of the Coffee Crop at IAC, Brazil

PÉTIARD Vincent

- PhD in plant breeding – Paris, France
- Biotechnology for Pharmaceutical Cosmetic and Food plant raw materials
- Head of plant sciences at Nestlé CRN, Tours, France
- Member of the French Academy for Agriculture

PFALZER Hans

- Graduate of the Tropical School – Swiss Tropical Institute, Basle, Switzerland
- Field manager in sugar cane
- Active in the Novartis Crop Protection Support Team
- Currently retired

RODRIGUES Júnior, Carlos José

- Agronomist – Technical University, Lisbon, Portugal
- PhD in plant pathology – University of Wisconsin, USA
- Coordinating researcher of the Tropical Scientific Research Institute (IICT)
- Director of the Coffee Rust Research Center (CIFC) – Oeiras, Portugal
- Currently retired

RODRIGUEZ PADRON Benigno

- Ing. agrónomo specialized in agricultural economy – Autonomous University of Chapingo, Mexico
- Master in economics – Colegio de Postgraduados, Mexico
- Currently researcher, professor and assistant director – CRUO, Autonomous University of Chapingo, Mexico

ROJAS FLORES Jesus

- Agricultural technologist CBTA7
- Ing. agrónomo specialized in production – ITESM, Querétaro, Mexico
- Manager of coffee and cocoa supply department and responsible for technical assistance to coffee producers with Nestlé, Mexico

SAITO Moeko

- BSc economics – Keio University, Japan
- Master of environmental management – Duke University, USA
- Worked for projects linked to sustainable coffee production, The Nature Conservancy (INC), international co-operation and marketing analysis
- Currently working at the World Bank as Junior Professional Associate

SNOECK Didier

- Ing. agronome, specialized in rural engineering – University of Louvain, Belgium
- Doctor ès sciences (microbial ecology) – University Claude Bernard, Lyon, France
- Coffee agronomist – CIRAD-CP, Cameroon

SNOECK Jacques

- Ing. agronome specialized in tropical regions – University of Louvain, Belgium
- INEAC – former Belgian Congo and Rwanda/Burundi
- IFCC/IRCC France and Africa
- Recently deceased

SOZA MALDONADO Lucino

- Ing. agrónomo specialized in phytotechnology – Autonomous University of Chapingo, Mexico
- Master in rural development sciences
- Researcher and professor for CRUS, Autonomous University of Chapingo Oaxaca, Mexico
- Executive director of CERTIMEX, Mexico
- Inspector of organic procedures and certification for CERTIMEX, Mexico

VAAST Philippe

- PhD of soil sciences and plant nutrition – University of California, Davis, USA
- Coffee ecophysiologist at CIRAD; currently posted at CATIE, Turrialba, Costa Rica
- Coordinator of the coffee agroforestry project in Central America (www.CASCA.project.com)
- Researcher and professor in eco-physiology, CATIE, Turrialba, Costa Rica

VASQUEZ MORENO Martin

- Ing. agrónomo specialized in soil sciences – Autonomous University of Chapingo, Mexico
- Collaborator with INMECAFE
- Agricultural advisor for the unit of coffee and cocoa supply department – Nestlé, Mexico

WALDBURGER Heinz

- Dipl. Ing. – ETH, Switzerland
- Formerly head of Nestlé computer services worldwide
- Since 1982 entrepreneur and consultant

WANGAI KIMENJU John

- PhD in plant nematology
- Section head of crop protection – University of Nairobi, Kenya

ZAMARRIPA COLMENERO Alfredo

- Agronomist – University of Chiapas, Mexico
- PhD of plant breeding – Ecole Nationale Supérieure Agronomique, Rennes, France
- Research INIFAP – SAGAR, Mexico (coffee breeding)
- National leader of research on coffee – Mexico
- Professor in genetics and plant breeding
- Pioneer of plant biotechnology in Mexico

Outline

Contents

Part III: Harvesting & Processing

Part IV: Storage, Shipment, Quality

Part V: Economics

Part VI: Data & Information

**Part 1
Growing**

1
The Coffee Plant

J. N. Wintgens

1.1
Taxonomy

The genus *Coffea* belongs to the family Rubiaceae. This family comprises many genera including *Gardenia, Ixora, Cinchona* (quinine) and *Rubia*. The latter includes *Rubia tinctoria* (Turkey Red), from which the name of the family Rubiaceae was derived. The genus *Coffea* covers approximately 70 species.

The two main species of coffee tree cultivated on a worldwide scale are *Coffea arabica* and *C. canephora* var. *robusta*. Minor cultivated species include *C. liberica* and *C. excelsa*, which are mainly restricted to West Africa and Asia, and account for only 1–2 % of global production.

In the past, coffee classification has undergone frequent alterations and the present system of classification is not yet the final version. Specialists are well aware that further native species of *Coffea* are likely to be discovered in Africa and possibly elsewhere. In addition, genome studies involving cellular studies and molecular chemistry, will undoubtedly highlight factors that will refine and simplify present day coffee classification.

1.2
Origins

The centre of origin of *C. arabica* is in Ethiopia (Abyssinia) in high plateaux areas at altitudes between 1300 and 2000 m. On the other hand, the origins of *C. canephora* are more widely dispersed in tropical Africa at altitudes below 1000 m. The species *C. liberica* originates from lowland habitats in West Africa, often coastal. The original identification of this species was made in Liberia, hence its name. A further species, *C. excelsa*, closely related to *C. liberica*, is also a native of lowland forest habitats in West and Central Africa.

1.3
Areas of Cultivation

C. arabica

Africa	On the high plateaux regions of the continent, Madagascar and also on the West Coast
Asia	At the higher altitudes across the continent from Arabia to the Philippines, including Yemen, India, Papua New Guinea, Mauritius, Reunion, New Caledonia, Vietnam and Hawaii

Coffee: Growing, Processing, Sustainable Production, Second Edition. Edited by J. N. Wintgens.
© 2012 WILEY-VCH Verlag GmbH & Co. KGaA. Published 2012 by Wiley-VCH Verlag GmbH & Co. KGaA.

Americas The high plateaux of the Tropical Americas; mid-altitude regions of South America; the mountainous lands of the Caribbean Islands

C. canephora

Africa The lowlands of west and central Africa and mid-altitude zones in the East

Asia Low and mid-altitude regions (India, Indonesia, Philippines, Malaysia, Thailand, China, etc.)

Americas Humid, tropical regions in the Northeast of Brazil (Conillon), Ecuador, Guyana, Mexico, Trinidad and Tobago, etc.

C. liberica

Asia Low-altitude regions, mainly in Malaysia, but also in Indonesia, the Philippines, Vietnam and Thailand

Africa The West coast, Equatorial Africa and Liberia

Americas Guyana and Surinam

C. excelsa

Asia Mainly in Vietnam, but also in Indonesia and the Philippines

Africa Central and West Africa, Chad, south Sudan, Madagascar, Mauritania, and others.

Americas Chiefly Puerto Rico

1.4
Botany

The coffee plant takes approximately 3 years to develop from seed germination to first fruit production. The fruit of the coffee tree is known as a cherry, and the beans which develop inside the cherry are used as the basic element for producing roast and ground coffee, soluble coffee powders, and coffee liquor.

A well-managed coffee tree can be productive for up to 80 years or more, but the economic lifespan of a coffee plantation is rarely more than 30 years.

1.4.1
Seed and Germination

The seed consists of a horny endosperm containing an embryo, which is wrapped in two husks: the outer parchment and the silverskin (or integument) just underneath (Fig. 1.1). The embryo, about 3–4 mm long, is composed of the hypocotyl (embryo axis) and two cotyledons (Fig. 1.2).

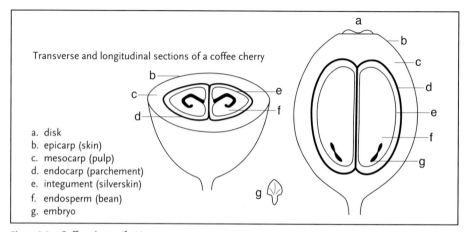

Transverse and longitudinal sections of a coffee cherry

a. disk
b. epicarp (skin)
c. mesocarp (pulp)
d. endocarp (parchement)
e. integument (silverskin)
f. endosperm (bean)
g. embryo

Figure 1.1 Coffee cherry (fruit).

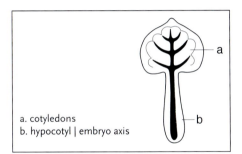

a. cotyledons
b. hypocotyl | embryo axis

Figure 1.2 The embryo.

Table 1.1 Germination rate of *C. arabica*, var. *caturra* at 12–13 % MC dried at different temperatures

Drying temperatures (°C)	Germination rate %
Ambiental temperature Artificial drying (oven)	95
40	95
45	95
50	80
55	45
60	4
70	0

From: J. Arcila – Cenicafe (1976).

The different varieties of coffee show differences in the size and shape of the coffee bean (seed) but, on average, beans are approximately 10 mm long and 6 mm wide. The weight of a parchment seed at 18 % moisture content is 0.45–0.50 g for Arabica and 0.37–0.40 g for Robusta. The average weight of hulled beans of both species at 12–13 % moisture content (MC) varies from 0.17 to 0.40 g.

Coffee seeds do not require a period of dormancy. Consequently seeds should be sown as soon as possible after ripening, when their moisture content is over 50 %. At this stage their germination rate is over 90 %. Viability decreases rapidly after 2 months (for Robusta) and 6 months (for Arabica) when stored at ambient temperature. If seeds have to be stored, they should be stored as parchment seeds which have to be dried slowly at a low temperature (not above 40°C) down to 12–13 % MC. The dried parchment seed stored at 10°C will maintain a high germination capacity for several months. Indeed, it has been demonstrated that Arabica seeds with 10–11 % MC stored at 15°C can be maintained for approximately 2 years.

The devastating effects of drying parchment seed at temperatures above 50°C is demonstrated by the data contained in Tab. 1.1.

It has also been shown that it is possible to extend the viability of coffee seeds with other methods of seed storage and two examples are shown below:

- Arabica parchment seeds (40–45 % MC) stored in humidified charcoal in a cool place kept their viability for 1 year.
- High viability can be maintained for some 30 months by storing humid Arabica parchment seeds with 40–41 % MC (on a fresh-weight bases) in air-tight polythene bags at a constant temperature of 15°C (Tab. 1.2) [10].

Attempts at chilling seeds to temperatures below 10°C and drying below 9 % MC have both resulted in a rapid loss of viability. Temperatures of below −15°C are lethal for coffee seeds. When storing coffee seeds, the parchment should not be removed.

The germination of coffee seeds is slow, taking 30–60 days under the most favorable conditions, which are:

- A high rate of ambient humidity
- An air temperature of 30–35°C
- A soil temperature of 28–30°C

Table 1.2 Viability of arabica coffee seeds in storage under optimum conditions

Species	Temp	Months of storage						Reference
	°C	0	8	15	24	30	36	
C. arabica	15	95	95	90	91	80	–	Van der Vossen (1979) [10]
C. arabica	19	95	94	95	90	80	60	Couturon (1980)
C. canephora	19	95	91	64	0	0	0	Couturon (1980)
C. stenophylla	19	90	80	50	0	0	0	Couturon (1980)

Note: Seeds stored at 40–41 % moisture content in air-tight polyethylene bags

A fall in air temperature slows down germination. This means that during the dry season, germination may be delayed by up to 90 days. If temperatures remain below 10°C for long periods, germination may not occur at all. The rate of germination can be increased by 6–10 days by removing the parchment, preferably by hand, before sowing. Hulled Catuai seeds have been known to germinate after 15 days at a temperature of 32°C.

Germination can also be accelerated by soaking the seeds in water for about 24 h before sowing.

As the hypocotyl axis grows, it lifts the seed out of the ground, this is known as an epigeous germination (Pict. 1.1).

Picture 1.1 Germination of a coffee seed 1 month after sowing. From: Nestlé.

Picture 1.2 Coffee plantlet 3 months after sowing. From: Nestlé.

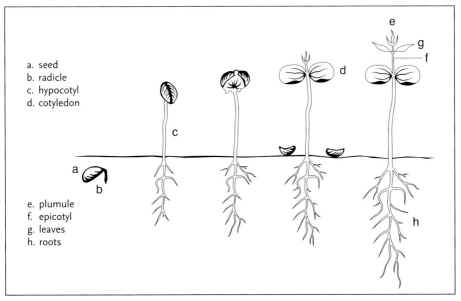

a. seed
b. radicle
c. hypocotyl
d. cotyledon

e. plumule
f. epicotyl
g. leaves
h. roots

Figure 1.3 Germination of a coffee plant.

Ten weeks after sowing, the parchment splits and the two cotyledons unfurl into two circular leaf-like structures. At this stage the root system consists only of a tap root with laterals (Fig. 1.3).

Three months after sowing, the apical bud lodged between the two cotyledons develops the first true leaves (Pict. 1.2).

1.4.2
Root System

At full development the coffee tree root system consists of five types of roots:

- The central tap root, often multiple, which generally terminates at a depth of approximately 0.45–1 m.
- The axial roots, generally four to eight, which can penetrate to a depth of up to 3 m and branch out in all directions.
- The lateral roots that run more or less horizontally, parallel to the soil surface at a distance of 1.2–1.8 m from the trunk.

- The feeder bearers of various lengths, distributed on the tap, axial and lateral roots.
- The root hairs that grow from the feeder roots are found at all depths, but are more numerous at the surface of the soil; the root hairs (feeders) are the main providers of mineral nutrition to the coffee plant.

In favorable soils coffee tree roots can explore up to 15 m³ of earth. In humid, heavy soils, the superficial roots concentrate mostly in the upper layers. In dry, sun-exposed soils, the root system is less superficial.

The removal of big lateral roots has a depressive effect on the uptake of nutrients by the tree. The vibrations of harvesting machines shaking the stems can also disturb nutrition in certain cases.

If the tap root is twisted or bent, or shows other malformations, nutritional uptake is decreased and the lifespan of the tree is shortened. Arabica coffee trees root deeper than Robusta coffee trees and

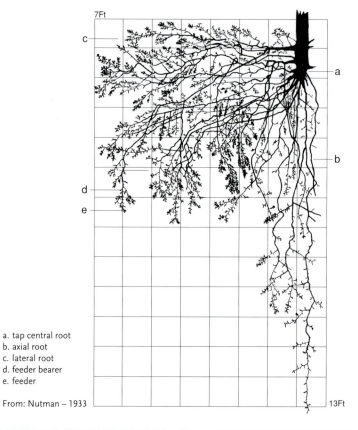

7Ft

c

a

b

d

e

a. tap central root
b. axial root
c. lateral root
d. feeder bearer
e. feeder

From: Nutman – 1933

13Ft

Figure 1.4
Typical root
system of a
6-year-old
C. arabica tree.

Picture 1.3 Young Robusta coffee plant root system. From: J. N. Wintgens.

Picture 1.4 Adult Arabica coffee tree root system. From: Ch. Lambot.

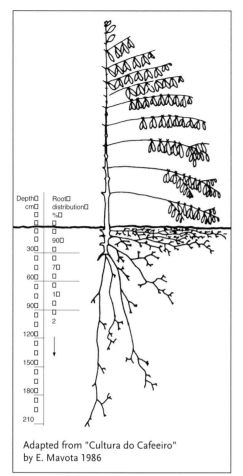

Adapted from "Cultura do Cafeeiro"
by E. Mavota 1986

Figure 1.5 Distribution of the root system of an adult coffee tree.

Optimum temperatures of the soil for root development and effective functioning of the rooting system are around 26°C during the day and no less than 20°C during the night.

1.4.3
Framework of the Tree

The development of the aerial parts of the coffee plant entails the lengthening of the vertical (orthotropic) main stem and the successive growth of pairs of opposite leaves at each node. Serial buds, headed by a "head of series" bud, are located in the leaf axil.

The "head of series" bud is the only bud able to generate a primary branch whereas the four to six serial buds can generate either new suckers (orthotropic stems) or flowers (Fig. 1.6, 1.7). Serial buds can remain dormant for a long time.

Primary branches grow off the main stem. They develop secondary branches that, in turn, give rise to tertiary and quaternary ramifications. Serial buds and "head of series" buds on horizontal (plagiotropic) branches are also located in the leaf axils (Fig. 1.8, Pict. 1.5).

therefore show a higher resistance to drought (Fig. 1.4).

Coffee trees can root deeply in a normal soil although about 90 % of their roots develop in the upper 30 cm layer (Fig. 1.5). This part of the root system is sensitive to climatic variations (temperature, drought, moisture). Protection can be provided by mulch that maintains the humidity of the topsoil and provides added mineral nutrition to the coffee plant. Under mulch protection, the density of the superficial root system can triplicate (Pict. 1.3, Pict. 1.4).

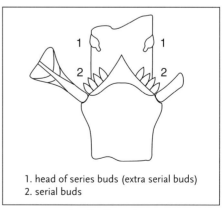

1. head of series buds (extra serial buds)
2. serial buds

Figure 1.6 Node of coffee Robusta.

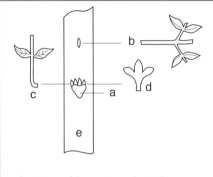

a. front view of the sectionned petiole
b. plagiotropic branch grown from a
   "head of series" bud
c. orthotropic shoot grown from a serial bud
d. flower grown from a serial bud
e. stem

Figure 1.7 Bud differentiation.

Picture 1.5 Coffee vegetative node with opposite leaves and axillary plagiotropic axis.
From: A. Charrier.

tree reaches full maturity and begins to yield a normal crop.

1.4.4
Leaves

Depending on the species and environmental conditions, a 1-year-old coffee plant develops approximately six to ten levels of plagiotropic branches. After 2 years, the coffee plant can reach a height of 1.5–2 m and the first flowers appear. After approximately 3 years, the coffee

Coffee tree leaves grow on a petiole in opposite pairs on the sides of the main stem and branches. They are dark green in color, shiny and waxed. Leaf shape is elliptical and the veins are mostly conspicuous.

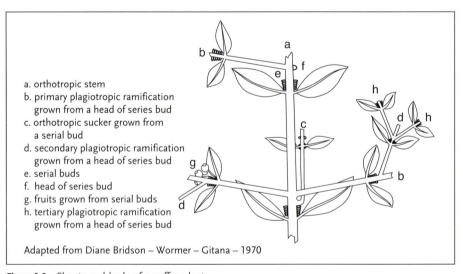

a. orthotropic stem
b. primary plagiotropic ramification
   grown from a head of series bud
c. orthotropic sucker grown from
   a serial bud
d. secondary plagiotropic ramification
   grown from a head of series bud
e. serial buds
f. head of series bud
g. fruits grown from serial buds
h. tertiary plagiotropic ramification
   grown from a head of series bud

Adapted from Diane Bridson – Wormer – Gitana – 1970

Figure 1.8 Shoots and buds of a coffee plant.

The leaves of the Arabica plant are slender and more delicate than those of either Robusta or Liberica species. The younger leaves of the Arabica plant are either light green or bronze depending on whether the plant is of Bourbon or Typica origin. The bronze color of the Arabica Typica fades with age.

The foliar surface of an adult coffee tree varies according to the species, its state of health, light, etc. Its average spread ranges from 22 to 45 m^2 and the life span of coffee leaves is 7–10 months for *C. canephora*. Small cavities called domatia are to be found on the lower epidermis of coffee leaves (Fig. 1.9).

The localization, shape, size and constitution, as well as the presence or absence of hairs around the opening of the domatia and the presence or absence of stomata on the outermost cell layer of the domatia, have been used to distinguish *Coffea* species and varieties.

In *C. arabica*, domatia occur at the insertion of the lateral veins and give a slight protuberance to the upper surface of the leaf, whereas no similar protuberance has been observed on the upper leaf surface of *C. canephora*.

The function of domatia is still a controversial subject and, to date, no specific function has been determined for this phenomenon. Leaf domatia, either of pit or pocket type, may contain bacteria, fungi or blue-green algae. These are generally washed in by rain. Similarly, mites, acaria or other small insects may use domatia as a shelter.

In those species where the stomata have the capacity to close rapidly, evaporation is better regulated and, therefore, resistance to drought is higher. The quantity of water evaporated per day by a non-shaded coffee tree is approximately 6 g/dm^2.

1.4.5
Fruit Formation

1.4.5.1   The Flower

The coffee flower consists of a white five-lobed corolla, a calyx, five stamens and the pistil. The ovary is at the base of the corolla and contains two ovules that, if duly fertilized, produce two coffee beans (Fig. 1.10).

Flowers open in the early morning and remain open throughout the day. In the afternoon of the same day, once fertiliza-

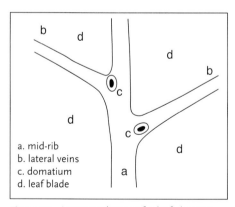

a. mid-rib
b. lateral veins
c. domatium
d. leaf blade

Figure 1.9   Lower epidermis of a leaf showing domatia in the angle formed by the mid-rib and a lateral vein.

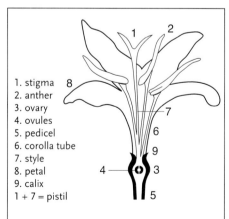

1. stigma
2. anther
3. ovary
4. ovules
5. pedicel
6. corolla tube
7. style
8. petal
9. calix
1 + 7 = pistil

Figure 1.10   Vertical section of a coffee flower.

tion has taken place, the anthers turn brown. Two days later the white corolla withers and the floral parts fall away leaving the ovary to develop. If fertilization has failed, the stigmas and the corolla remain bound to the ovary. The stigmas of the Robusta flowers remain receptive up to 6 days after flowering.

Coffee pollen is light in weight and is easily carried by the wind. Wind pollination may be of prime importance. However the sweet-smelling flowers also attract insects and it would seem very likely that insects more contribute to the pollination process than previously supposed.

In *C. arabica*, 90–95 % of the fertilization is carried out by wind. Foreign vegetation which harbours insects indirectly favorizes the pollination and the production of the coffee trees. In *C. canephora*, fertilization can only be carried out by the pollen of another tree (cross-pollination). In both cases, successful fertilization mainly depends on meteorological conditions. Strong rains or violent winds hinder pollination and lead to a greater proportion of non-fertilized flowers.

1.4.5.2 The Flowering Process

Normally, flowering starts during the second to third year. Flower buds develop in the leaf axils of all forms of plagiotropic branches and, more rarely, on orthotropic wood.

The development of a serial bud into a flower bud is largely controlled by plant hormones. Hormones are chiefly activated by photoperiodism (changes in day length) and by a drop in temperature (cool season).

Flat and triangular buds may or may not be florally determined, but if the buds appear to be thick and cased in rudimentary bracts covered with gum, they are florally determined. These buds can grow to a

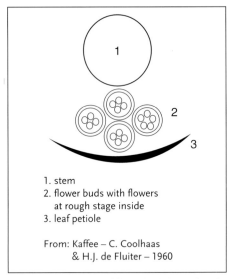

1. stem
2. flower buds with flowers
   at rough stage inside
3. leaf petiole

From: Kaffee – C. Coolhaas
& H.J. de Fluiter – 1960

Figure 1.11   Diagram of flower buds located in the leaf axil.

length of 4–6 mm and enter into a dormant phase until stimulated into flowering.

A normal inflorescence consists of four floral buds on a short stalk that grows off the main stem, generally known as cyme, floral cluster, anthela or glomerule. In most cases the cyme consists of four flowers, but this is not an absolute rule. For Arabica, the number of flowers per inflorescence can range from two to nine (Fig. 1.11).

This means four to twelve inflorescences per node that finally generate 16–48 flowers. On Robusta, flowers are more abundant (30–100 per node) and also larger (Fig. 1.12, 1.13, 1.14, 1.15).

Dormancy is usually broken by a sudden relief of water stress (rehydration) in the buds and/or a drastic fall in temperature. In the field, these stimuli often occur simultaneously with "blossom showers" at the end of a dry season. For best results, at least 10 mm of rainfall is necessary (Fig. 1.16).

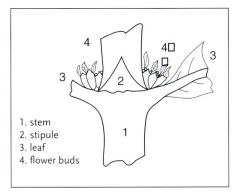

Figure 1.12 Robusta flower buds in the leaf axils.

1. stem
2. stipule
3. leaf
4. flower buds

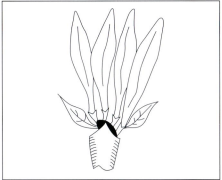

Figure 1.15 Inflorescences close to blossoming.

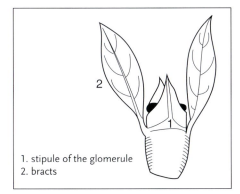

1. stipule of the glomerule
2. bracts

Figure 1.13 Detail of a Robusta flower bud.

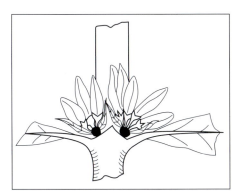

Figure 1.14 Coffee inflorescences (glomerules) close to blossoming.

The more severe the dry season, the more intense the flowering. In equatorial climates where the dry season is not shar-ply defined, coffee trees bloom during the whole year producing 25–50 flowering periods. Irrigation of any type, even mists, can break the dormancy period.

A flowering period usually starts 3–10 days after the activating stimulus occurs. Some dormant or undeveloped buds may not flower at this time but are likely to do so at a later stage. Unfortunately, staggered flowering of this type encourages the development of parasites like the Coffee Berry Borer. It also makes harvesting more difficult and more expensive (Pict. 1.6).

An acute form of floral atrophy termed "star flowering" for Arabica and "pink flowering" for Robusta can occur when the coffee tree is under stress. Stressful conditions are brought about by a lack of humidity after flowering, high temperatures or extended, heavy rainfall for Arabica and

Picture 1.6 Staggered uneven blossom on *C. arabica*. From: J. N. Wintgens.

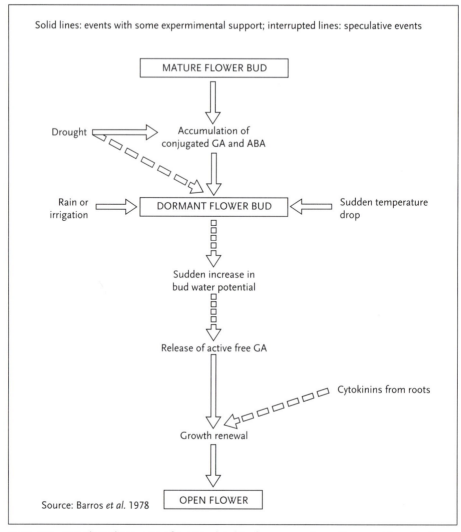

Figure 1.16 Hypothetical sequence of events related to dormancy and growth renewal in coffee tree flower buds.

Picture 1.7 Star flowers on *C. arabica* due to drought. From: J. N. Wintgens.

Picture 1.8 Pink flowers on Robusta due to drought. From: J. N. Wintgens.

Picture 1.9 Full-blooming Robusta coffee tree. From: GENAGRO.

Picture 1.10 Arabusta buds at "candle" state. From: J. N. Wintgens.

Picture 1.11 Blossoming Robusta coffee branch. Most of the flowers are concentrated on the 1-year-old section. From: J. N. Wintgens.

Picture 1.12 Clusters of coffee flowers in full blossom. From: GENAGRO.

excessive moisture stress for Robusta (Pict. 1.7, 1.8, 1.9, 1.10, 1.11, 1.12).

1.4.5.3 Pollination

In Arabica, pollen grains drop by gravity on to the lower layers of the coffee branches but at higher levels the flowers are pollinated by pollen carried on the wind and, to a lesser extent, by insects (5–10 %).

The quantity of pollen produced by an adult coffee plant is impressive, approximately 2.5 million grains per tree. Such a quantity is more than enough to fertilize the 20–30 000 flowers of the plant. Pollen can be wind-borne over a distance of approximately 100 m. Fertilization occurs when a pollen grain falls on the stigma and develops a pollen tube that grows down into the style and fuses with the ovary. Under favorable conditions pollen germination is fast and fusion takes place 1–2 h after the pollen grain has settled on the stigma.

1.4.5.4 The Fruit

The ovary, which contains the two fertilized ovules, starts to develop immediately following fertilization. During the first 2 months, however, the ovary grows very slowly, but eventually becomes definitely

Picture 1.13 Coffee fruit 3 months after fecundation. Right: external view of the coffee berry. Left: view of the ovule showing the embryonic sac that contains the zygotic embryo. From: Nestlé.

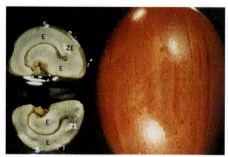

Picture 1.16 Coffee fruit 7 months after fecundation. Right: mature fruit (red skin). Left: transversal cut through each separated bean. Endosperm (E) reveals the zygotic embryo (ZE). From: Nestlé.

Picture 1.14 Excised green cherry showing the zygotic embryo. From: J. Berthaud.

Picture 1.17 Coffee fruit 7 months after fecundation. Right: mature fruit. Left: extract bean wrapped in silverskin (S). Endosperm (E) and embryo (ZE) are fully developed. From: Nestlé.

Picture 1.15 Coffee fruit 5 months after fecundation. Right: immature fruit (green skin). Left: transversal cut through the previously split fruit. Endosperm (E) is forcing back the integument (I). From: Nestlé.

visible in a dormant pinhead stage. From the second to third month of development, the ovary increases in size more rapidly and the integument occupies almost the entire space in each ovule. The embryonic sac grows and fills with endosperm (Pict. 1.13, 1.14).

From the third to the fifth month after fertilization, the fruit increases significantly in weight and volume. The endosperm slowly replaces the integument that is forced back to the periphery of the ovule (Pict. 1.15).

Between the sixth and the eighth month after fertilization, the fruit reaches maturity. The integument is now only represented by the silverskin. In the endosperm that fills the whole grain, the zygotic embryo has evolved to the "two-cotyledon" stage (Pict. 1.16, 1.17).

During the last month of maturation, the fruit completes its growth, and, depending on the variety, acquires a red or yellow color. Each fruit contains two "beans" coated in mucilage.

The ripe fruit is commonly referred to as a coffee berry, but in correct botanical terminology it would be described as a drupe or cherry (Fig. 1.17).

The time taken from flowering until the maturation of the coffee berries varies according to the variety, climatic conditions, agricultural practices and various other factors. As a general rule:

- *C. arabica* takes 6–9 months
- *C. canephora* takes 9–11 months
- *C. excelsa* takes 11–12 months
- *C. liberica* takes 12–14 months

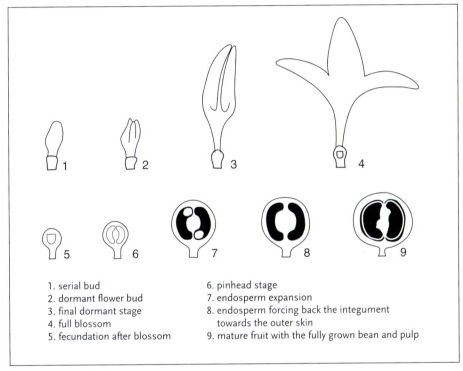

1. serial bud
2. dormant flower bud
3. final dormant stage
4. full blossom
5. fecundation after blossom
6. pinhead stage
7. endosperm expansion
8. endosperm forcing back the integument
   towards the outer skin
9. mature fruit with the fully grown bean and pulp

Figure 1.17 From bud to bean.

Picture 1.18 Heavily laden Robusta tree with even mature fruits. From: J. N. Wintgens.

Picture 1.19 Ripe Robusta cherries. From: J. N. Wintgens.

Picture 1.20 Left: Robusta ripe cherries. Right: fresh parchment. From: Nestlé.

Picture 1.21 Ripe cherries (whole and transversal section). From: Nestlé.

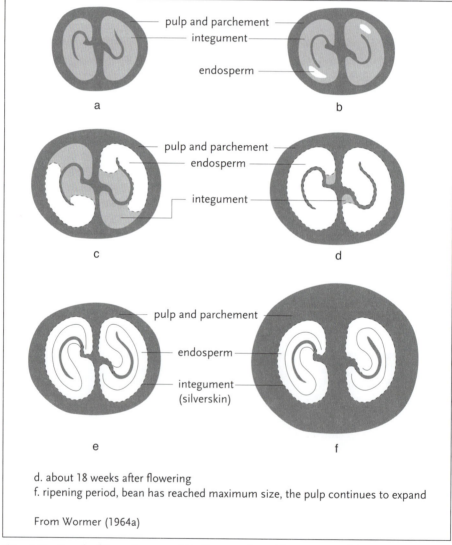

d. about 18 weeks after flowering
f. ripening period, bean has reached maximum size, the pulp continues to expand

From Wormer (1964a)

Figure 1.18 Growth of the endosperm within the space formed by expansion of the integuments.

Picture 1.22 Longitudinal section through a ripe cherry showing the two beans. From: Nestlé.

Maturation is slower at higher altitudes since the air temperature drops by 1°C per 180 m of elevation (Fig. 1.18, 1.19, 1.20, 1.21, 1.22).

1.5
Harvesting Strategies and Crop Management

Once the fruit is ripe, Arabica and Robusta plants show a fundamental difference in fruit bearing. Ripe fruit on the Arabica plant drop fairly soon after full maturation, whereas ripe fruit on the Robusta plant remain fixed to the branches for several weeks.

Unlike many other fruit tree species, the coffee tree sheds little excess fruit once the "pinhead" stage is over. In years of heavy flowering, this can lead to overbearing and, where the leaf coverage is insufficient to ensure the required level of photosynthesis, it can lead to die-back.

Thus, the coffee grower, especially the grower of Robusta, is faced with two options: he can either aim for a lower yield per plant which reduces the risk of damaging "die-back" or he can aim for high yields through more intensive crop management which involves an increased risk of damage to the plant caused by overbearing and "die-

back". The "low-yield option" or extensive management, which results in a reduction of floral initiation, means increased shading, minimal pruning and a minimum input of fertilizers, and, consequently, a reduced investment in labor and other inputs. On the other hand, the "high-yield option" entails severe pruning to regulate the leaf: fruit ratio, heavy fertilization and intensive crop management to ensure the trees carry their full crop with a minimum of dieback.

1.6
Photosynthesis

Wild coffee trees grow naturally under subdued lighting in the lower levels of the forests. This explains why, originally, coffee trees were planted under shade trees. The practice of planting under shade trees became more popular after it had been proven than on a single leaf, photosynthesis (respiration and carbonic nutrition) was more intense under reduced luminosity (10–60 %) than under full sunshine. However, later studies demonstrated that, only the peripheral leaves of the coffee tree are exposed to full sunshine and that most of the foliage is protected by the peripheral leaves. It was, therefore, concluded that artificial shading is superfluous and even harmful to productivity, especially in equatorial and sub-equatorial areas with fewer hours of sunshine. Nevertheless, it is important to realize that non-shaded coffee trees, with relatively higher rates of photosynthesis and a higher productive capacity, require more nutrients and care (intensive management). If they are not carefully husbanded, coffee trees with insufficient shade protection are rapidly exhausted and their production drops significantly.

Extreme temperature changes bring about leaf burn, frost damage, etc., and chlorosis (yellowing of the leaf) occurs when temperatures drop below 10°C, which kills the chloroplasts. The reduction in active chlorophyll levels results in lowered levels of photosynthesis.

Other benefits of planting of shade trees for coffee in dangerously exposed areas include the following:

- Reduction of topsoil erosion on steeper slopes
- The production of organic matter (litter-fall) which greatly improves the physical characteristics of the topsoil and generates nutrients
- The reduction in weed growth which avoids competition between weeds and the coffee crop for moisture and nutrients
- The curbing of the biennial bearing pattern and last, but not least, an improvement in the bean size
- The formation of alkaloids and aromatic compounds in the beans; these volatile substances contribute to the production of good quality coffee and are particularly important in Arabica coffee [5]

1.6.1
The Biennial Pattern

Another well-known phenomenon to coffee growers is biennial bearing (high yields in alternate years). Yield differences between years in the range of 5 to 10 times have been noted [11] but, as a general rule, differences are not quite so marked. In a high-yielding year, the tree sacrifices its new growth of wood to bear a heavy crop of fruit which results in a small crop of fruit the following year. When carrying a small crop of fruit, the production of new wood is more vigorous, resulting in a heavy crop of fruit the following year.

This biennial bearing phenomenon is more frequent in unshaded, mature Arabica trees. It is not related to climatic conditions and is more pronounced where management is deficient, Well-managed plantations, where proper pruning techniques and adequate fertilization promote vigorous growth, are less prone to biennial bearing.

1.6.2
The Leaf Area Index (LAI)

The LAI is the ratio of the total surface area of a plant's leaves to the ground area available to the plant [9]. The production capacity of any crop depends, among other factors, on the LAI. Experimental coffee farming in Colombia concluded that, for the Caturra cultivar of Arabica coffee, a LAI of 7.97 represented an optimal value. This corresponds to a density of 5000 4-year old trees/ha. A LAI of over 6 or 10 is considered necessary for the maximum production of dry matter in Arabica coffee.

1.7
Abnormalities in Beans

1.7.1
Polyembryony (Elephant Beans)

Two or more embryos develop in a single cavity of the ovary and form two or more abnormally shaped separate beans enclosed in a common parchment (endocarp) with each one enveloped in an individual silver skin (integument). (Fig. 1.19) Elephant beans are usually damaged by traditional pulping machines (drums, disks) and broken up by hulling.

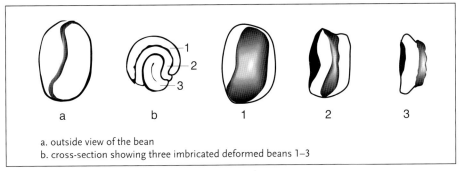

a. outside view of the bean
b. cross-section showing three imbricated deformed beans 1–3

Figure 1.19 Polyembryonic (elephant) beans in *C. canephora*.

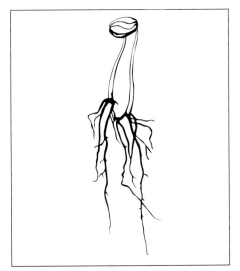

Figure 1.20 Germination of a polyembryonic seed of *C. arabusta* containing two endosperms with fertile embryos.

Polyembryonic seeds can produce as many plants as fertile seeds are united, since each seed contains a fertile embryo. Though the plants from these polyembryonic seeds are as good as the plants from normal seeds, there is no advantage in using them because of the complications in nursery practice (Fig. 1.20).

1.7.2
Monospermy – Peaberry or Caracoli

Normally the coffee cherry contains two beans (seeds), but it can happen that one of the seeds does not develop normally; in this case the other occupies the whole space and forms a larger berry called a peaberry or, better still, peabean. This phenomenon occurs when one of the two ovules aborts due to poor pollination. It can also be caused by incomplete endosperm development or infilling. *C. arabica*, of the *monosperma* variety, has a strong tendency to produce peaberries. This is also the case for coffee hybrids, where the tendency is even more marked. The presence of a peaberry is considered to be a defect in coffee production and is used as a criterion in coffee breeding. The peaberry seed germinates in the same way as normal beans. In an average crop, the peaberry yield ranges from 10 to 30 %, but in certain cases can exceed 50 %. An increase of 1 % of peaberry beans in a crop reduces production by 0.75 % [7].

As to quality, the peaberry bean is neither superior nor inferior to normal beans, but, in some places, the peaberry

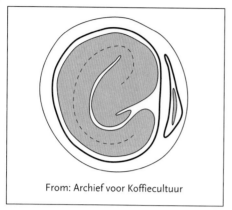

From: Archief voor Koffiecultuur

Figure 1.21   Transversal section through a pea-berry (caracoli).

is sold as a special grade. This may be justified as the shape of the peaberry lends itself better to roasting, but, on the other hand, it is also known to have a lower extraction rate for soluble coffee (personal comment by Dr F. Martinez, Nestlé, Mexico). Peaberries are often damaged in traditional crushing/shearing or hulling procedures (Fig. 1.21).

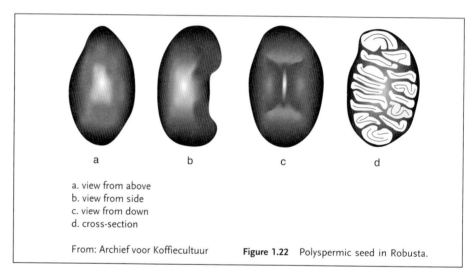

a. view from above
b. view from side
c. view from down
d. cross-section

From: Archief voor Koffiecultuur

Figure 1.22   Polyspermic seed in Robusta.

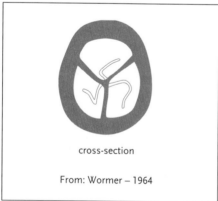

cross-section

From: Wormer – 1964

Figure 1.23   Polyspermic triangular seed in *C. arabica*.

1.7.3
Polyspermy

This occurs when the ovary has three or more cavities and generates one true seed per cavity. Each endosperm is wrapped in an integument (silverskin) and endocarp (Fig. 1.22, 1.23).

In certain varieties of coffee, polyspermy is a normal phenomenon as is the case for *C. arabica* var. *polysperma* (Menado Coffee) that can develop six to eight beans per cherry.

1.7.4
Empty Beans (Vanos)

When the endosperm fails to develop properly inside a normally developed endocarp, the result is an empty bean. The phenomenon can occur either on one bean only or on both beans and the causes can be either genetic or physiological. Whatever the causes, a high percentage of empty beans is seen as a defect in planting material (Fig. 1.24, 1.25).

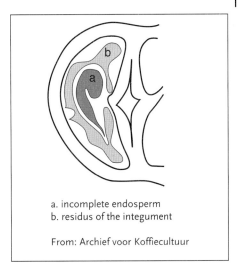

a. incomplete endosperm
b. residus of the integument

From: Archief voor Koffiecultuur

Figure 1.24 Empty beans – cross-section.

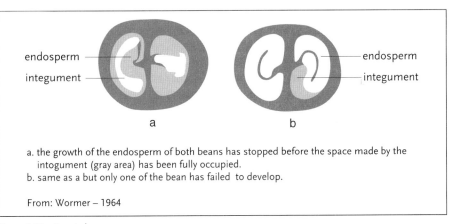

a. the growth of the endosperm of both beans has stopped before the space made by the intogument (gray area) has been fully occupied.
b. same as a but only one of the bean has failed to develop.

From: Wormer – 1964

Figure 1.25 Empty beans.

1.7.5
Misshapen Beans

Beans develop an unusual shape when the integument is misshapen or when the endosperm does not follow the shape of the integument.

The botanical origin of the coffee tree can cause this phenomenon but it can also be caused by the position of the fruit on the tree and the age of the plant.

Older branches tend to carry more misshapen beans.

1.7.6
Bean Size

Differences in the size of beans are, in part, determined genetically, but they are also influenced by environmental conditions and husbandry (nutrition, moisture, care, inputs, etc.).

Bibliography

[1] Coste, R. and Cambrony, H. *Caféiers et Cafés.* Maisonneuve et Larose, Paris, 1989.

[2] Cliffort, M. N. and Willson, K. C. *Coffee.* Croom Helm, New York, 1985.

[3] Clarke, R. and Macrae R. *Coffee. Vol. 4: Agronomy.* Elsevier, London, 1988.

[4] Coolhaas, C., de Fluiter H. J. and Koenig, H. P. *Kaffee.* Ferdinand Enke, Stuttgart, 1960.

[5] Op De Laak, J. *Arabica Coffee Cultivation and HCRDC.* Chiang Mai – Extension Manual for the Highlands of Northern Thailand, 1992.

[6] Gaie, W. and Flémal, J. *La Culture du Caféier d'Arabie au Burundi.* Administration Générale de la Coopération au Développement, Bruxelles, 1988.

[7] INIFAP. *Tecnologia para la Produccion de Café en Mexico.* INIFAP, Mexico, 1977.

[8] Cambrony, H. R. *Le Caféier.* Maisonneuve et Larose, Paris, 1987.

[9] Wrigley, G. *Coffee.* Longman, London, 1988.

[10] Van der Vossen H. A. M. – Methods of preserving the viability of coffee seeds in storage – *Seed Science and Technology,* Switzerland 7(1), 65–74, 1979.

[11] Gilbert S. M. – Variability in field of *Coffea arabica* E. Afr. agric. J3-131-39, 1938.

2
Botany, Genetics and Genomics of Coffee

A. Charrier, P. Lashermes and A. B. Eskes

2.1
Introduction

Documents dating back to the 10th century reveal that coffee, served as a beverage, originated in the Arab world, but the true origins of coffee consumption are lost in the mists of unwritten history and have only been transmitted by word of mouth as, for instance, the legend of the Ethiopian shepherd, Kaldi, or by various travellers.

On the other hand, Abyssinia has been officially recognized as the cradle of *Coffea arabica* since coffee surveys carried out in the 20th century revealed specimens growing wild in the highland forests of southwest Ethiopia.

The history of the coffee tree and of coffee as a beverage is closely linked to the growth of great empires and trade, first under the influence of the Arabs at the end of the first millennium, then the Turks in the 15th century and, finally, the European colonizers since the 18th century [37].

Currently, commercial green coffee production relies on two main coffee species: *C. arabica* and *C. canephora*. In actual fact, however, botanists regard all tropical plants of the Rubiaceae family which produce coffee beans as coffee trees. Hence, since the 16th century over 100 sponta-neous coffee species have been described. The taxonomic classification of the genus *Coffea* has become increasingly complex due to the many new species discribed during the 20th century in West Africa [15], Central Africa [27, 40], Madagascar [28, 29] and East Africa [7] and particularly in the last ten years (see Davis *et al.* for a synthesis) [VIII].

This chapter reviews the botanical description and taxonomy of coffee trees, the phylogeny, ploidy and reproductive systems of the coffee species, the cultivated coffee populations, and the conservation of coffee genetic resources in the light of the most recent scientific advances in coffee genetics and genomics.

2.2
Botanical Description and Taxonomy

2.2.1
Botany

Coffee plants are tree or shrubs characterized by:

- Plant architecture with single or multiple vertical main stems which carry primary horizontal branches at each internode to form secondary horizontal branches.

Coffee: Growing, Processing, Sustainable Production, Second Edition. Edited by J. N. Wintgens.
© 2012 WILEY-VCH Verlag GmbH & Co. KGaA. Published 2012 by Wiley-VCH Verlag GmbH & Co. KGaA.

Picture 2.1 Floral structure of the *Coffea* genus (i.e. *C. sessiliflora*). From: A. Charrier.

Picture 2.2 Floral structure of the *Psilanthus* genus (i.e. *P. mannii*). From: A. Charrier.

- Inflorescences in both the axillary position (monopodial development) and the terminal position (sympodial development).
- Opposite leaves with one leaf blade only, more or less leathery and thick.
- Domatia of variable shapes and positions.
- Stipulae which are often well developed.
- Nectar producing flowers of white or cream color with a corolla which is attractive to insects.
- Inflorescences that usually bear three to 10 flowers, although occasionally only one.
- An ovary with two carpels, each with one ovule.
- Cherries with sweet pulp containing two beans, each with a longitudinal slit.

The classification system into two genera is based on their floral structure:

1) The *Coffea* genus is characterized by a long style, medium corolla tube with anthers protruding from the corolla tube (Pict. 2.1).
2) The *Psilanthus* genus is characterized by a short style, long corolla tube with anthers encased in the corolla tube (Pict. 2.2).

Subgenera can be identified in each of these by the growth habit and the position of the inflorescence: monopodial with axillary flowers versus sympodial with termi-

nal flowers. The *Psilanthus* genus, which includes approximately 20 species, has a wide distribution covering the tropical humid regions of Africa, India, South-East Asia and the Pacific.

On the other hand, the geographic distribution of the *Coffea* genus is restricted to tropical humid regions of Africa and islands in the West Indian Ocean. See Fig. 2.1, Fig. 2.2 and Fig. 2.3.

2.2.2
Botanical Information

Botanical information on wild and cultivated coffee species is related to their habitats, geographical distribution and taxonomic characteristics. General sources of information are to be found in herbaria and museum libraries, travel reports of explorers, and in botanical gardens or in the field collections of coffee research stations.

2.2.2.1 Herbaria
Due to the historical ties with Africa, most European botanists and herbaria have a high degree of geographical specialization:

- The Royal Botanical Gardens at Kew and the British Museum (UK), with collections for eastern and southern African

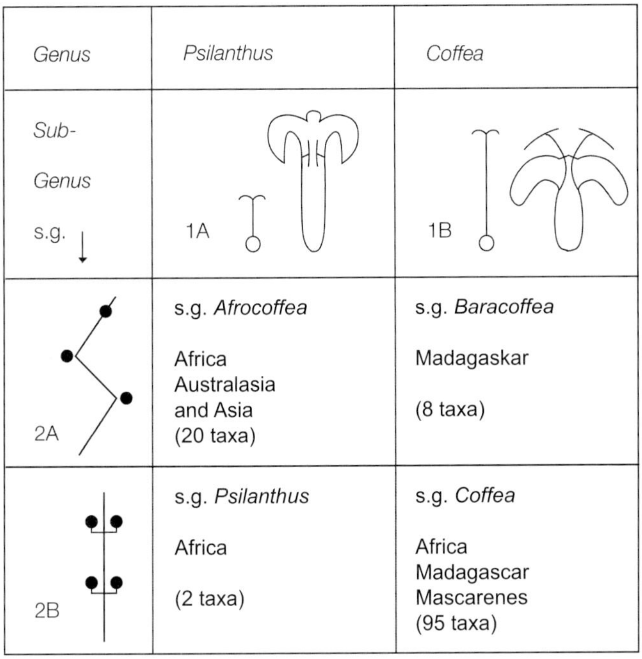

Genus	Psilanthus	Coffea
Sub-Genus s.g. ↓	1A	1B
2A	s.g. *Afrocoffea* Africa Australasia and Asia (20 taxa)	s.g. *Baracoffea* Madagaskar (8 taxa)
2B	s.g. *Psilanthus* Africa (2 taxa)	s.g. *Coffea* Africa Madagascar Mascarenes (95 taxa)

1A Long corolla tube, anthers not exserted, short style
1B Short corolla tube, anthers exserted, long style
2A Terminal flowers, predominantly sympodial development
2B Axillary flowers, monopodial development

Figure 2.1 Keys for coffee trees classification after Leroy [30], Bridson [6] and Davis [VIII].

countries, as well as from English-speaking West Africa.

- Jardin Botanique National de Belgique at Meise, Belgium, with emphasis on Central Africa.
- Museum National d'Histoire Naturelle at Paris, France, with collections from West and Central Africa, and the Malagasy region.
- Botanischer Garten und Botanischer Museum at Berlin-Dahlen, Germany which was almost completely destroyed in 1943.
- Herbarium Vadense of the Agricultural University of Wageningen and the Rijksherbarium of the University of Leiden (The Netherlands), with coffee specimens from Cameroon and Indonesia.
- Erbario Tropicale di Firenze, Florence (Italy), with collections from Ethiopia and Somalia.
- Botanical Institute of the University of Coimbra and Centro de Botanica da Junta de Investigaçoes Cientificas do Ultramar at Lisbon (Portugal), with collections from Angola and Mozambique.

The importance of national herbaria in African countries should not be underestimated; they often represent part of the duplicate herbarium specimens sent to Europe but they have also extended their collections through personal research.

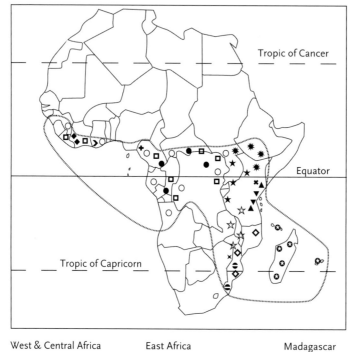

Figure 2.2 Distribution of *Coffea* species by geographic region [1].

West & Central Africa

+ *C. brevipes*
O *C. canephora*
● *C. congensis*
◆ *C. humilis*
□ *C. liberica*
➢ *C. stenophylla*

East Africa

✳ *C. arabica*
★ *C. eugenioides*
✖ *C. fadenii*
☆ *C. mufindiensis*
▼ *C. pseudo-*
   zanguebariae

◒ *C. racemosa*
✕ *C. salvatrix*
▲ *C. sessiliflora*
◇ *C. zanguebariae*

Madagascar

◉ *Mascarenes*
   60 taxa

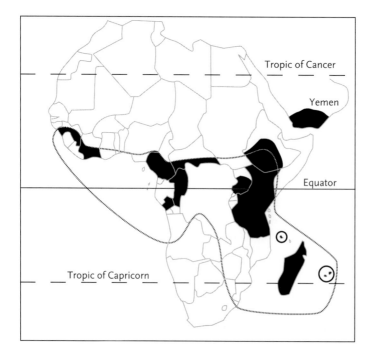

Figure 2.3 Collecting coffee missions in Africa since 1960.

Table 2.1 Coffee surveys and genebank locations

Years	Countries surveyed	Institutions[a]	Location of live collections
1964	Ethiopia	FAO	Tanzania, Brazil, India
1966	Ethiopia	ORSTOM	Ethiopia, Cameroon, Ivory coast, Madagascar
1960–74	Madagascar	MNHN, IRCC	Madagascar
1975	Central Africa	IRCC, ORSTOM	Central Africa, Ivory Coast
1975–81	Ivory Coast	ORSTOM	Ivory Coast
1977	Kenya	IRCC, ORSTOM	Kenya, Ivory Coast
1982	Tanzania	IRCC, ORSTOM	Tanzania, Ivory Coast
1983	Cameroon	IBPGR, IRCC, ORSTOM	Cameroon, Ivory Coast
1985	Congo	IBGPR, IRCC, ORSTOM	Congo, Ivory Coast
1989	Yemen	FAO, IPGRI, CIRAD	Brazil, Costa Rica
1987–91	New Caledonia	CIRAD, ORSTOM	New Caledonia

[a] ORSTOM now IRD; IRCC now CIRAD; IBPGR now Bioversity international.

2.2.2.2 Recent Collecting Missions

Exploration for wild coffee species was particularly intense in Africa at the end of the 19th century and during the 20th century. Evidence from that era is mostly to be found in botanical gardens. All coffee research centers maintain living collections of the two cultivated species used for breeding purposes, but very few other wild coffee species. The FAO has made an inventory of the existing coffee collections, which is regularly updated (Directory).

In order to fill the gaps in the various national inventories, FAO-IBPGR and French Institutes have intensified collections of wild coffee species in different African countries since 1960 (for a summary, see [5]). See Tab. 2.1.

Since 1967, 47 other collecting missions within Ethiopia have been conducted by JARC researchers in the main coffee production systems (wild coffee trees in forest and semi-forest systems; landraces in garden coffee system). 4,593 accessions of

C. arabica were collected, representing mainly landraces from gardens and homestead farms [XV]. Recently, some wild and cultivated C. canephora populations in Uganda have been sampled for studying their diversity and tolerance to coffee wilt disease [XXIV].

2.2.3
Taxonomy

Before Linnaeus' publication (1753) on the binomial species C. arabica, the coffee type from Arabia was the only one known and described by a short sentence in Latin (Jussieu, 1713): "Jasminum arabicum, lauri folio, cujus femen apudnos coffee deciur" ("Arab jasmine, with laurel type leaves, the beans of which we call coffee").

Other wild coffee species, listed below, were discovered during scientific surveys carried out since the 18th century in tropical regions of Africa and on the route to the East Indies. Initially these surveys were made on the islands (Bourbon, Zanzibar) and the more easily accessible coastal areas (Gold Coast, Mozambique). Later, as exploration developed, they were extended to the more central regions of Africa.

- The *C. mauritiana* species, on the Bourbon island (La Réunion), described by Lamarck (1783).
- The *C. racemosa* species collected in Mozambique by de Loureiro (1790).
- The *C. liberica* species found in Sierra Leone in 1792 and in Liberia in 1841.
- The Rio Nunez coffee type (*C. stenophylla*) and *C. liberica* found in Guinea Conakry in 1840.
- The Robusta coffee type and *C. congensis*, found in the basin of the Congo river between 1880 and 1900.

The first tropical flora published by Kew Gardens contains about 10 coffee species described by Hiern (1877). Botanical description of new coffee species, collected during the colonization of the African countries by the Europeans were made by:

- Froehner (1898) in a monograph on coffee and related species.
- Chevalier (1938, 1947) in a glossary on world coffee taxonomy.
- Lebrun (1941) on the coffee types from the former Belgian Congo (Democratic Republic of Congo).

The taxonomy of coffee species is based on the morphological description of the specimens held by the different herbaria. The different classifications made in the past are incomplete, sometimes confusing and partly obsolete as, for instance, "*Les Caféiers du Globe*" [15].

This is further illustrated by the recent reviews of coffee species taxonomy:

- Reclassification of the specimens held by the Museum National d'Histoire Naturelle (Paris) native species from Madagascar and Mauritius by Leroy [28, 29].
- Description of about 20 new taxa among specimens in the Kew Herbarium (London) from East Africa by Bridson [7].
- Description of new taxa native to the Congolese zone by Stoffelen [40].

The last and extensive synthesis made by Davis *et al.* [VIII] provides us an updated taxonomic conspectus of the genus *Coffea* with 103 species listed.

A few species of wild coffee are shown in Pict. 2.3 to Pict. 2.15.

Collectors have discovered new original taxa like *C. pseudozanguebariae* and *C. anthonyi*. The former was collected from four populations on the frontier region between Kenya and Tanzania [1].

The morphological characteristics of *C. pseudozanguebariae* are: thin leaf, short stipule, big domatia, six or seven lobed corolla, long fruit stalk and a purple black fruit color.

Its main originality is due to the absence of caffeine in the green coffee, a major trait specific to the wild species of Madagascar, Mascarenes, and comoros islands in fact, this caffeine-free species from continental East Africa as well as three species with caffeine from Madagascar were discovered only recently [35].

The second taxon was collected from populations on the border zone between south-east Cameroon and the North Congo.

Coffee trees of *C. anthonyi* are shrubs characterized by small leaves, red cherries and flowers like *C. eugenioides* with the same biochemical composition of green coffee (0.62 % DMB of caffeine and 4.65 % DMB of chlorogenic acids). The

Picture 2.3 *C. congensis* (Central African Republic). From: A. Charrier.

Picture 2.6 Fruit-bearing *C. liberica* var. *dewevrei*. From: A. Charrier.

Picture 2.4 *C. liberica* (Ivory Coast). From: A. Charrier.

Picture 2.7 *C. stenophylla* (Ivory Coast). From: A. Charrier.

Picture 2.8 *C. humilis* (Ivory Coast). From: A. Charrier.

Picture 2.5 *C. liberica* var. *dewevrei* (Central African Republic). From: A. Charrier.

Picture 2.9 *C. eugenioides* (Kenya). From: M. Lourd.

Picture 2.12 *C. racemosa* (Mozambique).
From: D. Le Pierres.

Picture 2.10 *C. sessiliflora* (Kenya, Tanzania).
From: A. Charrier.

Picture 2.13 *C. kapakata* (Angola).
From: J. Berthaud.

Picture 2.14 *C. farafanganensis* (Madagascar).
From: J. Vianey-Liaud.

Picture 2.11 *C. pseudozanguebariae*
(Kenya, Tanzania). From: A. Charrier.

Picture 2.15 *P. mannii* (Ivory Coast).
From: A. Charrier.

most remarkable characteristic of this new taxon is its self-compatibility.

In short, the number of inventoried *Coffea* species has increased regularly during the last two centuries to reach 103 species [VII]. Botanical descriptions of the existing specimens have yet to be completed on the basis of material collected during recent coffee surveys.

2.3
Genetic Diversity and Phylogeny

According to Davis *et al.* [VIII], their use should be avoided because subgeneric groups are based on weak morphological characterizations, *Coffea* sect. *Eucoffea* K. Schum. is an illegitimate name, *Coffea* sect. *Mascarocoffea* is invalid, as are all the series and subsections of Chevalier's classification without Latin diagnosis. We recommend now the current subgeneric classification in two subgenera.

The taxonomy identification should also be considered from the biological point of view. Genetic relationships between the coffee species have been established by the success rate of interspecific crosses, the meiotic behavior and the fertility of their interspecific hybrids [5, 9, 13, 31].

These studies lead to the major conclusion that all *Coffea* species share a common base genome, although noticeable differences exist between the various *Coffea* groups.

2.3.1
Genetic Diversity

In addition, genetic evaluation at intra-and interspecific levels has been carried out by using complementary methods such as enzymatic [1, 4] and biochemical markers [34, 35, IV]. In particular, a whole range of dif-ferent techniques has been used to detect polymorphism at the DNA level (2, 23, 25), including randomly amplified poly-morphic DNA (RAPD), cleaved amplified polymorphisms (CAP), restriction fragment length polymorphisms (RFLP), amplified fragment length polymorphism (AFLP), inverse sequence-tagged repeat (ISR) and simple sequence repeats or microsatellites (SSR) [XVIII]. During the last few years, the number of co-dominant markers has been considerably increased by SSR mining in coffee EST databases [I–III–XXVIII] offering new possibilities for genetic analysis thanks to their transfer-ability [VII–XXVII].

Single Nucleotide Polymorphisms (SNP) were analyzed in a few genes of two different biosynthetic pathways- sucrose metabolism in coffee fruits and lipid metabolism in coffee bean [XIX].

The classification of coffee trees and their structuring into genetic groups has been established after cluster and discriminate analysis of the DNA marker data and comparison with taxonomic groups.

Some species, like *C. canephora* (see below) or *C. stenophylla*, are polymorphic with discontinuities of morphological types and specific alleles without genetic barriers between groups. But in the case of *C. liberica*, morphological and molecular differentiation in two types (var. liberica and var dewevrei) is associated with a partial reproductive barrier [XXV]. Other coffee species, like ones from Madagascar or *C. pseudozanguebariae*, are monomorphic because of their low genetic variability. These genetic structures would appear to be associated to their modes of reproduction and the dynamics of the wild coffee populations, with a low number of reproductive individuals and spatial isolation in small groups scattered in the forest zones.

Self-incompatibility is normally observed in the diploid coffee species, in relation with a gametophytic system of incompatibility controlled by one multiallelic gene S for *C. canephora* [4]. However, notable exceptions, which have led to self-compatibility, should be noticed:

1) first of all, in the unique and well known tetraploid species, *C. arabica*
2) in two of the diploid species belonging to the sub-genus *Coffea, C. anthonyi* and *C. heterocalyx*
3) in the diploid species belonging to the Psilanthus genus, which have a floral structure favorable to self-pollination

Due to these reproductive mating conditions, the effects of inbreeding and genetic drift have led to the fixation of characteristics used to describe typical populations, without any genetic barriers.

2.3.2
Phylogenies

Phylogenies can be built from both chloroplast and nuclear genome markers. The low frequency of structural changes in the chloroplast DNA (cpDNA) has generated numerous conservative markers. In addition, its exclusively maternal inheritance has been observed in different coffee progenies. cpDNA variation present in the subgenus *Coffea* has been assessed by restriction fragment length polymorphism (RFLP) on both the total chloroplast genome and the atpB–rbcL intergenic region and by sequencing the trnL–trnF intergenic spacer. Phylogenetic relationships of *Coffea* species inferred from cpDNA variation are presented in Fig. 2.4 [18].

Results have confirmed a monophyletic origin of the subgenus *Coffea* species; several clades have been revealed according to the classical biogeographical distribution (West and Central Africa, Central Africa, East Africa and Madagascar, respectively). The low level of cpDNA variation exhibited by coffee species is probably related to the recent origin of the genus *Coffea*. Furthermore, the low number of characteristics supporting the respective branching and the low rate of homoplasy (i.e. the same characteristics not inherited from a common ancestor) suggest a rapid and radial mode of speciation for the primary clades.

The nuclear ribosomal DNA (rDNA) units consist of the three coding regions separated by intergenic spacers. These internal transcribed spacers (ITS) evolve more rapidly than the coding regions, and appear useful for assessing relationships among the closely related and highly diversified taxa in the genus *Coffea*. The sequences of the ITS2 region for 26 *Coffea* taxa and three *Psilanthus* species have been established [26]. The rDNA sequence divergence reveals a number of genetic groups with the same and consistent eco-geographical distribution.

Moreover, a recent study [XXII] based on a larger data set confirmed the different lineages within *Coffea* and their consistency with major biogeographical regions. It also supported the recognition of a single genus (i.e. *Coffea*) involving both *Coffea* and *Psilanthus* species, with a morphological and molecular support for *Coffea* plus *Psilanthus* and low sequence diversity between the two genera, but failed to resolve the issue of paraphyly vs monophyly for *Coffea* [IX].

Finally, these phylogenetic trees do not support the present classification of coffee tree taxa into two separate genera – *Coffea* and *Psilanthus*. Recently, intergeneric hybridization has even been achieved.

Further critical work is still needed to resolve this problem of generic delimitation at the *Coffaeae* tribe level [IX].

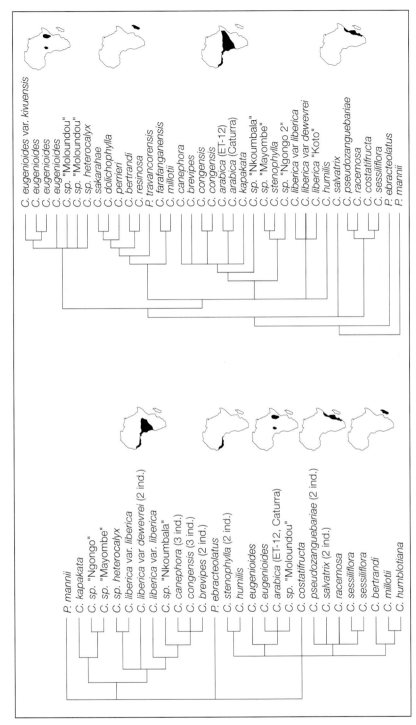

Figure 2.4 Phylogenetic classification of coffee taxa based on chloroplast DNA variation [17] and on nuclear ribosomal DNA [25], and on geographical distribution of the main clusters.

2.3.3
Origin of C. arabica

Earlier attempts to determine the genetic origin of *C. arabica* by cytogenetic studies and interspecific crosses have revealed a segmental allotetraploid origin; *C. eugenioides* and *C. canephora* have often been assumed to be the ancestral diploid parents of *C. arabica*.

Research into the origins of *C. arabica* can be now based on the results of the different DNA sequence evolution studies. cpDNA from *C. arabica* appears to be similar to cpDNA from *C. eugenioides* and *C. anthonyi* [18]. On the contrary, the ITS2 region from the ribosomal nuclear DNA of *C. arabica* diverged markedly from the sequences of those two taxa, but appeared to be almost identical to the sequences of *C. congensis* and *C. canephora* [26]. The present results strongly support the hypothesis of an allotetraploid origin of *C. arabica* involving a species close to *C. eugenioides* as the maternal progenitor and a species close to the canephoroid group as the paternal progenitor [XVI]. In addition, the very low divergence between the DNA sequences of the putative progenitors and *C. arabica* clearly indicate that the formation and speciation of *C. arabica* are recent events which are very likely to have occurred during the late Quaternary period.

2.4
Coffee genome

2.4.1
Cytotaxonomy

The basic chromosome number for the genus Coffea is considered to be n = 11, which is typical for most genera of the family *Rubiaceae*. Coffee somatic chromosomes are relatively small (1.5 to 3 µm) and morphologically similar to each other [V]. Observations at meiotic pachytene phase provided a significantly better chromosomal characterisation and allowed the identification of most bivalents of *C. Arabica*. The characterisation of the longitudinal differentiation of the mitotic chromosomes of coffee has progressed using other techniques such as fluorescent banding [XX].

In addition, the fluorescence in situ hybridization (FISH) method that opened up new perspectives. In particular, genomic In situ hybridization (GISH) was successfully applied for characterization of genomes and chromosomes in polyploid, hybrid plants and recombinant breeding lines. The genome organization of *C. arabica* was confirmed by GISH using simultaneously labeled total genomic DNA from the two putative genome donor species as probes [XVI]. Furthermore, FISH method was used to study the presence of alien chromatin in interspecific hybrids and plants derived from interspecific hybrids between coffee species [XIII]. More recently, BAC-FISH procedure was used to rapidly localize a given introgression on a specific chromosome [XIII].

As coffee trees belonging to the genus *Coffea* that contains approximately 100 taxa have the same number of chromosomes, variation in nuclear DNA content (qDNA) is another approach to estimate the genome size. Here, again, *C. arabica* is the exception. Flow cytometry, applied with success to coffee species [17, XXVI], is a rapid technique to estimate qDNA. Nuclei are extracted from leaves and nuclear DNA labeled by fluorochromes (propidium iodine). The genome size of the unique tetraploid species *C. arabica*, 2C = 2.61 pg, is approximately double that of the other coffee species. For the 14 diploid African species, qDNA ranges from 2C = 0.95 pg in *C. racemosa* to 1.78 pg in *C. humilis*.

Their genome sizes are distributed in three groups linked to the eco-geographic distribution and the phenology of coffee species (Fig 2.5):

- Group 1 (0.90±1.30 pg): *C. sessiliflora, C. racemosa* and *C. pseudozanguebariae* from East Africa have a short fruiting cycle.
- Group 2 (1.30±1.60 pg): *C. congensis, C. canephora, C. brevipes, C. eugenioides* and *C. liberica* from West and Central Africa.
- Group 3 (up to 1.60 pg): *C. humilis* from Ivory Coast.

2.4.2
Genetic maps

Several genetic maps have been constructed mainly for the 2 cultivated species [XXIII] in relation with the QTL heredity and breeding programmes (see chapters 3).

The low polymorphism has been a major drawback for developing genetic maps of the *C. arabica* genome. Nevertheless, a genetic map from a cross between Catimor and Mokka cultivars has been obtained in Hawaÿ. Furthermore, to overcome this limitation, efforts were directed to the development of genetic maps in *C. canephora* or interspecific crosses.

The earliest attempt to develop a linkage map was based on *C. canephora* doubled haploid (DH) segregating populations [XVII] and comprised more than 40 specific sequence-tagged site markers, either single-copy RFLP probes or microsatellites that are distributed on the eleven linkage groups. These markers constituted an initial set of standard landmarkers of the coffee genome which has been used as anchor points for map comparison and coverage analysis of bacterial artificial chromosome (BAC) libraries.

Species	DNA content (pg)
C. racemosa	1.035 A
C. pocsii	1.083 B
C. sessiliflora	1.109 C
C. pseudozanguebariae	1.131 CD
C. costatifructa	1.150 D
C. salvatrix	1.221 E
C. stenophylla	1.286 F
C. kapakata[*]	1.323
C. eugenioides	1.364 G
C. liberica var. Dewevrei	1.406 H
C. canephora RCA	1.440 I
C. congensis CR	1.478 J
C. liberica Koto	1.511 K
C. brevipes	1.523 K
C. heterocalyx	1.737 L
C. humilis	1.764 L

Letters indicate multiple mean comparison results using the Newman-Keuls test
[*] Represented by only one tree

Figure 2.5 Nuclear DNA content in African diploid coffee species. (From Noirot *et al.* 2003).

More recently, Crouzillat *et al.* [VI] reported the development of a *C. canephora* consensus genetic map based on a segregating population from the cross of two highly heterozygous parents (BP409 and Q121). More than 453 molecular markers such as RFLP and microsatellites were mapped covering a genome of 1258 cM. Recently, this map was used to map COS markers and perform comparative mapping between coffee and tomato [XXXIV].

In parallel several diploid interspecific maps were built. Those maps are based on either F_1 hybrid population resulting from cross between coffee diploid species *C. liberica x C. eugenioides* or progenies obtained by backcrossing of hybrid plants to one of the parental species *C. heterocalyx x C. canephora* and *C liberica x C. canephora* [XXIII].

2.4.3
Genomic resources

Recent advances in coffee genomics consist of bacterial artificial chromosome (BAC) libraries and a huge collection of expressed sequence tags (ESTs) [XXIII].

An Arabica BAC library using the cultivar IAPAR 59 was successfully constructed and validated. In total, the library contains 88,813 clones with an average insert size of 130 kb, and represents approximately 8 *C. arabica* dihaploid genome equivalents.

A Canephora BAC library was developed on a genotype (i.e. clone 126). The library contains 55,296 clones, with an average insert size of 135 Kb, therefore representing almost 9 haploid genome equivalents.

Recently, a large set (i.e. 46,914) of coffee expressed sequence tag (EST) with a special focus on developing seeds of *C. canephora* has also been released. In addition, two other *C. canephora* EST sequence sets were developed from mRNA isolated from leaves and fruits at different development and maturation stages with a total of 8,778 valid EST sequences. All together, a significant set of 55,692 ESTs mainly from fruits is already publicly available for *C. canephora*.

Regarding *C. arabica*, public accessibility to EST collections is limited to 1,226 ESTs. In total, there are less than 60 thousand entries of coffee ESTs, publicly available, deposited at the dbEST database.

However, other research groups involved in molecular genetics and genomics of *Coffea sp.* have also generated EST data that will become available to the coffee scientific community in the near future. The Brazilian Coffee EST Project is such an example, which generated single-pass sequences of a total of 214,964 randomly picked clones from 37 cDNA libraries of *C. arabica*, *C. canephora* and *C. racemosa*, representing specific stages of cells and plant development that after trimming resulted in 130,792, 12,805 and 10,510 good quality sequences for each species, respectively.

Coffee EST resources have also been developed by the Cenicafe research group in Colombia, which have in their database to date, 32,000 coffee EST sequences from 22 libraries organized in 9,257 *C. arabica* and 1,239 *C. liberica* unigenes.

Aiming at the development of EST-SSR markers for coffee an Indian research group reported an interim set of 2,092 ESTs of coffee generated at CCMB, Hyderabad, India.

EST sequences of two cDNA libraries from leaves and embryonic roots of *C. arabica* were also produced by an Italian group, to develop a cDNA microarray based on 1,587 non-redundant sequences.

2.5
Cultivated Coffee Populations

2.5.1
Cultivated Species

The *C. arabica* species is generally adapted to the tropical highlands (600–1500 m above sea level), whereas *C. canephora* is adapted to tropical lowlands (0–800 m above sea level). Arabica coffee is milder, more aromatic and contains less caffeine than Robusta coffee. Due to its autogamous nature, diversity within lines or within varieties of *C. arabica* is often limited. However, considerable diversity can be observed between *C. arabica* varieties, such as differences in adaptation capacity, plant height and shape, leaf size and shape, internode length, fruit color or shape, disease resistance, and yielding capacity. This diversity naturally represents factors of significant economic importance.

C. canephora is, genetically, a highly diverse species. In nature it is widely distributed in the wet lowland areas of West and Central Africa. This factor has favored differences in adaptation capacity and diversification of the species. Due to its strict allogamous nature (only outcrossing occurs), each plant can be considered as a unique genotype. These differences, as will be described later, have a direct bearing on the breeding methods, which would be suitable to these two species.

C. liberica is only cultivated in certain areas in Africa (Guinea) and South-East Asia (Malaysia, the Philippines and Vietnam). This species is highly diverse, and includes many varieties like Dewevrei, Excelsa, Liberica and Dybowskii. Some of these varieties are very drought resistant. The outstanding size of the trees (5–10 m) makes cultivation difficult and the large quantities of pulp cause processing problems, added to which its cup quality is generally considered inferior to Robusta.

2.5.2
C. arabica Populations

Pictures 2.17 to 2.26 illustrate a number of typical *Coffea arabica* populations.

2.5.2.1 Spontaneous and Subspontaneous Accessions

These accessions, originating from Ethiopia, Sudan or northern Kenya, are seed progenies collected from one or more trees. Considerable variations in characteristics can be found between but also within these accessions. Accessions from Ethiopia, like Kaffa, Geisha, Tafari-Kela, Ennarea, Gimma, Dalle and Dilla, introduced into Kenya, Kivu, Tanzania and India between

1930 and 1955, were considered to be quite productive and/or resistant to Coffee Leaf Rust (CLR). Other collections in Ethiopia were made by individual scientists, like Archer (1951), Lejeune (1954–56) and Sylvain (1952–54).

The early introductions from Ethiopia have been assessed chiefly in collections and have been little used in variety trials. Some exceptions are the Tafari-Kela variety, selected in India as being quite tolerant to coffee leaf rust, Harar accessions, which although generally susceptible to diseases, have proved to be quite productive in Brazil and in East Africa as well as Geisha (Picture 2.23), which appeared quite to be productive in Costa Rica and Tanzania.

Some materials were also collected on the Boma Plateau in Sudan, by Thomas in the 1940's (Rume Sudan, Barbuk Sudan) and on Mount Marsabit in Kenya, by IRD in 1977. The material from Sudan

Picture 2.16 Dispersion of *C. arabica* from Ethiopia and Yemen. From: J. Berthaud.

Picture 2.17 *C. arabica* cv. Ayasa (Yemen). From: A. B. Eskes.

Picture 2.18 *C. arabica* cv. Essaii (Yemen). From: A. B. Eskes.

Picture 2.21 Purpurascens mutant (purple leaves). From: A. Charrier.

Picture 2.19 CBD susceptible Caturra (Cameroon). From: A. Charrier.

Picture 2.22 Goiaba mutant. From: A. Charrier.

Picture 2.20 Caturra mutant (short internodes) used in a high-density plantation (Costa Rica). From: A. Charrier.

Picture 2.23 Geisha cultivar. From: J. P. Labouisse.

Picture 2.24 Hibrido de Timor resistant to *Hemileia vastatrix. From:* A. Charrier.

Picture 2.25 Cherries of *C. arabusta. From:* J. Berthaud.

Picture 2.26 Variety "Colombia" derived from Catimor population (Colombia). *From:* A. Charrier.

(Rume Sudan) appears to offer significant resistance to CBD and also contains an unidentified source of resistance to leaf rust.

After the FAO expedition (1964), a large collection (about 400 accessions) is maintained in Brazil, Costa,Rica, India and Tanzania. Field-testing of 200 accessions from this collection in Brazil generally revealed bad adaptation (especially Tepi origins) and productivity levels far below that of Brazilian cultivars. A major discovery made by IAC (Brazil) concerns the screening of a natural caffeine free coffee in the progenies of the FAO germplasm. Three plants almost completely free of caffeine of 3,000 Ethiopian *C. arabica* plants have been characterized; this trait seems to be recessive and open the door to this transfer to highly productive cultivars [XXX].

The IRD/IFCC collection (1967) of approximately 70 accessions has been introduced in Madagascar, Kenya, Ivory Coast, Cameroon and Costa Rica (CATIE). Its characteristics were first described in field collections at Ilaka-Est (Madagascar), Foumbot (Cameroon) and Mont Tonkoui (Ivory Coast) [13] Recent studies carried out in Cameroon, France (CIRAD), Kenya and Central America showed that this collection contains good sources of resistance to nematodes (*Meloidogyne incognita*), CBD and CLR. Morphological and agronomical traits of 148 accessions studied in Cameroon led to the differentiation of two main pheneotypic groups in relation with their geographical origins from Kaffa and Illubabor (south-west) vs Sidamo (south) and Harerge (east) in Ethiopia [XXIII] in accordance with genetic structuration.

In the 1970's, a mass selection programme was carried out in Ethiopian coffee populations, with a view to discovering CBD resistant accessions. The large number of accessions in the JARC field gene bank is being under observation for agronomic characteristics (diseases, yield, drought tolerance, coffee quality) Promising landraces selections and resistant lines are currently evaluated [II].

2.5.2.2 Accessions from Yemen

Early accessions from Yemen have become commercial varieties all around the world (Typica and Bourbon types) see picture 2.16. The relatively valid agronomic traits (yield potential, growth habit) of this material can be explained by the selection carried out by farmers. However, these varieties, with the exception of a few Typica accessions that contain some resistance to CBD, have proved to be generally susceptible to diseases and pests. Fixed Typica varieties generally have bronze-colored young leaves and an almost horizontal branching pattern, whereas the young branches of Bourbon are at an angle of about 60° to the main stem. Bourbon types may have green or bronze-colored young leaves. Ethiopian accessions have a more variable branching pattern and color in young leaves.

The leaves of Bourbon are larger, the cherries more round, and the branches thicker and less flexible than those of Typica. The accessions from Yemen normally display a single stem growth habit, whereas many Ethiopian lines have a tendency to form multiple stems. A recent survey carried out by IPGRI/CIRAD in South Yemen identified six local varieties, which often showed Typica characteristics. Similarly, however, many intermediate types were also found. These accessions were introduced into collections in Brazil, Colombia and Costa Rica.

2.5.2.3 Genetic Structure (Fig. 2.6)

As previously reported, low enzymatic variability has been detected in *C. arabica*. Genetic diversity for RAPD, AFLP and SSR markers between cultivated and wild accessions of *C. arabica* appeared to be effective in resolving the genetic variation in *C. arabica* species [2, 25 XXXI]. These studies indicated a relatively large genetic diversity within the *C. arabica* germplasm collection as with the studies carried out for botanical and morphological characteristics. As could be expected from the historical origin of *C. arabica* cultivars, Bourbon and Typica types showed important differences.

In addition, comparison of *C. arabica* populations throughout forests of Ethiopia using ISSRs [XXXII] reveals complex geographical patterns of genetic diversity and are genetically different from semi-domesticated plants (landraces). At a fine-scale, in forests of Berhane Kontir and Yayu (Geba Dogi), they have studied the spatial distribution of ISSR genotypes. Microsatellites are needed to assess heterozygosity and gene flow in these populations.

However, it has not been possible to distinguish the cultivars belonging to the same type, both Bourbon and Typica. A clear separation was observed between the Ethiopian germplasm collected in the south-west highlands of Ethiopia (Illubabor and Kaffa provinces) and the *C. arabica* cultivars spread worldwide from Yemen. This result supports the hypothesis that the *C. arabica* plants transferred to Yemen and selected in Yemen for cultivation by the Arabs could have originated in different regions of Ethiopia. Such differentiation can be estimated by molecular marker-based genetic distance measures and may explain the large heterosis effect which has been noted in F_1 hybrids resulting from crosses between Ethiopian acces-

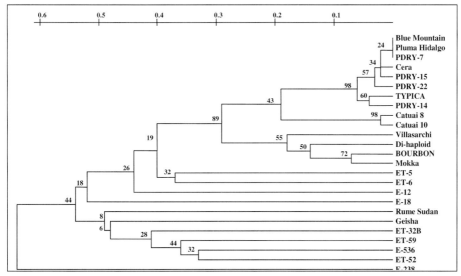

Figure 2.6 Dendrogram generated after UPGMA using AFLP-based genetic distance between 25 accessions of *C. arabica* [2].

sions and various cultivars in Brazil, Costa Rica and Ethiopia.

2.5.2.4 Mutants of Commercial Interest

Within the widely planted homogeneous Arabica varieties, variants have appeared which have attracted the attention of growers and researchers. Genetic analyses, mainly carried out in Brazil [11], have indicated that most of these variants were based on the mutation of one gene. Although over 40 mutants have been found, only a few are of commercial interest. The most important mutants are those affecting plant size, fruit shape and color, and leaf color. The color of young leaf tips may vary from deep bronze to light green. The bronze condition is due to an incomplete dominant gene. As observed in East Africa, bronze leaves may condition higher resistance to "dieback" and cold damage.

"Caturra" was found in 1937 as a mutant of the Bourbon variety in Brazil. It showed high early yield potential, but suffers from severe overbearing under Brazilian growing conditions. The short internodes of Caturra are based on a single dominant gene. There are yellow-fruited and red-fruited variants of Caturra (Caturra Amarelo and Caturra Vermelho), which are similar in yield capacity [12]. The Caturra variety proved to be well adapted to Colombia and Central America, where it has been used to develop high-density coffee-growing practices.

The "Caturra" gene for dwarfness was also identified in the Villa Sarchi variety from Costa Rica and the Pacas variety from El Salvador. Other mutants for dwarfness have been identified, each carrying different dominant genes: San Bernardo (or Pache) from Guatemala, and Villa Lobos and San Ramon from Costa Rica [12]. Recently, mutants for extreme dwarfness were identified in Guatemala (Pache Enano) and Mexico.

"Yellow"-fruited mutants, like Yellow Caturra, have been observed in a number of countries, the yellow color being due to one recessive gene. The color of the fruit does not affect yield in itself, but yellow-fruited varieties may tend to ripen earlier [11]. In some countries yellow-fruited varieties are avoided as they reputedly influence quality (the degree of ripeness is less well identified by the pickers) and because ripe yellow cherries tend to drop earlier than red cherries.

"Maragogipe" is a mutant detected in Brazil within the Typica variety. It has large cherries, long, slightly twisted seeds, long internodes and large leaves, often with twisted points. It has been widely distributed to many countries. The mutation is due to one dominant gene, which apparently has an indirect depressive effect on yield. Inevitably, it has been little used commercially [11]. In Central America, crosses between Maragogipe and Caturra are presently being selected. There appears to be a market for the large beans, which is related to appearance, rather than to flavor.

The "Laurina" or "Bourbon Pointu" variety originates from the Reunion Islands. Laurina plants have a dwarf conical plant shape with dense lateral branching, short internodes, short elliptic leaves and seeds which are pointed at the top. This mutant is due to one recessive gene in a Bourbon genetic background. This factor furthermore induces low caffeine content (about 0.6%). The yield of Laurina is low and the plant is highly susceptible to CLR. Laurina segregants obtained from crosses of Laurina with other varieties can be more productive [11].

The "Mocha" variety of Arabica, which also apparently originated from the Reunion Islands, has a dwarf conical plant shape with small internodes and leaves, and round cherries containing small round seeds. This type of Mocha variety is rather unproductive and very susceptible to CLR. Its characteristics are due to two genes – the recessive "Laurina" gene and the incomplete recessive "Mocha" gene. The mocha characteristic has been tentatively related to a special flavor quality, a trait that needs further confirmation. In germplasm collections, other varieties exist with similar names, like Erythrean Moka (or Mocha). It has not yet been ascertained whether these varieties contain the mocha gene.

"Semperflorens" is a recessive mutant, with a Bourbon genetic background, which flowers throughout the year (insensitive to photoperiodism). It was found in Brazil in 1934, but the same characteristic is present in Erythrean Moka [11]. The distinct flowering habit may be of some commercial value in marginal growing areas. Many other mutants have been observed in *C. arabica* generally affecting the leaf or fruit shape or color [11]. The best known are possibly "Purpurascens" (purple-colored leaves) and "Goiaba" (calyx retained until 5–7 months after flowering). The latter trait may reveal a promising tendency to delay attacks by the berry borer (Decazy, personal observations). A recently detected mutant in Costa Rica proves to have a longer maturation time, but a poor cup quality (Caturra "Lerdo").

Another economically important trait is genetic "male sterility". A few male sterile plants in Arabica have been found in the collection of the Instituto Agronomico of Campinas in Brazil around 1985. These plants showed absence of pollen grains, due to the early degeneration of the tapetum. This characteristic has appeared to be stable over the years. Crosses with local cultivars have resulted in normal male fertile plants. Consequently, the male sterile condition would seem to be recessive. Recently, a few male sterile plants have also been identified in Ethiopian accessions in the CATIE collection in Costa Rica. This trait may be exploited in the production of F_1 hybrid varieties in *C. arabica*.

2.5.2.5 Natural Crosses with Diploid Species

After the epidemic of orange leaf rust in the *C. arabica* plantations of Ceylon and of Java in the second part of 19th century, other coffee species showing higher tolerance to the disease were introduced. These introductions led to the appearance of natural interspecific hybrids between local cultivars of *C. arabica* and imported diploid coffee species growing together in the same botanical garden or plantation. Despite the cytological barriers, some of the hybrids were quite valuable at the tetraploid level and resistant to leaf rust.

The most famous examples were [16]:

1) Hybrids between *C. liberica* and *C. arabica*, i.e. Kalimas and Kawisari in Indonesia and S26 from the Doobla Estate in India. Some derivatives were the basis of the breeding programme of CRI of Balehonnur, India [39]. By hybridization with a Kent Arabica cultivar, Indian breeders obtained the well-known S795 family, which also carried the SH3 resistant gene from *C. liberica* and was developed in cultivation in 1947 [36]. A further hybrid found in Brazil gave origin to the population C387.

2) Hybrids between *C. canephora* and *C. arabica* named Arla in Indonesia, Devamachy in India and Hibrido de Timor (HdT) in Timor. HdT that has shown resistance to all known races of *H. vastatrix* began to replace the local Arabica in Timor island in the 1950s [38].

Crosses between *C. arabica* var. *caturra* and HdT generated a population called Catimor. These hybrids were first created by CIFC (Oeiras, Portugal) and selected in different American countries (Brazil, Colombia, Costa Rica, etc.) as well as in Kenya and in India.

Recently, other hybrid populations between *C. arabica* and *C. canephora* have been discovered and collected in mixed plantations on the island of New Caledonia. Investigation of these spontaneous interspecific hybrids by AFLP and SSR revealed an exceptional genetic diversity in that sympatric zone in New Caledonia [XXI].

Intense studies on artificial tetraploid hybrids between *C. arabica* and induced autotetraploid *C. canephora* or other diploid coffee species have been carried out to transfer traits of interest into Arabica cultivars. The standard type of first generation hybrid was designated *C. arabusta* [8], The agronomic value of the Arabusta hybrids is still limited in fertility and productivity. On the other hand, successive generations obtained by repeated backcrossing or by open pollination have generated different introgressed Arabica populations as:

- Icatu population (*C. canephora x C. arabica*) in Brazil and Colombia variety in Colombia with resistance to CLR;
- GCA population in Madagascar (*C. eugenioides x C. canephora*) x *C. arabica* a new low-caffeine hybrid.
- *C. racemosa x arabica* exhibiting different resistance levels to leaf miner in Brazil.

2.5.3 *C. canephora* Populations

Coffea canephora and *Congusta hybrides* population are shown in Pict. 2.26 to Pict. 2.32.

2.5.3.1 Spontaneous Populations

Spontaneous populations of *C. canephora* can be found in humid tropical lowland areas in West and Central Africa. They occur as rather small isolated groups of plants of varying age presenting relatively little variation within the groups [4]. In order to avoid the loss of original populations due to advanced deforestation, collections were organized by the French institutes ORSTOM and CIRAD in the 1970s and 1980s in Ivory Coast, Guinea, Cameroon, Congo and the Central African Republic. During these surveys, subspontaneous populations, like the Nana coffee type from the Central African Republic, were also collected. The collected material consists of more than 1000 genotypes, which were planted in observation plots in their countries of origin as well as in living collections at Man and Divo in Ivory Coast. Most of these populations present little direct agronomic value, but can be of great interest to the breeder for specific characteristics, like disease or pest resistance, technological qualities, tree shape and genetic origin.

Studies based on isozyme profiles carried out on wild *C. canephora* coffees revealed that these populations can be divided into two main groups – the "Guinean" group from West Africa (Ivory Coast and Guinea) and the "Congolese" group from Central African countries. These two groups have following phenotypic characteristics shown in Fig. 2.7a, 2.7b and 2.7c.

• Guinean genotypes possess relatively small, elongated leaves, rather short internodes, small beans (often less than 11 g/100 dry beans), high caffeine content (superior to 2.7 %) and a strong secondary branching pattern. The majority of plants is susceptible to local races of the coffee rust fungus (*H. vastatrix*) but many Guinean genotypes present tolerance to drought.

• Congolese genotypes have larger and broader leaves, long internodes, generally larger bean weight (often heavier than 13 g/100 beans), contain less caffeine (average of 2.5 %), and show little or no secondary branching and dying-off of older branches. Most plants have high resistance to CLR, but are susceptible to drought.

Congolese coffee types have been planted since the beginning of *C. canephora* cultivation and generally show better agronomic value than Guinean types. Both types are well presented in the collections in Ivory Coast; however, other countries possess nearly exclusively Congolese coffee types in their collections. An extensive study of the *C. canephora* populations in field collections in Ivory Coast based on morphological characteristics and molecular markers have led to the recognition in the Congolese genotypes of two and then four (SG1, SG2, B and C) genetic subgroups (23). Ugandan *C. canephora* genotypes of forest and cultivated origins were recently analysed using 24 SSR, in comparison with known diversity groups. A new group (Uganda) was identified, constituting the most eastern located genetic group [XXIV].

Global work on the genetic diversity of *C. canephora* including Ugandan populations based on SNP and SSR markers pointed out genetic relationships between subgroups [XIX]. Fst values for SSR's markers are indicators of their differentiation (fig Leroy). Apart for Guinean genotypes, clearly separated for both markers, all the other groups, including Ugandan group, have the same genetic background from which they discriminated since the past glaciations period, from SG1 and SG2 re-

Picture 2.27 Guinean coffee trees (Ivory Coast). From: A. Charrier.

Picture 2.28 Guinean type: small cherries with high caffeine content. From: A. Charrier.

Picture 2.30 Selected clone originated from hybrid progeny (Congolese × Guinean) with fruit bearing. From: A. Charrier.

Picture 2.29 Congolese coffee tree. From: J. Berthaud.

Picture 2.31 Purpurascens mutant. From: A. Charrier.

Picture 2.32 Vigorous and productive Congusta hybrids. From: A. Charrier.

Picture 2.33 Fruit-bearing Congusta: good bean size and cup quality. From: A. Charrier.

gions. Differentiation between groups is then mainly based on differences in allelic frequencies.

Genetic diversity among 40 accessions of robusta coffee genepool available in India was determined in comparison with 14 representative samples from a *C. canephora* core collection, using AFLP and 12 SSR markers [XXIX]. Accessions from the Indian gene bank tend to show good amount of genetic diversity but this diversity is incomplete present in comparison with core samples.

The value of Guinean genotypes for *C. canephora* breeding was stressed by Berthaud [4], who demonstrated that the most productive clones of *C. canephora* selected in Ivory Coast in the 1960s were in fact hybrids between Congolese and Guinean types. Hence, the breeding of *C. canephora* in Ivory Coast since 1984 has chiefly been based on a reciprocal recurrent selection strategy by hybridization between Congolese and Guinean genotypes [33].

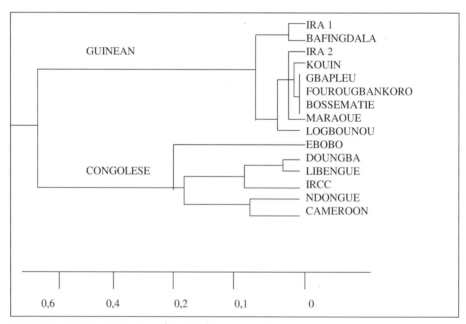

Figure 2.7a Genetic organization of *C. canephora* species [4].

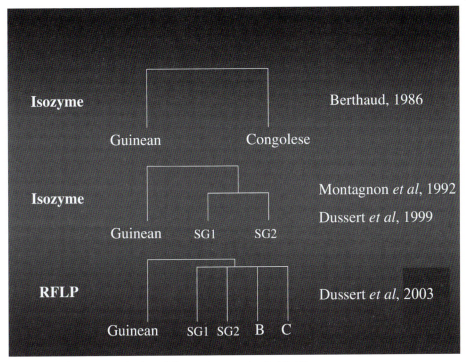

Figure 2.7b Genetic organization of C. *canephora* species [4, 22, 32].

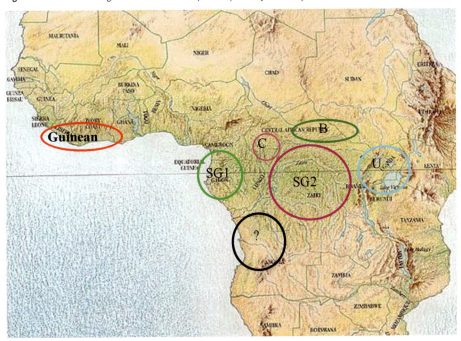

Figure 2.7c Geographical distribution of the C. *canephora* gene pools ([23, XXIV], from T. Leroy).

2.5.3.2 **Traditionally Cultivated Populations**
Over 80 % of the world's Robusta coffee is
still produced by traditional open-pollinated
varieties. The significant diversity present
in such varieties has been successfully
used for clonal selection by coffee breeders.

C. canephora populations from different
origins were introduced into Indonesia
around 1900. The locally selected Robusta
variety (or Laurentii) apparently originates
from a small plantation in the Democratic
Republic of Congo. This population displayed good yielding capacity, relatively
large beans and good resistance to leaf
rust. It was, therefore, selected as the
main variety planted in Indonesia [16].
Other C. canephora types introduced from
Africa are "Quillou" from Gabon, "Canephora" from different places in Central
Africa and "Ugandae" or "Bukobensis"
from Uganda. In Indonesia, the Quillou
(or Kouillou) as well as "Canephora" types
displayed greater susceptibility to leaf rust
than Robusta and often have bronze-colored
leaf tips, whereas Robusta only has green-
colored leaf tips. The Ugandae type shows
a more open growth habit with little secondary branching and short internodes [16].

At present, cultivated non-selected *C.*
canephora populations in Africa can be
mainly classified as Robusta and Kouillou
types However, local, spontaneous coffees
like "Maclaudi" and "Gamé" in Guinea,
"Niaouli" in Togo and Benin, the "Nana"
coffee type in the Central African Republic
and several local types (Nganda, Erect) in
Uganda have also been used commercially.
Local varieties have been intercrossed with
introduced varieties with the result that,
nowadays, it is often difficult to identify distinct varieties. However, some specific features of the traditional local varieties, like
the reputation of quality of the Gamé variety
as well as the drought resistance and the
short internodes of the Nana variety, can
be of interest.

In Brazil, the main cultivated C. canephora
variety introduced from Africa in 1900 is
called "Conillon" (Kouillou type, ortho-
graphic distortion). This variety displays
considerable diversity in growth habits,
bean size and susceptibility to leaf rust [10].

In C. canephora only a few mutants have
been observed so far, e.g. a mocha type and
purpurascens [16]. It may be of interest to
search for dwarf types in C. canephora
that would lend themselves to more intensive cultivation systems.

2.5.3.3 **Natural crosses with diploid species**
When coffee breeding started in Java,
Dutch agronomists studied several natural
interspecific hybrids [16]. A spontaneous
diploid hybrid between C. canephora var.
ugandae and C. congensis, called Congusta
or Conuga, had a certain amount of success in cultivation. These fertile hybrids
possess some interesting agronomic characteristics like vigor and productivity, adaptation to sandy soils, tolerance to temporary water-logging, and good bean size and
cup quality. Breeding of Congusta hybrids
has produced clonal varieties in Madagascar (hybrids HA, HB, H865, etc.) and in
India that are as productive as those of C.
canephora.

Recent plant collecting carried out by
ORSTOM in Africa has recreated a source
of C. congensis germplasm from Cameroon,
Congo and Central African Republic. A
program of controlled crosses between C.
canephora and C. *congensis* has been initiated in Ivory Coast to renew Congusta
populations.

2.6
Conservation

Different conservation means are shown in Pict. 2.34 to 2.41.

Coffee seeds are characterized by their very short life span in the hydrated state (1–6 months). *C. arabica* seeds can withstand desiccation down to 0.06–0.08 g H_2O/g dry weight, but the various coffee species show differences in desiccation sensitivity [22]. Coffee seeds cannot be considered to be standard and stored in seed banks because they remain cold sensitive, and desiccation does not increase their longevity. *C. liberica* seeds show typical recalcitrant storage behavior, whereas *C. arabica* seeds show "intermediate" storage behavior. See a recent review concerning coffee seed physiology providing information on recent findings on seed development, germination, storage and longevity [XI].

2.6.1
Coffee Gene Banks

In such cases, *ex situ* conservation is usually achieved in field gene banks:

- Important collections of *C. arabica* with material from the Ethiopian centre of genetic Diversity can be found at Jimma and Choche (Ethiopia), Turrialba (Costa Rica), Campinas (Brazil), Chinchina (Colombia), Lyamungu (Tanzania), Ruiru (Kenya), Foumbot (Cameroon), Man (Ivory Coast) and Ilaka-Est (Madagascar).
- A unique collection of over 50 wild species from Madagascar, Comoro, La Réunion and Mauritius islands is maintained at Kianjavato (Madagascar) with a representation of no less than 700 accessions.
- The main African coffee species are kept in the field collections located near Divo

and Man in Ivory Coast, with a representation of 8000 accessions of over 20 species collected from natural populations; moreover, working collections of *C. canephora* (and *C. arabica*) contain very valuable material in Ivory Coast, Togo, Cameroon, Angola, democratic Republic of Congo, India and certain countries of South-East Asia.

This type of conservation of a part of the biological diversity outside of their natural

Picture 2.34 Field gene bank of *C. arabica* on Foumbot (Cameroun). From: A. Charrier.

Picture 2.35 Field gene bank of wild coffee under natural forest in Divo (Ivory Coast). From: A. Charrier.

Picture 2.36 *In vitro* plantlets conservation *C. canephora* (IRD). From: A. Rival.

Picture 2.38 Cryopreservation techniques applied to zygotic embryos (IRD). From: A. Rival.

Picture 2.39 On farm conservation of traditional varieties by farmers in their home garden in Harar, Ethiopia. From: J. P. Labouisse.

Picture 2.37 *In vitro* plantlets conservation *C. pervilleana* (IRD). From: A. Rival.

Picture 2.40 Natural forest threatened by human activities in Ivory Coast. From: J. Berthaud.

Picture 2.41 Semi-forest coffee blossom near Jimma, Ethiopia. From: J. P. Labouisse.

habitats offers an easy access to the plant material for characterization, evaluation and subsequent utilization. Such collections are expensive to maintain, labor consuming and subject to genetic erosion [19]. For instance, accession losses in the *C. arabica* collection established at CATIE in Costa Rica were estimated to be 3.6% over ten years [X]. In the JARC collections, the erosion rate has reached 21% over 40 years, i.e. approximately 0.6% per year [XV].

Some accessions are susceptible to pests and diseases or not adapted to the local environment. Such problems can be limited by grafting on well-adapted rootstocks, duplication and planting in different ecological sites (9 locations in Ethiopia), as well as good agronomic practices.

As regards the documentation of *Coffea* germplasm collections, a descriptive list has been developed by IPGRI [24] in collaboration with crop experts. The documentation deck involves a wider application of the software BASECAFE developed by IRD [1] to handle data from the characterization and management in the basic coffee collections.

Such a database under Microsoft® Access software records the main passport data for all the JARC accessions [XV].

The construction of core collections provide large diversity to gene bank managers, research scientists and breeders with a reduced number of accessions. This strategy has been developed at the *Coffea* genus level [XII] and applied to *C. arabica* collection in CATIE represented by a sub-sample of 74 accessions [X] and to a representative sample of coffee species and scientific material maintained by IRD Montpellier in greenhouses.

2.6.2
In Vitro Techniques and Cryopreservation

Research into alternative conservation methods has recently been developed, particularly by means of *in vitro* techniques and cryopreservation. *In vitro* collections of plantlets, multiplied by micro-cuttings and maintained under slow growth, exist for coffee in a few research institutes. At IRD (Montpellier, France), the *in vitro* collection established by Bertrand-Desbrunais in 1991 has highlighted the limits of this technique, particularly with regard to genotypic selection and contamination. The maintenance of *in vitro* collections allows short- and medium-term storage for working collections. Slow growth techniques used at IRD with very low benzyladenine content of the culture medium permitted a subculturing interval of 6 months, and have led to the successful storage of germplasm of *C. congensis*, *C. canephora*, *C. liberica* and *C. racemosa* [21]. *In vitro* methods should improve coffee multiplication and the production of healthy plants for the safe transfer of *Coffea* germplasm.

Only cryopreservation (liquid nitrogen −196°C) [22] presently offers a safe long-term option. The cryopreservation of coffee material is not yet a routine technique for organized organs such as apices, embryos or seeds. The encapsulation/dehydration technique has been applied to apices of *C. racemosa* and *C. sessiliflora*. To date the maximal survival rates obtained, after apices cryopreservation, are 38%.

The cryopreservation technique developed for zygotic embryos is carried out by their excision from seeds before desiccation and freezing. These protocols were successful for *C. arabica*, *C. canephora* and *C. liberica*, but they present several constraints.

Figure 2.8 Repartition and main hotspots coffee species in Africa (From F. Anthony).

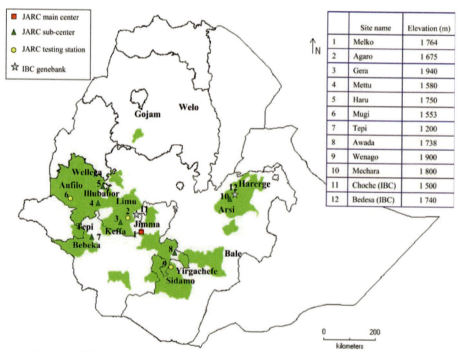

	Site name	Elevation (m)
1	Melko	1 764
2	Agaro	1 675
3	Gera	1 940
4	Mettu	1 580
5	Haru	1 750
6	Mugi	1 553
7	Tepi	1 200
8	Awada	1 738
9	Wenago	1 900
10	Mechara	1 800
11	Choche (IBC)	1 500
12	Bedesa (IBC)	1 740

Figure 2.9 Map of Ehtiopia showing the location of the conservation sites of JARC and IBC (Labouisse *et al.* 2008)

Dussert et al. [22] propose a new and simple approach for cryopreserving seeds from various coffee populations. This consists of drying the seeds under controlled relative humidity prior to cryopreservation lots of 50 seeds sealed in 10 ml polypropylene tubes. When seeds of *C. arabica* at 0.20 g H_2O/g dry weight are slowly precooled (1°C/min) to –50°C prior to immersion into liquid nitrogen, 30 % of them developed into normal seedlings. Higher survival rates (70 %) are obtained with zygotic embryos extracted from cryopreserved seeds. In contrast, pre-cooling is not needed for *C. costatifructa*, *C. racemosa* and *C. sessiliflora*. Refinements of various steps of the cryopreservation have improved results, and led to the development of a cryobank for *C. arabica* germplasm at CATIE (Costa Rica) [X].

Pollen storage at low humidity under a temperature of 5 or –18°C is currently used in coffee breeding for *C. arabica* and *C. canephora*. Pollen viability can be maintained from 1–2 months and up to 2 years. Pollen storage can be an effective tool for making crosses between coffee trees and species which do not flower simultaneously or which grow far apart.

2.6.3
In Situ Conservation

From a long-term point of view, *in situ* conservation in the natural habitat where *Coffea* species are found has also to be considered. To date there are few *in situ* reserves that have been set up specifically for the conservation of wild germplasm of *Coffea* spp. Protected areas in Africa do hold that goal in mind, in relation with the list of conservation priority species for *Coffea* established by IUCN [XIV]: 50 % are considered endangered, mainly in the African hotspots coffee regions (Fig. 2.8).

Consequently, the amount of data and the diversity of its content remains unknown, except for wild populations in Kenya [5], Mauritius [20] and Ethiopia [XXXII].

Up to now, conservation of a few remnants forest areas in Ethiopia are under development projects funded by international agencies. In 1998, three sites, namely Kontir-Berhan in Bench-Maji zone, Boginda-Yeba in Keffa and Geba-Dogi River in Illubabor had been identified for forest coffee conservation in a European Project. Since, a lot of integrated projects are mainly oriented towards the promotion of participative forest management and non timber products exploitation such as coffee [XXXV].

In addition to the conservation of coffee forests, it seems of the same importance to develop some on-farm conservation methods for landraces. It applies to *C. arabica* in its native zone of origin, Ethiopia, where indigenous people have developed a method of extractive picking of coffee cherries in national forests, and set up traditional varieties in home gardens and farms [XV].

The sustainability of in situ/on-farm conservation systems seems to be guaranteed only if farmers can benefit from the whole agroforestry system or garden system through coffee certification labels and ecotourism.

2.7
Future Outlook

In tropical humid Africa, where natural forests are being threatened by human activities, it is urgent to identify protected areas and to continue with the collection of coffee germplasm, particularly in countries such as Gabon, Democratic Republic of Congo and Uganda that have not recently been visited by collectors.

Basically, it must be remembered that the improved utilization of *Coffea* collections by coffee breeders depends on the documentation of the base collections in Africa (Ethiopia, Madagascar and the Ivory Coast) and the transfer of core collections for the two cultivated species, *C. arabica* and *C. canephora*, as well as for other species of wild coffee. While the main coffee field collections in Africa are not, generally speaking, in immediate danger, they are likely to be at risk in the near future because of lack of funding for day-to-day maintenance. Their potential value is also under-studied and under-used because only a limited number of researchers, mostly on part-time schedules, are working on them.

As a result, a multidisciplinary approach to *Coffea* taxonomy and the dynamics of populations in coffee trees is now required, involving a combination of botanical characteristics, genetic diversity of the biochemical and molecular markers, and biological studies. This approach would also be useful for the rational management and use of coffee germplasm.

2.8
Acknowledgments

The authors acknowledge the review assistance of F. Anthony and the illustration figures and slides of CIRAD and IRD researchers.

Bibliography

[1] Anthony, F. Les ressources génétiques pour l'amélioration des caféiers: collecte, gestion d'un conservatoire et évaluation de la diversité génétique. *TDM 81.* ORSTOM, Paris, 1992.

[2] Anthony, F., Combes, M. C., Astorga, C., Bertrand, B., Graziosi, G. and Lashermes, P. The origin of cultivated *Coffea arabica* L. varieties revealed by AFLP and SSR markers. *Theor. Appl. Genet.* **2002**, *104*, 894–900.

[3] Anthony F., Bertrand B., Quiros O., Wilches A., Lashermes P., Berthaud J., and Charrier A. Genetic diversity of wild coffee (*Coffea arabica* L.) using molecular markers. *Euphytica* **2001** *118*, 53–65.

[4] Berthaud, J. Les ressources génétiques pour l'amélioration des caféiers africains diploïdes: évaluation de la richesse génétique des populations sylvestres et de ses mécanismes organisateurs. *TD 188.* ORSTOM, Paris, 1986.

[5] Berthaud, J. and Charrier, A. Genetic resources of *Coffea*. In: *Coffee. Vol. 4: Agronomy.* Clarke, R. J. and Macrae, R. (eds). Elsevier, London, 1988, 1–42.

[6] Bridson, D. Nomenclatural notes on *Psilanthus*, including *Coffea* sect. *Paracoffea*. Kew Bulletin, **1987**, *42*, 453–460.

[7] Bridson, D. and Verdcourt, B. *Coffea*. In: *Flora of Tropical East Africa.* Rubiaceae *(Part 2).* Polhill, R. M. (ed.). Balkema, Rotterdam, 1988, 703–723.

[8] Capot, J. L'amélioration du caféier en Côte d'Ivoire. Les hybrides Arabusta. *Café Cacao Thé* **1972**, *16*, 3–16.

[9] Carvalho, A. and Monaco, L. C. Relaciones geneticas de especies seleccionadas de *Coffea. Café (Campinas)* 1968, **9**(4).

[10] Carvalho, A., Ferwerda, F. P., Frahm-Leliveld, J. A., Medina, D. M., Mendes, A. J. T. and Monaco, L. C. Coffee. In: *Outlines of Perennial Crop Breeding in the*

Tropics. Ferwerda, F. P. and Wit, F. (eds). Veenman, Wageningen, 1969, 189–241.

[11] Carvalho, A., Medina Filho, H. P., Fazuoli, L. C., Guerreiro Filho, O. and Lima, M. M. A. Aspectos genéticos do cafeiro. *Rev. Brasil. Genet.* **1991**, *14*, 135–183.

[12] Carvalho, A. and Fazuoli, L. C. Café. In: *O Melhoramento de Plantas no Instituto Agronomico. Vol. 1.* Furlani, A. M. C. and Viegas, G. P. (eds). Instituto Agronomico, Campinas, SP, 1993, 29–76.

[13] Charrier A. La structure génétique des caféiers spontanés de la région malgache (*Mascarocoffea*). Leurs relations avec les caféiers d'origine africaine (*Eucoffea*). *Mémoire ORSTOM 87.* ORSTOM, Paris, 1978.

[14] Charrier, A. (ed.). Etude de la structure et de la variabilité génétique des caféiers: résultats des études et des expérimentations réalisées au Cameroun, en Côte d'Ivoire et à Madagascar sur l'espèce *C. arabica* collectée en Ethiopie par une mission ORSTOM en 1966. *Bulletin 14.* IFCC, Paris, 1978.

[15] Chevalier, A. *Les Caféiers du Globe III. Systématique des Caféiers et faux Caféiers. Maladies et Insectes Nuisibles. Encyclopédie Biologique.* Lechevalier, Paris, 1947.

[16] Cramer, P. J. S. *Review of Literature of Coffee Research in Indonesia.* Inter-American Institute of Agricultural Sciences, Turrialba, Costa Rica, 1957.

[17] Cros, J., Combes, M. C., Chabrillange, N., Duperray, C., Monnot Des Angles, A. and Hamon, S. Nuclear DNA content in the subgenus *Coffea* (*Rubiaceae*), inter and intra-specific variation in African species. *Can. J. Bot.* **1995**, *73*, 14–20.

[18] Cros, J., Combes, M. C., Trouslot, P., Anthony, F., Hamon, S., Charrier, A. and Lashermes, P. Phylogenetic analysis of chloroplast DNA variation in *Coffea. Mol. Phylogenet. Evolution* **1998**, *9*, 109–117.

[19] Dulloo, M. E., Guarino, L., Engelmann, F., Maxted, N., Newbury, H. J., Attere, F. and Ford Lloyd, B. V. Complementary conservation strategies for the genus *Coffea*: a case study of Mascarene *Coffea* species. *Genet. Resources Crop Evolution* **1998**, *45*, 565–579.

[20] Dulloo M. E., Maxted N., Newbury H. J., Florens D. and Ford Lloyd B. V. Ecogeo-graphic survey of the genus *Coffea* in the Mascarene Islands. *Bot. J. Linnaean Soc.* **1999**, *131*, 263–284.

[21] Dussert, S., Chabrillange, N., Anthony, F., Engelmann, F., Recalt, C. and Hamon, S. Variability response within a coffee (*Coffea* spp.), core collection under slow growth conditions. *Plant Cell Rep.* **1997**, *16*, 344–348.

[22] Dussert, S., Chabrillange, N., Engelmann, F., Anthony, F., Vasquez, N. and Hamon, S. Cryopreservation of *Coffea*. In: *Biotechnology in Agriculture and Forestry. Vol. 50: Cryopreservation of Plant Germplasm.* Towill, L. E. and Bajaj, Y. P. S. (eds). Springer, Berlin, 2002, 220–233.

[23] Davis A. P., Chester M., Maurin O., Fay M. 2007. Searching for the relatives of Coffea (Rubiaceae, Ixoroideae): the circumscription and phylogeny of *Coffeeae* based on plastid sequence data and morphology. *American Journal of Botany* **94**: 313-329.

[24] IPGRI. *Descriptors for Coffee (*Coffea *spp. and* Psilanthus *spp.).* IPGRI, Rome, 1997.

[25] Lashermes, P., Trouslot, P., Cros, J., Combes, M. C. and Charrier, A. Genetic diversity for RAPD markers between cultivated and wild accessions of *Coffea arabica. Euphytica* **1996**, *87*, 59–64.

[26] Lashermes, P., Combes, M. C., Trouslot, P. and Charrier, A. Phylogenetic relationships of coffee-tree species as inferred from ITS sequences of nuclear ribosomal DNA. *Theor. Appl. Genet.* **1997**, *94*, 947–955.

[27] Lebrun, J. Recherches morphologiques et systématiques sur les caféiers du Congo. *Mem. Inst. Roy. Colon. Belge, Sci. Nat. Med.* **1941**, *II* (3), 1–186.

[28] Leroy, J. F. Coffeae novae madagascarienses. *J. Agric. Trop. Bot. Appl.* **1961**, *VIII* (1–3), 1–20.

[29] Leroy J. F. Coffeae novae madagascariensis et mauritianae. *J. Agric. Trop. Bot. Appl.* **1962**, *IX* (11/12), 525–530.

[30] Leroy J. F. Evolution et taxogénèse chez les caféiers. Hypothèse sur leur origine. Comptes rendus de l'Académie des Sciences (Paris), **1980**, *291*, 593–596.

[31] Louarn, J. La fertilité des hybrides interspécifiques entre caféiers diploïdes d'origine africaine (genre *Coffea*). *Thèse.* Universite Paris XI, Orsay, 1992.

[32] Montagnon, C., Leroy, T., Yapo, A. B. Etude complémentaire de la diversité génotypique et phénotypique des caféiers de l'espèce *Coffea canephora* en collection en Côte d'Ivoire. In: 14ème Conférence ASIC, San Francisco 1991, 444–450.

[33] Montagnon, C., Leroy, T., Charmetant, P., Yapo, A., Legnate, H., Berthaud, J. and Charrier A. Outcome of two decades of reciprocal recurrent selection applied to *C. canephora* in Côte d'Ivoire: new outstanding hybrids available for growers. In: *19ème Conférence ASIC*, Trieste, 2001.

[34] Ky, C. L., Louarn, J., Dussert, S., Guyot, B., Hamon, S. and Noirot, M. Caffeine, Trigonelline, Chlorogenic acids and sucrose diversity in wild *Coffea arabica* and *C. canephora* accessions. *Food Chem.* **2001**, *75*, 223–230.

[35] Rakotomalala, J. J., Cros, E., Clifford, M. and Charrier, A. Caffeine and theobromine in green beans from *Mascarocoffea*. *Phytochemistry* **1992**, *31*, 1271–1272.

[36] Prakash, N. S., Combes, M. C., Somanna, N. and Lashermes, P. AFLP analysis of introgression in coffee cultivars (*Coffea arabica* L.) derived from a natural interspecific hybrid. *Euphytica* **2002**, *124*, 265–271.

[37] Robbrecht, E. Un cadeau de l'Afrique. *L'Origine du Café*. Bibliothéque Royale, Bruxelles, 1996, 7–35.

[38] Rodriguez, C. J. Jr, Bettencourt, A. J. and Rijo, L. Races of the pathogen and resistance to coffee rust. *Annu. Rev. Phytopathol.* **1975**, *13*, 49–70.

[39] Sreenivasan, M. S. Breeding coffee for leaf rust resistance in India. In: *Coffee Rust: Epidemiology, Resistance and Management*. Kushalappa, A. C. and Eskes, A. B. (eds). CRC Press, Boca Raton, FL, 1989, 316–323.

[40] Stoffelen, P. *Coffea* and *Psilanthus (Rubiaceae)* in Tropical Africa: a systematic and palynological study, including a revision of the west and central African species. *PhD Thesis.*Katholieke Universiteit Leuven, Belgium, 1998.

Additional references

I   Aggarwal, R.K., Hendre, P.S., Varshney, R.K., Bhat, P.R., Krishnakumar, V., and Singh, L. (2007) Identification, characterization and utilization of EST-derived genetic microsatellite markers for genome analyses of coffee and related species. Theor Appl Genet 114: 359–372.

II   Bellachew B (1997) Arabica coffee breeding in Ethiopia: a review. In: 17th International Coffee Science Conference, ASIC, Nairobi, Kenya, 20–25 July 1997, pp 406–414.

III Bhat PR, Krishnakumar V, Hendre PS, Rajendrakumar P,Varshney RK, and Aggarwal RK (2005) Identification and characterization of gene-derived EST-SSR markers from robusta coffee variety 'CxR' (an interspecific hybrid of *Coffea canephora* and *Coffea congensis*). Mol Ecol Notes 5:80-83.

IV   Campa, C., Doulbeau, S., Dussert, S., Hamon, S. & Noirot, M. 2005. Diversity in bean caffeine content among wild *Coffea* species: evidence of a discontinuous distribution. Food Chemistry 91: 633–637.

V   Clarindo WR and Carvalho CR 2008. First *Coffea Arabica* karyogram showing that this specis is a true allotetraploid. *Plant. Syst. Evol.* (in press).

VI   Crouzillat D, Rigoreau M, Bellanger L, Priyono, S. Mawardi, Syahrudi, J. McCarthy, S. Tanksley, I. Zaenudin, and V. Pétiard (2004) A Robusta consensus map using RFLP and microsatellites markers for the detection of QTL. In: Proc 20th International Scientific Colloquium on Coffee, Bangalore (India). ASIC (http://www.asic-cafe.org).

VII Cubry P, Musoli P, Legnate H, Pot D, De Bellis F, Poncet V, Anthony F, Dufour M, Leroy T, 2008. Diversity in coffee assessed with SRR markers : structure of the genus Coffea and perspectives for breeding. Genome 51: 50–63.

VIII   Davis, A. P., Govaerts, R., Bridson, D. M. & Stoffelen, P. 2006. An annotated taxonomic conspectus of the genus Coffea (Rubiaceae). Botanical Journal of the Linnean Society 152: 465–512.

IX   Davis A. P., Chester M., Maurin O., Fay M. 2007. Searching for the relatives of *Coffea (Rubiaceae, Ixoroideae)*: the circumscription and phylogeny of Coffeeae based on plastid sequence data and morphology. *American Journal of Botany* 94: 313–329.

X   F Engelmann, E Dulloo, C Astorga, S, Dussert, F Anthony eds. 2007. Complementary strategies for ex situ conservation of coffee (Coffea arabica L.) genetic resources. A case study in CATIE, Costa Rica. Topical reviews in Agricultural Biodiversity, Bioversity International, Rome.

XI   Eira M.T.S., Amaral da Silva E.A., de Castro R.D., Dussert S., Walters C., Bewley D.J. and Hillhorst H.W.M. (2006) Coffee seed physiology. Brazilian J Plant Physiol 18:149–163.

XII   Hamon S, Noirot M, Anthony F (1995). Developing a coffee core collection using the Principal Component Score strategy with quantitative data. In: "Core collections of plant genetic resources". T Hodgkin, AHD Brown, ThJL van Hintum & EAV Morales éds. Wiley-Sayce Publication, Londres, pp. 117–126.

XIII   Herrera J.C., D'Hont A., and Lashermes P. (2006) Utilization of chromosome painting as a complementary tool for introgression analysis and chromosome identification in coffee (*Coffea arabica L.*) In: Proc 21th International Scientific Colloquium on Coffee, Montpellier (France). ASIC (http://www.asic-cafe.org).

XIV   IUCN. 2004. 2004 IUCN red list of threatened species. URL: http://www.iucnredlist.org

XV   Labouisse, JP.,Bellachew, B., Kotecha, S., Bertrand, B. Current status of coffee (*Coffea arabica L.*) genetic resources in Ethiopia: implications for conservation. *Genet. Resour. Crop Evol.*, 2008 (in press).

XVI   Lashermes P, Combes MC, Robert J, Trouslot P, D'Hont A, Anthony F, and Charrier A (1999) Molecular characterisation and origin of the Coffea arabica L. genome. Mol Gen Genet 261:259–266.

XVII   Lashermes P, Combes MC, Prakash NS, Trouslot P, Lorieux M, and Charrier A (2001) Genetic linkage map of *Coffea canephora*: effect of segregation distortion and analysis of recombination rate in male and female meioses. Genome 44:589–595

XVIII   Lashermes P, Andrade AC, Etienne H (2008). Genomics of Coffee, One of the World's Largest Traded Commodities. In "Genomics of Tropical crop plants" PH Moore and R. Ming (Eds), Springer, pp 203–226.

XIX   Leroy T., Cubry P, Durand N , Dufour F, De Bellis F , Jourdan I, Vieira LGE, Musoli P, Aluka P, Legnate H, Marraccini P, Pot D, 2006. *Coffea* spp and *C. canephora* diversity evaluated with micosatellites and SNPs. Lessons from comparative analysis. In: Proc 21th International Scientific Colloquium on Coffee, Montpellier (France). ASIC (http://www.asic-cafe.org).

XX   Lombello, RA and Pinto-Maglio CAF (2004) Cytogenetic studies in *Coffea* L. and Psilanthus Hook. Using CMA/DAPI and FISH. Cytologia 69:85–91.

XXI   Mahé L, Le Pierrés D, Combes M-C, Lashermes P (2008) Introgressive hybridization between the allotetraploid *Coffea* arabica and one of its diploid ancestors *C. canephora* in an exceptional sympatric zone in New Caledonia. Genome (in press)

XXII   Maurin O, Davis A, Chester M, Mvungi E.F. Towards a Phylogeny for *Coffea (Rubiaceae)*: Identifying Well-supported Lineages Based on Nuclear and Plastid DNA Sequences. Annals of Botany, 2007, 100:1565–1583.

XXIII   Montagnon C, Bouharmont P (1996) Multivariate analysis of phenotype diversity of Coffea arabica. Genet Resour Crop Evol 43(3):221–227.

XXIV   Musoli P. 2007. Recherche de sources de résistance à la trachéomycose du caféier Coffea canephora Pierre due à *Fusarium xylarioides* Steyart en Ouganda. Thése Montpellier SupAgro.

XXV   N'Diaye, A., Poncet, V., Louarn, J., Hamon, S. & Noirot, M. 2005. Genetic differentiation between *C. liberica* var.

liberica and *C. liberica* var. Dewevrei and comparison with *C. canephora*. Plant Systematics and Evolution 253: 95–104.

XXVI Noirot, M., Poncet, V., Barre, P., Hamon, P., Hamon, S. & Kochko (de), A. 2003. Genome size variations in diploid African Coffea species. Annals of Botany 92: 709–714.

XXVII Poncet V, Hamon P, Minier J, Carasco C, Hamon S, and Noirot M (2004) SSR cross-amplification and variation within coffee trees (Coffea spp.). Genome 47:1071–1081.

XXVIII Poncet, V., Rondeau, M., Tranchant, C., Cayrel, A., Hamon, S., de Kochko, A., and Hamon, P. (2006) SSR mining in coffee tree EST databases: potential use of EST-SSRs as markers for the Coffea genus. Mol Genet Genomics 276:436–449.

XXIX Prakash NS, Combes MC, Dussert S, Naveen S, and Lashermes P (2005) Analysis of genetic diversity in Indian robusta coffee genepool (*Coffea canephora*) in comparison with a representative core collection using SSRs and AFLPs. Genetic Res Crop Evol 52: 333–343.

XXX Silvarolla MB, Mazzafera P and Fazuoli LC 2004. A naturally decaffeinated coffee (*Coffea arabica*). Nature, 429, 826

XXXI Silvestrini M, Junqueira MG, Favarin AC, Guerreiro-Filho O, Maluf MP, Silvarolla MB et al (2007) Genetic diversity and structure of Ethiopian, Yemen and Brazilian *Coffea arabica* L. accessions using microsatellites markers. Genet Resour Crop Evol 54(6):1367–1379

XXXII Tesfaye GK, Govers K, Oljira T, Bekele E, Borsch T (2007) Genetic diversity of wild Coffea arabica in Ethiopia: analyses based on plastid, ISSR and microsatellite markers. In: 21st International Coffee Science Conference, Montpellier, 11–15 September 2006, [CD-ROM], pp 802-810

XXXIII Vieira, L.G.E., Andrade, A.C., Colombo, C.A., et al. (2006) Brazilian coffee genome project: an EST-based genomic resource. Braz J Plant Physiol 18:95–108.

XXXIV Wu, F., Mueller, L.A., Crouzillat, D., Pétiard, V., and Tanksley, S.D., (2006) Combining bioinformatics and phylogenetics to identify large sets of single-copy orthologous genes (COSII) for comparative, evolutionary and systematic studies: A test case in the euasterid plant clade. Genetics 174:1407–1420.

XXXV ZEF (2007) Conservation and use of wild populations of coffee arabica in the montane rainforests of Ethiopia. http://www.coffee.uni-bonn.de/project-overview.html

3
Coffee Selection and Breeding

A. B. Eskes and Th. Leroy

3.1
Introduction

3.1.1
General context

To obtain high yields of good-quality coffee, the choice of variety is of fundamental importance. Coffee breeding, i.e. the creation and development of new varieties, has already provided farmers with high-yielding cultivars which are adapted to different cultivation systems and show increased resistance to the major diseases – Coffee Leaf Rust (CLR) and Coffee Berry Disease (CBD). Further progress can be expected within the next 10–15 years by increasing resistance to different nematode species and to insects (particularly Coffee Leaf Miner and Coffee Berry Borer) by exploiting the vigor of hybrid varieties of *C. arabica* and *C. canephora*, and by improving bean characteristics which influence liquor quality (especially for *C. canephora*).

Future progress in coffee breeding will depend not only on the genetic variation present in breeding populations but also on the efficiency with which the potentially useful genetic variation can be identified and incorporated into newly improved varieties. Because coffee is a perennial crop, breeding programmes take many years to produce results and therefore require long-term commitments in terms of funding and human resources. In many of the major coffee-producing countries, the declining prices paid for coffee in recent years has restricted the allocation of funds and of human resources to coffee breeding and coffee research in general. However, it is significant that, today, the majority of coffee produced for world markets comes from plantations using selected varieties – this is especially true for *C. arabica*, although less so for *C. canephora*. In the recent past the contribution of improved, high-yielding varieties of coffee to maintaining profitability during a period of declining prices has been considerable. The viability of the coffee industry in the future will depend, to a large extent, on the continuing development of new varieties to overcome changing circumstances and new factors limiting yield levels.

Coffee-breeding techniques and the characteristics of existing varieties have been reviewed, relatively recently, in articles or books written by Carvalho [15], Van der Vossen [51], Charrier and Berthaud [23] and by Wrigley [52]. A review on breeding for rust resistance has been prepared by Kushalappa and Eskes [34]. Coffee taxonomy, in-

cluding a description of the several cultivated species of *Coffea* sp. and types, is presented in previous chapters, together with a
summary of the distribution of these species in the main geographical areas of the
world where coffee is produced.

This chapter is meant to provide background information for agronomists, giving an understanding of coffee-breeding
techniques and the various phases in the
development of higher-yielding varieties.

3.1.2
Characteristics of Coffee Species related to Breeding

The two main cultivated species of coffee,
C. arabica and *C. canephora*, are botanically
distinct. Arabica is a tetraploid autogamous
species, i.e. reproduced mainly by self-pollination. On the other hand, *C. canephora*
is diploid and strictly allogamous, due to
the presence of self-incompatibility alleles.
Consequently, every seed is obtained
through cross-pollination by neighboring
plants. Despite this difference between
the two species, interspecific crosses are
possible, although plants derived from
such crosses are generally less fertile [4].

Cultivars are cultivated varieties which
are more or less fixed genetically. In the
case of *C. arabica*, most cultivars are homozygous; they represent one genotype that is
reproduced by seed obtained by self-pollination (e.g. internationally recognized varieties such as Mundo Novo, and Caturra).
Crosses between different varieties are
called "F_1" hybrids. Within the breeding
process, "lines" are considered to be advanced progenies obtained by successive
cycles of self-pollination, but are not yet
necessarily homozygous; "pure lines" are
homozygous. The selection of lines in segregating populations, with the aim of
obtaining fixed homozygous varieties, is

called line or pedigree selection.
Populations derived from F_1 by selfing are
called $F_2 \cdot F_3$ are derived from F_2, etc. F_6
or more advanced populations are gener-
ally considered to be fixed varieties.

In the case of *C canephora*, varieties have
been developed within the species through
the selection of clones and of hybrid progenies (crosses between clones). The Canephora hybrid varieties are reproduced in
biclonal or polyclonal seed gardens; clones
are reproduced by vegetative means.

3.2
History of Coffee Selection

3.2.1
C. arabica

Arabica coffee originates from Ethiopia,
where it can still be found growing wild
or semi-wild in the undergrowth of tropical
highland forests. Early domestication and
selection of the species was carried out by
Arabs, who introduced it into Yemen probably in the 13th or 14th century. The local
varieties selected in Yemen have been at
the basis of coffee cultivation in all other
coffee-growing areas, with the probable
exception of Ethiopia.

Around 1700, coffee plants from Yemen
were introduced by the Dutch into Indonesia, and from there, some decades later,
into Central and South America. This latter
coffee material descended from one tree,
held by the Dutch in the botanical garden
of Amsterdam. It became named as
"Typica" or "Arabica" and has been widely
grown in the Americas for about two centuries, where it is still grown in some countries [15]. This variety is very homogeneous.

Coffee-planting material was also taken
from Yemen and introduced into Bourbon

(Reunion Island). From Bourbon, coffee was introduced into a number of countries in Africa and South and Central America. These Bourbon-type varieties proved to be generally higher yielding and genetically more variable than Typica. For example, in Brazil, the different Bourbon lines selected in individual fields revealed considerable variations in yield capacity [15]. The same appears to have been the case in East Africa, where Bourbon-type varieties were selected and are still widely grown.

Because Bourbon and Typica varieties were grown side by side in the same coffee regions, some degree of natural out-crossing has occurred. For example, in Brazil, it is supposed that the selected higher-yielding Yellow Bourbon variety originated from a natural cross between Red Bourbon and Amarelo de Botocatu, a Typica variety. The famous Mundo Novo variety was obtained in a similar way (selection from a natural cross between Bourbon and a Typica variety called Sumatra) [16, 18]. It is likely that similar situations have occurred in East Africa. This may explain why some of the varieties, which are held in collections, display characteristics related to both Bourbon and Typica, and could, in fact, be segregants from natural out-crossing.

The original introductions of C. arabica into India are thought to have also come from Yemen [49]. The traditional farmers' varieties cultivated post-1820 were called "Old Chiks". Following the increasing severity of coffee rust attack, they were replaced in the 1870s by the "Coorg" variety, which initially showed a higher resistance to rust. However, a few decades later, the Coorg variety was considered highly susceptible to the disease and was largely replaced from 1920 onwards by the offspring of a rust-resistant single-tree selection made by a planter called Kent. The "Kent"

variety lost its resistance within 10 years and was later proved to possess resistance only to specific types ("races") of the rust fungus, as shown by Mayne in 1932 [28]. Resistance to the new rust race that attacked Kent's variety was found in derivatives of spontaneous crosses between C. arabica and C. liberica, as in the S228 selection. A cross between S288 and Kent's variety gave rise to the well-known S795 variety, released for cultivation from 1946 onward. This variety retained considerable field resistance until the 1960s, although races capable of infecting it were traced by Mayne as early as 1939. The pioneer work done in India on different rust types has been the basis for rust resistance studies carried out in Portugal, at the Coffee Rusts Research Center (CIFC), since 1950 (see Section II.6).

3.2.2
Diploid Coffee Species

Presently, the main diploid coffee species grown is C. canephora, which is responsible for 30 % of the world coffee production. Cultivation of C. liberica used to be quite important in West Africa and South East Asia, but has now almost completely disappeared.

Liberica coffee was introduced by the Dutch into Indonesia around 1875. It soon became popular due to its better adaptation to the lower coffee belt and its higher resistance to leaf rust. However, in 1895 it was apparently severely attacked by the rust fungus. The solution for rust resistance was found by using C. canephora, first introduced into Java in 1900 [25]. Among the several Canephora types introduced, the "Robusta" type was selected due to its improved productivity and high level of rust resistance. From 1910 onwards, it was used on a large scale to replace Arabica and Liberica coffee in the lower coffee belt of Indonesia. Selections of Robusta from Java were re-introduced into Africa during the first part of the 20th century.

The substitution of *C. arabica* by *C. liberica* and later by *C. canephora* has also occurred in India, ever since the beginning of the 20th century. This substitution was motivated by the better natural adaptation of *C. liberica* and *C. canephora* to tropical lowlands and the better rust resistance of these species. Canephora coffee later replaced Liberica, not only because of its higher resistance to rust, but also because of the higher quality of coffee produced [25].

Ancient coffee cultivation in Africa occurred in Uganda, using local Canephora types for specific purposes (chewing of boiled beans) [41]. The cultivation of Canephora, as well as of other wild coffees, was sporadically attempted on the Atlantic Coast during the 19th century. Liberica coffee types as well as some local Canephora types were wiped out by the coffee wilt disease *Fusarium xylarioides* in the 1940s and 1950s. They were replaced mainly by Robusta types introduced from Java and the Democratic Republic of Congo and by "Kouillou" types from Gabon [23]. The "Kouillou" coffee type was also introduced into Brazil, where it became cultivated under the name "Conillon" (the change in name was due to a mistranslation).

3.2.3
Interspecific Crosses

Recent history of Arabica coffee breeding is largely related to the introgression of resistance genes from *C. canephora*, through interspecific hybridization. Spontaneous interspecific crosses have been found on several occasions in countries where diploid coffee species were cultivated besides *C. arabica*. Early evidence of the occurrence of such hybrids goes back to the end of the 19th century and the early 20th century. In Indonesia, the "Kawisari" (*C. arabica* × *C. liberica*) and the "Bogor Prada" (*C. arabica* × *C. canephora*) hybrids appeared to be quite productive and resistant to rust but their progenies were considered useless [25]. In India, the Devamachy hybrid (*C. arabica* × *C. canephora*), discovered around 1930, has been further used in breeding [49].

The most important spontaneous hybrid between *C. arabica* and *C. canephora* is the Hybrid of Timor (HdT), found on the island of Timor around 1920. Derivatives of this hybrid were selected locally for their rust resistance and vigor. Seeds were taken in the 1950s for resistance studies to the CIFC in Portugal and also to many other countries [28]. Two plants with resistance to all known rust types were used by CIFC in crosses with Caturra resulting in the well-known breeding population called "Catimor".

Other spontaneous hybrids between *C. canephora* and *C. arabica* have been identified more recently, around 1990 in New Caledonia by Mr Picot, a CIRAD researcher. The characteristics of the plants encountered there indicate the presence of tetraploid and triploid spontaneous hybrids as well as back-cross populations with Arabica phenotype (Eskes, personal observation).

Artificial hybrids have been created between *C. arabica* and *C. canephora*, with two main objectives. A tetraploid hybrid, produced by using a *C. canephora* parent with artificially doubled number of chromosomes, was created in Brazil in 1950. This hybrid has been back-crossed with Brazilian Arabica cultivars, resulting in the Icatu breeding population. A second approach was the direct use of the tetraploid hybrids between *C. arabica* and

C. canephora, called "Arabusta" [13]. Since the 1960's many Arabusta hybrids have been created and evaluated in Ivory Coast, Cameroon and also in Ghana with the objective of creating a better quality coffee adapted to tropical lowlands. The same objective was followed by crossing *C. canephora* with *C. congensis* (Congusta hybrid).

In Colombia, triploid hybrids between *C. arabica* and *C. canephora* have been created that have been used for further selection since the 1970s. In Brazil, back-cross generations from *C. arabica* × *C. racemosa* are being selected for resistance to Coffee Leaf Miner.

3.3
Selection Criteria

3.3.1
Agronomic Traits

3.3.1.1 Yield
Coffee yield is generally measured as fresh cherry weight per plant or per plot and converted into green coffee weight by applying a conversion ratio. This conversion ratio, or "out-turn", should therefore be calculated for each genotype. The average out-turn for Arabica varieties is around 18–20 %, for Canephora 20–25 % and for Arabusta 10–15 %. Depending on the genotype, coffee yields show wide variations with differences of two to four times between lowest and highest yields. Favorable growing conditions reduce the differences between genotypes. The observation of coffee yields over the first 4–5 years in production is usually sufficient to assess long-term yield potential. However, genotypes prone to "over-bearing" are less predictable and yields may be erratic throughout the life cycle of the plants. This is, typically, the case for Caturra and many Catimor lines under Brazilian growing conditions.

3.3.1.2 Vigor
Early plant vigor may be measured by the stem diameter of 1-year-old plants in the field or by the increase in stem diameter between the first and second year. This characteristic can be a good indicator for yield, for example in Canephora genetic coefficients of correlation of 0.70–0.93 have been found in experiments carried out in Ivory Coast [38]. Adult plant vigor is generally measured by plant height and/or diameter of the canopy that is often also correlated to yield. However, it may not necessarily be of interest to select the most vigorous plants. The yield:vigor ratio ("yield efficiency") has been applied in Canephora selection to obtain productive plants that occupy relatively little space [36].

3.3.1.3 Visual Breeders Score
In some countries a very useful scoring system is used to assess the performance of individual coffee trees. The score covers actual yield, yield expectation for the following year, vigor and plant shape. In Brazil, where a 10-point scale is used, the scoring carried out during the first three productive years in a variety trial has shown to be well correlated to subsequent yields over 15–20 year periods (Eskes, personal observations). This system may be favorably applied by the coffee breeder for pre-selection in large amounts of germplasm, for example in segregating F_2 or F_3 populations, without having to establish long term observations of yield and vigor.

3.3.1.4 Growth Habit
An increased interest prevails for the selection of dwarf coffee varieties that allow high density planting and easier harvesting. Dwarfness is generally related to dominant genes. This means that heterozygous

dwarf plants will give segregating progeny. Selection for dwarfness is possible at the nursery stage. In the field, it is not recommended to carry out a selection of dwarf and tall varieties in mixed stands because of the unequal competition that results between tall and dwarf plants.

3.3.1.5 Yield Stability

Multilocational variety trials are used to evaluate yield stability in different environments. As a result, variety recommendations may vary between regions in the same country. For example, the Canephora selection programme in Cameroon identified clones with differential adaptation to growing conditions in the western and eastern regions of that country.

3.3.2
Resistances

3.3.2.1 Fungal Diseases

Resistance to CLR (*Hemileia vastatrix*) can be assessed through subjective observations under field conditions. A five-point scale to score the degree of susceptibility is commonly used to identify susceptible (S) and resistant (R) plants. The level of field resistance to leaf rust may be affected by yield; rust incidence is generally higher in high yielding plants. To evaluate resistance under controlled conditions, quite simple tests have been developed using inoculations of detached leaves or of leaf disks [28].

Resistance to CBD, caused by *Colletotrichum kahawae*, can be observed in the field by estimating the percentage of black berries or the percentage of berry drop measured over the infectious period. In laboratory tests, inoculations of hypocotyls can be used for early screening and confirmation of resistance of the mother plants in the field. For the latter, detached berries have also been used. For more information on CLR and CBD resistance tests, the reader should refer to the chapter dedicated to these diseases (Section II.6).

Resistance to other diseases of local importance should be observed during epidemics both in the field and in the nursery, preferably by grading on a five-point scale for disease incidence (% infected organs) or disease severity (% of tissue destroyed by the disease). Genetic variation in resistance has been observed in Arabica for locally important diseases like Fusarium (*Fusarium* spp), Bacterial Blight (*Pseudomonas syringae*), Brown Eye Spot (*Cercospora coffeicola*) and, more recently, for the American Leaf Disease (*Mycena citricolor*). With Canephora, the same applies for resistance to Fusarium, Leaf Anthracnose (*Colletotrichum* spp.) and Pink Disease (*Corticium* spp.). Resistance can only be observed in environments conducive to the disease. For example, Brown Eye Spot is mainly due to an imbalance between nutrition, yield and shade. In this case, plants prone to over-bearing may show a higher degree of susceptibility. This has been observed in some countries with Catimor lines.

3.3.2.2 Physiological Disorders

Variation for drought resistance has been demonstrated between varieties of *C. canephora*. However, within *C. arabica* little variation for either drought resistance or frost resistance seems to occur. In both Arabica and Canephora, some varieties are better adapted to acid soils where there is aluminum toxicity and phosphate deficiency.

3.3.2.3 Resistance to Nematodes

The two major nematode types, which affect the coffee plant, are *Meloidogyne* spp. and *Pratylenchus* spp. Resistance can be observed by artificial inoculation of seedlings with a fixed number of nematodes (eggs, larvae and/or adults) carried out either in a laboratory or in a greenhouse. As nematode populations vary from one place to another, confirmation of resistance under field conditions would be required. Large variation in resistance, from near immunity to high susceptibility, is found in Arabica and Canephora in relation to *Meloigogyne* spp., but this is not the case in relation to *Pratylenchus*. However, Canephora as well as other diploid coffee species appear to be more tolerant than Arabica to the latter nematode type.

3.3.2.4 Resistance to Insects

Variation for resistance to major insect pests, such as Coffee Berry Borer (*Hypotenemus hampei*) and Coffee Leaf Miner (*Leucoptera* spp.), is rather limited for Arabica and Canephora. However, the Arabica mutant "Goiaba" seems potentially able to delay attacks of the Coffee Berry Borer. Resistance to the Coffee Leaf Miner can be found in several wild diploid species and has been introgressed into Arabica through crosses with *C. racemosa* in Brazil. An efficient early screening test has been developed for leaf miner using detached leaves or leaf disks [31]. Genetic variation for resistance to the twig borer has been observed in Canephora [42].

3.3.3
Quality

3.3.3.1 Technological Features

Important technological features are bean size (or weight), percentage of floating berries and percentage of seed defects (peaberries, elephant beans). Bean size is evaluated either by grading on sieves or by calcu-

lating the average weight of 100 berries. The percentage of floating berries is related to the presence of one or two empty locules, generally a manifestation of a genetic defect (early aborted seed or badly developed endosperm) [15]. Arabica varieties typically have less than 10 % empty locules, whereas interspecific hybrids may show up to 50 % empty locules.

Peaberries are roundish seeds. They result either from deficient pollination, unfavorable environmental conditions or genetic defects. Typically, selected Arabica varieties contain about 10 % peaberries, this percentage is substantially higher for Canephora, probably due to its allogamic nature. The origin of elephant beans (more than one seed per locule) is not well understood, but genetic and environmental effects are involved. Normally, in Arabica, less than 2–3 % of the seeds are elephant beans but some varieties may show 20–50 % under certain conditions (as is the case for S795, observed in Indonesia). The percentage of empty locules and other seed defects can be observed by cutting 200 berries per genotype, when they are 4–5 months old, just before the hardening of the seeds. This method was suggested by De Reffye [26] to determine the degree of fertility in coffee. In Arabica, seed defects are generally measured by counting the percentage of floating berries and peaberries after harvesting. Due to the seasonal and environmental influence, the technological quality features should be observed over at least two different years, at main harvesting periods. If high percentages are found in both years, one may conclude that the seed defects are due to genetic abnormalities.

Caffeine content varies amongst Arabica (0.6 for some mutants, normally 1.0–1.4 %) and Canephora genotypes (1.8–3.5 %). Low caffeine content can be an important selection criterion, especially in Canephora, as the consumption of coffee with a high caffeine content is limited. Presently, caffeine content can be estimated by extraction or by automated procedures. For instant coffee production, the percentage of soluble solids is one of the most important features. Canephora has a higher percentage of soluble solids than Arabica.

3.3.3.2 Flavor Features

Coffee flavor may be affected by environment, genotype, growing conditions and post harvest handling. With *C. arabica*, common cultivars seem to vary relatively little in flavor, i.e. a Brazilian variety grown in Costa Rica will produce Costa Rican quality coffee. High-quality coffee can be produced with Typica varieties (e.g. Blue Mountain, from Jamaica) as well as with Bourbon varieties (e.g. SL28, from Kenya). Some differences between dwarf breeding lines in relation to standard tall varieties have been observed recently in West Africa [51]. It is not yet known if these differences are directly related to the genotype or indirectly associated to the growth habit and physiologic conditions of the tree. It is generally admitted that progenies derived from interspecific crosses with Canephora may still carry less desirable quality characteristics from this species, and should therefore be selected essentially for flavor (e.g. Catimor).

Significant differences in flavor characteristics have also been demonstrated among Canephora genotypes [45].

3.4
Breeding Methods and Techniques

3.4.1
Generalities

Coffee breeding has benefited by selection within existing populations (farmers' varieties, germplasm collections) and by selection in artificially created crosses between varieties or species. The choice of selection methods to be used depends on the breeding behavior of the coffee species, on selection objectives (e.g. disease resistance, quality) and on the type of varieties to be selected (clone, pure lines, hybrid varieties).

Due to the self-pollinating nature of Arabica, the selection method mostly used for this species is line selection in segregating natural or artificially created populations (F_2, F_3, etc.), whereas Canephora breeding has mainly involved the selection of clones and of hybrids between clones.

Breeding methods have been reviewed elsewhere [15, 51, 23]. Arabica coffee breeding is historically related to selection for yield and rust resistance [27]. More recently, Arabica breeding objectives have diversified to include other important breeding goals, like CBD and nematode resistance, insect resistance, hybrid vigor, and quality. Canephora coffee breeding has mainly aimed at high yield and, to a lesser extent, rust resistance. Recently, the development of hybrid vigor and of increased quality have been added to these objectives.

Breeding techniques are the tools that the breeder uses in his breeding program for manipulating genetic variation. Pollination techniques, pollen and seed storage, the use of male sterility as well as vegetative propagation and grafting techniques will be briefly reviewed.

3.4.2
Selection Methods Applied to *C. arabica*

Early attempts to improve the performance of *C. arabica* were confined to farmer selections of individual trees as a source of seed. However, the results of farmer selections were erratic – they produced little benefit in terms of improved production when applied to the Typica varieties in South America but gave satisfactory results when used on different types of Bourbon varieties [15]. Important improvements in yield potential have been obtained in South America using trees derived from natural out-crossing of Bourbon and Typica varieties. For example, the famous Mundo Novo variety was selected in Brazil in the 1950s in a population derived from a natural cross between Sumatra (Typica) and Red Bourbon. The apparent vigor present in the early generations (F_2, F_3) could be fixed in more advanced generations, resulting in final selections being 60–120 % more productive than the parental varieties [15].

In more recent times the main breeding methods applied to *C. arabica* have followed a similar path but with the con-

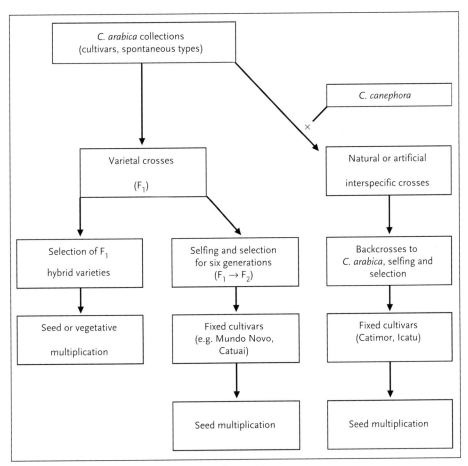

Figure 3.1 Breeding scheme for improvement of *C. arabica*.

trolled production of varietal and/or inter-specific crosses followed by selection from segregating populations (Fig. 3.1).

3.4.2.1 Yield

Hybrid vigor (heterosis) has only recently been exploited as a means of improving the yields of *C. arabica*. F_1 crosses of un-related genotypes of *C. arabica* [6, 22, 51] have produced significant increases in yields of coffee in a number of countries. In Cameroon, F_1 crosses between Ethiopian improved varieties and cultivated varieties [8] have given significant gains in yield over the parental varieties under less favorable growing conditions. In Central America, a breeding programme based on hybridization between Ethiopian improved varieties and local dwarf varieties (Catimor, Caturra) was initiated in 1990. This program has been conducted in several participating countries through a number of different institutions, including PROMECAFE, CATIE, CIRAD and IRD. The results obtained under this breeding program confirm that heterosis in most F_1 hybrids increases yield potential by 20–40 % when compared with the best parental control varieties [6].

3.4.2.2 Disease Resistance

Resistance to coffee rust has been developed by farmers and found in Ethiopian improved varieties. The Indian variety Kent, which contains the dominant SH2 resistance gene was, for example, selected by a farmer. Other dominant resistance genes (SH1, SH4) were identified later in Ethiopian improved varieties. The SH3 resistance factor was derived from a natural cross between *C. arabica* and *C. liberica*. These dominant genes have been of relatively little value in breeding as it was found in many countries that new rust types developed immunity to these resistance factors within 5–15 years [28]. Since the 1970s, efforts to obtain rust resistance have involved the selection of interspecific cross populations followed by back-crossing to *C. arabica* cultivars and line selection. Resistant varieties of this type have been released since 1985 (Catimor-type varieties, Icatu).

Resistance to CBD has been obtained through the development of natural resistance factors found in indigenous species, in cultivated varieties or by varietal crossing and back-crossing. In Ethiopia, a mass-selection procedure was applied in the 1970s to find resistance in wild or semi-wild populations [50]. In Cameroon, one of the varieties introduced called Java was selected and released in 1980 with both high yield and CBD resistance [8]. In Kenya, the hybrid variety Ruiru 11 was selected in the 1970s and 1980s. This variety combines three to four resistance genes obtained from different sources [51]. Further information on resistance to CLR and CBD is given in a separate chapter (Section II.6).

3.4.2.3 Nematode Resistance

Common cultivars are highly susceptible to destructive populations of nematodes (*Meloidogyne* spp. and *Pratylenchus* spp.). Variable levels of resistance have been found in diploid species (e.g. *C. canephora*, *C. liberica* and *C. congensis*) [15]. Unselected Canephora plants have been used for a long time as root-stocks for Arabica in Guatemala. In Brazil and in Central America, Canephora mother trees with high resistance to destructive local populations of *Meloidogyne* spp. were selected. Seed populations obtained from crosses between these mother trees resulted in the so-called

Apoatà and Nemaya root-stock varieties [2, 18]. The identification of high resistance to destructive species of *Meloidogyne* in Ethiopian improved varieties and of resistance to less destructive species in Catimor lines opens the way to obtaining hybrid varieties of Arabica with a broad spectrum of resistance to *Meloidogyne* spp. [5].

3.4.2.4 Insect Resistance

Resistance to Coffee Leaf Miner has been found in many diploid coffee species. In Brazil, a breeding programme is underway which is based on the introgression of resistance from *C. racemosa* [15]. Several back-cross generations to Arabica varieties show that high levels of resistance can still be found. Therefore, it can be assumed that this resistance is likely to be based on a few specific genes and could be fixed by further pedigree selection carried out over subsequent generations [30].

3.4.3
Selection Methods Applied to *C. canephora*

Due to the open-pollinated nature of the Canephora species, selection methods differ from those used for Arabica. Hybrids between clones have been selected to obtain improved seed varieties, which are more easily distributed to farmers than cuttings. However, hybrid varieties still display large variations because the parental clones are heterozygous. Selection of clones (rooted cuttings) has been undertaken in many countries in order to fix desirable characteristics and to maximize yield. Recently, improved knowledge of the genetic make-up of the Canephora species has allowed for the initiation of reciprocal recurrent selection (Fig. 3.2).

3.4.3.1 Conventional Hybrid Selection

The principle of hybrid selection is based on the testing of progenies from interesting mother trees, selected either in fields or amongst breeding plants or in collections. Mother tree selection based on the characteristics of individual trees can be effective for qualitative features (like disease resistance or caffeine content), but is less effective for yield [25, 23]. However, Bouharmont *et al.* [11] showed that the yield of clones evaluated in collections or in clonal trials can be correlated to the mean value of their progenies.

Progeny testing has been carried out by using open-pollinated progenies, by crossing with specific tester plants or by carrying out "factorial" crossing schemes (controlled crosses between different mother and father plants). The latter method allows the identification of genotypes that display good "general combining ability" (i.e. give good hybrids with a large range of other genotypes) or which display "specific combining ability" (i.e. give good hybrids only with some genotypes, not with others). With *canephora* coffee, it has been demonstrated that for yield characteristics the general combining ability predominates [23, 36].

Early stages of Canephora breeding programmes have assessed open-pollinated progenies to obtain valuable parents for hybrid production. In Indonesia this method was applied between 1915 and 1930, but only a small percentage of open-pollinated progenies proved to be outstanding. Later, the test-cross method or factorial crossing schemes were successfully used to select outstanding cross-combinations not only in Indonesia but also in India and in several African countries [17, 23]. The yield of hybrids selected by this method proved to be significantly higher than the yield of unselected local plants. In Cameroon, for

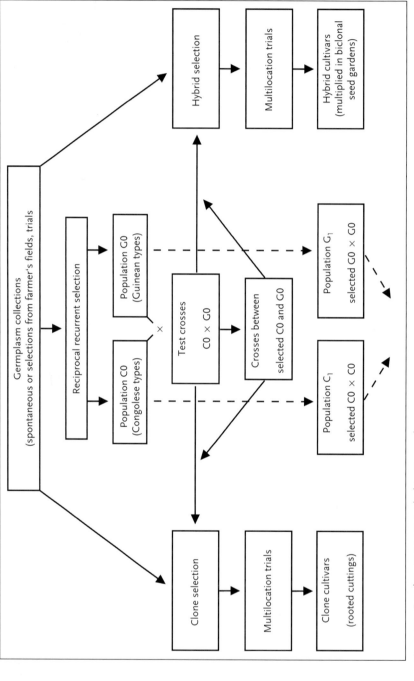

Figure 3.2 Breeding strategies applied to *C. canephora*.

example, the best hybrid combinations yielded about 50% more than open-pollinated progenies from unselected mother plants [11]. However, the yield potential of the best hybrids were generally lower than the yield potential of the best clones [14], except in recent developments of advanced hybrid selection in Ivory Coast [43].

3.4.3.2 Conventional Clone Selection

Clone selection has been carried out with success in most Canephora-growing countries from the 1930s to the 1950s onwards. The first step in clone selection is the constitution of a large germplasm collection containing plants selected by local farmers or introduced from other countries. These collections allow for a first evaluation of the value of a clone for interesting features like yield, vigor, bean size, growth habits, drought resistance, caffeine content, resistance to diseases and cup quality. The most interesting plants are multiplied as rooted cuttings and evaluated in clone trials together with standard varieties (seed or clones). It has been demonstrated that interactions exist between clones and the environment. This means that the best clones selected in one country are not necessarily suitable for other countries. In Cameroon, clones were selected with emphasis on specific adaptation to different ecological zones [10]. Potential yields of selected clones in Africa vary from 1.5 to 2.5 tons of green coffee/ha, whereas yield potential of unselected local seed populations is generally lower than 1 ton/ha.

Clone selection and hybrid selection can be related breeding strategies. After the selection of the best clones, these can be used to produce new hybrid combinations, which can then be used to select new promising clones. The relationship between parental behavior and offspring, justifies such an approach. If more than one cycle is applied, this strategy is called recurrent selection [17, 37].

3.4.3.3 Reciprocal Recurrent Selection

Reciprocal recurrent selection is a type of recurrent selection where progeny testing is carried out by crossing genotypes belonging to different populations. In other words, the value of the parental genotypes is tested by progenies obtained by crossing with testers from other genetic groups. The best parental clones belonging to the same group will then be crossed amongst each other to form a new population to be used for initiating the next selection cycle. This type of breeding scheme has been successfully used for maize crops and also for other perennial crops, such as palm oil. The advantage is that genetic variation can be maintained at a high level, allowing for continuous progress by using successive selection cycles. A favorable condition for initiating reciprocal recurrent selection is the existence of well-identified genetic groups, which reveal complementary features, and hybrid vigor in intergroup crosses.

For Canephora, the identification of the two main genetic groups, Guinean and Congolese [3] combined with the demonstration of the value of hybrids between these groups has justified the implementation of a reciprocal recurrent selection programme in Ivory Coast, initiated in 1985. About 100 representative plants of each group were crossed with a few testers from the other group [37]. More than 70 outstanding plants have already been selected within the best cross progenies. These can be expected to out-yield standard commercial clones by 20–50%. A few highly productive hybrids that yielded more than the average of the standard clones were also identified [43]. These candidate clonal and hybrid varieties now need to be tested in multilocational variety trials. The best genotypes of each group have been crossed to obtain improved Guinean and Congolese populations to be used in the second selection cycle [36].

This ambitious programme has already demonstrated its value by improving yield and yield efficiency. This programme is also expected to generate significant progress for other selection features like bean size or cup quality [36, 45]. Selections obtained by this programme hold great potential for future national or regional selection programmes in Africa and elsewhere.

3.4.3.4 Interspecific Hybridization

All coffee species can be crossed relatively easily. Two types of crosses have been used for improvement of lowland coffees:

- Arabusta hybrids (F_1 crosses between tetraploid Canephora clones and Arabica varieties) [13]
- Crosses of Canephora with other diploid species, mainly *C. congensis* (Congusta) and *C. liberica* (Ligusta).

The Arabusta hybrids were created and selected between the 1960s and the 1980s in Ivory Coast, Cameroon and Ghana in order to improve the quality of lowland coffees. Arabusta hybrids have responded well, obtaining increased quality features (large bean size, lower caffeine content and acceptable cup quality). However, these hybrids have so far failed to give satisfactory yields at low altitude. This is chiefly due to the low fertility of the flowers, irregular yields, a weakness of the main stems which causes them to bend and a high susceptibility to insect pests (Antestia bug, stem borers). An additional problem is the low out-turn ratio of Arabusta (10–

12%), due to the presence of empty locules. The Arabusta breeding strategy has been re-oriented towards the selection of more advanced generations (F_2–F_4) and of back-crosses to *C. arabica*, followed by intercrossing the best genotypes in each generation. This strategy can only be expected to give results in the long term and through the evaluation of many thousands of plants in each generation [21].

Crosses of *C. canephora* with *C. congensis* are fully fertile. *C. congensis* possesses interesting features such as good bean size, quality, plant shape and resistance to flooding, which could be exploited in F_1 hybrids or by further back-crossing with *C. canephora*. A few commercial Congusta clones have been selected in Madagascar [47]. Back-crosses of Congusta to *C. canephora* have appeared to show relatively low vigor and low productivity, as observed in Ivory Coast [53]. It would therefore be necessary to carry out further breeding. In India some interesting Congusta hybrid populations have been released.

F_1 crosses of *C. canephora* with *C. liberica* present a certain degree of sterility. Back-crosses were made in Ivory Coast and many plants presented interesting features, but yield has not yet been satisfactory [54].

3.4.4
Breeding Techniques and Tools

Pollination techniques have been described in detail by Carvalho *et al.* [17] and by Van der Vossen [51]. To avoid selfing, Arabica flowers need to be emasculated by removing petals and anthers 1–2 days before anthesis. This can be done manually or by using suitably adapted scissors. With Canephora, emasculation is often carried out as well because it facilitates hand-pollination. However, due to the strict self-incompatibility of this species emasculation is not really necessary. Branches with "female" flowers to be used for crosses should be covered with a paper bag or with a tightly woven cotton sleeve, supported by a light iron frame and attached to a bamboo stick in order to avoid the cloth contacting the flowers. This technique is better adapted to Canephora as it uses all the flowers on long branches. Pollen is obtained by collecting closed flower buds just before anthesis. These buds, when stored in plastic bags or in Petri dishes with a small piece of damp cotton wool, will open and produce viable pollen for 1–2 days. Pollination can be made by gently rubbing the anthers upon the stigmas of the emasculated flowers in the field. When using stored pollen, pollination can be carried out with a small brush dipped into the pollen. Branches with pollinated flowers need to be covered for several days, because unpollinated stigmas can stay fertile for about 10 days. If the cover is removed before 10 days, the stigmas should be cut off to avoid undesirable pollination.

Pollen storage is useful when the flowering of the parent plants is not synchronized or when the parent plants grow in different sites. Pollen can be collected from the flowers by gently rubbing the anthers with a small brush. Fresh and dry pollen can be correctly stored for several months in the refrigerator in tubes inside containers with silica gel. For Canephora, pollen can be successfully diluted in neutral talc powder. This facilitates the pollination process. High pollen viability can be maintained by storing under a vacuum at –18 °C for more than 2 years [51]. In order to be able to carry out more than one pollination campaign per year, coffee trees can be induced to flower during the dry season by watering. This induction of flowering by watering is more efficient towards the end of the dry season when flower buds are completely initiated.

Seed storage is practiced by breeders when sowing dates do not coincide with harvesting or when seeds, cumulated from different crossing campaigns, need to be sown together. Exposure to the sun may kill the seed, so coffee seeds to be used for sowing should be dried in the shade. The coffee seed does not have a dormancy period but the germination potential of dry parchment seeds (10–12 % humidity) usually drops after 6 months storage in paper bags at room temperature. Arabica seeds can be conserved up to 30 months at cooler temperatures (15–19 °C) in sealed plastic bags where the humidity of the seed is at 40–41 % water content [51].

Male sterility has been identified in some improved varieties of *C. arabica* in Brazil [40] and in Costa Rica. The use of male sterility can be an important tool for the breeder to produce hybrid seeds by natural pollination, using the male sterile variety as a female parent in seed gardens. Male sterility can be used directly if crosses obtained through naturally occurring male sterile plants prove to be agronomically interesting. The breeder could also introduce the male-sterile characteristic into commercial cultivars by back-crossing.

The *vegetative propagation* of coffee species is an important tool for reproducing clonal varieties. Rooted cuttings have been used on a large scale to reproduce Canephora clones as well as Arabusta hybrids. With Arabica, mass propagation methods for rooted cuttings have been studied in detail in Kenya [51], but have so far not been used on a commercial basis. Developments of vegetative propagation by somatic embryogenesis show that this *in vitro* method is a potentially valuable tool for large-scale multiplication of heterozygous Arabica varieties (e.g. F_1 hybrids) and Canephora clones.

Grafting of Arabica varieties on root stock varieties of Canephora with resistance or tolerance to nematodes is used commercially in Brazil and Guatemala. The method used is hypocotyledonary grafting of seedlings in the pinhead stage, as described first by Reyna [46]. By using this method, Arabica appears to be graft-compatible to many diploid coffee species [24]. However, earlier reports have indicated a degree of graft-incompatibility between Arabica and some *C. liberica* or Canephora genotypes [25, 33, 35]. This is probably related to the fact that, in these early studies, older plants were used for grafting.

Progress with horticultural vegetative propagation as well as micropropagation is described elsewhere in this manual.

3.5
Variety Trials

3.5.1
Statistical Layout

The statistical layout of breeding plots in the field depends on the genetic material and the objective of the trials. Germplasm collections are generally constituted by small unreplicated plots with five to 10 plants per variety or clone. However, it can be an advantage to use two replications and interplant the improved varieties at regular intervals with standard varieties. In this way, quantitative features could also be more reliably assessed when observing these collections.

Selection trials for *C. arabica* may contain genetically homogeneous varieties (F_6 or more advanced generations considered as pure lines, or F_1 hybrid varieties) or segregating progenies (F_2–F_5, back-crosses). Highly segregating populations (F_2, F_3, back-crosses) can be evaluated in observation plots, using standard commercial varieties as controls. The objective is generally to select individual plants in these populations, therefore repetitions are not necessarily useful. More advanced populations and fixed varieties are usually tested in randomized block designs with six to eight replicates and eight to 10 plants per plot. This type of design only needs lateral border rows and no separation between blocks if the varieties are all of the same size. However, spacing between blocks is useful for an easier identification of the plots, facilitating observations and avoiding errors. In general it is not recommended to compare varieties of different sized plants (tall and dwarf) within the same experiment. If this needs to be done, it is better to plant the tall varieties side by side within the blocks and to adjust spacing according to the requirements of the different plant sizes.

In breeding experiments high variations between plants can be observed due to the heterogeneous nature of the field (soil fertility, slopes, effect of shade trees). The most efficient statistical design to cope with field heterogeneity is complete randomization of single-tree plots. In this way, each tree represents a replicate, therefore the number of plants and hence the size of the experiment, can be reduced. Such designs can be efficiently used even where unequal numbers of plants are available. While only 25 plants per variety are recommended in such experiments, the minimum number of recommended plants in randomized block design is about 50. These advantages have caused many coffee breeders to adopt completely randomized single-tree plot designs. However, this type of experiment is mostly carried out on experimental stations where highly qualified staff is available as mistakes are more easily made.

Statistical designs of clonal trials of Canephora are much the same as those used for fixed varieties of Arabica. However as genetic variation within hybrid progenies of Canephora is generally higher than that of Arabica, it is recommended to use a larger number of plants for seed progeny testing (at least 40 for completely randomized single tree plot designs and 60–80 for randomized block designs).

Preliminary variety trials are generally carried out on one specific site and with a limited number of plants per variety. Multilocational confirmation trials are needed to evaluate the development of pre-selected varieties in a range of different environments. For preliminary experiments it is often preferable to use less than optimal plant husbandry methods (average fertilization regime, absence of phytosanitary treatments) in order to be able to select genotypes that have a good adaptation to average agronomic conditions and prevailing diseases and/or pests. Confirmation trials should be carried out under prevailing recommended agronomic conditions (spacing, fertilizer and phytosanitary treatments). For preliminary experiments, wide spacing would be more beneficial to ascertain individual plant behavior. This is especially important for dwarf varieties, which are commercially planted at 5000–10 000 plants/ha. Efficient selection for important individual plant features, like yield, vigor and resistance to over-bearing would be impossible with such a high density of plantation.

3.5.2
Observations to be Carried Out

The breeder should carry out all observations related to his breeding objectives. Vigor is generally observed in 1 year-old plants in the field and in adult plants. Yield should be observed for one growing cycle or at least for four consecutive harvesting years. For individual plant selection in segregating populations, the use of breeders scores (see Section 3.3.1.) can be a useful substitute for laborious yield determinations, especially in countries where coffee is harvested many times per year. Coffee, as a biennial crop plant, shows alternative high- and low-yielding years. In Brazil, it has been observed that selection in high-yielding years can be more efficient than selection carried out in low-yielding years. Whenever epidemics occur, observations on the reaction of plants to prevailing diseases and pests should be made. Quality and presence of seed defects in the varieties should be determined twice during the growing cycle. As these observations are labor intensive, they are generally only carried out on genotypes that are already pre-selected for yield.

3.5.3
Statistical Analyses

Generally, standard "analyses of variance" are used to evaluate the significance of genotype effects in variety experiments. In completely randomized single-tree plot designs, analyses of co-variance have been used to take into account the micro-heterogeneity within the experimental plots, with the average value of neighboring plants being used as the co-variable. This method, proposed initially by Papadakis, has been successfully applied in cocoa and coffee breeding trials in Ivory Coast [39, 19]. Co-variance can explain part of the residual variation, which improves the estimation of the variables due to genotypes, and thus the precision of the analysis.

The analysis of co-variance can also be applied to estimate the effect of vegetative vigor on plant yield. By using the adult plant stem diameter as co-variable, yield data can be adjusted for vigor, which may help to select plants that have a better yield-efficiency and may, therefore, be better suited to high-density planting [19] (see also Section 3.3.1.).

3.6
Description of Main Cultivated Varieties

3.6.1
Selected *C. arabica* Varieties

3.6.1.1 Typica-type Varieties
These varieties present main characteristics which are similar: low- to average-yielding capacity, large elongated cherries and beans, relatively small elongated leaves, good cup quality, flexible main stems, rather thin primaries, a nearly horizontal branching pattern and bronze-tipped young leaves. With the exception of some lines like "Guatemala" or "Blue Mountain" which appear to have some resistance to CBD, Typica varieties are susceptible to all main diseases, pests and nematodes.

Typica varieties are still grown in Colombia, Central America, the Caribbean region ("Blue Mountain"), Papua New Guinea ("Blue Mountain"), Pacific, Indonesia ("Java Typica", "Blawan Pasumah" or "BLP", "Bergendal") and Cameroon ("Jamaique"). The traditional Typica or Arabica variety grown in South and Central America is probably genetically very similar to Blue Mountain, Java Typica and Jamaique. It could be that the more vigorous and more productive Typica variety called "Sumatra" in Brazil is related to Bergendal, still grown in Northern Sumatra, and to the "Blawan Pasumah" variety selected in East Java which also originates from Sumatra [29]. Some of the most expensive commercial coffee brands are produced with Typica varieties (e.g. Blue Mountain coffee). Dwarf mutants of the Typica variety were detected in Central America that are called "Villa Lobos", "San Ramon" and "Pache".

3.6.1.2 Bourbon-type Varieties
In most countries, where Typica varieties were grown together with Bourbon varieties, farmers tend to prefer the latter because of its generally higher vigor and yielding capacity. Bourbon-type varieties are still grown in Colombia, Central America and West Africa. In Brazil, the Yellow Bourbon variety appears to be more productive than the Red Bourbon variety, but Yellow Bourbon may have been derived from a natural cross between Red Bourbon and "Amarelo de Botucatu", a yellow-fruited Typica variety [18]. Bourbon varieties have broader leaves with rounder cherries and beans than Typica varieties, the main stem and primaries are stronger and stiffer and the branching pattern is more erect. In Central America, the "Tekisic" variety is a Bourbon-type variety still grown in El Salvador. Bourbon varieties are still widely grown in West Africa. Selections like "Mibirizi" "Jackson" or "Bourbon Mayaguez", cultivated in Rwanda and Burundi, as well as several SL selections in Kenya (e.g. "SL28") or "N39" in Tanzania can be considered as Bourbon-type varieties [32]. SL28 presents some drought resistance under Kenyan conditions. The "Arusha" variety in Papua New Guinea is probably related to Bourbon varieties selected in Tanzania. Bourbon varieties can give excellent cup quality (e.g. SL28 in Kenya), but also show susceptibility to all major coffee diseases and pests.

3.6.1.3 S795 Variety
S795, released in the 1940s in India, was created by crossing the Kent variety and the S288 selection [49]. It combines two resistance factors (SH2 and SH3) to CLR.

This hardy variety is still widely grown in India and also in Indonesia, although it has largely lost its rust resistance. Introductions of this variety in other countries often show segregation for rust resistance and seed characteristics (Eskes, personal observation).

3.6.1.4 Java Variety

This variety was selected in Cameroon, among many other Arabica varieties, and has been released for cultivation since 1980 [8]. It was originally introduced from Java, and resembles certain "Abyssinia (AB)" numbers still to be found in collections in East Java (Eskes, personal observations). Abyssinia is the name of an introduction of Arabica coffee from Ethiopia into Java, made by Cramer in 1928 [25]. Java is a vigorous variety and displays good resistance to CBD in Cameroon. In years of high yield it shows high susceptibility to rust; however, due to its vigor, it is able to recuperate its foliage in low-yielding years. Its yield potential in Cameroon can be estimated to be 1.5–2 tons/ha, whereas other known varieties like Caturra and Mundo Novo yield less than 1 ton under the same conditions [8, 9]. The fruit and seed of Java are elongated and young leaf types are bronze colored.

3.6.1.5 Maragogipe

"Maragogipe" is a mutant variety (from Typica) that appeared in Brazil [16]. It has long internodes, large leaves and large cherries and beans, but relatively low yield. There is a certain amount of commercial interest for the large bean size. Therefore, this variety is grown to some extent in Central America and in Mexico. More recently, derivatives of crosses of Maragogipe with Pacas ("Pacamara") and with Caturra ("Maracatu") are being selected and are now also grown.

3.6.1.6 Caturra

Caturra is a dwarf variety with short internodes, it was discovered as a mutant of the Red Bourbon variety in Brazil in 1937. In Central America, similar mutant varieties were detected ("Vila Sarchi" and "Pacas"), but are not cultivated to any extent. Caturra has a high yield potential (over 2000 kg/ha under good growing conditions), but was never recommended in Brazil due to lack of hardness and severe over-bearing observed after three to four production cycles [18]. However, it seems well adapted to the growing conditions that prevail in Colombia and Costa Rica, where it has served as a basis for high-density planting (5000–10000 plants/ha). Like Bourbon, this variety is susceptible to all main coffee diseases and pests. Leaf and fruit characteristics are similar to those of the Bourbon varieties.

3.6.1.7 Mundo Novo

This variety still occupies about 40% of the coffee area in Brazil, where it is well appreciated because of its high-yielding capacity and hardness. It was selected by the Instituto Agronômico of Campinas in the 1940s and 1950s, and is derived from a natural cross between Sumatra and Red Bourbon. The tall-growing Mundo Novo variety yields about 30% more than the best Bourbon lines in Brazil [18]. Although it is normally planted at low density (1200–1600 plants/ha), due to its open growth, it can also be successfully grown at higher planting densities (3000–4000 plants/ha). Some of the best lines of Mundo Novo are designated by the codes LCMP 376-4, 388-6, 379-19, 464-12, 467-11, 501-5, 502-1 and

515-20. A selection of Mundo Novo with larger beans is named "Acaià" (LCP 474). Although Mundo Novo is very well adapted to the Brazilian climate and cultivation practices, this variety has shown variable adaptation to other countries and is little used elsewhere. It is susceptible to all main coffee diseases and pests. Leaf and fruit characteristics are intermediate between Typica and Bourbon.

3.6.1.8 Catuai

This dwarf variety selected by the Instituto Agronômico of Campinas in Brazil in the 1950s and 1960s also occupies about 50% of the coffee acreage in that country, and is widely used in Central America as well. It was selected as F_5 and F_6 lines derived from a cross between Yellow Caturra and Mundo Novo. It has the dwarf stature of Caturra, but has inherited the higher vigor and hardness of Mundo Novo. Some of the best red-fruited selections ("Red Catuai") are designated as LCH 2077-2-5-81, 2-5-99, 2-5-44, 2-5-72, 2-5-46 and 2-5-144, and yellow-fruited selections ("Yellow Catuai") as LCH 2077-2-5-62, 2-5-100, 2-5-86, 2-5-66 and 2-5-32 [18]. Catuai is susceptible to main coffee diseases and pests. In Mexico, a variety similar to Catuai was selected from Mundo Novo × Yellow Caturra cross progenies that is called "Garnica". Another similar variety, derived from a Red Catuai × Mundo Novo cross, but with more vigor than Red Catuai, was released in 2000 in Brazil under the name "Ouro Verde".

3.6.1.9 Catimor- and Sarchimor-type Varieties

This group of similar varieties is derived from crosses between the HdT and Caturra (Catimor) or Vila Sarchi (Sarchimor). The first cross of this type was made at the CIFC in Portugal in the 1960s. F_2 populations were tested in Angola, and F_3 and F_4 populations in Brazil, in the 1970s. More advanced populations were reselected for rust resistance at CIFC and sent to many countries as F_5 or F_6 populations. Two different plants of HdT were used in Portugal for these crosses: CIFC 832/1 and 832/2. In Colombia, another origin of HdT was used the CIFC 1343 population. Agronomic features of selections derived from Catimor Sarchimor and Cavimor (HdT × Catuai) have been described by Bettencourt [7]. In Brazil, several early introductions of Catimor have shown over-bearing which is why they have not yet been used commercially in that country. In many other countries, the agronomic behavior of Catimor is quite similar to that of Caturra.

The most important characteristic of Catimor varieties is their resistance to prevailing races of CLR, although in Indonesia, Thailand and India several Catimor lines prove presently to be more or less susceptible. Derivatives of HdT may also have resistance to CBD or to certain populations of root-knot nematodes, as found in some Colombian Catimor lines and in Sarchimor progenies.

Catimor-type varieties have been distributed to farmers from the 1980s onward. One of the first commercial Catimor varieties is the "Colombia" variety, released around 1985, is presently widely grown in that country. This variety consists of a mixture of selected F_5 or F_6 lines, either with yellow- or red-colored fruits [44]. In Central America, Catimor varieties originating from Portugal or Brazil have been selected and were released to farmers at the end of the 1980s. Two origins were mainly selected, i.e. T5175 and T8667. More or less advanced populations of T5175 have been selected in many Central American countries and were released under different names (e.g. "IHCAFE 90" in Honduras, "Catrenic" in Nicaragua). The cup quality of this variety is now considered inferior to other Arabica varieties (Bertrand, personal communication). T8667 was introduced from Brazil as an F_6 population and has been selected mainly in Costa Rica, were it was released in 1995 under the name "ICAFE 95". Similar varieties were released under different names in Mexico ("Oro Azteca"), Honduras ("Lempira") and El Salvador ("Catisic"). In India, advanced populations of Catimor, received from CIFC, were released to farmers under the name "Cauvery" in the late 1980s.

Sarchimor varieties were released in Brazil by IAPAR ("IAPAR59", released in 1993) and by IAC ("Obata" and "Tupi", released in 2000). These have good resistance to local races of CLR, and Tupi and IAPAR59 also have resistance to certain nematode populations (*Meloidogyne exigua*).

3.6.1.10 Icatu

The Icatu variety, selected by the Instituto Agronômico de Campinas in Brazil, was released officially in 1993. This variety consists of advanced breeding lines derived from a *C. arabica* × *C. canephora* hybrid, back-crossed to Brazilian Arabica cultivars. It is a tall-growing variety that has many characteristics in common with Mundo Novo. However, it presents resistance to common races of leaf rust and may yield 30–50 % more than Mundo Novo. Red- and yellow-fruited progenies have been selected. Some selections of Icatu were released for commercial use in Brazil by the end of the 1990s. This population is still being selected in order to obtain lines

which also show resistance to root-knot nematodes (*Meloidogyne* spp.) [18].

3.6.1.11 Ruiru 11

This dwarf variety was selected as a result of a breeding programme carried out in Kenya in the 1970s and 1980s. The main objective of this program was to obtain a high-yielding variety with resistance to CLR and CBD [51]. The Ruiru 11 variety is a composite hybrid variety consisting of crosses between Catimor lines and selected breeding progenies. The breeding progenies were obtained by intercrossing different CBD-resistant varieties (Rume Sudan, K7, HdT) with the local susceptible SL28 cultivar, followed by back-crossing to SL28. Individual plants that present high resistance to CBD are used for crosses with CBD and rust-resistant Catimor lines, thus obtaining the Ruiru 11 variety. This variety, despite its heterogeneous genetic background, presents a reasonable phenotypic uniformity in the field due to the dominant "Caturra" gene inherited from Catimor. The variety Ruiri 11 is reproduced by hand-pollination [1].

3.6.1.12 Root-stock Varieties for Nematode Resistance

So far, two Canephora seed varieties have been selected for use as root-stock for Arabica with resistance to nematodes. In Brazil, the "Apoatã" variety was released in the late 1980s. This variety consists of seed obtained on selected mother trees planted in a polyclonal seed garden [18]. In Central America, in 1995, the "Nemaya" root-stock variety was selected, which shows a high percentage of resistant plants to prevailing populations of *Meloidogyne* spp. The Nemaya variety is made up of a cross between two mother plants selected in the CATIE collection at Turrialba [2].

3.6.2
Selected *C. canephora* Varieties

As described earlier, the majority of cultivated Canephora is still made up of unselected open-pollinated seed varieties. In many countries, however, a selection of clones and hybrids has been successfully carried out.

In Indonesia, early selections for yield in Canephora were based on open-pollinated seed progenies obtained from selected clones. It appeared that clones, like BP39 and 42, gave progenies that were widely adapted and productive. The use of hybrid seed production in biclonal seed orchards planted with these clones was recommended at the end of the 1930s. Selection of clonal varieties was also carried out at the same time, resulting in recommendations of "polyclonal varieties". However, some clones were well adapted to certain places but less adapted to others [25]. The Indonesian coffee-breeding programme was interrupted during the World War II and many selections were apparently lost. The pre-war selections of Canephora are apparently still being recommended for commercial use in Indonesia.

In several African countries, Canephora breeding was initiated at the end of the 1950s, with support from the French Coffee and Cocoa Research Institute (IFCC, now part of CIRAD-CP). The selection of hybrids as well as of clonal varieties was carried out, aiming at yield, disease or pest resistance, bean size and adaptability. In the 1960s and 1970s, seven clones were selected in the Ivory Coast [14], 10 in the Central African Republic [27], 7 in Uganda [41], 8 in Madagascar [47] and 12 in Cameroon [10]. In Cameroon, it was shown that four of the recommended clones had wide adaptability, whereas the others were more specifically adapted to one of the two production zones (eastern or western area). The above-mentioned clones have a production capacity of over 2 tons/ha, average to good bean size and variable caffeine content (1.8–3.2 %). At the same time hybrid varieties were also selected, for example, in Ivory Coast, where six hybrids were recommended. Although less productive than clones, these hybrids present more adaptability, especially in marginal growing areas [14].

In Brazil, clones were recently selected from the heterogeneous Conillon variety, which is quite susceptible to leaf rust, although more-or-less resistant to drought. Based on the evaluation of 77 clones for yield, maturation time, bean size and resistance to leaf rust, three clonal varieties were recommended by INCAPER Institute: INCAPER 8111, 8121 and 8131. These varieties consist of 10, 15 and 14 clones, which present respectively early, normal and late fruit maturation and an average yield about 30 % higher than a standard seed mixture [12].

In Mexico, five clonal selections were made in progenies from Turrialba, Costa Rica. They were released as a polyclonal variety under the name "ROMEX".

3.7
Multiplication of Selected Varieties

With *C. arabica*, most selected varieties consist of pure lines. This means that they are nearly homozygous and can be reproduced in separate seed gardens. Generally, the breeders provide basic seed of fixed breeding lines (F_6, F_7) to specialized farmers or seed production organizations. Inspection should be carried out to eliminate possible off-types in seed production gardens. The seed gardens of individual varieties need to be separated from other coffee varieties, in order to avoid out-crossing. No information is available on the exact distance needed to be completely sure that no out-crossing will occur, but a minimum distance of about 25 m between plots seems to be safe. Wind-breaks between the seed production plots are recommended in order to provide an additional barrier against out-crossing.

As a result of new breeding methods, hybrid varieties are now being selected in a few countries (Kenya, Cameroon, Central America). The Ruiru 11 variety selected in Kenya is reproduced by controlled hand-pollination. A pollination team induces flowering by irrigation of the male parents, collects pollen from these flowers and uses this pollen to fertilize female parents. This procedure is quite laborious but with good organization several millions of hybrid seeds can be produced annually (Agwanda, personal information). The cost of such hybrid seed is estimated to be less than US$50/kg (around 3500 seeds), which appears to be an acceptable price to the farmers. Alternatively, hybrid Arabica varieties may be reproduced *in vitro* by somatic embryogenesis. The high potential of this method for the rapid multiplication of Arabica is now well established (see Section 3.4.4.). Research carried out by Sondhal [48] indicates that only a low frequency of deleterious mutants appears when applying somatic embryogenesis to multiply fixed Arabica varieties. Further studies on the genetic stability of hybrid Arabica genotypes are being undertaken in Central America.

For *C. canephora*, polyclonal varieties can easily be reproduced by traditional horticultural methods. Rooted cuttings can be obtained from clonal cutting gardens, consisting of orthotropic bent stems planted at high density that continuously produce orthotropic shoots. Such gardens can produce about 1 million cuttings/ha/year. It is to be noted that, due to self-incompatibility, several intercompatible clones are required to be part of one polyclonal variety of Canephora in order to obtain a good seed set. Commercial multiplication by somatic embryogenesis of selected Canephora clones has also been carried out in an EU-funded project in Uganda (Berthouly, personal communication).

In order to reproduce hybrid varieties of *C. canephora*, biclonal or polyclonal seed gardens have been used. The six selected hybrids in Ivory Coast are now reproduced in biclonal seed gardens. In order to overcome pollination barriers, due to non-synchronous flowering, triclonal seed gardens were initially suggested [14]. However, this method resulted in lower-quality planting material due to a presence of less productive cross-combinations. Therefore, these seed gardens have recently been transformed into biclonal seed gardens [20], in spite of a certain degree of under-pollination. Canephora seed gardens, like for Arabica, should be isolated but, due to the allogamic nature of Canephora, extra care should be taken to ensure correct isolation. Wherever possible, large intercropping strips with another tree crop should be implemented to prevent unwanted out-crossing.

3.8
Conclusions and Perspectives

The breeding procedures for coffee are time-consuming. Traditional Arabica breeding has relied mainly on varietal crosses or on interspecific crosses, followed by backcrossing and line selection. With this procedure, 20 to 35 years are necessary to produce new fixed seed varieties. New breeding methods and tools may help to reduce the time needed for variety selection. Recent developments in coffee micropropagation make it feasible to rapidly select heterozygous clonal Arabica varieties (F_1 hybrid varieties or plants selected in any segregating population). Once research on the stability of micro-propagated coffee in the field is completed, such clonal varieties can be reproduced by somatic embryogenesis or by traditional horticultural multiplication methods. F_1 hybrids may also be reproduced by hand-pollination, as already practiced in Kenya, or in future, by using male sterility.

It can be expected that Arabica breeding will continue to develop the hybrid vigor observed in crosses among Ethiopian varieties and cultivated varieties. This breeding method also allows for the combination of the dominant features present in wild coffee types (like nematode and rust resistance) with resistance already present in available cultivars or breeding lines (Catimor, Sarchimor), thus cumulating resistance genes. These advantages have given incentive to a hybrid selection programme initiated in Central America since 1992 [6]. The best selections of vegetatively propagated F_1 trees are being tested in a large regional trial set-up in Central America since 2001.

Breeding objectives of *C. arabica* have mainly involved yield potential, plant stature (dwarf varieties) and rust resistance. Current breeding objectives include also nematode, CBD and insect resistance (Coffee Leaf Miner, Coffee Berry Borer), as well as quality. Resistance to root-knot nematodes, caused by *Meloidogyne* spp., has been obtained by using selected Canephora root-stock varieties. Further development can be expected by selecting diploid coffee species for improved resistance or for tolerance to the *Pratylenchus* nematode and/or to other root parasites (scale insects). Breeding for Coffee Leaf Miner resistance is in an advanced stage in Brazil, where progenies derived from a *C. arabica* × *C. racemosa* cross are being selected. Resistance to Coffee Berry Borer may be very difficult to obtain by traditional methods, as the genetic variability for resistance is limited. Attempts to use genetic transformation in order to obtain resistance to this important pest are therefore justified.

Breeding for quality in *C. arabica* is expected to mainly involve the maintenance of quality features in traditional varieties. Derivatives from interspecific crosses may present less desirable quality features (seed defects, higher caffeine or less refined cup quality) that have to be counter-selected. However, such interspecific cross progenies, may also lead to new desirable quality features like lower caffeine content. Some peculiar Arabica varieties like "Mokka" are also of interest for further investigation in relation to cup-quality. Cup-quality research will have to take into account the significance of environmental impact, post-harvest technologies and the physiological conditions of the plants.

For *C. canephora*, breeding has been effective for yield, disease resistance and adaptation. However, due to socio-economic conditions selected clonal or hybrid varieties are, as yet, little used by farmers. For instance it is estimated that only 10–20% of the area is planted with selected varieties in Ivory Coast. In other countries, like Brazil, Canephora selection has only recently been initiated. Further progress in Canephora breeding can be expected by the development of hybrid vigor, as is the case for the reciprocal recurrent selection programme carried out in Ivory Coast. Several hybrids were recently selected that out-yielded the commercial standard clone varieties [43]. Improved quality features (larger bean size, lower caffeine content, better cup quality, percentage of extractable soluble solids) have become important breeding objectives. Progress can be expected as the Canephora species reveals a broad genetic variation to develop these features.

Derivatives of interspecific crosses have been used for the improvement of *C. arabica*. However, so far the selection of crosses between *C. arabica* and *C. canephora* (Arabusta) have not yet successfully improved lowland coffee types. Only the long-term selection of large numbers of recombinant genotypes, selected in more advanced Arabusta generations and/or in back-crosses with *C. arabica*, may open the way to generating productive and suitable lowland coffee types with improved quality.

Following pictures (Pict. 3.1 to 3.7) show coffee plants from different origins.

Picture 3.1 *C. arabica.*
(A) Katii single stem Yemen. From: A. B. Eskes.
(B) Catuai Costa-Rica. From: A. B. Eskes.
(C) SL28 Kenya. From: A. B. Eskes.
(D) Mokka Kenya. From: A. B. Eskes.

Picture 3.2 *C. canephora.*
(A) Robusta (single stem) Thailand. From: J. N. Wingtens.
(B) Robusta (free growing) Mexico. From: J. N. Wintgens.

Picture 3.3 *C. liberica.*
(A) Liberica Malaysia. From: J. N. Wintgens.
(B) Liberica single stem (front with robusta behind). From: J. N. Wintgens.
(C) Liberica (cherries) Malaysia. From: J. N. Wintgens.

Picture 3.4 *C. excelsa.*
(A) Excelsa (tall plants) with Robusta Vietnam. From: J. N. Wintgens.
(B) Excelsa (cherries) Vietnam. From: J. N. Wintgens.

Picture 3.5 *C. arabica* hybrids.
(A) Catimor T 5175 line (vigorous) Guatemala. From: A. B. Eskes.
(B) Catimor in fructification Thailand. From: J. N. Wintgens.

Picture 3.7 Kapakata and Timor hybrid Columbia. From: J. N. Wintgens.

Picture 3.6 *C. canephora* hybrids.
Arabusta (Ecuador). From: J. N. Wintgens.

Bibliography

[1] Agwanda, C. O. Hybrid seed production in arabica coffee: consequences of isolation techniques in preventing alien pollen contamination. In: *15th International Scientific Colloquium on Coffee*. ASIC, Paris, 1993, 180–182.

[2] Anzueto, F., Bertrand, B. and Dufour, M. Nemaya, desarollo de una variedad portainjerto resistente a los principales nematodos de America Central. *Bol. Promecafé* **1995**, *66/67*, 13–15.

[3] Berthaud, J. Propositions pour une nouvelle stratégie d'amélioration des caféiers de l'espèce *C. canephora*, basée sur les résultats de l'analyse des populations sylvestres. In: *11th International Scientific Colloquium on Coffee*. ASIC, Paris, 1985, 445–452.

[4] Berthaud, J. and Charrier, A. Genetic resources of *Coffea*. In: *Coffee. Vol. 4: Agronomy*. Clarke, R. J. and Macrae, R. (eds). Elsevier, London, 1985, 1–42.

[5] Bertrand, B., Anzueto, F., Pena, M. Y., Anthony, F. and Eskes, A. B. Genetic improvement of coffee for resistance to root-knot nematodes (*Meloidogyne* spp.) in Central America. In: *16th International Scientific Colloquium on Coffee*. ASIC, Paris, 1995, 630–636.

[6] Bertrand, B., Aguilar, G., Santacreo, R., Anthony, F., Etienne, H., Eskes, A. B. and Charrier, A. Comportement d'hybrides F_1 de *Coffea arabica* pour la vigueur, la production et la fertilité en Amérique centrale. In: *17th International Scientific Colloquium on Coffee*. ASIC, Paris, 1997, 415–423.

[7] Bettencourt, A. J. Características agronómicas de seleções derivados de cruzamentos entre Híbrido de Timor e as variedades Caturra, Villa Sarchi e Catuaí. In: *Communicações Simposio sobre Ferrugens do Cafeeiro*. CIFC, Portugal, 1983, 351–374.

[8] Bouharmont, P. Sélection de la variété Java et son utilisation pour la régénération de la caféière Arabica au Cameroun. *Café, Cacao, Thé* 1992, *36*, 247–262.

[9] Bouharmont, P. La sélection du caféier Arabica au Cameroun (1964–1991). *Document CIRAD 1-95*. CIRAD, Montpellier, France, 1995.

[10] Bouharmont, P. and Awemo, J. La sélection végétative du caféier Robusta au Cameroun. Première partie: programme de sélection. *Café, Cacao, Thé* **1979**, *23*, 227–254.

[11] Bouharmont, P., Lotode, R., Awemo, J. and Castaing, X. La sélection générative du caféier Robusta au Cameroun. Analyse des résultats d'un essai d'hybrides diallèle partiel implanté en 1973. *Café, Cacao, Thé* **1986**, *30*, 93–112.

[12] Braganca, S. M., De Carvalho, C. H. S., Da Fonseca, A. F. A., Ferrao, R. G. and Silveira, J. S. M. Emcapa 8111, Emcapa 8121, Emcapa 8131: primeiras variedades clonais de café Conilon lançadas para o Espirito Santo. *Com. Tecn. 68*. Empresa Capixaba de Pesq. Agr., EMCAPA, Vitória, ES, Brazil, 1993.

[13] Cambrony, H. R. Arabusta and other interspecific fertile hybrids. In: *Coffee. Vol. 4: Agronomy*. Clarke, R. J. and Macrae, R. (eds). Elsevier, London, 1985, p. 263–290.

[14] Capot, J. L'amélioration du caféier Robusta en Côte d'Ivoire. *Café, Cacao, Thé* **1977**, *21*, 233–244.

[15] Carvalho, A. Principles and practice of coffee plant breeding for productivity and quality factors: *Coffea arabica*. In: *Coffee. Vol. 4: Agronomy*. Clarke, R. J. and Macrae, R. (eds). Elsevier, London, 1985, 129–166.

[16] Carvalho, A. Histórico do desenvolvimento do cultivo do café no Brazil. *Documentos IAC 37*. Instituto Agronómico, Campinas, SP, Brazil, 1993.

[17] Carvalho, A., Ferwerda, F. P., Frahm-Leliveld, J. A., Medina, D. M., Mendes, A. J. T. and Monaco, L. C. Coffee. In: *Outlines of Perennial Crop Breeding in the Tropics*. Ferwerda, F. P. and Wit, F. (eds). Veenman, Wageningen, 1969, 189–241.

[18] Carvalho, A. and Fazuoli, L. C. Café. In: *O Melhoramento de Plantas no Instituto Agronómico, Vol. 1*. Furlani, A. M. C. and Viegas, G. P. (eds). Instituto Agronómico, Campinas, SP, Brazil, 1993, 29–76.

[19] Charmetant, P. and Leroy, T. Méthodologie de la sélection caféière en Côte d'Ivoire. In: *13th International Scientific Colloquium on Coffee*. ASIC, Paris, 1989, 496–501.

[20] Charmetant, P., Leroy, T., Bontems, S. and Delsol, E. Evaluation d'hybrides de *Coffea canephora* produits en champs semenciers en Côte d'Ivoire. *Café, Cacao, Thé* 1990, *34*, 257–264.

[21] Charmetant, P., Le Pierres, D. and Yapo, A. Evaluation d'hybrides Arabusta F_1 (caféiers diploïdes doublés × *Coffea arabica*) en Côte d'Ivoire de 1982 à 1989. In: *14th International Scientific Colloquium on Coffee*. ASIC, Paris, 1991, 422–430.

[22] Charrier, A. (ed.). Etude de la structure et de la variabilité génétique des caféiers. Résultats des études et des expérimentations réalisées au Cameroun, en Côte d'Ivoire et à Madagascar sur l'espèce de *Coffea arabica* L. collectée en Ethiopie par une mission ORSTOM en 1996. *Bulletin IFCC 14*. IFCC, Paris, 1978.

[23] Charrier, A. and Berthaud, J. Principles and methods in coffee plant breeding: *Coffea canephora* Pierre. In: *Coffee. Vol. 4: Agronomy*. Clarke, R. J. and Macrae, R. (eds). Elsevier, London, 1985, 167–198.

[24] Couturon, E. Mise en évidence de différents niveaux d'affinité des greffes interspécifiques chez les caféiers. In: *15th International Scientific Colloquium on Coffee*. ASIC, Paris, 1993, 209–217.

[25] Cramer, P. J. S. *Review of Literature of Coffee Research in Indonesia*. Inter-American Institute of Agricultural Sciences, Turrialba, Costa Rica, 1957.

[26] De Reffye, P. Le contrôle de la fructification et de ses anomalies chez les *Coffea arabica*, robusta et leurs hybrides "Arabusta". *Café, Cacao, Thé* 1974, *18*, 237–254.

[27] Dublin, P. L'amélioration du caféier Robusta en République Centrafricaine. Dix années de sélection clonale. *Café, Cacao, Thé* 1967, *11*, 101–138.

[28] Eskes, A. B. Résistance. In: *Coffee Rust: Epidemiology, Resistance and Management*. Kushalappa, A. C. and Eskes, A. B. (eds). CRC Press, Boca raton, FL, 1989, 171–292.

[29] Eskes, A. B. Breeding for durable resistance of arabica coffee to coffee rust (*Hemileia vastatrix*). *Final Report on FAO Consultancy in Indonesia*. CIRAD, Montpellier, France, 1991.

[30] Guerreiro Filho, O. Identification de gènes des résistance à *Perileucoptera coffeella* en vue de l'amélioration de *Coffea arabica*. PhD Thesis. Ecole Nationale Supérieure Agronomique de Montpellier, 1994.

[31] Guerreiro Filho, O., Medina Filho, H. P. and Carvalho, A. Método para seleção precoce de cafeiros resistentes ao bicho mineiro, *Perileucoptera coffeella*. *Turrialba* 1992, *42*, 348–358.

[32] Jones, P. A. Notes on the varieties of *Coffea arabica* in Kenya. *Monthly Bull. Coffee Board of Kenya* 1956, *251*, 305–309.

[33] Kumar, A. C. Nematode Problems of Coffee. *J. Coffee Res.* 1986, *16* (suppl.)

[34] Kushalappa, A. C. and Eskes, A. B. (eds). *Coffee Rust: Epidemiology, Resistance and Management*. CRC Press, Boca Raton, FL, 1989.

[35] Le Pierres, D. Considérations sur les incompatibilités de greffe pour la culture du caféier. In: *12th International Scientific Colloquium on Coffee*. ASIC, Paris, 1987, 783–790.

[36] Leroy, T. Diversité, paramètres génétiques et amélioration par sélection récurrente réciproque du caféier *Coffea canephora* P.

PhD Thesis. Ecole Nationale Supérieure Agronomique de Rennes, France, 1993.

[37] Leroy, T., Montagnon, C., Charrier, A. and Eskes, A. B. Reciprocal recurrent selection applied to *Coffea canephora* Pierre. I. Characterization and evaluation of breeding populations and value of intergroup hybrids. *Euphytica* **1993**, *67*, 113–125.

[38] Leroy, T., Montagnon, C., Cilas, C., Charrier, A. and Eskes, A. B. Reciprocal recurrent selection applied to *Coffea canephora* Pierre. II. Estimation of genetic parameters. *Euphytica* **1994**, *74*, 121–128.

[39] Lotode, R. and Lachenaud, P. Méthodologie destinée aux essais de sélection du cacaoyer. *Café, Cacao, Thé* **1988**, *32*, 275–292.

[40] Mazzafera, P., Eskes, A. B., Parvais, J. P. and Carvalho, A. Stérilité mâle détectée chez *Coffea arabica* et *Coffea canephora* au Brésil. In: *13th International Scientific Colloquium on Coffee.* ASIC, Paris, 1989, 466–473.

[41] Millot, F. Recherches en sélection caféière en Ouganda. In: *Contribution à l'Etude de la Cacaoculture et de la Caféiculture en Ouganda. IFCC Bull.* **1974**, *12*, 47–66.

[42] Montagnon, C., Leroy, T., Cilas, C. and Eskes, A. B. Differences among clones of *Coffea canephora* in resistance to the scolytid coffee twig-borer. *Int. J. Pest Management* **1993**, *39*, 204–209.

[43] Montagnon, C. Optimisation des gains génétiques dans le schéma de sélection récurrente réciproque de *Coffea canephora* Pierre. *PhD Thesis.* Ecole Nationale Supérieure Agronomique de Montpellier, France, 2000.

[44] Moreno, R. G. and Castillo-Z, J. La variedad Colombia. Cenicafé, Chinchina, Caldas, Colombia. *Bol. Tecn. 9.* CENICAFE, Colombia, 1984.

[45] Moschetto, D., Montagnon, C., Guyot, B., Perriot, J. J., Leroy, T. and Eskes, A. B. Studies on the effect of genotype on cup quality of *Coffea canephora. Trop. Sci.* **1996**, *36*, 18–31.

[46] Reyna, E. H. Un nuevo método de injertacíon en café. *Bol. Tecn. 21.* Dir. Gen. de Investigacion y Control Agropécuario, Min. de Agr., Guatemala, 1966.

[47] Snoeck, J. La rénovation de la caféiculture malgache à partir de clones sélectionnés. *Café, Cacao, Thé* **1968**, *12*, 223–235.

[48] Sondhal, M. R. and Bragin, A. Somaclonal variation as a breeding tool for coffee improvement. In: *14th International Scientific Colloquium on Coffee.* ASIC, Paris, 1991, 701–710.

[49] Sreenivasan, M. S. Breeding coffee for leaf rust resistance in India. In: *Coffee Rust: Epidemiology, Resistance and Management.* Kushalappa A. C. and Eskes A. B. (eds). CRC Press, Boca Raton, FL, 316–323.

[50] Van Der Graaff, N. A. Selection of arabica coffee types resistant to coffee berry disease in Ethiopia. *PhD Thesis.* University of Wageningen, 1981.

[51] Van Der Vossen, H. A. M. Coffee selection and breeding. In: *Coffee, Botany, Biochemistry and Production of Beans and Beverage.* Clifford, M. N. and Willson, K. C. (eds). Croom Helm, London, 1985, 48–96.

[52] Wrigley, G. *Coffee.* Longman, New York, 1988.

[53] Yapo, A., Charmetant, P., Leroy, T., Le Pierres, D. and Berthaud, J., Les hybrides Congusta (*C. canephora* × *C. congensis*): comportement dans les conditions de Côte d'Ivoire. In: *13th International Scientific Colloquium on Coffee.* ASIC, Paris, 1989, 448–456.

[54] Yapo, A., Leroy, T. and Louarn, J. Contribution à l'amélioration de *Coffea canephora* par hybridation interspécifique avec *Coffea liberica.* In: *14th International Scientific Colloquium on Coffee.* ASIC, Paris, 1991, 403–411.

4
Coffee Propagation

J. N. Wintgens and A. Zamarripa C.

4.1
Background

Prior to planting coffee trees with a view to either setting-up, extending or renewing a plantation, due consideration should be given first to the location of the site and then to the selection of suitable stock. The variety of coffee or clone should not only be selected with a view to its productivity, but should also take into account values like coffee quality and production costs. The potential long-term income generated by the plantation is built up on the following inter-related factors.

4.1.1
Productivity

Initially, potential productivity depends on the variety or the clone selected. Development, however, can only be favorable if the new plants adapt well to local agroclimatic conditions, if they are resistant to local parasites and if the cultivation methods chosen are adapted to the variety in question. There is not much point in selecting a productive variety of coffee if the pruning methods implemented do not encourage the renewed growth of annual fruit-bearing branches. Where pruning

methods produce the desired results, fertilization should be calculated with a view to support the ensuing crop. Again, adapted pruning and fertilization methods will not suffice if diseases, parasites and other pests are not correctly controlled.

Hence, productivity can only be ensured by the choice of a suitable variety and adapted crop management practices.

4.1.2
Coffee Quality

The variety of coffee chosen is a basic factor which conditions the ultimate quality of the product. On the other hand, it has also been established that harvesting methods and, more importantly, post-harvesting processing methods have a considerable impact on coffee quality.

4.1.3
Production Costs

Production costs are directly linked to the choice of a cultivation system – either intensive, semi-intensive or extensive.

In other words, once the coffee planter has decided on the cultivation system he plans to implement, and the type and quality of the coffee he plans to produce, he will

Coffee: Growing, Processing, Sustainable Production, Second Edition. Edited by J. N. Wintgens.
© 2012 WILEY-VCH Verlag GmbH & Co. KGaA. Published 2012 by WILEY-VCH Verlag GmbH & Co. KGaA.

have to study how to adapt to the following factors:

Local climate	Temperature, average and extremes
	Total annual rainfall and its distribution
	Length of the dry season (or seasons)
	Seasonal winds, their strength and points of impact
Cultivation system	Under shade or sun-exposed
	Manual or mechanized with or without irrigation
Cultivation techniques	Free growth or capping
	Soil cover, mulch, cover plants or clean weeding
	With or without fertilization
Diseases and pests	Genetic tolerance

Once these multiple factors have been identified and studied, the planter must contact local research centers, specialized local work groups and advisory centers who can provide suitable advice and support when it comes to making essential decisions.

4.2
Propagation Methods

Propagation by seed is the cheapest and easiest method for coffee propagation. Coffee can also be vegetatively propagated either by grafting, stem-cutting or *in vitro* propagation.

Robusta and Arabica plantations have been established from seed for many years. Seeds issued from auto-fertile Arabica mostly produce identical plants to the mother plant, except in the case of hybrids (genetically non-stabilized crosses).

On the other hand, seeds issued from self-sterile Robusta, Excelsa and Liberica result from a cross between an identified mother plant and a non-identified male parent. These seeds produce genetically different plants to the mother tree and may even be inferior. The plantation generated by these seeds is heterogeneous and variability can be observed in leaves, fruit size, vigor productivity, etc.

4.3
The Choice of a Propagation System

The choice of a propagation technique is influenced by two essential factors:

- The coffee species to be propagated
- The ultimate goals of the project, either for research and experimentation or for commercial purposes.

4.3.1
Arabica

Thanks to its high grade of autogamy, the most appropriate and cheapest propagation method is by seeding. This maintains a high proportion of the characteristics of the parental material (91–93 % according to Krug and Carvalho [7A]).

The techniques of cutting and graft can be applied to a few plants to rescue worthy material. *In vitro* systems, particularly somatic embryogenesis, are applicable to the propagation of F_1 hybrids for generating a rapid production of reliable planting material for extended coffee areas.

4.3.2
Coffea canephora (Robusta, Conillon) and other Allogamic Species (Excelsa, Liberica)

Due to the fact that the seeds are issued from cross-pollination, the progeny from this material is variable. Growers, therefore, prefer vegetative propagation for these species. The classic root-cutting or grafting methods are easy to implement and does not cost much. They are, however, limited by the fact that the root-stock orchard (propagation plot) can only produce a limited number of cuttings or scions. For the quick introduction of new clonal varieties in a large area, somatic embryogenesis is the most appropriate method. Decision making also depends on factors like the skillfulness of local labor, the costs involved and national phytosanitary regulations.

4.4
Propagation by Seeds

4.4.1
The Selection and Preparation of Seeds

Seeds produced from any part of the coffee tree as well as abnormal seeds (peabean, elephant bean, etc.) are genetically identical. Nevertheless, it has been observed that larger seeds are more precocious and seedlings are better developed.

Although plants produced from poly-embryonic seeds (elephant) are as good as seedlings from normal seeds, they add unnecessary complications to the nursery practice because of the multiple sprouts emerging from one same seed.

Seedlings produced by peabeans are the same as those grown from normal seeds.

The sources of seeds for propagation should be restricted to disease-free areas. Zones with endemic Coffee Berry Disease, Coffee Leaf Rust and, especially, seed-borne diseases like tracheomycosis should be discarded. As a measure against seed-borne diseases, dried seeds should be treated with a seed dressing fungicide.

Cherries harvested at full maturity provide optimal germination conditions. Cherries harvested at the beginning or the end of the harvesting season should not be used for germination purposes. After harvesting, the lighter cherries are eliminated by floatation. The heavy fruits are depulped with care in the hours that follow to avoid fatal fermentation. If there is over 5% of floating cherries in the batch, they should be discarded. Where only a small quantity of seeds is required, it is preferable to depulp manually. On the other hand, where important quantities are required, depulping can be carried out mechanically by pulpers with widely adjusted channels.

After pulping all the abnormal beans should be discarded. The beans retained should be fermented for a short time (maximum 10 h) and afterwards the degraded mucilage should be washed away.

During this operation any remaining floater or damaged beans should be removed. The seeds are then slowly dried on mesh trays under shade cover to avoid overheating which might impair their germination capacity and split their parchment. The time required for drying to bring the moisture content (MC) of the seeds to 35–40% is 12–15 days depending on the atmospheric conditions. It is not essential to ferment and wash the pulped parchment beans. After depulping, the sticky parchment beans can be sown immediately. They can also be mixed with charcoal powder or wood ash as this combines well with the mucilage to form a coating which prevents the beans from sticking together and helps drying.

When the seeds are to be used on site, the fresh parchment beans can be mixed with wood ash and sown immediately, but when they have to be stored or transported they must be mixed with charcoal and packed in aerated bags. Normally, by the

Table 4.1 The viability of coffee in storage under optimal conditions (in air-tight bags with 40–41 % MC)

Species	Temperature	Months of storage						Reference
		0	8	15	24	30	36	
C. arabica	19°C	95	94	95	90	80	60	Couturon [4B]
C. canephora	19°C	95	91	64	0	0	0	Couturon [4B]
C. stenophyla	19°C	90	80	50	0	0	0	Couturon [4B]

third month, their germination capacity drops from 95 to 70–75 %, by the ninth month germination capacity is only 20–25 % and after 15 months of storage germination capacity is completely lost. On the other hand coffee seeds dried at 40 % MC and kept in air-tight bags at 15°C can keep their full germination capacity for about 6 months (Tab. 4.1).

The number of seeds per kilogram varies according to their size and density. Table 4.2 provides average values.

Table 4.2 Number of good seeds per kilogram of dry parchment coffee

Species	Coste (1968)	Wellman [9A]
C. arabica	2000–2200	2100
C. canephora	2500–2700	2200–2660
C. excelsa		2100
C. liberica		1540

When calculating the number of seedlings required for planting, a margin of 25 % should be added for replacing useless or dead seedlings. It has also been noted that seeds with diminished germination power generate stunted seedlings.

The time required for seeds to germinate varies according to their preparation, their age, and the temperature and moisture of the soil. The most appropriate soil temperature for germination is 28–30°C.

Normally Arabica seeds germinate as per the following criteria:
• Fresh harvested seeds   after 32 days
• Seeds stored for   after 50 days
  8 weeks after harvesting
• Older seeds   after 42–70 days

The speed of germination can be accelerated if the parchment is removed or by soaking the seeds in water for 24 h, but these operations require special care and attention. If there is the slightest doubt as to the quality of the seeds (certification, long conservation period, etc.), they should be pre-germinated. This can be done by spreading them between wet sacks or on damp sand covered with hessian or on a hessian envelope between layers of vermiculite.

When the root tip breaks through the parchment the seeds are ready for sowing. The entire batch of germinated seeds should be planted simultaneously and not selected with the most advanced ones for planting first, because this can result in a high proportion of broken or twisted roots.

Pre-germination offers certain advantages. It reduces of the amount of water and attention required to bring the seeds through their first stage and saves at least 3 weeks in nursery care. It also facilitates the screening of non-viable seeds and enables a rapid identification of seed viability. The major disadvantage of pre-germination is the fact that the seeds can be lost if the new root tip is damaged. Planting pre-germinated seeds in polybags appears to be the best method. Generally speaking, however, pre-germination is regarded as an exceptional practice.

In any case, in order to avoid too high a percentage of seed failure, it is always advisable to carry out a germination test with 100 seeds before planting.

Parchment provides the seeds with a substantial protection. Therefore, it must not be removed and it is advisable to discard seeds with split parchment.

Coffee seedlings are nurtured either by sowing directly in the fields or by sowing in the nursery. The choice of system depends chiefly on local traditions and farming practices.

4.4.2
Sowing Directly in the Field

This system consists of planting several seeds in holes which are pre-prepared in the plantation. Each hole receives three to five seeds which are covered with a fine layer of topsoil. Where the land is steep the seeds are protected from water run-off by a small crescent-shaped mound of earth located on the upper side of the hole.

The seeds are sheltered by a few branches placed on each side of the hole or stacked, pyramidically above it. When the seedlings have reached a certain size, those that have developed best are kept and the others are eliminated. This method avoids the need to set up a nursery and bypasses the transplantation risks, but it is little used because of several major disadvantages:

• The need to store seeds until the beginning of the wet season when sowing can begin
• The large quantities of seeds required

• The maintenance of a large area of land during the germination stage and the initial growth period.

In Brazil, the traditional system was to clear land and plant a number of seeds together in "covas" at the usual planting distance. This system is interesting because it induces self-shading, an improved distribution of branching and generates potentially good-yielding plants. It has now been introduced into other of countries with satisfactory results.

4.4.3
Sowing in Nurseries

4.4.3.1 The Nursery Site
In order to ensure regular irrigation, the nursery should be located close to a source of good quality water, with low salinity and free from any plant parasites including nematodes. It should also be sheltered from prevailing winds which may dry or break young plants. Accessibility should be easy for transport. A gentle slope is the best option to reduce frost risks and allow for a good drainage of cold air and excess water. Valley bottoms and hollows that can become frost pockets should be avoided. If the slope is over 8%, the land should be protected against erosion.

The soil should present the characteristics of a good coffee soil, i.e. deep, light textured, friable, fertile, free-draining, not capped and, above all, free of nematodes. Soils which are alkaline, saline, waterlogged and heavy textured should be avoided (see also Section I.6). A soil analysis should be carried out to confirm that the site is acceptable. In certain cases defects like heavy texture or low fertility can be corrected.

Trees are not desirable on the nursery site because their shade cannot be regulated. As a result, it is better to rely on artificial shade. The nursery should not be located above 1800 m because temperatures below 10°C slow down the growth of the seedlings. In order to avoid transport problems, the nursery should be located as close as possible to the plantation area. The nursery should never be located on old plantation site nor should it be used several years consecutively because of the threat of soil-borne parasites (nematodes in particular).

Picture 4.1 Shaded nursery roof built of palm leaves.

4.4.3.2 **Layout of the Nursery**

The nursery includes the seedbeds or germination beds where the seeds are sown to germinate and to produce seedlings, as well as the nursery beds where the seedlings are nurtured until they are ready for planting out in the field. If the nursery is located on a slope, cold air should be prevented from moving down the slope into the nursery by transversal vegetal or artificial barriers placed above the nursery and the lower part should be left sufficiently open to allow cold air to flow out.

In order to prevent water run-off into the nursery area, a storm drain should be dug on the upper side of the site. It is also sometimes necessary to fence the nursery against stray animals and thieres.

4.4.3.2.1 **Shade**

Traditional shade consists of a roof placed on wooden posts. It should be about 2–2.5 m high in order to allow people to enter without stooping. The ties for the posts and the grills for the roof are made with plain galvanized wire.

It should provide about 50 % shade and be covered with grass thatching or with long reeds made of available local material (palm leaves, fern, etc.). An even better covering can be made with a closely woven plastic net (Pict. 4.1, 4.2).

Picture 4.2 Shaded nursery roof built with plastic mesh.

In order to protect the nursery itself from wind and to maintain the humidity level, the sides of the nursery enclosure should be fenced in with the same material that is used for the roof (Fig. 4.1).

An alternative shade protection is to build low individual shade protection over each bed. This need not be more than 1 m high and leaves the pathways free. The vegetal matting that provides the shade can be rolled back to give access to the beds and to provide more light to the plants. If the nursery, has a certain size, high shade roof is more appropriate.

A more permanent shade roof entails a greater investment and is chiefly recommended for commercial nurseries with a continuous program of plant production; it is also better for the vegetative production

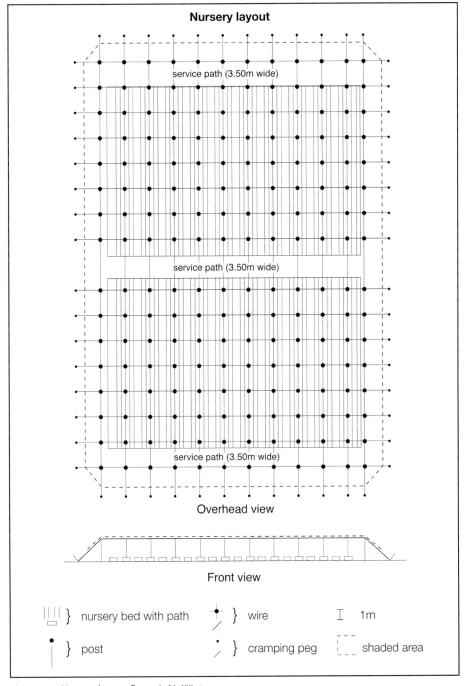

Figure 4.1 Nursery layout. From: J. N. Wintgens.

of the plants. In this case, shade is provided by plastic nets that exist in different mesh sizes and which ensure a light penetration of 40–70 %. Thick-caliber black mesh lasts, on average, 8 years. If this option is chosen, the posts should be made of concrete.

4.4.3.2.2 The Seedbeds

The seedbeds should be installed about 1 month before sowing. Normally, the number of seedlings obtained per square meter of seedbed varies from 600 (in row sowing) to 1500 (broadcast sowing).

To begin with, the ground where the seedbeds are to be installed should be completely cleared and cleaned. The beds are then marked out at a width of 1.2–1.4 m and separated by circulation paths of 0.5–0.6 m width. The soil should be dug to a depth of about 30 cm, broken up and finally the top layer should be sifted. The beds must be raised about 25 cm above the path level to ensure adequate drainage of excess rain or irrigation water. Once sifted, the topsoil removed from the paths can be used as the top layer for the seedbeds. The length of the beds depends on the size and the topography of the site and the importance of the nursery. Normally it varies between 10 and 40 m.

If the slope is over 8 %, the beds should be built to follow the contour lines and even be terraced where drainage is poor. Successful germination has been achieved by placing the seeds in germination beds which contain washed river sand, vermiculite and pearlite in equal parts. Where only a small number of seedlings is required, any type of small tray, 15 cm deep which holds 200–500 seeds, can replace the seedbed.

Washed river sand alone constitutes a good substrate for coffee seed germination. Before sowing the germination beds must be sterilized in order to eliminate all potential damaging agents which could harm the future coffee plant, i.e. nematodes, insects, diseases and weeds. For this purpose chemicals like Dazomet (40 g/m^2), Trimaton 51, Basamid (30–40 g/m^2), Formol 40 %, methyl bromide (454 mg/10 m^2) are suitable for the specific control of nematodes, fungus, insects and small weed seeds. The most commonly used products are Banrot combined with a mix of Previcur and Derosal. For the control of nematodes and insects, products like Furadan 5G, Nemacur, Counter 10G, Disyston 12G are most recommended. Basamid is often preferred by growers because it is easy to use and efficient.

Before choosing a product it is wise to consult local agricultural services who will advise as to efficiency, costs and legal restrictions. It is advisable to ensure that the seeds have been treated against soilborne diseases and insects before sowing. Seeds can be sown in rows (5 cm × 2 cm) or broadcast (Pict. 4.3).

To mark out the row, strings can be stretched along the beds and secured by pegs at the end of each bed. Planting positions can be determined with a planting "comb". Other devices like a notched planting board can also be used. In spite of traditional allegations, the actual position of the seed in the soil is not important.

Picture 4.3 Sowing in rows.
From: Comité Français du Café.

Picture 4.4 Seedbeds which have been sown in rows. From: Comité Français du Café.

When broadcast sowing is practiced, the seeds should be spread as uniformly as possible at a density of about 2500 seeds/m², which corresponds to an average sowing distance of 2 cm × 2 cm (Pict. 4.5).

For improved distribution, the seeds can be mixed with sand or sieved earth. In both cases, after sowing the seeds should be covered with a 1-cm layer of washed sterilized river sand. Following this, the seedbeds should be covered with a mulch layer of chopped grass or coffee parchment about 1 cm thick. This precaution serves as a protection against temperature variations and helps to maintain the moisture within the beds.

During the whole germination period the young plants feed only on their own reserves and on water. Therefore, there is no

Picture 4.5 Seedbeds sown by broadcast.

need to add organic matter or fertilizer to the soil.

Sowing should take place at intervals of 1–2 weeks in order to stagger the subsequent transplanting operations.

Maintenance of Seedbeds

Germinating coffee seeds are very sensitive to water stress. In case of excessive drought the seeds die, but if the soil is too wet the seeds may rot and young seedlings can be killed by damping-off. The soil of the germination beds should never be allowed to dry out completely. To achieve this, irrigation should be regulated as shown in Tab. 4.3.

Table 4.3 The irrigation of seedbeds

Season	Period	Irrigation required
Hot (20–25°C)	rainy	after 2 days without rain
	dry	watering twice a day: morning and evening
Cold (15–20°C)	rainy	after 5 days without rain
	dry	watering twice a day: morning and evening

Infestations in Seedbeds

Even if the soil is disinfected prior to sowing, an outbreak of infection can occur after germination. The most destructive infection is damping-off caused by *Rhizoctonia* and *Fusarium* fungus. Another serious seed infection is often caused by nematodes which is revealed by flaccidity and yellowing of the leaves. The most harmful pest in seed beds are crickets that cut the tender stems of the coffee plantlets. Table 4.4 gives guidelines on how to control pests and diseases in seed beds.

If weeding is necessary it must be carried out by hand when the seedlings have emerged and are about 2 cm high with the cotyledons still enclosed in the parch-

Table 4.4 Phytosanitary problems in coffee seedbeds and their control

Problem	Cause	Parts of the plant affected	Pesticide	Doses (per ha) and application[a]
Damping off	*Rhizoctonia solani* and other fungi	stem and roots	Quintonzene PENTACHLOR 600F	22 kg. Apply to the soil and the stems of the plants.
Brown Eye Spot	*Cercospora coffeicola*	leaves	BENLATE or PROMYL	70 g. Apply to the foliage. Up to three applications at 2-week intervals.
Brown Leaf Spot	*Phoma costarricensis*	buds, stems and leaves	Fosetil-Al ALIETTE 80 % WP or Iprodione ROVRAL 50 WP	800 g of Fosetil or 200 g of Iprodione. Apply to the foliage. Up to three applications at 2-week intervals.
Crickets, ants, defoliation larvae, cutworms, scales and aphids	several genus and species	leaves, buds and stems	Diazinon BASUDIN 40	400 g. Apply to the foliage. It can be mixed with fungicides applied to leaves.
Nematodes/ white grubs	*Meliodogyne incognita*, *Pratylenchus coffeaea*, *Phyllophaga* spp.	roots	Aldicarb TEMIK 156 or Fenamifos NEMACUR 10G	7.9 kg of Aldicarb or 3 kg of Fenamifos 20 days after germination. Place between the furrows of the plants.

[a] Based on 200 l water/ha.

ment husk. At the latest weeding can take place when the cotyledons start to unfold at the "butterfly" stage.

Transplanting the Seedlings

Once the coffee plantlets have reached the stage known as "soldier" or "match-stick" or "butterfly" they must be dug up without damaging the roots (Pict. 4.6).

Plantlets can be transplanted directly in nursery beds or in containers like polybags. The preparation of the plantlet entails: dampening the seedbeds and lifting out the coffee plantlet, and eliminating any seedlings which present defects like:

- Misshapen growth
- Underdevelopment
- Chlorotic foliage
- Twisted or damaged roots (Pict. 4.7).

Picture 4.6 Seedlings at the "soldier" or "matchstick" stage. From: Comité Français du Café.

Picture 4.7 Seedlings with deformed roots.

If it is necessary to transport the plantlets for long distances, they should be protected against drying out by wrapping them in humid material. It is not recommended to dip the plantlets in water because of the possible transmission of pathogens. The plantlets should be exposed as little as possible to air and sun in order to avoid dehydration.

It is recommended to clip part of the tap root of the plantlets to avoid bending during transplanting and to encourage the development of lateral rootlets. If, for example, the root length of a plantlet is of 12 cm, this should be clipped to about 7 cm.

When transplanting, it is important to always plant the seedling with its collar at soil level. Transplanting is made by opening a hole of about 10–15 cm in depth with a dibble of 4 cm in diameter. The plantlet is inserted deeply into the hole and pulled back up so that the lateral roots drop down. The earth is then pressed laterally against the root with the dibble.

4.4.3.2.3 The Nursery Beds

Layout

Nursery beds are prepared almost in the same way as seedbeds. The soil must be broken up at a depth of 40–50 cm and all rubbish, like stones, pieces of roots, sticks, etc. must be removed without bringing up the subsoil. It is advisable to also incorporate 2–5 kg/m^2 of well-rotted organic material (manure, compost) at a depth of 15 cm together with 50 g phosphoric rocks or another fertilizer which is rich in phosphorus. It is also recommended to superficially incorporate 1–1.5 kg/m^2 of mycorhized soil or an equivalent commercial preparation.

The soil should be sterilized in the same way as the seedbeds, chiefly in the case of Arabica propagation. It is preferable not to use such areas which have previously been planted with coffee or used for nursery.

As a measure of protection against waterlogging, nursery beds should be built up to a height of 10–20 cm according to the soil texture.

The transplanting distance within the beds depends on the coffee variety and above all on the size of the seedlings which are to be transplanted in the field:

- Small size varieties   20 cm × 20 cm
- Large size varieties   25 cm × 25 cm

The length of the beds depends on the importance of the nursery but should not exceed 40 m for logistical reasons. Their width can vary from 1.2 to 1.5 m with separation paths of 50–60 cm wide.

Fertilization

Nursery beds need to be fertilized in staggered applications. The first application should take place when the plantlets have developed two or three pairs of leaves and the following at intervals of 75 days.

Fertilization is related to the real soil fertility. As a result, it is recommended to base the nutritional program on soil analysis. Nitrogen and phosphorus are the most important elements for the development of young coffee plants. The most appropriate fertilizer formulation is 10-30-10 (800 kg/ha) together with ammonium nitrate but, again, this needs to be confirmed by soil analysis.

Fertilizer must be spread between the coffee rows and buried. If the leaves of the plantlets present symptoms of deficiency in micronutrients, foliar sprays are more suitable.

Weed Control

If the nursery is small, weed control can be carried out manually. Herbicides for chemical control should be selected according to the type of weeds. There are several herbicides like oxyfluorfen (GOAL) combined with alaclor (LAZO) that can be used for pre-emergent weed prevention; however, once again, this has to be determined after consulting local advisory agencies.

To restrict the incidence of weeds and reduce soil evaporation the nursery beds should be mulched with coffee parchment, chopped grass or any equivalent material.

Pest and Disease Control

Table 4.5 describes the most common pests and diseases in coffee nurseries and their control.

Cutting the Roots

This operation consists of cutting back the tap root of the coffee plant to a length of

Table 4.5 The most common pests and disease in coffee nurseries and their control

Problem	Cause	Parts of the plant affected	Pesticide	Doses (per ha) and application[a]
Brown Eye Spot	*Cercospora coffeicola*	leaves	copper oxychloride at 50 % CUPRAVIT	1.2 kg. Apply to the foliage every 15 days. Between three and five applications are recommended.
Brown Leaf Spot	*Phoma costarricensis*	buds, stems and leaves	Clorotalonil or Fosetil-Al ALIETTE 80 % WP	1.5 kg of Clorotalonil, 1.6 kg of Fosetil. No more than four alternated applications of both products.
Fusariosis	*Fusarium* spp.	base of the stems and the roots	quintocene PENTACLOR 600	22.5 kg. Diluted in water and applied to the soil in one sole application.
Nematodes/ white grubs	*Meliodogyne incognita*, *Pratylenchus coffeaea*, *Phyllophaga* spp.	roots	Aldicarb TEMIK 156	7.9 kg of Aldicarb incorporated into the soil in one sole application.
Scales	*Saessetia* spp., *Coccus* spp.	stems and branches	Aldicarb TEMIK 156 or Diazinon BAZUDIN 40	7.9 kg of Aldicarb incorporated into the soil. 800 g of diazinon applied to the leaves in one sole application.
Aphids	*Toxoptera aurantii*	buds and leaves	Diazinon BAZUDIN 40	480 g mixed with water and applied to the foliage in one sole application.

[a] Based on 400 l water/ha.

12–15 cm at 2–3 months before transplanting it into the field. The objective is to encourage the growth of new lateral roots and to balance the distribution of the root system. This practice is not essential, but it is recommended.

To begin with, the plantlets should be abundantly watered. Then a narrow, sharp spade should be inserted into the soil at an angle of 45° and at a distance of about 12 cm from the base of the coffee plantlet. At a depth of about 15 cm, knock the spade energetically so that it cuts the tap root of the plants. After about 1–2 weeks, the plant should be sprayed with a mix of micronutrients, urea and VYDATE at low concentration (0.3–0.5%) in order to strengthen it and protect it against nematode attacks.

The period the coffee plant needs to remain in nursery beds varies according to the species, the time of plantation and local climatic conditions, etc. As a general rule, Arabica needs 9–15 months in the nursery beds, whereas Robusta, whose growth is faster and more vigorous, only needs 6–8 months. To "harden" the plantlets prior to transplanting in the field, shade should be removed gradually. After 6 months, it can be reduced by 30% and 3 months later, it can be reduced again by a further 30%.

Preparation for Planting Out

The beds should be watered a few hours before lifting the seedlings. This makes the operation easier. The practice of cutting back the leaves to reduce the loss of water through transpiration and spraying the leaves with a sugar solution is recommended by some authors, but the positive effects of interventions of this type have not been fully demonstrated to date.

Each plantlet is lifted out with a fork and closely inspected. Any "off type" should be destroyed and never kept for planting later.

Under extreme conditions the number of "discards" can be as high as 50%.

A good plantlet must be straight and strong with about six to 10 pairs of leaves. It should be 40–60 cm high and 3–4 pair of laterals branches should already have developed on the stem. The diameter of its stem should be about the size of a pencil at the base. If the transplantation is carried out with bare roots, these should be trimmed and wrapped in damp sacking as soon as the earth has been removed to prevent them from drying out. A good practice is to dip the roots in a thick slurry of clay which grants a protective coating against evaporation.

If the plant has grown too tall it should be stumped back to 25 cm from the collar. This is done with plants older than 18 months and over 2 m high. If the tap

Picture 4.8 Excessively developed seedlings.

root has grown too long or if it has been damaged or bent, it should be reduced (Pict. 4.8). Overgrown plants from the nursery can also be transplanted with a ball of soil around their roots but the most common practice is bare-root planting (see also Section I.7).

4.4.3.2.4 Planting in Containers

The most used containers are polybags. Generally, the bag is made of black diothene which is 0.05–0.08 mm thick. The size varies according to the time the seedling will remain in the bag and the size of the plant when transplanted.

The dimensions of the bags are given either as flat or as filled bags. The most common flat size for one plant can be from 15 cm × 25 cm to 18 cm × 30 cm. When filled, a bag of 17 cm × 30 cm has a diameter of 11 cm and holds about 3 kg of moist potting soil. The recommended size for filled bags is 30 cm × 12 cm in diameter for seedlings.

The bottom of the bag must be perforated and the lower half of the bag must have a double row of holes to ensure the drainage of excess water and a correct aeration of the roots.

Polyethylene sleeves can also be used instead of polybags. The sleeves are filled with a funnel fitted with a cylindrical spout corresponding to the size of the sleeve.

Biodegradable sleeves are also available now. They can either be filled by hand or mechanically. If the bags used are too small, there is a risk of "pot bound" roots and bent taproots. If the plant has to remain in the bag for longer than planned, the bottom of the bag should be cut off to allow the tap root to grow downwards rather than forcing it to bend and curl round in the bag.

The bag must be filled with a fertile, sterilized potsoil. The mix can be prepared in many different ways according to the availability of local materials but, basically the mixture must contain fine loam, sand and organic matter. A typical mixture should contain the following elements:

- 1 part of fine, dry farmyard manure which has been damped or coffee pulp which has fermented for at least 3 months,
- 1 part of well sieved loamy top soil or fine river alluvial deposit, and
- 1 part of coarse sand.

The mixture must be sieved to remove foreign bodies (stones, clods, etc.) and then sterilized as recommended for seedbeds, especially for Arabica coffee which is sensitive to nematodes.

Some planters like to incorporate a complete fertilizer in the mixture (2-1-1) at a rate of 4 to 5 kg/m^3 or diammonium phosphate at the rate of 2 kg/m^3. This practice is chiefly to be recommended if analysis has revealed serious deficiencies in macronutrients. Otherwise, it is preferable to apply fertilizers after transplanting, once the roots have reached their full absorption capacity or use slow release fertilizers.

Transplanting in bags should be carried out at the "soldier" stage as recommended for nursery beds. The transplanted bags should be placed in beds which are set up just like the nursery beds. They must be well drained and free from nematodes, pests and diseases. They can also be sterilized, especially in the case of Arabica.

The bags should be arranged in double rows and separated by a distance of 15–20 cm in order to ensure correct ventilation and avoid weak thread-like growth. Galvanized wire, bamboo or any other available material can be used to separate the double rows of polybags (Fig. 4.4, Pict. 4.9).

Fertilization starts once the plants have developed at least one pair of new leaves.

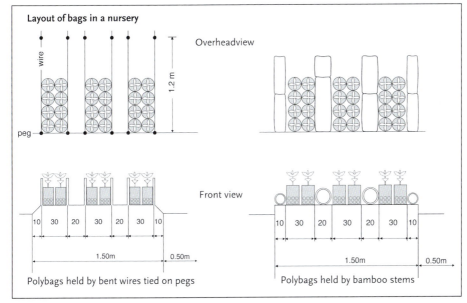

Figure 4.2 Layout of bags in a nursery. From: J. N. Wintgens.

Picture 4.9 Bags arranged in double rows and maintained with wire.

Picture 4.10 Seedlings at their butterfly stage, sown directly into bags.

This is usually about 3–4 weeks after transplanting. Diammonium phosphate in doses of 4–5 g/bag is recommended for the first application of granulate. This can be followed by repeated foliar spraying with the same fertilizer at a concentration of 0.3 %.

Deficiencies in micronutrients can be compensated by foliar spraying with added urea. Pest and disease control is similar to the practices recommended for nursery beds. To reduce evaporation, a fine layer of mulch can be spread over the bags.

4.4.3.2.5 Direct Sowing in Nursery Beds or in Containers

This technique consists of sowing the seeds either in nursery beds or directly in the bags. The seedlings are raised in the beds or bags until they are sufficiently developed for field planting. Sowing is similar to the practices used in seed beds. The same applies to treatment afterwards.

Sowing directly in nursery beds (20 cm × 20 cm) is reserved to bare-root planting for the production of plants of 30–40 cm with six or seven pairs of leaves. If the intention is to produce smaller plants (20 cm) the seeds can be spaced at 12–15 cm.

Sowing directly in bags ensures a better revival of the seedlings in the field. Coffee in bags can be handled in the same way as coffee transplanted in containers. Given the cost of the operation it is preferable to sow at least two seeds per bag and retain the best. Healthy seedlings which are weeded out can be used to replace missing seedlings in other bags or they can be re-planted in new bags (Pict. 4.10).

Comments

- The current tendency is to use pre-pre-pared commercial mixes to fill the poly-bags or the sleeves. This has the advantage of avoiding any pathogenic contamination of the substract and encouraging healthy growth.
- In large plantations where hundreds or thousands of plants are produced each year, seed-bed filling and coffee seed planting is entirely mechanized.

4.5
Vegetative Propagation

Vegetative propagation is only justified where seed propagation cannot produce plants which conform to the mother plant. This is the case for Robusta, Conillon, Liberica, Excelsa, mutants, non-stabilized hybrids (Arabusta, Congusta, etc.) or recent crosses which have not yet been selfed for a sufficient number of generations to genetically stabilize them. Plants developed vegetatively should bear an exact resemblance to their parents. This means that they breed true to type.

Vegetative propagation plays an important role in the selection of coffee, since it enables the propagation and the exploitation of genetic heterozygotic material like outstanding clones of C. canephora and interspecific hybrids like Congusta.

Vegetative propagation methods used in coffee include grafting, stem-cutting and in vitro propagation including microcutting and somatic embryogenesis.

4.5.1
Grafting

Grafting is a cheap and efficient method to upgrade coffee trees. Consequently, it should be encouraged wherever possible. The operation consists of joining two pieces of living tissue, a scion and a rootstock, in such a way that they will develop like an individual plant. The technique of grafting can be used in the following cases:

- To obtain a more resistant root system for trees which are facing soil-borne problems (nematodes, root scales, humid soils, etc.).
- To multiply trees with valuable characteristics which may not be transmitted by other methods of propagation.
- To top-work, without uprooting, existing plantations with inferior planting material (period and concentration of ripening, uniformity and size of the beans, yield, cup taste and bean composition) can all be improved by grafting.

Grafting can be undertaken at various stages or ages of the plant:

- "Soldier" or cotyledon stage in the nursery: in this case, the scion and the rootstock must be the same size – grafting Arabica on Robusta is used to improve nematode resistance or to secure a superior clone with a vigorous root system.

- One-year seedlings in the nursery: by grafting an improved scion on Arabica or Robusta rootstock.
- Mature trees in the plantation: by grafting an improved scion on existing trees of inferior quality.

Problems can occur if the rootstock and the graft are incompatible. As a result, it is important to determine the compatibility of the stock and the scion prior to the establishment of a large-scale propagation program.

The advantage of grafting is that one can combine the best characteristics of the rootstock, i.e. sound root systems which are resistant to soil parasites, with the desirable characteristics of a selected clone used as a scion. Grafting material can be imported as rods (small quantities), produced locally in special plots or taken from adult trees. Grafting adult plants in the plantation enables the introduction of material which can improve the performance of the existing trees. This operation is not very expensive and offers the added advantage of saving time as well a rejuvenating the plantation.

Whatever the grafting method, a certain heterogeneity in the plants cannot be avoided. This is due to the interaction between the scion and the root-stock which can influence plant vigor, resistance to pests, diseases or drought as well as bean quality, etc.

The genotype has a certain influence on the success of the grafting, e.g. clone IFCC 197 may reach a success rate of 85–90 % as opposed to 40 % for clone IFCC 126.

The physiological state of the coffee trees involved also has a great influence on the potential success of grafting. It is also an established fact that a sucker taken close to the soil presents better chances of success than one located higher up. Grafted plants should be frequently watered and

have adequate light. The grafting point itself should also be well protected.

Various types of grafts have been tested throughout the different stages of plant growth. The best results have been obtained with seedlings when the plants are at the "soldier" stage. As the plants mature, greater care must be taken and there is less chance of success.

4.5.1.1 Splice or Approach Grafting (Ingles Simple)

This method has been widely implemented and is considered to be the simplest. It consists of making a transverse bevel incision of about 2.5 cm in the stem of the rootstock. A similar incision is made in the scion and the two cuts are locked together and bound with a plastic tape 0.5-cm wide. Audio-cassette tape can also be used as binding material. The most efficient and least expensive tape for binding the graft is auto-adherable parafilm M which can be found in most local shops as it is extensively used in households.

After about 40 days, once the graft has "knit" satisfactorily, the tape should be removed. A satisfactory knit of the graft depends essentially on correct binding. It is, therefore, necessary to check that the binding is tight at the outset and remains tight while it is in place.

4.5.1.2 Wedge Grafting or Reyna Method (Injerto de Incrustación)

Wedge or cleft grafting is another method which is easy to perform and has produced good knitting results. Grafting is undertaken when the root-stock and the scion seedlings are at the "soldier" stage. The stock is cut crosswise at 5 cm above the collar and a longitudinal incision of 2.5 cm is made in the stem. Two cuts are then imme-

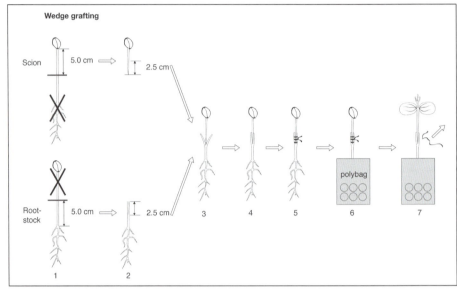

Figure 4.3 Wedge grafting at the "soldier" stage.
Phase 1: Decapitate the future scion 5 cm below the cotyledon and eliminate the rooted part. Cut the stem of the future stock 5 cm above the root collar and eliminate the upper part. Phase 2: A slanting 2.5-cm bevel cut is made in the scion. A 2.5 cm longitudinal incision is made in the upper part of the root-stock. Phase 3 + 4: The wedge-shaped scion is inserted into the slit of the root-stock so that the scion and the stock lock together tightly. Phase 5 + 6: The graft is bound with par-afilm (0.5 cm × 35 cm). Once the graft is well bound, it is pricked out in the propagation box or directly into a polybag. Phase 7: The binding is removed after 40–45 days. From: J. N. Wintgens.

Picture 4.11 Wedge grafting – seedling before grafting.

Picture 4.12 Binding the grafted stock.

Picture 4.13 Wedge grafted plant. From: Nestlé.

Picture 4.14 Grafted seedlings before pricking out.

Picture 4.15 Grafted plants which have just been pricked out in the nursery.

Picture 4.16 Grafted plants in polybags.

diately made in the scion, first a cross-cut at 5 cm below the cotyledon. This part is slant-cut to form a 2.5 cm long wedge which corresponds to the incision in the stock.

The wedge shaped cut of the scion is then placed in the cleft of the stock and the graft is fitted so that the incisions correspond exactly. They are bound with a narrow tape

Picture 4.17 Grafted stock. The biggest plant is 5 months old. From: Nestlé.

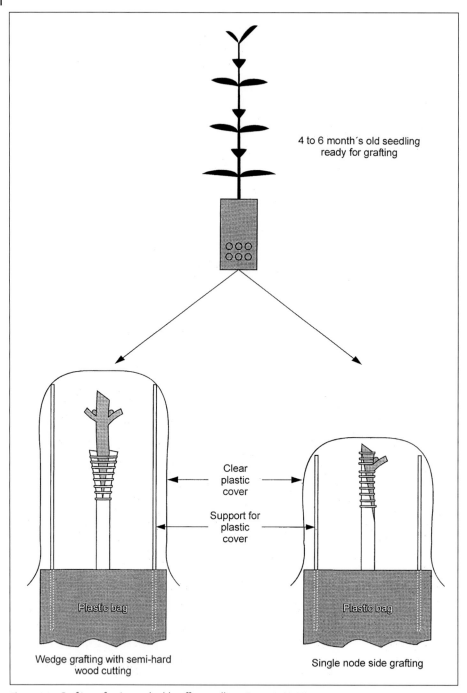

4 to 6 month's old seedling
ready for grafting

Clear
plastic
cover

Support for
plastic
cover

Plastic bag

Plastic bag

Wedge grafting with semi-hard
wood cutting

Single node side grafting

Figure 4.4 Grafting of a 6-month-old coffee seedling. From: J. N. Wintgens.

Picture 4.20 Growing phases of a side-grafted seedling at 2, 3 and 5 months. From: Nestlé.

measuring 0.5 cm in width (Fig. 4.3, Pict. 4.11, 4.12, 4.13, 4.14, 4.15, 4.16, 4.17).

4.5.1.3 Grafting Transplanted Seedlings

The 4- to 5-month-old seedling is stumped at a height of approximately 8–10 cm. A 3–4 cm longitudinal incision is then made along the stem for wedge grafting with a 3- to 4-month-old semi-hard green wood scion (Fig. 4.4, Pict. 4.18, 4.19, 4.20).

A parafilm is used to bind the graft and the grafted stock is covered with a translucid plastic cap. The same technique can also be applied on 3- to 4-month-old bare-root seedlings (Fig. 4.21A).

It is important to note that when using polythene caps to cover the polybags, the cap should be removed gradually. The first stage is when the young shoot starts to develop. This is when the cap should be lifted just above the surface of the bag. Complete removal can be made 3–4 weeks later.

4.5.1.4 Hypocotyledonary Grafting with Somatic Embryos and Microcuttings

This is a combination of *in vitro* and *in vivo* practices. It consists of grafting either somatic embryos or microcuttings on *in vivo* grown seedlings at the cotyledon stage. To date, this method has been essentially

Picture 4.18 Wedge grafting of a 6-month-old coffee seedling. From: B. Bertrand.

Picture 4.19 Wedge grafted 6-month-old coffee seedling. From: Nestlé.

Picture 4.21
(A) Grafted coffee plant protected with a translucid plastic cap.
(B) Grafted embryo on a seedling grown *in vivo*.
(C) Grafted microcutting on seedling grown *in vivo*.
From: B. Bertrand.

used for experimental purposes (Pict. 4.21 B, 4.21 C).

4.5.1.5 Coffee Rehabilitation

Rehabilitation is used for restoring or improving yield. This exercise is now an important factor in the search for higher productivity; it is generally recommended for trees that are low yielding and disease prone.

In mature grafting, adult plants are maintained as root-stocks. The grafted stock or the crown normally matures after 18 months as compared to the 24–30 months required for new plants.

Once the new stem is established it is important to regularly remove the sprouts emerging from the original stem. This will avoid mistaking them for the grafted stem.

There are many different opinions as to the best system to adopt for the rehabilitation and upgrading of unproductive older trees. The methods discussed provide an overview of some of the options available.

4.5.1.5.1 Upgrading by Wedge Grafting

Old or degenerate trees are stumped to a height of 30–40 cm from the ground. Young sprouts are allowed to develop and are evenly spaced, leaving four to five sprouts per stand. After 3–4 months, three of the most vigorous sprouts are selected for top grafting.

The stock is decapitated at a height of about 10–15 cm. An incision of approximately 2 –3 cm is made through the stem to receive the wedge. The wedge (scion) is taken from the semi-hard green

Picture 4.22 Upgrading by means of wedge grafting/securing the union with parafilm binding. From: Nestlé.

Picture 4.23 Upgrading by means of wedge grafting new sprouts. The graft is protected with a translucid plastic cap. From: Nestlé.

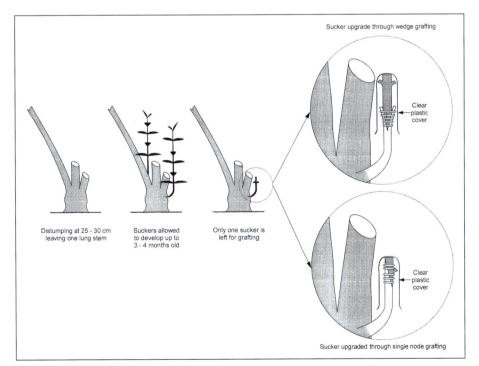

Figure 4.5 Upgrading by means of wedge and single node grafting. From: J. N. Wintgens.

Picture 4.24 Upgrading by means of wedge grafting new sprouts. The scions are beginning to develop.

Picture 4.26 Upgrading by means of patch grafting. A clear plastic sheet is used to cover the graft and both ends are tied down with string. From: Nestlé.

wood-cutting of a selected clone (Pict. 4.22).

The union is then secured firmly with a parafilm and covered with a clear plastic cap (Pict. 4.23, 4.24).

Grafting can also be successfully carried out with a single node side graft (Fig. 4.5).

Picture 4.25 Upgrading by means of patch grafting. A single node cutting is placed in the window of the stock and bound with a tape. From: Nestlé.

Picture 4.27 Upgrading by means of patch grafting. The grafted single node cutting is beginning to sprout. From: Nestlé.

4.5.1.5.2 Upgrading by Patch Grafting

Trees selected for this purpose must have all their primary branches cleared up to 1 m from the base. A small window of 1 cm × 4 cm to accommodate a single node cutting is opened on one of the

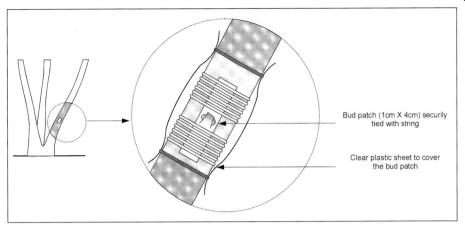

Figure 4.6 Upgrading by means of patch grafting (sketch).

stems at about 30–40 cm above ground level. The window should located on the upper side of the stem.

A single node cutting is firmly placed in the window and securely bound with tape. A clear plastic sheet tied down at both ends with string is used to cover the graft (Pict. 4.25, 4.26, 4.27).

Just after a shoot emerges the cover, should be slit with a small incision to allow air to flow into the pocket.

The plastic cover can be removed 2–3 weeks after slitting. The primary branches can be gradually removed to allow unhindered growth. Stumping of the old stems should also be carried out progressively as the new crown develops or when they obstruct the development of new stems (Fig. 4.6).

4.5.1.5.3 Upgrading by Shield Budding

Shield budding consists of taking a shield bud from a young orthotropic stem and inserting it into a T-cut window made in an old branch or stem. The green and still soft shield is removed in one shallow, slicing cut. The sliver of wood which still adheres to it after the cut must be carefully removed prior to inserting the shield bud into the T-cut window. The bud is taped into the window, but the green eye remains exposed and is covered with a protective sleeve of waxed paper. After 3–4 months, the grafted bud begins to develop and grow.

Grafting is generally more popular than shield budding because the callusing time is shorter, the union is stronger and the method requires less skilled labor. However, new and more interesting methods are now being developed and are likely to supplant these upgrading techniques (Fig. 4.7, 4.8).

On site grafting of mediocre adult or old plants is gradually being abandoned in flavor of replanting with new young stock. This is chiefly because on site grafting is both tricky and costly.

Replanting also enables improved laying out and correct spacing which was rarely the case in old plots, not to mention the obvious benefits of using young and vigorous stock.

4.5.1.6 Necessary Precautions to be taken when grafting or budding

• The seeds for stock growing must be taken from young, healthy, robust trees.

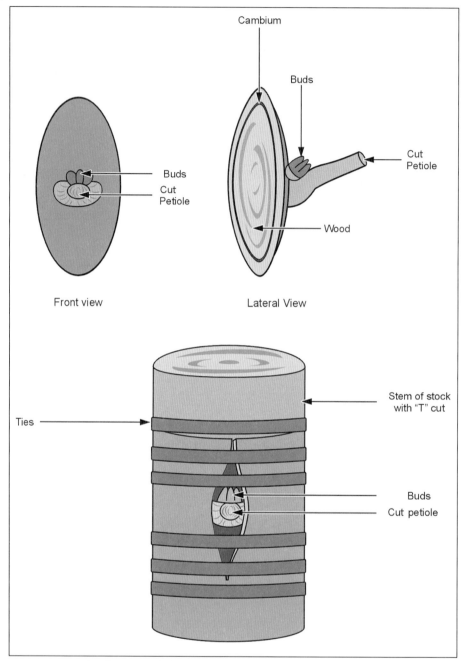

Figure 4.7 Upgrading by means of shield budding. From: J. N. Wintgens.

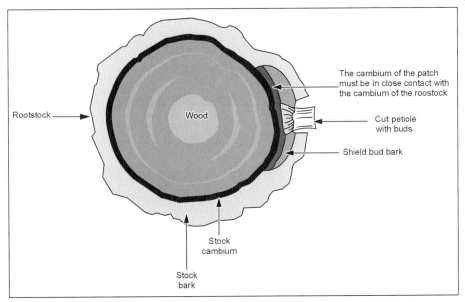

Figure 4.8 Upgrading by means of shield budding. Cross-cut at the level of the callus.

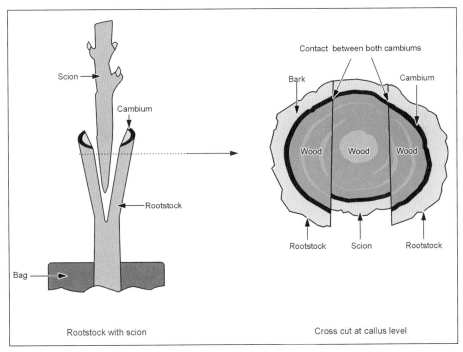

Figure 4.9 Wedge grafting/cambium contact.

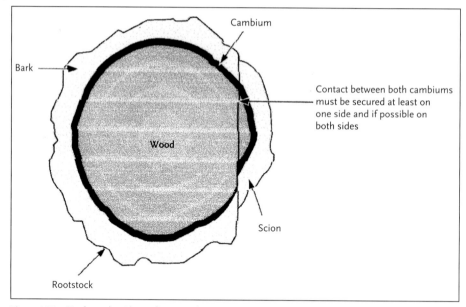

Cambium

Bark

Contact between both cambiums must be secured at least on one side and if possible on both sides

Wood

Scion

Rootstock

Figure 4.10   Single node side grafting/cambium contact.

- The nursery beds must be carefully labeled.
- Sowing the seeds both for the stock and for the scion should be staggered so that the material is constantly maintained at the desired growth stage for grafting.
- Seeds for stock should be sown 10–15 days before sowing the seeds to be used for scions.
- Ideally, grafting should be carried out in the spring season so that transplanting from the nursery to the coffee plantation is not delayed – the best results are obtained when grafting is practiced during the wet season.
- The objective of all grafting techniques should be to bring the cambiums of the rootstock and the scions into close contact so that they can fuse and build a callus which will make the circulation of sap possible from one to the other (see Figs. 4.9 and 4.10).

- A correct level of atmospheric humidity must be maintained during the grafting and curing period – grafted plants must be watered 2–3 times daily to maintain the required level of humidity, especially on sunny days.
- The graft should only be exposed to the sun gradually so that it can adapt more easily to its new environment.
- After grafting, the plants should be pricked out in nursery beds or in polybags.
- Grafts must not be buried as this will result in the scion taking root in the substrate.
- Scions, cuttings or patches for grafting or budding should only be taken from orthotropic stems – grafting material taken from plagiotropic wood retains its horizontal growth and will never generate a good yielder.
- The most appropriate clones to be used as rootstock are Nemaya (Guatemala) and Apoata (Brazil).

4.5.1.7 Advantages of Grafting Arabica on Canephora

- Grafting Arabica on Canephora offers the possibility of establishing Arabica in soils which have a high root parasite rating (nematode, root-scales, etc.).
- The maturation of the cherries will be more uniform because the plant will have a higher tolerance to water-deficient soil.
- The lifespan of the coffee plantation will be longer because physical defects and unproductive coffee trees will only appear at a later date.

4.5.1.8 Disadvantages of Grafting

- The coffee grower will have to take more care of the graft at the nursery stage.
- Grafting requires qualified and skilled labor in order to achieve the highest possible success rate.
- Grafting must take place at the optimal time, under favorable conditions so as to reduce the risks of knitting failure.
- Whatever the method used, a certain heterogeneity in the plants cannot be avoided when grafting – this is due to the interaction between the scion and the root-stock.

4.5.2
Horticultural Cutting

4.5.2.1 Choosing the Site for a Propagation Centre

A centre for propagation with cuttings needs to provide a root-stock orchard for producing the suckers required for cuttings and shelters for preparing them. It also needs to set aside specific areas for propagation and pricking out as well as specially shaded beds in which to grow young cuttings until they are planted out. The propagation centre should be located in a flat

Picture 4.28   Root-stock orchard with ground cover of *Flemingia congesta*.

area, preferably close to a source of water. Large trees should not be kept within the precincts of the propagation centre.

4.5.2.2 The Root-stock Orchard

The suckers, which ultimately become cuttings are grown in the root-stock orchard in which coffee trees are planted at a high density The layout of the root-stock orchard influences the density of the trees.

There are two major options:

- Planting in regular rows with a distance of 1 m between the rows and 75 cm within the rows (density = 13 333 plants/ha).
- Planting in twin rows with a wide inter-row of 1 m, a narrow inter-row of 0.5 m and a distance of 0.75 m within the rows (density = 17.777 plants/ha).

It is recommended to plant *Flemingia congesta* or any other appropriate cover plant in the wide inter-rows (Pict. 4.28).

In order to obtain a rapid and abundant growth of suckers for the production of cuttings, fertilization is important. Basically, the dominant element of the fertilizer should be nitrogen. Correct soil analysis will provide the quantities of nutrients to be supplied.

When planting out, 25–30 g of fertilizer should be placed in each hole. The

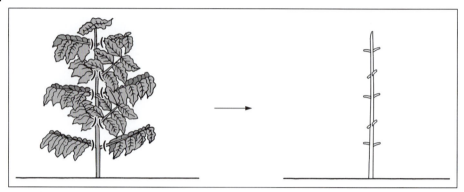

Figure 4.11 Clearing the primaries after 1.5 years of growth. From: P. Pochet.

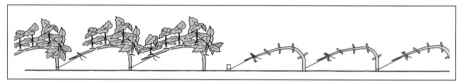

Figure 4.12 Bending to encourage the growth of shoots. From: P. Pochet.

same quantity of fertilizer should be spread in a circle around the foot of each tree every 3–4 months, preferably before a rainy period or after watering. Urea can be applied as a foliar spray after each sucker harvest at a rate of 25 g per coffee tree.

4.5.2.2.1 Maintenance of the Root-stock Orchard and Harvest

The root-stock orchard must be kept very clean. Pest and disease control should be strictly carried out as in a normal plantation.

The coffee plants are grown as single stem and all primaries should be removed after 12–18 months. New plagiotropic branches which may appear should also be removed (Fig. 4.11).

To encourage the formation of orthotropic shoots, the stem of the plant is bent along the axis of the row as soon as the plant is well anchored (Fig. 4.12).

Picture 4.29 Brick propagation box with cuttings.

A root-stock orchard should be able produce about 200 cuttings/m^2/year after 2 years. In general, suckers are removed 4 times a year, but production drops significantly in the dry season. Consequently, the root-stock orchard should be irrigated during this period to encourage the growth of suckers.

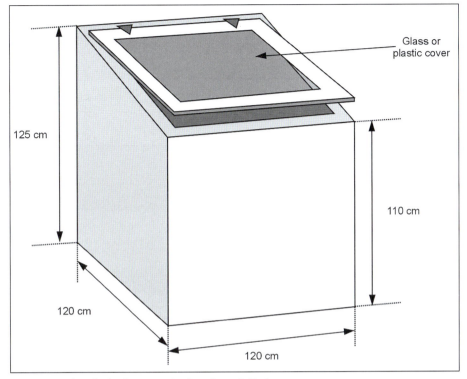

125 cm

Glass or plastic cover

110 cm

120 cm

120 cm

Figure 4.13 Plan of a brick propagation box. From: P. Pochet.

Picture 4.30 Large group of brick propagator boxes. From: Genagro.

4.5.2.2.2 **Propagator Beds**

The cutting production site is essentially a shaded area where the propagator beds are located. A propagator bed consists of 1 m × 1 m tub or box made of bricks. A layer of gravel should be used to line the bottom of the box which is then filled with a thick layer of adapted substrate. It is closed with a separate, removable cover made of glass or plastic (Pict. 4.29, Fig. 4.13, 4.14).

One square meter of propagator substrate can receive 400–500 cuttings. Hence, with two cycles per year, 1 m^2 can produce about 1000 cuttings annually.

The following figure shows two groups of six propagator boxes separated by a 2.5-m wide path to ensure easy access. The shade roof is about 2 m high and it juts out an extra 2.4 m laterally on the sun-lit side of the propagator boxes to offer added protection from direct sun (Fig. 4.15).

The shade supports should not obstruct access nor transit and the pathways should always remain clear. Protection along the

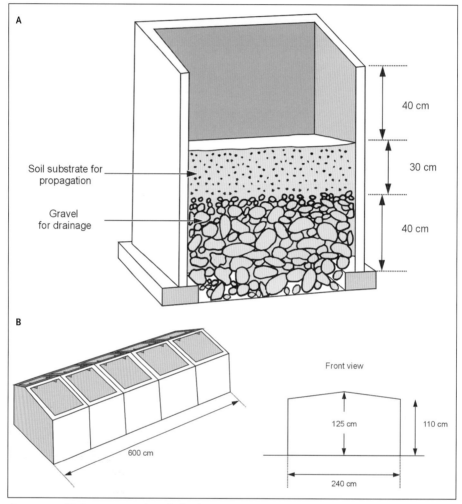

A

Soil substrate for propagation

Gravel for drainage

40 cm

30 cm

40 cm

B

Front view

125 cm

110 cm

600 cm

240 cm

Figure 4.14 (A) Cross-section plan of a brick propagation box showing the substrate and the drainage system. (B) Battery of 10 propagators. From: P. Pochet.

sides can be made of loose matting, palm leaves or specially adapted plastic netting.

The shade-roof should intercept about 65% of the sunlight. If it intercepts less, the temperature within the propagator boxes will be too high. On the other hand, if it intercepts more than 65% of the outside light, development and growth will not be optimal. Pict. 4.30 shows a large battery of propagators.

Other Devices for Developing Propagators

The two least costly devices for cutting production are to either prick out the cuttings in cutting pits or directly into polybags filled with organic earth. Cutting pits are dug directly into the soil and covered with a translucent polythene film. Cuttings in polybags and cutting pits are shaded and protected in the same way as those in propagator boxes (Fig. 4.16, 4.17, Pict. 4.31).

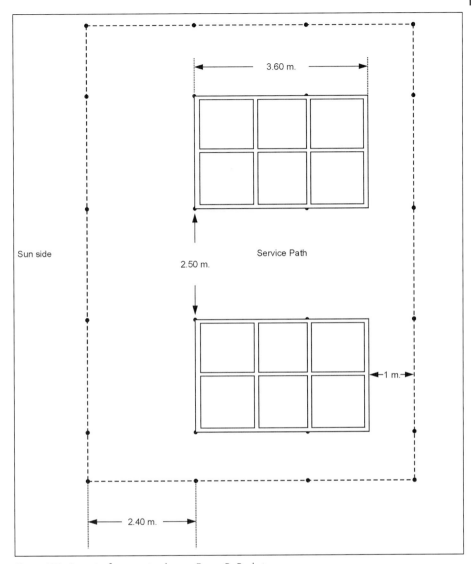

Figure 4.15 Layout of propagator boxes. From: P. Pochet.

The Substrate for Raising Cuttings

The substrate needs to be balanced between a capacity for filtering excess water and a capacity for retaining a suitable moisture level for the healthy development of the cutting. A number of substances can be used: fine river sand, peat, compost, sawdust, rice husks and coffee parchment, either pure or mixed. Before using an organic substrate it is useful to submit it to fermentation for several weeks. During this period, it should be kept damp by watering. After a period of 6–8 weeks, the temperature of the substrate will drop, this indicates that fermentation has been completed.

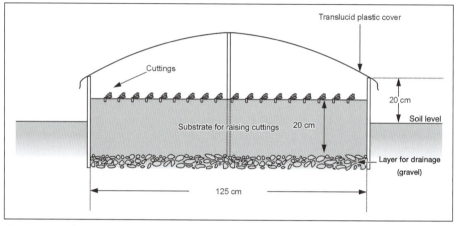

Figure 4.16 Rudimentary cutting pit.

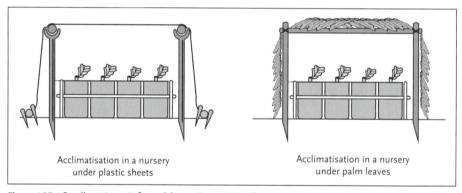

Acclimatisation in a nursery
under plastic sheets

Acclimatisation in a nursery
under palm leaves

Figure 4.17 Small cutting pit for polybags. From: P. Pochet.

Picture 4.31 Grafted stock planted in polybags.

Picture 4.32 Team of workers preparing cuttings.
From: O. Borbon.

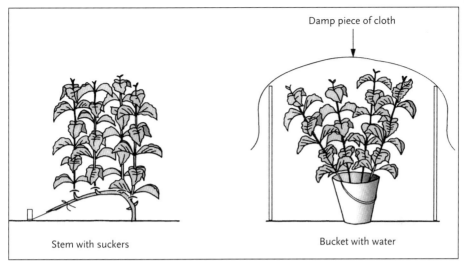

Figure 4.18 Conditioning shoots. From: P. Pochet.

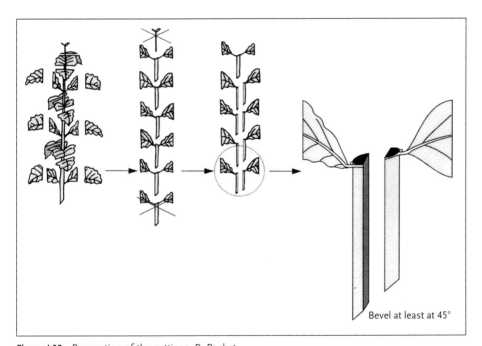

Figure 4.19 Preparation of the cuttings. P. Pochet.

The substrate can only be used for two or three cutting cycles because infection, especially fungal infection, sets in progressively. Spraying with fungicide can delay this process, but it is preferable to replace the substrate as soon as the infection shows significant development.

Picture 4.33   Cleaved and uncleaved cuttings. From: P. Charmentant.

Preparation of the Cuttings

Removing the shoots and trimming the cuttings must be carried out early in the morning while it is still cool and by 9 or 10 a.m., the operation should be completed.

Early in the morning, before 8 a.m., the suckers should be cut off the root-stock orchard with clippers or a pruning knife. They are immediately placed in a bucket partially filled with water. The top of the shoot must be protected from drying

out with a damp cloth (Fig. 4.18). Once the bucket is full of shoots, it is transported to the shaded area where trimming takes place (Pict. 4.32, Fig. 4.19).

Trimming consists of removing part of the leaves with either scissors, clippers or a sharp knife. Then the shoot is cut into sections just above each inter-node. The tender top and the hard base of the shoot are eliminated.

Once the sections have been cut, a bevel at a 45° angle is made at the base of the cutting and then it is cleaved. If the cuttings are too tender, they are not cleaved (Pict. 4.33).

All precautions must be taken to prevent the cutting from drying out. As soon as the cutting is cut and trimmed, it must be kept in water. The tools (scissors, clippers and knives) must be perfectly sharpened to avoid tumefaction of the plant tissue. Clean cuts are essential, particularly for the bevel and the cleavage, as the slightest crushing will later become a source of rot. When these operations are completed, the

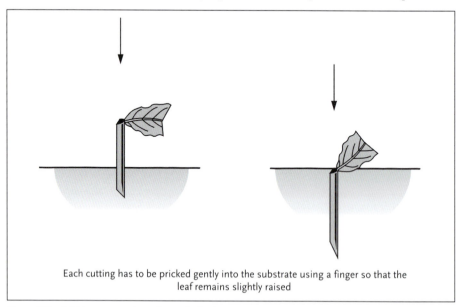

Each cutting has to be pricked gently into the substrate using a finger so that the leaf remains slightly raised

Figure 4.20   Pricking out cuttings. From: P. Pochet.

Picture 4.34 Planted cuttings in a propagator bed.

cuttings are ready to be pricked out in the propagator beds.

Certain authors recommend soaking the base of the cuttings in rhizogenic substances to encourage root development. These substances range from cattle urine (fed with fresh grass) to hormones of synthesis, such as indol butyric acid (solution of 500 ppm), naphtalenacetic acid β, dichlorophenoxyacetic acid, 2,4,β-indol propionic acid, etc. These products are chiefly recommended for Arabica cuttings which "take" less easily than Robusta.

4.5.2.2.3 Pricking Out the Cuttings

Once the cuttings are prepared, they are arranged in parallel rows by pushing them vertically into the humidified substrate until the petiole stump touches the surface and remains in contact with the damp substrate (Fig. 4.20).

Ideally, planting density should be between 400 and 500 cuttings/m^2 (5 cm × 5 cm and 4.5 cm × 4.5 cm). The cuttings are watered and covered as soon as a propagator box is completed (Pict. 4.34).

The cuttings should be humidified by misting with a fine spray of water until the microdroplets on the leaves begin to join together and flow (about 0.5–1 l/m^2). Watering maintains the ambient humidity at 100 % relative humidity and stabilizes the inside temperature at 25–30°C.

Picture 4.35 Rooted cuttings. From: CIRAD.

When the atmosphere is dry, watering should take place 2–3 times per day. It can be automated by installing fixed pipes fitted with several jets which are controlled either manually or by sensors. Perfect confinement is essential for the ultimate success of the operation and small propagator boxes with individual covers offer the best solution.

After 20 days, the healing cutting begins to bud and 2.5 months later the first roots appear. Where rooting has not taken place after 3.5 months, the cutting should be considered as defective and destroyed (Pict. 4.35).

After 6–8 weeks, rooted cuttings can be removed from the propagator box and slightly trimmed to shorten the tap root and remove twisted roots. Then they should be immediately drilled into a plastic bag or a nursery bed (Fig. 4.21).

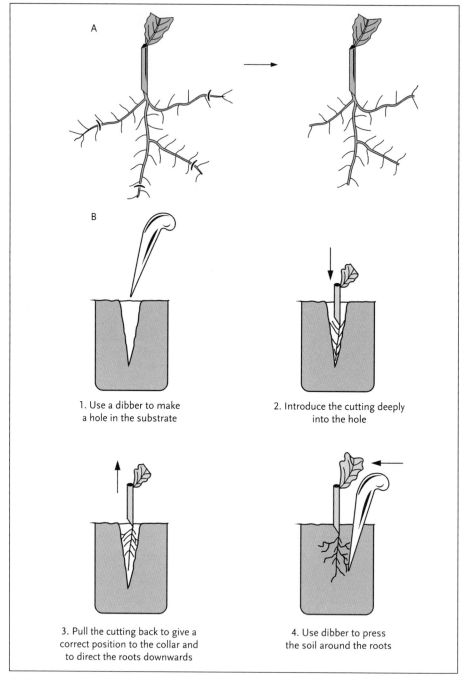

Figure 4.21 (A) Trimming the roots of cuttings. (B) Dibbing a cutting into a polybag. From: P. Pochet.

The success rate is normally 60–80 % depending on the variety, the quality of the cuttings and the care given to the operation.

The cuttings are pricked out in a standard nursery substrate consisting of sieved loam (20 %), organic soil (60 %) and sandy soil (20 %), all of which has been previously sieved. Where polybags are used, they should be perforated at the base and made of black polyethylene, 0.05 mm thick and treated against UV. The size of the bags should be 10.5 × 25 cm if the cuttings are to remain for eight months or 10.5 × 30 cm if they are to remain for a longer period.

The cuttings are inserted at mid-length with a dibble and the roots carefully placed downwards. The precautions to be respected are the same as those for transplanting seedlings at the "soldier" stage. The bags are placed on beds of 140–150 cm in width and separated by service pathways which are 50–60 cm wide. For the first 3–4 weeks, the cuttings are kept under a plastic tunnel in order to maintain a high level of atmospheric humidity identical to the level in the propagator boxes. Prior to the removal of the cutting and its final transplanting in the field, the shade is gradually reduced to accustom the young coffee plants to sunlight and minimize the physiological shock. Once the cuttings have hardened, they are exposed to normal sunlight and treated like nursery plants.

They can be planted out when they have six to eight pairs of leaves and one to three pairs of primaries. In the case of Robusta, this is generally 8–9 months after transplanting.

4.5.2.2.4 Data for Robusta

- 100 m^2 of root-stock orchard planted at 1 m × 0.75 m = 133 mother plants (minimum).

Table 4.6 Robusta cutting production

	Number of months	Accumulated months
Plantation of the root-stock orchard	0	0
Start production of cuttings from the root-stock orchard	18	18
Length of time the cuttings stay in the propagator boxes	3	21
Acclimatization of cuttings under shade	1	22
Length of time the cuttings stay in the nursery	8	30
Planting in open field		30

- Cuttings produced in 1 year per 100 m^2 of root-stock orchard with four harvests of 40 cuttings per tree = 133 × 160 = 20 000 cuttings.
- As the propagator can accept 500 cuttings/m^2 with two cycles per year, the capacity is 1000 cuttings/m^2/year.
- For 20 000 cuttings, the necessary propagator surface is 20 m^2.
- An average 60 % success rate means that there remain 12 000 usable cuttings (20 000 × 0.6) (for Arabica this percentage is lower because of the variety and problems linked to altitude).
- Nursery surface for 12 000 plants = 120 m^2 (net).

Table 4.6 gives an average timing on Robusta cutting production.

4.5.2.2.5 Distribution of the Rooted Cuttings

Transporting rooted cuttings in polybags filled with substrate directly to the field is costly and is only practical for short distances or for a small number of plants.

In general, for longer distances and large quantities of plants, the cuttings are removed from the propagators and transported with bare roots. In this case, a certain number of necessary precautions need to be implemented.

Transport

The rooted cuttings should be uprooted from the propagators in the evening. Then they should be tied together in small bunches and wrapped to retain their humidity. Wherever possible, the plantlets should be transported during night so that they arrive on site for transplanting in the early morning.

Transporting cuttings over long distances is risky. An overnight transport by truck is the maximum reasonable time for safety. Consequently, countries which use this propagation method must have a network of cutting stations available or a very efficient means of transport. The best solution is to set up several cutting stations which supply nearby planters or planters who produce their own cuttings in bags.

Pricking Out

As soon as the cuttings are delivered, the planter should prick them out in polybags filled with potsoil. This work must be carried out immediately on the morning of their arrival. This means that everything must be ready to receive them: this includes bags filled with earth, protection leaves, shade, etc. Added to which, the number of cuttings delivered to a planter must not exceed the amount he is capable of pricking out the same morning. Farmers often group together to ensure the handling of a large delivery of cuttings.

Acclimatization

Once the cuttings are pricked out, they should be stored under a shade roof. During the first 2 weeks they are protected by a plastic sheet which reproduces the atmospheric conditions of the propagator from which they have recently been removed.

Period of Transition in the Tree Nursery

The care of the young plants under shade in the nursery is the same as for cuttings which have been acclimatized in propagators. The best time for supplying cuttings to planters is between 9 and 12 months preceding the date scheduled for transplanting into the field.

4.6
Conditions for Successful Grafting and Horticultural Cutting

4.6.1
The Influence of the Vegetal Material

The genotype has a certain influence on the success of grafting and cutting. Arabica, for instance, produces hardly any roots after 4–6 months, whereas Robusta roots within 2–3 months and Arabusta in only 2 months.

The physiological stage of the graftwood or cutting also significantly impacts the success of propagation. This, in turn, is influenced by the care given to the mother plant i.e. fertilization, mulching, irrigation in case of drought, limitation of the number of suckers, etc.

There is also proof that suckers taken closely to the root system offer a better potential for propagation.

4.6.2
Environmental Conditions

- Cuttings of grafted plants should be placed in favorable conditions with respect to light.

- Frequent but light watering is recommended.
- The substrate must be sieved, disinfected and well aired.
- The graft point should be well protected.
- Phytosanitary treatment should be implemented where necessary.
- A constant temperature of 26–30°C accelerates the growing process.
- Ideally, propagation centers should be established in low areas; at higher altitudes, above 1000 m, the cuttings only root after 16–20 weeks, whereas in lower areas 8–12 weeks is sufficient.

4.7
In Vitro Propagation

4.7.1
Microcuttings

This technique, developed by Dublin and Custer in the early 1980s, is based on the same principles as that of horticultural cuttings, but it is carried out *in vitro*, i.e. by tissue cultures in aseptic conditions. The nodes of orthotropic stems, which must be neither too soft nor too woody, are removed and split lengthwise, then they are disinfected with an anti-fungal solution (Iprodione or Benomyl are more efficient) and calcium hypochloride (Pict. 4.36, 4.37).

Each node is placed into a solidified gelose medium containing cytokinin which favors cellular division. Microcuttings are cultivated in a special medium and kept at a temperature of 26°C for a photoperiod of 16 h.

After 3–5 months, the axillary buds of the nodes produce new shoots which have to be separated from the initial bud and then split into new segments called microcuttings (Pict. 4.38).

Both the axillary and the terminal buds of a microcutting are able to develop new

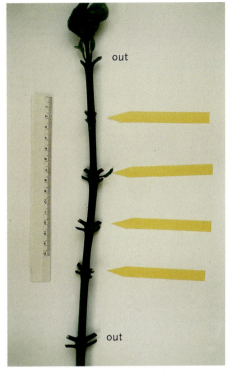

Picture 4.36 Choice of selected nodes on a green orthotropic stem The arrow indicates the selected nodes. From: Nestlé – CRN.

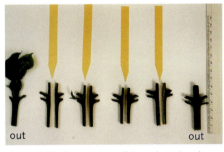

Picture 4.37 Preparation of the selected nodes. Each selected node indicated by an arrow is split into 2 cuttings. The nodes at the tip and the bottom are rejected. From: Nestlé – CRN.

microcuttings. This cycle is repeated until a sufficient number of shoots is produced. Depending on the genotype, the multiplication rate is usually two, sometimes

Picture 4.38 5 month old nodes after planting. Left: no response, the node dried up.
Right: The axillary bud of the node has developed and is ready to be subcultured to build up
new cuttings. From: Nestlé – CRN.

Picture 4.39 Newly generated buds ready to be subcultured. From: Nestlé – CRN.

Picture 4.40 Newly produced microcuttings with leaves and roots after a 2 week stay in the nursery. From: Nestlé – CRN.

Picture 4.41 New plantlets in the field nursery. From: Nestlé – CRN.

three and, occasionally, four cycles per month. In the best case, starting from one plantlet, approximately 20 000 coffee plantlets can be obtained in 1 year (Pict. 4.39, 4.40).

To harden the microcuttings, i.e. to transfer them into *in vivo* conditions, the lower leaves are removed and the basal cut is treated with a root-inducing hormone. They are then plugged into small blocks of peat. During the first 2–4 weeks of their acclimatization phase, the plantlets must stay under a plastic sheet in order to maintain the relative humidity at about 100 % and to protect them from intense light. Two weeks later, new pairs of leaves are formed and new roots begin to appear. At this stage, the coffee plantlets are ready for delivery to the nursery as clonal plantlets in plastic bags. The success rate for the acclimatization of microcuttings in field nurseries is generally about 90 % (Pict. 4.41).

This multiplication rate of microcuttings is distinctly superior to that achieved with horticultural cuttings but it has a major drawback which is a high risk of infection of the apex taken from the tree. Another drawback is the cost of the operation due to the specialized labor involved.

All in all, however, this technique shows a marked progress when compared to tra-

ditional techniques. On the other hand, its performance is not quite so impressive when compared to the very high multiplication rate achieved by somatic embryogenesis for mass propagation.

4.7.2
Somatic Embryogenesis

(see Section I.5)

4.8
Transfer of *In Vitro* Produced Material

In vitro produced material can be transferred in five different ways depending on the stage of the production chain.

4.8.1
Frozen or Dehydrated Cellular Mass (Calluses)

The calluses (undifferentiated cells) produced by somatic tissue, leading to the development of embryos, are frozen in aseptic vials for transfer. The recipient country needs basic laboratory equipment to produce and wean microplants (transfer to nursery). The equipment is similar to that of a quality control microbiology laboratory and includes a laminar flow hood, autoclave, shaking table and growth chamber. This equipment costs approximately US$ 75 000–150 000.

Once the calluses have been delivered to their country of destination, they are gradually defrozen until they reach a normal ambient temperature and multiplication can be initiated to produce the required number of embryos. These embryos need to be isolated, pricked into an appropriate solid medium, weaned and then transferred to a nursery until they are transplanted into the plantation.

The principal advantage of this technique is that most of the work is carried out by local labor.

4.8.2
Frozen or Dehydrated Embryos

This step follows the production of calluses. The calluses produce embryos which are isolated. The embryos are dispatched either fresh, frozen or dehydrated. Once they are coated, they receive the name of false or artificial seeds.

In this case the infrastructure required for processing is far less technical and costly because the embryos merely require "sowing" into a specific support to be weaned and then transferred to a nursery. Sowing is carried out by local labor, but the embryo sorting must be carried out in the central laboratory.

4.8.3
Sterile Microplants

This method has many disadvantages (volume, glass vessels, risk of transport, cost) and therefore tends to be abandoned.

The three techniques described above offer the obvious advantage of aseptic transport of material. On the other hand, at destination, competent staff is required to wean the plants and prepare their transfer to a normal nursery. To train a person in weaning techniques can take at least 2 weeks, but the cost of necessary material is very low (a plastic film tunnel and shading).

4.8.4
Rooted Microplants (Non-sterile)

This practice involves rooting microplants in an earthy support and transferring them to a nursery. This can entail difficulties with phy-

tosanitary authorities due to the presence of earth. It is a more expensive solution because drilling and weaning have to be carried out at the central laboratory and the total volume which has to be transported is considerable. At destination, however, the recovery of the plants is more rapid.

4.8.5
"Bare-root" Microplants (Non-sterile)

This method requires the central laboratory to take the 2-month-old plant from the solid medium and plant it in an earth support. The 3- to 4-month old plant is removed from the earth support after weaning. The roots are then washed with sterile water and the plants are transported to their destination. These plants can easily be subjected to 1–2 weeks of transport. Once received, the plants are transplanted into individual earth bags and grouped together under a plastic tunnel, simulating their transfer to a nursery. The plants can be transplanted into the plantation after 6 months.

This solution is of interest as it is the easiest method to implement. It is, however, less reliable from a phytosanitary point of view and also entails more handling in the central laboratory.

4.9
The Cost of Propagation Material (Tab. 4.7)

It is difficult to establish provisional costs, especially since they are largely dependent on the technique chosen as well as local conditions and labor costs. Cost comparison must be made at the stage of transferring the plants into the nursery because, in actual fact, the nursery and plantation costs will be similar whatever the origin of the plant.

Table 4.7 The cost of propagation material

Type of vegetal material to be transferred	Packing and transport	Work at receiving end	Time required before transferring to nursery	Estimated production cost of one plant ready for transferal to the nursery (in US$)	
				In Europe	In producing countries
Frozen cellular mass (callus)	frozen vial (liquid nitrogen)	multiplication, production, sorting and sowing, germination, weaning, development	1 year	0.06	0.04
Frozen or de-hydrated embryos	frozen vial (liquid nitrogen)	sowing, germination, weaning, development	32 weeks	0.06	0.04
Sterile microplants (ex embryos)	plants on aseptic medium	weaning, development	20 weeks	0.06	0.04
Microplants in earth support (ex embryos)	microplants in earth support (non-aseptic)	development	16 weeks	0.06	0.04
Bare-root microplants (ex embryos)	plants in damp boxes (non-aseptic)	development	18 weeks	0.06	0.04
Sterile microcuttings	plants on aseptic medium	weaning, development	20 weeks	0.80	0.30
Bare-root microcuttings	plants in damp boxes (non-aseptic)	development	18 weeks	1.00	0.30
Stem cuttings (in vivo)	local production	weaning, development	24 weeks	–	0.035–0.050
Grafted plant	local production	grafting on locally raised root-stock	18 weeks	–	0.25
Seeds (stabilized Arabica varieties)				0.003	

The less expensive multiplication method is, of course, seed propagation (US$ 0.003/plant). Asexual multiplication will therefore only be justified when seed propagation will not enable the production of a plant which is similar to the parent. (Robusta and intraspecific Arabic hybrids, interspecific hybrids Congusta, etc.)

For vegetative propagation, under present conditions, *in vivo* propagation by stem cuttings remains an interesting option because it is easy to perform. Smallholders can carry it out on a small scale by pricking out stem cuttings directly into earth bags. The cost per plant is very low: US$ 0.035–0.05/plant.

The production of plants by *in vitro* micro-cuttings is probably the safest method of *in vitro* propagation because the genetic uniformity and the identical type of the regenerated plant are guaranteed. It is nevertheless a more expensive method (the price of one plant produced by *in vitro* microcuttings is US$ 0.3–0.5) than propagation by somatic embryos and therefore should be used for limited quantities, e.g. when setting up propagation plots like a root-stock orchard.

Alternatively, *in vitro* propagation by somatic embryogenesis is cheaper. The plant produced has an estimated cost of US$0.04–0.08.

4.10
The Choice of a Propagation Method

Before taking an option on any particular propagation method, the first consideration should refer to the ultimate goal of the plantation. Once this is clearly defined, the type of coffee chosen, management systems, economic and environmental factors, among other elements, will inevitably indicate the most suitable propagation method for the plantation in question.

4.10.1
The Pros and Cons of Different Varieties of Coffee

Coffea arabica is the best suited to seed propagation. Seed generated progeny from seeds is rather uniform and a variety propagated by seed remains almost true. Hence, seed propagation is adequate for most Arabica varieties except for non-stabilized hybrids. In the latter case, seeds can only be used if they have been produced by intercrossing the two original parents. Somatic embryogenesis can be very useful to propagate Arabica hybrids.

In the case of Canephora, Excelsa and Liberica, however, seed-generated progeny is far less uniform because seeds are always produced by cross-pollination. Vegetative propagation is recommended for Robusta, Conillon, Liberica, Excelsa, mutants and non-stabilized interspecific or recent crosses that have not been selfed for a sufficient number of generations to induce genetic stability.

4.10.2
The Supply of Clonal Material

To date, the most interesting option for ensuring the supply of large quantities of clonal material (1 million plants or more) is still *in vivo* propagation by stem cuttings. It is easy to perform and inexpensive. On the other hand, it remains a relatively slow process because of the limited number of orthotropic shoots produced in a root-stock orchard.

Robusta can be propagated by stem cuttings or somatic embryogenesis depending on the production cost of plantlets in the country concerned and the quantities needed (smallholder, nursery, large estate). When large quantities of planting material are needed urgently, *in vitro* propagation by somatic embryogenesis can be a valuable alternative.

These two methods are interchangeable. For instance, somatic embryogenesis can be used for quick introduction of new varieties and then stem cuttings used for normal propagation.

Clonal material supplied by grafting is more limited in quantity than cuttings or somatic embryogenesis, but this method is practiced on a large scale in some countries like Brazil, El Salvador or Guatemala. Grafting can also be applied to obtain trees with a nematode-resistant root system (Arabica grafted at "soldier" or cotyledon stage on Robusta) and it can be used to top-work existing plantations with inferior planting material without uprooting them.

4.10.3
Requirements Influence the Alternative Choices

Supply of vegetal material for the establishment of propagation plots, seed gardens, local adaptation trials or small plantations. Small quantities required (up to 10 000 plants).
In this case, bare-root plants or microcuttings produced in a central laboratory are adequate. Graftwood could also be sent for local grafting. This would allow adult trees to be obtained faster for multiplication purposes. However, this solution is more limited (in quantity) and has a larger phytosanitary risk.

Supply of large quantities of vegetal material (up to 1 million plants a year)
Robusta 1 million plants (700–800 ha)
         = 400 –1600 tons green coffee

Arabica 1 million plants (200–400 ha)
         = 100–600 tons green coffee

The best and most economical solution for Arabica, is seed propagation. Asexual multiplication by somatic embryogenesis is only justified when propagating new hybrids that cannot be seed multiplied. The genetic improvement brought about by the use of hybrids (vigor, productivity, pests and diseases resistance) must compensate for the slight cost increase of the plantation when applying somatic embryogenesis compared to seed propagation.

Supply of massive quantities of vegetal material (over 1 million plants a year).
Seeding remains the cheapest solution for the propagation of Arabica. Robusta propa-

gation, however, entails the production of plants from embryos. These are either imported or produced *in vitro* in local laboratories or development units.

The development of plants by embryogenesis is the same wherever the embryos come from. It, therefore, makes no difference whether they are imported from a central laboratory in a foreign country or produced in the coffee-growing country where the embryo will, finally, take root and become a coffee tree.

As a result, the development of local production units for embryogenesis depends more on legal or political motives than economic reasons. The large-scale production of plantlets originated from somatic embryos will, undoubtedly, show extensive development in the foreseeable future.

4.11
How to Plant Clones

Individual plants issued from the same mother plant by vegetative propagation (stem-cuttings, somatic embryos, scions) constitute a clone. All trees belonging to the same clone are genetically and morphologically the same. Therefore, coffee trees of the same clone cannot fertilize each other. As a result, they become a sterile or non-pollinated species.

On the other hand, coffee trees issued from different clones are able to fertilize (pollinate) each other to a certain extent, depending on their grade of diversity. Investigation and research studies have now come to the conclusion that approximately five different clones of sufficient diversity should be planted together in production plots to obtain a good fertilization and rate of production.

Clones should be planted in single or twin rows. The preferred clone can be

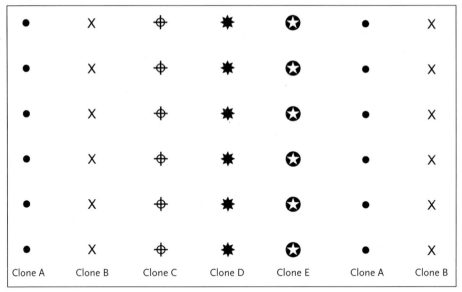

Figure 4.22   Homogenous distribution of clones in the fields. From: P. Pochet.

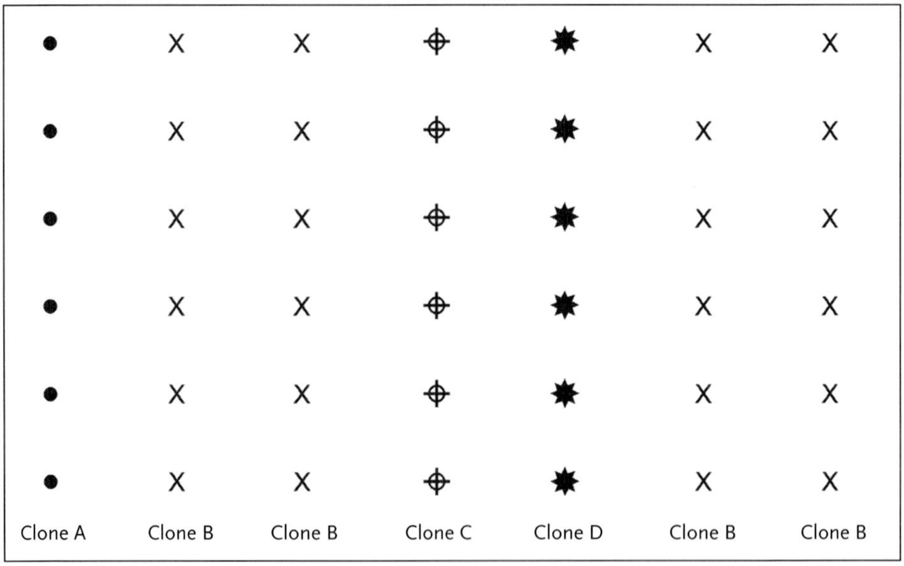

Clone A	Clone B	Clone B	Clone C	Clone D	Clone B	Clone B

Figure 4.23   Preference is given to clone B. From: P. Pochet.

planted in a greater number of lines. However, planting more than two adjoining lines with the same clone should be avoided.

The diversity of the clones, i.e. their mutual capacity to pollinate, can be assessed summarily by DNA fingerprinting which reveals the genetic distance of the clones.

In order to confirm this compatibility between clones, it is recommended to check the pollination and fruit-bearing rate of planted trees (Fig. 4.22, 4.23).

4.12
Expression of the Potential of Planting Material (Table 4.8)

The genetic factors of the coffee plant interact with environmental conditions and agricultural practices. As a result, the following comments should be taken into consideration:

- There is no point in planting superior genetic material in an unsuitable environment where agricultural practices are not up to the standards required for its optimal development.
- Improving agricultural practices (pruning, fertilization) on a weak genetic material does not provide a significant increase in productivity and could be a waste of time and money.
- In most cases, the use of selected clones rather than seeds from unselected seedlings can lead to a significant increase in the production potential (extensive cropping is the exception).

- In countries where extensive and semi-extensive cultivation prevails, it is better to use selected clonal material, or at least biclonal seeds, while progressively improving agricultural practices.
- Where an intensive cropping system is planned, the use of elite clonal material is fully justified; on the other hand, where extensive cropping systems with rudimentary agricultural practices prevail and no projects for improvement are underway, the use of elite clonal material or biclonal seeds can barely be justified.
- Consequently, the choice of planting material must take into account the cropping system to be implemented.
- Whatever the type of cropping, polyclonal material or seeds from unselected seedlings should not be used as they will not lead to an increase in crop performance even where improved agricultural practices are implemented.

Table 4.8 Expression of the potential of planting material according to environmental conditions and management practices

Type of planting material	Intensive cropping without shade 1500 to 3000 kg/ha	Intensive cropping with slight shade 1500–2000 kg/ha	Semi-intensive cropping with dense shade 500–1000 kg/ha	Extensive cropping with heavy shade 300–500 kg/ha
Elite clones selected for yield and quality[a]	100	80	50–60	10–20
Clonal seeds[b] issued from a clonal garden planted with two special clones selected for seed production (uniform population)	70	50–60	40	10–20
Polyclonal seeds[b] issued from a clonal garden planted with multiple clones selected for production (non-uniform population)	50–60	40–50	30–40	10–20
Seeds[b] harvested in commercial plantations from unselected seedlings	40–45	N/A.	30–40	10–20

[a]Applicable to Arabica and Robusta.
[b]Applicable to Robusta only.

Complementary explanations on Table 4.8

Intensive cropping without shade 1.5–3.0 tons/ha High planting density (1500–5000 plants/ha) No shade Pest and disease control easily monitored Good quality maintenance practices (cleaning, pruning, fertilizing) Irrigation when needed	**Intensive cropping with shade 1.5–2.0 tons/ha** Slightly lower planting density Slight shade Pest and disease control easily monitored Good quality maintenance practices (cleaning, pruning, fertilizing) Irrigation when needed
Semi-intensive cropping 0.5–1.0 tons/ha Lower planting density (1000–2500 plants/ha) Dense shade Pest and disease control sometimes neglected Less quality maintenance practices	**Extensive cropping 0.3–0.5 tons/ha** Very low planting density or mixed cropping Heavy shade Practically no pest and disease control Rudimentary maintenance practices

Bibliography

[1] ANACAFE. *Manual de Caficultura.* ANACAFE, Guatemala, 1998.

[2] Andrade, N. G. *Cafetales y Café.* Ministerio de Aqricultura y Cria, Caracas, 1988.

[3] Cambrony, H. R. *Le Caféier.* Maisonneuve & Larose, Paris, 1989.

[4] Coffee Growers Association. *Coffee Handbook.* Coffee Growers Association, Zimbabwe, 1987.

[4A] Coste, R. Le Caférier. Ed. Maisonneure & Larose, Paris 1968.

[5] Gaie, W. and Flemal, J. *La Culture du Cafeier d'Arabie au Burundi.* AGCD, Brussels, 1988.

[6] ICAFE. *Manual de Recomendaciones para el Cultivo del Café.* INCAFE, Costa Rica, 1998.

[7] INIFAP. *Tecnologia para la Production de Café en Mexico.* SAGAR, FPV, INIFAP, Xalapa, VE Mexico, 1997.

[7A] Krug, C.A. and Carvalho A. The Genetics of *Coffea.* In: Demerek, E.T. (ed.) *Advances in Genetics* 1951, 4, 127–158.

[8] Pochet, P. *Robusta Propagation by Cuttings.* AGCD, Brussels, 1987.

[9] PROCAFE. Manual del Caficultor Salvadoreño. PROCAFE, San Salvador, El Salvador, 1997

[9A] Wellman, F.L. Coffee, Botany, Cultivation and utilization. Pub. Leonard Hill, London 1961.

[10] Wintgens, J. N. and Petiard, V. *Coffee Propagation Policy.* NESTLÉ, Vevey, Switzerland.

5
Biotechnologies Applied to Coffee

A. Zamarripa C. and V. Pétiard

5.1
Definition and Biological Bases

Biotechnology is defined by the European Biotechnology Federation as the integrated use of engineering, biochemistry and microbiology to achieve the technological application of the capabilities of microorganisms, cultivated tissues cells and parts thereof.

Over the last two decades, biotechnology has developed significantly, modifying both plants as well as animal breeding strategies and industrial production systems. However, the first applications were on cheese, beer and wine.

Plant biotechnologies can be defined as the procedures for manipulations in artificial media set up with the intention of identifying/preserving and propagating as well as modifying plants to improve their use for mankind. For example, thanks to biotechnologies, it is now possible to obtain improved pest- and/or disease-resistant varieties in less time, and, by somatic embryogenesis, it is possible to speed up their diffusion at a lower cost.

Plant tissue culture uses the basic property of the plant kingdom known as totipotency (the power to express everything which is encoded in its genetic information including the regeneration of a whole plant or the production of high added value compounds). Totipotency, therefore, means that any plant cell, whatever its type/specialization, possesses all the necessary genetic information to develop another individual with similar characteristics to those of the plant from which it originates.

The above means that, theoretically, any species/variety of the plant kingdom may be cultivated and regenerated *in vitro* from any cell. However, the success of regeneration in a laboratory requires previous knowledge of various factors, such as growth conditions (temperature, light, etc.), culture media, type and dosage of growth hormones, as well as knowledge of the regenerative aptitude of the various plant organs/pieces (primary explant).

The applications of plant biotechnologies can be summarized as described in Fig. 5.1. One should note that some are geared to the modification of the genome to create more diversity while others are geard to the description, conservation, and fixation in the progenies or propagation of a given genotype they did not modify. Assimilating plant biotechnologies to genetically modified organisms (GMOs) is, therefore, a simplistic approach. Genetic engineering is a biotechnology. Biotechnologies are not genetic engineering.

Coffee: Growing, Processing, Sustainable Production, Second Edition. Edited by J. N. Wintgens.
© 2012 WILEY-VCH Verlag GmbH & Co. KGaA. Published 2012 by Wiley-VCH Verlag GmbH & Co. KGaA.

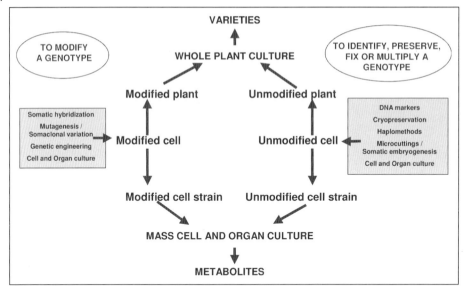

Figure 5.1 Techniques in plant biotechnologies.

One should also keep in mind that whatever their practical goal, plant biotechnologies are only tools to be used according to the species, breeding strategy and objectives, and have to be strongly tied to the work of breeders to be successful.

5.2
Markers for Identification/Genetic Mapping/Selection (Fig. 5.2)

5.2.1
Introduction

For the creation of varieties, the breeder requires the description of each genotype, e.g. its resistance to diseases, or its grain size or composition, in order to select the best parents for crosses and the promising progenies in any specific characteristic.

Normally, the description of varieties is based on morphological, physiological and biochemical criteria. However, the use of these criteria has its limitations – mainly the influence of the environment on the characteristics of the plant being observed.

Over the last 15 years, molecular biology has developed valuable tools called molecular markers, which are used to conduct faster, detailed and more reliable descriptions based on the plant genome, and not only on its visible expression. The information obtained from markers may be used for many purposes, such as the evaluation of the genetic diversity, the identification of the varieties, and plant breeding which can be guided by the markers and subsequent genetic maps. These markers can also be used for the identification of varieties in their derived products, including processed food products [45]. In the case of coffee, the use of molecular markers has already given rise to significant advances.

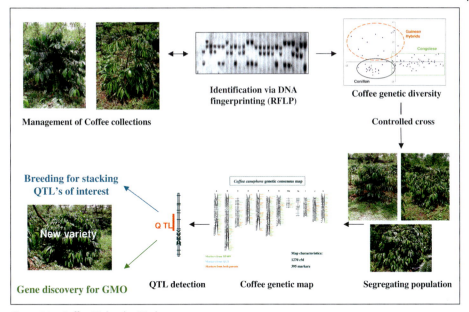

Figure 5.2 Coffee Molecular Markers.

5.2.2
Techniques

There are two types of molecular markers: proteins and the DNA markers, called random amplified DNA polymorphism (RAPD), restriction fragments length polymorphism (RFLP) and microsatellites [short sequence repeats (SSRs)].

5.2.2.1 Isozymes
In addition to DNA markers, which are the most common ones used today, proteins (isozymes) are the most widely used markers. They may be separated according to their molecular weight and electrical charge, in a gel placed under an electrical field. Then they can be visualized by dyeing the gel with specific reagents. However, their polymorphism in a given species may be limited and could also be influenced by the environment. They are, therefor, less reliable than DNA markers.

5.2.2.2 RAPD (Fig. 5.3)
RAPD analysis is conducted in three stages:

1) Heat denaturation of the DNA and separation of the two strands of the molecule
2) Attaching of the "primers" generally constituted by a short complementary sequence of the DNA
3) Synthesis of a specific DNA fragment delimited by the anchoring of the two primers by means of the action of the DNA polymerase of a bacteria named *Thermophilus aquaticus*, currently named Taq polymerase.

The three successive stages are repeated up to 35 times. This allows for an exponential amplification of the synthesized fragments. The DNA fragments generated are separated by electrophoresis, dyed (e.g. in an ethidium bromide solution) and observed under ultraviolet light.

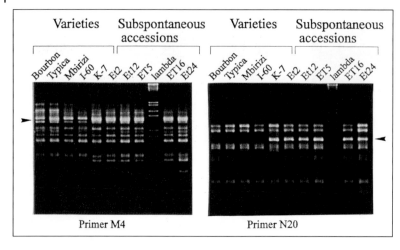

Figure 5.3
Examples of RAPD markers in *C. arabica*.

This RAPD technique is quite rapid and easy and does not require any previous knowledge of the genome. However, it is not as precise and specific as RFLP or SSRs, and therefore of less interest mainly because it does not allow the differentiation of homozygous and heterozygous markers.

5.2.2.3 RFLP

RFLPs are polymorphisms the size of DNA fragments generated by a treatment with restriction enzymes. RFLP analysis can be generally described in six stages (Fig. 5.4).

1) Extraction of crude DNA from a plant organ (e.g. leaves)
2) Specific fragmentation of the DNA by a restriction enzyme
3) Separation of the DNA fragments by electrophoresis in an agar gel
4) Transfer of the separated fragments to a nylon membrane (Southern blotting)
5) Specific hybridization of the homologous fragments with a given DNA sequence probe
6) Detection of the hybridization strips (probe/fragments) mainly using radio-labeling, but also chemical luminescence methods.

5.2.2.4 SSRs

Microsatellites are SSRs of DNA such as $(GA)_n$ or $(GT)_n$. These SSRs are highly polymorphic in plants due to the great level of variation in the number (n) of tandem repeats of two DNA "letters". As the DNA sequences flanking these SSRs are well conserved between different plants of a given species or between different species, the diversity of (n) allows the characterization of individual plant.

The development of these markers requires:

• DNA sequencing of each marker in order to define the primers corresponding to the sequence of both flanking sides of the repeat.
• Amplification of the marker using Taq polymerase as previously described for the RAPD technique.
• Characterization of the different SSR forms according to their size (n value) using fluorescent or radioactive labeling.

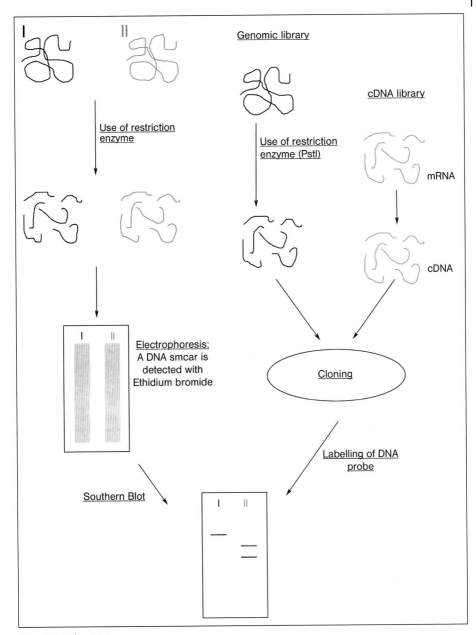

Figure 5.4 The RFLP process.

This technique is certainly the most powerful DNA marker technique in terms of polymerphism detection, but it is obviously dependent on the availability of microsatellite DNA sequences.

5.2.3
Applications to Coffee

An old study based on seven enzymatic systems led to the classification in two geographical groups of 471 Robustas from natural populations and from a collection managed by the Wood Institute Coffee and Cocoa Department (IDEFOR) of the Ivory Coast.

The first, named Guineans, consists of the populations of Western Africa, and the second, named Congolese, consists of the populations of Central Africa [7]. This work, based on enzymatic polymorphism revealed by electrophoresis, allowed the characterization of the diversity of the Robusta species and finally led to the initiation of a recurrent reciprocal breeding amongst the groups [33].

However, among the numerous applications of molecular markers, one of the most relevant is the preparation of genetic maps, which can be used to determine the chromosomic location of genes that affect characteristics of agronomic or industrial interest.

Paillard *et al.* [40] have used RFLP and RAPD markers to develop the genetic map of *Coffea canephora* Robusta. The genetic map, with an average distance of 10 centimorgan (cM) between two adjacent markers, located 47 RFLP loci and 100 RAPD loci, situated on 15 linkage groups.

Lashermes *et al.* [31] used RAPD markers for the analysis of 20 genotypes of *Coffea arabica*. The data obtained led to their classification into three groups: the Ethiopian wild genotypes, and the Bourbon and Typica groups. However, due to the small number of markers, characterization of the varieties within each group was not achieved.

Recent studies of phyllogenetic relations of the coffee species, by chloroplastic DNA analysis [13] and ribosomal DNA [32], have led to the identification of *C. brevipes,*

C. canephora, C. congensis, C. eugenoides and *C.* sp. "Moloundou" as the genetic species closest to *C. arabica.*

These studies reinforce the hypothesis of Lashermes *et al.* [31], which suggests that *C. arabica* may come from the *natural* hybridization of two *Coffea* species, close to *C. eugenoides* and *C. canephora.*

More recently, the use of microsatellites led to new phyllogenetic studies as well as to a new, more dense genetic map based on 350 markers and showing the number of linkage groups expected for Robusta when considering the chromosome number ($n = 11$) of this species (Crouzillat *et al.,* accepted for publication, 2003) (Fig. 5.5).

Molecular markers represent valuable tools to be used in a number of ways to support the breeding of new improved coffee varieties. For example, they can be used for the molecular fingerprinting and classification of germplasm. They enable the elimination of duplicates and mislabeled trees, the identification of varieties, and an estimation of the genetic diversity of the collections.

The use of these techniques may also lead to flagging a characteristic of agronomic/quality interest on the DNA and, therefore, enabling the early selection of trees which would have expressed the desired trait at the adult stage. It may also guide cross-breeding experiments by cumulating into a single plant different chromosome segments [currently called quantitative trait loci (QTL)] determining one or a number of desired traits. Considering that a coffee selection cycle requires 5 years, the use of molecular markers could be of even greater interest for coffee than for the main annual field crops, such as corn, where it is already very commonly used. It will speed up the breeding process, and make it more precise and cheaper to achieve desirable improvements.

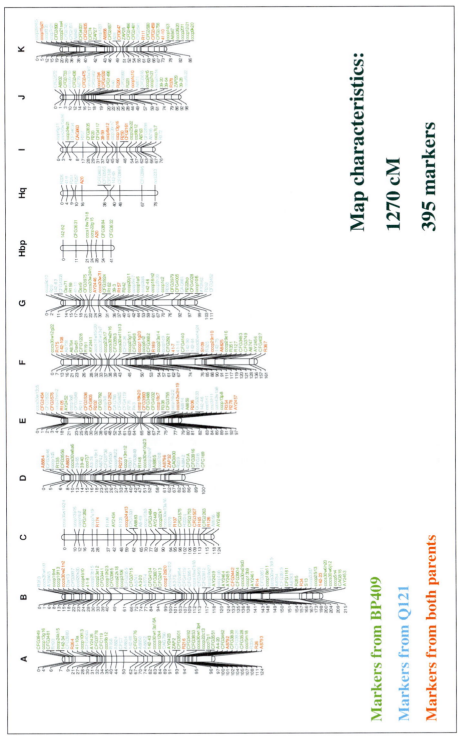

Figure 5.5 Robusta genetic map.

5.3
Cryopreservation (Fig. 5.6)

5.3.1
Introduction

Genetic diversity is a prerequisite to any plant-breeding program aimed at the creation of varieties with an improved agronomic, processing or qualitative value. It is, therefore, necessary to have in the collection the plants containing the genetic information required for the development of new varieties. The coffee plant is a species recalcitrant to conventional storage methods, since the seeds are sensitive to desiccation and to temperatures lower than 10°C [24].

Coffee seeds lose their viability in less than 6 months when stored at room temperatures. Hence, the conservation of genetic resources is conducted in field collections set from seeds or cuttings. Practically all coffee-growing countries have their own coffee collections in experimental centers. However, due to high maintenance costs, budgetary limitations may make it mandatory to concentrate efforts only on *C. arabica* and *C. canephora* materials, discarding the rest of the *Coffea* species.

Conservation in the field is risky, mainly due to exceptional climatic factors like tornadoes and drought which have recently affected several coffee-growing countries. Phytosanitary and even political risks can also jeopardize valuable breeding material. Therefore, several specialists agree that, for safety reasons, duplicates of collections should be kept.

Bearing these factors in mind, over the last 10 years various public and private entities, mainly European, have developed new techniques for the medium- and long-term conservation of coffee genetic resources in order to secure the preservation of genetic diversity.

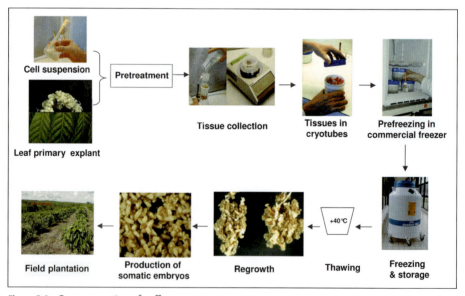

Figure 5.6 Cryopreservation of coffee.

5.3.2
Methods

5.3.2.1 Conventional Cryoconservation

This entails treatment and freezing with cryoprotectants, in order to reduce the intracellular water content and to increase the concentration of soluble compounds such as sugars. Freezing is conducted in a programmed freezer decreasing the temperature from 0.5°C/min down to −40°C before transferring the material to liquid nitrogen.

5.3.2.2 Simplified Method

This method is based on the induction of freezing tolerance during a pre-treatment phase, without any further addition of cryoprotectants. After this pre-treatment phase the samples are placed in a freezer at −20° C for 20 h before transfer to liquid nitrogen [57].

5.3.2.3 Pre-drying/Freezing

This entails partial tissue dehydration under controlled relative humidity as a pre-treatment phase before transfer to liquid nitrogen.

5.3.3
Applications to Coffee

The conservation of coffee zygotic embryos in liquid nitrogen has been reported by Abdelnour-Esquivel *et al.* [1]. Florin *et al.* [24] have also reported the conservation of coffee zygotic embryos. The embryos isolated from ripe coffee beans are placed on a filter paper, which is then introduced in a Petri dish either directly or in a medium supplemented with sucrose. After a 7-day incubation in the dark, the embryos are partially dehydrated in a container with a controlled relative humidity of 43 %. Five days later, the embryos are transferred to cryotubes and frozen directly in liquid nitrogen.

The best survival rate of 78 % was obtained with pre-treatment on a medium containing 0.6 M sucrose. The development rate of embryos into plantlets was comparable to that of non-frozen zygotic embryos [24].

However, the preservation of zygotic embryos is of no interest for cross-pollinated heterozygous plants such as Robusta since the genetic background of their seed embryos is variable and not similar to that of the mother plant one would like to preserve. On the other hand, it

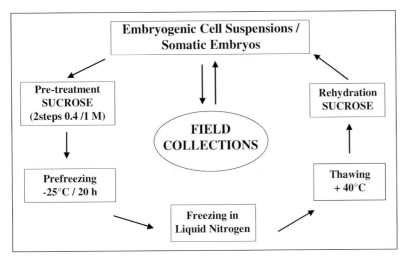

Figure 5.7 Simple freezing method for coffee collections.

may be of great interest for Arabica as long as self-pollination has been controlled.

Using the simplified method (Fig. 5.7), Tessereau [57, 58] reports the cryoconservation of cell suspensions of three coffee species which maintained their viability and embryogenic potential.

In the case of somatic embryos, freezing has also been possible with similar results for both conventional and simplified methods [24]. However, the simplified freezing method based on an inductive pre-treatment would be easier to perform for the conservation and distribution of germplasm.

Using this simplified method, Tessereau [57, 58] reports survival rates up to 90–100 %. After a lag phase of 3–5 weeks, the re-growth of frozen embryos takes place by the production of secondary embryos, which have been developed into plantlets similar to those obtained from the non-frozen embryos. However, the direct re-growth of embryos as reported by Mycock and Berjak [36] might be more desirable for the commercial diffusion of selected plants, since it would not require intermediate *in vitro* growth phases nor would it lead to multigerm-type "seeds" [44].

Hatanaka *et al.* [26] report the successful cryoconservation of somatic embryos of Robusta by dehydration of the embryos coated with alginate gel pearls before their immersion in liquid nitrogen. The maximum survival rate was 63 %. More recently, Dussert *et al.* [22] reported for the first time the cryoconservation of Arabica beans in liquid nitrogen, after drying and pre-freezing before their immersion in liquid nitrogen.

After field validation of these technologies in order to ensure that they do not induce unexpected variations/mutations, the above conservation methods will represent a significant advance – on one hand, to ensure the genetic patrimony of the *Coffea* genus and, on the other hand, for the industrial process of multiplication of improved varieties by somatic embryogenesis.

Cryoconservation should lead to the setting up of safer and less costly collections [44], and facilitate the exchange of genetic material through avoiding the risk of dissemination of phytopathogens such as *Colletotrichum kahawae* (Coffee Berry Disease) to coffee-growing areas which are still free of the disease.

5.4
Haplomethods (Fig. 5.8)

5.4.1
Introduction

Plant characteristics, are controlled by one or more genes located in the chromosomes. In cells, each one of the chromosomes is present in two copies, so the plant is called diploid (2n). However, for coffee, Robusta is diploid, but Arabica is tetraploid, which means that it may contain four instead of two copies of a given gene.

Haploidy is the natural genetic status of a cell or an individual which contains only one set of chromosomes instead of two ("n" status). Only the gametes, i.e. the male reproductive cells (pollen) and the female reproductive cells (embryonic sack), are haploids. At the time of fertilization their fusion generates a diploid zygotic embryo capable of forming a new individual.

Haplomethods is the term given to the regeneration of whole plants from haploid gametic cells. The development of pollen grains is called androgenesis (male) and that of the embryonic sack cells is called gynogenesis (female).

In autogamous (self-pollinated) species such as Arabica the plant breeder seeks pure breed lines. Their homogeneity and stability leads to their use either as com-

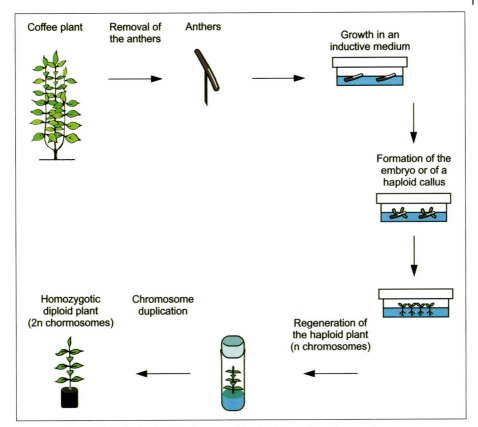

Coffee plant   Removal of   Anthers
             the anthers

Growth in an
inductive medium

Formation of the
embryo or of a
haploid callus

Regeneration of
the haploid plant
(n chromosomes)

Homozygotic   Chromosome
diploid plant   duplication
(2n chormosomes)

Figure 5.8 Protocol for the *in vitro* development of haploid plants by androgenesis.

mercial varieties or as parents of true hybrids. Their creation by successive self-pollination may require five to six generations, which for coffee means more than 20 years. The haplomethods, therefore, have attracted ever-increasing interest to speed up the plant breeding process.

5.4.2
Methods

In the case of the coffee plant, as far as we know, only androgenesis has apparently been successful. Anthers are grown in Petri dishes onto a specific medium containing growth hormones, mineral elements, vitamins and sugar to induce the division of the microspores. The Petri dishes are placed in culture chambers where the temperature and light are monitored. After a few weeks of cultivation, haploid embryos (gametic embryos) may be observed. These are then subcultured on an adequate medium for their development to the plantlet stage.

Once the haploid plantlets have been acclimatized, the number of chromosomes is doubled to restore the diploid status. An alkaloid named colchicine is generally applied on the buds to cause diploidy. The plants obtained in this way are obviously 100 % homozygous (pure breed line) since their two sets of chromosomes come from the duplication of a single one.

5.4.3
Applications to Coffee

There are relatively few papers reporting the *in vitro* development of coffee haploid embryos and subsequent plantlets. That might be due to the relatively easy development of haploid coffee plants from the natural occurrence of double embryos in the seeds. One of these so-called twin embryos may be haploid. These naturally occurring haploids have been used in various breeding projects. It was, therefore, not as important as for some other crops to develop *in vitro* haplomethods.

In various Forums, Carneiro [9–12] has reported success in growing isolated anthers and microspores of Arabica Catuai and different progenies of Catimor. Approximately 80 % of the microspores exhibited embryogenesis. The induction of embryos appeared directly from pollen grains through calli formation. The conversion of embryos into plantlets requires 6 months [12].

Ascanio and Arcia [3, 4] also reported the regeneration of haploid plants from Arabica Garnica anthers. In the case of Arabica, which is an autogamous species, haplomethods significantly reduce the time required to develop new homozygous lines to be used either directly as varieties or as parental lines for the production of homogeneous so-called F_1 true hybrid varieties.

In the case of Robusta, which is an allogamous species, haplomethods would allow the creation of pure breed lines (which could not be obtained in the conventional way: no self-pollination) to be used as parents of homogeneous hybrid varieties.

5.5
Somatic Embryogenesis (Fig. 5.9)

The last plant biotechnology techniques which do not modify the genotype are geared to the rapid vegetative propagation of selected plants. Both micropropagation

Figure 5.9 Coffee somatic embryogenesis.

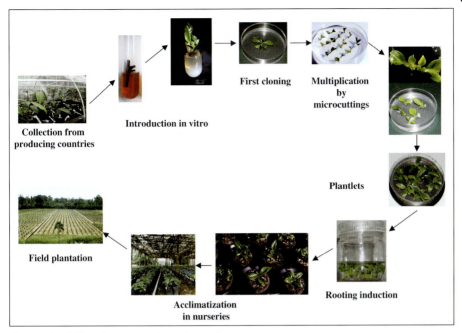

Figure 5.10 Coffee micropropagation: microcuttings.

techniques, microcutting (Fig. 5.10) and somatic embryogenesis, can be applied to coffee. However, we will focus in this chapter on somatic embryogenesis because it may be used for the commercial distribution of selected varieties, whereas microcutting is used only for laboratory or breeding purposes.

5.5.1
Introduction

A zygotic embryo is the natural result of fertilization during the course of sexual reproduction.

Somatic embryogenesis is defined as the development of embryos from somatic tissue or cells. Somatic embryos (soma = body) are developed by going through the globular, heart, torpedo and cotyledon successive stages, similar to those of zygotic embryo development. Somatic embryos have a bipolar structure (radicular and caulinar meristems).

Somatic embryogenesis is a type of vegetative (asexual) reproduction exploiting plant cell totipotency. It happens naturally in various species, but it has also been extensively studied by *in vitro* culture since the 1960s.

5.5.2
Methods

The whole process currently includes the main following steps:

- Induction of embryos or embryogenic tissues from primary explant (e.g. a piece of fresh leaf).
- Multiplication of embryogenic tissues (calli or cell suspension).

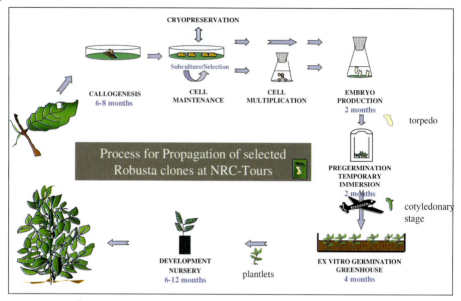

Figure 5.11 Coffee somatic embryogenesis: the whole process.

- Expression of somatic embryogenesis (formation of embryos).
- Maturation of embryos (development up to the green stage).
- Acclimatization to *ex vitro* conditions (development into plantlets in greenhouses or nurseries).

Figure 5.11 summarizes the whole process.

5.5.2.1 Induction

The formation of somatic embryos is the result of a close interaction of various factors which differ greatly according to the species, genotype, primary explant, nutrient medium, culture conditions, etc.

Regeneration of plants from cell cultures has been reported to depend on the *genotype* for various species, such as *Glycine max* L. [41] and *Pisum sativum* L. [29].

For coffee, Zamarripa [63] reported a strong genotypic effect finding 0–100 % of primary explants to be embryogenic according to the variety. Testing eight genotypes and three growing methods, he has shown a culture medium × genotype interaction for somatic embryogenesis. Thus, a genotype can be very reactive, semi-reactive or recalcitrant, according to the process or to the culture medium.

In theory, thanks to totipotency, any type of *plant organ* can be used for regeneration. However, some "young" tissues like immature zygotic embryos and seedling hypocotyls appear to be more often reactive than very differentiated old tissues.

With coffee, somatic embryogenesis has been induced from various *explants* such as stems [16, 47, 54], leaves [17, 27, 37, 46, 49, 51, 60, 62], ovules [9, 30], anthers [3, 9] and even leaf protoplasts [2, 52].

Studies chiefly targeted *culture medium* components; more specifically, the plant growth hormones used in each stage of the process. Natural or synthetic growth hormones can stimulate, inhibit or modify plant differentiation processes. They play a decisive role in the induction and expression of somatic embryogenesis.

For coffee, the auxin/cytokinin hormonal balance was originally used to induce and control the development of coffee somatic embryos on a semi-solid medium [15,18, 46, 50]. The development of somatic embryos with only one cytokinin (BAP) was also reported by Dublin [17] and Yasuda *et al.* [60].

In 1993, Zamarripa pointed out that embryogenesis can be observed on a medium containing both auxin and cytokinin for Arabica, while the addition of auxins is not required for Robusta and Arabusta. For example, Zamarripa [63] reports that embryogenesis of Arabusta is very limited or non-existent in a medium containing auxin and cytokinin, whereas it is highly productive in a medium containing only BAP. On the same species the addition of 80 mg/l of adenine, increase the embryo production from approximately 60 000 to 170 000 embryos/l [64].

5.5.2.2 Multiplication

Either on solid medium or liquid medium, the multiplication of embryogenic tissues is a key step because it greatly and rapidly scales up the number of potential embryos to be produced. However, it is clear that liquid medium is cheaper and more adapted to commercial production than solid medium.

Pétiard *et al.* [43] reported the use of somatic embryogenesis in a liquid medium as a mass multiplication technique. The embryogenic cell strains are maintained by regular subculture at high density in a well-defined liquid culture medium. The subculture frequency and the inoculation density vary according to the suspensions. For coffee, the frequency is 3 weeks and

the inoculation density is 10 g fresh weight (FW)/l [61].

One important possible limitation to the commercial application of this process is the loss of embryogenic potential observed in the course of regular subcultures. Cryoconservation of embryogenic cell suspensions may be integrated with the process [43] to overcome this limitation. Nevertheless, Zamarripa [63] reported that embryogenic suspensions of Robusta and Arabusta can maintain their potential for more than 2 years, after which it dwindles to the point of becoming practically unproductive in the third year.

5.5.2.3 Expression: Production of Somatic Embryos

Large-scale multiplication requires embryo production in a liquid medium. The most common culture method is the use of Erlenmeyer flasks agitated on a gyratory shaker. This method is very practical for the optimization of the process when many parameters have to be tested. For industrial production, the use of larger flasks or a bioreactor may be essential, since it makes the scaling-up easier. It also allows the measuring and control of different parameters, such as the oxygen and carbon dioxide concentrations, the pH, as well as an easier automation of the process. Continuous renewal of medium may also be applied [38, 43].

For coffee, the production of somatic embryos is obtained by the inoculation of previously multiplied embryogenic tissues or cell suspension at low densities in an expression medium containing BAP as the sole growth hormone [62]. In 1989, Zamarripa [61] reported that *the inoculation density* plays an essential role in leading cell suspension to produce embryos, showing that low densities of 0.5–1.0 g FW/l, are optimal. On the contrary high inoculation densities (above 3 g FW/l) strongly inhibit embryogenesis.

As to the oxygen demand during somatic embryogenesis, the various reported results indicate some differences according to the plant species. In the case of coffee, Zamarripa [63] has shown that embryo production in a bioreactor is limited at low oxygen concentrations. He, therefore, recommended maintaining the dissolved oxygen concentration above 60%, at least during embryo formation. Barbier [5] has reported similar results in Erlenmeyer flasks. She mentions that 80% of dissolved oxygen is necessary for good coffee somatic embryogenesis.

Ethylene has an auxin-type activity. It has, therefore, been described as an inhibitor of somatic embryogenesis in several species [25]. However, for coffee, Dao [14] has shown that ethylene has a favorable effect on Arabusta embryogenesis.

Figure 5.12 Large-scale temporary immersion for embryo maturation.

5.5.2.4 Maturation: Development up to the Green Stage

The last *in vitro* step of this propagation process is the development of embryos into small green plantlets suitable for *ex vitro* acclimatization in a greenhouse or even directly in a nursery. It is clear that the success rate and the simplicity of this step are key factors for the economical viability of the whole process.

In the case of coffee, the established procedure can be divided in two stages: maturing and germination [63]. The maturing of somatic embryos (acquisition of a green color, elongation of the hypocotyls, development of cotyledons) lasts 4 weeks on a solid medium containing BAP as the sole growth hormone. The germination stage is conducted in a similar medium, but without BAP. With this two-media sequence, up to 72% of embryos may be developed into plantlets. However, this two-stage procedure requires two hand-made subcultures in solid medium on an embryo per embryo basis. As embryo production run in liquid medium does not require as much labor, this maturation step constitutes a major source of costs which need to be reduced.

In order to do so, Berthouly *et al.* [8] and Etienne *et al.* [23] used temporary immersion growing systems for the development of somatic embryos up to a stage where they can be directly transferred for acclimatization. Using larger flasks which may contain up to 10 000 to 15 000 germinating embryos (Fig. 5.12), this system might cut down plantlet production costs thanks to the elimination of the two hand-made subcultures [21].

5.5.2.5 Acclimatization to *Ex Vitro* Conditions

Provided that the growing conditions (substratum, temperature and humidity) are adequate, the acclimatization of *ex vitro* coffee plantlets in greenhouses or nurseries is not critical when they already have three or four pairs of leaves. The success rate of the acclimatization of such somatic embryos produced in a liquid medium may reach

90–100 %. However, when using smaller germinated embryos or plantlets produced by temporary immersion, this acclimatization rate can be as low as 30–50 %. Loosing some germinated embryos during acclimatization may not be a big cost issue if they have been produced and germinated in liquid media on a large-scale basis [20]. However, their "sowing" in greenhouses may be time consuming and, therefore, might constitute a significant waste of money in the case of low acclimatization rates.

5.5.3
Applications to Coffee

The massive production of somatic embryos, based on the use of low inoculation densities and regular renewal of medium, permits the production of up to 400 000–500 000 Robusta and Arabusta embryos per liter of medium in 6–7 weeks [63, 65]. It is important to point out that the productivity attained by this process has not been exceeded to date. In other words, 1 l of embryo culture might be enough to plant 400–500 ha as long as all the produced embryos can be germinated as viable plantlets.

Ducos *et al.* [19] also reported the development of up to 500 000 embryos per liter of medium in 50 days of cultivation, thereby demonstrating the feasibility of scaling-up in a bioreactor the massive production of Robusta embryos.

Somatic embryogenesis can be integrated into the breeding scheme of all coffee species:

- For Robusta and Arabusta, it allows the rapid propagation of selected clones.
- For Arabica, its main application will be for F_1 hybrid propagation, thereby avoiding manual hybrid seed production. According to Kahia and Owuor [28], the manual production of Arabica hybrids

does not satisfy the demand of the farmers, due to the limited number of seeds produced. It may also lead to the production of numerous out-types by undesired self-pollination of the parents. At present, 1 ha of seed garden can produce 5 million hybrid seeds. With the protocol developed by Zamarripa [63], the same quantity could be produced in about 20 l of culture medium.

Coffee somatic embryogenesis is, therefore, of great practical interest for three main reasons:

- It allows us to speed up the introduction to the farmers of new improved varieties regardless of the method used for their creation.
- It also allows us to apply new breeding strategies leading to varieties which might not have been commercially propagated without this technology (e.g. F_1 Arabica hybrids).
- It opens the door for other biotechnological methods that require a regeneration protocol, such as cryopreservation of heterozygous selected clones, somaclonal variation, fusion of protoplasts and genetic engineering.

However, for large-scale practical use, this technology had to be validated by confirming the true-to-type status of the produced plants at the field level. It is clear that a high rate of out-type trees may stop the development of this technology, as already happened few years ago for other crops like oil palm. Multilocated, large-scale field trials with five Robusta clones have not shown major undesired somaclonal variation [21]. Moreover, even if it is not definitely proven, it seems that, according to different scientists, the first significant production of *in vitro* derived trees might be a few months earlier than that of the control.

If this is the case, the faster production of young trees could be an advantage which may compensate for the slightly higher cost of the planting material.

Currently, the Regional Cooperation Program for the Technological Development and Modernization of Coffee Growing (PROMECAFE) is successfully developing a project for the technical and economic validation of the somatic embryogenesis process for the massive propagation of F_1 hybrids of Arabica.

Similarly, INIFAP in association with Nestlé is currently producing 1.5 million Robusta plantlets from different selected clones to be distributed to growers in VeraCruz state. That project might constitute the first real commercial application of coffee somatic embryogenesis – a biotechnology which may well hold the promise of a great future for coffee.

Another way of in vitro propagation is the coffee micro-propagation by microcuttings. This technique is developed further in Section I.14 ("Coffee propagation").

5.6
Somaclonal Variation

After reviewing plant biotechnologies aimed at identification, preservation, fixation and propagation of a given genotype, but which do not modify it, we will now consider those aimed at genetic diversity enlargement and, therefore, modify the genotype of the considered plant.

5.6.1
Introduction

Somaclonal variation is the unexpected, possibly random, modification of *in vitro* regenerated plant characteristics. They may be due to genetic, epigenetic or even physiological modifications of the original plant. Due to its occurrence in somatic cells (non-reproductive cells), they are called somatic or somaclonal variations. It may lead to the isolation of new stable plant types.

When undifferentiated cells are grown *in vitro* for a long period of time, they may undergo variations in their genes, in the number of chromosomes and even alterations of the physiological mechanisms that control the expression of their genes. In other words, the cells may be either modified in their genetic patrimony and/ or lose the "memory of their genetic programming", therefore regenerating plants that are not true to the original type. In some cases, the newly generated characteristics may be transmitted by vegetative or sexual reproduction.

The variation may have two main origins: it may either be pre-existing in the somatic cells or induced by the *in vitro* culture process [35].

This phenomenon, which is uncontrolled and apparently appears randomly, is therefore of interest when considering the creation of diversity, but it becomes a very strong limitation when investigating the true-to-type propagation of elite varieties as described previously.

5.6.2
Methods

The methodology for the development of variant plants is very similar to *in vitro* vegetative multiplication, with the exception of certain factors that induce the likelihood of inducing variations in the cells. These favorable factors might be the number of subcultures before regeneration or the nature and dosage of growth hormones. In some cases, one may add some kind of mutagenesis treatment and then the technology is called *in vitro* mutagenesis.

For somaclonal variation, *in vitro* culture conditions are, therefore, designed to increase the variability, inversely to multiplication, where the objective is to minimize as much as possible the risk of creating out-type plants.

As for mutagenesis, the main limitation of somaclonal variation is the screening of the variants, which takes place in the field and requires the observation of thousands of plants to have one chance to be successful with regard to a given trait.

5.6.3
Applications to Coffee

Söndahl and Bragin [50] reported remarkable results for somaclonal variation applied to nine genotypes of Arabica. The study was conducted on adult plants regenerated by somatic embryogenesis. Taking all characteristics into consideration, the average variation frequency was about 10 % of evaluated clones.

They also pointed out that the rate of "variants" depends on the genotype. The lowest variant frequency was observed on Red Catuai and Aramosa, with variation rates of 3.3 and 3.1 %, respectively, while Yellow Bourbon, Caturra and Catuai presented variation rates of 30.6, 22.0 and 22.3 %, respectively.

As to the modified plant characteristics, the highest frequencies were found in fruit color, ranging from yellow to red (42.3 %), and in plant size, which changed from tall to short (3.8 %).

Somaclonal variation can be applied to commercial varieties, where it is desired to induce a given modification (e.g. a size reduction of the plant), without destabilizing the whole set of the genetic background. Due to the long breeding cycle,

the generation of coffee varieties requires a minimum of 25 years. Thus, somaclonal variation appears to be an *in vitro* method of interest to create new varieties with improved characteristics in a significantly shorter period of time [44].

It is interesting to note that, in the end, somaclonal variation has been the first commercial application of coffee biotechnologies thanks to the Söndahl project. One of the variants he detected is a better-yielding Arabica cv. Laurina which has been patented and is now grown in Brazil. The yield improvement has made this variety economically viable and led to the production of low caffeine beans with a superior cup-quality profile.

5.7
Genetic Engineering (Fig. 5.13)

Whatever the present discussion/opposition about GMOs, one cannot deny the potential of this technology aimed at precise modifications of the genetic background of a plant. Even if coffee is by far not the most important target for this technology at present, it has to be considered in the present review.

5.7.1
Introduction

The gene is the hereditary unit that is located in the chromosomes which contains the necessary information for the expression of a given characteristic in an organism. Contrary to hybridization, which leads to a mixed genetic information coming from the two progenitors of the created hybrid, genetic engineering is a method that permits the transfer of a single gene or several selected genes to the genetic patrimony of a living organism (bacteria,

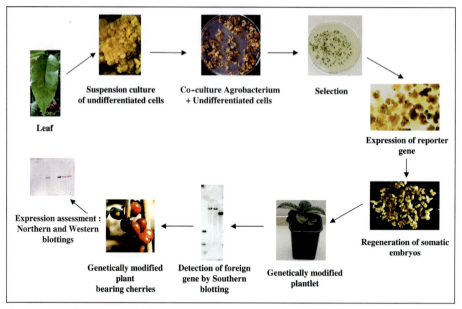

Leaf

Suspension culture of undifferentiated cells

Co-culture Agrobacterium + Undifferentiated cells

Selection

Expression of reporter gene

Expression assessment : Northern and Western blottings

Genetically modified plant bearing cherries

Detection of foreign gene by Southern blotting

Genetically modified plantlet

Regeneration of somatic embryos

Figure 5.13 Coffee genetic modification.

plant, animal, human). Genes may come from plants from the same species or from other species, from bacteria or even from an animal/human source.

Since the first reported plant transformation in 1983, genetic transformation methods have been rapidly developing and the first GM food crop was commercialized in 1996; furthermore over 120 millions acres of GM plants were grown in 2002 [42].

For agronomic purposes, the most widely transferred characteristics are resistance to herbicides, viral diseases and insect attacks. For qualitative purposes, fatty acid and sugar composition as well as ripening control have been given the most important emphasis. This is probably due to the fact that such characteristics are monogenic and/or determined by pathways and genes already known.

5.7.2
Methods

Three conditions are required for the efficient transfer of genes to a plant genome:

- The availability of a gene(s) of interest and a promoter(s) to express it when and where it is required.
- An efficient method for the introduction of genes into the plant cell (transformation).
- The development of a protocol for plant regeneration from the transformed cells.

5.7.2.1 Gene of Interest
The isolation (cloning) of a gene of interest requires:

- The extraction and fragmentation of the DNA by means of restriction enzymes, which operate as biological scissors and/or the isolation of transcribed DNA (cDNA).

- The identification of the gene, by means of the hybridization with a labeled DNA segment corresponding to the product of expression of the target gene.
- The multiplication of the gene into millions of copies, by means of their introduction into a so-called cloning bacteria (such as *Escherichia coli*).

5.7.2.2 Introduction of the Gene of Interest: Transformation

Several transformation methods exist, but only a few will be briefly described hereafter.

5.7.2.2.1 *Agrobacterium*

Agrobacterium (soil bacteria) can provoke a stable integration of a part of their DNA in the genetic material of the plant they attack, causing tumors *(A. tumefaciens)* or the anarchic formation of roots (*A. rhizogenes*). This "gene-transmitting" property has been used by scientists. However, in order to use *Agrobacterium* it is necessary to "disarm" the bacteria by elimination of its oncogenic DNA fragment inducing the tumors. Then the gene of interest can be introduced into the bacteria.

Cells or tissues from the receiving plant are co-cultivated with the transformed bacteria and the foreign gene of interest is integrated. This system is simple and cheap; it also has the advantage of transferring only the DNA from specific sequences. To date, over 100 species have been shown to be transformable by *Agrobacterium*.

In order to select the few plant cells which have integrated the foreign gene, a so-called marker (selectable) gene, e.g. an antibiotic or herbicide resistance gene, is associated with the gene of interest. On selective medium containing the corresponding antibiotic or herbicide, only transformed cells which carry the marker (se-

lectable) gene will develop and possibly regenerate. Untransformed cells will not grow.

5.7.2.2.2 Shot Gun Approach (Biolistics): Particle Bombardment

Biolistics is based on the acceleration of DNA-coated particles towards the plant cells with such force that they penetrate the cell wall and membrane. Once inside the cell, DNA dissociates from the particles and integrates in the plant genome.

This method is capable of overcoming the host-range restrictions of *Agrobacterium*, thus opening the way to the transformation of recalcitrant species.

5.7.2.2.3 Electroporation

This method consists of submitting a protoplast suspension (cells of which the cell wall has been enzymatically digested), plus the DNA to be transferred, to a low-intensity electrical field, for a very short period of time. Electric shocks open the membranes and help the entrance of the DNA into the cells where it might be integrated in the nuclear genome. The procedure is relatively simple and fast, but protoplasts are very often difficult to regenerate into a whole plantlet.

5.7.2.2.4 Microinjection

This is a direct physical method used to introduce DNA molecules inside a cell, under optical control. It is one of the most difficult and arduous methods, but it has the advantage of allowing the control of the quantity of transferred molecules as well as to target this introduction into the nucleus or cytoplasmic corpus such as chloroplasts. Nevertheless, it is more currently used for animal genetic modification than for plants.

5.7.2.3 Regeneration of Transformed Plants

The stable transformation of plant cells unable to regenerate into a whole plant would obviously be of no interest for the plant breeder. The biologist will, therefore, take full benefit for transformation of all the know-how developed about regeneration processes such as somatic embryogenesis. He will have to compromise between the selection of transformed cells and regeneration. Using too selective medium may often lead to the destruction of so many cells that the surviving ones are no longer able to regenerate due to the toxicity of compounds released by the dead cells.

5.7.2.4 Applications to Coffee

Coffee transformation has been attempted by different methods (Tab. 5.1). Several authors report the use of *A. rhizogenes* [53, 56], of *A. tumefaciens* [34, 39, 52], of electroporation [6, 59] and of biolistics [59].

In 1991, Spiral and Pétiard [52] reported the first coffee transformation through co-culture of protoplasts with *A. tumefaciens*, showing a transient expression of the glucuronidase gene.

The first electroporation of protoplasts was reported by Barton *et al.* [6]. They obtained transformed somatic embryos, but did not succeed in developing plantlets out of them. Van Boxtel [59] showed a transient expression in protoplasts treated by electroporation, but did not succeed in regenerating plantlets.

In 1993, Spiral and Pétiard [53] described the transformation of Robusta, Arabica and Arabusta plants using *A. rhizogenes*. The method integrated the GUS reporter gene with a transformation frequency of over 10 % for the three genotypes. Bioassay of glucuronidase showed the expression of the GUS gene. The polymerase chain reaction (PCR) and "Southern blotting" showed its integration, therefore constituting the first definitite proof of coffee transformation.

Sugiyama *et al.* [56] reported similar results. Van Boxtel [59] applied biolistics to various coffee tissues and revealed the transient expression of the GUS gene.

Table 5.1 List of methods used for coffee genetic modification

Authors	Transformation method	Source tissue	Species	Results
Spiral and Pétiard [52]	*A. tumefaciens*	protoplasts	*C. arabica*	transient expression
Barton *et al.* [6]	electroporation	protoplasts	*C. arabica*	transgenic plants
Spiral and Pétiard [53]	*A. rhizogenes*	somatic embryos	*C. canephora, C. arabica* Arabusta	transgenic plants
Van Boxtel [59]	electroporation	protoplasts	*C. arabica*	transient expression
Van Boxtel [59]	biolistics	various tissues	*C. arabica*	transient expression
Sugiyama *et al.* [56]	*A. rhizogenes*	various tissues	*C. arabica*	transgenic plants
Leroy *et al.* [34]	*A. tumefaciens*	somatic embryos	*C. canephora*	transgenic plants
Shinjiro *et al.* [48]	*A. tumefaciens*	cells	*C. canephora*	transgenic plants

More recently, Leroy *et al.* [34] reported the successful introduction of the *cry1ac* gene of *Bacillus thuringiensis* in Robusta and Arabica. The goal of this study was to confer resistance to the Coffee Leaf Miner *Perileucoptera Coffeela* and to other *Leucoptera* species. According to these authors, the somatic embryos were co-cultivated with an *A. tumefaciens* strain containing the *cry1ac* gene, the GUS reporting gene and a gene that conferred herbicide resistance. More than 100 different transformed plants (events) were obtained for each genotype. The gene expression was studied by "Western blotting" analysis. The insecticide protein was detected in most of the transgenic plants.

Some of these plants have been transferred to French Guyana for field evaluation and have, therefore, been the first field release of GMO Coffee.

One venture capital company from Hawaii named ICTI (Integrated Coffee Technology Inc.) has also claimed coffee genetic modifications for the development of caffeine-free coffee and synchronized ripening. The latest modification should allow the production of more homogeneous fully ripe beans by mechanical picking and, therefore, improve their quality.

Finally, Shinrijo *et al.* [48] have reported the development of Robusta plantlets transformed with a construct containing an RNA I of theobromine synthase (CaMXMTI) in order to turn off the caffeine pathway. Their leaf caffeine content is reduced by up to 70%, but beans have not been produced yet. Thus, the development of GM caffeine-free coffee looks feasible, but it will take time because additional breeding steps will still be required.

Generally speaking, and despite the present public acceptance issues, the use of genetic engineering opens the way to the improvement of the agronomic and technolo-

gical characteristics of many crops, thus constituting a great promise for the future. This is even more significant in the case of coffee, where traditional breeding methods may require more than 25 years for the development of a new improved variety.

Private and public entities are now developing research with the end goal of introducing genes of practical interest into the coffee varieties. Such genes might increase resistance to diseases and pests like the Coffee Berry Borer, and control fruit ripening or the caffeine and sugar content. However, the high deregulatory costs, and possible public acceptance issues and image consequences might discourage companies to move in this direction even if coffee would greatly benefit from this so-called modern biotechnology.

5.8
Conclusions

As seen from this review and summarized in Tab. 5.2, most current biotechnological tools are now available for coffee. They already allow:

- The identification, classification and conservation of the germplasm.
- The acceleration and guidance of selection methods.
- The random creation of diversity through somaclonal variation.
- The targeted creation of diversity through genetic engineering.
- Last, but not least, the mass propagation of improved varieties that would allow coffee growers to rapidly take advantage of the benefits of the genetic innovations.

While it is clear that a lot of research is still required to fully implement these tools (checking the stability and heredity of the modifications, evaluating the field confor-

Table 5.2 Biotechnological methods and their application to the genetic improvement of coffee

Methods	Implementation
Molecular markers	evaluation of the genetic diversity; construction of genetic charts
Cryocon-servation	preservation of germplasm and international exchange
Haplomethods	acceleration of selection methods
Somaclonal variation	random genetic modification
Genetic transformation	targeted genetic modification; increase in genetic variability
Somatic embryogenesis	acceleration of the diffusion of new varieties

mity of the *in vitro* produced plants, optimizing the processes to reduce the costs, etc.), it is important to emphasize the recent advances made in the field of biotechnology applied to coffee.

However, that should not lead to a too optimistic conclusion by hiding the orphan crop status of the coffee species. Compared to other main annual field crops of rich countries (e.g. corn), the efforts either by conventional techniques or in association with biotechnologies are still very limited. It is clear that all the required tools are, to a certain extent, available. Would they be intensively applied for the benefit of all actors along the coffee chain? This is a question the public institutes of producing countries through to the processors and consumers should consider if one does not want to further increase the gap between "northern" and "southern" agricultures.

Bibliography

[1] Abdelnour-Esquivel, A., Villalobos, V. and Engelmann, F. Cryoconservation of zygotic embryos of *Coffea* spp. *Cryo-Lett.* **1992**, *13*, 297–302.

[2] Acuna, J. R. and De Pena, M. Plant regeneration from protoplast of embryogenic cell suspension of *Coffea arabica* L. cv. Caturra. *Plant Cell Rep.* **1991**, *10*, 345–348.

[3] Ascanio, E. C. E. and Arcia, M. M. A. Haploids from anther culture in *Coffea arabica* L. In: *International Congress of Plant Tissue Culture, Tropical Species.* Bogota, 21–25 September, 1987, abstr. 68.

[4] Ascanio, E. C. E. and Arcia, M. M. A. Efecto del estado de desarrollo de las anteras y de un shock térmico sobre la androgénesis en *Coffea arabica* L. var. Garnica. *Café, Cacao, Thé* **1994**, *38* (2), 75–80.

[5] Barbier, V. Influence de l'environnement gazeux sur l'organogenèse; multiplication végétative de la chicorée (*Cicorium intybus*); embryogenèse somatique du caféier (*Coffea arabica*). *Memoire de DEA.* Université Pierre et Marie Curie (Paris VI), 1990.

[6] Barton, C. R., Adams, T. L. and Zarowitz, M. A. Stable transformation of foreign DNA into *Coffea arabica* plants. In: *14th International Scientific Colloquium on Coffee.* ASIC, Paris, 1991, 460–464.

[7] Bertaud, J. Les resources génétiques pour l'amélioration des caféiers africains diploïdes. Evaluation de la richesse génétique des populations sylvestres et de ses mécanismes organisateurs. Conséquences pour l'application. *Travaux et Documents de l'ORSTOM.* ORSTOM, Paris, 1986, 379.

[8] Berthouly, M., Dufour, M., Alvard, D., Carasco, C., Alemanno, L. and Teisson, C. Coffee micropropagation in a liquid medium using the temporary immersion technique. In: *16th International Scientific Colloquium on Coffee*. ASIC, Paris, 1995, 514–519.

[9] Carneiro, M. F. Androgenesis on cvs. *Coffea arabica* L. In: *Reproductive Biology and Plant Breeding. XIIIth Eucarpia Congress*, Angers, 5–11 July, 1992, 143–144.

[10] Carneiro, M. F. Induction of double haploids on *Coffea arabica* cultivars via anther or isolated microspores culture. In: *15th International Scientific Colloquium on Coffee*. ASIC, Paris, 1993, 133.

[11] Carneiro, M. F. Androgenesis in different progenies of Catimor. In: *16th International Scientific Colloquium on Coffee*. ASIC, Paris, 1995, AP7.

[12] Carneiro, M. F. Determination of ploidy level of coffee plants regenerated through anthers culture by using the chloroplast counting technique. In: *17th International Scientific Colloquium on Coffee*. ASIC, Paris, ASIC, Paris, 1997. p. 105.

[13] Cros, J., Combes, M. C., Trouslot, P., Anthony, F., Hamon, S., Charrier, A. and Lashermes, P. Phylogenetic analysis of chloroplast DNA variation in *Coffea* L. *Mol. Phylogenet. Evol.* **1997**, *9*, 109–117.

[14] Dao, M. Influence de l'éthylène sur l'embryogenèse somatique en medium liquide de la carotte et du caféier. *Mémoire de Maîtrise de Sciences et Techniques*. Université d'Orléans, 1992.

[15] De García, E. and Menendez, A. Embriogénesis somática a partir de explantes foliares del cafeto "catimor". *Café, Cacao, Thé* **1987**, *31* (1), 15–22.

[16] Dublin, P. Multiplication végétative *in vitro* de l'arabusta. *Café, Cacao, Thé*, **1980**, *24* (4), 281–289.

[17] Dublin, P. Embryogenèse somatique directe sur fragments de feuilles de caféier Arabusta. *Café, Cacao, Thé* **1981**, *25* (4), 237–242.

[18] Dublin, P. Techniques de reproduction végétative *in vitro* et amélioration génétique chez les caféiers cultivés. *Café, Cacao, Thé* **1984**, *28* (4), 231–244.

[19] Ducos, J. P., Zamarripa, A., Eskes, A. B. and Pétiard, V. Production of somatic embryos of coffee in a bioreactor. In: *15th International Scientific Colloquium on Coffee*. ASIC, Paris, **1993**, 89–96.

[20] Ducos, J. P., Gianforcaro, M., Florin, B., Pétiard, V. and Deshayes, A. A technically and economically attractive way to propagate elite *Coffea canephora* (Robusta) clones: *in vitro* somatic embryogenesis. In: *18th International Scientific Colloquium on Coffee*. ASIC, Paris, 1999, 295–301.

[21] Ducos, J. P. and Pétiard, V. Propagation de clones de Robusta (*Coffea canephora* P.) par embryogénèse somatique en milieu liquide. In *VIIIèmes Journées Scientifiques du Réseau Biotechnologies, Amélioration des Plantes et Sécurité Alimentaire*, Marrakech, Maroc, 2002, 142–159.

[22] Dussert, S., Chabrillange, N., Engelmann, F., Anthony, F., Hamon, S. and Lashermes, P. Cryopreservation of coffee (*Coffea arabica*) seeds. In: *17th International Scientific Colloquium on Coffee*. ASIC, Paris, 1997, 67.

[23] Etienne, H., Bertrand, B., Anthony, F., Cote, F. and Berthouly, M. L'embryogenèse somatique: un outil pour l'amélioration génétique du caféier. In: *17th International Scientific Colloquium on Coffee*. ASIC, Paris, 1997, p. 66.

[24] Florin, B., Tessereau, H. and Pétiard, V. Conservation à long terme des ressources génétiques de caféier par cryoconservation d'embryons zigotiques et somatiques et de cultures embryogènes. In: *15th International Scientific Colloquium on Coffee*. ASIC, Paris, 1993, 106–114.

[25] George, E. F. and Sherrington, P. D. Factors affecting growth and morphogenesis. In: *Plant Propagation by Tissue Culture*. George, E. F. and Sherrington, P. D. (eds). Exegetics, Eversley, 1984, 125–171.

[26] Hatanaka, T., Azuma, T., Uchida, N. and Yasuda, T. Effect of plant hormone on somatic embryogenesis of *Coffea canephora*. In: *16th International Scientific Colloquium on Coffee*. ASIC, Paris, 1995, 790–797.

[27] Herman, E. B. and Hass, G. J. Clonal propagation of *Coffea arabica* L. from callus culture. *Hot Science* **1975**, *10* (6), 588–589.

[28] Kahia, J. W. and Owuor, J. B. O. *In vitro* propagation of the disease resistant *Coffea arabica* L. cultivar Ruiru 11. *Kenya Coffee* **1990**, *55* (646), 901–905.

[29] Kysely, W. and Jacoben, H. J. Somatic embryogenesis from pea embryos and shoot apices. *Plant Cell Tissue and Organ Culture* **1990**, *20*, 7–14.

[30] Lanaud, C. Production de plantules de *C. canephora* par l'embryogenèse somatique réalisée à partir de culture *in vitro* d'ovules. *Café, Cacao, Thé* **1981**, *25* (4), 231–235.

[31] Lashermes, P., Trouslot, P., Anthony, F., Combes, M. C. and Charrier, A. Genetic diversity for RAPD markers between cultivated and wild accessions of *Coffea arabica. Euphytica* **1996**, *87*, 59–64.

[32] Lashermes, P., Combes, M. C., Trouslot, P. and Charrier, A. Phylogenetic relationships of coffee-tree species *Coffea* L. as inferred from ITS sequences of nuclear ribosomal DNA. *Theor. Appl. Genet.* **1997**, *94*, 947–955.

[33] Leroy, T., Montagnon, C., Charrier, A. and Eskes, A. B. Reciprocal recurrent selection applied to *Coffea canephora* Pierre 1. Characterization and evaluation of breeding population and value of intergroup hybrids. *Euphytica* **1993**, *67*, 113–125.

[34] Leroy, T., Henry, A. M., Royer, M., Altosaar, I., Frutos R., Duris, D., Phillipe, R. Genetically modified coffee plants expressing the *Bacillus thuringiensis* cry1Ac for resistance to leaf miner. Plant Cell Reports **2000**, *19*, 382–389.

[35] Mestre, J. C., and Pétiard, V. La nature de la variabilité des cellules végétales en culture; les diverses causes possibles de son expression. *Bull. Soc. Bot. Fr., Actaul. Bot.* **1985**, *132* (3), 67–68.

[36] Mycock, D. J. and Berjak, P. Cryostorage of somatic embryos of coffee. In: *Fourth International Workshop on Seeds, Basic and Applied Aspects of Seed Biology.* ASFIS, Paris, 1993, 879–884.

[37] Neuenschwander, B. and Bauman, T. W. A novel type of somatic embryogenesis in *Coffea arabica. Plant Cell Rep.* **1992**, *10* (12), 608–612.

[38] Noriega, C. and Söndahl, M. R. Arabica coffee micro-propagation through somatic embryogenesis via bioreactors.

In: *15th International Scientific Colloquium on Coffee.* ASIC, Paris, 1993, 73–81.

[39] Ocampo, C. A. and Manzanera, L. M. Advances in genetic manipulation of the coffee plant. In: *14th International Scientific Colloquium on Coffee.* ASIC, Paris, 1991, 378–382.

[40] Paillard, M., Lashermes, P. and Pétiard, V. Construction of a molecular linkage map in coffee. *Theor. Appl. Genet.* **1996**, *93*, 41–47.

[41] Parrott, W. A., Willians, E. G., Hildebrand, D. F. and Collins, G. B. Effect of genotype on somatic embryogenesis from immature cotyledons of soybeans. *Plant Cell Tissue and Organ Culture* **1989**, *16*, 15–21.

[42] Peerenboom, E. Une bactérie "passeur de gènes". In: *L'alimentation au Fil du Gène.* Fondation Alimentarium, Nestlé, 1998, 83–84.

[43] Pétiard, V., Ducos, J. P., Florin, B., Lecouteux, C., Tessereau, H. and Zamarripa, A. Mass somatic embryogenesis: a possible tool for large-scale propagation of selected plants. In: *Fourth International Workshop on Seeds, Basic and Applied Aspects of Seed Biology.* ASFIS, Paris, 1993, 178–191.

[44] Pétiard, V., Bollon H., Ducos, J. P., Florin, B., Paillard, M., Spiral, J. and Zamarripa, A. Biotechnologies appliquées au caféier. In: *15th International Scientific Colloquium on Coffee.* ASIC, Paris, 1993, 56–66.

[45] Pétiard, V. and Crouzillat, D. Use of DNA identification techniques for the determination of genetic material of cocoa in fermented or roasted beans and chocolate. *PCT/EP99/08268. WO 00/28078,* 2000.

[46] Pierson, E. S., Van Lammeren, A. A. M., Schell, J. H. N. and Staritsky, G. *In vitro* development of embryoids from punched leaf discs of *Coffea canephora. Protoplasma* **1983**, *115*, 208–216.

[47] Sharp, W. R., Caldas, L. S., Crocomo, O. J., Monaco, L. C. and Carvalho, A. Production of *Coffea arabica* callus of three ploidy levels and subsequent morphogenesis. *Phyton* **1973**, *31* (2), 67–74.

[48] Shinjiro, O., Hirotaka, U., Yube, Y., Nozomu, K. and Hiroshi, S. RNA interfer-

ence: Producing decoffeinated coffee plants. *Nature* **2003**, *423*, 823.

[49] Söndahl, M. R. and Sharp W. R. High frequency induction of somatic embryos in cultured leaf explants of *Coffea arabica* L. *Z. Pflanzenphysiol, Bd., S.* **1977**. *81*, 395–408.

[50] Söndahl, M. R. and Bragin, A. Somacional variation as a breeding tool for coffee improvement. In: *14th International Scientific Colloquium on Coffee*. ASIC, Paris, 1991, 701–710.

[51] Söndahl, M. R. and Noriega, C. Coffee somatic embryogenesis in liquid cultures. In: *World Congress on Cell and Tissue Culture*, Arlington, VA, 20–25 June 1992, abstr. 1010.

[52] Spiral, J. and Pétiard, V. Protoplast culture and regeneration in *Coffea* species. In: *14th International Scientific Colloquium on Coffee*. ASIC, Paris, 1991, 383–391.

[53] Spiral, J. and Pétiard, V. Développement d'une méthode de transformation appliquée à différentes espèces de caféier et régéneration de plantules transgéniques. In: *15th International Scientific Colloquium on Coffee*. ASIC, Paris, 1993, 106–114.

[54] Staritski, G. Embryoid formation in callus tissues of coffee. *Acta Bot. Neerl.* **1970**, *9 (4)*, 509–514.

[55] Staritski, G. and Van Hasselt, G. A. M. The synchronised mass propagation of *Coffea canephora in vitro.* In: *9th International Scientific Colloquium on Coffee*. ASIC, Paris, 1980, 597–602.

[56] Sugiyama, M., Matsuoka, C. and Takagi, T. Transformation of coffee with *Agrobacterium rhizogenes.* In: *16th International Scientific Colloquium on Coffee*. ASIC, Paris, 1995, 853–859.

[57] Tessereau, H., Lecouteux, C. Florin, B., Schlienger, C. and Pétiard, V. Use of a simplified freezing process and dehydration for the storage of embryogenic cell lines and somatic embryos. *Rev. Cytol. Biol. Végét. Bot.* **1991**, *14*, 297–310.

[58] Tessereau, H. Développement d'une méthode simplifiée de cryoconservation

de tissus et d'embryons somatiques végétaux et étude de l'acquisition de la tolérance à la congélation. *Thèse de Doctorat.* Université Pierre et Marie Curie, Paris VI, 1993.

[59] Van Boxtel, J. Studies on genetic transformation of coffee by using electroporation and the biolistic method. *PhD Thesis.* University of Wageningen, Holland, 1994.

[60] Yasuda, T., Fujii, Y. and Yamaguchi, T. Embryogenic callus induction from *Coffea arabica* leaf explants by benzyladenine. *Plant Cell Physiol.* **1985**, *26* (3), 595–597.

[61] Zamarripa, C. A. Production d'embryons somatiques de caféier (*Coffea arabica* L.) en milieu liquide. *Mémoire de DEA.* Ecole Nationale Supérieure Agronomique de Rennes, Université de Rennes I, 1989.

[62] Zamarripa, C. A., Ducos, J. P., Tessereau, H., Bollon, H., Eskes, A. B. and Pétiard, V. Développement d'un procédé de multiplication en masse du caféier par embryogenèse somatique en milieu liquide. In: *14th International Scientific Colloquium on Coffee*. ASIC, Paris, 1991, 392–402.

[63] Zamarripa, C. A. Etude et développement de l'embryogenèse somatique du caféier en milieu liquide (*Coffea canephora* P., *Coffea arabica* L. et l'hybride Arabusta). *Thèse de Doctorat.* Ecole Nationale Supérieure Agronomique de Rennes, France, 1993.

[64] Zamarripa, C. A. Influencia de la adenina sobre la producción de embriones somáticos de café Arabusta. In: *11th Congreso Latinoamericano de Genética (Area Vegetal) y XV Congreso de Fitogenética*, Nuevo León, 25–30 September, **1994**, 148.

[65] Zamarripa, C. A. Optimización de la embriogénesis somática del café Arabusta (*Coffea canephora* P × *Coffea arabica* L) a partir de una suspensión celular. *Agric. Téc. Méx.* **1994**, *20* (1), 27–41.

6

Environmental Factors Suitable for Coffee Cultivation

F. Descroix and J. Snoeck

Coffee-growing areas are situated approximately between latitude 22° N and latitude 26° S. Appropriate sites for coffee growing should be selected with respect to six basic environmental factors, i.e. temperature, water availability, sunshine intensity, wind, type of soil and topography of the land.

6.1
Temperature

Temperature values and their fluctuations have a significant impact on the behavior of coffee trees. No coffee species can survive in temperatures bordering on 0 °C. Sensitivity to cold as well as to high temperatures can vary between coffee species and even between individual plants. Both wind and air humidity can greatly influence the effect of air temperatures. Influence differences caused by altitude are calculated at 0.6 °C/100 m.

6.1.1
Arabica

The optimal mean temperature for this species which is native of the high Abyssinian plateaux, is given as 18 °C during the night and 22 °C during the day. Tolerated extremes extend to 15 °C during the night and 25–30 °C during the day. Temperatures higher than 25 °C cause reduced photosynthesis and prolonged exposure to temperatures above 30 °C incur leaf chlorosis and generate "star flowers", or blossom wilting, as well as defective fruit set. High temperatures also favor the development of Coffee Leaf Rust (*Hemileia vastatrix*) and fruit blight (Cercospora) in coffee plants and accelerate fruit maturation, whilst low temperatures favor Coffee Berry Disease.

Low temperatures also cause a white or yellow discoloration of the leaves. This discoloration is not uniform and the initial effect is often limited to the fringe of the leaves. More serious lesions occur on leaves and fruit exposed to temperatures below 4 °C. These appear reduced in size, often distorted and mottled, and may eventually scorch and fall. Symptoms are more severe when high daytime temperatures are followed by low temperatures at night. In severe cases, excessive branching to secondary and tertiary stems occurs, and short tips blacken, distort and shrivel. Exposure to temperatures below −2 °C for a period of over 6 h causes serious damage to the coffee plant and may even lead to its death, particularly if the temperature drops to −3 or −4 °C.

Coffee: Growing, Processing, Sustainable Production, Second Edition. Edited by J. N. Wintgens.
© 2012 WILEY-VCH Verlag GmbH & Co. KGaA. Published 2012 by Wiley-VCH Verlag GmbH & Co. KGaA.

Variations in temperature play a very important role in coffee behavior. For Arabica, the maximum tolerance is a range of 19 °C and for Robusta it is 12–14 °C. If, even for only a short period during the year, the diurnal temperature variation exceeds these tolerances, a poor coffee crop can ensue (Robinson, 1986, comments to G. Wrigley in [7]). The effect of wide temperature variations such as the sudden drop of temperature at sunrise is known as the "Hot and Cold Disease". One of the reasons for shading coffee is to reduce wide temperature variations.

6.1.2
Robusta

C. canephora Robusta originates in the low, hot zones of the Guinea/Congo forests. For this species, optimal average annual temperatures are given between 22 and 28 °C. Problems start when temperatures drop below 10 °C and the tree dies at around 4–5 °C. High temperatures are also hazardous for Robusta, especially in a dry atmosphere, as photosynthesis is reduced at temperatures above 30 °C.

When planning a new coffee plantation one needs to bear in mind the fact that areas swept by dry continental winds during the cold season have lower temperatures than zones exposed to humid maritime winds. Another element is the fact that thermal conditions are not always the same on the surface of the soil as they are just a little higher in the space occupied by the canopy of the coffee tree. Temperatures decrease with height and differences of 4–5 °C have been registered from between the soil surface and the canopy layer. Nurseries, in particular, should be wary of areas where nocturnal temperatures are extreme.

Unfavorable temperatures can be attenuated by selecting a suitable site for settling the plantation, by using appropriate cultural practices like shade trees or windbreaks and by implementing frost protection measures.

6.2
Water Availability

This factor includes both rainfall and atmospheric humidity.

Rainfall is the most important restrictive factor for coffee growing. Two inseparable elements should be taken into consideration: the total annual rainfall and its monthly or, better still, weekly distribution.

6.2.1
Rainfall

The rainfall pattern must include a few months with little or no rain as this period is necessary to induce flowering. A total annual rainfall between 1400 and 2000 mm is favorable for Arabica growing, whereas Robusta needs about 2000–2500 mm. Rates below 800–1000 mm for Arabica and 1200 mm for Robusta, even if they are well distributed, can be hazardous to the productivity of the coffee plantation, particularly if artificial irrigation is not possible.

An annual rainfall of over 2500–3000 mm is common in a number of coffee production areas and does not cause significant damage to the coffee plant provided both surface and vertical drainage is sufficient. On the other hand, excessive rains can generate other drawbacks like erosion and make it difficult to sun-dry the crop. Areas with an annual rainfall of over 3000 mm can be considered as less appropriate for economically valid coffee cultivation.

A period of drought favors flowering concentration and, as a result, limits harvest staggering. A dry season of 2–3 months which coincides with the harvest period is ideal.

Arabica coffee can withstand a dry period of 4–6 months provided the soil is deep and offers good water retention. In the case of Robusta, however, the dry season should not exceed 3–4 months because of high evapotranspiration due to more elevated temperatures.

A month where rainfall is less than twice the monthly average temperature is normally considered to be a dry month, e.g. for Arabica 36 mm (2 × 18) and for Robusta 50 mm (2 × 25).

Coffee species are evergreen so transpiration is continuous. The rate of water loss by evaporation from free water surface depends on the temperature of the air and the relative humidity (RH), and on air movement and radiant solar energy. In 1948, Penman [5] derived an equation giving the open-pan evaporation as E_0. Loss of water from vegetation E_1 (evapotranspiration) is always less than E_0 because of the closing of the stomata. The E_1 value of any plant can be determined by measuring the water balance with a lysimeter. The results are normally expressed as the annual E_1/E_0 ratio. The E_1/E_0 ratio for an evergreen mountain rain forest is equal to 0.95 and for pasture lands it is equal to 0.75. For Arabica in Kenya, the ratio varies between 0.5 in dry months and 0.8 in wet months.

The average annual water requirement for coffee was studied over a period of 12 years and was found to be 951 mm [2, 6]. Where the dry season is normal and soils provide a high water retention capacity, Arabica can be grown satisfactorily without irrigation where the annual rainfall is approximately 1100 mm [1].

Overall rainfall distribution throughout the year is a decisive factor for scheduling cultivation practices and harvesting. Where the rainfall distribution is unimodal, there is one major flowering period. In the case of a bi-modal rainfall pattern, there will be two periods of blossoming and, consequently, two harvest periods, known as the "early crop" and the "late crop". Coffee issued from the "late crop" is of better quality.

6.2.2
Atmospheric Humidity

The atmospheric humidity or RH of the air has a marked influence on the behavior of the coffee plant, particularly in the case of Robusta. The best RH level for Robusta is 70–75 % and for Arabica it should be around 60 %. A high level of atmospheric humidity will reduce water loss, whereas a low level will increase evapotranspiration. In the case of Arabica, persistent levels above 85 % may affect the quality of the coffee.

Air humidity is increased by cloud cover and mist. As a result, the lack of rainfall can be partially compensated in regions where cloud cover and mist prevail during the dry season. The presence of mist can also increase the dampness of the soil due to condensation on the leaves. In 1960, Parsons recorded 250 mm of drip water from a pine tree near San Francisco.

The RH level can also vary considerably between the interior and the fringe of the coffee plantation. On a clear morning, Kirkpatrick recorded a RH level of 56 % in the open and 95 % under the canopy of the coffee trees.

Morning dew in the mountains also provides a supply of water which compensates for the hardships of the dry season.

6.3
Sunlight and Shading

Due to their origins, all coffee plants are naturally heliophobes; however, in farming conditions, flowering is significantly stimulated by direct sunlight and, inasmuch as adequate fertilization, in particular nitrogen, is provided, sunshine can boost productivity.

Initially, coffee was farmed under natural or artificial shading conditions to recreate the original forest environment but now coffee plantations are often established in direct sunlight. Further observation has even led to the conclusion that the peripheral foliage of the coffee tree has the effect of auto-shading which makes further shading redundant, particularly in cloudy areas. Shade still remains useful and even necessary in certain conditions as it helps to attenuate the effects of extreme high and low temperatures. A further beneficial effect of shade has been revealed by recent studies which indicate that it improves the quality of coffee.

Shading also diminishes the risk of erosion, restricts weed growth and generates a mulch which protects and enriches the soil with organic matter.

The effects of shading on the intensity of the photosynthesis process in coffee trees has been abundantly researched; however, to date, the results of these studies have not produced conclusive evidence to establish basic guidelines for producers. In Madagascar, an interaction has been shown between shade and fertilization: optimal yield was obtained with a combination of light shade and fertilizers as compared to no shade and the same fertilizer formula. Increase of shade densities diminishes beneficial effect of fertilizers, probably because of lower photosynthetic activity.

Similarly, elsewhere it has been ascertained that a lack of luminosity can diminish productivity. In Mayumbe (Democratic Republic of Congo), for instance, surveys have concluded that Robusta coffee trees are practically unproductive because they lack sufficient luminosity even though the average soil fertility is perfectly adequate (Jurion [4]). In this area, exposure to sunshine is only 1330 h/year.

To ensure satisfactory results, exposure to sunlight should reach approximately 60 % of the potential during the rainy season and 60–75 % during the dry season. This is the case at Rubona in Rwanda where the average annual exposure to sunlight between 7 a.m. and 5 p.m. is 58 % during the rainy season and 75 % in the dry season (Jurion [4]).

Consequently, one can conclude that, for best results, coffee requires an average of 2200–2400 h of sunlight per year.

6.4
Wind

Strong winds do not suit coffee growing, although a light air movement sustains the gaseous exchange throughout the foliage. Tornadoes and cyclones, for instance, cause significant damage such as defoliation or branch breakage and even lodging, particularly in light or shallow soils. In areas where cyclones are known to be frequent like Madagascar, the Philippines, the Caribbean, Vietnam, Hawaii or Hainan, for instance, damage can be considerable. Coffee growing on the Island of Reunion was abandoned in the 19th century because of a serious outbreak of Coffee Leaf Rust combined with frequent, violent cyclones.

This factor must, therefore, be taken into consideration when establishing a plantation and basic precautions are required. These include avoiding known cyclone paths, using topographic protection and creating windbreaks. It is advisable to select low-growing coffee plants with sturdy, resistant branches for plantations in these areas.

Less powerful, but longer-lasting, winds like trade winds, sea or ground breezes and relief winds can also be harmful. They generate physiological drawbacks like excessive evapotranspiration which incurs branch die-back, and blemishes the flowers and fruit which are close to maturity. The use of shade trees and windbreaks is necessary in areas where these conditions prevail.

6.5
Soil Characteristics

Coffee plants prosper as well in alluvial and colluvial soils with a favorable texture as in volcanic formations. On the other hand, the depth of the soil above any obsta-cle like hard-pan, the water table or bed-rock is most important.

Soil depth should be at least 2 m to allow tap root development and ensure the necessary water supply to the plant during the dry spell of the year. In dryer areas, the soil should be even deeper, as deeper soil allows for the development of roots over a wider ground volume. Coffee roots can invest the soil extensively to a depth of over 3 m (Fig. 6.1).

A hard pan can have various causes, but most often it is due to the presence of rock or a lateritic shield. This type of formation can be neither by-passed nor penetrated, so it is advisable to avoid planting coffee where a hard-pan is known to be present. In cases where the lateritic formation has broken up, however, the roots of coffee trees are able to make their way through the fissures and clefts so that the plant prospers. This is particularly true for Robusta, which is the most productive in areas where the laterite shield is crumbling and disintegrating because the soil offers good permeability which ensures

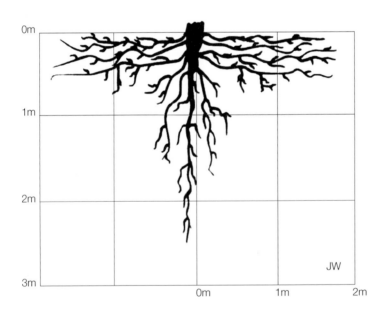

Figure 6.1 Normal root system of an adult Arabica coffee tree.

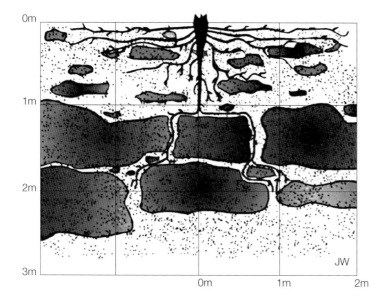

Figure 6.2 Root system of a Robusta coffee tree growing through a broken up laterite hard-pan.

favorable conditions for air and water infiltration. Coffee plantations have been installed in this type of soil by knocking out holes with a mine rod or dynamite and they have prospered more successfully than in lower-lying heavy clay soils (Fig. 6.2).

On the other hand, it is advisable to be wary of hard-pans formed by lava, calcareous tufa or mud-stone as their density or thickness may represent an impenetrable barrier for the roots of coffee trees. Indurated illuvial horizons are also indicative of an unfavorable area for installing coffee trees.

Gravel strata which consists of pebbles with no soil between them should not be over 15-cm thick for the roots to be able to penetrate the layer of gravel and reach the soil below (Fig. 6.3).

Soils where the water table is permanently high should never be planted with coffee. The water table should be at least 1.5 m below the soil surface. Where the water table is temporary, the duration of its presence is the decisive factor for installing cof-

fee trees, but these areas remain marginal (Fig. 6.4).

Soils which are subject to occasional flooding should also be avoided as coffee tree roots are easily asphyxiated. Badly drained soils or heavy soils with excessive clay content are not recommended. In regions with a low average annual rainfall or a long dry season, sandy, rocky or heavily laterized soils with poor water retention are also to be avoided.

As the main part of the radicular system of the coffee tree develops in 30 cm of the upper soil layer, the physical properties of the topsoil are more important to the coffee tree than those of the deeper subsoil.

Soil porosity, i.e. the percentage occupied by air and water, gives an indication of its drainage capacity, its permeability and easy rooting of the plant. Soils which are favorable to the coffee plant have a porosity of 50–60% (water + air), the mineral content (45%) and the organic content (2–5%) make up the remainder.

The physical properties of the soil, however, remain far more important because

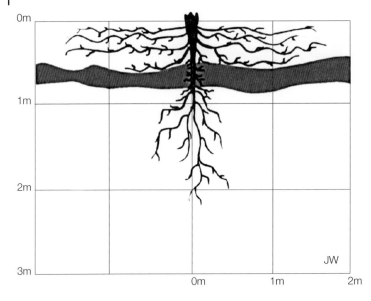

Figure 6.3 Root system of an adult Arabica coffee tree growing through an indurated illuvial horizon.

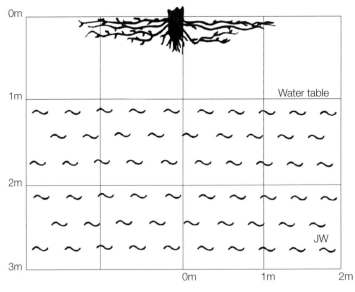

Figure 6.4 Root system of an adult Arabica coffee tree stopped by a water table at 1 m.

they cannot be modified, whereas its chemical properties can be corrected. The types of soil best suited for coffee are those that originate from lava, volcanic ash, basic rocks and alluvial deposits, which all provide a high cation-exchange capacity (CEC) and a favorable organic status. As a general rule, suitable soils for coffee should not contain more than 20–30 % of coarse sand (larger than 2 mm) and 70 % of clay in the upper layers of 30–50 cm.

Soils with a high percentage of organic materials are strongly recommended as they are more fertile, less prone to erosion, and offer better water and nutrient retention capacity. As a result, coffee planters prefer to settle their plantations in cleared forest soils where native fertility has been little or not at all affected by previous cultivation.

To sum up, one can conclude that the ideal type of soil to be recommended for coffee growing should be deep, permeable, slightly acid and porous, i.e. well-structured with a favorable texture. It should

also offer a high water retention capacity. A further point to be considered is the fact that coffee roots need a lot of oxygen. Consequently, heavy and poorly drained soils are unsuitable.

Where soils were formed on site, quality is determined by the characteristics of the parent rock. Soils issued from granite rocks, for instance, are acid with a poor cation exchange capacity, whereas basalt born soils are fertile and rich in nutrients.

Table 6.1 Typical coffee-growing soils throughout the world

Type of soil	Coffee species planted	Geographical areas
Volcanic ash (sometimes partially laterized)	Arabica	Central America, Colombia, Mexican highlands, Cameroon highlands, Malaysian highlands, Javanese mountain soils (Keloed volcano), Uganda (Mount Elgon)
Soils generated by rocks of a crystalline complex like diorite, gneiss, granite, mica, basalt, megaschist, i.e. arenitas, reddish-yellow podzolic soils, reddish-brown laterite soils, massapé, terra roxa (purple land), terra vermelha (degraded terra roxa)	Arabica (mostly)	Brazil
Heavy clay soils	Arabica	Andean zone, central Java
Laterite soils	Arabica and Robusta	India, West and Central Africa, Ethiopian plateau, Indonesia (Malang)
Recent volcanic soils (high fertility)	Arabica (mostly)	East Africa, Kivu, Hawaii, Central America
Red and Yellow latosols (acid with a low CEC)	Robusta	The lower lands of the Democratic Republic of Congo, Ivory Coast
Alluvial soils (high fertility)	Robusta and Arabica	several limited areas
Kikuyan red loam soils of volcanic origin	Arabica	Kenya
Red sandy clay and gravely loam	Robusta	Uganda
Podzolic soils near the sea	Liberica	Malaysia
Volcanic ash mixed with fragments of lava rocks	Arabica	Hawaii (Kona)
Black humid mountain soils (deep and fertile)	Arabica	Indonesia

6.6

Topography

Flat lands or slightly rolling hills are best suited to the installation of a coffee plantation. This type of topography generally offers deep soils with a good water retention capacity. Furthermore, farming and mechanization are also much easier.

On the other hand, coffee can be grown successfully on steep slopes. The steeper the slope, however, the more costly the farming as conservative measures need to be implemented to avoid erosion and mechanization is far more difficult. Slopes that are steeper than 20% are difficult to mechanize, although slopes with a gradient of 20–40% can still be farmed with tools drawn by animals. Steeper slopes need to be farmed manually.

Overviews of the environmental factors suitable for coffee cultivation are given in Tables 6.2 and 6.3.

Table 6.2 Environmental factors suitable for Arabica coffee cultivation

Levels of suitability		S1		S2	S3	N1	N2
Degree of limitation		0	1	2	3	4	4
Climatic characteristics							
Temperatures (°C)	mean annual	18–20	16–18	15–16	14–15		<14
	mean maximal annual	19–20	20–22	22–24	24–26		>26
		25–26	26–28	28–30	30–32		>32
	mean daily minimal	24–25	22–24	20–22	18–20		<18
		15–17	17–19	19–21	21–23		>23
	temperature of the coldest month	14–15	10–14	7–10	4–7		<4
Rainfall (mm)	annual	1600–1800	1800–2000	2000–2200	>2200		<800
		1400–1500	1200–1400	1000–1200	800–1000		
	length of dry season (months)	2.5–3	3–4	4–5	5–6		>6
		2–2.5	1–2	0–1			
Relative humidity (%)	mean RH of driest month	50–60	60–70	70–80	80–90		>90
		50–55	40–50	30–40	20–30		<20

S1 = Best suited units with very few (three or four) slight restrictive factors or none at all.

S2 = Average units with more than three or four slight restrictive factors and/or no more than two or three moderately restrictive factors.

S3 = Marginally suitable units with more that two or three moderately restrictive factors and/or no more than one serious restrictive factor provided it does not totally exclude farming.

N1 = Unsuitable units with one or more serious restrictive factor and no more than one very serious restrictive factor which totally excludes farming. Potentially suitable if improvements were implemented.

N2 = Totally and potentially unsuitable units.

Table 6.2 Continued

Levels of suitability		S1		S2	S3	N1	N2
Degree of limitation		0	1	2	3	4	4
Soil characteristics							
Slope (%)	without irrigation	0–4	4–8	8–16	16–30	30–50	>50
	with irrigation	0–2	2–4	4–8	8–16		>16
Hydrous conditions	Drainage	good	good	moderate	imperfect	poor, but drainable frequent	poor, not drainable frequent
	submersion	none	none	none	occasional	frequent	frequent
Physical characteristics of the soil	depth of soil (cm)	>200	150–200	100–150	100–50		<50
	texture	clay clayey-silty silty-clayey	loam silty-clayey-sandy	silty-sandy	sandy-silty	sand	sand
	% of coarse elements >2 mm	0–3	3–15	15–35	35–55		>55
Chemical characteristics of the soil	pH (H$_2$O)	5.5–6.0	5.3–5.5	5.0–5.3	4.5–5.0	<4.5	
	apparent CEC (meq/100 g clay) 0–15 cm (%)	>24	16–24	<16+	<16–		
	saturation in cations of layer 0–15 cm (%)	>80	50–80	35–50	20–35	<20	
	organic carbon of layer 0–15 cm (%)	>2.4	1.2–2.4	0.8–1.2	<0.8		

See legend p. 177.

Table 6.3 Environmental factors suitable for Robusta coffee cultivation

Levels of suitability		S1	S2	S3	N1	N2
Degree of limitation	*0*	*1*	*2*	*3*	*4*	*4*
Climatic characteristics						
Temperatures (°C)	mean annual	22–28	22–25	20–22	18–20		<18
	mean maximal annual	>29	27–29	24–27	22–24		>22
	mean daily minimal	>20	18–20	16–18	14–16		<14
Rainfall (mm)	annual	2000–2500	1600–2000	1400–1600	1200–1400		<1200
	length of dry season (months)	2–2.5	2.5–3	3–3.5	3.5–4		>4
Relative humidity (%)	mean RH of driest month	70–75	75–80	80–90	>90		
		60–65	45–60	35–45	30–35		<30

See legend p. 177.

Table 6.3 Continued

Levels of suitability		S1		S2	S3	N1	N2
Degree of limitation		0	1	2	3	4	4
Soil characteristics							
Slope (%)	without irrigation	0–4	4–8	8–16	16–30	30–50	>50
	with irrigation	0–2	2–4	4–8	8–16		>16
Hydrous conditions	drainage	good	good	moderate	imperfect	Imperfect, but drainable	Imperfect, not drainable
	submersion	none	none	none	occasional	frequent	frequent
Physical characteristics of the soil	depth of soil (cm)	>200	150–200	100–150	50–100		<50
	texture	clay, clayey-silty, silty-clayey	loam, silty-clayey-sand	silty-sandy	sandy-silty	sand	sand
	% of coarse elements > 2mm	0–3	3–15	15–35	35–55		>55
Chemical characteristics of the soil	pH (H₂O)	5.5–6.0	5.3–5.5	5.0–5.3	4.5–5.0	<4.5	
	apparent CEC (meq/100 g clay)	>16	<16+	<16–			
	saturation in cations of layer 0–15 cm (%)	>35	20–35	<20			
	organic carbon of layer 0–15 cm (%)	>1.5	0.8–1.5	<0.8			

See legend p. 177.

Bibliography

[1] Achternich, W. *Sprinkling in Coffee Cultivation*. Perrot Regnerbau, Calw/Württemberg, 1958.

[2] Blore, T. W. D. Further studies of water use by irrigation and unirrigated arabica coffee in Kenya. *J. Agric. Sci. Camb.* **1966**, *67*, 145–54.

[3] Cliffort, M. N. and Willson, K. C. *Coffee, Botany, Biochemistry and Production of Beans and Beverage*. Croom Helm, London, 1987.

[4] Jurion F. *Milieux écologiques et spécialisation agricole*; Semaine d'études des problèmes intertropicaux. Bulletins de recherches agronomiques de Gembloux, Hors série, 1972.

[5] Penman, H. L. Natural evaporation from open water, bare soil and grass. *Proc. R. Soc. Ser. A* **1948**, *193*, 120–45.

[6] Wallis, J. A. N. Water use by irrigated arabica coffee in Kenya. *J. Agric. Sci. Camb.* **1963**, *60*, 381–388.

[7] Wrigley, G. *Coffee*. Longman, London, 1988.

7
Establishing a Coffee Plantation

J. N. Wintgens and F. Descroix

7.1
Choice of Site

7.1.1
General Considerations

The establishment of a coffee plantation requires a high level of investment and the returns from such an investment are obtained over the productive life of a plantation, which usually lasts 20 to 30 years. Thus the selection of the site of a plantation at the outset is a very important decision. There are numerous factors which should be taken into account in choosing a site for a coffee plantation and these factors may be divided into two broad categories:

- Those environmental and physical factors which influence the potential productivity of a plantation and, therefore, its long-term profitability.
- Those socioeconomic factors which affect the costs of establishment and operation of the coffee plantation, and, therefore, the economic feasibility of the enterprise.

7.1.2
Environmental and Physical Factors

7.1.2.1 Macroclimatic Features

It can be assumed that the search for a suitable location for a new coffee plantation will only take place where the macroclimatic conditions are broadly suitable for the crop. However, the macroclimatic conditions in the coffee-producing countries of the world are not uniform, and have to be assessed particularly in relation to the effects of altitude and latitude on temperatures. Altitude has a major influence on average temperatures with a diminishing thermal gradient of 0.6 °C/100 m of elevation. Similarly for each degree of latitude away from the equator, the corresponding reduction in temperature is estimated at 0.5 °C. Thus, the combination of the effects of altitude and latitude on average temperatures has a major impact on the location of coffee-growing areas. For example, in equatorial Kenya and Cameroon the most favorable areas for Arabica cultivation are at altitudes of between 1200 and 2200 m as compared with Brazil where the main traditional coffee areas are between latitudes 15 and 25° S and altitude range is restricted to 600–1200 m by frost damage which occurs at higher altitudes. Similarly with

Coffee: Growing, Processing, Sustainable Production, Second Edition. Edited by J. N. Wintgens.
© 2012 WILEY-VCH Verlag GmbH & Co. KGaA. Published 2012 by Wiley-VCH Verlag GmbH & Co. KGaA.

Robusta, the normal altitude range at the equator is from sea level up to 1000–1200 m, but in New Caledonia in the latitude range 20–25° S Robusta is confined to only a few meters above sea level.

Macroclimatic conditions can be defined using climatic data from meteorological stations that are located in close proximity to the area in which the coffee plantation is to be sited. The data that should be collected and analyzed includes long-term records of rainfall, maximum and minimum temperatures, relative humidity, wind direction observations and wind speeds, and solar radiation. It is important that the position of the meteorological stations in terms of altitude and latitude are accurately defined, and that the effects of these variables on climatic factors in the locality being studied are clearly understood.

7.1.2.2 Topographic Features

Topographic features frequently modify macroclimatic conditions and within relatively small areas a number of microclimates may occur. For example, in hilly regions, slope aspect has an important influence on average temperatures – in subtropical regions of the northern hemisphere a south-facing slope has average temperatures which are 2–4 °C higher than on a north-facing slope. These higher temperatures, which are due to differing exposure to the sun, are a consequence of higher solar radiation so that coffee on slopes facing the direction of the sun benefit from greater photosynthetic activity and an increase in floral induction irrespective of altitude. Also, in humid zones, the coffee trees growing on slopes exposed to the morning sun are less susceptible to fungal diseases because the dew dries off much more rapidly, and the dry conditions and increased solar radiation are less favorable

to fungal growth. Slope aspect is also important in relation to the prevailing wind. In windy zones during the dry season, coffee trees grown on slopes exposed directly to the wind suffer more from water stress than those on more protected slopes because of the higher evapotranspiration rates that occur with higher wind speeds.

The topography also has a major impact on nighttime temperatures and the incidence of frost. This subject is dealt with in Section I.11. In general the air cools rapidly at ground level as heat loss by radiation occurs after sundown. On sloping land the cold air will flow down the slope and will accumulate in depressions sometimes causing severe ground frosts. This phenomenon must receive due consideration when choosing a site for a coffee plantation in frost prone areas.

7.1.2.3 Identification of Sites for more Detailed Study

It is assumed that the coffee plantation is to be located in an undeveloped area with few restrictions in terms of human occupation or land ownership. In these circumstances the first step in the selection of potential sites for the establishment of coffee plantations in a particular search area will be the assembly of all relevant maps covering that area. In some situations ordnance survey maps may be available and they are usually produced at scales ranging from 1 : 25 000 to 1 : 50 000. These maps contain contour lines at intervals between 5 and 50 m, major watercourses, natural vegetation and land use types, other major natural features, and details of human settlements, access roads, footpaths and other man-made structures. Such maps provide the basis for the approximate demarcation of potential sites for the establishment of coffee plantations for further in-depth studies.

In the absence of ready-made ordnance survey maps, the initial step in site selection could be an examination of satellite photographs that might be available. Satellite photographs with good resolution and that are cloud-free can provide much useful information on land forms and geology, vegetation, and land use. If appropriate equipment is available together with the necessary information to allow accurate field interpretation then detailed mapping is possible to scales of 1 : 5000. In some countries, government departments are able to assist with mapping based on satellite photography and, in others, private companies provide a mapping service based on such photography.

In the majority of situations in lesser developed countries the identification of potential sites for coffee plantations will have to depend on the use of up-to-date aerial photography to produce suitable maps for planning purposes. Thus, where up-to-date aerial photographs are not available, the first requirement is to procure new photographs covering the search area. Using photogrammetric techniques, the stereographic aerial photographs, usually at a scale of between 1 : 10 000 and 1 : 25 000, can be used to produce accurate contour maps. Following the field identification of features and recording of these features on the aerial photographs, maps may be produced with a level of detail which is similar to that which is found on published ordnance survey maps.

At the mapping stage, it is important to carry out a reconnaissance soil survey to identify the major soil types. At the reconnaissance level survey auger sampling should be carried out at a density of one pit sampling per soil group. Laboratory analyses to establish the physical and chemical characteristics of the soil should be carried out on a sample from each horizon in the profile of each pit. This soil information should be used to prepare a reconnaissance level soil map of the search area.

The soil map combined with other available maps of the search area, when used in conjunction with aerial photo interpretation verification, will usually be sufficient to identify potential sites for coffee plantations. However, the choice of site may be further narrowed by consideration of local features, which would involve the following:

- A review of all available macroclimatic data from within the search area which would include an assessment of rainfall characteristics and an examination of temperature fluctuations throughout the year.
- The collection of evidence to indicate the effects of topography on temperatures and the localized occurrence of frosts.
- The collection of information on the importance of slope aspect in moderating temperatures, radiation and evapotranspiration rates.
- The analysis of wind data to determine the likelihood of potentially damaging wind speeds and whether protection is required to compensate for any adverse effects of prevailing winds.
- The calculation of potential evapotranspiration rates and irrigation water need – where there is a large irrigation water need for coffee, the availability of irrigation water will be an important factor in plantation site selection.
- In addition factors affecting the costs of plantation development also have to be considered, e.g. the cost of land development (land clearance, erosion control measures, land drainage requirements, land cultivation, etc.), the cost of access (off-farm and on-farm roads), the costs of supply of services (water and electricity), etc.

7.1.2.4 Detailed Studies of Selected Sites

It is desirable that aerial photography at a 1 : 10 000 scale be available for the detailed studies required on potential coffee plantation sites since the optimum scale of mapping required for detailed planning of a plantation is 1 : 2000. The maps to be produced include a contour map with a contour interval of 5 m, a vegetation and land-use map, a map showing infrastructural features (settlements, roads, paths, water sources, power lines, etc.), and a soil map.

A very detailed soil survey should be carried out, involving an auger sampling density of about 200 observations/km². The actual number of observations made will depend on the judgment of the soil surveyor as to the degree of variation in the soil type and conditions across the site. Observations are usually made on a grid system and traces are cut at appropriate intervals. A typical example of a survey traverse along a trace is shown in Fig. 7.1.

Existing lines of access (roads and footpaths) or natural lines (watercourses or ridges), which are easily identifiable on the aerial photographs, are used as baselines for the trace cutting. The sampling sites should be at regular intervals and at points where a marked change is observed in either geology, soil type or land use. Each sampling site is marked on the aerial photographs. The surveyor's records should consist of rapid descriptions as outlined below:

- Type of topsoil
- Type of vegetation (forest, savannah, fallow or cultivated land)
- Topography
- Other particularities.

This data is used to delineate the boundaries of the different soil types and vegetation and land use categories (Fig. 7.2).

Auger sampling is used to identify the main soil groups or soil types in the survey area. Subsequently, each soil type or soil group represented within the survey area should be further investigated. A pit should be dug to either a depth of 2 m or to the depth of the soil profile to bedrock and

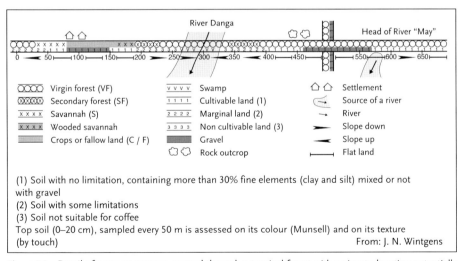

	Virgin forest (VF)		Swamp		Settlement
	Secondary forest (SF)	1 1 1 1	Cultivable land (1)		Source of a river
x x x x	Savannah (S)	2 2 2 2	Marginal land (2)		River
x x x x	Wooded savannah	3 3 3 3	Non cultivable land (3)		Slope down
	Crops or fallow land (C / F)		Gravel		Slope up
			Rock outcrop		Flat land

(1) Soil with no limitation, containing more than 30% fine elements (clay and silt) mixed or not with gravel
(2) Soil with some limitations
(3) Soil not suitable for coffee
Top soil (0–20 cm), sampled every 50 m is assessed on its colour (Munsell) and on its texture (by touch)   From: J. N. Wintgens

Figure 7.1 Detail of a survey traverse opened through a tropical forest with a view to locating potentially suitable "coffee land".

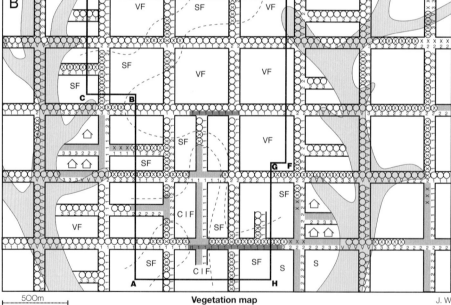

Figure 7.2 Delimitation of a coffee plantation on the basis of survey elements: (A) vegetation map and (B) soil map.

this provides the opportunity for detailed examination of the profile. Soil samples should be taken from each pedological horizon for physical and chemical analysis. The profile description should include the usual observations and particular attention should be paid to the drainage condition. Where inadequate drainage of the soil is revealed by the presence of mottling or light-colored, white or pale yellow clay at a depth of less than 1.5 m from the surface, the soil quality can be considered as unsuitable for coffee growing. In addition the depth and lateral spread of the roots of the natural vegetation should be recorded.

The results of the physical and chemical analyses of the soil from each of the several horizons in the profile provide essential guidelines as to the future potential of land for coffee production. The main selection criteria for coffee plantation land are summarized in Tab. 7.1.

Pedological studies of existing mature coffee plantations have revealed that up to 40 % of coffee tree populations produce below average yields because they had been established on unsuitable soils. Thus, it is important that the soil map should be used in planning the layout of a new plantation so as to avoid the areas of unsuitable soils. The avoidance of the unsuitable soils will significantly raise average coffee yield levels over the life of the plantation.

7.1.3
The Socioeconomic Environment

The socioeconomic environment must be studied before establishing a coffee plantation as this can influence the choice of production systems and farming practices. Under traditional coffee production systems, manual labor is used for the vast majority of tasks and the total annual labor requirement is between 250–300 man days/ha. The peak seasonal demand for labor occurs during the harvesting season and this

Table 7.1 Summary of the major factors influencing site selection for coffee plantation development

Soil characteristics for successful coffee sites	Lowland areas for Robusta <900 m	Highland areas for Arabica >900 m
Slope	<50 %	<100 %
Clay + silt	>30 %	>30 %
Profile 2 m		
Root system	Deep (>1.5 m) and well distributed	
Accidents	Good drainage (no mottling above 1.5 m)	
	No hard-pan (laterite, tuff, rock) above 2 m	
Organic Matter	>2.5 %	>5.0 %
Nutrients		
CEC (meq%)	>5	>10
P (ppm) Truog	>15	>20
pH (H$_2$O)	5.5	5.5–6.5
Toxic elements	Fe, Mn and Al should not exceed tolerated levels of toxicity	
Flooding	Historical confirmation should be recorded confirming that the area is not subject to flooding	

operation accounts for approximately 60–65 % of the total annual labor requirement of the plantation. In order to meet this high seasonal demand for labor a large source of casual labor must reside within easy reach of the plantation. Thus, when choosing a site for a new coffee plantation to be operated with traditional harvesting practices, potential sources of casual labor must be identified.

The labor requirement for the coffee production tasks other than harvesting will vary depending on the natural conditions prevailing, and on the crop husbandry practices and levels of mechanization employed. The natural conditions, which influence labor demand include the ecological factors which favor or limit weed growth, the landforms which require soil conservation practices, the climatic conditions which determine irrigation need, etc.

Crop husbandry practices also influence labor requirements as, for instance, mulching requires more labor than clean weeding and stumping is more labor intensive than topping. In general, extensive methods of cultivation of coffee (with shade trees, without irrigation, irregular pruning, low levels of agrochemical inputs and low yielding) will require less labor than intensive cultivation (no shade trees, with irrigation, regular pruning, high levels of agrochemical inputs and high yields). The labor requirement for a coffee plantation can be reduced by the mechanization of a number of tasks including weed control, the application of agrochemicals and harvesting. Thus, the size of the permanent labor force will be determined by the system of coffee crop management adopted.

The study of the social environment aims to assess the availability of manpower within the local community and the capacity of that manpower to meet the labor requirements for the development and subsequent operation of a new coffee plantation. The work capacity of locally available manual labor can vary considerably between countries and zones, depending on the level and quality of nutrition of the population and on the general health of the people. Thus, in certain areas where the standard of living is low, the workers' food supply can restrict work capacity to no more than 4–5 h/day. In other areas high levels of disease, e.g. AIDS, malaria and filariasis, may result in high levels of absenteeism in the workforce and a restricted ability to carry out strenuous work such as holing or stumping.

The period of peak casual labor requirement for harvesting coffee may coincide with a high labor need on surrounding smallholdings or on local commercial enterprises which are large employers of labor. Thus there are conflicting demands for the available manual labor which must be taken into account when assessing the availability of manpower for the coffee harvest. While such conflicting demands frequently create more problems for smallholding coffee production, it is also valid for larger plantations which employ a temporary labor force drawn from smallholdings. Other competing demands for the limited supply of manual labor, which may or may not be seasonal, should also be taken into account, e.g. local manufacturing and processing industries, other agricultural businesses or estates, nearby large construction projects, nearby conurbations, etc. Competition in the labor market is likely to result in higher labor costs which is an important consideration when making the choice of site for a coffee plantation.

The permanent labor force for a coffee plantation should come from local communities, but it is unlikely that sufficient numbers of people can be recruited within easy traveling distance of a new large coffee plantation. Where sufficient numbers of people are not available from convenient local sources, provision will have to be made for housing part of the permanent labor force on site. However, the costs of providing even low-cost accommodation for workers are considerable and have to be kept to a minimum. Thus, the proximity of local settlements should be a factor in site selection.

7.2
Planning the Plantation Infrastructure

7.2.1
Overall Layout of the Plantation

The basic soil map of the plantation at a scale of 1:2500 with contours and other main features of the selected site will be used for planning purposes. The areas of soils designated for coffee production will be demarcated and the infrastructure layouts should be superimposed on this map. The main infrastructures (housing, offices, processing facilities, workshop, storage sheds, warehouses, etc.) should, as far as possible, be located in the centre of the plantation site in order to limit operational movements to a minimum. However, the location of these facilities should also take account of water availability and power supplies, and access to other public services and local communities. Where appropriate these buildings should be located on marginal land rather than on prime coffee land and the area should be flat – building construction costs will be significantly higher on sloping ground.

The main access roads should follow the central axes of the plantation, run in straight lines and should be designed to carry relatively heavy traffic loads over pro-

longed periods. Secondary roads, which are necessary for the movement of labor, vehicles and equipment for strategic tasks, should serve the individual fields. The numbers and size of fields within a new plantation will depend on topography. Where the land is relatively flat, individual fields may be in the range 2–20 ha. On steeply sloping ground, field sizes will be restricted and secondary road densities will be higher. Where an irrigation system is to be installed, field sizes may be influenced by the irrigation layout requirements. The field sizes with sprinkler and drip irrigation systems will normally be restricted to less than 10 ha, whereas with centre pivot systems of irrigation the field size may be 100 ha.

Ideally, an outside ring road should be constructed to permit alternative access for fire-control, supervision and security. Road construction and maintenance costs should be kept to a minimum. All roads should be planned and designed to minimize the risk of erosion. Road layouts should aim to limit the number of bridges and avoid poorly drained lands.

7.2.2
Housing and Social Infrastructure

The amount of housing required on the plantation will depend on staffing levels and on the housing required to ensure that adequate security of supply of permanent manual labor will be maintained. Staffing levels will be determined by the size of the plantation and by the coffee production system to be adopted. More senior staff are likely to be required on plantations which are managed intensively with modern technology than on traditional extensively managed plantations using large amounts of unskilled labor. Conversely the requirement for manual labor will be less on the intensively managed plantation than on the traditional extensively managed plantation.

The need for social infrastructure will depend to a large extent on the proximity of existing social facilities. Where the new plantation is located several miles from existing facilities, then such facilities should be provided on the plantation. Essential social facilities that need to be provided on a large plantation include a health clinic, school, community centre, religious facility (church, mosque, temple), and shops and/or a market area.

7.2.3
Office Accommodation

The number of office buildings and the floor space required will depend on the size of the plantation and on the administrative tasks that are to be undertaken on site. The more intensive management systems using the latest in modern technology are likely to need more office space than the more traditional systems. More intensive management with high levels of technology implies the use of more mechanization, the application of larger quantities of a greater variety of agrochemicals, the use of irrigation and much higher levels of production. All of which add to the administrative burden.

Office space will be required for the managerial team and for their administrative support. Other senior administrators requiring office space include a cashier/accountant to deal with salaries and wages, a personnel officer responsible for hiring and firing of manual labor, a procurement officer responsible for crop inputs, a maintenance officer responsible for all the buildings on site, and a security officer responsible for all aspects of security on site.

7.2.4
Processing and Storage Facilities

Pulping equipment is required together with an ample supply of clean water. The floor space required for the processing facilities depends on the size of the plantation and on the peak rate of harvesting. The rate of harvesting is governed by the uniformity of flowering, which in turn is influenced by the climatic conditions, which stimulate the flowering.

Dried coffee beans have a long storage life. Consequently, the storage capacity required on a coffee plantation is influenced by the size of operation, by transportation factors and by marketing strategies.

Storage capacity is required for agrochemical inputs since because of the dangers of contamination of the coffee, they cannot be mixed in storage. It is advantageous to have separate storage facilities for bulky fertilizers and for insecticides and fungicides which are in relatively small containers.

7.2.5
Workshops and Machinery Sheds

The type and size of the workshops will be determined by the extent of the plantation, the level of mechanization and the proximity and availability of machinery services. On a large plantation in a remote location a well-equipped workshop will be essential, particularly if there is a high level of mechanization.

7.2.6
Other Special Purpose Land Areas

In addition to the land areas which are required for housing and buildings, where appropriate, land should be allocated in close proximity to the housing area for the manual labor for allotments, where the permanent workforce may cultivate some of their basic food supplies and for orchards that will provide all the plantation employees with a supply of fruit.

Land within the plantation area which is unsuitable for coffee or other forms of agriculture should be developed for tree species which can be used as fuel supply for heating water and cooking food, and/or as a source of timber for use on the plantation for fencing, gates, foot bridges, etc.

7.2.7
Preparation of Plantation Area Map

The plantation layout should be outlined on the combined soil and contour map of the selected area. Obviously, prior to the preparation of the plantation layout decisions will have to be made on the target size of the coffee area, the production systems to be followed, the role of irrigation, the size of the permanent manual labor force to be housed on site, the number of permanent technical and management staff to be housed, and floor area of office space required and the public services required. Many of these decisions would be assisted by a full feasibility study of the plantation development designed to establish optimum conditions in relation to initial investment requirements and the financial viability of the plantation in the long term. A schematic illustration of the layout of a Robusta coffee plantation is shown in Fig. 7.3.

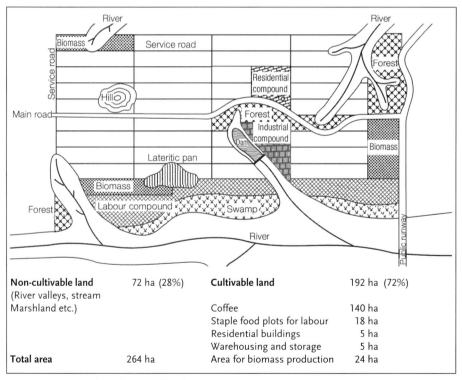

Non-cultivable land (River valleys, stream Marshland etc.)	72 ha (28%)	Cultivable land	192 ha (72%)
		Coffee	140 ha
		Staple food plots for labour	18 ha
		Residential buildings	5 ha
		Warehousing and storage	5 ha
Total area	264 ha	Area for biomass production	24 ha

Figure 7.3 Layout of a Robusta coffee plantation.

7.3
Land Development

7.3.1
Land Clearance

Where natural primary and/or secondary forests are present on the selected area of land for a new coffee plantation the first stage of land development will be the clearance of all trees and bush vegetation. Normally the first operation in land clearance is the felling of trees and bush using chain saws or suitable manual saws and/or chains, pulled by two crawler tractors.

After felling, any useful timber is extracted. The remaining aerial parts of the trees and bush vegetation are then placed in windrows. Any tall trees should be felled at such an orientation to simplify the task of windrowing (Fig. 7.4). The windrows

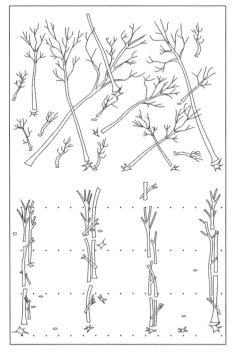

Figure 7.4 Windrowed trunks and branches.

should be lined up as inter-spacing rows between the future rows of coffee plants.

The second land clearance operation is the removal of tree stumps and of major woody roots of the natural vegetation from the upper soil profile. This operation is undertaken by bulldozers equipped with large and powerful root rakes or, on smaller plantations, stationary winches and wrenches may be used. The stumps and major roots are then collected and either removed and deposited in a suitable location outside the development area or placed in windrows for burning.

As a general rule, the larger tree stumps are left *in situ* because their extraction is expensive and time consuming. Only stumps that may obstruct roads or other infrastructure need to be removed immediately. Regrowth of stumps left in the ground may be controlled by the use of herbicides. A number of product are available for this purpose including sodium arsenite, potassium nitrate, ammoniun sulfamate, glyphosate or 245-T. In order to improve the penetration of these herbicides they should be poured into a circular groove cut around the base of the stump.

In circumstances where there are no time constraints on the development of the coffee plantation, a longer-term approach to land clearance may be adopted. For example, a cheaper method of land clearance consists of removing the undergrowth, felling the tall trees, leaving the fallen vegetation *in situ* for a few months and planting the coffee before the removal of the vegetation. The trunks and branches are left to decompose naturally, and so enrich the soil of the coffee plantation with organic matter. In another approach in regions where root diseases may threaten the plantation, trees may be killed by ring barking at least 2 years prior to planting.

Ring-barking depletes the root starch, which acts as a substrate to the fungus which generates the root disease of coffee (*Armillaria mellea*). Once the substrate is exhausted, the trees can be felled and the stumps removed. However, ring-barking of trees has no effect on the occurrence of the soil-borne *fungi Ganoderma* spp., *Fusarium solani* nor *Fusarium stilboides*.

The herbicides, ammonium sulfamate, 245-T and glyphosate, may be used to kill trees more quickly than ring-barking and without depletion of the root starch. The application of these herbicides can be made to the chipped bark at the base of the trunk. The trees die within approximately 1 year after the application and can either be removed by traditional felling methods or by means of tree-dozers or winches. These herbicides may be applied either by a knapsack sprayer or applied directly to exposed surfaces with a brush.

The least expensive method for clearing consists of total incineration of all natural vegetation on the site and planting without windrowing. This method, however, is contrary to present-day concepts of farming as it affects the organic matter of the topsoil, and paves the way for soil erosion and possible leaching (Pict. 7.1, 7.2, 7.3). In cases where this method is chosen for clearing, incineration should, at least, be carried out when the soil is damp, so as to preserve the humus as far as possible. In the Ivory Coast it has been observed that coffee planted after incineration of the forest yields less than coffee planted without incineration because of excessive accumulation of K and Ca in the upper layers of the soil.

In Brazil land clearance may be carried out using only manual labor under a system of sharecropping. The land owner enters into an agreement with a person, known as an "impetreiro" extending over a period of 4–6 years. Under the agree-

Picture 7.1 A tract of primary forest land cleared by incineration for coffee planting. Note that sheet erosion has already started because the land was left without protection. From: J. Berthaud.

Picture 7.2 A tract of secondary forest land cleared by incineration for coffee planting. Note that superficial erosion has already started because the land was left without protection. From: J. N. Wintgens.

Picture 7.3 Young Robusta plantation in secondary forest land cleared by incineration. Erosion is minor because the slope is gentle. From: GENAGRO – Belgium.

ment the impetreiro will undertake the work of land clearance on a plot of land,

and will be responsible for the establishment of a coffee plantation on that land and for the subsequent maintenance of the coffee until the end of the period of the agreement. In return the impetreiro will receive a share of the value of the timber extracted from the area cleared, will be allowed to cultivate annual food crops for his own benefit on the land being developed for coffee and will receive a share of the coffee harvest over the period of the agreement. In some cases the impetreiro may receive a token salary. The major drawback to this system is that there is no guarantee as to soil protection during the period the land is contracted out to the impetreiro.

7.3.2
Land Improvement Measures

In most tropical soils, where coffee is likely to be planted on newly cleared ground, a relatively thin layer of topsoil contains a high proportion of the organic matter and available plant nutrients contained in the soil profile. Thus, it is important that any disturbance of the topsoil during the land development process is kept to a minimum. Heavy mechanical operations, such as the felling of trees and shrubs by bulldozing, the use of a heavy root rake for root removal, deep moldboard ploughing to incorporate trash to depth in the profile and subsoiling, all of which bring infertile subsoil to the surface, should be avoided where possible. However, where the use of heavy machinery for land clearance is unavoidable the variable fertility levels resulting from the infertile subsoil brought to the surface should be rectified through land leveling and the planting of leguminous green manure crops over a period of 1–2 years after land clearance. The green manure crops will raise organic matter levels in

the topsoil and recycle plant nutrients from depth in the profile to the topsoil.

In former coffee plantations, it is advisable to uproot and remove all remaining coffee and shade trees in order to avoid the propagation of root diseases. A leguminous cover plant like *Mucuna utilis*, a great producer of organic material, can be established to raise soil fertility levels. The new coffee plantation should not be established until several growths of leguminous cover plants have been ploughed back into the soil.

On coffee plantations where the planting of shade trees is planned, specially adapted and non-competitive shade trees should be planted as soon as the clearing has been completed. The majority of indigenous naturally occurring trees compete aggressively with the coffee for water and minerals, and so should not be allowed to remain on the coffee plantation area as shade trees.

After land clearance an assessment should be made of the potential weed problems on the proposed coffee areas since there are a number of weed species that threaten the success of the coffee crop. These weeds should be significantly reduced, if not eliminated, before the coffee is planted. Perennial grasses and sedges present the most serious weed problems since they are much more difficult to control and offer much greater competition to the coffee than the dicotyledonous weeds which grow mostly in the rainy season. Perennial weeds are not only highly competitive in terms of nutrients and water but some species also secrete toxic substances through their roots, which block the mineral assimilation in the coffee plant. The most common pernicious perennial weeds which can be expected in typical coffee environments, are *Imperata cylindrica* (Cogon grass) (Pict. 7.4, 7.5), *Cyperus rotondus* (nutgrass, cebollin), *Cynodon dactylon* (Bermuda grass, chiendent), *Digi-*

Picture 7.4 Perennial grasses: *Imperata cyindrica* (Alang-alang, Cogon grass). From: J. N. Wintgens.

Picture 7.5 Young coffee trees invaded by *Imperata cylindrica*. From: J. N. Wintgens.

taria scalarum (red millet) and various ferns.

The mechanical control of these pernicious weed species is largely ineffective unless carried out under the most favorable conditions of soil moisture. Manual and mechanical cultivation inevitably results in the production of numerous splits of rooting material (rhizomes, stolons, corms) of the perennial weeds and their dispersion in the cultivated layer of soil. Under favorable moisture conditions much of this material is able, through vegetative reproduction, to produce a new generation of perennial weeds. For manual hoeing or mechanical cultivation to successfully reduce populations of perennial weeds, the operation must be carried out when the soil is moist and has to be followed by a period of drought during which the rhizomes, stolons and corms are completely desiccated. The change in conditions from a moist soil to a completely dry soil may not occur in the coffee-growing areas with well-distributed year-round rainfall and even in those coffee-growing areas with a dry season such changing conditions are not predictable. Consequently, chemical methods of weed control are generally preferred to mechanical methods in the modern coffee plantation.

The use of non-selective herbicides, such as Round-up (glyphosate), provides excellent blanket control of weeds on land prior to planting the coffee. However, more than one application may be required to satisfactorily control populations of the major perennial weeds. In most cases, the long-term control of these weeds may be more effectively obtained with the use of selective systemic herbicides. Selective herbicides are available for the control of most grasses and for sedges. It is important that the herbicides used to control weed growth prior to planting the coffee should not have any long-term residual effects that adversely affect the performance of the coffee.

Annual weeds generally present fewer problems of control than perennial weeds. There are effective herbicides, both contact and systemic, for the control of dicotyledonous weeds. Prior to planting the coffee, the herbicides should be applied to prevent the annual weeds seeding and herbicide use should cover two seasons to effectively reduce annual weed populations. Annual weed control can also be achieved by manual cutting or hoeing of the weeds which needs to be implemented before the seeds have time to mature. Left too late, slashing only helps to propagate the weed even further.

Though efficient weed control prior to planting the coffee is necessary, the resulting bare soil does present some drawbacks by increasing exposure to erosion by rainfall and run-off, a loss of organic matter from the topsoil and a loss of soil humidity. If there is a prolonged period between effective elimination of weed growth and the establishment of the coffee then the ground should be covered with a mulch or a suitable cover plant as soon as possible. Fast-growing leguminous creeping cover plants are not recommended since they may be difficult to control once the coffee trees are planted. Fast-growing legumes which are not creepers, are best suited for the purpose.

Where innocuous, non-invasive broad-leaf plants such as *Alingsaga* sp., *Ageratum mexicanum*, *Bidens spinosa*, *Salvia obscura*, *Salvia privoides*, *Commelina diffusa*, *Pseudochinoleana polystaquia*, *Hyptis atrorubens*, *Tradescantia* sp. and *Impatiens balsamina* are present, they should be retained as a soil protection measure.

7.3.3
Land Preparation before Planting the Coffee

7.3.3.1 Soil Conservation Needs
Most coffee-growing areas are located in undulating and often mountainous terrain with high seasonal rainfall. High seasonal rainfall invariably includes some high-intensity storms. The natural vegetation found in these areas is commonly dense tropical forest. The development of a coffee estate in this environment involves the clearance of the forest and the exposure of the soil to the rainfall. Inevitably, the water-borne erosion hazard is very high.

Water-borne erosion is caused by the impact of raindrops on the soil resulting in the detachment of soil aggregates and the transport of those aggregates in the surface run-off from the land to another place. Run-off is that part of the rainfall which does not infiltrate into the soil profile, but flows down the slope. The rate of erosion

or soil loss is determined by the amount and intensity of rainfall, and the speed of run-off. The faster the speed of run-off, the higher the volume of soil aggregates (or silt load) that can be carried and the higher the rate of erosion. The speed of run-off is determined by slope and the condition of the soil surface, and can be controlled to some extent. Soil conservation measures, which are designed to reduce the amount and speed of run-off by maximizing the infiltration rate and by creating obstacles to reduce the speed of run-off, aim to minimize the erosion rate.

The potential benefits in terms of reducing the rates of erosion is illustrated by the results of a trial carried out in West Africa which are shown in the Tab. 7.2. The highest rate of soil loss of almost 1500 tons of soil over 3 years is equivalent to a loss of 15 mm depth of topsoil per annum. This rate of topsoil depletion is well in excess of any replacement that may occur through soil weathering and organic matter accumulation, and would undoubtedly lead to a catastrophic decline in soil fertility in the short term. In such circumstances there is an urgent need for soil conservation measures to be installed.

Soil conservation measures should be planned using a 1:2500 scale contour map combined with frequent references to the soil map of similar scale. On slopes of more than 5%, most soil conservation measures are based on contour cultivation. The appropriate contour interval should be delineated on the contour map at the outset. In addition, the disposal of run-off from the cultivated areas will require carefully designed storm drains and waterways, most of which will follow natural gulleys and drainage channels. Also, devices that will encourage the infiltration of water (inter-row ridges and silt pits) may be included in the plans where appropriate,

Table 7.2 Annual soil loss from observation plots under three types of land use

Land use	Anti-erosions measures	Annual Soil Loss over 3 years (MT/ha)
Coffee planted in rows and clean weeded	No erosion control measures; bare soil within the coffee areas	1494.7
Annual food crops; crop residues allowed to remain *in situ*	Rows planted along the contour; closed contour ditches	169.6
Coffee trees planted along the contour	Mulch covering the soil and additions made at regular intervals	0.4

and on very steep land the construction of bench terraces and/or individual terraces will be needed. A detailed review of the practical implementation of the planned soil conservation measures (to be installed before planting the coffee) is contained in the following sections.

7.3.3.2 Contour Planting

Where a coffee plantation is to be installed on land with a slope of more than 5 % the coffee trees should be planted along the contour, i.e. the rows of coffee should run at right angles to the direction of the slope. Anti-erosion measures such as contour ridges, contour bunds, contour ditches and vegetative measures of erosion control also run along the contour between the coffee rows, and are designed to disrupt the downward flow of rainfall run-off and to collect and convey the run-off to major waterways that carry the water to a drainage channel. The selected contour interval at which the rows of coffee are to be planted will be determined by an examination of the contour map and the preparation of outline plans for the erosion control measures required (Pict. 7.6).

The initial demarcation of the field layouts and contour alignments for planting the coffee trees will be made with conventional survey equipment. However, for the marking out of rows of planting holes in the field, the U-framed ridge or "caballete" is specially recommended (Fig. 7.5).

The standard size of 3 m × 1 m can be adapted as required. The level line is indicated by a simple water gage used by carpenters which is fixed to the centre of the main bar of the frame. The frame can be made of wood or hollow metal. The adjustable graduated device is used to give the line a pre-defined incline. To trace a slope

Picture 7.6 Contour planting. From: Nestlé.

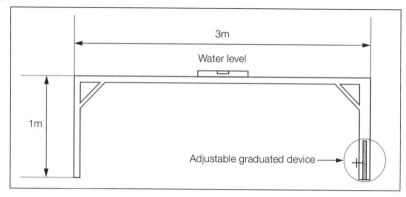

Figure 7.5
U-framed
ridge.

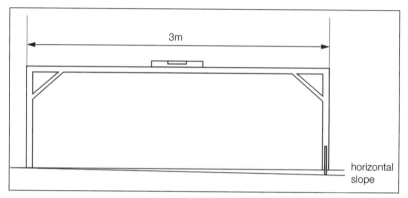

Figure 7.6
Adjustable
graduated
U-framed
ridge.

with a gradient of 1%, for example, the graduated device will be stepped up by 3 cm (1% of the length of the 3 m ridge bar) (Fig. 7.6).

7.3.3.2.1 Defining the Slope of the Land

Knowledge of the average slope of the land chosen for the plantation is indicative and will help to decide on the most suitable anti-erosion measures to be implemented as well as the density of the plantation (Fig. 7.7).

7.3.3.2.2 Pegging the Contours

Pegging the base lines

From the highest point of the slope the base-lines are marked from A to B. In practice, it is advisable to peg as follows:

Slope (%)	Distance between master rows (m)
<15	>30
15–20	15–30
20–50	6–15
>50	<6

The steeper the slope, the closer the spacing between the master rows (Fig. 7.8).

Pegging the Master Rows

The master rows are also pegged out by using the U-framed ridge. To trace the line as precisely as possible, the person in charge of the pegging out will start out with one end of the U-framed ridge placed against the peg of the baseline. He then needs to adjust the entire frame until the water gage on the ridge of the frame is horizontal. At this point, he can stake out the

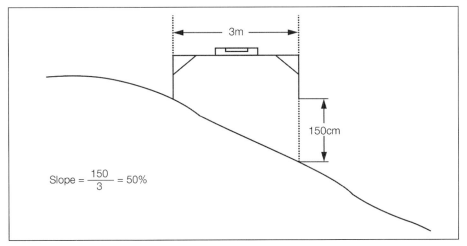

$$\text{Slope} = \frac{150}{3} = 50\%$$

Figure 7.7 Slope evaluation.

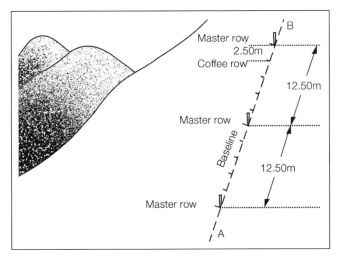

Figure 7.8
Pegging the base lines.

next peg for the master row. The operation should be repeated until the end of the row (Fig. 7.9).

Various uneven natural features of the land may lead to a sinuous course for the master row. This can be adjusted by moving stakes which stray too far from the general course (Fig. 7.10).

Marking Out Planting Holes on the Master Rows

Once the master rows have been laid out and staked, the planting holes need to be marked. Once the planting hole sites are pegged, the master row stakes can be removed and the hole sites marked with shorter sticks (Fig. 7.11).

Pegging the Contours

Once the planting hole sites are pegged out along the master rows, the contours still need to be completed. This can be done either by means of two wooden sticks with forked ends or by a cord with three fixed metal rings (Fig. 7.12).

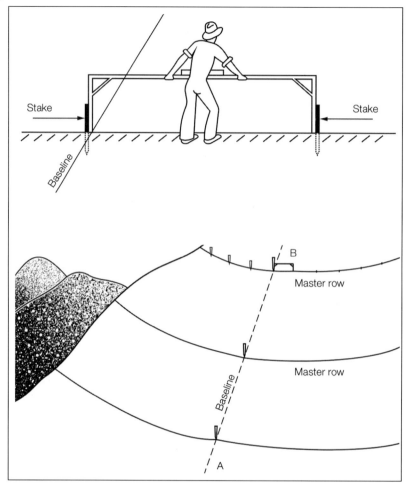

Figure 7.9
Pegging
the master
rows.

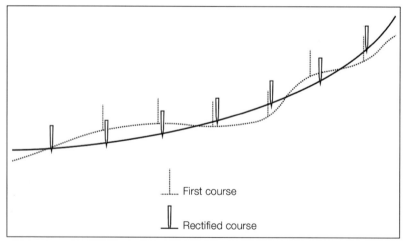

Figure 7.10
Master
row course
rectification.

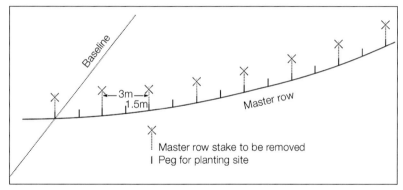

Figure 7.11
Marking out the planting sites.

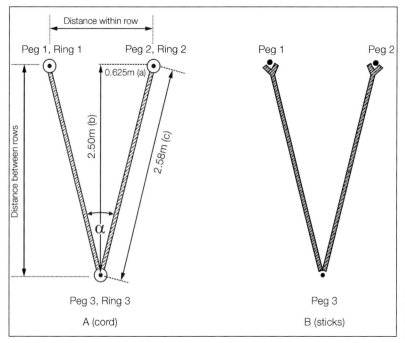

Figure 7.12
Devices for contour pegging.

The length of the cords or wooden sticks, i.e. the distance "c" between pegs 2 and 3 is always longer than distance "b" (between the rows). In the above example, "c" (2.58 m) is approximately 3% longer than "b" (2.50 m). The wider the angle, the greater the difference. In cases where the plantation is equidistant, the difference will be about 15%. For each particular planting layout, these distances should, therefore, be calculated separately, according to the Pythagoras formula: $c^2 = a^2 + b^2$.

Rings 1 and 2 of the cord should be slipped on to pegs 1 and 2 respectively. Ring 3 should then be pulled out until the cord is taut. At that point, peg 3 should be placed in ring 3 and driven into the soil.

When using wooden sticks with forked ends, the forked ends should be placed against pegs 1 and 2 and swiveled until

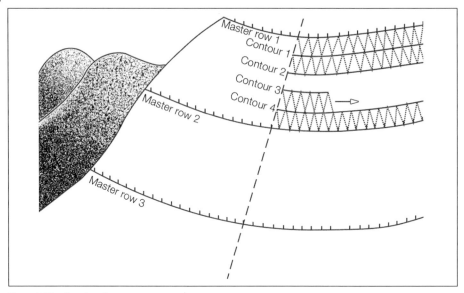

Figure 7.13 Pegging the contours.

the two ends meet. At that point, peg 3 should be driven into the soil.

This operation should be repeated along the entire contour of the first master row, then again along the second contour until mid-distance between master rows 1 and 2, and then repeat the operation from master row 2 (Fig. 7.13).

On coming to the final contours, the width of the inter-rows are frequently modified because the master rows are not strictly parallel. In this case, it is necessary to alter the spacing of several lines. In some cases it may even be necessary to delete or add row sections (Fig. 7.14).

7.3.3.3 Transversal Planting

Though contour planting remains the most efficient method for erosion protection of the soil, this system often complicates work in the fields because the rows are frequently irregular. For this reason, many farmers prefer the method of transversal planting (Fig. 7.15).

7.3.3.4 Measures to Control Removal of Rainfall Run-off

Where a plantation area is located in mid-slope it is important that rainfall run-off entering the plantation from higher areas should not cause any damaging erosion in the coffee areas. Thus, the run-off from the higher ground either should be diverted to specially constructed storm drains, which pass around the plantation, or should be collected and contained in well-protected, natural waterways or drainage channels that pass through the plantation.

The size of the storm drains should be calculated to evacuate the excess water produced from the higher ground taking into account rainfall intensity probabilities, and the area and condition of the higher ground which influence run-off rates.

Where natural waterways are insufficient to cope with the evacuation of excess water due to heavy rains or storms, artificial water ways can be installed. To avoid undue erosion, artificial waterways should be shallow and wide, and always protected

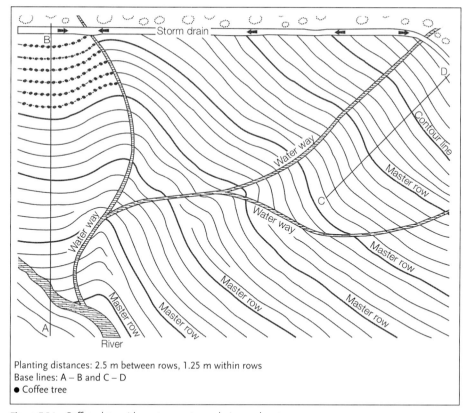

Figure 7.14 Coffee plots with contours, storm drains and waterways.

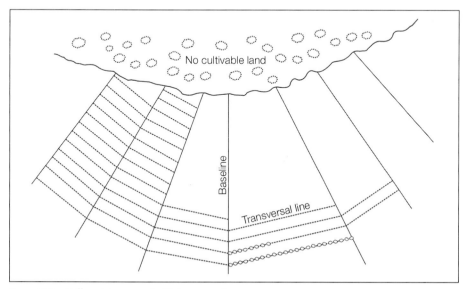

Figure 7.15 Coffee plot prepared with transversal rows.

with well-rooted grassy vegetation like *Paspalum* spp. The depth and width of an artificial waterway depends on the soil infiltration capacity, the slope and the catchment area, but as a general rule they should be 0.9–1.5 m wide and 20–30 cm deep (Fig. 7.16).

Measures to control the removal of rainfall run-off from the coffee plantations on sloping land should aim to slow down the surface flow of water during heavy rainfall so that erosion and topsoil removal are avoided, and so that there is maximum infiltration into the soil. Inter-row contour ditches may be constructed at suitable intervals to intercept the run-off flowing down the slope and through the rows of coffee, and to convey that water to the natural waterways. The slope of the contour ditches should be minimized to reduce the rate of flow and avoid erosion in the channels.

Inter-row ridges may be used instead of contour ditches to retain run-off and allow seepage into the soil. Inter-row ridges are not recommended in plots where the slope is steeper than 40%. Inter-row ridges can be dug out either manually or, where conditions permit, with an animal or tractor-drawn ridger. The edges of the ridges should be stabilized with Vetiver plants. Like the contour ditches, the inter-row ridges divert excess run-off into the natural waterways for evacuation from the hillside (Fig. 7.17).

Distance between the inter-row ridges calculated according to the gradient of the slope:

Gradient of the slope (%)	Distance between the ridges (m)
10	40
20	25
30	20
40	15

The waterways which are used for the evacuation of the run-off from the plantation

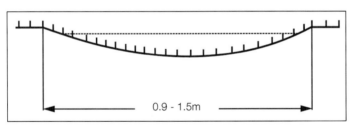

0.9 - 1.5m

Figure 7.16
Artificial waterway.

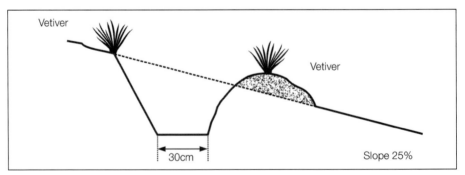

Vetiver

Vetiver

30cm

Slope 25%

Figure 7.17 Inter-row ridge.

resulting from high-intensity rainfall need to be protected to avoid problems of erosion. This protection may be provided by establishing vegetation in the bottom and along the edges of the channels. The aerial parts of the vegetation serve to slow down the flow of water along the channel, whilst the roots bind the soil. Many types of perennial vegetation may serve to provide channel protection including grasses, shrubs and small trees. Where the naturally occurring vegetation does not provide a satisfactory ground cover in the channels the introduction of dense growing grasses such as *Paspalum* spp. is recommended.

The erosive capacity of the flow of water in the natural waterways may be reduced by the installation of drop structures. Water is dropped several feet into a stilling basin or on to rocks and the energy of the flowing water is dissipated. Drop structures are an expensive solution to the problem of channel erosion, but in unstable soils they may be the only solution.

The system of controlling the removal of excess run-off from a new plantation should be carefully planned before the coffee is planted. Much of the implementation of the plan should also be completed before planting. The layout of the contour ditches, the construction of the ditches, and the development of natural drainage channels like grassed waterways should be undertaken simultaneously with the planning and installation of the access roads and irrigation systems.

7.3.3.5 Limitation of Topsoil Removal by Absorption Pits

On average slopes with a gradient between 6 and 20%, the anti-erosion effect of the vegetation on the site combined with mulching can be reinforced by opening a network of silt pits. The silt pits will catch run-off surface water, and impound soil particles and fragments of vegetation. Once they have filled up, they should either be cleaned out or simply abandoned. In this case, new silt pits can then be opened in nearby inter-rows. Silts pits can also be filled with organic fertilizers like compost or manure. This will offer nitrogen compensation and ensure an easily accessible reserve of nutrients to the coffee trees (Fig. 7.18).

The silt pits should be placed in alternance from one row to the other and their size can vary according to conditions on the site. As a general rule, they will be 0.6–3 m in length, 0.4 m wide and 0.3 m deep. In some cases it may be advisable to locate silt pits in the coffee rows themselves in order to avoid hampering transit along the inter-rows.

The density of the pits will be calculated according to the gradient of the slope:

Gradient of the slope (%)	Spacing of the silt pits
7	every 5 inter-rows
10	every 4 inter-rows
15	every 3 inter-rows
20	every 2 inter-rows

From: Gaie and Flémal

7.3.3.6 Other Soil Conservation Measures

An efficient and relatively cheap operation for reducing erosion on slopes of 10–50% is to install individual terraces. These consist of small circular or oval-shaped platforms of approximately 40–60 cm in width with a backslope of 5% towards the hill. The slope will facilitate the infiltration of water and prevent fertilizers from being carried away downhill by rainwater. Individual terraces should be dug out prior to holing (Fig. 7.19).

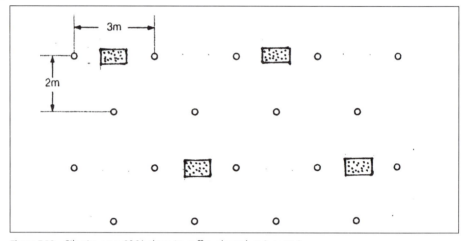

Figure 7.18 Silt pits on a 20% slope in coffee planted at 3 m × 2 m.

Where the slope is steeper than 30% and where all other measures have proved unable to prevent erosion, it will be necessary to build bench terraces. This is a costly operation which involves 500–800 man days/ha.

Bench terraces must follow the contours of the terrain and, like individual terraces, they should backslope at a degree of 5% towards the hill. The width of the bench terrace should be 1–1.5 m and the coffee tree placed slightly more than halfway out from the hill in order to avoid suffocation of the coffee roots after heavy rainfall (Pict. 7.7, 7.8, Fig. 7.20A).

Building bench terraces entails considerable movement of the soil which can im-

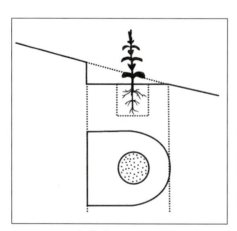

Figure 7.19 Individual terrace on a 25% slope.

Picture 7.7 Arabica planted on bench terraces in China. From: J. N. Wintgens.

Picture 7.8 Arabica planted on bench terraces in Yemen. From: A. Eskes.

pair fertility if due precautions are not taken. It is therefore advisable to remove the topsoil prior to digging out the terrace and to restore it in the hole once the terrace is completed. Vetiver plants should also be installed to anchor the soil (Fig. 7.20B).

In steeply sloped areas, regular anti-erosive measures can be reinforced by installing small palisades between the coffee rows. Palisades can be made of wood or bamboo stakes planted at intervals 0.5–1 m. Ligneous debris can also be stacked against them.

These are small structures built across gullies or other rainwater evacuation waterways to reduce the speed of the water-flow and to diminish gully erosion. They are generally made of split bamboo stalks planted in the soil at strategic points where risks of gully erosion are high. They are frequently used in Latin America where they are known as "palos piques". A similar result can be achieved by placing large stones in the waterways.

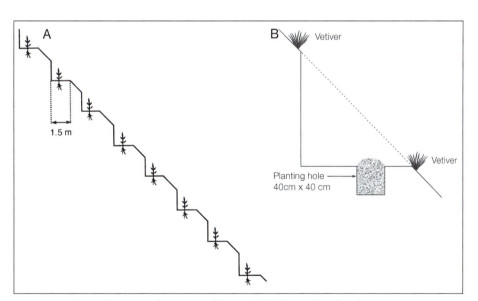

Figure 7.20 (A) Bench terraces of 1.5 m in width on a 100% slope with coffee planted at 2.5 m × 2.5 m. (B) Bench terrace with 100% slope showing Vetiver protection and planting hole refilled with topsoil.

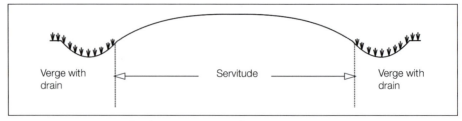

Figure 7.21 Cross-section of a plantation service road.

Wherever possible, roads should follow the contour lines so as to minimize their erosion and the costs of maintenance. Major roads require a width or servitude of 6–8 m, whereas secondary service roads only need to be 3–4 m. wide. Road verges should always be protected with appropriate grass like *Paspalum* spp. and should be properly maintained (Fig. 7.21).

Where necessary, water evacuation drains should be protected with either stone or wood facines or natural vegetation.

The least costly and most efficient protection of road banks is the Vetiver plant. Cementing is not advisable as it is costly and does not last long. It is also not very aesthetic or environmentally friendly and does not allow for water infiltration.

7.4
The Interplanting of Coffee

7.4.1
The Use of Cover Plants in Coffee

Clean weeding is rarely practiced, except in regions like Yemen or Zimbabwe, where rainfall is scarce and weed growth is eliminated for the purpose of soil moisture conservation, and in countries like Brazil, where coffee cherries are gathered from the ground and any weed growth inhibits cherry collection (Pict. 7.9).

Instead of clean weeding, in many situations, a cover crop is planted in the coffee. The cover crop is generally planted with the aim of covering all the ground space between the coffee trees leaving only a small "saucer" of bare weed-free space at the base of each tree.

Most cover crops are perennial, low-growing leguminous species which protect the soil against erosion, compete well with weeds susceptible to shading and contribute some nitrogen to the soil reservoir which becomes available to the coffee. However, in typical areas where Arabica coffee is cultivated, there is a distinct divide between dry and rainy seasons, and many otherwise suitable cover crop plants compete with the coffee trees for water during the dry season. On the other hand, in typi-

Picture 7.9 Clean weeding on red latosol. The top layer has been washed away and the soil is dangerously exposed to sun-baking and erosion by rain or wind. From: J. N. Wintgens.

cal Robusta areas, where the dry season is much less severe, a well adapted cover crop is normally very beneficial to the coffee trees as it not only protects the soil but also ensures an addition of organic matter to the topsoil and discourages the growth of weeds.

Two types of cover crop are recognized – the natural cover crop which consists of allowing spontaneous vegetation to develop and the artificial cover crop which is seeded either before or after the planting of the coffee trees. Many different leguminous species have been experimented over the years as artificial cover crops which include the following:

- *Pueraria javanica, P. thunbergiana*
- *Calopogonium mucunoïdes*
- *Vigna sinensis*
- *Dolichus lablab*
- *Mimosa invisa* var. *inermis*
- *Desmodium ovalifolium, D. intortum*
- *Arachis pintoi*
- *Stylosanthes* spp.
- *Crotalaria* spp.
- *Tephrosia* spp.
- *Flemingia congesta*
- *Glycine wightii*
- *Canavalia ensiformis*
- *Indigofera suffructicosa*
- *Leucaena* hybrids
- *Centrosoma macrocarpum.*

Unfortunately, no ideal cover crop exists. The choice of the artificial cover crop must be made on the basis of species adaptation to the environment, morphology or growth habit, rooting depths, phenology, response to dry conditions, etc., but also taking into account type of coffee (variety, growing habits, etc.) and cultivation practices. A cover crop should be easy to establish and should neither invade the coffee trees nor offer strong competition for water or plant nutrients. On the other hand, a cover crop should be vigorous and, ideally, should be able to compete strongly with the most pernicious weeds. In most situations the growth of a leguminous cover crop can be stimulated by applications of phosphate fertilizers and dolomite.

Vigorous climbing and twining plants like *Pueraria*, *Calopogonium* and *Mucuna* are costly to control since they require frequent cutting back to prevent them from invading the coffee rows and covering the trees (Pict. 7.10).

Other plants like *Stylosanthes* spp., which are upright in growth habit and do not produce tendrils, will compete strongly with the coffee trees for water if the season is particularly dry (Pict. 7.11).

Calopogonium spp. and *Mimosa invisa* tend to dry up when water is scarce. This response to dry conditions means that these two species do not compete with the coffee trees for water in the dry season. However, both species produce large quantities of dry leaf litter, which adds to the danger of fire in the dry season. Shallow

Picture 7.10 Pueraria cover plants in Robusta coffee. From: G. Wrigley.

Picture 7.11 Stylosanthes cover plants in coffee. From: J. N. Wintgens.

Picture 7.12 Soil invaded by perennial grass (*Cynodon dactylon* and *Eleusine indica*). From: J. N. Wintgens.

rooting cover crops will gradually decline if organic matter levels in the soil are allowed to diminish (increase of C:N ratio) and are likely to be replaced by acid tolerant, perennial grasses which compete very strongly with the coffee plants both for water and for nutrients (Pict. 7.12).

On the basis of acquired knowledge and experimentation to date, the following choices appear to be the most advisable:

- For Robusta plantations: *Pueraria, Mucuna* (chiefly recommended for new plantations), *Flemingia congesta, Mimosa invisa* var. *inermis* (spineless), *Arachis pintoi, Leucaena* (seedless).
- For Arabica plantations: *Arachis pintoi, Desmodim intortum* (Greenleaf desmodium), *Mimosa invisa* var. *inermis, Indigofera suffricticosa* (Pict. 7.13, 7.14, 7.15, 7.16).

Picture 7.13 *Arachis pintoi* cover plants. From: J. N. Wintgens.

Flemingia congesta is a tall erect plant with roots several meters deep. It does not compete with coffee trees when regularly cut back as soon as it reaches a height of about 1.6 m. Its roots are associated with *Rhizobium* which enrich the soil with nitrogen. Installation is somewhat delicate be-

Picture 7.14 *Mimosa invisa* var. *inermis* cover plants. From: J. N. Wintgens.

Picture 7.16 *Leucaena leucocephala* cover plants. From: GENAGRO – Belgium.

Picture 7.15 *Flemingia congesta* cover plants. From: GENAGRO – Belgium.

Picture 7.17 Typical leaves of *Flemingia congesta*. From: J. N. Wintgens.

cause its seeds are protected by a tough envelope. Scarification or, better still, soaking at a temperature of 80 °C is advisable prior to sowing. It is also advisable to treat the seeds with a powder insecticide as a preventive measure against ants, in particular. A triple row should be sown in the centre of the inter-row. To limit competition, the outside rows of *Flemingia* should be distanced by at least 1 m. from the coffee trees. A slight nitrogen fertilizer is recommended at the outset to provide the necessary input until the *Rhizobium* develop.

When *Flemingia* is cut back, 2–3 times annually, the fallen vegetation constitutes abundant mulch with high protein content. The leafy stalks of *Flemingia* can also be used as fodder for livestock. Rabbits, in particular, are very partial to this type of fodder. *Flemingia* also has the added advantage of rapid re-growth, even when cut down to

ground level or after a fire, because of the buds located under its collar. It attracts no known parasites with the possible exception of ants who remove the seeds (Pict. 7.17).

Mimosa invisa var. *inermis* is easy to install and provides reasonable protection to the soil.

Arachis pintoi does not prosper well under heavy shade.

The seedless *Leucaena* is a valid cover plant for Robusta, but, as it becomes woody when it grows older, it hampers circulation between the coffee rows. Therefore, it is only worthwhile if it can be kept in check with a rotary cutter because manual lopping is both costly and tiresome.

Mixed stands of cover crops are sometimes used, e.g. *Calopogonium mucunoides* is planted together with *Desmodium ovalifolium* and similar plants.

7.4.2
Strip Cropping for Erosion Control

The practice of strip cropping is used as an erosion control measure. The rows of coffee, which are planted on the contour, alternate with parallel strips of natural or specially planted herbaceous vegetation which also follow the contour. The functions of the belts of vegetation between the coffee are to stabilize the hillside, to prevent erosion (sheet, rill and gulley erosion) developing and to filter out the silt load in the run-off (Fig. 7.22).

In most cases the anti-erosive strip is made up of two to four rows of well-rooted perennial plants. For this purpose *Vetiveria zizanoides* (Vetiver), *Leucaena diversifolia* (K-156), *Flemingia congesta*, *Tripsacum* (Guatemala grass), *Cymbopogon citratus* (Lemon grass), *Crotallaria*, *Calliandra calothyrsus*, *Tephrosia vogelii* (Fish poison bean) and *Indigofera endecaphylla* (Creeping indigo) are best suited. Most of these suggested species can be periodically cut or lopped to provide mulching material for the coffee. The leguminous species in the list are particularly useful for mulching and will also provide some additional nitro-

gen to the soil some of which may be picked up by the coffee. Depending on local conditions, the anti-erosive strips can be made up either of four rows of graminaceous plants or two rows of graminaceous plants alternating with a row of leguminous plants.

The distance between the anti-erosive strips is calculated according to the gradient of the slope:

Slope (%)	Spacing of the anti-erosive strips (m)
5	20–25
10	15–20
15	12–15
20–30	10–12
30–40	7–10
40–50	5–7
>50	<5

The establishment of strips of grasses and the legumes with a more prostrate growth habit will act as a barrier to the movement of topsoil down the slope and could result in the progressive building up of terraces. Consequently, the plant species used for hedging should be well-rooted in order to ensure a solid cramping and avoid the

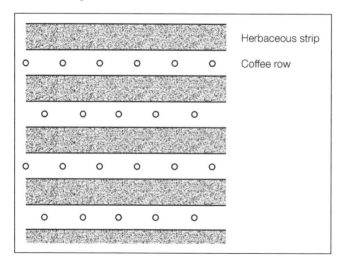

Herbaceous strip

Coffee row

Figure 7.22
Natural herbaceous
anti-erosive strips.

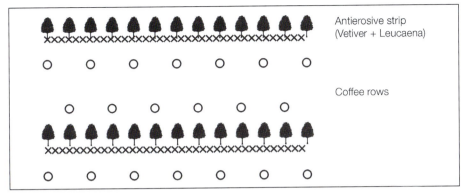

Antierosive strip
(Vetiver + Leucaena)

Coffee rows

Figure 7.23 Planted anti-erosive strip on a slope with a 50 % gradient.

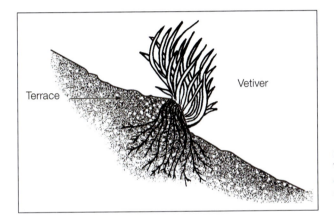

Vetiver

Terrace

Figure 7.24 Cross-section of a Vetiver hedge showing an accumulation of soil on the upper side of the slope and the formation of a small terrace.

plants being carried away by the weight of accumulated soil on the upper side of the slope. Vetiver alternated with *Leucaena* is a potentially effective combination (Fig. 7.23).

To install a Vetiver strip, shoots of Vetiver should be planted at intervals of 40–50 cm in a row running parallel to the coffee rows. Leguminous shrubs are planted 50 cm above the Vetiver hedge at intervals of 1 m. To prevent competition for sunlight, the shrubs should be cut down to 25 cm as soon as they reach a height of 1.6 m. The resulting fallen vegetation can be used as mulch for the coffee plants. To avoid the suffocation of Vetiver plants, the mass of topsoil trapped in their roots should be periodically loosened up and dry stalks removed. This will also encourage the growth of young shoots (Fig. 7.24, Pict. 7.18, 7.19, 7.20).

Picture 7.18 Slips of Vetiver in polybags ready for planting. From: J. N. Wintgens.

Picture 7.19 Vetiver anti-erosive strip in young Arabica coffee planted in contour in Zimbabwe. From: J. N. Wintgens.

Picture 7.20 Vetiver anti-erosive strip in an old rejuvenated Robusta plantation in Thailand. From: J. N. Wintgens.

A belt of coffee with an equal width of an alternating strip of fruit trees (such as bananas, citrus, guava, etc.) or an alternating strip of short-growing forest trees is referred to as alley-cropping. This system can be a good anti-erosive measure, but the trees in the alternative strip must not affect the performance of the coffee. The choice of tree is, therefore, limited.

7.4.3
Shade Trees in Coffee

7.4.3.1 Advantages and Disadvantages of Shade and Shade Trees

The species of coffee which have given rise to the cultivated coffee varieties of today are indigenous to the lower storey vegetation of tropical forests. Thus, the cultivated varieties of coffee have inherent characteristics which ensure good adaptation to growing in shaded conditions. Until around 40 years ago coffee was grown under shade trees in the vast majority of plantations. The advantages claimed for shade and shade trees in traditional coffee management systems are as follows:

- Shade limits weed growth, particularly of grasses and sedges, and consequently weeding costs are reduced.
- Shade reduces oxidation and rate of decay of organic matter in the soil.
- Shade reduces plant metabolism which results in more regular flowering.
- Shade trees enrich the soil through organic material from twigs and leaf fall.
- Leguminous shade trees fix atmospheric nitrogen in the soil some of which reaches the coffee trees.
- Shade trees protect against extremes in atmospheric and soil temperatures.
- Shade trees provide increased protection against frost and strong winds.
- Shade trees provide improved soil protection against erosion.
- Shade trees provide a reduction of *Cercospora* spp. and of attacks by the White Stem Borer.

The disadvantages of shade and shade trees in modern intensive management systems are as follows:

- Shade trees compete with the coffee for plant nutrients and water particularly when the shade tree root systems are superficial or shallow and extensive exploiting a large area of ground.

- Shade trees require regular lopping and thinning which is labor-intensive and, therefore, costly.
- Shade tree branches which fall due to lopping or strong winds can damage the coffee trees, causing a reduction in productivity.
- Shade reduces photosynthetic activity in the leaves of the coffee trees and causes the elongation of inter-nodes of new shoots as they seek sunlight – both the reduction in photosynthesis and the heliotropic response of new shoots lead to lower yields (this yield limitation is especially marked in cloudy areas).

The disadvantages of shade and shade trees listed above have been generally seen to outweigh the advantages in the modern plantation. Consequently, the vast majority of coffee plantations established in the last 30–40 years have been developed with no shade. However, there are a number of situations, usually defined by agroclimatic conditions, where shade may still be preferred. Shade trees may still be preferred in zones where temperature fluctuations are unfavorable and shade trees have a moderating effect on extreme events, in zones where high winds occur (constant and/or strong winds) shade trees provide sheltered conditions preventing damage to the coffee trees, and in zones where there is a risk of damage from tornadoes and violent storms (including hail) shade trees may reduce physical coffee tree damage. In areas with higher than optimum temperatures, the presence of shade trees will protect the coffee from high solar radiation, will help to moderate the soil temperatures and will limit evapotranspiration rates. On poor soils, shade trees reduce the

rate of growth of the coffee and lower yields, but at the same time the requirements for plant nutrients is reduced and the risk of nutrient deficiencies developing in the coffee will be largely avoided.

There is no doubt that, given favorable soil and farming conditions, potential coffee yields are far higher without shade than with shade. Under full sunlight, the faster rate of growth of the coffee has to be supported by more intensive crop husbandry including significantly increased applications of inorganic fertilizers, higher rates of pesticide use and the use of irrigation. Also, without shade there will be a need for increased action to control weed growth, pruning activities will be more onerous and longevity of the plantation may be significantly reduced. The physical sustainability of highly intensive coffee production without shade in the long term is uncertain. The long-term prospects for the economic viability of highly intensive coffee production is also questionable, particularly in view of the experiences of world coffee production during the last decade. The intensive systems of coffee production without shade are generally expensive, and are more susceptible to changes in production costs and coffee prices.

Organic coffee is already a rapidly expanding marketing line for coffee. The production of organic coffee requires abandoning all practices in which agro-chemical products are applied to the crop. The prohibition of the use of inorganic fertilizers in itself precludes the production of coffee in the without-shade situation for the organic market. Thus, the organic coffee grown under shade receives no costly agro-chemical inputs and is, therefore, a low-cost operation. At the same time, organic coffee is making significantly higher prices on the world markets than coffee produced on intensively managed plantations. Because of the low intensity of the management system, organic coffee produces only relatively low yields in environmentally benign conditions, but the outlook for long-term sustainability is excellent.

7.4.3.2 Choice of Shade Trees

Ideally shade trees should provide a homogenous canopy with a minimum area of clear spots and no undue leaf density. Shade trees should intercept some 20–45 % of the total light falling on the plantation. The shade trees must be deeply-rooted to avoid competing with the coffee trees for water in the upper 0.75 m of the soil profile. The selected species should be well adapted to local environmental conditions and they should be fast growing, long lived, deep rooted and wind resistant. Evergreen species that resist defoliation under extreme climatic conditions and maintain constant and adequate shade protection throughout the year should be preferred. A further selection criterion is that the shade trees are able to withstand intensive pruning and to produce substantial quantities of litter that can be used as mulching material for the coffee. Care should be taken to ensure that the selected shade trees are not alternative host plants for any insect pests of coffee.

7.4.3.3 Tree Species used as Shade Trees

A wide range of tree species are used as shade trees in coffee and the most commonly used are listed and are briefly described in the following paragraphs together with illustrations. The tree species used vary according to continent and agroclimatic conditions.

7.4.3.3.1 The *Leucaena* Genus

There are 10 main species of the *Leucaena* genus of which *Leucaena leucocephala* is the best known, particularly the K8, K28, K67 and the Peru strains.

The *Leucaena* spp. are fast-growing leguminous trees which can reach a height of 10 m in 2 years. They produce high quality fire wood (4500 kcal/kg) and their leaves, rich in proteins (25 % DM) can be used as fodder for most herbivores, with the exception of equines who react negatively to their toxic mimosine content. They are also very resistant to drought and offer no competition to coffee trees.

Due to these exceptional qualities, trees of the *Leucaena* spp. have been called "wonder trees", and they are used in coffee plantations as shade trees, cover plants and for erosion control.

The major drawback for *Leucaena* was their massive seed production and their sensitivity to attacks of the psyllid *Heterophylla cubana*.

Currently, hybrid varieties produced by crossing *L. leucocephala* with *L. glabrata*, *L. pulverulenta*, *L. diversifolia* (K.743), *L. palida* (KX12) etc. are used wherever possible as shade trees in coffee plantations as they have developed a resistance to psyllids and produce less or no seeds at all. The hybrid KX2 (K8 × KX12) is one of them.

Leucaenas chosen as shade trees in coffee plantations should be:

• Resistant to the psillid bug
• Seedless
• Adapted to the altitude
• Tolerant to acid soils (pH < 5.5) (Pict. 7.2, 7.22).

Further information on *Leucaena* shade trees can be obtained from: NFTA, PO Box 680, Waimanalo, HI 96795, USA.

Picture 7.21 Three year-old *Leucaena leucocephala* in lowland Panama. From: J. N. Wintgens.

Picture 7.23 Aerial view of *Mimosa scabrella* on Arabica in Mexico. This is an environmentally friendly vegetative association. From: J. N. Wintgens.

Picture 7.22 *Leucaena leucocephala* as a shade tree for Robusta in Malaysia. From: J. N. Wintgens.

7.4.3.3.2 *Calliandra calothyrsus*

This shade tree is best suited for alley-cropping. It produces a high quality firewood, its foliage can be used as animal fodder, and it thrives in a wide range of soils and climatic conditions.

7.4.3.3.3 *Mimosa scabrella*

This shade tree comes from Brazil and has been widely used with good results for many years (Pict. 7.23, 7.24).

Picture 7.24 *Mimosa scabrella* shading Arabica in Mexico, seen from the inside. From: J. N. Wintgens.

7.4.3.3.4 *Gliricidia sepium*

This plant is popular in Latin American cocoa plantations, hence its Spanish name of "Madre de cacao". As a shade plant it presents the drawback of generating too many shoots at the base of the trunk and not spreading enough canopy for adequate shade. It also undergoes severe seasonal defoliation leaving both the coffee trees and the soil exposed to full sunlight. Apart from that, it prospers under poor conditions, is drought resistant, and adapts to a wide range of soil conditions and altitudes (from 0 to 1500 m).

Picture 7.25 *Glyricidia sepium* shading Robusta in Malaysia. From: J. N. Wintgens.

In Latin America it is also known under the name of "mata raton" because it contains a toxic substance used for baiting rodents. In spite of that, its leaves can be used as fodder for ruminants. It is, however, more suitable when used for alley-cropping or fencing (Pict. 7.25).

7.4.3.3.5 Ingas

These are chiefly used in Latin America. There are a number of varieties of *Inga* which differ widely. The most commonly used are *I. edulis* (Colombia, the West Indies and the Far East), *I. sparita, I. spectabilis* (Central America) and *I. deniflora*.

When choosing a variety of *Inga* it is advisable to examine its specific behavior in a given area as it is receptive to attacks of pests and diseases which can also attack coffee trees. Ingas withstand intensive lopping or thinning and resist defoliation (Pict. 7.26).

7.4.3.3.6 *Erythrina* spp.

These are chiefly used in Latin America, but also in Asia, for shading both coffee and cocoa. A number of different varieties exist but the most commonly used are *E. poeppigiana* (Central America) and *E. ombrosa*. In Indonesia, the seedless *E. lithosperma* (Dapdap) is often affected by parasites. *Erythrina* withstands intensive lopping but does not resist defoliation (Pict. 7.27).

Picture 7.27 Erythrina shade trees after severe trimming. From: J. Snoeck.

7.4.3.3.7 *Grevillea robusta* (Silver oak)

Though *Grevillea robusta* is not a leguminous plant, it is greatly appreciated as a shade tree in many countries from India and Indonesia to Eastern Africa and Central America for its deep rooting which offers no competition to coffee trees and its widespread loosening action on the soil. It should be planted at 8 m × 8 m (Pict. 7.28).

Picture 7.26 Inga (Chalum) shading Arabica in Mexico. From: J. N. Wintgens.

Picture 7.28 *Grevillea robusta* (Silver oak) shading Arabica in Mexico. From: J. N. Wintgens.

7.4.3.3.8 Minor Tree Species used for Shade

- *Pithecolobium saman* (Rain tree). This was tried out in tropical America, but was not a success because of the size of its canopy and the large surface occupied by its trunk.
- *Croton mubango*. An appreciated shade tree in Central Africa where it is widely used.
- *Cordia alliadora*. This shade tree can be successfully felled as commercial timber.
- *Cassia siamea*. This shade tree is not recommended because its canopy is too dense.
- *Parkia roxburghii*. The use of this shade tree is limited.
- Albizias. A considerable number of varieties exist but the most interesting ones are: *A. lebbek* (Madagascar and Central America), *A. malacocarpa* (Cameroon, Co-lombia, San Salvador), *A. procera* (Africa), *A. stipulata* (Malaysia, Madagascar, Uganda, Congo and Zimbabwe), *A. moluccana* (Indonesia, Cameroon), this variety has reputedly fragile branches, and *A. falcataria*, a very fast-grower reaching a height of 7 m in under a year. As a general rule this species prefers rainy zones (above 1500 mm/year) and lower reaches (below 1500 m). Albizia is difficult to trim as its branches are fragile. Its root system is also superficial, but it is excellent for renewing forest growth and generates considerable green fertilizing materials (Pict. 7.29, 7.30).
- *Sesbania sesban*. Generally discarded as a permanent shade tree because of its relatively short life cycle (7–10 years).
- *Casuarina oligodon* (Yar) is used as a shade tree in Papua New Guinea and is planted at 7 m × 6 m.
- *Casuarina junghuhniana* (Jemara) is used as a shade tree in Northern Thailand.
- *Alnus* spp. are adapted to high and humid areas. Although they are not leguminous trees, they are able to fix atmospheric nitrogen.

Picture 7.29 *Albizia falcataria* shading Arabica in Malaysia. From: J. N. Wintgens.

Picture 7.30 *Albizia stipulata* shade. From: Ch. Lambot.

7.4.3.4 Temporary Shade

Under unfavorable conditions like high winds or excessive sunlight, it may be necessary to protect young coffee trees in the early stages of their development with temporary shade trees. One way of doing this is to maintain the original low forest strata and gradually remove it as the coffee trees grow; an alternative solution is to plant fast-growing plants between the coffee rows. These plants will also be removed as the coffee trees and the permanent shade trees develop.

The plants most widely used for temporary shade are bushy leguminous plants like: *Tephrosia candida, Cajanus cajan, Crotal-*

Picture 7.31 Hedge of hybrid Sorghum providing temporary shade and wind break to a young Arabica plantation in Hawaii. From: J. N. Wintgens.

Picture 7.32 Coffee under its native forest shade. From: A. Charrier.

laria, Flemingia congesta, Zebrina pendula or even a sterile Sorghum hybrid (Pict. 7.31).

If it has been decided to install temporary shade, this should be done prior to the plantation of coffee as it will ensure early soil protection and shade to the young coffee plants.

When planting shade trees, spacing will differ from 8 m × 8 m, to 6 m × 10 m or 12 × 12 m according to their development capacity and the density of the shade required. These distances also have to be adapted to the spacing of the coffee trees. Slower-developing shade trees should be initially planted at high density and then weeded out as they develop so as to ensure constant, but not too dense, shade. Shade trees can also be planted within the coffee rows themselves so as to avoid hampering transit along the inter-rows.

7.4.3.5 **Other Types of Shade Trees**

In days gone by, planters in India maintained a number of original forest trees like *Acrocarpus, Adenanthera, Dalbergia, Ficus* or *Artocarpus heterophyllus*, to mention a few, as shade trees for their coffee plantation.

While this technique most certainly represented solid soil preservation qualities,

it did not necessarily generate high-yield production. On the other hand, productive coffee trees of over 100 years old can still be found in India, maintained in this type of plantation (Pict. 7.32).

Modern coffee cultivation practices, however, do not advise such techniques as original forest trees do not ensure homogenous shading, and could well compete with the coffee trees for nutrients and moisture. The same advice also applies to the use of fruit trees or decorative trees for shading. Citrus, mangoes, guayaba, pommerose, aguacate, breadfruit, ficus, banana trees or pines, for instance, frequently offer inadequate shade (either too dense or too sparse) and, in most cases, compete severely with the coffee trees. The same applies to the rubber tree *Hevea brasiliensis*, which has never been a successful shade tree for coffee (Pict. 7.33, 7.34, 7.35).

Picture 7.33 Coffee under banana shade in Mexico. From: J. N. Wintgens.

Picture 7.34 *Hevea brasiliensis* (rubber tree) shading young Robusta in Thailand. The shade does not yet affect the coffee trees. From: J. N. Wintgens.

Picture 7.35 Robusta coffee suffering under dense *Hevea* shade in Thailand. From: J. N. Wintgens.

7.4.4
Wind Breaks in Coffee

7.4.4.1 The Need for Wind Breaks

Exposure to constant wind can lead to significant damage to the coffee tree, particularly if combined with low air humidity. Such conditions cause excessive evapotranspiration which induces drying out and may cause extensive leaf scorch which ultimately results in stunted growth. Also, violent gusts of wind can break branches and stems or even uproot coffee trees with detrimental effects on yields. Where damaging wind conditions exist it is highly advisable to set up wind breaks especially in exposed areas.

The benefits of wind breaks may be defined as follows:

- Evapotranspiration losses can be reduced by up to 33 % in coffee which is sheltered from a prevailing constant wind of low humidity.
- Large fluctuations in temperature are buffered by wind breaks, which is particularly important in areas subject to the movements of cold air from higher latitudes.
- Behind wind breaks there is a lower rate of air mixing and the relative humidity of the air within the coffee trees is maintained at a slightly higher level than would be the case without a wind break.

- During drier years or when high winds prevail, the protection afforded by wind breaks avoids leaf scorch and physical damage to the leaves, and can improve yields.

The disadvantages of wind breaks include the following:

- The coffee rows directly adjacent to wind breaks may suffer from competition for light, moisture and nutrients, but any production loss in these rows is generally compensated for by higher yields in the unaffected but protected rows.
- The area given over to the development of wind breaks is lost for coffee production, but some compensatory income may be generated by the wind break trees, e.g. they may be ultimately harvested for commercial timber or they may produce fruit or fodder or mulch, all of which has some value.
- Wind-breaks appear to have a negative effect on the air temperature within the coffee plantation as the difference between night and day temperatures can be increased by as much as 3 % – this is caused by the fact that wind breaks effectively reduce evapotranspiration during the day and, as a result, the daytime air temperature is higher in the protected areas.
- High and dense wind breaks may become an inconvenient obstacle should aerial spraying be used in the plantation.

7.4.4.2 The Design of Wind Breaks

Wind-breaks should be planted across the slope and against the path of prevailing winds. The density of the wind break plantation increases in direct ratio to the gradient of the slope. Wind-breaks can be planted either parallel or perpendicular to the coffee rows. In cases where damaging prevailing winds may blow from different directions during the year it is advisable to plant perpendicular secondary wind breaks which end up by forming a protective square around the exposed area of the coffee plantation. It is generally recognized that wind breaks protect a distance which corresponds to 7 times their height on the leeward side and twice their height on the windward side.

To be effective, wind break trees on a coffee plantation should grow higher than shade trees. One or two rows of high trees and one row of shorter trees should be planted 2–3 years prior to the installation of the coffee plantation. The density of the wind breaks requires careful consideration. Where density is sparse, too much air filters through and efficiency is reduced; on the other hand, where the wind break is too dense, the air stream bounces off and flows over the top, generating damaging turbulence on the leeward side. Therefore, the density of the wind breaks must be calculated so as to allow air to filter through softly and create an air cushion on the leeward side. This air cushion will protect the coffee trees from the damaging effects of strong winds.

It is better to plant wider hedges in several thin rows rather than one tightly planted, continuous row of a single type of tree. The trees selected for wind breaks should have a hardy root system and offer strong resistance to wind.

The distance between the wind breaks should be 7 times the height of the fully grown trees. However, as wind break trees can take 5–10 years to reach full development, alternative intermediary solutions should be implemented. To accomplish this, one can either plant wind breaks at mid-distance and remove every other row as they reach full height or plant temporary wind breaks between the coffee rows.

On slopes, the base of the wind break trees should be stripped of branches up to a level of approximately 3 m to allow for air drainage in case of frost. Where overhead irrigation is practiced, the position of the sprinklers should be taken into account prior to installing wind breaks.

7.4.4.3 Plant Species for use as Wind Breaks

Plants which are recommended as temporary wind breaks include Giant Sorghum, *Flemingia congesta*, *Sesbania* sp., Pigeon pea, *Crotalaria* or *Tephrosia*. All these species are easy to establish and are fast growing biennials or perennials.

The most appropriate species for wind breaks are *Grevillea robusta* (Silver oak), *Acrocarpus* spp. *Eugenia jambos*, *Croton glabelus* and *Croton reflexifolius*, *Hakea saligna*, *Juniperus procera*, *Podocarpus milanjianus*, *Casuarina equisetifolia*, *Artocarpus integrifolia* (Jackfuit tree), *Terminalia* spp., *Cupressus betami* and *C. brachitanga* spp. These trees should be planted at 3–4 m intervals along the row and at a minimum distance of 6 m from the coffee trees.

Highly competitive trees like *Eucalyptus* spp., *Pinus* spp. and *Bambusa* spp. can only be used outside the coffee plantation. Shrubs like *Leucaena*, *Calliandra* and *Eleocarpus* should be planted at 1 m intervals along the row (Fig. 7.25).

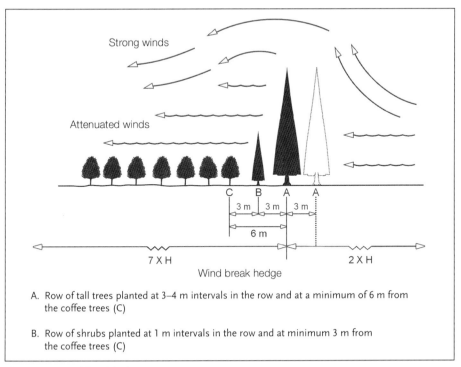

A. Row of tall trees planted at 3–4 m intervals in the row and at a minimum of 6 m from the coffee trees (C)

B. Row of shrubs planted at 1 m intervals in the row and at minimum 3 m from the coffee trees (C)

Figure 7.25 Wind break hedge.

7.4.5
Intercropping in Coffee

7.4.5.1 **Temporary Intercropping or Catch Cropping**

Intercropping during the 2–3 years that precede the first coffee harvest is an efficient means of increasing land productivity, optimizing the use of labor and improving the farmer's income. Furthermore, the care directed to the intercrops (fertilization, weeding, etc.) will ultimately benefit the coffee crop as well. This practice, however, requires a certain number of precautions to avoid damaging the coffee plantation itself.

- To avoid erosion, intercropping should not be carried out on slopes steeper than 15 % unless the entire area is completely mulched.
- Soils with low fertility and/or a poor water retention capacity should not be intercropped.
- Crops which could host coffee parasites should not be used.
- To avoid undue lateral shading of the young coffee plants, crops with a high, profuse aerial development should not be used.

- Sufficient spacing must be allowed between the coffee plants and the intercrop.
- Weeds should be severely kept under control.
- Soil nutrients extracted by the intercrop should be compensated by adequate fertilization particularly in acidic soils (pH < 4.5) where a support of dolomite lime or calcium ammonium nitrate at a rate of 300 kg/ha should be provided.

A number of intercrops have been tried out, but many have been discarded because of their adverse effect on the growth of the coffee trees. As a general rule, cassava, maize and potatoes should not be used because of their high nutrient removal. On the other hand, beans, soya, cowpea, peanuts, yam and mountain rice can be used as they offer relatively less competition to the coffee plants.

Intercropping in adult coffee plantations is not recommended but it is possible during the 2 years which follow the stumping of old coffee trees. In Kenya, experiments along these lines have shown that it is possible to grow four successive harvests of beans planted between the coffee rows without harming the subsequent coffee yield. See Tab. 7.3 (Fig. 7.26, Pict. 7.36).

Table 7.3 Density and Spacing of the intercrop

Crop	Distance from the coffee trees (m)		Spacing (cm)	
	Year 1	*Year 2*	*Between rows*	*Within the row*
Dwarf bean	0.5	0.8	25–30	20
Runner bean	1.0	1.5	40–50	15
Soya	0.6	1.0	30	15
Peanut	0.6	1.0	25–30	15
Wild rice	0.9	1.2	20	10
Potato	1.1	1.5	80	40
Yam	1.0		100	100

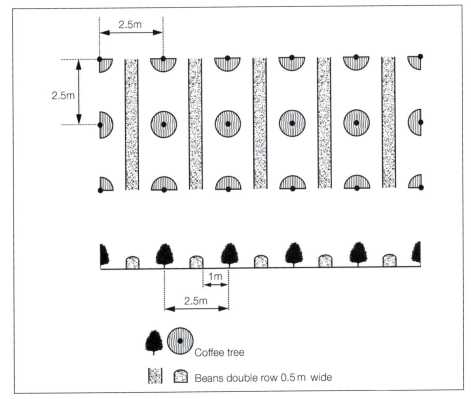

Figure 7.26 Intercropping with beans.

Picture 7.36 Intercropping beans in a young, mulched coffee plantation.

7.4.5.2 Permanent Intercropping

Permanent intercropping is practiced in certain regions chiefly by smallholders attempting to make the best possible use of their land and their labor resources. It affords a diversification of production and thus distributes the risks over several crops while eliminating the disadvantages of revenue dependence on a single crop. The choice of intercrop is made according to on-site conditions, the size of the farm, water availability and market outlets.

In Malaysia and Sri Lanka, the tendency to associate coconut palms and coffee trees on plantations is rapidly gaining in popularity as it appears to be beneficial to both plants because palm trees break down immobilized phosphorus in the soil which makes it available to the coffee plants. The planting distance of coconut trees is generally 8–9 m. The foliage of the coconut palms intercepts approximately 40% of the sunlight and the annual coffee yield averages at 450–600 kg/ha. In this case, the coconut production is also higher than in coconut monoculture (Pict. 7.37).

In India, intercropping coffee trees with black pepper (*Piper nigrum*) can be highly profitable. Shade trees are used to support the pepper and, technically speaking, it appears to be an excellent association (Pict. 7.38).

It should be pointed out, however, that these beneficial types of association are the exception rather than the rule. In the majority of cases, intercropping with coffee trees is not the best way to maximize yields as the different crops often compete for soil nutrients, water and light. This generally results in lower yields for each crop when compared with monoculture practices.

Nevertheless, permanent intercropping is a means of securing a more regular in-come from the plot, growing produce for domestic needs and ensuring year-round work for the labor force. More importantly, it also reduces the paralyzing risks of heavy financial losses caused by the undue fluctuations in coffee prices (Pict. 7.39, 7.40, 7.41, 7.42). This practice is now quite widespread in organic coffee cultivation.

Picture 7.39 Intercropping Betel nut palm (*Arec/catachu*) with Robusta in Thailand.

Picture 7.37 Intercropping coconut palms with coffee Robusta in Malaysia. From: J. N. Wintgens.

Picture 7.38 Intercropping black pepper (*Piper nigrum*) with coffee in India. The shade trees are used to support the climbing pepper plant. From: J. N. Wintgens.

Picture 7.40 Intercropping Durian (*Durio zibetinus*) with Robusta in Thailand. From: J. N. Wintgens.

Picture 7.41 Intercropping banana and *Parkia* sp. in Thailand. From: J. N. Wintgens.

Picture 7.43 Strip cropping banana with coffee. From: Nestlé.

Picture 7.42 Intercropping Cardamon (*Amomum* spp.) with Arabica in India. From: Nestlé.

Picture 7.44 Strip cropping pineapple with coffee. From: J. N. Wintgens.

7.4.5.2.1 Strip Cropping

This practice consists of growing the intercrop in strips which run parallel to the coffee rows. The strips can either be planted with fruit trees like banana, avocado, citrus, papaya, pineapple, mangosteen, rambotan, durian, cloves, etc., or with other trees. The use of other perennial trees like oil palm and hevea should be avoided because they compete too severely with the coffee trees for soil, nutrients, water and light (Pict. 7.43, 7.44, Fig. 7.27).

7.4.5.2.2 Agroforestry

In this case the intercrop consists of forest trees planted in between the coffee trees and in alternate rows to the shade trees. Forest trees are considered to be more suitable than fruit trees for intercropping as they require less maintenance, and offer less competition for soil, water and nutrients (Fig. 7.28).

7.4.5.2.3 Protection against Cold

Cold air, which is heavier than hot air, flows from higher to lower levels. Consequently, it tends to accumulate in depressions, dips or hollows. Topographical zones like frost pockets, valley bottoms or down slopes, which lend themselves to the accumulation of cold air, should not be planted with coffee. It is also important to plan corridors for channeling cold air through the plantation down to cleared zones at lower levels. For more precise information about local frost exposed sites, official meteorological services and neighboring farmers should be consulted.

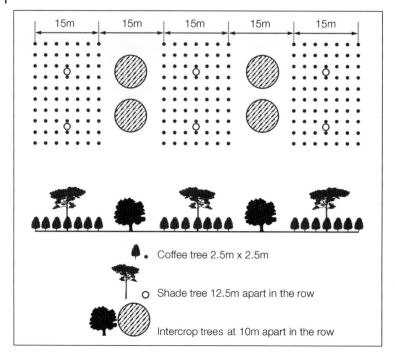

Coffee tree 2.5m x 2.5m

Shade tree 12.5m apart in the row

Intercrop trees at 10m apart in the row

Figure 7.27
Strip cropping.

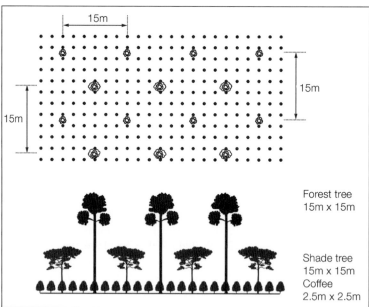

Forest tree
15m x 15m

Shade tree
15m x 15m
Coffee
2.5m x 2.5m

Figure 7.28
Agroforestry.

7.5
Planting Practices

7.5.1
Density, Layout and Spacing

7.5.1.1 General Considerations
The notion of density in a plantation is identified by the number of holes per unit of surface and not by the number of coffee trees planted.

The density in a plantation depends on the species, Robusta or Arabica, the growth habits of the plant, normal or dwarf, the cultivar, the fertility of the soil, the level of fertilization and the cultivation methods used. The more abundant the development of the plant and the more fertile the soil, the lower the density required in the plantation. An overly high density will reduce exposure to light and lower fruit yields. Density also depends on the pruning system. Topping the trees encourages a greater elongation of lateral branches than free growth with periodic stumping. As a result, topping requires increased spacing between the trees. This initial choice of farming practice, combined with other choices like manual or motorized farming, monoculture or intercropping, irrigation, anti-erosive measures, etc., will ultimately determine the spacing between the rows.

Density also depends on the gradient of the slope – spacing should be greater on steep slopes to ensure adequate exposure to light for the lower parts of the coffee trees. Where the slope is moderately steep (above 10 %), it is advisable to plant the coffee trees in contour.

Finally, in zones which are regularly threatened by attacks of parasites, the inter-rows should be wide enough to allow for the passage of fumigation machines. Where harvesting is carried out with self-propelled harvesters, the inter-rows should be about 4 m wide.

It should be noted that the plantation densities discussed above are purely indicative as each site on each plantation is different and only local trials will provide reliable information as to the relevant density and spacing required.

In conclusion, ideal plantation density should be calculated to generate long-term, optimal yields without causing undue soil depletion. It should also afford easy access to the coffee plants for adequate maintenance and harvesting. The final decision is not easy to make, all the more so as the end results will only come to light a few years later.

7.5.1.2 Robusta
Robusta coffee trees are vigorous so only one tree is necessary per hole. The density of the plantation can range from 1250 to 2220 coffee trees/ha. In full exposure to sunlight, the optimal density is approximately 2000 plants/ha (either seedlings or clones). Under shade, the density will be lower and will vary according to the density of the shade. The denser the shade, the lower the density of the coffee trees (1250–1660 trees/ha). See Tab. 7.4.

The number of stems varies from 5000 to 7500/ha according to the density.

Whilst the cultivation of coffee at four stems per tree is current for densities of less than 1700 plants/ha, it is advisable to work with three stems per tree for densities of more than 1700 plants/ha.

7.5.1.3 Arabica with Normal Growth Habit
Recent studies have shown that optimal density in high-altitude equatorial zones is between 2500 and 3300 trees/ha. In lower areas or those closer to the tropics, the den-

Table 7.4   Density, layout and spacing for Robusta

Farming practices	Layout	Good to very good soil fertility or fertilized coffee trees		Medium to poor soil fertility or non-fertilized coffee trees	
		Spacing (m)	Trees (/ha)	Spacing (m)	Trees (/ha)
Manual maintenance	square	2.8 × 2.8	1275	2.4 × 2.4	1735
	rectangle	3.0 × 2.0	1666	3.0 × 1.5	2222
	triangle	3.0 × 3.0	1280	2.6 × 2.6	1710
Mechanized maintenance	rectangle	4.0 × 2.0	1250	3.0 × 1.8	1850

Table 7.5   Density, layout and spacing for Arabica

Farming practices	Altitude	Layout	Good to very good soil fertility or fertilized coffee trees		Medium to poor soil fertility or non-fertilized coffee trees	
			Spacing (m)	Trees (/ha)	Spacing (m)	Trees (/ha)
Manual maintenance	high	rectangle	2.5 × 1.5	2660	2.5 × 1.2	3330
		square	2.0 × 2.0	2500	1.75 × 1.75	3265
	low	rectangle	3.0 × 2.5	1330	3.0 × 2.0	1666
		square	2.75 × 2.75	1320	2.5 × 2.5	1600
Mechanized maintenance		rectangle	3.0 × 2.5	1330	2.75 × 1.25	2430
			2.74 × 2.74	1330	2.75 × 1.5	2430

sity generally recommended varies between 1100 and 1600 trees/ha. See Tab. 7.5.

Where mechanized maintenance is practiced, experimental strip planting in several more closely planted rows, separated by wide inter-rows, has been attempted. This has proven unsatisfactory as productivity in triple rows turned out to be less than productivity in double rows and less again than productivity in single equidistant rows. Reducing space between the coffee trees leads to increased competition and, to a certain extent, can also foster the spread of diseases like Coffee Berry Disease or Coffee Leaf Rust.

According to the density decided for the plantation, the choice for growing the coffee trees on two, three or four stems can be made on the basis of the following:

Trees/ha	Number of stems	Total Stems/ha
1000–1300	4	4000–5200
1400–2000	3	4200–6000
2100–3300	2	4200–6600

7.5.1.4   Dwarf Arabica

In several countries, the best productivity results have been obtained with densities of 10 000 to 12 500 coffee trees/ha spaced at 1 m × 1 m and 2 m × 0.40 m, especially in the first planting cycle.

Nevertheless to secure sustainable, long-term production and to allow access to the plots for maintenance purposes, producers prefer to plant 4000–7000 trees/ha. This corresponds to a spacing between 2 m × 1.2 m and 1.7 m × 0.8 m.

Dwarf Arabica is usually grown on a single stem.

In Costa Rica, where high-density farming is perfectly mastered, producing the world's highest yields, farmers have adopted a spacing of 1.6 m × 0.8 m with two plants per hole. Due to the rapid exhaustion of the plant, pruning is carried out by cyclical stumping back ("rock'n roll").

Highly intensive methods of this type are costly and rapidly deplete the nutrients in the soil which, in turn, entails intensive fertilization (2–5 tons/ha) and, finally, accelerates acidification. They are therefore incompatible with sustainable agricultural philosophies and practices.

7.5.1.5 Other Layouts

"Cova planting" consists of planting several plants per hole. This practice is popular in Brazil where two to three plants per hole are the general rule. Similar experiments have been relatively successful in Zimbabwe, but met with very little success in Kenya, and, as mentioned above, experiments with Robusta have not been successful at all.

"Double- or twin-row planting" consists of planting two relatively close rows of coffee with a wide inter-row between them. This facilitates mechanized maintenance work and allows for a larger mulch covering in the wider inter-rows. A typical example of spacing twin rows could be:

2 m within the rows

2.5 m for narrow inter-row

3.5 m for the wide inter-row.

This would give a planting density of 1666 plants/ha.

"Hedge planting" consists of planting trees much closer within the rows and increasing the spacing between the rows. A typical example of hedge planting would be:

2.25 m within the rows

3.7 m for a wide inter-row.

This would give a planting density of 1200 plants/ha.

7.5.1.6 Calculation of Planting Density

To calculate the planting density, a hectare, or 10 000 m², must be divided by the area needed for one coffee plant. Consequently, where the spacing of the coffee plant re-

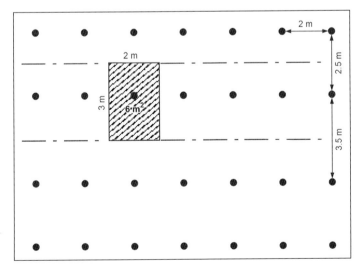

Figure 7.29
Planting in twin rows.

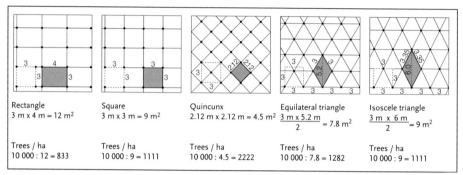

Rectangle	Square	Quincunx	Equilateral triangle	Isoscele triangle
3 m x 4 m = 12 m²	3 m x 3 m = 9 m²	2.12 m x 2.12 m = 4.5 m²	$\frac{3 \text{ m} \times 5.2 \text{ m}}{2} = 7.8 \text{ m}^2$	$\frac{3 \text{ m} \times 6 \text{ m}}{2} = 9 \text{ m}^2$
Trees / ha	Trees / ha	Trees / ha	Trees / ha	Trees / ha
10 000 : 12 = 833	10 000 : 9 = 1111	10 000 : 4.5 = 2222	10 000 : 7.8 = 1282	10 000 : 9 = 1111

Figure 7.30 Planting layout.

quires 3 m × 3 m, planting density is 1111 plants per hectare or 10 000/9.

The same result can be reached by multiplying the number of rows in 100 m by the number of coffee trees in a 100 m row; hence, where the spacing of the coffee plant requires 3 m × 3 m, planting density is 1111 plants/ha or 33 333 × 33.333.

When it comes to twin planting, the same method of calculation applies, but it is based on the axis of the inter-rows (Fig. 7.29).

Geometrical layouts can be set up in squares or rectangles, but these layouts are being progressively abandoned in favor of triangle and quincunx arrangements which offer improved territorial occupation and better protection against erosion (Fig. 7.30).

7.5.2
Staking, Holing, Refilling, Planting

7.5.2.1 Staking Out Planting Sites

The best way to ensure regular staking is to begin the line up from the axis of the roadways as shown on the following figure. To obtain straight coffee rows, surveyor's markers should be used for staking out. The site for each hole for planting the coffee trees should then be staked out with pegs of 50 cm long (Fig. 7.31).

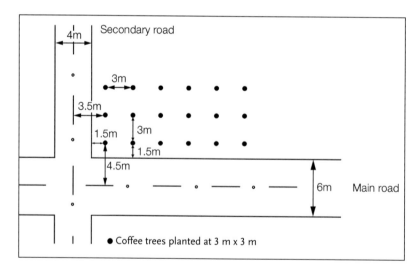

Figure 7.31 Staking out planting sites.

7.5.2.2 Holing and Refilling

Once the planting sites are staked out, the holes can be dug. Ideally, the holes should be opened a few weeks prior to planting. The young plants need to be installed as soon as possible after the first rains have well started. On the other hand, if the soil is too wet, the holes should be neither dug nor filled to avoid the formation of puddles and muddy depressions. Where the earth is compact, the soil should be loosened and aired by subsoiling to a depth of 70 cm along the future planting rows before digging the holes.

Prior to holing, a ring of approximately 1 m in diameter should be cleared around the peg. To foster initial growth, the topsoil of the cleared ring should be used for infilling the hole.

The size of the hole will condition the root development and the growth of the young plant. In very favorable farming conditions, holes of 30 cm diameter and 30 cm in depth are sufficient, but where the earth is heavy or contains coarse elements, larger holes should be dug, i.e. 40 or 60 cm diameter.

Before refilling, the surveyor in charge of the operation should use a gauging rod to check the size of the holes (Fig. 7.32). Refilling should take place with the least possible delay after holing in order to prevent humus loss and compacting caused by rain or sun-baking. If the holes have been drilled with a mechanized drill, the edges should be broken up prior to refilling. Industrialized plantations generally use portable mechanical drills or drills set up on a tractor which can drill holes of 30–70 cm in diameter to a depth of 50–80 cm.

The soil used for infilling should be compacted by trampling each layer to avoid the formation of air pockets which can be detrimental to the development of the root system of the coffee plant. Similarly, un-

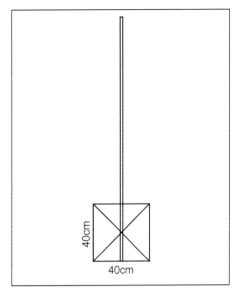

Figure 7.32 Gauging rod.

wanted matter like stones, scraps of plastic, glass or tinfoil, etc., should be removed from the hole. Once the hole has been filled, it should be capped with a small mound of topsoil to allow for further settling caused by rainfall. The pegs used for staking out are then planted in the centre of the hole to mark the planting site.

It is advisable to wait at least 2 weeks before planting so that the rain can suitably dampen the site and complete the settling of the earth.

7.5.2.3 Fertilization at Planting

In soils with low fertility, planting without prior fertilization can jeopardize the initial development of the plant. Soil analysis prior to planting will provide the guidelines for suitable fertilization. The action of the fertilizing material is more beneficial when it is spread at the bottom of the hole. Mixing fertilizing material with the refill soil has given poor results. For sandy soils or soils with poor CEC, organic fertilization composed

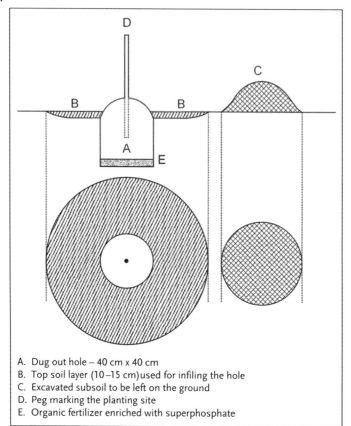

A. Dug out hole – 40 cm x 40 cm
B. Top soil layer (10–15 cm)used for infiling the hole
C. Excavated subsoil to be left on the ground
D. Peg marking the planting site
E. Organic fertilizer enriched with superphosphate

Figure 7.33
Dug out planting hole.

of manure, compost or well-decayed coffee pulp, enriched with 200 g of superphosphate at a rate of 10 l per hole is recommended. Single superphosphate is recommended for soils with a pH above 5.0 and treble superphosphate for soils with a pH under 5.0. This re-establishes the balance of nutrients in the soil and improves its water retention capacity (Fig. 7.33).

7.5.2.4 Planting Out

Planting is a very important operation which will not only influence the future development of the coffee tree but also its life-long performance and the mortality rate after planting.

The best period for planting is at the beginning of the rainy season when the rains have set in and the soil is humid down to a level of 50 cm below the surface. The best day for planting is a damp, cloudy day. Before planting, the coffee plants should be gradually adapted to direct sunlight by progressive removal of the shade of the nursery. Planting material should be sorted before planting and only vigorous and well developed plants retained. Seedlings should bear six to ten pairs of leaves and cuttings two to three pairs of primary branches (crosses). Leaf trimming is not advisable nor is the practice of spraying with a sugary solution or a sticker as practices of this type only delay the recovery of the plant.

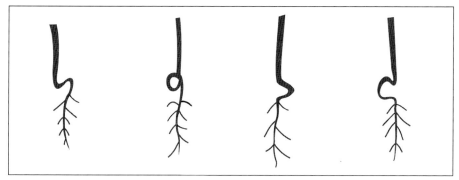

Figure 7.34 Twisted tap roots.

The age of the plant when it is installed largely depends on the variety and prevailing agroclimatic conditions. Where polybags are used, which is very frequently the case, the period spent in the nursery would be as follows:

Robusta 6–9 months
Arabica 12 months

At this point, the plants are generally about 40–60 cm high and have 8–12 pairs of leaves and sometimes one or even several pairs of primary branches.

The most common malformations to be found in nursery plants are twisted taproots. Seedlings which present this malformation should be discarded as they will never develop properly and will always be poor yielders (Fig. 7.34).

Approximately 1 month after planting out, young plants which have not recovered correctly should be removed and replaced. Average plant loss after planting out amounts to approximately 5 %.

Slanted planting at 30° reduces root development and can lead to slow degeneration or even cause the coffee tree to keel over in periods of high production or strong winds.

7.5.2.5 Planting Materials and Planting Methods

A wide variety of planting materials and planting methods are used to establish a coffee plantation – each has been devised for specific situations. A brief description of the combinations of planting materials and methods is presented.

7.5.2.5.1 Direct Seeding

Seed at stake consists of sowing several coffee seeds on the surface of the plant hole. Once the seedlings have developed, the four to six most vigorous plantlets should be maintained and the remainder weeded out. This method is chiefly used by planters who practice "Cova planting" in Brazil and in Zimbabwe. It can also be carried out in polybags or plastic sleeves.

Direct drilling is more labor-intensive and consequently more costly, added to which it does not allow for examination of the tap roots prior to final installation on the plant site.

7.5.2.5.2 Bare Root Plants

This system can be used for plants which have not yet developed primary branches. The uprooted plantlets are freed of soil and examined. After trimming, the roots are tied into bunches and transplanted to the planting site. At this stage the average length ratio between the aerial part of the plant and its roots is about 1.8–2.5.

7.5.2.5.3   Bare Root Stumps

This system is used for transplanting plants that have developed in the nursery for more than 18 months. They are stumped to a height of 25 cm at 2 weeks prior to planting, their lateral roots and their tap root are trimmed just before transplanting.

In both bare root planting systems, the roots of the plantlets are coated with clayed mud and wrapped in leaves, grasses or pieces of hessian before their transport to the field for planting.

Bare root planting is a less costly system, but plant recovery can be precarious. As a result, the stumps need little hats made of straw, branches or palm leaves to protect them from sunburn and withering. As the plant stumps are more developed, they are

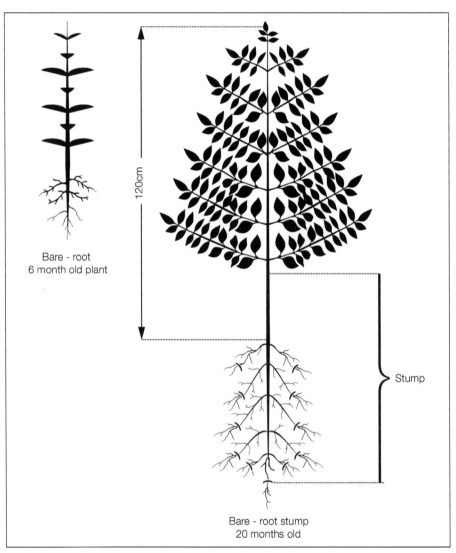

Bare - root
6 month old plant

120cm

Stump

Bare - root stump
20 months old

Figure 7.35   Bare root plants.

preferred for replacing missing plants in adult plots (Fig. 7.35).

7.5.2.5.4 Earth Blocks

Pots and woven baskets

Little pots made of bamboo sections or baskets woven with locally grown materials are filled with soil. The coffee plantlets develop in these containers until they are transplanted. The major drawback is that these containers are limited in size so planting has to take place early. The system is, however, still practiced by small producers with limited budgets (Fig. 7.36).

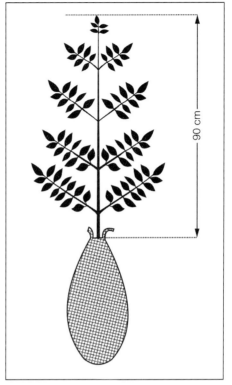

Figure 7.37 Wrapped earth block containing a 15-month-old plant.

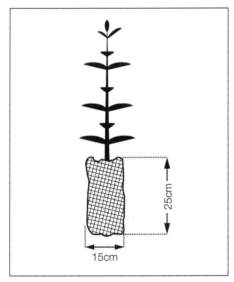

Figure 7.36 Woven basket.

Wrapped Earth Blocks

When the young coffee plants are removed from the nursery, the protective ball of earth around their roots is wrapped in broad leaves or a piece of hessian. The wrapping is removed immediately prior to planting. This system is mostly used for transplanting coffee trees that have been kept in the nursery for too long and whose size exceeds the normal size for transplanting (Fig. 7.37).

Java Planter

The young coffee plant is dug out of the soil by means of a 30-cm long metal cylinder known as a Java planter. The cylinder has a longitudinal slit which opens easily. One side of the cylinder has a sharp edge which is used to penetrate the earth and cut out the soil around the roots. When installing the plant on site the metal cylinder is opened and the plant slipped into the planting hole.

The various earth block systems mentioned above offer the advantage of protecting the roots when transplanting, but they also entail cumbersome and costly transport problems. As a result, they have mostly been discarded (Fig. 7.38).

Figure 7.38 Java planter.

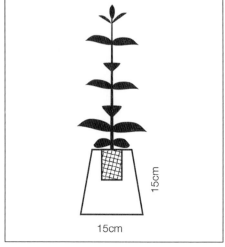

Figure 7.39 Molded block.

7.5.2.5.5 **Molded Blocks**

Small blocks of earth are molded with a mixture of clay and compost. A considerable number of different tools has been developed for this purpose ranging from a small portable tool for two molds to sophisticated mechanized systems. Given the small size of the molds, transplanting and installing the young coffee plants has to be carried out before the trees are 6 months old. In spite of this slight drawback, the system is quite popular as it is relatively cheap and easy to transport the plantlets (Fig. 7.39).

7.5.2.5.6 **Transplanting in Plastic Wrappers**

Plastic wrappers are usually made of 200 gage black diothene film. This category includes polybags and plastic sleeves.

Polybags

Polybags are very frequently used in arboriculture and their sizes vary from case to case. For coffee trees, flat polybags of

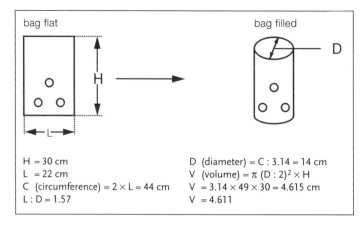

H = 30 cm
L = 22 cm
C (circumference) = 2 × L = 44 cm
L : D = 1.57

D (diameter) = C : 3.14 = 14 cm
V (volume) = π (D : 2)^2 × H
V = 3.14 × 49 × 30 = 4.615 cm
V = 4.611

Figure 7.40 Polybag.

30 cm × 22 cm are recommended. When they are unfolded and filled with earth, they are cylindrical and have a diameter of approximately 14 cm (Fig. 7.40).

Polybags are perforated with small holes to allow excess water to trickle out. When filled the 30 cm × 22 cm polybag weight approximately 5–6 kg according to the density of the mix used for filling

Planting procedures:

- From the center of the planting hole, extract a volume of soil which roughly corresponds to the volume of the filled polybag.
- Slice the base of the polybag at 2–3 cm from the bottom with a sharp machete – this operation will cut off the end of the tap root if it has grown to the bottom of the polybag and avoids planting the coffee tree with twisted, curled or bent tap roots (Pict. 7.45).
- Place the bottomless bag in the planting hole to check if it fits and that the collar of the plant appears at ground level.
- Slice the bag, remove it and insert the cylinder of earth into the planting hole.
- Fill the remaining space with topsoil.
- Compact the earth around the coffee plant by hand and feet to ensure good contact with the roots and avoid air pockets.
- Build a little mound of earth 5–8 cm high around the base of the coffee plant

Picture 7.45 Slicing the bottom of the polybag prior to planting. From: Ch. Lambot.

to allow for settling and to avoid water collecting at the stem.

If plants grown in polybags are not installed before they are 1 year old, the bottom of the bag should be opened so that the tap root can continue to grow without curling round inside the bag.

The use of polybags is very widespread at present and has superseded practically every other method of growing and transplanting coffee trees (Fig. 7.41).

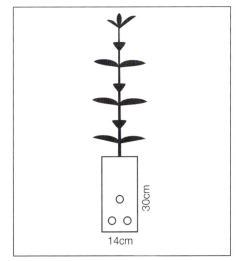

Figure 7.41 Polybag with coffee plant.

Sleeves

The sleeves are generally 25–30 cm long by 20.5 cm when flat. Once they are unfolded and filled, their diameter is 13 cm (20.5 : 1.57). They are filled by means of a cylinder which is inserted into the sleeve. Sleeves should be used in the same way as polybags except that they are not sealed at the bottom (Fig. 7.42).

A new type of biodegradable sleeve composed of vegetal material is also available now. These sleeves can either be filled by hand or cut in sections and filled mechanically.

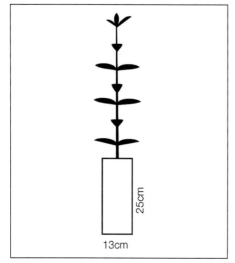

Figure 7.42 Sleeve.

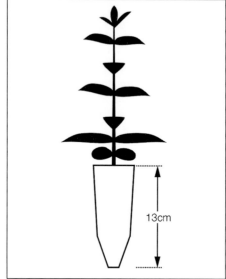

Figure 7.43 Planting cone.

7.5.2.5.7 **Other Devices**

Planting cones

This system is already successfully used in both horticulture and forestry, particularly for growing Eucalyptus and Pine trees. It consists of a small cone made of polypropylene which is 13 cm long by 5 cm in diameter and contains 150 cm^3 of growing substrate. The bottom of the cone is perforated. The cones are assembled on trays specially adapted to their size which hold 256 units/m^2. They are filled with a mixture of 60% earth and 40% organic material. The mixture is well sieved and previously treated with BASAMID G. Coffee plantlets are placed in the cones at the "soldier" stage. Special care should be taken to ensure that the tap root remains straight. Direct seeding is possible but not advisable. Field planting is carried out 4–5 months after transplanting the seedlings into the cones when the plantlet is 18–20 cm high and has grown four to six pairs of true leaves. This system is not labor intensive nor is it costly in nutrient input or transport. It also offers the added advantage of protection against infections, but the initial costs need to be taken into consideration (Fig. 7.43).

Fiber Pots and Paper Bags

Small pots or blocks made of fibrous vegetal material of the Jiffy pot type and reinforced paper bags are alternative solutions which are very similar to planting cones. They are managed in the same way and, like for the cones, the time the plantlets can remain in the nursery is limited to 4–5 months.

The installation of coffee plants in the field is generally carried out manually. Only in certain specific cases where labor costs are high, like in Australia or Hawaii or where frequent renewal of coffee plants in the rows is necessary, can plantation be carried out with planting machines. Mechanized planting is, however, limited to bare root, molded block, fiber pot and paper bag planting systems.

7.6
Mulching as a Post-planting Soil Management Tool

7.6.1
Functions of Mulching

Ground cover in the form of an organic mulching material is used widely to protect the soil against erosion, to reduce soil moisture losses by evaporation from the soil surface and to inhibit weed growth. Most organic or vegetative mulching materials are subject to decay, and the products of decay are slowly incorporated in the topsoil through the activities of the soil fauna and natural weathering processes, thus raising the organic matter levels contained in the topsoil. The increase in organic matter levels has a beneficial effect on the physical and chemical properties of the topsoil, improving soil structure and porosity, and raising fertility levels through an increase in plant nutrient availability; CEC is greatly enhanced by high humus levels. Mulch also increases water infiltration and soil moisture-holding capacity as well as regulating the soil temperature.

7.6.2
Mulching as an Erosion Control Measure

The mulching of coffee is an effective erosion control measure. The mulching material protects the soil so that raindrops do not impact on the soil surface and dislodge soil particles which can be transported down slope in run-off. Instead, the raindrops fall on to the mulching material and the water then filters through the mulch on to the soil surface. The run-off generated then moves down slope on a broad front with the mulching material obstructing the flow and restricting the rate of run-off so as to prevent the formation of rills and gulleys. The restriction in rate of run-off provides more time for the infiltration of the rainfall, which also serves to reduce soil losses due to erosion.

Mulching material should be laid out parallel to the contour lines and remain available to protect the soil during the entire rainy season. As a result, mulching material should be chosen from vegetation which decays slowly like stems of *Argrostis, Hyparrhenia, Papyrus, Pennisetum, Panicum, Tripsacum, Setaria,* Vetiver, *Leucaena,*

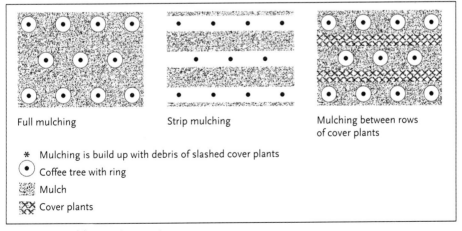

Full mulching   Strip mulching   Mulching between rows of cover plants

* Mulching is build up with debris of slashed cover plants
(•) Coffee tree with ring
▓ Mulch
▓ Cover plants

Figure 7.44 Mulching against erosion.

Flemingia, Calliandra, Tephrosia, Lemon grass, etc.

In order to avoid the propagation of mulching plants inside the coffee plots, they should not be harvested when they carry seeds.

Three methods of mulch spreading are commonly used: full mulching, strip mulching and mulching between rows of cover plants (Fig. 7.44). With the exception of strip mulching, where mulch is not spread in between the coffee trees along the rows, a ring with a radius of approximately 30 cm should be left free of mulch around the base of the coffee plant so as to avoid attacks by parasites.

7.6.3
Mulching as a Soil Moisture Conservation Measure

Where the dry season is very severe and long, the supply of stored soil moisture available to the coffee is a critical factor in determining potential productivity, and in some cases even for the survival of the coffee. Losses of stored soil moisture to the atmosphere occur either through evapotranspiration from the coffee trees or through direct evaporation from the soil surface. Losses of soil moisture through the coffee trees cannot be controlled, but losses from the soil surface can be very much reduced, if not eliminated, by mulching. The mulch not only limits the rate of evaporation, but also cools the surface soil. The amount of water saved in a coffee plantation by mulching during the dry season depends on the temperature and the amount of mulch applied. The layer of mulch spread should be approximately 10 cm thick at the outset to allow for packing.

In young plantations, 1–2 years old, where the root system of the coffee plants

Picture 7.46 Fresh grass mulch in a young Arabica plantation. From: GENAGRO – Belgium.

is little developed, mulch can be spread in strips of about 1.2 m wide (Pict. 7.46).

In adult coffee plantations, mulch is spread in strips of 1–1.5 m wide on or between the coffee rows. If there is not enough mulch material available, it can be spread every other row. To reduce the risk of attacks by parasites, the 30-cm radius ring around the coffee tree trunks should not be neglected (Fig. 7.45).

7.6.4
Mulching as a Weed Control Measure

Mulching controls weed growth by preventing the germination of seeds or by smothering germinated seedlings or young shoots emanating from adventitious roots, rhizomes, stolons, corms, etc. In

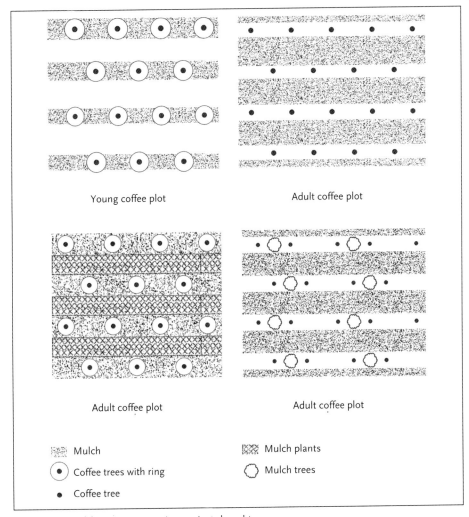

Figure 7.45 Mulching as a protection against drought.

order to achieve effective long-term weed control by mulching, the quantity of mulch required is considerable for not only must the vast majority of the ground surface be covered but a depth of mulch of at least 15 cm is required to prevent some of the more vigorous weeds reaching the daylight. Even if such quantities of mulch could be made available, it would be prohibitively expensive. Consequently the use of mulching to control weed growth will have to be combined with other methods of weed control.

7.6.5
Sources of Mulching Materials

Mulching material can be generated from several sources, which include the by-products of coffee (hulls, parchment, etc.)

and harvest residues such as corn and sorghum stover; bean and soya haulms also constitute excellent mulching material. The use of harvest residues, however, may not be advantageous since it represents a drain on the fertility of the land used for staple-food production and such materials should be ploughed back into these lands rather than transferring their fertility to the coffee plantation.

Picture 7.47 *Mucuna puriens*, a high producer of biomass, in Ecuador. From: J. N. Wintgens.

The production of mulching material (biomass) outside the coffee plantation requires additional land which is not always available. The average area of land required outside the coffee plantation for the production of the annual mulch requirement for 1 ha of coffee is approximately 1–1.5 ha for plots planted with mulching grass or 3–5 ha for natural regeneration on fallow plots, depending on the local conditions and the plant species involved.

Heavily fertilized Napier grass (*Pennisetum purpureum*) produces 50–70 tons/ha of dry grass per annum. This is enough to ensure complete mulch coverage for 1–2 ha of coffee or 2–4 ha of mulching on alternate rows, whereas a sterile hybrid of Napier known as "King Grass" ensures a greater production and does not present the drawback of propagation by seeding. In damp tropical climates, *Mucuna utilis*, *Mucuna pruriens* and *Tripsacum fasciculatum = T. laxum* (Guatemala grass) generate considerable quantities of biomass. The same applies to the *Crotalaria*, *Indigofera* and *Tephrosia* genera, *Flemingia congesta*, *Zebrina pendula*, *Crotalaria* spp., and *Cajanus cajan*. The local research centre will provide information as to the species of mulch plants best suited to local conditions (Pict. 7.47, 7.48).

It should also be noted that mulching material needs to be changed at regular intervals in order to avoid degrading the balance of nutrients in the soil. Mulching out-

Picture 7.48 *Tripsacum* spp. (Guatemala grass), a high producer of biomass. From: GENAGRO – Belgium.

side the plantation not only entails the need for extra cultivation areas, it also requires considerable extra labor for production and harvesting as well as added expenses for transport. The resulting costs often limit the amount of mulch applied annually which does not always ensure sufficient protection to the coffee plantation. Consequently, the present tendency is to grow mulching plants and produce mulch within the coffee plots and/or the plantation itself.

7.6.6
Multi-purpose Plants used for the Production of Mulch

The association of multi-purpose mulching trees with coffee trees on-site allows the production of low-cost mulch as well as providing protection against erosion, drought and weed proliferation. The multi-purpose species cultivated within the plantation must offer the following characteristics:

- Little or no competition to the coffee trees for water and nutrients
- Resistance to slashing
- Resistance to prolonged dry spells
- Easy installation and rapid growth
- Considerable production of green material (biomass).

Besides offering protection to both the ground and the coffee trees, mulch trees should also enrich the soil in organic material and nutrients.

Leguminous trees are the preferred multi-purpose mulching plants because they fix atmospheric nitrogen if the appropriate *Rhizobium* nodules are present; where not present, inoculation of seed or soil with *Rhizobium* spp. will be necessary in the initial phase of development. These leguminous trees produce mulching material with relatively high nitrogen levels.

Two different techniques can be used to manage leguminous mulch plants depending on whether they are herbaceous, bushy or arborescent. Herbaceous or bushy plants should be planted in double or triple rows spaced at a distance of 30–50 cm and at a minimum distance of 1 m from the coffee trees. Mulch plants are regularly cut back to 20–25 cm from the ground as soon as they reach 1.5–1.6 m in height and the material resulting from slashing is spread throughout the plot. In favorable conditions, mulch plants can produce 10–15 tons of green material/ha, which can be immediately used for covering the coffee rows and inter-rows.

Mulch plants which grow in the inter-rows of the coffee plantation also provide protection against cold or drying winds and regulate the soil temperature at certain times of the year.

Leguminous mulch plants require space to develop so the inter-rows should be wide enough to accommodate them. This, inevitably, reduces the density of the coffee trees and also means that mulch plants of this type are incompatible with the high density farming practices in Arabica plantations. To avoid this drawback, coffee can be planted in double or twin rows which would leave a wider space for mulching plants while reducing potential competition to the coffee trees (Fig. 7.46).

Wherever possible, however, mulch trees are preferred as they take up less ground area and form their foliage above the coffee trees. Mulch trees are conducted on a single stem, pruned at the time of planting and maintained at a height of about 3 m to prevent them from shading the coffee trees. They are severely thinned leaving only a few branch stumps. *Erythrina* trees withstand this drastic treatment without undue discomfort and are used successfully in an ever-increasing number of plantations in Central America. They also produce a considerable amount of plant matter which covers and enriches the soil.

Forest trees can also be planted in monoculture plots on the marginal soils of a plantation which are not suited to coffee trees. Teak (*Tectona grandis*), eucalyptus, acacia, pine and cypress trees are well adapted to poor soils. Forest trees are also useful when planted along road verges, as wind breaks or for marking the limits of the concession. All these trees may be

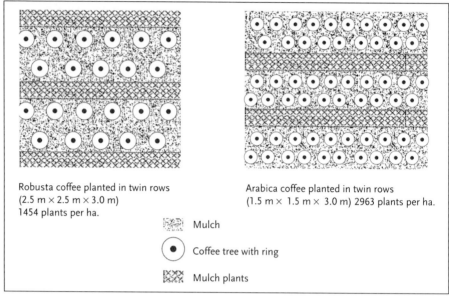

Robusta coffee planted in twin rows
(2.5 m × 2.5 m × 3.0 m)
1454 plants per ha.

Arabica coffee planted in twin rows
(1.5 m × 1.5 m × 3.0 m) 2963 plants per ha.

Mulch

Coffee tree with ring

Mulch plants

Figure 7.46   Twin row plantation.

regarded as dual purpose and a source of mulching material as well as being used for other purposes.

As a general rule, spacing between special purpose mulch trees is from 6 m × 6 m to 6 m × 10 m depending on the slope, the type of tree and the spacing of the coffee trees. Mulching material may be obtained from many species of trees on the coffee plantation which are multi-purpose for in addition to producing mulching materials these trees also per-

form functions such as anti-erosive hedges, cover plants, wind breaks, and shade trees.

Also, at the end of the life of the plantation these trees may be used for timber. Some of the multi-purpose forest tree species which are recommended are shown in Tab. 7.6.

The planting distances given in are those applied in monoculture plantations. Where individual trees are inset in a coffee plantation, planting distances will vary according to on-site conditions and intended use.

Table 7.6 Multi-purpose forest tree species

	Planting distance (m)	*Use(s)*
Low altitude (500–800 m a.s.l.)		
Balsam of Tolu (*Myroxylon balsamum*)	10 × 10	commercial, medicinal, cosmetic
Red Cedar, shingle tree (*Acrocarpus fraxinifolius*)	10 × 10	timber, construction
Mexican mahogany (*Swietenia humilis*)	10 × 10	timber, furniture
Casuarina or Bull oak (*Casuarina equisetifolia*)	10 × 10	firewood, roofing, posting
Laurel (*Cordia alliodora*) cannot be trimmed	10 × 10	timber, construction, medicinal
Albizia (*Albizia saman*)	10 × 10	construction
Tabebuia chryzanta	10 × 10	construction
Colubrina ferruginosa	6 × 6	carpentry, firewood
Medium altitude (800–1200 m a.s.l.)		
Mahogany (*Swietenia humilis*)	10 × 10	timber, furniture
Neem tree (*Azadirachta indica*)	6 × 6	construction, insecticide, firewood
Tabebuia rosea	10 × 10	construction
Cedro or Spanish cedar (*Cedrela odorata*) can be trimmed	10 × 10	timber, furniture and cigar boxes
Yellow olivier (*Terminalia obovata*)	10 × 10	timber, construction
Walnut tree (*Juglans olonchana*)	10 × 10	timber, fruit
High altitude (>1200 m a.s.l.)		
Cedrela (*Cedrela tendusi*)	10 × 10	timber
Silver oak (*Grevillea robusta*)	10 × 10	construction and firewood

a.s.l. = above sea level.

7.7

Conclusions and Prospects

Given the current unstable situation in the coffee trade, growers no longer have much lee-way for "trial and error". Mistakes are too costly and wrong choices made at the outset of the project generally lead to very serious financial problems. Considering that the average lifespan of a coffee plantation is approximately 30 years, it is of paramount importance to make the right decisions from the very beginning.

Choosing a suitable location to set up the coffee plantation is the first and most important decision that must be made. This entails an in-depth study of the selected site with emphasis on the quality of the soil and the immediate environmental conditions, both climatic and those impacted by human settlements. These are the basic factors which are essential to the development of a healthy, long-lasting plantation.

The next factor which also requires careful consideration and study is the choice of the planting material. Here, significant progress has been made thanks to the development of new technologies for coffee breeding and selection. The range of varieties available is vast and the choice of a variety must be based on knowledge of which material is best suited to local agro-environmental conditions. A variety which produces outstanding results in one area may be totally unsuitable to another because it cannot adapt to local conditions.

Cloned planting material offers a number of advantages, particularly for allogamous species like Robusta. It should be noted, however, that new breeds are often developed on the basis of their performance (yield, resistance to disease, etc.), as opposed to finer choices like quality and, more particularly, the cup-quality potential. This also means that the initial choice of a coffee variety should be made with respect to the final goal the planter wishes to achieve.

When deciding on the coffee planting material to install, one must take into account which cultivation practices have been chosen. Again, this choice needs to be made on the basis of the surrounding agro-environmental conditions and the desired end results.

Density in coffee planting, for instance, has already been researched, and it is now known that the ideal density of a plantation is determined by the variety and the type of location. This is a factor that will greatly influence the performance of the future plantation.

Similarly, the considerable advantages of combining shade plants and cover plants are also well known nowadays. Combined cultivation practices of this type are more "environmentally friendly". They also ensure a better protection of the soil, reduce the costs of fertilization and have a beneficial effect on coffee quality.

The points discussed in this chapter are basic, broad guidelines to indicate the importance of defining goals and the means to attain them, and the aim of this chapter is to provide general, all-round advice. However, when it comes to a particular situation, local advisors and promotion centers are the people to consult as they can offer their specific knowledge.

Bibliography

[1] ANACAFE – *Manual de Caficultura*, Guatemala, 1998.

[2] Bertrand B. & Rapidel B. – *Desafios de la Caficultura en Centroamérica*, CIRAD & ICCA, San José Costa Rica, 1999.

[3] Cambrony H.R. – *Le caféier*, Maisonneuve et Larose et A. C.C. T., 1989.

[4] Carvajal J. F. – *Café, Cultivo y Fertilización*, Instituto Internacional de la Potasa, Berna, 1984.

[5] Clarke R. J. & Macrea R. – *Coffee, Agronomy*, vol. 4, Ed. Elsevier Applied Science, London, 1988.

[6] Coste R. & Cambrony H. – *Caféiers et Cafés*, Ed. Maisonneuve & Larose, Paris, 1989.

[7] FEDECAFE – *Manual del Cafetero Colombiano*, Ed. CENICAFE, Colombia, 1979.

[8] Gaie W. & Flémal J. – *La culture du café d'Arabie au Burundi*, ISABU, Ed. AGCD. Bruxelles, 1988.

[9] Haarer A. E. – *Modern Coffee Production*, Ed. Leonard Hill Ltd. London, 1958.

[10] IBC-DIPRO – *Cultura de Café no Brazil*, Pequeno manual de recomendacões, 1986.

[11] ICAFE – *Manual de recomendaciones para el cultivo del café*, CICAFE Costa Rica, 1998.

[12] PROCAFE – *Manual del Caficultor Salvadoreño*, San Salvador, 1977.

[13] Wrigley G. – *Coffee*, Ed. Longman, Wiley & Sons NY. USA, 1988.

8
Crop Maintenance

8.1
Fertilization

J. Snoeck and Ch. Lambot

8.1.1
Guidelines and Targets

The soil can be compared to a reservoir which can only provide what it contains and whose ingredients need to be replaced as soon as they are depleted, if it is to continue to nourish the plants which live on it. Here, we focus essentially on fertilization as a means of providing and maintaining optimal quantities and combinations of ingredients within the soil so that depletion caused by the activity of coffee trees does not unduly impoverish it or turn it into sterile land which can no longer be used for cultivation purposes.

Plant tissues are composed of a number of elements of which 16 are essential to its physiological development. These are carbon, oxygen, hydrogen, macronutrients like nitrogen, phosphorus, potassium, sulfur, calcium and magnesium, and micronutrients like manganese, boron, iron, zinc, copper, aluminum and molybdenum.

Carbon, oxygen and hydrogen are supplied by air and water, whereas the other 13 elements are directly absorbed as mineral nutrients taken from the soil by the roots of the plants.

When fertilizing coffee trees, two distinct aspects should be considered: (1) the compensation of the actual deficiencies and (2) the necessary inputs to replace depletion caused by the coffee trees themselves and by lixiviation. Fertilization should also cater for the needs of the various microorganisms which play an active part in the quality of the topsoil.

Replacing mineral depletion in the topsoil and neutralizing the effects of toxic elements are basic prerequisites to ensure production in the coffee plantation. The need for parallel fertilization should, normally, be significantly reduced if the soil has been well selected. Frequently, however, it is necessary to correct the fertility of the new land to adapt it to the requirements of the coffee trees. This generally means increasing the organic matter (OM) in the soil, and eliminating the toxic effects of manganese, aluminum and iron. It also entails increasing the nutrient content and correcting the balance between the principal cations. The more economical way to achieve this result is either to fertilize each individual planting hole before planting, or to fertilize a ring around each tree or to fertilize strips along the rows in high-density plantations. Parallel fertilization is the last factor of

Coffee: Growing, Processing, Sustainable Production, Second Edition. Edited by J. N. Wintgens.
© 2012 WILEY-VCH Verlag GmbH & Co. KGaA. Published 2012 by Wiley-VCH Verlag GmbH & Co. KGaA.

intensification and should only be used if all other cultural practices are well applied. The coffee varieties might also be able to valorize the investment placed in fertilization.

8.1.2
Interaction with other Cultivation Practices

Fertilization can interact favorably with other cultivation practices. Shading, for instance, can reduce the need for mineral nutrients in coffee trees by about one-third. Coffee grown under shade responds less well to applications of fertilizer.

Cover plants and shade trees, particularly leguminous, also reduce the need for intense fertilization as they generate a high nitrogen input. When deeply rooted, they absorb mineral nutrients at lower levels and bring them up into the topsoil. On the other hand, these plants may compete with the coffee trees for nutrients other than nitrogen, especially in impoverished soils.

Mulching the soil significantly reduces the need for chemical fertilization and ensures a considerable input of OM which enriches the soil with nutrients, essentially potassium. Mulching, however, requires regular checks of the cation ratio to avoid an imbalance between magnesium and potassium. An excess of potassium shortens leaf life and accelerates leaf fall. It also decreases the efficiency of nitrogen fertilizers.

In all cases, mulching enhances the efficiency of mineral fertilizers and the water-retention capacity of the soil.

8.1.3
Fertilization: Organic Fertilization

It has been established that coffee trees prosper well in soils which are rich in OM and the productivity of a coffee planta-

tion is directly related to its level in the soil. The optimal OM level varies between 2 and 5 % depending on the texture of the soil.

OM plays an important part in the productivity of coffee trees because of its influence on the physiological, chemical and biological characteristics of the soil. It also airs the soil, favors water infiltration, reduces erosion and activates animal life.

OM also considerably improves the cation-exchange capacity (CEC) in tropical soils. OM also helps to restrain the acidity induced by certain nitrogen fertilizers. This is important because a high level of acidity in the soil reduces microbial activity, and furthers the development of toxicity caused by the presence of aluminum and manganese.

It is also considered very likely that the presence of OM in the soil encourages the activity of various microorganisms like mycorrhisa, *Rhizobium* and other organisms which antagonize root diseases in coffee trees. Consequently, it is very important to preserve the native humus level of the soil. This must be taken into consideration right from the start, particularly when preparing the land prior to planting the coffee.

The OM in the soil should be preserved and, where possible, improved by good cultivation practices. The soil should be protected by shading, cover plants and/or organic mulch. It should be noted, however, that mulching with grass or manure can have a negative influence on the quality of the coffee produced by inducing the development of undesirable beans with a raw brown appearance and a poorer roast. This is brought about by an imbalance in magnesium because of the high potassium content of the grass and high levels of potassium and calcium in the manure.

Manure should be applied at the beginning of the rainy season at the rate of

10–15 kg for each tree. It should be lightly covered with either topsoil or straw and should not touch the trunk as this could damage the coffee tree.

8.1.4
Fertilization: Mineral Fertilization

8.1.4.1 Requirements of Coffee Trees
Arabica coffee prospers in slightly acid soils with a pH ($CaCl_2$) of 4.5–5.5 and a pH (H_2O) of 5.5–6.5.

Where the pH level is less than 4.0, the levels of aluminum and manganese should be ascertained and possibly corrected by liming.

Requirements of coffee trees can be considered in relation to the soil texture assessed according to the clay and silt content.

The ratio between the principal cations also needs to be carefully considered and should be contained within given limits:

	Lower limit	Upper limit
Mg:K	<2	>5
Ca:K	<3	>14
Mg:Ca	<0.2	>0.8

The optimum K:Ca:Mg ratio is approximately 6:76:18% of the sum of the exchangeable bases. For further details, see Tables 8.1.1 and 8.1.2.

8.1.4.2 Depletion of Nutrients
In 1984, Carvajal [1] surveyed a 3-year-old coffee plantation which produced 1 ton of marketable coffee and ascertained the nutrient requirements of the coffee trees (Tab. 8.1.3)

For Robusta coffee trees, the production of 1 ton of marketable coffee leads to the following rate of nutrient depletion:

- N   33.4 kg
- P_2O_5 6.1 kg
- K_2O 44.0 kg
- CaO   5.4 kg
- MgO   4.2 kg.

Table 8.1.1 Fertility requirements of coffee trees

Fertility parameters	Levels		
	Low	Medium	High
pH (H_2O)	4.5	5.5	6.5
Total phosphorus content			
P (ppm)	220	440	660
P_2O_5 (%)	0.5	1.0	1.5
Assimilable phosphorus			
P (ppm) Olsen-Dabin	30	45	60
P (Troug) ppm	15	22	30
CEC (meq/100 g)	5	10	25
Base saturation percentage (BSP)	40	60	100

Table 8.1.2 Guidelines for reading the results of soil analysis

Fertility parameters	Levels	Percentage of clay and silt (fine elements)					
		13	18	23	28	34	41
OM (%)	low	1.23	1.43	1.64	2.17	2.53	3.09
	medium	1.90	2.40	3.00	3.50	3.80	4.50
	high	2.96	3.61	6.27	7.17	7.44	9.81
N (%)	low	0.88	1.11	1.25	1.58	1.67	1.86
	medium	1.04	1.27	1.56	1.74	1.85	2.12
	high	1.55	1.89	3.29	3.73	3.92	5.14
P_2O_5 (Troug) ppm	low	4.2	3.8	4.0	3.8	4.2	5.4
	medium	15.0	13.4	13.0	13.0	12.5	12.1
	high	27.0	32.6	28.0	36.0	30.6	27.0
K (meq%)	low	0.07	0.10	0.11	0.14	0.16	0.24
	medium	0.13	0.18	0.28	0.35	0.42	0.42
	high	0.25	0.46	0.70	0.85	1.03	1.08
Ca (meq%)	low	1.89	2.08	2.29	2.61	2.59	3.03
	medium	3.22	4.09	4.62	5.35	5.36	5.88
	high	5.93	7.91	11.30	12.57	13.84	18.67
Mg (meq%)	low	0.34	0.37	0.40	0.45	0.50	0.68
	medium	0.55	0.63	0.80	0.95	1.07	1.20
	high	0.82	1.03	1.50	1.58	2.13	2.25
Total cations (meq%)	low	2.30	2.55	2.80	3.20	3.25	3.95
	medium	3.90	4.90	5.70	6.65	6.85	7.50
	high	7.00	9.40	13.50	15.00	17.00	22.00

Source: Forestier [3].

Table 8.1.3 Nutrient requirements of 3-year-old Arabica coffee trees (kg/ha/year)

Part of the tree	N	P_2O_5	K_2O	CaO	MgO
Roots and trunk	15.4	5.0	31.4	13.2	3.6
Branches	14.3	4.6	22.8	8.4	5.5
Leaves	52.9	22.9	54.2	26.3	11.3
Ripe fruit	29.5	5.9	41.4	4.6	5.5
Total	112.1	38.4	149.8	52.5	25.9

Source: Carvajal [1].

If one includes the requirements for growth, the estimated annual rate of nutrient depletion for a good yielding Robusta coffee tree could be 135 kg of N, 34 kg of P_2O_5 and 145 kg of K_2O, this represents a ratio of 4.0:1.0:4.3.

Mineral depletion is significant, particularly for nitrogen and potassium. The tissues of the coffee tree fix nutrients, although some of them are returned to the soil during the natural process of decaying. Nitrogen and bases are partially lost through leaching, particularly in areas of heavy rainfall and where the soil offers a low retention capacity. This can be solved simply by compensating for the nutrient depletion entirely on the basis of the analysis of the soil and the leaves. Correct readings of soil analysis compared to expected productivity will provide the basis for a balanced fertilization program.

8.1.4.3 The Fertilization Program

The results of soil analysis, calculation of the cation levels and the content of assimilable phosphorus can provide the necessary indications for establishing a suitable fertilization program. See Tab. 8.1.4.

Table 8.1.4 Suggested fertilization formulas $N:P_2O_5:K$ for coffee under varying conditions of soil, phosphorous and potassium, and optimum soil reaction (pH 4.4–5.4)

Potassium (K) soil test ratings and/or (Ca + Mg):K ratios	Soil test ratings for phosphorus (P Truog)		
	Low (<15 ppm)	Medium (15–30 ppm)	High (>30 ppm)
Low: <0.2 meq% K or 0.2–0.4 meq% K	15:15:15	16:8:16	18:0:18
and (Ca + Mg)/K = 10 or more	17:17:17	12:12:17+2	CAN, ASN
	19:19:19	18:9:18	KCl,K₂SO₄
	22:21:17	14:12:14ᵃ	
		15:15:15:6ᵃ	
Medium: 0.2–0.4 meq% K	18:18:9	20:10:10	20:0:0
and (Ca + Mg):K = 10 or less	15:15:6:4	25:10:10	CAN, ASN
	10:10:7.5ᵃ	32:10:11ᵃ	KCl, K₂SO₄
or >0.4 meq% K and (Ca + Mg)/K = 10 or more	11:8:6ᵃ	25:5:5	
High: >0.4 meq% K and (Ca + Mg):K = 10 or less	25:25:0	24:10:0	straight nitrogen fertilizers
	20:20:0	24:12:0	CAN, ASN
	22:22:0	20:10:0	Urea (etc.)
	24:23:0	30:10:0	

ᵃNormally used as foliar feed.
CAN = calcium ammonium nitrate, ASN = ammonium sulfate nitrate, KCl = potassium chloride, K₂SO₄ = potassium sulfate.
Source: Kenya Coffee [4].

Table 8.1.5 Results of soil analyses

Parameters	Soil analyses/diagnosis	Optimum contents	Required inputs
% clay and silt	37.1 clay loam		
pH (H$_2$O)	4.15 low	5.5	
Assimilable P (Troug) ppm	17.5 medium	22.5	
Ca (meq/100 g)	0.25 low	5.75	1.30 + 1.65
Mg (meq/100 g)	0.15 low	1.26	1.11
K (meq/100 g)	0.27 low	0.42	0.15
CEC (meq/100 g)	24.8 high	10.00	
Al (meq/100 g)	3.45 excess	0.04	
Ratio Mg/K	0.56 low	2–5	
Ratio (Ca + Mg):K	1.50 low	10–20	

An excess of aluminum or manganese in the part of exchangeable bases in the soil combined with a very low pH level presents a risk of toxicity for coffee trees. Hence, the soil needs to be neutralized by liming. Similarly, where deficiencies or imbalances between the various elements are noted, these should be corrected by an input of chemical fertilizers or lime. Table 8.1.5 provides an example for calculating the required rates to correct deficiencies according to the results of soil analysis.

Calculation of Inputs of Potassium
To reach the optimal requirement of 0.42 meq/100 g, the soil needs an input of 0.15 meq/100 g of K, which must be doubled to compensate losses by leaching, immobilization and retrogradation in the soil.

For a soil density of 1.2, to correct the content at a depth of 20 cm the requirements are:

$2 \times 10\,000$ (m^2) $\times 0.2$ (m) $\times 1200$ (kg) $\times$ 0.15 (meq)/0.1 (kg) = 7 200 000 meq of K/ha

Given that 78 mg of KCl at 60 % K$_2$O are needed for 1 meq of K the dose should be:

7 200 000/1000 $\times$ 78 = 561 600 g KCl or approximately 560 kg of KCl/ha

Calculation of Inputs of Magnesium
To obtain a ratio of Mg:K equal to 3, the required Mg^{2+} level equals 1.26 meq/100 g and therefore 1.11 meq/100 g must be added to the soil. As a result, the requirements are:

10 000 (m^2) $\times$ 0.2 (m) $\times$ 1200 (kg) $\times$ 1.11 (meq)/0.1 (kg) = 26 640 000 meq of Mg/ha.

With a dolomite containing 28.3 % of CaO and 17.0 % of MgO, the requirements are:

26 640 000 $\times$ 20.16/1000 = 537 062 g of MgO or 3122 kg of dolomite.

Calculation of Inputs of Calcium
The 3122 kg of dolomite will provide the soil with 883 kg/ha of CaO, equivalent to 1.30 meq/100 g of Ca. This will not be enough to bring the (Ca + Mg):K ratio up to 10.

Consequently it will have to be completed by an input of lime. A similar calculation will indicate that an input of 1500 kg of slaked lime will provide the soil with 1.65 meq/100 g of Ca. This liming will have to be renewed every 3 years to try to neutralize the excess aluminum and to reach the optimum of 5.75 meq/100 g Ca.

The quantities of dolomite and slaked lime inputs can be reduced by applying the material either in circles round the coffee trees or in strips along the rows.

8.1.4.3.1 Macronutrients

Nitrogen

Nitrogen is essential for the vegetative growth of the trees. It boosts the development of branches and leaves. It also greatly influences the production of coffee by increasing the number of flowers and of fruits per clusters, and by extending leaf life. It helps to combat die-back. It is chiefly absorbed in the form of NO_3 and also NH_4.

Trial and experience in fertilization have revealed that nitrogen fertilization produces excellent results provided the cation balance is appropriate and the coffee plantation well maintained. The annual nitrogen input generally varies between 60 and 200 kg/ha, although it can sometimes be as high as 400 kg/ha according to the expected yield. Robusta trees need less nitrogen than Arabica.

Nitrogen deficiencies are more frequent in sun-exposed coffee plantations, especially after a high yield. This is also the case during the dry season as dry topsoil hinders root function. In such cases, mulching is advisable because it protects the soil from excessive drying out.

Nitrogen inputs can be made in different ways depending on the soil acidity, cost-efficiency and the form of nitrogen. Urea is, of course, the cheapest nitrogen input, but in soils with a high water content part of the nitrogen will be wasted as ammonium. Ammonium sulfate contains sulfur which can have a positive effect on certain types of soils, but it increases soil acidity. Sodium nitrate should not be used as a fertilizer in coffee plantations because of the release of sodium.

Unshaded coffee trees respond better to applications of nitrogen than shaded trees. Nitrogen inputs at the beginning of the dry season strengthen coffee tree resistance to drought which, in turn, generates a positive effect on the yield (Tab. 8.1.6).

Potassium

Potassium is of significant importance to the physiological development of the coffee tree, particularly for the development and the maturation of the fruit. A high content of potassium can be found in soils of volcanic origin and in soils which are regularly mulched.

As potassium depletion of the soil is more rapid than nitrogen depletion, potassium deficiency can become a factor which limits production.

Potassium is an antagonist of magnesium and calcium, which means that

Table 8.1.6 Form of nitrogen and phosphate fertilizers in relation to soil pH

Type of fertilizer	pH in the soil		
	<4.4	4.4–5.4	>5.4
Nitrogen	CAN or ASN or urea	CAN or ASN or urea	AS or ASN
Phosphates	rock phosphate or SSP or DSP	SSP or DSP	DAP or DSP

From: CRF, Ruiru, Kenya.
CAN = calcium ammonium nitrate, ASN = ammonium sulfate nitrate, AS = ammonium sulfate, SSP = single superphosphate, DSP = double superphosphate, DAP = diammonium phosphate.

soils with a high content of potassium frequently show deficiencies in magnesium and calcium, and, conversely, a high percentage of magnesium and calcium in the soil can lead to a potassium deficiency. Excess of potassium inhibits the positive effects of nitrogen.

As a result, the ratio of K:Mg:Ca needs to be seriously considered when calculating the input of fertilizers. The level of potassium in the soil should not fall below 156 ppm or 0.4 meq/100 g with an optimal ratio of K:Ca:Mg at $6:76:18\%$ of exchangeable bases. Potassium is absorbed by the roots as a monovalent ion of K^+.

The most common types of potassium fertilizers are K_2SO_4 (48% of K_2O) and KCl (60% of K_2O). According to soil analyses results, if potassium is needed, an average input of not more than 150–200 kg of K_2O/ha in two applications per year is recommended. Practical experience shows that applications of over 330 kg of K_2O/ha/year are counter-productive and have practically no effect on the yield. It also appears that K_2SO_4 provides better results than KCl because an accumulation of chloride in the plant has a negative effect on production.

Phosphorus

Phosphorus is necessary for the development of roots, wood and young buds. It is absorbed in the form of ions of $H_2PO_4^-$ and HPO_4^{2-} to produce organic components. Phosphorus depletion after yielding is low. Experiments with phosphate fertilizers in coffee plantations have not generated significant results except when the fertilizer is incorporated into the planting hole or applied in large quantities. It is, however, very useful to apply phosphorus at the bottom of the planting hole to stimulate root formation.

In adult plantations, phosphate fertilizers which encourage the development of the root system should be buried in the topsoil. The doses of phosphate fertilizers should be calculated in relation to the soil content of P_2O_5 and its acidity. For acid or very acid soils, the recommended form is tricalcic phosphate or rock phosphate: 200 g per hole. For soils of medium acidity, the recommended rate of single or triple superphosphate is 100 g per planting hole.

Phosphorus and zinc have an antagonist interaction, consequently an excess of phosphate inputs can bring about a zinc deficiency.

For adult coffee trees, phosphorus inputs, if their necessity is indicated by soil analyses, should be made in the form of a compound fertilizer at the rate of 50–60 kg of P_2O_5/ha/year. As phosphorus tends to be immobilized in the soil, compound fertilizers should be lightly worked into the topsoil. In acid soils, iron and aluminum immobilize phosphorus as tricalcic phosphates.

OM in the soil is an important reservoir of phosphorus, nitrogen and sulfur. Consequently, wherever possible, it is advisable to maintain the phosphorus level of the soil by the application of manure, compost mulch, etc.

Calcium and Magnesium

Calcium is important to the development of terminal buds and flowers. Magnesium is one of the constituents of chlorophyll and therefore important for photosynthesis. Magnesium deficiency is frequent and easily identified because of its typical symptoms on the leaves. Magnesium is absorbed through the roots as the bivalent ion Mg^{2+}.

Calcium and magnesium deficiencies can have a negative effect on the bean quality. Magnesium deficiency shortens leaf life and is therefore detrimental to yield.

Some compound fertilizers contain 3–6 % of MgO. Soil acidity will determine the form of magnesium input: for a pH of over 5.0, magnesium sulfate is recommended (Kieserite, Epsom salt), and for a pH below 5.0, calcined magnesite or dolomite should be applied. Rates of 125 kg of MgO/ha/year are recommended.

Recommendations for calcium and magnesium applications should be based on soil analysis, taking the K:Ca:Mg ratio into account as well as the degree of exchangeable acidity (H_p).

If the H_p value is above 0.5 meq%, lime should be applied to bring the pH to a level of 4.5–5.5 ($CaCl_2$). For the correction of soil acidity, calcium and magnesium are applied as ground limestone ($CaCO_3$), dolomite ($CaCO_3 + MgCO_3$) or calcined magnesite (MgO) at rates of 250–700 kg/ha according to the results of the soil tests. Where soil acidity is normal, the magnesium deficiency can be corrected by a soil application of Kieserite ($MgSO_4$) or by foliar sprays of Epsom salt, magnesium chloride or magnesium nitrate in an aqueous solution at 1 %.

Liming

Many crops react positively to massive initial liming, but subsequent liming generates a negative effect. This has given rise to a popular proverb which says that "liming enriches the father and ruins the son".

Apart from one or two exceptions, however, liming does not appear to have a significant effect on the production of coffee trees. Nevertheless, it should be noted that high quantities of lime applied without taking soil analysis into account can generate a significant imbalance in the adsorption complex since calcium cations may dislodge the cations of potassium, magnesium and other elements, and finally bring about marked deficiencies.

Liming is a practice which should chiefly be used for correcting toxicity and soil acidity, particularly in the case of an excess of manganese and aluminum in the soil.

8.1.4.3.2 Micronutrients

Due to their visible symptoms, deficiencies and toxicity are relatively easy to identify on the coffee leaves. On the other hand, soil analysis does not necessarily highlight defi-

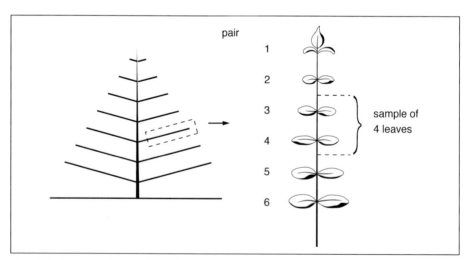

Figure 8.1 Sampling of leaves for analysis.

Table 8.1.7 Optimum leaf nutrient levels

Nutrient	Deficiency	Optimum range	Excess
Nitrogen (%)	<2.0	2.5–3.0	>3.5
Phosphorus (%)	<0.1	0.15–0.2	>0.2
Potassium (%)	<1.2	1.5–2.6	>2.6
Sulfur (%)	<0.05	0.10–0.20	>0.25
Calcium (%)	<0.5	0.7–1.3	>1.5
Magnesium (%)	<0.15	0.2–0.4	>0.5
Iron (ppm)	<50	50–150	>220
Manganese (ppm)	<20	75–150	>400
Copper (ppm)	<3.0	6–15	>25
Zinc (ppm)	<7.0	10–15	>20
Molybdenum (ppm)	<0.10	0.15–0.20	>0.30
Boron (ppm)	<30.0	40–100	>100

ciencies. Leaf analysis, completed by visual observations, will often confirm the deficiencies or toxicities (Fig. 8.1, Tab. 8.1.7).

Zinc

Zinc is absorbed through the roots as a bivalent ion Zn^{2+} or as a chelate. Its deficiency is frequent in acid or exhausted soils. It can be compensated by foliar spraying with either zinc sulfate ($ZnSO_4$) or zinc oxide (ZnO) in concentrations of 1%. It should be noted, however, that compensating a zinc deficiency by foliar spraying can aggravate damage caused by *Colletotrichum coffeanum* if suitable sanitary precautions are not taken. Soil applications of zinc are not recommended because zinc is strongly immobilized in the upper 0.5–1 cm of the soil.

Aluminum

At a level situated between 0.5 and 1.0 ppm, the presence of aluminum in the soil is relatively favorable. A concentra-tion of over 2 ppm of Aluminum hinders the absorption of calcium, magnesium, ammoniac nitrogen, iron and manganese.

At low soil pH combined with low rates of cations, particularly calcium, aluminum toxicity is most likely to occur. Analysis of the aluminum content in the soil should be done to calculate requirement of calcium fertilizer to correct toxicity by following formula:

$$CaO \text{ (kg/ha)} = 1.5 \text{ meq Ca} \times (28.1 \times 1.4 \times 20 \text{ cm}) \times \text{meq\% Al}$$

Where:

1.5 meq Ca	= quantity required to neutralize 85–90% of 1 meq exchangeable Al in a soil with 2–7% OM
28.1	= mg of CaO oxide in 1 meq% Ca
1.4	= soil density in g/cm^3
20	= corrected soil depth in cm
meq% Al	= from the soil analysis

Iron

A lack of Iron is frequently noted in alkaline soils where it generates the typical symptom of chlorosis in young leaves. Iron chelate used to compensate this deficiency can be applied in a foliar spray or directly onto the soil. The latter treatment is more expensive. A cheaper way of solving the problem is to reduce the alkalinity of the soil by applying sulfate fertilizers such as ammonium sulfate.

Manganese

Excessively alkaline soils are frequently prone to manganese deficiency. This can be compensated by spraying a manganese sulfate solution at 1 % but, again, it is simpler to previously reduce the alkalinity of the soil. Manganese toxicity in the soil is generally associated with an aluminum toxicity in very acid soils. Calcareous applications will reduce soil acidity and correct the toxicity.

Copper

The spraying of cupric fungicides has considerably reduced copper deficiencies in the soil. Toxic effects of an excess of copper are more frequently noted because of repeated spraying for disease control. The famous "tonic effect" of cupric fungicides is not due to the nutrient input of copper, but rather to the fungicide effect which reduces the pressure of fungi and other microorganisms on coffee plants.

Sulfur

A sulfur deficiency is occasionally noted in sandy soils. It can be compensated by an application of sulfate of ammonia or any other fertilizer which contains sulfur. Sulfur is absorbed through the roots as SO_4^{2-}.

Boron

Boron deficiencies are frequent. They can be compensated by applying 30–60 g of borax per tree directly into the soil or by foliar spraying with a borax solution at 0.5 %. Boron is absorbed through the roots as non-dissociated boric acid.

8.1.4.4 Practical Indications for Applying Fertilizers

8.1.4.4.1 Scheduling

Normally coffee fertilization is applied in four stages: in the nursery, in the planting hole, on young, growing plants and on adult trees in production.

In the nursery, fertilization can be applied to young plants from the time they leave the seed bed until they are transplanted into the field. This period lasts from 6 to 12 months. Fertilization can begin when the plantlets have two or three pairs of leaves and should take place approximately every 2 months. Fertilizer in a solution or by foliar spraying is recommended.

In the planting hole, it is usually recommended to apply well rotted cattle manure or compost (about 20 l) and phosphate. For phosphate choice and rates see Section 8.1.4.3.1.

On young trees growing in the field, three applications of fertilizer can be made during the rainy season. The formulas should be based on the soil analyses. Yielding coffee trees can be fertilized at the same time as the growing trees, but the formulas and the rates of the fertilizer should be adapted to the soil analysis and expected yields.

In the case of stumping for rejuvenating, fertilization should be interrupted, or at least significantly reduced, during the unproductive period.

8.1.4.4.2 Soil Applications

For a correct input into the soil, the fertilizer should be placed either in a ring around the base of the trunk or in strips along the rows. Application in a ring around the base of the trunk is used for young trees. The diameter of the ring should be directly related to the foliar spread as in Fig. 8.2.

Fertilizer applied in strips along the rows is a method generally used for high density adult plantations (Fig. 8.3). When applying fertilizer to the soil, remove coverage like plastic protection sheets prior to application.

Lime inputs should be dug into the soil to prevent the formation of a hardened crust on the surface. Calcareous and nitrogen inputs should be separated by a period of at least 2 months or by spacing the rings. To avoid a toxic concentration, chemical fertilizers should not be applied to young coffee trees until at least 3–4 weeks after their transplantation.

Nitrogen fertilizers can be applied on the surface of the soil, but their application should be spaced out in order to avoid losses through rainfall or flooding and to ensure the regular feeding of the coffee plant. As a general rule, three or four applications per year during the rainy season are enough. The first two applications can be combined with other elements like potassium, phosphorus, magnesium and, where necessary, boron.

Potassium inputs should be made in two applications per year and phosphorus, being less mobile, in one application at the beginning of the rainy season. Phosphorus dislocates the nitrogen and the potassium in the adsorption complex. As a result, high quantities of phosphorus fertilizers in a sole application should be avoided.

Deficiencies in trace elements are generally compensated by foliar sprays. Soil ap-

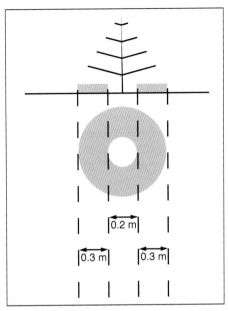

Figure 8.2 Application of fertilizer in a ring around a young coffee tree.

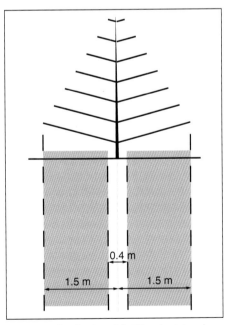

Figure 8.3 Application of fertilizer in strips along the rows in an adult coffee plantation.

plications of trace elements should be considered as a preventive measure, whereas foliar sprays are more of a curative measure.

8.1.4.4.3 Foliar sprays

Foliar sprays offer the advantage of rapid absorption through the stomata located on the underside of the leaves. To obtain best effects, foliar sprays need to be repeated frequently and special care must be taken to ensure that the spray reaches the underside of the leaves.

Their major advantage resides in the fact that, inasmuch as the chemical compositions are compatible, fertilization by foliar sprays can be combined with phytosanitary treatments. It also has the added advantage of avoiding any negative interaction between the different elements in the soil. The concentrations for foliage applications should comply with the directions for use issued by the manufacturer and should never exceed 1%.

Foliar sprays are generally used to compensate a lack in microelements but can also be used for nitrogen, phosphorus and magnesium. If urea is used, its biuret content should not exceed 0.5%. Urea may be added to all cases of micronutrient applications at not more than 0.5%. For an input of phosphorus by spraying, diammonium phosphate can be used. For magnesium spraying, Epsom salt (magnesium sulfate), magnesium chloride or magnesium nitrates are recommended.

8.1.4.4.4 Fertilization Deficiencies and their Effects on Coffee Quality

Optimal cup quality is directly dependent on a correct ion balance in the soil:

- A lack of zinc will lead to the production of small, light grey-colored beans which produce a poor liquor.

- Excessive nitrogen inputs will affect the quality of the beans and increase their caffeine content.
- A lack of iron will encourage the production of amber beans.

Generally speaking, overproduction in coffee trees leads to a decrease in cup quality. This is accentuated in the case of marked deficiencies in mineral nutrients.

8.1.5
Conclusions

To summarize the targets and guidelines in this chapter, it can be said that fertilization should have two main targets:

1) To correct any native imbalances in the soil prior to setting up the coffee plantation, and
2) To ensure a regular and sustained nutrient input for the coffee tree during its various stages of development and yield.

The recommendations included in this chapter should be considered in the light of general "rule of thumb" advice because each coffee plantation or, for that matter, even each individual plot, has its own special needs which are conditioned by external factors specific to the area. Coffee growers should, therefore, not hesitate to consult official regional centers for more detailed local advice and, more importantly, rely on personal, on site, recorded observations.

Frequently, the implementation of coffee fertilization is still inadequate and, as a result, inefficient. In many places it is either excessive, deficient or unbalanced. A fertilization program can only be successful and cost-effective if it is adapted to the prevailing conditions of the land and the specific nutrient requirements of the plant. Consequently, to ensure optimal results, fertilization needs to be carefully planned and regularly supervised according to the local situation, soil analysis and personal observations.

In view of the present economic fragility of the coffee industry, the entire production system needs to be reviewed and distinctly improved. Fertilization, in particular, needs to be reconsidered and approached from a more scientific angle. It should be based on data generated by soil and leaf analysis and, possibly, enzyme fluctuations. A study of the lifecycle of the coffee tree can also reveal cyclic physiological needs which could be met by more suitable fertilization.

Generally speaking, from both an economic and an environmental point of view, nutrients supplied to coffee trees by organic fertilization of the soil (cover plants, shading, compost, manure and the residues of coffee processing) offer the most beneficial option for coffee growers.

8.1.6
Annexes

8.1.6.1 **Physical Soil Composition**
Cultivable soil is composed of mineral and organic matter, water and air, as is illustrated in Fig. 8.4.

Air encourages oxidation, and caters to the needs of animal, vegetal and microbial life in the soil.

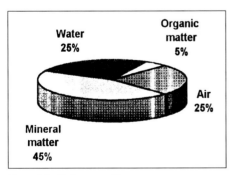

Figure 8.4 Soil composition.

Water dissolves the nutrients and facilitates their absorption through the roots of plants. The movements of the roots are conditioned by physical characteristics like density, compactness and porosity, among others.

Soil fertility is generated by the efficiency of the clay and humus adsorption complex which binds the nutrient components.

Since texture is the most permanent characteristic of the soil, it influences a number of other soil properties such as: structure, consistency, soil-moisture regime, permeability, infiltration rate, runoff rate, erodibility, workability, root penetration and fertility.

The international system used for defining soil separates is:

Clay	0–2 µm
Fine silt	2–20 µm
Coarse silt	20–50 µm
Fine sand	50–200 µm
Coarse sand	200–2000 µm
Coarse elements	over 2 mm (2000 µm)

FAO soil classification for textural quality is:

Coarse:	<15 % clay and >70 % sand (50–2000 µm)
Medium	<35 % clay and <70 % sand (50–2000 µm)
Fine	>35 % clay

Soil texture is measured in laboratories by separating the solid components of the soil. In the field, different practical methods can be used like, for instance, kneading a damp sample of earth to assess its texture by feel.

The percentage of sand, silt and clay can be plotted on a diagram (Soil textural triangle; Fig. 8.5) to determine the textural class of the soil. A loamy soil texture would, for example, have a 13 % clay, 41 % silt and 46 % sand content.

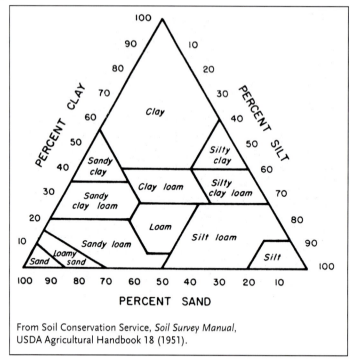

From Soil Conservation Service, *Soil Survey Manual*,
USDA Agricultural Handbook 18 (1951).

Figure 8.5
Soil textural triangle.

8.1.6.2 The Soil as a Reservoir of Nutrients

The mineral nutrients of the soil are generated by the natural meteorization of the parent rock, the mineralization of OM and the input of fertilizers. Within the soil, nutrient molecules are split up into anions, which are negative, and cations, which are positive, e.g.

- KCl splits up into 1 cation K^+ and 1 anion Cl^-
- $Ca(NO_3)^2$ splits up into 1 cation Ca^{2+} and 2 anions NO_3^-.

Table 8.1.8 gives further examples of nutrient molecules. The interaction between these ions is conditioned by the acidity of the soil and the intensity of the electric charges of the ions.

The most mobile elements are potassium, nitrates, borates, molybdates, chlorides and sodium. Sulfates, calcium and

Table 8.1.8 Cations and anions in the soil

Cations		Anions	
Ca^{2+}	Calcium	PO_4^{3-}	Phosphate
Mg^{2+}	Magnesium	SO_4^{2-}	Sulfate
K^+	Potassium	CO_3^{2-}	Carbonate
NH_4^+	Ammonium	NO_3^-	Nitrate
Na^+	Sodium	Cl^-	Chloride
H^+	Hydrogen	OH^-	
Cu^{2+}	Copper	$H_2PO_4^-$	
Zn^{2+}	Zinc	HCO_3^-	
Al^{3+}	Aluminum	HPO_4^{2-}	
Fe^{2+}	Iron	$B_4O_7^{2-}$	
Fe^{3+}	Iron	MoO_4^{2-}	
Mn^{2+}	Manganese		

magnesium are less mobile, and heavy metals like iron, copper, manganese and zinc are virtually static.

Positive cations are fixed by the negative clay–humus complex in the soil and this creates a food store for vegetation. The fixa-

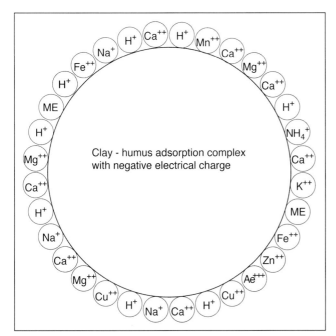

Figure 8.6 Clay–humus adsorption complex. From: USDA Agricultural Handbook 18 (1951).

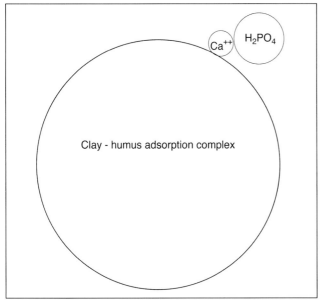

Figure 8.7 Anion fixation of phosphates on clay–humus complex.

tion potency of the cations in the soil complex varies according to their individual characteristics. The hydrogen cation is the least volatile, followed by the trace elements, and the more volatile calcium, magnesium, ammonium, potassium and sodium (Fig. 8.6).

The PO_4 anion uses the advantages of the bridge formed by calcium anions which enables it to be fixed into the soil complex (Fig. 8.7).

The Cation-exchange Capacity (CEC)

The total capacity of the soil for cation exchange (cation adsorption) is expressed by the CEC and is measured in milliequivalents per 100 g of dry soil (meq%).

The meq% is the ratio between the quantity of one element (expressed in mg) and the equivalent weight of the element.

Example for calcium:

Atomic weight $= 40$, Valence $= 2$
Equivalent weight $= 40:2 = 20$
Milliequivalent weight $= 20:1000 = 0.020$
1 meq% Ca $= 0.020g:100$
$= 200:1\,000\,000$ or 200 ppm

To convert from meq% to ppm, the equivalent weights (Tab. 8.1.9) should be multiplied by 10, e.g. 1 meq% of Ca $= 20 \times 10 = 200$ ppm.

Effect of pH

The effect of pH on the availability of common elements in soil is illustrated in Fig. 8.8.

8.1.6.3 The C:N Ratio

The C:N ratio gives an indication of the type of OM present in the soil and, in particular, the degree of humification. For temperate soils, the ratio is high, whereas in tropical soils, it is lower due to higher temperatures and more intense microbial activity.

The incorporation of partially decomposed organic residues can greatly affect the C:N value. Non-decomposed straw residues tend to increase the ratio whilst leguminous residues, high in nitrogen content, tend to reduce it. This factor should be taken into consideration when applying organic fertilizers because it could generate a nitrogen deficiency.

The C:N ratios of several materials are:

Straw	40
Acid humus	>20
Good soil	12
Impoverished soil	<10 (low OM content)

Table 8.1.9 Main equivalent weights used for soil analysis (molecular weight/valence)

Na^+	23	NaHO	31	CO_3Na_2	53	SO_4H_2	44.5
K^+	39	K_2O	47	CO_3K_2	69	CO_3NaH	84
Ca^{2+}	20	CaO	28	CO_3Ca	50	CO_3KH	100
Mg^{2+}	12	MgO	20	CO_3Mg	42	$(CO_3)_2CaH_2$	81
Cl^-	35.5	Br	80	SO_4Na_2	71	$(CO_3)_2MgH_2$	73
SO_4^{2-}	48	SO_3	40	SO_4K_2	87	NaCl	58.5
CO_3^{2-}	30	CO_2	22	SO_4Ca	68	KCl	74.5
CO_3H^-	61	NO_3	62	SO_4Mg	60	$CaCl_2$	55.5
S^{2-}	16	SO_2	32	$SO_4Fe.7H_2O$	139	$MgCl_2$	47.5

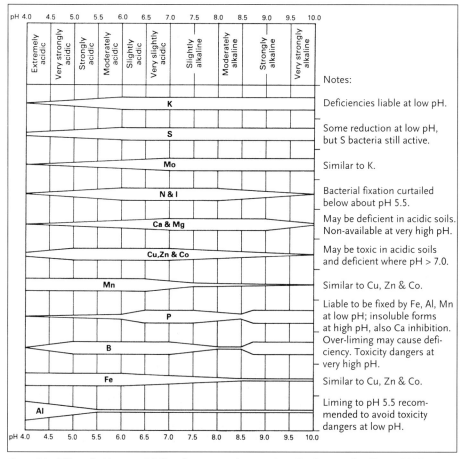

Figure 8.8 Effect of pH on availability of common elements in soils. Source: after Truog (1948)

8.1.6.4 Procedures for Leaf and Soil Sampling

8.1.6.4.1 Leaf Sampling

Leaf samples should be made up of 32 leaves taken from eight coffee trees, four leaves from each tree (see also Fig. 8.1).

- Collect the third and fourth pairs of leaves from branches located half-way up the tree.
- All dust must be removed from the leaf samples.

- The samples should be oven-dried prior to analysis.
- Samples collected at the beginning of the rainy season, before fertilization will enable the detection of mineral deficiencies.
- Samples collected 4–6 weeks after foliar spraying will enable the assessment of the efficiency of compensation by fertilization.
- Samples should be collected regularly, every year so as to keep a check on the evolution of the chemical condition of the coffee trees.

8.1.6.4.2 Soil Sampling

- Soil samples should be taken at the beginning of the rainy season when OM is actively in the process of mineralization and nitrates are generated – it also corresponds to the period of vegetative and fruit development.
- Soil samples should be taken from areas which normally receive fertilizers – in young coffee trees, this corresponds to the ring related to the spread of the foliage; in adult plantations, samples should be taken from the strips.
- Soil samples should be taken with a gauge at depths of 0–20 cm, after removing the superficial undecayed debris.
- Composite samples for analysis are made up of 20 random samples which are mixed.
- Regular soil analysis should be carried out every 4–5 years to keep a check on the evolution of the fertility of the plantation.

8.1.6.5 Mixing Fertilizers

For both logistic and economic reasons, it can be advisable to mix fertilizers before their application. This should, however, be carried out after taking into account certain restrictions as indicated in Figure 8.9.

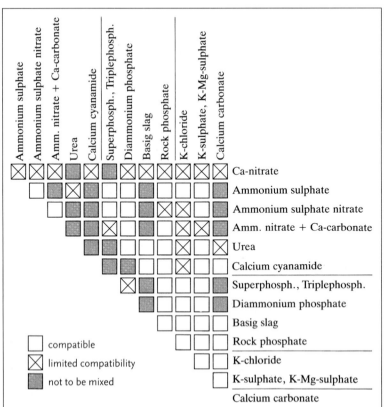

Figure 8.9 Possibilities of mixing fertilizers (limited compatibility is mainly due to hygroscopicity). From: IFA – World Fertilizer Manual.

8.1.6.6 Mineral Deficiency Symptoms on Coffee Leaves

Nitrogen

The typical symptom revealing a lack of nitrogen is the general chlorosis of young leaves. Later, the mature leaves also become chlorotic and, as a result, the entire plant turns yellow or almost white.

The leaves are very sensitive to sun and often exhibit severe scorching. Affected leaves are dull and lack-luster; whitish veins occasionally appear in the older leaves. The leaves fall and the branches gradually wither from tip to base (dieback). Growth slows down.

In nurseries, seedlings in the center of the beds frequently lack nitrogen. Nitrogen deficiency strongly affects plant growth (Pictures 8.1, 8.2, 8.3 and 8.4).

Picture 8.3 Nitrogen deficiency. Coffee seedlings affected by nitrogen deficiency. From: J. N. Wintgens (2002).

Picture 8.4 Nitrogen deficiency. Effect on plant growth. Right: deficient plant (1.46–1.92 % N). Left: control plant (3.00–3.72 % N). From: HITAHR Hawaii (1986).

Phosphorus

A phosphorus deficiency is revealed by a slight mottled chlorosis of the older leaves with a faint interveinal yellowing. At advanced stages, the yellow patches develop into a reddish or dark red autumn tint.

Younger leaves turn a dark blue–green and hang downwards and backwards. The older leaves soon fall off. Phosphorus deficiency also reduces the root development of the coffee plant (Pict. 8.5 and 8.6).

Potassium

A lack of potassium can be noted by the apparition of a chlorotic band along the margins of older leaves. Later, dark brown ne-

Picture 8.1 Nitrogen deficiency. Young leaves affected by chlorosis. From: J. N. Wintgens.

Picture 8.2 Nitrogen deficiency. The entire plant is affected by chlorosis. From: J. N. Wintgens (2002).

Picture 8.5 Phosphorus deficiency. Advanced stage on older leaves.

Picture 8.8 Magnesium deficiency. "Herring-bone" chlorosis.

Picture 8.6 Phosphorus deficiency. Effect on root development reduction. Right: control plant (0.14–0.15 % P). Left: deficient plant (0.05–0.08 % P). From: HITAHR Hawaii (1986).

Picture 8.7 Potassium deficiency. Evolution of symptoms on older leaves.

crotic spots appear in the same place. The spots enlarge until the entire margins of the leaves are necrotic, whereas the central portion of the blade remains green. These symptoms are similar to salinity damage.

Young leaves appear to be unaffected by a lack of potassium, but the older affected leaves fall off and die-back is the final stage. Tests revealed that the leaves of deficient plants contained 0.36–1.07 % leaf potassium compared to the leaves of the control plant which contained 2.62–3.86 % (Pict. 8.7).

Magnesium

Initially the symptoms of a magnesium deficiency reveal a slight chlorosis along the margins of the older leaves but later the leaves sink and become olive-green. Only the mid-rib and the veins retain a healthy dark green color. The chlorotic areas take the shape of a parallelogram. Leaves on a coffee tree suffering from a lack of magnesium have often been described as having "herring-bone" chlorosis.

A prolonged period of Magnesium deficiency will cause older leaves to abscise whereas young leaves remain unaffected. Tests revealed that the leaves of deficient plants had magnesium levels of 0.04–0.11 % compared to the leaves of the control plant which had levels of 0.3–0.35 % (Pict. 8.8 and 8.9).

Calcium

Typical symptoms of a calcium deficiency affect young leaves which turn to a bronze color, particularly along the margins. The mid-rib remains dark green leaving an

Picture 8.9 Magnesium deficiency on older leaves. Younger leaves are unaffected.
From: J. N. Wintgens (2002).

Picture 8.10 Calcium deficiency.
Saw-like edges on both sides the main leaves.

Picture 8.11 Calcium deficiency. Effect on root development reduction. Right: control plant (0.94–1.16% Ca). Left: deficient plant (0.36–0.55% Ca).
From: HITAHR Hawaii (1986).

area of saw-like edges on both sides of the main veins. The leaf blades are cupped downwards.

In advanced stages, emerging leaves are already necrotic and finally the entire apical bud suffers from die-back. Older leaves show signs of chlorosis, generally along the margins, or necrotic patches on the blade. Poor root development is also observed in calcium-deficient plants (Pict. 8.10 and 8.11).

Boron

A Boron deficiency affects the youngest leaves which are light green, smaller than normal, curved, twisted and mottled with small necrotic spots. The young leaves are unable to unfurl normally, which results in irregular margins and a rough surface and suberization or corking of the underside of the leaf. The death of the terminal branch causes branching in a fan-like pattern which is similar to the pattern caused by Antestia damage.

The concentration of boron in the leaves of deficient plants was 9 ppm against 31–52 ppm in the control plant (Pict. 8.12).

Picture 8.12 Boron deficiency. Advanced symptoms. Irregular margins of young leaves and abnormal growth of apical buds.
From: HITAHR Hawaii (1986).

Zinc

A zinc deficiency also affects young leaves, which remain small, narrow and brittle. They are shaped like a spear-head and snap easily. Interveinal chlorosis is also visible on the leaf blade revealing the fine

Picture 8.13 Zinc deficiency. Effect on young leaves. From: HITAHR Hawaii (1986).

Picture 8.14 Symptoms of zinc deficiency on a coffee tree branch. Older leaves remain unaffected. From: W. Gaie (1988).

Picture 8.15 Iron deficiency. Advanced stage. From: J. N. Wintgens (2002).

veinal network. This is similar to the aspect of the leaf in the case of an iron deficiency.

A zinc deficiency is further typified by a shortening of the internodes, excess sprouting and die-back of the branches in the final stage. Older leaves remain unaffected. Zinc levels of less than 10–15 ppm have been reported in deficient plants (Pict. 8.13 and 8.14).

Iron

A lack of iron is revealed by interveinal chlorosis in young leaves. During the early stages, the interveinal area of the leaves is light green, later it turns yellow or creamy-white. At a later stage, the dark green of the veins contrasts strongly with the pale mesophyle tissue and appears as a delicate network pattern (Pict. 8.15).

Sulfur

Here the symptoms of the deficiency are only marked by a general chlorosis of the younger leaves and a faint, diffused interveinal yellowing particularly along the midrib. This symptom helps to distinguish a sulfur deficiency from a nitrogen deficiency.

Another factor which distinguishes a sulfur deficiency from a nitrogen deficiency is that the leaves do not lose their luster and the color of the underside of the leaf is much lighter than that of the upper side. Shoot growth remains uninhibited.

Tests revealed that leaf sulfur concentrations in deficient plants were 0.04–0.05 % against concentrations in the control plant which were 0.12–0.17 % (Pict. 8.16).

Manganese

A manganese deficiency can usually be noted in the younger leaves, from the first to the third pair. They turn from a pale olive-green to a bright lemon-yellow

and show a number of small yellowish dots. The fourth pair is generally a normal green. There is no reduction in leaf size and no fall of leaves (Pict. 8.17).

Picture 8.16 Sulfur deficiency on young coffee leaves. From: HITAHR Hawaii (1986).

Picture 8.17 Manganese deficiency on young coffee leaves.

Bibliography

[1] Carvajal, J. F. *Cafeto – Cultivo y Fertilizacion.* Instituto Internacional de la Potasa, Bern, 1984.

[2] Coussement, S., Gaie, W., Warnier, F. and Sottiaux, G. *La Fumure Minérale du Caféier d'Arabie au Burundi.* ISABU, OCD, 1971.

[3] Forestier, J. Relations entre l'alimentation du caféier robusta et les caractéristiques analytiques du sol. *Café, Cacao, Thé* **1964**, *VIII* (2), 89–112.

[4] Kenya Coffee. Technical circular 59. Standard recommendations for fertilizer. *Kenya Coffee* **1989**, *54* (628), 511–517.

[5] Landon, J. R. *Booker Tropical Soil Manual.* Booker Agriculture International/Longman, London, 1984.

[6] Malavolta, E. Nutrition, fertilization et chaulage pour le caféier. In: *Cultura do Cafeeiro*, Piracicaba (SP), 1986.

[7] Snoeck, D., Bitoga, J. P. and Barantwaririje, C. Avantages et inconvénients des divers modes de couverture dans les caféiers au Burundi. *Café, Cacao, Thé* **1994**, *XXXVIII* (1), 41–48.

[8] Snoeck, D. and Snoeck, J. Programme informatisé pour la détermination des besoins en engrais minéraux pour les caféiers arabica et robusta à partir des analyses du sol [with English translation]. *Café, Cacao, Thé* **1988**, *XXXII* (3), 201–218.

[9] Snoeck, J. and Jadin, P. Mode de calcul pour l'étude de la fertilisation minérale des caféiers basés sur l'analyse du sol [with English translation]. *Café, Cacao, Thé* **1990**, *XXXIV* (1), 3–21.

8.2
Soil Protection

Ch. Lambot and P. Bouharmont

8.2.1
Introduction

The absence of international agreements between coffee-producing countries has generated a surplus in coffee production, registered on a worldwide scale. This, in turn, has lead to significant fluctuations in the price of coffee, already greatly influenced by market laws and speculation. The general trend to lower the price is dangerous as coffee producers lose the incentive to maintain the quality of their product. However, this trend will, undoubtedly, lead to a re-evaluation of the price of quality coffee as consumers learn to appreciate good coffee. It is therefore of paramount importance that agronomists in charge of the upkeep of coffee plantations master the techniques required for the production of quality coffee.

The influence of cultivation techniques on the quality of the coffee produced is a subject that has not been fully researched to date. Isolated observations have been made during experiments and, although these should be further developed, they already point to the main factors which need to be taken into consideration.

In the following survey of maintenance techniques we will try to highlight the effects these techniques can have on the quality of the coffee produced with a view to guiding agronomists in the difficult choice they will have to make between cost-effective quantity and/or quality production.

A considerable number of maintenance techniques are known to date. They range from relatively simple farming practices to sophisticated technologies and include practices like shading, mulching, chemical weed control and multicropping. The choice of a specific method will initially depend on the goals the coffee grower would like to achieve but available resources and environmental conditions will also influence it.

A total inventory of ways and means would scarcely be suitable here, and may well prove to be extremely difficult to establish, but we will attempt to provide the basic agronomic principles which can influence the farmer's choices, and help him to set up appropriate and adapted farming practices.

8.2.2
Weed Control

8.2.2.1 Principles and Objectives
Natural vegetation consists of a mixture of dicotyledons and monocotyledons among which grasses generally compete with the coffee trees. Grasses are great consumers of nitrogen, the most important nutritive element in coffee growing. They often weave a net of rhizomes, intertwining with the roots of the coffee trees, which slows down their development. Weeds compete with coffee trees for moisture and nutrients, and sometimes impair allelopathic effects. Their action is detrimental to coffee growth (Pict. 8.17a).

Ideally, weeding should aim at limiting the growth of harmful vegetation and, as far as possible, to permanently eliminate noxious vegetation, particularly grasses.

The negative effects of weeds depend mostly on their type, the fertility of the soil and the hydrous status of the ground. Perennial weeds and sedge grass are the most competitive and the most difficult to eliminate. Annual weeds which develop

Picture 8.17a Effect of noxious grass on coffee development. Left: Coffee grown with leguminous cover. Right: Coffee grown with allelopathic noxious grass. Experiment at University of Hawaii - Manoa. From: J. N. Wintgens.

only during the rainy season are less damaging.

It is important to be able to recognize the types of weeds in order to assess their negative effects on the coffee trees and determine the most suitable methods for their elimination. Perennial grasses such as *Digitaria scalarum*, *Cynodon dactylon* and *Imperata cylindrica* present the worst problem. These grasses grow very easily in soil exposed to sunlight and tend to spread if the dicotyledons are eliminated.

Traditionally, weed control is brought about by mulching, by intercropping, by ground cover vegetation (cover crops), by slashing back and by weeding. Certain cultivating techniques ensure ground protection by diminishing the amount of direct sunlight and improving the balance between dicotyledons and grasses:

- Shading, though it often entails a decrease in production
- Fertilizing, which also increases the foliage density of the coffee trees
- Pruning by topping the coffee trees which generates improved ground cover by the foliage
- Increasing the density of the plantation
- Planting ground cover vegetation, although certain species tend to die out after a few years.

8.2.2.2 Ground Cover Vegetation

8.2.2.2.1 Advantages
Ground cover vegetation is useful for a number of reasons but, first and foremost, the aim is to totally or, at least, partially eliminate noxious adventice plants. This can be achieved either by installing competitive plants or by replacing noxious plants with plants which are beneficial to coffee. Leguminous plants, for instance, enrich the soil with nitrogen absorbed from the atmosphere and erect plants contribute mulching material which helps to maintain soil moisture. The use of ground cover vegetation can significantly reduce maintenance costs in a plantation.

The most frequently used cover plants are *Pueraria* sp., Greenleaf *Desmodium*, Seedless *Mimosa*, *Stylosanthes* sp., *Flemingia congesta* and *Arachis pintoi*.

8.2.2.2.2 Maintenance of Ground Cover Vegetation
The maintenance of ground cover vegetation depends on the type of cover plant used. This choice is, in turn, determined by growth habits and the reproduction cycle of the cover plants. Weeds need to be kept in check and prevented from competing with the ground cover vegetation as this can present a maintenance problem within the coffee plantation.

Growth Habit
Cover plants should be slashed at the end of the rainy season, just before the long dry season. This is necessary to prevent them from competing with the coffee trees for moisture.

Picture 8.18 Cover plant, *Flemingia congesta*, having reached the height for cutting down. From: C. Lambot.

Creeping and climbing plants such as *Pueraria* sp. should be kept under control regularly either by slashing or simply by flattening them down. Protection of coffee trees against climbing cover plants is important. This entails cleaning a protective circle or strip under the coffee plants and ensuring frequent maintenance rounds. Depending on the growth rate of the ground cover vegetation, a minimum of three to four annual rounds should be scheduled. Slashed residues are generally left on the spot and do not hinder plant re-growth. Collecting them and using them as mulch under the coffee trees is both difficult and time-consuming.

Erect cover plants like *Flemingia congesta* or *Leucaena leucocephala* are generally sown in two or three rows between the rows of coffee trees. Their growth must be controlled so as not to hinder the development of the crop trees. Once they are approxi-

mately 1–1.5 m high, they should be cut down, leaving a short stump of 20 cm (Pict. 8.18). The cuttings are then spread under the coffee trees where they disintegrate into quality mulch.

Reproduction

Ground cover vegetation can be annual or perennial.

In the case of annual plants like *Calopogonium mucunoides* it is necessary to re-sow the plant or to let it go to seed before the last cutting. This is a major drawback, often increased by the need to slash back the cover plant before the beginning of the dry season. The seeds produced will, however, germinate when the rains set in and ensure the continuity of the ground cover vegetation. The spineless *Mimosa invisa* presents this drawback though plants can survive for several years if the dry season is not too severe.

Perennial plants like *Pueraria* sp. need to be replanted after a few years because as they get older, their vegetation thins out. Replanting should take place at the beginning of the rainy season.

Occasionally, replanting causes problems due to the need to avoid repeated leguminous cultivation in the same soil. An alternative technique is to plant *Pueraria* sp. at the beginning of the plantation and main-

Picture 8.19 Ground cover composed of mulch and *Leucaena leucocephala*. From: C. Lambot.

tain it as ground cover vegetation. Then, planting *Pueraria* is easy in direct sunlight, especially after the original vegetation has been burnt down. Once the *Pueraria* degenerates, soil maintenance can be ensured by herbicides which present less of a threat to adult coffee trees than to a young plantation.

Perennial species have the advantage of being more or less permanent. This could nevertheless present a drawback when changes in ground cover are decided and the perennial plant needs to be uprooted. This operation can be difficult and expensive especially with *Leucaena* because of its deep rooting system and lignified stems (Pict. 8.19).

The initial choice of species therefore needs to take potential uprooting difficulties into consideration.

Weeding

Whatever the species chosen, ground cover vegetation is never perfect and there are always weeds which need to be eliminated. Selective weeding is done by hand by uprooting the weeds or with the help of tools to pull out weeds with rhizomes.

Localized treatments can be made with selective herbicides for leguminous cover plants. In the case of erected cover plants, paraquat treatments can be sprayed on the leafless stumps after pruning, thus only weeds which have developed between the rows and on the sides will be affected by the herbicide treatment.

Impact on Quality

Soil protection with creeping ground cover vegetation can lead to hydrous and mineral competition with the coffee trees as is the case with *Stylosanthes*. This is not favorable to a high production of quality coffee, although it is less damaging than spontaneous weed growth, especially grasses. Where there is hydrous competition, the

cherries produced by the coffee trees are smaller.

On the other hand, mulch supplied by the cuttings of erect ground cover vegetation such as *Flemingia*, ensures protection against evaporation and supplies nutrients to the coffee trees. Furthermore, *Flemingia* has very deep taproots so it does not compete with coffee for soil moisture.

8.2.2.3 Mulching

8.2.2.3.1 Goals

Robusta coffee requires an annual rainfall of over 1200 mm and a dry season which does not exceed 3 months. Arabica coffee requires average temperatures between 15 and 24 °C, and an annual rainfall between 1000 and 3000 mm, but with an optimum of 1500 mm. The length of the dry season should not exceed 4 months [2]. When these conditions are not fulfilled, either due to a lack of annual rainfall or an excessively long dry season, mulching is the best way to limit soil evaporation and compensate for the lack of water.

Test results on the Arabica coffee trees in the Rubona Station in Rwanda show that in rainfall conditions of 1100–1200 mm and a dry season of 3.5 months, permanent mulching is the best technique for soil maintenance. Coffee trees under permanent mulch double their production when compared to coffee trees maintained by clean weeding [10]. In Kenya, Arabica coffee trees under permanent mulch produce 20 % more than those maintained by clean weeding and 44 % more than those with *Centrosema plumerii* ground cover vegetation [6]. A new experiment made at Rubona has recently confirmed the superior efficiency of permanent mulch when compared to leguminous ground cover as shown in Tab. 8.2.1.

Table 8.2.1 Comparative production of Arabica coffee under mulch and under living ground cover at Rubona in kg of green coffee per hectare per year [7]

Type of ground cover	Production	Percentage in comparison
Permanent mulch	1417	100
Desmodium sp.	1152	81
Stylosanthes sp.	1140	80
Mucuna sp.	1040	73
Glycine max (soja)	995	70

Picture 8.20 Mulched young coffee trees.

In Cameroon, Bouharmont [1] made a study on the impact of mulch compared to that of ground cover vegetation on the husbandry of groundwater. Mulch consisting of grasses was the most efficient method of maintaining a satisfactory rate of humidity in the soil and compensating the effects of drought. Mulch reduces the effect of soil evaporation on the entire surface it covers [1] and Pict. 8.20.

Limiting soil evaporation and compensating the severity of the dry season are really necessary in regions where rainfall conditions are not ideal to coffee. Further benefits stem either directly or indirectly from the use of mulch. Cultivation on a very steep slope, as practiced in certain regions of Central Africa, entails reverting to anti-erosive methods as shown in Tab. 8.2.2.

Mulch provides other significant advantages, which Gaie and Flémal [3] describe as follows:

- It eliminates the effects of direct impact of rain on the ground, protecting the surface area and preventing compacted topsoil.
- It regulates the soil temperature by forming a screen which reduces temperature variations at ground level.
- It reduces the development of natural vegetation growth and its negative effects on the coffee trees.
- It improves the pH and soil structure by providing humus in the layers explored by the coffee tree roots and by increasing water retention and the adsorption capacity of the exchange complex.
- It provides nutrients such as nitrogen, potassium, phosphorus, magnesium and calcium, as well as trace elements.

The quantities of elements provided vary according to the quantity and the nature of the mulch (see Tab. 8.2.3). An annual

Table 8.2.2 Total earth loss over 3 years of various anti-erosion methods installed on a slope of 45 % in the Mumirwa (Burundi) [5]

Ground cover	Anti-erosion measure	Total earth loss (tons/ha)
None, ground constantly cultivated	none, Wieschmeier plot	1494.7
Subsistence crops	Silt pits	169.6
Coffee	permanent mulching	0.4

Table 8.2.3 Quantity of nutritive elements (kg/ton of green material)

Source of mulch	N	K₂O	P₂O₅	CaO	MgO
Tripsacum laxum	2.9	11.0	1.1	0.4	0.6
Setaria sphacelata	2.7	6.4	0.8	0.3	0.4
Eragrostis olivacea	3.5	2.1	0.5	0.6	0.4
Banana (stem and leaf)	6.8	10.3	1.1	7.3	2.6

mulching of 40 tons of green matter per hectare of *Eragrostis* theoretically covers a large part of the nutritive needs of the coffee tree. Where banana leaves and stems are used, those needs are, in theory, totally covered. Banana tree residue has been defined as high-performance mulch material for coffee tree production and soil enriching [2].

8.2.2.3.2 Methods

Mulching material is derived from natural areas like pastures or fallow marshland. It can also be obtained from specially cultivated land or from crop residue. Annual mulching needs are generally estimated to be between 20 and 40 tons of dry material per hectare. This entails cutting grass on 3–6 ha of fallow land or using 0.5–1.0 ha to grow plants for mulching. Mulch plants, like *Themeda gigantea* or *Tripsacum laxum*, also require considerable fertilization to compensate for the depletion of mineral elements.

Mulch should be spread thick enough to limit ground evaporation. In Burundi, the recommendded thickness is 10 cm. Mulch should be spread perpendicular to the main slope so as to allow maximum rainfall penetration into the ground and to avoid water run-off [3]. In Kenya, alternate row mulching has been found to yield better results [6]. In Burundi, this technique has resulted in a production equivalent to total ground mulching, with low water stress for the coffee trees.

8.2.2.3.3 Problems Linked to Mulching

Correct mulch spreading is not always easy and entails compliance with a number of basic rules.

The most favorable period for mulching is at the end of the rainy season in order to minimize losses due to soil evaporation during the dry season.

Mulching material is less available in highly populated areas. This shortage is increased by the expansion of coffee plantations with high mulch requirements. The need for large quantities of vegetal matter for mulch competes with other farming activities like livestock or composting. The use of harvest residues, leaves and banana stems leads to a transfer of fertilizing material from staple food plots and banana plantations to coffee plantations.

Mulching also requires a significant amount of labor. This aspect is accentuated if the areas for growing mulching material are far from the coffee plantations. The average labor required to mulch a hectare of adult coffee trees varies from 450–600 man days/year. The exclusive use of agricultural soil for the production of mulching material, and the high labor requirements necessary for the cutting, transport and spreading of mulch, significantly increase production costs. The use of mulch is therefore more suitable for specific cases where its production costs are not exorbitant and where marginal rainfall conditions make it absolutely necessary.

Ideally, to counteract these drawbacks, sufficient mulching material should be produced either on the plantation itself (cover plants, shade tree pruning residue) or in the immediate vicinity of the coffee plots.

8.2.2.3.4 Impact on Coffee Quality

Ground mulch generates bolder coffee beans and an improved rate of green coffee per weight of fruit [10]. This improvement is probably due to a better hydrous exchange within the coffee plantation. Nevertheless, according to Mitchell [9] mulch can have an unfavorable influence on the quality of the bean and, as a result, on the liquor because it generates a soil imbalance of K/Mg brought about by a higher supply of potassium and a relative lack of magnesium. Roasted coffee sometimes has a duller aspect and gives a less acid liquor. It is most important to monitor the effect of cationic imbalances in the soil brought about by mulching because of their impact on both the quality and the production of coffee. See also Section I.8.1.

Soil sampling should take place every 5 years so that necessary corrections by means of calcareous magnesia or a magnesia fertilizer can be implemented.

8.2.2.4 Plastic Sheets

Ground protection can also be ensured with black plastic film. This is sold either in rolls or in rectangles which are adapted to the normal width of the alleys between coffee rows. The black plastic film is rolled out along the rows once the soil has been prepared and prior to planting. The coffee trees are planted in slits made in the film. Rectangles of black plastic film are easier to handle than rolls. They also have an added advantage because they can be used in plots where the trees are already planted.

The beneficial effects of plastic film covering are similar to those of mulch as it also prevents moisture evaporation, eliminates weed competition and protects the soil against erosion. The end result leads to significant increases in yield. In 1988, Snoeck noted a 35 % increase in yield over 3 years in an Arabusta plantation in the Ivory Coast where a plastic film cover had been used for the first time in preference to natural ground cover.

In Brazil, yields were doubled in plots where rectangles of plastic film of 1 m × 1.2 m were used as a cover over clean weeding.

As a result, it can be concluded that black plastic film could be a worthwhile substitute for mulch, particularly where mulching material is not easily obtainable and where labor is expensive.

Plastic film should be chosen on the basis of its resistance, i.e. its thickness and the level of the UV protection provided.

8.2.2.5 Manual Soil Maintenance

As mentioned previously in the paragraph about cover crops, natural vegetation which consists of a mixture of dicotyledons and monocotyledons competes with coffee trees for moisture and nutrients and sometimes impairs allelopathic effects. Grasses are the most noxious because they are great consumers of nitrogen, the most important nutritive element in coffee growing, added to which, they often weave a net of rhizomes, intertwining with the roots of the coffee trees, which slows down their development.

Manual and mechanical soil maintenance entails the slashing of natural vegetation when it grows too high and the elimination of noxious species wherever possible. Weeds can be eliminated by traditional methods or with herbicides but careless uprooting can cause erosion and often exposes the soil to the adverse effects of direct sunlight.

Certain perennial weeds like *Digitaria scalarum*, *Cynodon dactylon* or *Imperata cylindrica* flourish in direct sunlight and propagate by means of an underground network of rhizomes which have to be pulled out in order to eliminate the weed. When this is done manually or with tools, it can often damage the roots of coffee trees. Systemic herbicides are useful in such cases.

Sedge grasses also present a threat to coffee cultivation, especially in irrigated conditions or without shading. *Cyperus rotundus* and *Cyperus esculentus*, in particular, have roots which produce substances that are toxic to coffee trees. Grass roots also intoxicate the soil but to a lesser degree. Dicotyledons offer less competition to the coffee trees for nutrients and moisture.

Slashing can also reduce the negative effect of weeds but if repeated eliminates dicotyledons and favors the growth of sedge and grasses which are not desirable. Where water and nutrients are not too limited, slashing is preferable to total weeding. As a general rule, weeds should be regularly slashed during the rainy season and uprooted at the beginning of the dry season to produce mulch.

Clean weeding is commonly practiced for ground maintenance on plantations. It is generally believed to be beneficial to coffee production but leaves the bare soil very vulnerable. Where rainfall conditions and soil fertility are favorable, it is preferable to practice selective weeding. Dicotyledons are thus maintained, whereas grasses are regularly uprooted.

Frequently, soaring labor costs no longer allow for slashing or uprooting. It has become necessary to revert to mechanized soil maintenance or the use of herbicides.

8.2.2.6 Mechanized Soil Maintenance

Mechanization requires easy, unhindered access to the coffee plot and a slope which does not exceed 10 %. The use of a rotary cutter is not advisable because it encourages the spread of creeper grasses. It is preferable to use a low speed rotavator which cultivates the soil to a depth of 2–3 cm without altering its structure [11].

The use of mechanical tools should be alternated with the herbicides in order to avoid compacting the soil.

8.2.2.7 Chemical Soil Maintenance – Herbicides

Herbicides are used to replace manual and mechanical weeding, either to compensate for the lack of labor or to lower maintenance costs. It can also be useful to destroy certain particularly noxious or resistant hardy weeds like *Imperata cylindrica* and sedge grasses.

The herbicides currently used in coffee cultivating can be subdivided into two main groups: contact herbicides and systemic herbicides.

8.2.2.7.1 Contact Herbicides

Gramoxome (paraquat) is the most currently used contact herbicide in coffee plantations. It acts rapidly on natural vegetation by destroying the aerial organs of the plants without seriously affecting the root system. Its action is noticeably equivalent to cutting back vegetation practically to ground level. It is most efficient against annual dicotyledons, whose stumps generate less second growth than most grasses. Its action is more long-lasting at the beginning of the dry season than in the rainy season. Gramoxone does not alter the ratio of ground coverage between dicotyledons and grasses [1].

Gramoxone must be applied with shields to reduce the risk of spray dispersion and possible contact of the product with the coffee trees. Damage can be caused if the product directly contacts the young tissues of the coffee tree. In the case of plantation maintenance with herbicides, eight treatments are necessary in the first year, with 6–8 l/ha with an active ingredient content of 200 g/l. During the following years, the number of treatments can be reduced to four or five localized applications using no more than 2–3 l/ha. Additional treatments can occasionally be made with glyphosate or dalapon to eliminate perennial grasses.

8.2.2.7.2 Systemic Herbicides

The leaves absorb systemic herbicides without destroying them. The sap then carries the herbicide to the weed organs and blocks the production of substances necessary for photosynthesis. This leads to the death of the aerial organs and the root system. The effects of systemic herbicides are all-round and more permanent than those obtained with contact herbicides. Later regrowth will mostly consist of random natural weed association. The main beneficial effect of systemic herbicides is their anti-grass action. As a result, dominant grass is progressively replaced by dicotyledons.

Glyphosate (Roundup) based herbicides are the most commonly used systemic herbicides in coffee cultivation. They have no residual effect on the soil. They are more expensive than contact herbicides but they are better for controlling sedge grasses. They are most effective when used at the flowering stage of the plants. Regular application doses correspond to 3–5 kg/ha of a solution at 33%. Glyphosate is generally used as an additional treatment to paraquat to eliminate more tenacious weeds. To ensure best results the herbicide requires 6 h without rainfall after spraying. It can be applied with a low volume spray system. See Section II.7.

Other herbicides can sometimes be used in special cases. See Tab. 8.2.4 and Tab. 8.2.5.

8.2.3
Shade Trees

8.2.3.1 Goals

Shading has a direct influence on the composition of the vegetal ground cover as it fa-

Table 8.2.4 Main herbicides used in coffee cultivation

Technical name	Market name	Rate (kg or l/ha)	Observations
Herbicides applied to the soil (prior to emergence)			
Diuron	Diuron	1.25–3.75	not for young trees under 2 years old
Simazine	Simazine	3.10–5.00	any time after transplanting
Oxyfluorflen	Goal	3.00–6.00	any time after transplanting
Herbicides applied on vegetation			
Dalapon	Basfapon	5.50–7.80	directed spray at any time
Paraquat	Gramoxone	1.50–4.00	directed spray at any time
Glyphosate	Roundup	1.00–6.00	directed spray at any time
Fluazifop butyl	Fusilade	1.00–8.00	can be used at any time, including in the nursery

Table 8.2.5 Examples of herbicide mixes used in coffee cultivation

Mix	Commercial name	Rate (kg or l/ha)	Observations
2,4-D + diuron	Tufordon	3.00–5.00	applied on vegetation
2,4-D + glyphosate	Comand	2.50	applied on vegetation
Ametrine + simazine	Gesatop Z	3.00–4.00	applied to the soil prior to emergence

vors dicotyledons at the expense of grasses. Shade also reduces soil and leaf temperature, and maintains more regular levels of temperature. At the same time, it also reduces vegetative growth and flower induction in coffee trees which can, in turn, lower production levels.

On the other hand, by cushioning the impact of atmospheric conditions, shade also has a positive effect, particularly when leguminous shade trees are used because they enrich the soil with OM and nitrogen. Shade influences cultivation practices like pruning, fertilization, choice of OM, etc. Shade trees are also an efficient anti-erosion measure.

Coffee grown under shade will need to be rejuvenated by stumping more frequently which means that there will be more years with reduced yields, but this drawback is less of an issue with dwarf varieties like Caturra or Catimor. This problem can be totally avoided by topping the coffee trees.

In cases where the soil needs a nitrogen supply, shading is beneficial and reduces the need for excessive nitrogen inputs. In plantations where *Leucaena malacocarpa* was used as a shade tree, it has been noted that a nitrogen fertilizer was less necessary than in sun-exposed plots on the same soil.

Shade trees are recommended as a protective measure when environmental conditions can be difficult for coffee, particularly in areas which are exposed to high temperatures, heavy rainfall, frost or hail and, more importantly, when intensive agricultural practices are not acceptable.

Once again, the choice resides in balancing the costs related to fertilization (purchase and spreading) against the obtainable market prices for coffee.

8.2.3.2 Choice of Shade Trees

Suitable shade trees (Tab. 8.2.6, Pict. 8.21) should fulfill the following criteria:

- Rapid growth
- A deeply imbedded root system that will not compete with the coffee trees

Table 8.2.6 Leguminous shade trees most frequently used in coffee plantations

Name	Plant spacing (m)
Inga densiflora	10–12
Inga spectabilis	10–12
Inga edulis	10–12
Erythrina poeppigiana	10–25
Erythrina fusca	6–20
Erythrina edulis	5–10
Ibizzia carbonaria	12–20
Pithecollobium saman	15–25
Enterolobium cyclocarpum	15–20
Leucaena leucocephala	6–10
Gliricidia sepium	6–12
Albizzia stipulata	15–20

Picture 8.21 Shade composed of *Albizzia stipulata* trees. From: J. Flémal [3].

- Abundant leaf growth which is easy to control
- Constant leaf growth which persists during the dry season
- Resistance to wind
- Preferably leguminous
- Not be a potential host for pests and diseases affecting coffee trees.

8.2.3.3 Maintenance of Shade Trees

The shading of coffee trees should never be too dense as this greatly reduces production. Ideally, screening should only exclude between 25 and 40 % of potential light, and certainly no more than 40 %. This entails adequate pruning and lopping to limit the growth of shade trees. In some respects, Arabica appears to be better suited to cultivation under shade than Robusta.

The size of the shade trees must take into account particulars linked to the individual species and their growth, as well as climatic conditions which determine the pruning period of the trees.

Certain trees, like *Erythrina* sp., defoliate during the dry season while others, like *Inga* sp., for instance, keep their leaves.

There are many pruning methods for shade trees but two major techniques can be highlighted:

- Trees that generate stump shoots easily are controlled by a cyclic renewal of the stem – stems are cut when they become too big and are replaced by new stems selected amongst the suckers on the stump (e.g. *Leucaena leucocephala*).
- Species that do not sprout shoots easily are managed to generate a single main stem that branches out 4–5 m from the ground – single-stem shade trees are preferred as they compete less with the coffee trees and allow better light control.

Erythrina species allow tipping to a level of approximately 3 m from the ground on a main trunk. The branches are regularly pruned and lopped in order to control their screen effect against sunlight.

The management of shade trees should aim at creating a sufficiently high canopy so as not to hinder the growth of the coffee trees. Trees which generate shoots easily can be pruned either at stump level for a complete tree renewal or to a level of approximately 3 m above ground on a main trunk.

Eliminating shade allows for more intensive coffee cultivation. This makes it possible to adjust the intensity of coffee management to the variations of economic factors such as the ratio between the sale value of the coffee and the production costs for intensive coffee growing.

The residues of shade tree pruning can be used as firewood or as mulch (leaves and young twigs only). Certain species such as *Leucaena leucocephala* can be used as animal fodder although with certain limitations because of its mimosine content (Pict. 8.22).

In Latin America, it is generally recommended to prune shade trees twice a year, the first one after main flowering or harvest and the second one just before the start of the second period of heavy rains.

Picture 8.22 Shade composed of *Leucaena leucocephala* trees. From: W. Gaie [3].

However, pruning must always be adapted to the prevailing local environmental conditions. The guiding principle is to avoid excessive hydrous competition between the shade trees and the coffee trees during the dry season. This entails removing part of the foliage of the shade trees at the beginning of the dry season. Pruning residues spread on the ground also reduce evaporation. A further reason for pruning the shade trees at the beginning of the dry season is to allow sufficient sunlight for the blossom initiation of the coffee trees.

8.2.3.4 Temporary Shade and the Production of Timber

Temporary shade trees can be installed to protect young coffee trees for 1–2 years. They generally consist of shrubs like *Cajanus cajan*, *Crotalaria anagyroïdes* or *Flemingia congesta* which efficiently protect the young trees against overexposure to sunlight.

Forest trees for timber are sometimes used as shade trees in coffee. This is one way of diversifying plantation production and income. The main problem is the felling of the forest trees. This has to be done carefully to avoid damaging the coffee trees and, preferably, during the period when pruning and lopping is scheduled. Forest

trees used as shade for coffee should be carefully selected so that they do not compete unduly. *Cordia alliodora* or *Grevillea robusta* are possibly the best choice for Arabica and *Terminalia ivorensis* for Robusta coffee.

8.2.3.5 Impact on Quality

Shading coffee trees generally leads to the production of larger fruit. This may well be due to a lower yield of the trees and to a slower ripening period.

Shading maintains a microclimate in coffee plantations which sometimes generates a higher incidence of insects and parasites, detrimental to the quality of the coffee. On Robusta coffee trees, the damage due to the twig borer is significantly greater under shade than in direct sunlight but, on the other hand, shade distinctly reduces the development of the White Stem Borer.

8.2.4 Wind breaks

The main purpose of wind breaks is to reduce evapotranspiration in coffee leaves.

Wind breaks should be placed perpendicular to the mainstream of dominant winds. Due consideration should be given to height and to width. The zone protected by the wind break is 5–7 times its height and wide wind breaks that are not too densely planted are preferable because they offer adequate protection while still allowing wind to filter through.

The most frequently used plants for wind-breaks are: *Grevillea robusta*, *Casuarina* sp. (Filao), *Erythrina* (poplar), *Pinus patula*, *Flemingia congesta* and *Leucaena leucocephala*.

As it sometimes takes up to 10 years for the trees of permanent wind breaks to reach the desired height, it is often advisable to install temporary wind breaks which can be progressively removed as the coffee trees grow. Vegetation used for this purpose can be *Cajanus cajan*, *Crotalaria* sp., *Flemingia congesta*, *Sesbania* sp. or *Tephrosia* sp.

Branches below three meters must be cut back to avoid stunting the growth of the coffee trees.

Ideally, wind breaks should be high enough to protect the coffee trees from strong winds but sufficiently sparse to allow some wind to filter through at a lower level. See also Section I.7.

8.2.5
Infilling

Gaps due to coffee trees missing in a plantation can be caused by several factors, i.e. death following transplantation, death after an accidental incident such as lightening or localized fire, death due to old age, death due to the absence of shoots following severe pruning or death due to pests and diseases.

In order to maintain productivity, missing trees should be replanted otherwise returns will decrease. It is unfortunately very difficult to replant coffee trees on an existing plot, due to the strong competition prevalent amongst coffee trees. This competition is in direct proportion to the density of the plantation. Therefore, rigorous supervision is required at the time of planting to minimize the mortality rate.

When mortality sets in on young coffee trees, rapid replanting must be ensured as soon as climatic conditions are favorable. In adult plantations, replanting will be made according to the causes of death and the number of missing trees.

If a bacterial or fungal disease causes the decline of coffee trees, strict measures must be taken to avoid contamination of the rest of the plantation. It is often advisable to allow several years between death and replanting in order to avoid further onslaughts of the disease. If the dead coffee trees form an extensive area across the plantation, for example after lightening, fire or an attack by parasites, they will be relatively easy to replant. Competition is then limited to the coffee trees on the fringe of the area. For isolated dead coffee trees, the replanting problem is much more delicate. Root and aerial competition from the older coffee trees is difficult to restrain.

In this case, it is of paramount importance to take necessary precautions before replanting. The planting hole must be large enough to interrupt the development of roots from the neighboring trees. It must then be well fertilized to encourage rapid growth. The most favorable period should also be selected for replanting.

Infilling must take place when aerial competition is diminished, i.e. when the adult coffee trees have been stumped for rejuvenation. Every precaution must be taken to encourage rapid growth for the young coffee trees. As the new, young coffee plants are frequently scattered throughout the plantation, there is a risk that supervision may be neglected. Their development should, therefore, be closely monitored on a regular basis. Thick mulch is necessary to slow down the growth of vegetation in the immediate vicinity and to accelerate the growth of the young coffee trees. It is essential to use well-nurtured nursery plants belonging to a vigorous coffee material. In Cameroon, the frequent failure in Robusta infilling only became successful when the healthiest clones were used. The average rate of infilling in a new plantation after 1 year is approximately 5 %.

The trees selected for transplanting should already be trained in the nursery where any necessary bending or tipping has already been done.

Despite all precautions, successful replacement remains risky and it is not uncommon to see new coffee trees stunted or dying. Where the percentage of missing coffee trees is too high, it may even be necessary to replant the entire plot.

8.2.6
Conclusions and Future Options

The maintenance of coffee trees is a technique which is the result of many years of practice, and calls on the know-how and experience of farmers and agronomists alike. Coffee growing is a wonderful example of the integration of practical experience and agronomical technology.

The level of intensive management used depends on environmental conditions, available resources and the farmer's knowledge and training. These various factors inevitably lead to a great variety of types of farming but the end result is the sustainable production of coffee.

It would be unwise to advise farming practices which lead to soil depletion and the deterioration of the plantation. Associating coffee growing with shade trees or cover plants is an effective way of developing an environmentally friendly coffee cultivation system where the coffee tree grows and develops in soils which maintain their initial fertility.

Bibliography

[1] Bouharmont, P. L'utilisation des plantes de couverture et du paillage dans la culture du caféier arabica au Cameroun. *Café, Cacao, Thé* **1979**, *XXIII* (2), 75–102.

[2] Carvajal, J. F. *Cafeto – Cultivo y Fertilización.* Instituto Internacional de la Potasa, Bern/Switzerland, 1984, 254.

[3] Flémal, G. and Gaie, W. *La Culture du Caféier d'Arabie au Burundi.* Publication du Service Agricole 14. AGCD, Brussels, 1988, 198.

[4] Guizol, P. *L'érosion des Sols au Burundi. Résultats après Trois Années de mesures sur les Parcelles Expérimentales de Rushubi.* ISABU, 1983.

[5] Hömberg, B. F. and Ripken, R. R. *Guia para la Cafeicultura Ecologica.* GTZ, 2001, 153.

[6] Haarer, A. *Modern Coffee Production.* Leonard Hill, London, 1962.

[7] ISAR. *Plantes Economiques – Le Caféier d'Arabie.* Rapport Annuel 1968. **ISAR**, **1968**, 183–204.

[8] Mitchell, W. Grasses for mulching coffee. *Kenya Coffee* **1968**, *XXXIII*, 327–335.

[9] Mitchell, H. W. Cultivation and harvesting of the arabica coffee tree. In: *Coffee. Vol. 4. Agronomy.* Clarke, R. J. and Macrae, R. (eds). Elsevier, London, 1988, 43–90.

[10] Snoeck, J. Le caféier d'Arabie à Rubona. *Bull. d'Inf. de l'INEAC* 1959, *VIII* (2), 69–99.

[11] Snoeck, J. Cultivation and harvesting of the robusta coffee tree. In: *Coffee. Vol. 4. Agronomy.* Clarke, R. J. and Macrae, R. (eds). Elsevier, London, 1988, 91–127.

8.3

Pruning

Ch. Lambot and P. Bouharmont

8.3.1

Introduction

Correct pruning requires detailed knowledge of the morphology of the aerial system of coffee trees.

Arabica and Robusta coffee trees are characterized by the dimorphism of vegetative stems. These are divided into vertical growth (orthotropic) and horizontal growth (plagiotropic). Vertical growth forms the trunk of the coffee tree and horizontal growth forms the branches.

The activity of the apex bud, located at the tip of the orthotropic shoot, ensures height growth. At each leaf axil there are two types of buds: serial buds located close to the leaf insertion and an extra-axillary bud. Serial buds and offshoots from the stump can develop into suckers or orthotropic shoots. The extra-axillary bud, on the other hand, develops only one horizontal branch, which is called the primary if it starts from the trunk. Primary branches also provide the leaf axils with serial and extra-axillary buds. These buds can develop leafy sprouts to produce the secondary plagiotropic branches, or floral buds. The development of floral buds calls for floral initiation procedures.

An illustration of the morphology of the coffee plant can be found in Section I.1. These physiological characteristics are the unavoidable factors which determine pruning techniques. When pruning, one needs to bear in mind the following basic principles:

- The elimination of a primary branch is irreparable as it can never be replaced.

- The elimination of the apex bud leads to the growth of a number of suckers or orthotropic shoots.
- The elimination of the top of a primary branch leads to the development of a number of secondary branches.

Furthermore, one must remember that:

- The development of floral buds is encouraged by exposure to direct light and inhibited by shading.
- Floral buds on Robusta can develop on wood that is less than 1 year old (Fig. 8.10).
- Arabica trees also start floral development on wood that is less than 1 year old but also bear flower buds on older wood of 1–2 years old.
- Coffee trees are often prone to a biennial bearing pattern where a high production year generally alternates with a year of low production.
- Overproduction consumes reserves in the tree and leads to the decay of the fruit-bearing branches (die-back) which can even extend to the roots of the coffee tree.

After a few years of non-restricted, natural growth coffee trees develop their characteristic habit. Its single main stem reaches full height and generates the primary branches from which the secondary and tertiary develop. After several years, the growth of the coffee tree slows down and there is virtually no further development of new primary branches. If left to itself, the coffee tree enters a long period of reduced vegetative and generative activity with little development. As berries are borne by young plagiotropic branches, production decreases. See [3] and Fig. 8.11, Pict. 8.23, Pict. 8.24 and Pict. 8.25.

Pruning should be geared to achieve the following results:

- Maintain the morphology and the balance of the coffee tree in order to facili-

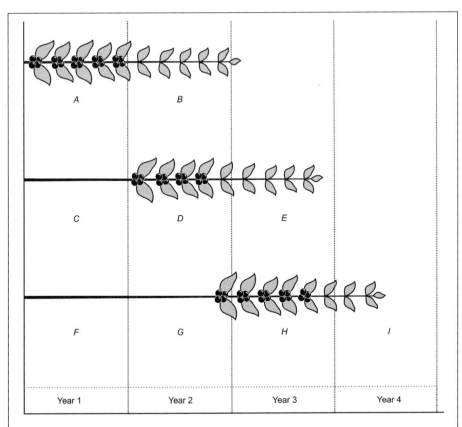

A: One year old wood in fructification
B: Young wood, less then one year in vegetative development
C: Two year old unproductive wood
D: One year old wood in fructification
E: Young wood, less then one year in vegetative development
F: Three year old unproductive wood
G: Two year old unproductive wood
H: One year old wood in fructification
I: Young wood, less then one year in vegetative development

Observation: Each year the productive length of the branch is getting shorter – therefore it is advisable to remove the branch after two yields.

Figure 8.10 Robusta coffee fructification cycle on a freely growing multiple-stem tree.

tate harvesting and phytosanitary treatments.
- Eliminate all dead wood, unproductive branches and suckers.
- Encourage the growth of new stems and fruit-bearing branches.

- Ensure correct ventilation and allow sunlight to penetrate throughout the coffee tree in order to diminish the impact of pests and disease and encourage floral induction.

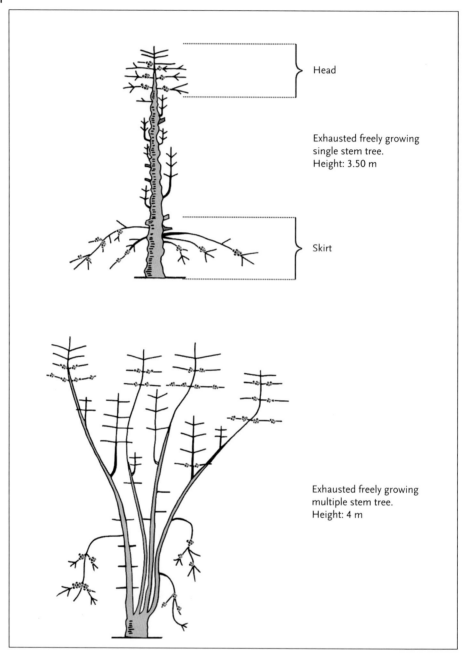

Figure 8.11 Exhausted coffee trees.

Picture 8.23 Over 80-year-old single-stem Robusta coffee tree. From: J. N. Wintgens.

Picture 8.24 Old Robusta multiple-stem tree almost unproductive because of mismanagement and lack of pruning. From: J. N. Wintgens.

Picture 8.25 Abandoned unproductive multiple-stem freely growing Robusta trees. From: J. N. Wintgens.

- Avoid overproduction which generates a risk of die-back.
- Reduce biennial bearing.

8.3.2
Pruning Systems

Several systems exist for pruning. Each system has its advantages and disadvantages which are often the subject of animated discussions amongst planters and agronomists, each extolling the virtues of the system they prefer. Generally speaking, however, the choice of a system depends on the major differences which exist between the physiological reactions of coffee trees and their immediate, specific environmental conditions.

When determining the most suitable pruning system, there are four basic factors which need to be taken into consideration:

- The choice between single-stem and multiple stem systems
- The choice of natural, non-restricted growth or growth controlled by capping
- Methods for formative and maintenance pruning
- Methods for rejuvenation.

On the other hand, the choice of pruning methods or practices is determined by the species, plantation density, the morphology of the tree and its growth pattern, shading, essential phytosanitary treatments, and to what degree intensive farming practices are applied. Conversely, where pruning practices have been decided before installing the plantation, their choice will determine the density of the plantation, the use of shade, fertilization practices, etc. The availability and cost of qualified labor is also a determining factor.

8.3.2.1 Pruning of Freely Growing Coffee Trees

8.3.2.1.1 Pruning of Freely Growing Multiple Stems

Methods encouraging multiple stem growth without capping are the most common in coffee production. The system first entails preliminary pruning to train the coffee tree on the basis of several vertical stems.

Various methods are used:

- Capping the plants either in the nursery or in the field
- Bending the plants either in the nursery or in the field (Agobadia)
- Inclined planting in the field.

See Pict. 8.26, Pict. 8.27 and Pict. 8.28.

Capping involves nipping the terminal bud of the plantlet so that the extra-axillary buds can develop. This operation can be

Picture 8.26 Trained coffee tree by Agobadia – From: J. N. Wintgens.

Picture 8.27 Trained coffee tree by Agobadia showing emerging suckers. From: Ch. Lambot.

Picture 8.28 Robusta coffee tree trained on three stems by capping – young stems were bent by hand. From: J. N. Wintgens.

carried out on very young plants in the nursery, as soon as three pairs of leaves are well developed. Capping generally allows the growth of only two suckers. If more suckers are required, it is necessary to repeat the process on the newly developed suckers.

Bending the plantlets involves keeping the principal stem inclined horizontally by using a wooden hook fixed into the soil or a rope as shown in Fig. 8.12. Bending the plants can also be carried out in the nursery, by blocking the stem of the plantlets between two bags. This practice gives rise to several suckers, which is useful when it comes to selection.

When plants are bent or capped in the field, it is important to carry out these operations during the hot, humid season when growth is intense and suckers develop faster. Formative pruning carried out in the nursery means that the seedlings need to remain in the nursery for a longer period of time, i.e. 3–6 months more, because growth has been slowed down.

Planting young trees at an angle of 30° also generates spontaneous sucker growth. In this case there is less spontaneous growth than with bending so it is sometimes necessary to bend the coffee trees when suckers do not appear, added to which this system can also create an imbalance in the root system of the tree. Inclined planting is chiefly used for Robusta in a hot, humid climate. We are not much in favor of this method, because experiments have shown lower yields during the first years.

The number of shoots selected after bending in inclined planting varies from two to four for Arabica and from two to five for Robusta depending on environmental conditions and the density of the plantation. In the case of bending and inclined planting, the main stem of the coffee tree should be cut off. The suckers to retain are selected according to their distribution and vegetative growth. They should stem off at a level between 5 and 20 cm from the ground.

The sooner formative pruning is carried out, the sooner the coffee trees will yield. However, yields which are too precocious lead to a risk of die-back. This can have a devastating effect on the future development of the tree. Therefore, early formative pruning can only be carried out when all the environmental conditions are favorable.

After formative pruning, primary brunches and, occasionally, some secondary branches, are kept for growth and production. Maintenance pruning only involves regular de-suckering and the removal of lower primary branches which have produced twice. Dead wood should also be removed.

The pruning cycle of coffee trees depends on the speed of growth, which in turn depends on local environmental conditions and the type of planting material used (dwarf varieties or regular sized Arabica trees). The usual cycle is 5–7 years.

The freely grown multiple-stem technique is recommended in the following cases.

- For Robusta coffee trees:
  - Under high temperatures at low altitudes
  - In favorable environmental conditions
  - With varieties adapted to this technique
  - On condition that the optimal length of production cycles is strictly respected. This entails frequent stumping.
- For Arabica coffee trees:
  - In favorable environmental conditions
  - Where labor availability is limited
  - Where phytosanitary treatments are unnecessary or where the equipment needed to treat high coffee trees is available
  - Under careful management where there is no, or very little, shade
  - On condition that the optimal length of production cycles is strictly respected, i.e. the stumping frequency.

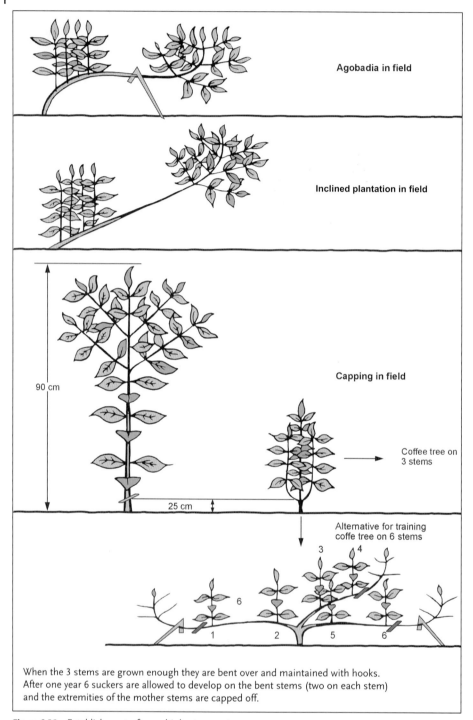

Agobadia in field

Inclined plantation in field

Capping in field

90 cm

25 cm

Coffee tree on 3 stems

Alternative for training coffe tree on 6 stems

3   4

6

1   2   5   6

When the 3 stems are grown enough they are bent over and maintained with hooks. After one year 6 suckers are allowed to develop on the bent stems (two on each stem) and the extremities of the mother stems are capped off.

Figure 8.12 Establishment of a multiple-stem system.

8.3.2.1.2 Pruning of Single Stems

Pruning single stems without capping is seldom practiced. It is used for "Cova" planting when several trees are placed in the same hole which is a variant of the multiple-stem system where formative pruning is not necessary and for "hedge planting" at 4 m × 0.80 m. In this case, the single-stem trees are pruned, but not capped, in order to allow for easy mechanical maintenance of the inter-rows and to avoid intertwining branches and stems in the rows.

In these cases maintenance pruning only consists of regular de-suckering, the removal of dead branches and the lower primaries which have already fructified twice. In other words, the same type of maintenance pruning as is applied to freely grown multiple stems.

Pruning systems for freely grown single-stem trees are also especially advisable for compact coffee trees like Caturra, Catuai or Catimor planted at high density.

8.3.2.2 Pruning System with Capping

The vertical stem is capped at a height of 1.6–1.8 m above ground to keep the trees low to facilitate picking and spraying. The crop is borne initially on primaries and then on secondary and tertiary branches that are replaced progressively as they become old and unproductive.

8.3.2.2.1 Single Stems

Apart from the capping, this system requires no formative pruning because the coffee trees are managed on a single stem.

When the trunk reaches a height of 1.6–1.8 m the terminal bud is nipped or capped at the end of the stem. This is to ensure that production continues by means of the annual renewal of horizontal branches which have fructified.

Depending on the age of the tree, capping can be staggered at increasing height. This method has been used for Robusta, though with mitigated success, with a view to strengthening the lower primary branches which tend to disappear, as the tree grows older. The staggered capping is sometimes considered to be formative pruning. This practice is of little interest except that it reduces the risk of overproduction at a young age. It also reduces yields significantly (Fig. 8.13).

Certain basic rules should be observed when cap cutting coffee trees as shown in Fig. 8.14.

The Costa Rica or double-lyre formative pruning is an alternative between single and multiple stem training. The original stem is capped and after each capping two suckers are allowed to develop into orthotropic stems until about four to eight are formed. One reason for using this method is to keep the fruiting layer as low as possible. When the stems are exhausted they are cut off and a new sucker is trained to restore the structure (Fig. 8.15).

Maintenance pruning entails the elimination of certain secondary or tertiary fruit-bearing branches in order to maintain beneficial aeration and light throughout the coffee tree. This practice is specially recommended for Arabica, which produces more secondary and tertiary branches than Robusta, but it requires a considerable amount of well-trained labor, which is not always easy to come by and which significantly increases production costs.

On the other hand, it is essential to regularly remove suckers, which develop along the main stem, particularly at its top to the detriment of fruit-bearing branches. In this way Robusta coffee trees can be exploited for up to thirty years or more without further intervention.

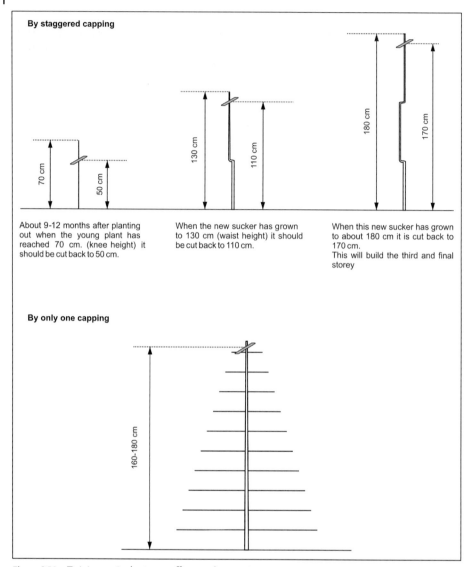

Figure 8.13 Training a single-stem coffee tree by capping.

As a general rule, pruning with capping is better suited to the Robusta tree, which tends to develop fewer secondary and tertiary branches than the Arabica. Indeed, it has been shown in different trials in Cameroon and in the Ivory Coast, that some Robusta coffee clones give better yields when topped on one or two stems, because these clones develop secondary and tertiary branches naturally. After a few years only the upper main primary branches remain, all lower primaries die off, resulting in an enormous umbrella which goes down to the soil. No pruning was applied during 15 years of the experiment except regular de-suckering. This simplified pruning sys-

The cutting must be made 5 cm above a node. The short peg remaining above the node is to reduce the risk of the stem splitting.
At the same time one or two primaries should be cut back just beyond their first leaves, otherwise the weight of these two side branches may split the main stem.

Figure 8.14 Diagram of cap cuts.

tem should be well accepted by traditional farmers who are loath to regular rejuvenating of their coffee trees because they always see a few flowers or fruits that will be sacrificed. Table 8.3.1 shows the yields of the Ivory Coast trial planted in 1973.

Multiple stems were rejuvenated in 1980 and 1985 and harvest was from the lung stems.

The Arabica coffee tree shows a greater development of secondary and tertiary plagiotropic branches. Selective pruning is therefore necessary to prevent stifling by surplus vegetative growth.

The capped single stem pruning system is advisable in the following cases:

• For Robusta:
  – Where plantation is not too dense
  – For varieties or clones which generate enough secondary and tertiary branches
  – For cultivation under shade
  – For cultivation at high altitude with low temperatures
  – Where the soil is of poor fertility
  – Where stem borers are not a major problem because of their difficult control in older stems.
• For Arabica:
  – Where the varieties planted are of a tall morphology and require repeated phytosanitary treatments

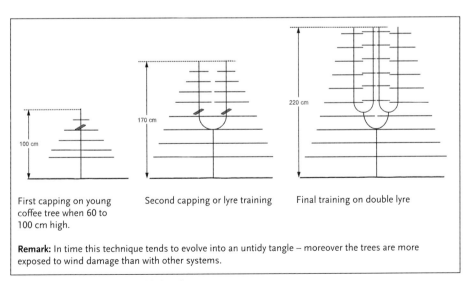

First capping on young coffee tree when 60 to 100 cm high.

Second capping or lyre training

Final training on double lyre

Remark: In time this technique tends to evolve into an untidy tangle – moreover the trees are more exposed to wind damage than with other systems.

Figure 8.15 Costa Rica or double-lyre formative pruning.

Table 8.3.1 Pruning trial on Robusta coffee planted in 1973 at IRCC, Divo, Ivory Coast (yields in kg/ha clean coffee – cumulative per period)

Years	Cumulated mean for six clones			Cumulated mean for clone 126		
	One topped stem	Two topped stems	Multiple stems rejuvenated	One topped stem	Two topped stems	Multiple stems rejuvenated
1975–1979	4054	4229	6707	4958	5059	5551
1980–1984	6879	7048	7851	10952	10340	9073
1985	2522	2670	796	3309	3724	1481
1986	1267	1411	947	no data	no data	no data
Totals	14722	15448	16301	19219	19123	16105

– Where husbandry is under constant control
– Where unfavorable agroclimatic conditions prevent healthy regeneration of new orthotropic shoots
– Under shade where the freely growing system would require frequent regeneration.

8.3.2.2.2 Multiple-stem Pruning with Capping

Pruning capped multiple-stem trees can be carried out on both Robusta and Arabica. In the case of Robusta, it is a variant of the capped single stem pruning but practiced on two, three or even four stems. The stems are developed by formative pruning in the same way as they are in freely grown conditions and then they are capped. Both multiple-stem and single-stem trees require similar maintenance pruning and their development is more or less the same.

For Arabica coffee tree farming, a relatively recent variant of this system has been developed to facilitate phytosanitary treatments. After formative pruning, the stems are left to grow freely until a height of 2.5 m. They are then capped to a height of approximately 2 m. This height is maintained during the remainder of the farming cycle.

Maintenance pruning must strive to keep the number of secondary branches under control and eliminate certain primary branches in order to aerate and allow light to penetrate to the center of the canopy [2].

This system entails intensive labor input for the selection of the secondary branches and frequent de-suckering. The lower primary branches also tend to disappear and the coffee tree becomes umbrella shaped with the fruit yield concentrated at the top of the tree. To avoid this type of development, the farming cycle is limited to a 3–5 year period.

The exploitation of capped multiple-stem coffee trees is advisable in the same conditions as in the case of capped single-stem trees (see above).

8.3.2.2.3 Advantages and Disadvantages of Free Growing versus Capping

The advantages and drawbacks of capping or not capping coffee trees can be summed up as follows:

• Capping tends to stabilize production in cases where marginal conditions could lead to biennial coffee crops. The crop develops at an easily accessible height which facilitates both harvesting and phytosanitary treatments. On the other hand, particularly for Arabica, maintenance pruning is more difficult to carry out and requires a considerable number of experienced workers.

- The freely growing system is simpler and quicker to manage. Its major drawback is that it can lead to significant biennial crops. Harvesting and phytosanitary treatments are also more difficult when the stems grow above 2.5 m.

Problems linked to biennial production caused by the pruning cycle can be partially reduced by plantation management. Where the size of the plantation is large enough, it is advisable to rejuvenate a different sector each year. If, for instance, the production cycle runs on a 5-year basis, one-fifth of the plantation should be rejuvenated each year. In this way, problems linked to biennial crops on the plantation as a whole are avoided even if they remain valid for individual trees.

8.3.3
The Rejuvenation of Coffee Trees

8.3.3.1 Principles and Objectives

Whatever pruning system is adopted for coffee management, it is frequently necessary, or at least advisable, to renew the fruit yielding branches during the course of their productive life. The timing and methods used for rejuvenation depend on the system chosen for coffee tree management.

When the freely growing system is practiced, rejuvenation of the stems must take place regularly and be scheduled systematically. Postponing rejuvenation leads to the development of tall stems with reduced production and which are difficult to harvest.

On the other hand, in the case of capped stems, rejuvenation is seldom necessary and sometimes even useless. It can, how-

ever, be justified after die-back due to over production or when the trunk has been badly damaged by attacks of borer insects on one or several stems.

8.3.3.2 Rejuvenation of Freely Growing Coffee Trees

8.3.3.2.1 Rejuvenation of Freely Growing Multiple-stem Coffee Trees

For non-capped coffee trees, most of the berries develop on the primary plagiotropic branches. As the stems grow, the berry bearing branches get higher. The orthotropic stems are periodically replaced when they are exhausted. The lower primary plagiotropic branches must be removed when they have produced twice or when they begin to droop.

The annual production curve of multiple-stem coffee trees shows a decrease in yield after a few years. Ageing or exhaustion is characterized by a slow down in growth of the vertical axes, which flatten out at their tips, and by a significant reduction in the number of new productive branches.

The timely renewal of exhausted vertical stems with new shoots growing from the stump will enable production to return to its normal level. This practice also avoids harvesting difficulties due to the height of the stems [3]. Cutting back the old stems leaving only a lung stem to be removed after the next harvest rejuvenates the stems. Shoots then develop at the base of the coffee tree and, after selection, the most healthy and better placed shoots become the replacement stems (Pict. 8.29 and Fig. 8.16).

Total removal of all the old stems at once by stumping can lead to a physiological imbalance in the coffee tree. The trees are unable to survive on the nutrient reserves of

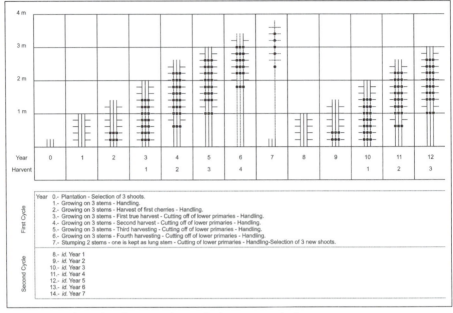

Figure 8.16 Robusta fructification cycle on a freely growing multiple-stem tree.

Picture 8.29 Multiple-stem Robusta tree with one lung stem, 5 months after stumping. From: Genagro.

the roots and stump alone. This imbalance may create deformed shoots or even, in the case of a high production year, cause the coffee tree to die of exhaustion [4]. If rainfall conditions are unfavorable or the soil is not very fertile, blank stumping of the coffee trees can endanger the next harvest or even the life of the trees (Pict. 8.30, Pict. 8.31, Pict. 8.32, Pict. 8.33).

One thing is for sure, this type of stumping deprives the coffee farmer of any return on his coffee trees for 1 or 2 consecutive years and causes a severe physiological shock to the coffee trees.

Classical rejuvenation pruning methods present several difficulties. To begin with the rejuvenation of old stems by stumping on a breather curtails production for a year. Coffee growers tend to prolong production cycles for fear of losing the harvest on the old stems. Biennial production remains frequent with this method though, as mentioned earlier, when managed correctly by

Picture 8.30 Stumped multiple-stem Robusta tree with its first shoots (with no breather stem left). From: J. N. Wintgens.

Picture 8.32 Stumped Arabica multiple-stem coffee tree trained on 3 stems. From: Genagro.

Picture 8.31 One row of stumped multiple-stem Robusta trees with no breather left. From: J. N. Wintgens.

Picture 8.33 Stumped Arabica multiple-stem coffee tree trained on 3 stems (after first partial harvest). From: Genagro.

sectors, it should not necessarily impact the plantation as a whole.

Where cultivation takes place on steep slopes, coffee trees tend to shade each other. This makes it more difficult for vigorous suckers to grow during rejuvenation pruning. Self-shading results in a lower exposure to sunlight and is often accentuated by inadequate spacing between the coffee trees as they need adapt to the slope of the land. It can also lead to excessive density in the plantation.

Although coffee trees seem able to survive severe pruning under excellent agroclimatic conditions, this is not the case in regions with low rainfall, little shade and poor soil fertility.

Arabica trees are not as sturdy as Robusta. As a result, young replacement stems grow less rapidly and produce less quickly. These drawbacks can be partially avoided by adopting slower and less physiologically stressful methods of rejuvenation as explained below:

During the first year, two stems are retained on the coffee tree and the level of productive storeys is raised to 1.7 m from the ground. As a rule, it is preferable to remove the stems located in the center of the coffee tree, unless their production is clearly superior to that of the external stems. Stumping the stems enables sunlight to reach the center of the coffee tree and encourages the growth of suckers, thus counteracting the effects of self-shading.

The stump shoots are selected during the first year.

During the second year, only one stem is retained. The choice of the stem to be eliminated should bear in mind the creation of lanes with lung stems and lanes without. These lanes should be oriented east-west in order to ensure maximum light penetration inside the coffee trees.

This technique described by Finney [2] is practiced in Kenya to ensure a maximum space and light for the development of new suckers. Favorable conditions for the growth of new shoots can be maintained despite the presence of residual old stems. The plagiotropic branches of the lung branches that do not bear any fruit must be removed to reduce shading on the suckers.

During the third year, the residual old lung stem is stumped (Fig. 8.17).

This pruning technique can be varied where environmental conditions are particularly favorable (i.e. sufficient rainfall and a short dry season), or in high density plantations with 3000 or more coffee trees per hectare. In this precise case, only one stem will be retained in the first year and selected to form lanes with lung stems on an east–west axis. These lung stems will be retained for two years.

8.3.3.2.2 Rejuvenation of Freely Growing Single-stem Trees

Smaller varieties of Arabica and Catimors are generally grown on a freely growing single stem. Depending on environmental conditions and cultivation practices like shading, rejuvenation becomes necessary after 5–8 years because the lower plagiotropic branches become unproductive, growth slows down and yields drop significantly.

Various techniques like integral stumping, stumping to 50 cm above the ground or stumping after selecting a replacement shoot are used for rejuvenation. Integral stumping can only be implemented without risk where the agroclimatic conditions are most favorable otherwise it is not recommended, as it could lead to serious physiological problems (Fig. 8.18).

Stumping to 50 cm above the ground must preserve the primary plagiotropic

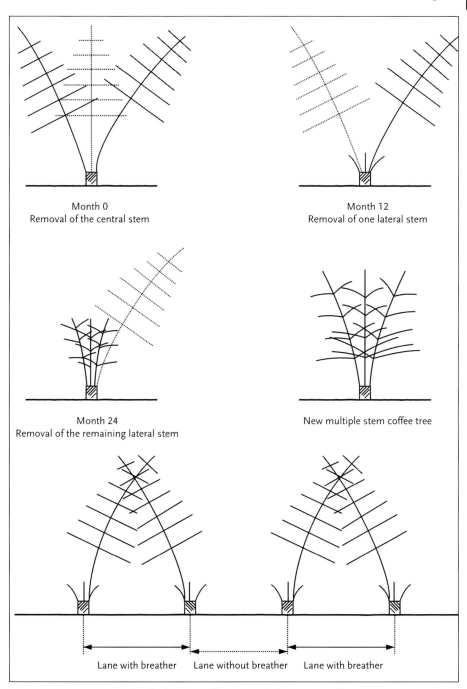

Month 0
Removal of the central stem

Month 12
Removal of one lateral stem

Month 24
Removal of the remaining lateral stem

New multiple stem coffee tree

Lane with breather   Lane without breather   Lane with breather

Figure 8.17   Rejuvenation of multiple-stem coffee trees with the creation of lans.

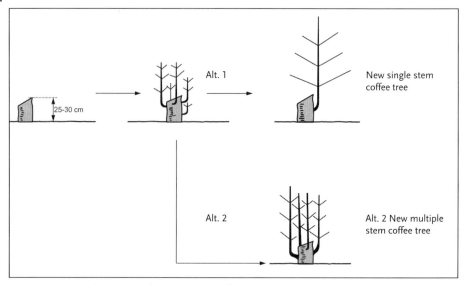

Figure 8.18   Integral stumping of a single-stem coffee tree with conversion option in single- or multiple-stem tree.

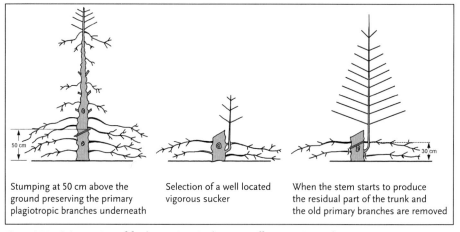

Stumping at 50 cm above the ground preserving the primary plagiotropic branches underneath

Selection of a well located vigorous sucker

When the stem starts to produce the residual part of the trunk and the old primary branches are removed

Figure 8.19   Rejuvenation of freely growing single-stem coffee trees (smooth system).

branches which maintain sap circulation in the coffee trees. These branches should occasionally be lopped so as to allow sunlight to reach the stumps. Stumping is followed by the selection of a replacement shoot which must stem off as low as possible. When this shoot starts producing, but not before, the stump can be cleared by cutting off the residual parts of the old stem and the old primary branches (Fig. 8.19).

Another technique has been tried with success on Catuai in Burundi. It consists of removing the plagiotropic branches below a height of 1 m to encourage the growth of a number of shoots at the base of the coffee trees. The future orthotropic

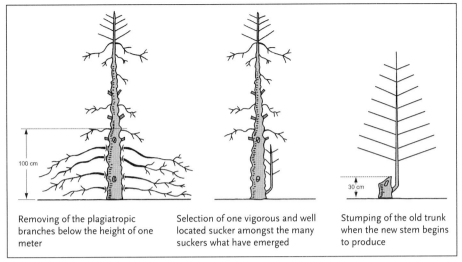

| Removing of the plagiatropic branches below the height of one meter | Selection of one vigorous and well located sucker amongst the many suckers what have emerged | Stumping of the old trunk when the new stem begins to produce |

Figure 8.20 Rejuvenation of freely growing single-stem coffee tree (smooth systems).

stem is then selected from these shoots. The old stem is removed when the new stem begins producing, generally after about 2 years. This technique avoids interrupting production and prevents a physiological shock to the coffee trees (Fig. 8.20).

High-density plantation with small sized coffee trees hinders the growth of new shoots because of excessive shade. This drawback can be partially overcome by pruning the coffee trees in alternate order either 1/4/2/5/3 or 1/3/5/2/4 for a 5-year cycle or 1/3/2/4 for a 4-year cycle. The "rock n roll" pruning method used for smaller sized coffee trees, more specifically for Catimor, seems well adapted. In this case capping and stumping are carried out alternately on single-stem trees, row by row in alternate order. Management in rows allows easier access for phytosanitary treatments and harvesting. It also improves wind protection (Fig. 8.21).

8.3.3.3 Rejuvenation of Capped Coffee Trees

8.3.3.3.1 Rejuvenation of Capped, Single-Stem Trees (Classical Single Stem)

Capped coffee trees grown on a single stem, become umbrella-shaped after a few years. Their productivity starts to decrease and rejuvenation becomes necessary. Several techniques exist for rejuvenating capped, single-stem trees.

Rejuvenation by stumping generally entails cutting down the trunk to a height of 30–40 cm and keeping, wherever possible, several primary inferior branches to ensure the circulation of sap in the plant. Shoots then develop at the base of the coffee tree and the number of shoots retained depends on choice of farming methods as this type of rejuvenation can also lead to a multi-stem system (see Fig. 8.22A).

When the coffee tree has no lower branches, the lung stem can be the stem on which a primary branch was maintained. If there is too much growth, a section of the branch may also be used.

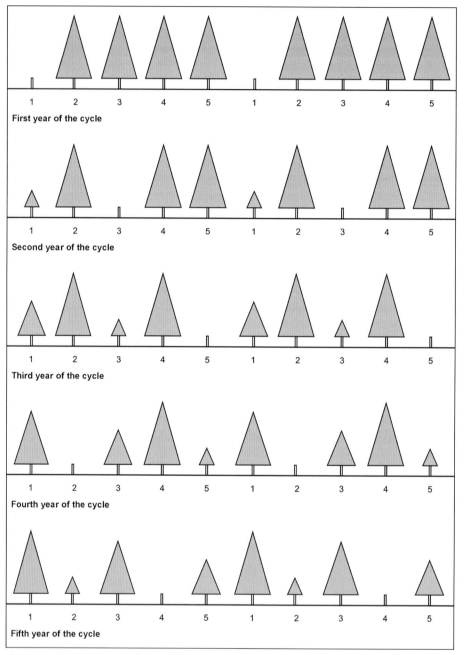

Figure 8.21 Rejuvenation row by row in alternative order (small size coffee trees).

When this method is implemented, it is then necessary to systematically remove all the suckers which sprout off the trunk during the period of re-growth. Only the sucker selected for rejuvenation should be maintained. Stumping of the trunk needs to be carefully managed because it can cause a breakage of the replacement shoot (see Fig. 8.22B).

When the trunks are very old, the "hinge" system is advisable. This consists of notching the trunk just enough to fell it without totally stumping it. The sap can still circulate and generate new suckers. The felled tree is finally stumped and removed once the new shoots have developed. Another alternative is to remove the umbrella and stump the tree at mid-height (see Fig. 8.22C).

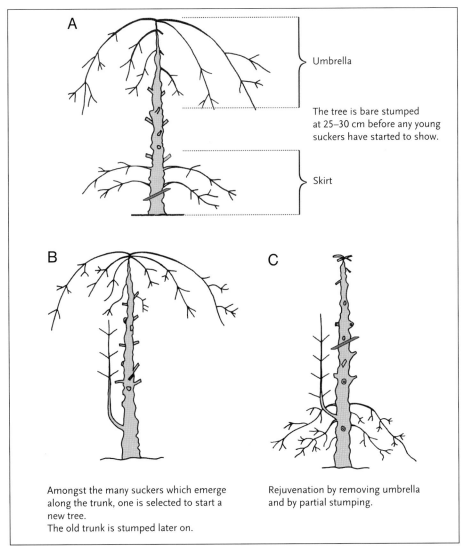

A

Umbrella

The tree is bare stumped at 25–30 cm before any young suckers have started to show.

Skirt

B

Amongst the many suckers which emerge along the trunk, one is selected to start a new tree.
The old trunk is stumped later on.

C

Rejuvenation by removing umbrella and by partial stumping.

Figure 8.22 (A)–(C) Rejuvenation of capped single-stem coffee trees.

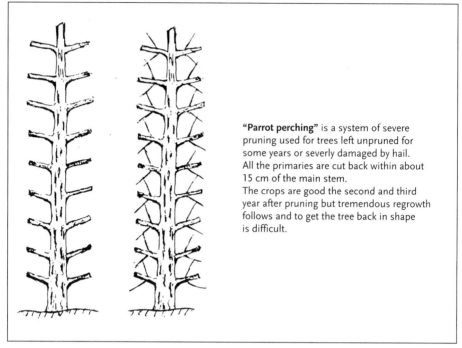

"Parrot perching" is a system of severe pruning used for trees left unpruned for some years or severly damaged by hail.
All the primaries are cut back within about 15 cm of the main stem.
The crops are good the second and third year after pruning but tremendous regrowth follows and to get the tree back in shape is difficult.

Figure 8.23 Parrot perching pruning.

Rejuvenation by "parrot" pruning is carried out mostly on Arabica coffee trees. All the primary branches are cut back to 10–20 cm from the trunk so as to allow the development of new secondary branches (Fig. 8.23).

On Robusta, if the canopy is thinning, a spontaneous shoot growth on the stump can be retained and capped to the desired height. This allows for either refilling the entire height or only the diminished part of the canopy.

The different methods described above highlight the fact that rejuvenation for capped, single-stem trees requires case-to-case study. Each tree has particular needs and the system chosen for rejuvenation must meet these needs. This implies that well-trained, experienced workers with the necessary know-how are required to manage plantations of this type.

This is not the case for the rejuvenating of freely growing single-stem trees where all the trees on a given plot can be rejuvenated at the same time and in the same way.

8.3.3.3.2 Rejuvenation of Capped Multiple-Stem Trees

When this pruning system has been chosen, it is necessary to carry out stem rejuvenation because the lower primary branches tend to disappear as the tree ages causing it to gradually become umbrella-shaped. This is frequently the case for Robusta.

Rejuvenation by stumping can be achieved by using the same methods as those practiced on non-capped multiple-stem coffee trees if all the stems need to be rejuvenated. The same drawbacks apply to the physiology of the coffee trees where agroclimatic conditions are unfavorable.

Partial rejuvenating can also be carried out when one or two stems have been damaged by accident or by an attack of borer insects. The damaged stems are stumped and replaced by new shoots while the healthy stems are maintained. The healthy stems act as lung stems and ensure production.

The rejuvenation of capped multiple-stem trees also requires case-to-case studies as well as the selection of rejuvenating techniques which are best adapted to individual needs. Similarly, this also requires experienced, well-trained workers with know-how.

8.3.3.4 Criteria for Rejuvenation

Objective guidelines, like the height of the stems and the years of production as well as evaluation guidelines based on the observations and the experience of agronomists, need to be taken into account prior to taking any decisions related to the rejuvenation of coffee trees.

Coffee trees can only be stumped when their stems are over 3 m high and their growth has slowed down. Excessive height significantly increases the risk of stem breakage at the time of harvest. It also leads to a reduced yield. If the stems have already produced for over 5 years and the predictions for the following year are poor, rejuvenation should be considered since maintaining the stems would only pointlessly set the rejuvenation of the coffee plants back for a year during which growth will be more vegetative than productive. It is therefore advisable to start rejuvenation when productivity reaches a low ebb. This represents an investment in the future, which is all the more beneficial as production is already low.

On larger scale plantations, it is advisable to schedule rejuvenating by sectors each year. This method helps to maintain a regular annual yield (Fig. 8.24).

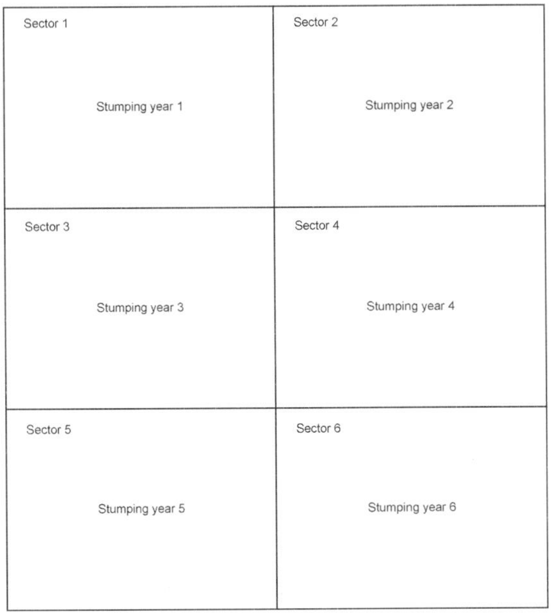

Sector 1	Sector 2
Stumping year 1	Stumping year 2
Sector 3	Sector 4
Stumping year 3	Stumping year 4
Sector 5	Sector 6
Stumping year 5	Stumping year 6

Figure 8.24 Rejuvenation by sector.

It is advisable to accompany the rejuvenation operation of the coffee trees by a severe thinning of the shade trees and a regeneration of the cover plant (plowing).

8.3.4
Methods and Tools

Whatever the methods chosen for pruning and rejuvenation, satisfactory results can only be ensured by strictly respecting basic rules when it comes to the actual implementation.

Pruning must be performed with a perfectly sharpened compass saw, and cutting must start from the exterior and progress towards the interior of the coffee trees. This prevents the stems from damaging the bark when they fall. The sawn surface must be at an angle of 30° so that water will run off and avoid rotting the base of the stumps. The stumps must be perfectly cleaned to encourage the growth of shoots and some branches must be lopped to allow adequate sunlight to reach the stumps.

If there is a risk of diseases developing, like bacteria spread by *Fusarium lateritum* in Arabica trees or *Fusarium xylarioides,* responsible for tracheomycosis on canephora, or even *Xyllella fastidiosa,* preventive measures must be taken when pruning the coffee trees. All tools should be disinfected between work on each coffee tree and pruned branches must be burnt, not removed to other parts of the plantation.

Rejuvenation pruning should never take place during a severe dry season. This would accentuate the stress suffered by the trees. It is therefore advisable to wait until the return of the first rains.

It is imperative to consider the tree in its entirety, bearing in mind that any intervention in the aerial part of the tree will affect the root system. Pruning which is either too severe or performed during the dry season or after a year of high yield, will inevitably generate a physiological shock to the tree. This, in turn, will force the coffee tree to draw on its reserves to the detriment of the root system.

Using gradual pruning methods and managing the plantation in sections where rejuvenation is scheduled in different sectors at different stages, will ensure a more regular production. This, in turn, will avoid the insecurity of low-income returns in years where production is low because of rejuvenation or pruning.

8.3.5
How Pruning Influences Coffee Quality

Little information exists on the influence of pruning methods on the quality of coffee. Nevertheless, it is a well-known fact that the finest fruits are harvested from the primary branches, whereas fruit produced on secondary and tertiary branches is smaller. Therefore, bean size will be improved by pruning without capping, where the primary branches of the tree yield the best berries.

However, this observation can be belied by the impact of disease and parasites on quality. Capping facilitates phytosanitary treatments and more efficient health control, as the entire coffee tree is easier to reach. This is of paramount importance to coffee quality. Atomizer spraying could, however, solve this problem for non-topped coffee trees.

8.3.6
Conclusions

This chapter has undoubtedly highlighted the complexity of pruning and rejuvenating methods, but it has also shown that the different methods can be grouped into systems. The choice of a system, adapted to each plantation, will be conditioned by factors like the variety of coffee, the on-site agro-environmental conditions and the availability of experienced workers.

Pruning and rejuvenating systems should be implemented with a view to ensuring regular production without weakening the coffee trees. The system chosen should also take into account the work force required and the costs of labor involved.

This means that the best system chosen in each case should balance an optimal management of the coffee trees against the prevailing agroclimatic conditions and the resources available for production. It is very likely that the best system and specific methods will only be found after a period of trial and error, as close observation of the trees and their reactions to different pruning methods is necessary to determine which methods are most suitable.

A rational, discerning choice of pruning and rejuvenating systems will make all the difference between a balanced, sustainable and well-adapted plantation and a badly managed plantation with perpetual production problems and ongoing costs.

Controlled observation carried out over recent years has highlighted the fact that rejuvenation by means of repeated drastic methods like total stumping, generally exhausts coffee plants prematurely. This is partially due to the depletion of reserves stored in aerial tissues but it is also caused by the gradual die-back of the root system.

As a result, partial stumping leaving a breather branch, is now considered to be a more advisable rejuvenating method, as well as parrot perching and other well-tested, less aggressive methods.

Bibliography

[1] Anon. Pruning. *Kenya Coffee* **1984**, *49* (572), 9–13.

[2] Finney, A. Le système de taille en tiges multiples écimées pour le caféier Arabica. *Café, Cacao, Thé* **1988**, *XXXII* (1), 33–43.

[3] Flémal, G. and Gaie, W. *La Culture du Caféier d'Arabie au Burundi.* Publication du Service Agricole 14. AGCD, Brussels, 1988.

[4] Mitchell, H. W. Cultivation and harvesting of the arabica coffee tree. In: *Coffee. Vol. 4. Agronomy.* Clarke, R. J. and Macrae, R. (eds). Elsevier, London, 1988, 43–90.

[5] Wrigley, G. *Coffee.* Longman, London, 1988.

[6] Coste, R. *Caféiers et Cafés.* Maisonneuve and Larose, Paris, 1989.

8.4
Irrigation

R. Goodyear

8.4.1
Some Climatic Characteristics of Coffee-growing Countries

Coffee is grown in tropical rainy climates (Koppen's classification) within the tropics (i.e. between latitudes 25° N and 25°S) where average temperatures do not fall below 18 °C. Within the tropical belt, the two main species of coffee are adapted to tropical rainy climates as modified by altitude, which reflect conditions at their centers of origin. Arabica coffee is grown in the altitude range 500–2000 m above sea level (m.a.s.l.); Robusta coffee is invariably found in the altitude range 200–800 m, whilst Liberica coffee is a product of a tropical lowland or coastal environment. Optimum altitudes for bean quality of both Arabica and Robusta coffees are highest at the equator.

Within the areas of the globe with a tropical rainy climate, rainfall conditions are variable. In general tropical rainfall follows the movement of the sun so that at the equator the sun passes overhead twice a year and rainfall distribution tends to be bimodal and equinoctial. At the tropics of Cancer and Capricorn rainfall is monomodal and occurs in the months around the summer solstice – June in the north and December in the south.

For much of the equatorial belt rainfall with bimodal distribution significant rainfall occurs in every month and average temperatures show little variation through the year. With increasing distance from the equatorial belt seasonal differences become increasingly apparent – the length of the rainy season decreases, the number of dry months increases and seasonal differences in temperatures are more extreme. In other words, there is a distinct separation of wet and dry seasons, which may also be referred to as summer and winter seasons. Large parts of Brazil provide an exception to this generalized description of the rainfall pattern related to latitude since even at latitudes of more than 20° S the dry season months may be ill-defined covering only 2–3 months at most.

Average annual rainfall totals for some of the major coffee producing countries in the world are given in Table 8.4.1 together with the number of dry months and length of the dry season (a dry month is defined arbitrarily as a month in which the long-term average rainfall is less than 50 mm) From this data it must be concluded that there will be a need for rain water conservation measures to overcome or minimize the soil moisture deficits during the dry months and/or a need for the application of irrigation water to avoid such deficits.

8.4.2
Effective Rainfall

In the tropical rainy climates of the major coffee-producing countries a high proportion of the rainfall is seasonal and occurs in high intensity storms. Thus, during the wet season the rainfall exceeds the level of evapotranspiration and evaporation occurring from the soil and there are seasonal surpluses of water. On the other hand, during the dry season coffee consumptive use of water exceeds rainfall and seasonal deficits occur.

Irrigation needs in coffee is obviously depend on the amount of rainfall which can be utilized by the coffee crop, commonly referred to as the effective rainfall. The factors which determine effective rainfall in a perennial crop, such as coffee, include the

Table 8.4.1 Typical rainfall patterns in some major coffee-growing countries in Africa and South America

| Location | | | | Average annual rainfall (mm) | Dry season characteristics. | | Average annual tempera-ture (°C) |
Country	Place	Latitude	Altitude (m.a.s.l.)		Months with < 50 mm average an-nual rainfall	Total rain-fall in dry months (mm)	
Uganda	Kampala	00 32° S	1144	1244	none	–	21.2
Tanzania	Lyamungu	03 20° S	1470	1545	Aug–Oct (3)	102	20.2
Ethiopia	Jimma	07 39° N	1750	1490	Nov–Feb (4)	63	19.0
Angola	Carmona	07 58° S	829	1627	Jun–Sep (4)	50	19.8
Zimbabwe	St Pauls	17.80° S	1330	950	Apr–Oct (7)	102	20.5
Columbia	Chinchina	04 58° N	1360	2536	None	–	20.0
Costa Rica	San Jose	09 56° N	1170	1850	Dec–Apr (5)	70	21.0
Brazil	Santa Teresa	21 10° S	620	1410	Aug–Sep (2)	75	22.0

rainfall characteristics (seasonality or distribution and intensity) and land characteristics (topography, soil textures and soil moisture status).

During the wet season, rainfall surpluses on a coffee plantation are lost to run-off and deep percolation. The losses to run-off are influenced by a number of factors including topography or slope, the condition of the soil surface and the soil textures in the profile. Coffee is a crop of hilly and mountainous regions and so is frequently grown on sloping land. Obviously, any gradient will increase potential rates of run-off and higher rates of run-off results in reduced infiltration into the soil profile. The condition of the soil surface also influences rate of run-off. A bare soil will give rise to higher rates of run-off than from land which has some form of obstruction to surface flows of water down the slope such as a cover crop, a mulch, belts of vegetation on contour bunds and contour ditches designed to hold water for infiltration.

Soil textures, by governing the rates of infiltration and permeability of a soil, not only influence the losses of rainfall surpluses to run-off but they are the main factors determining the losses to deep percolation. Heavy soil textures (clay) have low infiltration rates (1.0–10.0 mm/h) and so, with high intensity rainfall, (at least 50 % of rain in the humid tropics falls at an intensity of more than 10.0 mm/h) losses to run-off are substantial. Coarse soil textures have high infiltration rates, e.g. with a sandy loam soil infiltration rates are within the range 12.5–75 mm/hour and losses to deep percolation are commonly more significant than losses to run-off.

The moisture status of the soil profile when rainfall occurs has an important influence on run-off and deep percolation losses. Coffee is grown where rainfall exceeds an annual average of around 1500 mm per annum and the wet season is usually around 7–9 months. Such a seasonal rainfall pattern results inevitably in prolonged periods of saturated soil profiles. Thus, when high intensity rainfall occurs on a saturated profile the total rainfall may be lost to run-off and deep percolation.

8.4.3
Coffee Cultivation Practices to Maximize Effective Rainfall

8.4.3.1 Introduction

The climatic conditions under which much of the world's coffee is grown include a dry season. Inevitably during the dry season the coffee crop is subject to moisture stress which restricts yield levels and in some cases the quality of coffee produced. Where the use of irrigation is not an option, there are a number of cultivation practices which can be employed to more effectively utilize the rainfall through improvements in the replenishment of the soil water reservoir and in the conservation and use of stored soil moisture, and so increase the productive potential of the coffee. These practices are discussed below.

8.4.3.2 Soil and Water Conservation Measures

Coffee plantations are frequently located in hilly, sometimes mountainous, areas which are susceptible to soil erosion. Thus, coffee lands are often sloping and, since a large proportion of the rainfall is of high intensity, are susceptible to soil erosion. In these circumstances soil conservation measures are necessary to avoid serious losses of topsoil to erosion by rainfall run-off. If topsoil losses are allowed to continue unchecked over long periods the depth of useful soil profile shrinks, the volume of soil that can be exploited by the coffee plants declines and the soil water reservoir is reduced. Much of the root development in coffee is concentrated in the upper 25 cm of the profile so that a reduction in depth of topsoil has serious adverse effects not only on water availability but also on plant nutrient levels in the profile.

The speed of flow of rainfall run-off over the soil surface determines the quantity of sediment that can be transported and therefore the erosion rate. Thus, the main objective of soil conservation measures is to reduce the speed of flow of rainfall run-off. At the same time the slower rate of flow of run-off allows for higher infiltration rates and so assists with water conservation by improving the rate of replenishment of the soil water reservoir. The rates of rainfall run-off can be significantly lowered with the manipulation of the soil surface using techniques involving soil movements and/or vegetation either in the form of cover crops or in barrier rows along the contour.

Soil conservation techniques that are designed to reduce the speed of flow of rainfall run-off by changes in the land surface include the formation of contour bunds and the construction of bench or individual terraces. In addition, structures such as contour ditches and grassed waterways may be installed which are designed to conduct excessive rainfall from the sloping land avoiding erosion and soil losses whilst at the same time encouraging infiltration. The establishment of vegetative cover of the ground between the coffee plants is often used to control erosion.

Any vegetation growing amongst the coffee trees will compete with these trees for the available supplies of soil moisture. Consequently care has to be taken to ensure that the advantages of erosion control that arise from a cover crop or established barrier vegetation are not outweighed by the adverse effects of increased moisture stress on coffee yields. In most areas with a long dry season vegetative measures to control soil erosion will only be appropriate if the species of vegetation used has a natural dormancy period during the dry season and so does not compete for moisture with the coffee at a critical time.

8.4.3.3 Use of Artificial Ground Cover

The use of artificial ground cover for the reduction of losses of stored soil moisture by evaporation from the soil surface is practiced in some countries. The materials used for such ground cover also control weed growth and, thus, prevent losses of stored soil water through the unwanted vegetation.

The beneficial effects of the use of black plastic sheeting as ground cover were amply demonstrated at the Abengourou Research Station in the Ivory Coast. The annual average rainfall at Abengourou is 1370 mm, with a dry season of 6 months, November to March (two dry months with less than 50 mm and four lean months with 50–100 mm average). The estimated average accumulated deficit in adult coffee is around 400 mm at the end of the dry season. In a trial begun in 1974 various clones of Arabusta coffee (a hybrid of Arabica × Robusta) were planted in the center of bands of black plastic sheeting 1.20 m in width and compared with the same clones planted in the traditional manner. The overall yields of coffee obtained from harvests in the first three years are shown in Tab. 8.4.2.

The Abengourou trial illustrates that the use of plastic sheeting does not prevent catastrophic yield reductions in dry years as for example in 1977. However, in more normal years the yields of coffee were between 25–35 % higher under black plastic sheeting than in traditionally managed coffee having a mainly bare soil with minimal weed growth.

8.4.3.4 Use of Natural Organic Mulches

The use of mulch as a means of slowing down the flow of water over the soil surface and therefore aiding infiltration and reducing run-off has been referred to earlier. A cover of mulch also reduces evaporation from the soil surface and prevents the growth of weeds which otherwise would also be responsible for accelerated soil moisture losses if allowed to become established. Thus, mulches have an important role in moisture conservation. In addition it has been demonstrated that the practice of mulching may effectively eliminate soil erosion, improve soil structure and availability of nutrients in the topsoil and reduce the range of soil temperatures. All

Table 8.4.2 Results of a trial comparing the production of coffee grown with black plastic sheet ground cover with traditionally managed coffee

	Mortality rate of transplanted seedlings (%)	Yields of marketable coffee (kg/ha)			
	1975	1976	1977	1978	Cumulative yield 1976–78
Coffee under traditional management	35	1650	284	1918	3852
Coffee trees within 1.2-m bands of black plastic sheeting	10	2220	438	2551	5209
Benefits of using black plastic sheeting	−25	+570	+154	+633	+1357

Source: J. Snoeck (2003).

these benefits of mulching are not only reflected in increased yields in the long term but also in an improved quality of coffee produced.

The practice of mulching in coffee is widely used in central and eastern Africa, particularly in Kenya, Tanzania, Burundi and Rwanda. A wide variety of mulching materials are used including grasses, arable crop residues, banana trash, leguminous tree loppings, animal manure and composted vegetable waste of all kinds. The main disadvantage of mulching is that it has high costs in terms of labor requirements for collection, preparation and application. With the present low prices being paid for coffee, mulching is not economic on commercially viable coffee plantations, but is worthwhile on smallholdings where the cost of labor is very low.

8.4.3.5 Use of Stones as a Ground Cover

In the Yemen a unique system of water conservation is used in coffee plantations to overcome the climatic conditions in which average annual rainfall is only between 300 and 500 mm. In the absence of any organic mulching materials, the ground is covered with a layer of stones after establishment of the coffee. The thick layer of stones not only minimizes losses of rainfall to run-off by slowing the movement of water and maximizing infiltration but also reduces topsoil losses to erosion. Also, in these low rainfall conditions a low planting density is used as a means of improving water use efficiency. The concentration of limited water supplies on fewer plants is more efficient in terms of the production of coffee beans than would be the case with a higher planting density.

8.4.4 Coffee Water Use

Coffee is an evergreen perennial crop and for optimum growth requires a plentiful supply of moisture through the year. In the natural situation coffee is found growing in tropical climates with several dry months and under shade conditions in tropical forests where it has to compete for moisture with other associated vegetation. Thus, it may be assumed that coffee has evolved in situations where periodic moisture stress occurs and that all coffee varieties have some genetically linked drought tolerance. Indeed, it is generally accepted that with both Arabica and Robusta coffee, one or two periods of moisture stress are desirable in the year in order to stimulate flower development.

Coffee requires water from the soil for photosynthesis, for the translocation of nutrients from soil to plant tissues and for transpiration from the leaf surface as a means of controlling leaf temperature. The rate of water use by the coffee tree is dependent on the climatic conditions and the response of the plant tissues to these conditions. Transpiration takes place through the leaf stomata. During the day the stomata are open, except when the leaf wilts and they close at night.

In a coffee plantation water is removed from the soil not only through the coffee trees, but also by evaporation from the soil surface and by transpiration through other surrounding vegetation, such as weeds, cover plants and shade trees. Thus, the total water lost to the atmosphere from an area of coffee is referred to as the evapotranspiration. Penman developed a formula, which is now widely used for estimating reference evapotranspiration rates. The formula incorporates measurements of climatic factors, including temperature, solar radiation (or sunshine hours), wind velocity, humidity and vapor pressure, to derive reference evapotranspiration rates (ET_0) for particular situations.

In tropical rainy climates Penman's calculated reference evapotranspiration rate is generally between 4–5 mm/day or 1400–1800 mm/year. This calculation is based on total vegetative cover of the ground surface and assumes an absence of water stress. In coffee plantations the tree cover varies from the initial planting to maturity and with the spacing used. Also, the coffee evapotranspiration rate varies through the year since it is dependent on the level of soil moisture. As moisture stress increases, so the stomatal mechanisms on the coffee leaves reduce transpiration rates. Thus, in calculating actual evapotranspiration of coffee an adjustment has to be applied to the reference evapotranspiration rate made to take account of the actual surface area covered by actively transpiring coffee foliage and of the degree of moisture stress. The adjustment is made by applying a crop coefficient which has been devised through lysimeter studies in which actual crop water use by the coffee tree (E_t) has been accurately measured and compared with calculated potential evapotranspiration (ET_0). It has been demonstrated in a number of countries that under conditions of ample soil moisture supplies and at the closest tree spacings used in commercial production, the coffee crop coefficient is of the order of 0.8 whilst at the widest spacing used with moisture conservation measures in place, the coffee crop coefficient may be reduced to as little as 0.65.

Direct measurement of evaporation from an open water surface (E_o) has also been widely used as a measure for estimating crop evapotranspiration rates and, indeed, is still used today in many countries for the routine determination of coffee irrigation schedules. Studies at Ruiru in Kenya in 1963 demonstrated that the overall annual water loss from the soil, or actual evapotranspiration, in a standard unirrigated coffee plantation was 864 mm which was equivalent to 57 % of measured open water surface evaporation which compared with actual evapotranspiration of 1067 mm or 69 % E_o, from irrigated coffee. At the same station it was estimated that, with maximum coffee planting densities and minimal moisture stress, the actual evapotranspiration could be expected to rise to 1219 mm or 80 % of open water surface evaporation.

8.4.5
The Need for Irrigation in Selected Coffee-growing Countries

The need for irrigation can be illustrated by comparing potential evapotranspiration with rainfall. Generally reference crop evapotranspiration is between 120 and 150 mm/month so actual coffee evapotranspiration is between 84 and 105 mm/month. The dry season has been defined arbitrarily as including those months with less than 50 mm of rainfall. In the equatorial belt where rainfall is well distributed through the year, as exemplified by the rainfall data for Kampala in Uganda and for Chinchina in Colombia (see Tab. 8.4.1), average monthly rainfall is more than 50 mm for each month, so that accumulated water deficits are likely to be small and there is little need for irrigation.

In the middle tropical belt, as exemplified by data rainfall from Jimma in Ethiopia, Carmona in Angola and San Jose in Costa Rica (see Tab. 8.4.1), there are 4 or 5 months in which the average monthly rainfall is less than 50 mm. Under these conditions the accumulated soil moisture deficit in coffee can be expected to be in the range 200–500 mm towards the end of the dry season in an average year. Coffee will survive limited periods of soil moisture deficits of 200–300 mm and produce acceptable yields, but with more severe deficits high yields can only be attained with irrigation.

In areas nearer the tropics of Cancer and Capricorn the dry season may extend to 6 or 7 months, e.g. at St Pauls in Zimbabwe (see Tab. 8.4.1) and in Yemen. Severe moisture deficits in excess of 300 mm can be expected to develop under rainfed conditions and coffee cultivation is unlikely to be viable without the use of irrigation.

From 1990 through 2002, during a period of steadily declining prices, ever-increasing efforts have been made in some countries to extract maximum yield levels on their coffee plantations through the use of high levels of input. In these situations the yield responses to applications of irrigation water have increased. For example, with irrigation, hybrid varieties of coffee are already able to produce high yields of coffee in the second and third years after planting. Also, when such varieties receive heavy applications of nitrogenous and phosphate fertilizers the yield response to water is significantly enhanced. The two countries which have been highly successful in promoting massive increases in coffee exports in the period 1990–2002 are Brazil (a rise of 6.5 million bags or 37 % increase) and Vietnam (a rise of 13 million bags or a 10-fold increase). Although a high proportion of these production increases has been due to the development of new land, both these countries have relied heavily on the use of irrigation in the development of this new land.

8.4.6
Factors affecting Irrigation Water Management

8.4.6.1 The Soil Water Reservoir

In assessing irrigation water requirements, the soil water reservoir in relation to the coffee plant must be defined. The soil water reservoir of a coffee tree is provided by the volume of soil which can be exploited by the roots of the tree. Root development is determined mainly by the physical characteristics of the soil, but in certain rare situations rooting depths may be restricted by conditions such as a high water table or by unfavorable chemical conditions in the soil.

Rooting depth is the main factor determining the volume of soil which constitutes the soil water reservoir. Coffee rooting depths are variable and depend on soil conditions.

In many situations coffee is grown in hilly areas where there is high erosion and the A horizons of the soil profile are shallow, sometimes not exceeding 50 cm. whilst in other areas of volcanic soils the depth of the A horizons of the soil are more than 2 m. These two contrasting situations give rise to wide differences in the amount of available soil water. In very deep red loam soils in Kenya, coffee removed the water from a 3 m depth of profile during a prolonged drought period. On the other hand, coffee roots do not grow within 50 cm of a permanent water table nor can they grow through dry soil. Also, in conditions of a plentiful water supply, coffee root growth is restricted to surface soil layers at a depth of 25–40 cm. The elimination of any moisture stress with irrigation results in limited depth of penetration of the tap root and poor development of primary and secondary lateral roots to the deeper layers of the soil profile. In these circumstances, if irrigation is withheld the coffee will quickly succumb to drought.

Soil textures restrict the penetration of coffee roots. Heavy clay soils are poorly aerated and so may restrict the depth of rooting to little more than 50 cm so that the A horizon is not fully exploited. On the other hand, coarser textured soils permit the coffee roots to penetrate to very much deeper levels.

8.4.6.2 Soil Moisture-holding Characteristics

The amount of water which can be held in the volume of soil exploited by the roots of the coffee plant is determined by the volume of pore space between the soil particles. In general, coarse-textured, gravelly or sandy soils have a smaller percentage of total pore space than fine-textured, loam and clay soils and the expected range on typical coffee soils is between 40 and 50 %. The total pore space of a soil on a percentage basis is given by the equation:

$$n = 100 \ (1 - A_s/R_s)$$

where n is the percentage pore space, A_s is the apparent specific gravity and R_s the real specific gravity, approximately 2.65 for most soils.

Soil water that occupies the pore spaces may be classified as hygroscopic, capillary and gravitational. Hygroscopic water is attached to soil particles and conglomerates and is largely immoveable except by excess heat. Capillary water is that which is retained in the pore spaces against the forces of gravity in a soil with unobstructed drainage. Also, under free-draining conditions, the water that is that removed by gravity from a saturated soil is known as gravitational water.

The vast majority of gravitational water drains through the soil before it can be used by the plant. After the removal of gravitational water the moisture status of the soil is described as being at field capacity. At field capacity, soil moisture is between 0.10 and 0.33 atm. It has been demonstrated that plants reach permanent wilting point when soil moisture tension rises to around 15 atm. Thus, the generally accepted level of total available moisture is that which may be extracted from the soil by increasing the soil moisture tension at field capacity to 15 atm. The total available moisture can be expressed as percentage moisture, percentage volume or depth.

The moisture holding characteristics of a range of typical coffee soils are shown in Tab. 8.4.3. The total available moisture ranges from 116 mm/m in a sandy loam to 225 mm/m in a clay.

Coffee tree growth is adversely affected by moisture stress at lower soil moisture tensions than the 15 atm of permanent wilting point. Consequently, the concept of readily available moisture, which is generally accepted as an arbitrary 75 % of the total available moisture, may be used to provide guidelines for determining irrigation schedules. The frequency of irrigation required on a coarse textured sandy loam soil is much higher than on clay soils.

8.4.6.3 Irrigation and the Production Cycle

Irrigation has become a vital component of modern intensive coffee production systems. Without irrigation much of the rapid expansion of coffee areas which has taken place in recent years, e.g. in Brazil and Vietnam, would not have been possible. In young plantations the use of irrigation to supply ample water to permit maximum rate of tree growth and development enables the coffee to become productive at

Table 8.4.3 Moisture-holding characteristics of typical soils

Soil texture	Total pore space (% of volume)	Apparent specific gravity	Field capacity moisture (% dry weight)	Permanent wilting point (% dry weight)	Total available moisture (% dry weight)	Total available moisture (mm/m)
Sandy loam	43 (40–47)	1.50 (1.40–1.60)	14 (10–18)	6 (4–8)	5 (4–6)	116 (92–150)
Loam	47 (43–49)	1.40 (1.35–1.50)	22 (18–26)	10 (8–12)	12 (10–14)	167 (141–191)
Clay loam	49 (47–51)	1.35 (1.30–1.40)	27 (23–31)	13 (11–15)	14 (12–16)	192 (167–216)
Silty clay	51 (49–53)	1.30 (1.25–1.35)	31 (27–35)	15 (13–17)	16 (14–18)	208 (183–233)
Clay	53 (51–55)	1.25 (1.20–1.30)	35 (31–39)	17 (15–19)	18 (16–20)	225 (200–250)

Source: *Irrigation Practices and Principles*. O. W. Israelsen and V. E. Hansen.

a much earlier stage than would be the case if it were reliant solely on rainfall. This advantage is greatly magnified with the use of high levels of fertilizers – there is an additional response in growth and early production of young coffee due to the interaction of ample supplies of water and nutrients. A significant yield may be obtained in the second year after planting – in Western Bahia in Brazil yields of 30–40 sacks (60 kg sack)/ha are reported only 18 months after planting. The early production and the subsequent high yields make the investment in irrigation attractive even in the present unfavorable economic environment.

Irrigation water management is not simply a matter of meeting the full water demand of the coffee by eliminating any soil moisture stress. Over-irrigation that causes waterlogged conditions has to be avoided. Also the supply of irrigation water should not be maintained at a level that will result in a restricted depth of rooting particularly during the early stages of tree development. If soil moisture deficits do not occur, the roots will be concentrated inevitably in a shallow surface layer of soil. Thus, in the early stages of development of coffee trees the frequency of irrigation

should aim to allow quite severe soil moisture deficits to develop so that the roots respond by growing deep into the soil profile in the search for water. In general, increasing the rooting depth of coffee increases the volume of soil being exploited and therefore increases the availability of stored soil water and nutrients.

Also, the application of water has to take into account the role of the moisture status of the soil in relation to the flowering physiology and fruit development in coffee. In countries with a marked dry season the phenology of coffee is clearly different to that in equatorial countries where significant rainfall occurs in every month. There is a clear link between the water stress, which occurs in those coffee growing areas with a well-defined dry season and the growth and flowering cycle of the coffee. A period of moisture stress appears to be essential for the normal development of flowers and if water stress does not occur, the buds do not develop or develop abnormally. Under natural conditions, the flowering of coffee takes place some days after the period of moisture stress is ended by the first showers that signify the coming of the wet season – often referred to as "blossom showers". In Hawaii, it is reported that the flowers open almost exactly 8 days after the first "blossom showers", whilst in East Africa flowering is usually between 9 and 15 days after the first shower. In Viscosa, Brazil, the flowering of coffee followed the first rains by 12 and 18 days in successive years; the difference between years may be explained by lower temperatures in the second year. A fall in temperature associated with the "blossom showers" has also been identified as an important stimulus of flowering in a number of countries.

It has long been recognized that the application of irrigation water can be used instead of rainfall to break the drought period and, therefore, to stimulate flowering. However, there is some dispute about the effectiveness of surface and drip irrigation as a means of stimulating flowering, and overhead systems of irrigation are generally recognized as being more effective in breaking flower dormancy. The clearer flowering response obtained with overhead systems of application of irrigation water may be partly explained by the drop in the temperature of the aerial parts of the coffee trees, caused by the cooling effect of the water, whilst the wetting of the flower buds may also be a factor.

In order to maximize labor use efficiency for the harvesting operation, it is desirable to concentrate the maturation of the fruits in the shortest possible period. Thus, irrigation water management practices can be used to control the timing of flowering and the subsequent period of harvesting. In Australia, irrigation and stressing trials have been used to develop a strategy to synchronize flowering with the adjustment of frequency of application of water. In Zimbabwe, irrigation is withheld towards the end of the dry season and then commenced again with the first showers at the beginning of the wet season. In Hawaii, where mechanical harvesting of coffee is being introduced on a large scale, irrigation practices designed to stimulate and synchronize anthesis in coffee are being attempted. However, this approach is being hampered by irregular rainfall, which both interferes with the intensity of drought and causes irregular flowering. Consequently, research is being carried out investigating the effectiveness of the use of giberellin for synchronizing anthesis and ethylene for synchronizing fruit ripening to facilitate the use of mechanical harvesting.

8.4.6.4 Irrigation Water Demand

It has been noted earlier that coffee is grown in conditions of relatively high average rainfall. However, in most situations the rainfall (incoming water supply) is seasonal and there is a dry season of several months duration. Also, in tropical areas, dry periods of several weeks in duration can occur in the wet season. On the other hand, the water demand of the perennial coffee (outgoing water losses) is influenced by other climatic variables, including day length, temperatures, humidity, wind speed and radiation, and is relatively constant throughout the year. In the absence of rainfall, the water demand of coffee must be met from the stored soil moisture. Prolonged periods of depletion of stored soil moisture result in soil moisture deficits. Increasing soil moisture deficits eventually adversely affect the growth and productivity of the coffee. Thus, the use of irrigation in coffee is to supplement the water supplied by rainfall and so avoid any development of damaging soil moisture deficits.

The application of irrigation water based on calculated water demand is important for a number of reasons. Optimum rates of application of water are cost-effective and important for crop health – they maximize productivity, facilitate the production of a high quality of coffee and make the best use of the (usually scarce) available resources of water. Obviously rainfall is the key factor influencing the frequency of application of irrigation water and, because rainfall is erratic for much of the year, the timing of applications are inevitably irregular. There are a number of different instruments and techniques that may be used on a coffee plantation to assist with making the practical decisions of when to irrigate and of how much water to be applied.

The simplest instrument that is used for monitoring crop water needs and for the determination of irrigation water demand is the evaporation pan. The pan dimensions vary; the US Weather Bureau Class A Pan is 10 inches deep and 47.5 in diameter whilst that being used in Hawaii is 12 inches deep and 24 in diameter. The pan is half filled with water. The water level is checked once a week and the amount by which the water level has declined is multiplied by 0.60 (crop coefficient) in the case of young coffee and 0.75–0.80 in the case of fully developed coffee, to provide estimates of the consumptive use of water by the coffee. Irrigation is carried out when the cumulative consumptive use of the coffee exceeds 30–40 % of the calculated stored soil moisture within the root zone.

The soil tensiometer is an instrument that monitors the level of soil moisture by measuring the suction pressure needed to extract the water from the soil. Tensiometers are placed in the active root zone where the most rapid changes in soil moisture take place, say at levels of 15, 30 and 45 cm. The instrument is only effective on soils of lighter textures and should be maintained at 20–30 cbar on sandy soils and 30–40 cbar on loams. Commercially manufactured versions of the tensiometer are now available which are calibrated to give a direct indication of irrigation need to the user.

A more modern, sophisticated and accurate system of estimating irrigation water demand in coffee is through computerized irrigation telemetry. An automatic weather station measures temperature, relative humidity, solar radiation, wind and rainfall, and computes evapotranspiration at 5-min intervals or any interval specified. At the same time, soil moisture sensors located within the rows of coffee trees provide measurements of the soil moisture status of the soil which are transmitted to the weather station and incorporated in the calculations of irrigation water requirements.

The peak irrigation water requirement is a critical factor in the design of an irrigation system. The most important factor influencing peak irrigation water requirement in the majority of coffee areas is the peak evaporation rate which will rarely exceed 5.5 mm/day or, say, 40 mm/week, and may be significantly less further away from the equator in the dry season where dry season days are shorter and temperatures lower. The timing and period over which the peak in irrigation water demand may occur in an average year must also be taken into account when designing the irrigation system. If the peak evaporation rates occur over a relatively short period it may be better in cost benefit terms to permit some moisture stress and reduce peak capacity of the irrigation system.

The overall irrigation water demand of coffee varies from situation to situation depending on the relationship between rainfall and evapotranspiration and this relationship will vary from year to year. In general, the differences between annual rainfall and evapotranspiration are not important since the overall water balance does not reflect the seasonal and short-term soil moisture deficits that occur. It is the short term soil moisture deficits that determine the overall irrigation water demand of coffee. In most situations the short-term soil moisture deficits vary widely from year to year and rarely result in tree mortality. However, these deficits are responsible for wide fluctuations in coffee yields and for wide variations in coffee quality. Thus, the benefits from irrigation arise from consistently high yield levels and from the reliability of the quality of coffee produced.

The estimation of optimum quantities of irrigation water to be provided for a coffee plantation is difficult because it involves many variables including the consumptive use of the coffee, rainfall probabilities through the year, the volume and moisture-holding capacity of the soil being exploited by the coffee roots and the incremental responses of coffee to applications of irrigation water, in terms of yield and quality, for particular sets of circumstances. In general terms, the total annual demand for irrigation water on a plantation may be based on the sum of the average total dry season deficit in the coffee and the losses that occur in delivering this quantity of water to the trees (i.e. the irrigation efficiency). The latter is dependent on the method of irrigation being used.

8.4.6.5   Irrigation Water Quality

Surface water taken from a river or a dam rarely causes any problems in coffee since the rainfall is invariably sufficient to prevent any accumulation of salts in catchment areas. On the other hand, groundwater can contain high levels of salts which may cause damage to the coffee. Consequently, the quality of water supplied from wells or boreholes should always be checked before proceeding with its use on coffee. The most common salt hazard is high levels of calcium and magnesium salts which raise the pH of the soil to a level at which critical elements such as phosphorus become unavailable. Excess sodium in the water leads to deterioration in soil structure which leads to impeded infiltration and drainage. High levels of boron and bicarbonate have direct effects on the coffee plant.

8.4.7
Methods of Irrigation

8.4.7.1   Surface Irrigation

Surface Flooding
Surface flooding is rarely used for coffee partly because coffee is not normally grown on land which is topographically suitable for this type of irrigation system. However, surface flooding causes serious erosion problems even on flat land and for that reason is ruled out. The erosion leads to exposure and ultimate destruction of the surface layer of roots of the coffee which results in very restricted tree growth and low productivity.

Basin Irrigation
Basin irrigation of individual trees is sometimes used for coffee but, like surface flooding, is not satisfactory. The construction of the bunds which are formed around each tree causes damage to the coffee roots, particularly on sloping ground, resulting in reduced productivity. Water is conveyed in light aluminum or plastic pipes to individual trees and although this method is water efficient, it has a high labor requirement. The main advantage of surface irrigation is that the quality of water in terms of silt load and other pollutant materials is not a limiting factor.

8.4.7.2 Overhead Irrigation

8.4.7.2.1 Sprinkler Irrigation

The typical sprinkler system includes the pumping plant, the main line pipe, the lateral distribution pipes, the riser pipes and the sprinkler head. Sprinkler systems may be classified as semi-permanent or portable. The semi-permanent system has a permanent pump, buried main lines, and the lateral lines and sprinklers remain fixed during the irrigation season. The portable system is fully transportable including the pumping plant. The labor requirement to operate a semi-permanent system are about half that required for a fully portable system. The layout of sprinklers requires careful planning, particularly on sloping land to avoid an irregular distribution of the applied water with over-watering in some areas whilst other parts remain dry.

Sprinkler systems of irrigation were the main form of overhead irrigation, but in recent years the popularity of this form of irrigation has declined and it has been replaced by more efficient systems of under tree irrigation. Sprinkler systems, using rotating nozzles, are suitable for all areas of coffee plantation, and they can be used on any terrain. Although sprinkler application rates can be precisely controlled, the water is not directed on to the soil but on to the foliage of the coffee trees and the bare soil between rows. Consequently, losses due to evaporation are high, particularly in hot dry conditions, and irrigation efficiency is often not much better than with surface methods of application. Irrigation efficiency is further lowered in persistent windy conditions, which cause an irregular distribution of the water as well as causing higher evaporation rates.

Silt and debris are very detrimental to sprinkler equipment. The abrasive action of silt causes excessive wear on pump impellers and sprinkler nozzles and bearings. Ideally the source of water should be debris-free from either groundwater or reservoirs or lakes. Power is required for pumping the water and operating the sprinkler. Medium-pressure sprinklers operate at 30–50 psi and can apply 6–9 mm/h, whilst high-pressure sprinklers operate at 80 psi and can apply 13 mm/h. However, the rate of application is limited by droplet size – large droplets cause a breakdown in the structure of the topsoil, which drastically reduces infiltration rates, leading to excessive losses of water as run-off and a reduction in irrigation efficiency. Application rates should be restricted to less than 10 mm/h.

Overhead irrigation using sprinklers removes any protective fungicides that have been applied and so increases the susceptibility of the coffee to leaf diseases and Coffee Berry Disease or leads to an increased use of fungicides. It also encourages the development of most leaf and fruit diseases.

8.4.7.2.2 Center Pivot

Center pivot irrigation has been developed on a large scale in several countries for a range of crops but it is only in recent years that the technology has been applied to coffee. The center pivots being used for the irrigation of coffee cover areas ranging from 50 to more than 100 ha. The conditions required to allow the adoption of center pivot irrigation include access to a large contiguous land area with relatively flat topography and a convenient source of plentiful clean water on site.

In the development of an area for the center pivot irrigation of coffee the water supply system is constructed first, followed by the installation of the center pivot pumping plant and the wheeled booms. This is followed by the preparation of the land for planting coffee which can be planted in straight rows. This is also valid for delineating access roads and service areas.

The cost of installing a center pivot irrigation system is relatively low compared with other overhead and drip irrigation systems. The cost per ha is currently 1000 USD for center pivot vs. 2500 to 3000 for drip irrigation. The labor required to operate the center pivot system is very low, system maintenance costs are low and there are additional labor saving benefits in that fertilizers, fungicides and other chemicals may be applied in the irrigation water. With the center pivot system of irrigation, application rates can be precisely controlled. Because 24-h operation is possible, the windy periods of the day may be avoided so guaranteeing the even application of water and minimizing losses due to evaporation. Thus, the water application efficiency with the center pivot is relatively high.

It has been amply demonstrated in Brazil and Zambia that the potential production benefits that may be obtained with center pivot irrigation are considerable. Maximum responses to applied fertilizer can be obtained from the beginning of the life of the plantation. It is reported that the first harvest of pivot irrigated coffee in western Bahia in Brazil takes place about 18 months after planting with a yield of 30–40 sacks of green coffee (60 kg sack)/ha. At 42 months, when the coffee is considered mature, yields are in the range 70–100 sacks/ha (4.2–6.0 MT/ha). The only draw back with center pivot irrigation is that it cannot operate in plots planted with shade trees.

8.4.7.3 Ground-level Irrigation

8.4.7.3.1 General Advantages of Ground-level Irrigation

Systems of ground-level irrigation have a number of common advantages over systems of overhead irrigation. With under-tree irrigation at ground level the water is more effectively distributed as, with the several methods of application, the water is directly targeted on to the root zone of the coffee. As a result, evaporation losses are kept to a minimum and the limited area of wetted ground restricts weed growth, which also serves to reduce unnecessary evapotranspiration losses. Thus, the total losses of water to the atmosphere are much less with ground-level irrigation than with overhead systems of irrigation. Also, the adverse effects of the impact of droplets on the soil surface that occur with overhead irrigation – topsoil compaction, soil capping, increased losses to run-off – are avoided when using ground level systems of application of water. It may be concluded that the water use efficiency of ground-level irrigation systems is much better than for overhead systems due to the lower evaporation losses and reduced run-off losses.

In general, the ground level methods of irrigation do not wet the foliage of the coffee so that crop protection chemicals are not removed from the leaves by washing as they are with overhead systems of irrigation. Leaf diseases that thrive in humid conditions, such as Coffee Leaf Rust and Coffee Berry Disease, are encouraged by overhead irrigation but not by ground-level irrigation. Also the risk of leaf scorch is much reduced under systems of ground-level irrigation. Because of the more efficient delivery of water to the coffee trees, a poorer quality of water may be used with ground-level irrigation since it is easier to manage a leaching programme to prevent the accumulation of salts in the surface layers of the soil.

Fertilizers can be applied to the coffee trees in the water through all ground-level irrigation systems. The soluble plant nutrients can be metered into the main pipe at a constant dilution rate, provided that the concentrations of chemicals are not at levels which cause the precipitation of any of the soluble nutrients. It is also possible to apply certain herbicides, insecticides and fungicides in the water.

The capital costs of ground level systems of irrigation are somewhere between the costs of center pivot systems and those for traditional overhead systems. On the other hand, the operational power requirements for ground-level irrigation systems are much less than for overhead irrigation and so pumping costs are lower.

8.4.7.3.2 Mini-sprinklers

Mini-sprinklers are interesting for tree crops such as coffee, as not only can application rates be closely controlled, but the water can be restricted to the area around the individual trees and the diameter of wetting may be adjusted as the tree grows. The layout involves the use of plastic piping with a main pipe, laterals and sublaterals. As the system is designed to work at low pressure of 1.5–2 atm, not more than 12 sprinklers can be fed· from a sublateral otherwise the pressure falls too low. On each sprinkler the nozzle is mounted on the riser at approximately 20 cm from the ground. By interchanging parts in the sprinkler head the radius covered may be varied from 0.5 to 3 m with a delivery rate of between 40 and 100 l/h. The mini-sprinkler system is not very popular because the sprinklers are easily stolen.

8.4.7.3.3 Trickle or drip Irrigation

Trickle irrigation involves slow and low-volume applications of water to the coffee through low-pressure pipelines. Usually a fixed distribution system is used and irrigation is carried out continuously over long periods. Water use efficiency is high. Other irrigation systems which use a similar concept, but with variations in equipment, are referred to as drip irrigation and a microjet system of irrigation. These variations are less suitable for coffee than the conventional trickle irrigation. The engineering designs for trickle irrigation systems are straightforward with easy installation, even on land with irregular topography.

A trickle irrigation system consists of a trickle head unit, where the incoming water flow is filtered and regulated, and a plastic or polyethylene header pipeline that supplies lateral pipelines which are fitted with emitters. The pressure in the lateral line is usually about 1 atm and this is reduced in the emitter so that the water leaves the emitter as a drip. A typical flow rate is between 2 and 8 l/h, depending on the type of emitter. The emitters should be spaced at closer intervals on light textured soils than on heavy soils. Because of the normally low application rate, trickle irrigation is particularly well suited to sandy soils and heavy textured clay soils. Water use efficiency is very high with trickle irrigation and as much as 90% of the water applied may be used by the coffee. The high water use efficiency of trickle irrigation systems combined with low working pressures ensure that pumping costs are minimal.

The trickle head unit filters out the coarse suspended solids in the water but the colloidal and dissolved salts remain in the water, and may cause problems referred to earlier. In addition, the trickle head unit is designed to allow the controlled addition of fertilizers and other agrichemicals into the irrigation water. This method of applying fertilizer is sometimes referred to as fertigation and greatly increases the efficiency of nutrient uptake from the applied fertilizer.

The labor requirement for a trickle irrigation system is quite low with a high percentage of labor being used to maintain the emitters. However because of the complexities of the trickle head unit and emitters and the intricacies of operation and maintenance of the system a high level of skill is required in the workforce. Thus, some individual labor costs will be substantial, but because of low pumping costs the overall recurrent costs of operation and maintenance are relatively low.

8.4.8
Conclusion:

In the present circumstances recommendations for coffee irrigation are the following:

- Big estates with gentle slopes and water availability: center pivot
- Coffee plantations with hilly topography and water availability: overhead irrigation
- Coffee plantations with gently or hilly topography but water limitations: drip irrigation

Draw backs of irrigation systems
Drip
- Needs very clean water
- Clogging problems
- Limited capacity to trigger the flowering
- Hoses prone to damage and vandalism (rodent, people)
- Difficulties to balance the outflow of the drippers (topography, distances etc.)

- Faintly short period of usefulness (approx. 6 years)

Overhead
- Wasteful use of water
- Efficiency affected by wind
- Creates favorable microclimate for diseases and weed proliferation

Mini-sprinkler
- Expensive (more than drip)
- Very prone to vandalism and animal damage

Picture 8.34 Center Pivot.

Picture 8.35 Center Pivot (observe the straight coffee rows).

Picture 8.36 Operating Center Pivot.

9
Vermicomposting in Coffee Cultivation

E. Aranda D., L. Duran O. and E. Escamilla P.

9.1
Introduction

In 1985, Mexico was the first country to propose and use vermicomposting techniques to transform coffee pulp into vermicompost. With the aid of earthworms, like *Eisenia andrei*, *E. fetida* and *Perionyx excavatus*, coffee pulp and many other waste materials can be transformed into a material rich in available nutrients to plants, which can either be added to the soil to improve its structure and fertility or sold as marketable potting soil or plant growth media [1].

Earthworms can be reared in a variety of ways to accelerate the breakdown of various organic wastes. The case study of coffee pulp vermicomposting in Mexico demonstrates the interest of this technique. The practice of vermicomposting results in an effective and efficient recycling of organic material and a better utilization of organic residues. The benefits and advantages of vermicompost for agricultural production are obvious.

Vermicomposting as a practice stemmed from the knowledge that certain species of worms consume organic waste very rapidly and convert them into a beneficial growing substrate. Organic waste materials are consumed, ground and digested by the worms with the help of aerobic microflora. They are then naturally converted into much finer particles which contain plant nutrients in a form which is more easy to assimilate than those contained in the parent compound.

The end-product, a broken down fecal material, known as worm feces or "castings", is a very finely structured, uniform, stable and humic organic material with excellent porosity and water-retention capacity, rich in nutrients, hormones, enzymes and microbial fauna.

Coffee pulp, the main byproduct in the coffee industry, represents around 40 % of the weight of the fresh cherries. It was, and still is, dumped in open piles next to rivers where most of the coffee processing plants are commonly located. Although major efforts have been made to use fresh coffee pulp as food for ruminants, the caffeine, polyphenols, tannins, chlorogenic and caffeic acids contained in the pulp limit its use in this way.

Where coffee pulp is dumped on open land, in a very few days there is a self-inhibited proliferation of thermophilic microorganisms which ferment the coffee pulp generating high temperatures, a low pH, a high demand for biologic oxygen and turning the material into a sluggish mass.

Coffee: Growing, Processing, Sustainable Production, Second Edition. Edited by J. N. Wintgens.
© 2012 WILEY-VCH Verlag GmbH & Co. KGaA. Published 2012 by Wiley-VCH Verlag GmbH & Co. KGaA.

Picture 9.1 Pile of coffee pulp in a dump showing the dark upper layers and the poorly fermented, yellowish mass below. From: E. Escamilla P. CRUO-UACH, Mexico.

The lack of oxygen makes the pulp go from bright red to yellow and, finally, dark brown when aerated (Pict. 9.1).

Obviously, the microfauna proliferation in open dumps will only affect the superficial layers, no deeper than 20–30 cm. If the pulp is left where it was dumped for 8–10 months, the inner layers of the pile remain untouched by the fermentation and the next production of coffee pulp, dumped on top of the old pile, will cover everything again, killing the microfauna and further delaying organic transformation. Hence the conclusion that coffee pulp requires a more effective and efficient aerobic treatment to promote the total transformation of the material.

9.2
Vermicomposting Studies on Coffee Pulp

9.2.1
In Vitro Earthworm Studies

In 1985, observations of *E. fetida* and *P. excavatus* confirmed their potential use for recycling organic waste [3].

The rearing of *E. andrei*, *E. fetida* and *P. excavatus* under laboratory conditions has shown that the three earthworm species

perform fairly well in coffee pulp, growing, reproducing and transforming organic matter into an earthy, stable casting. *E. andrei* were able to produce more biomass and casts. They have a lower mortality rate and a higher carrying capacity in coffee pulp [5].

When comparing the average maximum individual weight of *E. andrei* in different population densities, the results for coffee pulp were more uniform than in the case of cow manure. Cast production was over 4 times its own live weight per day.

P. excavatus were more density dependent, grew more rapidly, reached sexual maturity earlier and produced a larger number of cocoons in coffee pulp than *E. andrei*. Consequently, *P. excavatus* also present an interesting prospect for vermicomposting in tropical conditions, added to which it tolerates higher temperatures and a lower pH.

9.2.2
Comparative Vermicomposting Development

An initial population of 500 g of *P. excavatus, E. andrei* and *E. fetida* was reared separately in three sheltered wooden boxes 1.0 m x 1.0 m x 0.4 m deep, in order to compare their development, using coffee pulp as a substrate [14]. Additionally, a fourth treatment box was inoculated with 166.7 g of the three species together, to compare poly- versus mono-specific cultures. Two controls were used; one with a composting treatment (ploughed over with a shovel every 15th day) and the other with a single initial aeration. From beginning to the end of the coffee pulp stabilization process, the earthworm population was counted and weighed to measure growth and reproduction rates.

The final biomass tripled for *E. andrei* (1691.6 g) and almost tripled in the case of *P. excavatus* and *E. fetida* (1440.9 and

Table 9.1 Ecological and biological parameters of *P. excavatus* and *E. andrei* grown in coffee pulp substrate, soil plant litter and cattle manure [5]

Observations	Perionyx excavatus		Eisenia andrei	
	Coffee pulp	Soil + plant litter	Coffee pulp	Cattle manure
Individual maximum weight (g)	0.705 g	0.211 g	0.816 g	0.946 g
Theoretical maximum growth rate (mg/earthworm/week)	80.58	19.07	60.22	98.01
Estimated carrying capacity (g earthworm/m^2)	532.44	74.22	3371.11	926.66
Apparition of clitellum (days)	15–30	15–30	30–45	15–30
Average weight of mature individuals (mg)	393	158	646	526
Cocoon distribution	aggregated in coffee pulp fibers	–	uniform throughout the substrate	uniform throughout the substrate
Fecundity rate (cocoons/worm/week)	7	–	1	3
Cocoon incubation (days)	15–22	–	14–24	14–24
Cocoons hatched (%)	58	–	85	98
Number of hatchlings (earthworms/cocoon)	2	–	3	3
Fertility rate (earthworms/adult/week)	8	–	3	9
Estimated maximum cast production (casts/g earthworm/day)	5.4	1.96	3.72	0.69
Biomass production (mg biomass/g earthworm/day)	15.59	9.09	28.18	14.59
Mortality rate (%)	2.4	18	4.4	33.4

1346.6 g respectively). The treatment with the three species together reached the lowest biomass increase (1089.0 g); however, this was still twice the initial biomass.

Table 9.1 presents the most relevant chemical and biological changes seen at the beginning of the experiment and at its end. The final worm population quantities showed a higher value with *E. andrei* and *E. fetida* (26 834 and 23 503 individuals respectively) and a lower value with *P. excavatus* (5828 individuals), which presented some escape behavior after the first genera-

tion. The quantity of specimens of both *Eisenia* species quintupled. In addition, the poly-cultural treatment (with 11 205 individuals) was dominated by the *Eisenia* species but the treatment as a whole was less efficient in growth and reproduction than the same species growing in monocultures.

9.2.3
Vermicomposting as a Plant Growth Medium

At the end of the stabilization treatments previously mentioned, the resulting vermi-

Table 9.2 Chemical characteristics of vermicompost obtained from coffee pulp [12]

Analysis	Treatments						
	Initial coffee pulp	E. andrei	E. fetida	P. exca-vatus	Three species together	Composted pulp	Control (single aerated pulp)
Water content (%)	87.30	64.66	65.61	68.92	60.64	63.09	66.99
Total solids (%)	13.70	35.34	34.39	31.08	39.36	36.91	34.68
pH	7.08	6.51	6.51	6.72	6.29	6.02	6.09
Total nitrogen (%)	2.00	4.00	4.23	4.07	4.00	3.30	3.99
$N-NO_3$ (ppm)	–	1137.82	53.60	1775.01	1130.26	1806.57	18.45
$N-NH_4$ (ppm)	–	829.45	538.54	86.58	846.00	652.85	748.89
Mineral nitrogen (%)	–	0.20	0.06	0.26	0.20	0.25	0.08
Organic nitrogen (%)	–	3.80	4.17	3.80	3.80	3.05	3.91
Total phosphorus (%)	0.11	0.25	0.24	0.23	0.22	0.25	0.20
Inorganic phosphorus (%)	–	0.14	0.13	0.14	0.15	0.14	0.19
Organic phosphorus (%)	–	0.11	0.11	0.09	0.07	0.11	0.015
Calcium (%)	0.75	1.72	1.73	1.65	1.50	1.60	1.40
Magnesium (%)	0.36	0.80	0.82	0.78	0.69	0.75	0.73
Sodium (%)	0.03	0.07	0.07	0.07	0.07	0.07	0.06
Potassium (%)	0.35	0.73	0.78	0.73	0.70	0.53	0.52
Copper (ppm)	1.00	19.68	19.01	19.92	19.79	19.67	19.93
Iron (ppm)	589.91	5310.23	2203.80	562.70	536.63	9250.73	3406.43
Manganese (ppm)	257.73	722.39	643.32	745.52	748.70	982.81	571.47
Zinc (ppm)	147.89	662.55	1287.76	103.63	596.31	886.46	374.12
Ash (%)	4.00	15.49	12.70	17.29	14.04	28.21	16.08
Organic matter (%)	96.00	84.05	87.30	82.71	85.96	71.79	83.92
Carbon (%)	55.68	48.84	50.29	48.36	49.86	41.64	48.68
C:N ratio	27.66	12.22	11.88	11.89	12.47	12.62	12.20
Caffeine (%)	0.022	0.004	0.007	0.006	0.008	0.008	0.012
Tannins (%)	0.16	0.010	0.010	0.009	0.012	0.007	0.014

Table 9.2 Continued

Analysis	Treatments						
	Initial coffee pulp	E. andrei	E. fetida	P. exca-vatus	Three species together	Composted pulp	Control (single aerated pulp)
Organic compounds							
Humin (%)	–	74.30	84.22	70.34	73.89	83.95	67.50
Humic acids (%)	–	15.11	11.00	22.08	13.76	10.45	22.68
Fulvic acids (%)	–	10.60	4.78	7.58	12.35	5.60	9.82
Humic acid/fulvic acid ratio	–	1.43	2.32	2.93	1.11	1.88	2.32
Enzyme assays							
Acid phosphatases IU/g	–	9.59	9.14	8.42	6.99	8.49	7.83
Alkaline phos-phatases IU/g	–	6.74	8.08	5.71	4.68	3.88	2.14
Urease IU/g	–	0.92	1.29	1.08	0.66	0.17	0.60

compost was collected for physical and chemical evaluation (see Tab. 9.2). In the composted control, a less efficient transformation took place. The resulting material was hard, dry and moldy with lumps of unprocessed coffee pulp. Vermicompost is undeniably more uniform and finely structured. After drying in the shade, the end product presents a more attractive brownish-black color and texture, with a 55–60% water content and a bulk density close to 500 g/l.

As a rule, vermicomposts double the nutrient content of organic wastes, passing from 28 to 12 of the C:N rate, with a well-balanced chemical content, a high nitrogen level, unusual for such a vegetal substrate. Earthworm activity appears to have stimulated ammonifying bacteria and reduced denitrifying ones, when compared with the compost control. In addition, enzyme activity in the vermicomposts is significantly higher than in the controls.

Harmful compounds, such as caffeine and polyphenols, were almost completely decomposed during the process in all six treatments, but less so in the controls [12]. The high percentage of humins in the vermicompost indicates higher microbial activity than in the controls. The total phosphorus content of the initial coffee pulp is low, but after decomposition it almost doubled, while the control showed the lowest concentration. Potassium is the most sensitive mineral, because it can either be preserved in the finished product or drained out if the substrates are excessively soaked during the stabilization process.

9.2.4
The Effect of Coffee Pulp Vermicompost on Coffee Plants

Experience indicates that for best results, vermicompost should be generated from the crop where it is to be used. Most of the coffee-processing plants owners not only process their own coffee, but also that of many other coffee growers in their area. This represents a double advantage when setting up vermicomposting production. They have the waste materials of many other growers, as well as their own, to fertilize the coffee crop. It is also less expensive for them to use the byproduct generated by the plantation than to dispose of it and then have to buy chemical fertilizers. Most vermicompost producers use vermicompost to prepare nurseries and to fertilize adult coffee trees [13].

9.2.5
The Effect of Coffee Pulp Vermicompost on other Plants

Currently, little scientific work is being conducted to test the effects of coffee vermicompost on other crops, although a lot of practical experiments are underway. Vermicompost and peat-moss have different, but complementary properties [14]. Castings can compensate many of the inherent deficiencies of peat-moss.

Castings increase nutrient contents, otherwise provided by chemical "starter nutrients" (calcium nitrate, potassium nitrate, phosphoric acid, iron and micronutrients) and also increase nutrient absorption, otherwise boosted by horticultural perlite. They increase the water-retention capacity, otherwise provided by "wetting agents", curb evaporation and, incidentally, reduce the need for adding superficial vermiculite. Due to their buffering capacity, castings reduce the acidity of peat-moss, otherwise performed by dolomite and limestone. They also regulate the pH values and reduce the risk of attacks by plant pathogens.

When compared with other organic substrates, the success with coffee pulp castings appears to be related to purity, texture and their sustained nutrient content.

Coffee pulp has proved to be a reliable and predictable material for producing high quality castings. Since the vermicomposting method also generates an excellent product in terms of visual appearance, touch and smell, garden centers are increasingly interested and marketing techniques are being implemented to extend sales to urban areas.

9.3
The History of Vermicomposting

9.3.1
Vermicomposting throughout the World

As presented in detail by Bouché [6], vermiculture throughout the world has a long history. Due mainly to the popularity of residential "buy-back" growing programs, there was a rapid expansion in the earthworm business in the US. An estimated 100 000 people were involved in projects of this type in the late 1970s. Then, in 1978, the *Wall Street Journal* published a feature article exposing the pyramid-type system which had invaded the booming earthworm growing business. By the spring of 1980, the boom had gone bust and almost overnight this fledgling industry disappeared [7].

In the 1980s, parts of central Europe, as well as Spain, Italy and France, underwent the same opportunistic wave, with an initial introduction of presumably "Californian Hybrid" worms. Propaganda publications and pamphlets also resulted in fast expansion and rapid fall. Despite the booming earthworm business, the movement did not spread to developing countries in the tropics, probably because, at the time, the intensive high-input agricultural systems (like the "Green Revolution") were gaining popularity in these countries following the commercial expansion of international equipment and agrochemical companies.

In recent times, with the emergence of the internationally recognized concept of sustainable agriculture, vermicomposting and other soil management methods have had a better chance to help reduce the problem of waste management. One clear message is that the worm business does not mean easy money and that worm breeders should be warned to avoid falling into traps or illegitimate practices similar to those of the 1970s [6].

Scientific communities and research programs have given a more serious, critical, scientific measure to vermiculture, highlighting the effective potential tool that compost worms represent. Documented knowledge has now been assembled on the possibilities offered by worms in organic waste management, their breeding needs, their structural composition, their organic matter quality as well as their impact and potential uses in the soil, on plants and on the environment as a whole.

Since 1978, there has been an increasing scientific interest in the assessment and development of the potentials of processing large quantities of waste by using worms.

The global environmental crisis has turned the attention of developed countries to the less privileged third world countries, facilitating the assessment and support of sustainable soil management practices.

Third world countries discard a considerable amount of available, often unused, vegetable or animal matter, produced by agriculture and domesticated animals. Land in tropical regions is more fragile and exposed to degradation than in temperate areas [3].

In under-privileged countries, there are strong population pressures and risk factors associated with the degradation of natural resources. At the same time, urbanization continues, individual consumption grows and the need for waste processing is ever more present. Tropical countries have better and more uniform environmental conditions for vermiculture than those found in temperate climate countries. Labor is also cheaper and more available.

Generally, lower amounts of metals, pesticides and contaminants are present in their waste. The breakdown of organic waste by worms can convert these overlooked and neglected assets into profitable and useful enterprise, while decreasing environmental pollution. Worm breeding can be used to process organic waste matter in the tropics and to convert it into a valuable solution to a major environmental problem.

9.3.2
Vermicomposting in Mexico

9.3.2.1 Earthworms

The most common earthworm species used world-wide to process organic matter is *E. fetida*, "the tiger or branding worm", and the closely related *E. andrei*, "the red worm". Both have a wide range of temperature tolerance and can live within an ample range of moisture content; both are tough worms, easily handled and usually become dominant when mixed with other species [8] (Pict. 9.2).

Picture 9.2 Earthworm *Eisenia andrei*, also known as the "Red Worm". From: Aranda D. Terranova, Mexico.

Some 50 000 individuals/m^2 (with 3540 g/m^2) of *E. andrei* are commonly obtained from coffee pulp under controlled conditions. For practical purposes, it is assumed that a beneficial population density is about 15 000–25 000 worms/m^2 in field conditions. The equivalent of this quantity of worm biomass is about 2000 g/m^2 or an average of about 0.100 g/worm.

Worms can also be used as a protein source for animals. Worthwhile experiments have been conducted in Mexico, using worms to feed captive animals [2] and salmon trout, *Oncorhynchus mykiss*, as an alternative protein source (Toscano, in preparation). At the present time, the high-altitude coffee area in Veracruz State also has trout farms, which are potential worm consumers.

9.3.2.2 Substrates

The most widely known organic waste used to produce vermicompost in Mexico is coffee pulp, but cattle manure, sugarcane filtercake ("cachaza") and domestic residues are also commonly used. Worms can also be found in paper pulp solids, spent mushroom substrates, sheep, rabbit, horse and pig manure, trimmings from floriculture industries, banana stems, su-

garcane tips, water hyacinth and other related water plants, ruminal contents, municipal solid wastes, and anaerobically digested effluents [9].

Many other organic producers are also increasingly interested in finding similar solutions for their material, including organic waste water, which can be treated and cleansed by cheap and easy vermifiltration systems [11].

Additional studies are being made to conduct and standardize the silo processing of organic materials (in this case coffee pulp), in order to maintain the quality of the substrates and, consequently, of the products [10].

A wide range of organic remains are still available. There are animal, vegetal or urban and industrial residues which are still to be tested for their potential in vermiculture. Although many vermicomposting farms work with a single waste source, others use several, generally with a view to improving a specific nutrient content or to take advantage of seasonal waste availability.

9.3.2.3 Concepts

The first vermicomposting concepts in Mexico were made of building blocks. Set on a bed of gravel, they were about 0.80 m wide by 0.60 m high and of various lengths (Pict. 9.3).

Several designs of this model were also made with local or readily available materials like wood, wood planks, bamboo, plastic or metal grids and even with rubber conveyor belts. In these cases, it is possible to find different kinds of shelters using black plastic film, fern leaves (*Pteridium aquilinum*), as well as living creeper plants (Pict. 9.4).

In larger farms, the most common models used are simple windrows on open land, shaded by native or planted trees.

Picture 9.3 A battery of vermicomposting tubs made of building blocks set on a bed of gravel. Cuatzalan, Puebla. From: CRUO – Mexico.

Picture 9.4 Vermicomposting structure made of bamboo at "La Catalina" CENICAFE, Colombia, 1999. From: E. Escamilla P. CRUO-UACH, Mexico.

Picture 9.5a Vermicomposting tray with coffe pulp. From: J. N. Wintgens.

Picture 9.5b Vermicomposting windrows set up between coffee, rows and covered with sugarcane tips. From: E. Escamilla P. CRUO-UACH, Mexico.

Other farms cover the beds with sugar cane tips or spent jute sacks from the coffee industry. In time, these materials are also decomposed and incorporated into the product (Pict. 9.5a and b).

9.3.2.4 Equipment

Machinery is not commonly used in vermicomposting farms. Only the larger ones have their own trucks, tractors or towed vehicles to move the waste and vermicompost. The windrows are watered in the field by irrigation or with trucks but some still have garden hoses and the smaller farms simply use watering cans. No sophisticated machines are used to

screen or crumble the castings, most of the equipment is simply adapted from coffee industry or agricultural all-purpose tools. Efforts need to be made to design cheap, rustic and efficient vermicomposting equipment and tools.

On most small farms, the only equipment to be found is gardening tools like-shovels, rakes, pitchforks, wheelbarrows and wooden bricklayer-type screens.

9.3.2.5 Human Resources

The low-input technology of vermicomposting is characterized by relatively low labor costs, a large labor force and extensive land surfaces. Generally speaking, for field activities, one man is needed per 500 m^2 of vermicomposting beds.

A coffee expert once said: "due to seasonal coffee cropping, it was difficult to keep good workers throughout the year, but with vermicomposting sites, we can now maintain a group of reliable workers, working mainly at the industrial coffee processing units and, for the rest of the year, at the vermicomposting plant".

9.3.2.6 Operational Methods

Prior to using coffee pulp for vermicomposting, it is advisable to stir it up for a few days to reduce its acidity and cool it down. When preparing the windrow, spread an initial 5 cm layer of pulp at the bottom. Then spread a layer of material containing worms on top of this layer. Leave the two layers for 8–10 days and then add another 5 cm layer of coffee pulp or other organic material (Pict. 9.6).

Traditional vermicomposting operating methods are mostly based on the addition of successive thin layers of food to the surface of the beds. Stirring the food together with the castings in the beds increases the worm colonization and reduces the heat. In this basic model, earthworms will always move upwards in search of food. About 3–4 months later, when the windrows are full, the upper layers of material, where

Picture 9.6 Feeding earthworms with dry grass. From: J. N. Wintgens.

the majority of the worms are, can be lifted to harvest the layers of inner castings. When the castings are harvested, the upper layer of food and worms is shifted to an empty windrow, to restart the cycle with a new supply of food [9].

In some cases, where the main objective is not to sell the vermicompost, but to use the material directly on the coffee plantation, a different, even simpler method, has been successfully implemented in Mexico. A single harvest of castings is planned to coincide with the coffee-planting season. The coffee pulp is initially placed in lines of successive mounds deposited on the ground by trucks and thereafter shaped as windrows. The worms are inoculated approximately 2–3 weeks after the thermophilic phase of the coffee pulp.

Only occasional stirring of the upper layers is required to make the worms move to lower layers for food. When the whole substrate is close to being consumed, another row of mounds is built in the spaces left with the next coffee pulp generation. The escape behavior of *P. excavatus* and *E. andrei* on rainy nights causes them to migrate gradually to the neighboring windrows and the worm-worked windrow can be harvested once they have all migrated.

By this method, more than 2500 trucks filled with approximately 10 000 tons of coffee pulp from a single coffee factory have been successfully transformed from the last five coffee crops [3]. The vermicompost produced has already been distributed in thousands of holes for new coffee plants. Also, instead of the costly treatment by anaerobic digestion of washing waste water from the same coffee factory, successful experiments have been carried out by adding the waste water to the vermicomposting beds in the dry season and using the beds as "vermifilters".

9.3.2.7 **Pulp–Vermicompost Conversion**

Through processing organic residues, whether by composting, by vermicomposting or by any other organic method, the quantity of commercial fertilizer is inevitably reduced, both in volume and in weight. An approximate, simplified description of organic matter decomposition could compare the process to that of respiration, i.e. the consumption of oxygen and the release of carbon dioxide. At the end of the process, the quantity of carbon present is significantly reduced, so the matter that initially showed a 30:1 C:N ratio, would show approximately 12:1 in the finished vermicompost. With the exception of carbon, other chemicals retain almost their total quantities concentrated in the vermicompost.

When comparing the nutrient content in the fresh pulp with the vermicomposted product, the latter generally contains approximately twice the amount of each chemical element (except for carbon). This is due to the material reduction, as only 30% of the initial fresh weight remains. In a dry base, vermicompost represents approximately 40% of the total solids in the initial coffee pulp [12]. For practical purposes and not taking into account the possible variations in the humidity contents of the coffee pulp, or even of the fertilizer obtained, it is estimated that each ton of coffee pulp yields approximately 250 kg of finished organic product.

9.3.2.8 **Vermicompost Harvesting**

Prior to their use either as a market product or by the producers themselves, the worm worked castings need to be harvested, dried, sieved and also bagged.

Harvesting is commonly done manually, 3 or 4 days after a thin layer of feeding is placed to attract most of the worms to the fresh waste in the upper layer. Some producers repeat the method twice or even three times to ensure that the cast below is almost free of worms. The harvested castings are still muddy and difficult to sieve or handle, therefore drying is needed.

One of the most difficult technical obstacles to overcome in the production of vermicompost is to dry the end-product, reducing the initial moisture content from 75–80 to about 55% because of the high water-holding capacity and also the high humidity of the air in the coffee grown areas. Content of less than 45% decreases the efficiency of the casts and diminishes the microbial activity, it also changes their attractive black color to a dusty, grey, disintegrated material, which is difficult to rewet. Most of the farms which sell castings simply sun-dry the material in their yards. This system is costly and slow.

Simple wooden or mechanical sieves (with a 0.5–0.6-cm net) can be used to screen the casts to remove stones, debris, metal, wood or even glass, in order to produce a more uniform, smooth and high-quality product.

Local sales are usually made in bulk, while bagging is done for more distant markets. The most common way of bagging products is in second-hand raffia (polypropylene) 40 kg sacks used for agri-

Picture 9.7 Bagged vermicompost ready for sale. From: CRUO-UACH, Mexico.

cultural purposes or in 10 and 20 kg colored plastic bags for retail stores (Pict. 9.7).

Freight and trucking are limiting factors in marketing, but in some cases they are necessary to build up acceptance and reach broader, expanding markets.

9.4
Vermicompost Qualities

9.4.1
Earthworm Fertilizer

Vermicompost is recommended and can be widely used as an organic fertilizer in any agricultural growth. It is an efficient soil improver, a biological reactivator of fertility and a nutrient regulator in the soil. As an organic fertilizer, it contributes an important quantity of soil nutrients, such as nitrogen, phosphorus and potassium, which are readily available for the plants, and micronutrients, such as Fe, Ca, Mg, Mn, Zn and Cu.

Vermicompost is a stabilized product with a neutral pH and is naturally free of seeds, pathogens, pesticides and heavy metals.

It comes in the form of cylindrical aggregates 1–2 mm long and has a pleasant fresh-earthy smell. It is dark-brown, almost black in color and presents a uniformity, lightness and porosity that give it exceptional visually appealing characteristics.

Humification implies the generation of relatively complex, high molecular weight, organomineral substances, such as humic acid, fulvic acid and humines, which act as virtual "magnets" to attract, hold and regulate the release of mineral nutrients and organic compounds. One of the most important aspects of humus is that each negatively charged particle is capable of retaining, by adsorption, an endless quantity of cations, which are measured by the cationic-exchange capacity (CEC).

It should also be noted that a very important quality of the humus is what is known as the buffer capacity, which acts as the resistance to abrupt changes in the soil acidity or the pH.

Some of the nutrients contained in the vermicompost are immediately absorbed by the plants, while others are only released at slow rates, ensuring long lasting nutrition. This aspect is a very interesting synchrony between nutrient retention and actual nutrient requirements of the plants.

Vermicompost offers better seed germination and improved, faster plant growth. At the same time, the soils to which it is applied improve their structure, fertility and boost the efficiency of mineral fertilizers. As an additional benefit, the use of vermicompost implies a wiser and more harmonious use of natural resources.

9.4.2
Organic Composts or Fertilizers

It is worthwhile to stress the specificity and the distinctive characteristics of organic fertilizers as opposed to chemical fertilizers, since this comparison is frequently made when choosing between the two products. While both products are useful, it is important to assess their respective characteristics in order to determine in which cases each of them should be used.

- Vermicompost and compost are natural and organic, whereas fertilizers are chemical or synthetic.
- Fertilizers normally contain a reduced number of chemicals, basically nitrogen, phosphorus and potassium in high concentration, while vermicompost contains more nutritional elements, in smaller concentrations.

- In vermicompost, chemical elements are found in the form of complex organic compounds, while in fertilizers the elements are generally in the form of simpler compounds, such as minerals salts.
- Fertilizers cannot be used in high concentrations because they may "burn" the plants, while vermicompost can be used without any restriction.
- With fertilizers, rapid and immediate responses in production or growth are sought, without regard to the medium or long term effect on the soils, while vermicompost seeks both to improve production and to preserve and improve soil fertility.
- Vermicompost is both a nutrient and a growth plant medium, while fertilizers are only chemical concentrates.
- Fertilizers tend to dissolve rapidly in water, while vermicompost does not dilute easily, it maintains its structure and has a high water- and air-retention capacity.
- When manufactured, fertilizers consume energy and create pollution, while vermicompost uses organic waste matter. Not only is used the solid fraction or vermicompost, but it is also possible to take advantage of the liquid fraction called thea that is produced during the process of lixiviation.
- Contrary to the apparent meaning of their name, high rates of fertilizers frequently tend to salinize or deteriorate soil fertility, while vermicompost enriches the soil making it more fertile and productive.
- Vermicompost contains active substances and live microorganisms, while fertilizers are a sterile, inactive mixture without any regulating effect.

Nevertheless, in spite of all the above-mentioned advantages and comments, it would be practically impossible to depend solely and exclusively on organic fertilizers at a national or planetary level since the volumes that can be generated worldwide would never be sufficient to cover the present demand.

Considering, too, that a great proportion of organic matter is still wasted and dumped without any use, significant efforts would still have to be made before attempting to dispense with chemical fertilizers.

9.5
Perspectives

The rapid development of this technology indisputably demonstrates that vermicomposting is an effective, efficient and economically viable conversion method of organic waste and that its application can contribute considerably to the productive and economical use of organic wastes for a wise, environmentally friendly preservation program.

- At present, coffee pulp is no longer considered to be a polluting second rate product and its value as a raw material for the preparation of organic fertilizers is beginning to be widely recognized. Thus, more coffee industries are interested in processing their own coffee pulp to produce earthworm fertilizer both for their own consumption and for retailing.
- In comparison to other substrates, the earthworm compost obtained from coffee pulp presents exceptional properties in structure, humidity retention, nutrient content, uniformity and consistency in its results.
- The properties of vermicompost depend chiefly of the parent organic matter used to generate it. Organic waste of dif-

ferent sources cannot be expected to produce similar material.

- To evaluate the potential quality of vermicompost, one should not only consider its N-P-K nutrient content, but also its biochemical and biological properties.

- If vermicompost producers are to maintain the interest of the agricultural world, they must establish and maintain rigorous standards and scientific back up. They should also lend an attentive ear to the comments and opinions of the users of their products.

- Great efforts have been made in Mexico, where after two and a half years of meetings held between vermiculture producers and agricultural, and standards institutions, an official publication of the Mexican Quality Standard of Worm Casting ("Humus de Lombriz", NMX-FF-109-SCFI-2008) was published in June 2008, setting out the quality specifications required for worm castings which are produced or marketed in this county. This Mexican standard (of a voluntary nature) considers quality specifications for testing methods for attributes such as granulometry (particle size), maturity, type of substrate, moisture content, presence of living seeds, worms, impurities and pollutants.

- Standards without supervision are not standards. If earthworm fertilizer is to gain the reputation it deserves, strict monitoring and control systems need to be implemented. Research institutions or state agencies should not be the only organizations to assume the responsibility of deciding which standards should be attained to ensure quality and consistency, the recognized professional input of vermicompost producers is also indispensable.

- Although vermicomposting has spread significantly in the coffee-growing sector

of Mexico, it is necessary to increase recognition and knowledge of this environmentally friendly technology in order use coffee pulp efficiently and to avoid the serious pollution problems generated by dumping it.

- Ultimately, organic waste, at each level of production in the various sectors of agricultural production, (including crop farming, animal husbandry and fisheries) should be transformed by earthworms and recycled as vermicompost.

- So far, most of the research has concentrated on the transformation of coffee pulp. The results of this research and its promotion, together with the demand for the development of similar applications to other types of organic waste, demonstrate that the potentials of this technology are gradually gaining recognition.

- A close partnership between vermicompost producers and research institutions must be promoted and maintained with a view to conduct and develop studies in connection with earthworms, vermicompost technology and final products, to improve technology and to consolidate its use. Research into the impact on plants should also be implemented.

- Training courses, therefore, require improved support and back-up, with professional, scientific and updated information.

The objectives and essential goal of modern vermicomposting are to secure and control new economic alternatives, better recycling options and the productive utilization of organic waste, as well as to ensure significant improvements in environmental preservation and soil fertility. Recognition and attention should, therefore, be granted to its study, development and consolidation.

Bibliography

[1] Aranda, D. E. La utilización de lombrices en la transformación de la pulpa de café Abono Orgánico. *Acta Zoológica Mexicana* **1988**, *27*, 21–23.

[2] Aranda, D. E. and Aguilar, R. S. H. Tasa de extracción de lombrices de *Eisenia andrei* para su utilización como proteína animal. In: *Utilización de Lombrices en la Pulpa de Café en Abono Orgánico. CONACYT (Clave: 045-N9108) Final Report.* Barois, I. and Aranda E. (eds). CONACYT, 1995, 47–54.

[3] Aranda, D. E., I. Barois, P. Arellano, S. Irisson, T. Salazar, J. Rodriguez and J. C. Parrón, Vermicomposting in the tropics. In: Earthworm Managaement in Tropical Agroecosystems. Lavelle P., L. Brussaard and P. Hendrix (eds.) CABI Publishing, London, 1999, 253–287.

[4] Aranda, D.E. and Barois I. Coffee Pulp Vermicomposting Treatment. In: Coffee Biotechnology and Quality. Sera, T., C.R. Soccol, A. Pandex and S. Roussos (eds.) Kluwer Academic Publishers, The Netherlands. 2000, 489–506.

[5] Arellano, C. R. P. Descomposición de la pulpa de café por *Eisenia andrei* (Bouché, 1972) y *Perionyx excavatus* (Perrier, 1872) (Annelida, Oligochaeta). *Tesis de licenciatura.* Facultad de Biología, Universidad Veracruzana, 1997.

[6] Bouché, M. B. Emergence and development of vermiculture and vermicomposting: from a hobby to an industry, from marketing to a biotechnology, from irrational to credible practises. *On Earthworms. Selected Symposia and Monographs UZI.* Bonvicini Pagliai, A. M. and Omodeo, P. (eds). Mucchi, Modena, 1987, 519–532.

[7] Carmody, F. Hooked on worms! A 20 year commitment to the environment. *Worm Digest Mag.* **1996**, *14*, 3–9.

[8] Edwards, C. A. and Bohlen, P. J. *The Biology and Ecology of Earthworms.* Chapman & Hall, London, 1996.

[9] Duran, O. L. Estudio comparativo del lombircompostaje de tres residuos orgánicos y de dos métodos de adición del sustrato. *Tesis Profesional.* Ingeniero Agrónomo Zonas Tropicales, Universidad Autónoma Chapingo, 2000.

[10] Gaime-Perraud, I. Cultures mixtes en milieu solide de bactéries lactiques et de champignons filamenteux pour la conservation et la décaféination de la pulpe de café. *Thèse de Doctorat.* Université Montpellier II, 1995.

[11] http://www.recyclaqua.agropolis.fr/default.html

[12] Irisson, N. S. Calidad del abono y de la lombriz de tierra resultante del lombricompostaje de la pulpa de café. *Tesis de Licenciatura.* Universidad Veracruzana, 1995.

[13] Rodríguez, H. J. Respuesta de cafetos (*Coffea arabica* L) en semillero-vivero a la aplicación de abono orgánico de pulpa de café y fertilizante químico. In: *Utilización de Lombrices en la Pulpa de Café en Abono Orgánico. CONACYT (Clave: 045-N9108) Final Report.* Barois, I. and Aranda E. (eds). CONACYT, 1995, 1–15.

[14] Salazar, T. C., Aranda, D. E. and Barois, I. Estudio comparativo del lombricompostaje de la pulpa de café por *Eisenia andrei, Eisenia fetida* y *Perionyx excavatus* en condiciones de campo. In: *Utilización de Lombrices en la Pulpa de Café en Abono Orgánico. CONACYT (Clave: 045-N9108) Final Report.* Barois, I. and Aranda E. (eds). CONACYT, 1995, 28–41.

10

Organic Coffee

L. Sosa M., E. Escamilla P. and S. Díaz C.

10.1
Introduction

Many of the synthetic chemicals or agri-chemicals which are now widely and routinely used in modern agriculture are toxic both to humans and to many living creatures in the natural environment. Consequently, over the last decade, consumer demands for foods free of synthetic chemicals and a growing awareness of the adverse effects of these chemicals on the environment have resulted in new trends in farming. Farming systems are being developed which aim to eliminate synthetic chemicals and to minimize environmental disruption caused by certain farming activities. These new farming systems are referred to as organic agriculture or organic sustainable farming. The extent of organic agriculture is still relatively insignificant (1–2 % in developed countries), but there is a growing interest in natural farming systems and consumers are willing to pay higher prices for organically produced food. As a result, the adoption of organic agricultural production can be expected to intensify rapidly.

10.1.1
Organic Agriculture

The IFOAM (International Federation of Organic Agricultural Movement) defines organic agriculture as a holistic farming system which has the following objectives:

- To enhance the exploitation of biological cycles in the farming system
- To work with renewable resources and within a closed system with regard to organic matter and nutrient elements
- To maintain genetic diversity
- To provide animals with natural living conditions
- To minimize all forms of pollution which could be generated by farming
- To consider the social and economical impacts of farming systems.

Organic agriculture does not allow the use of synthetic chemicals either for fertilization or for plant protection. It also outlaws practices such as irradiation for pest control and the use of genetic engineering in plant and animal breeding.

Coffee: Growing, Processing, Sustainable Production, Second Edition. Edited by J. N. Wintgens.
© 2012 WILEY-VCH Verlag GmbH & Co. KGaA. Published 2012 by WILEY-VCH Verlag GmbH & Co. KGaA.

10.1.2
Regulations

National and international regulations by which organic products are identified are already being implemented. These regu-la-tions cover crop- and animal-breeding methods, all the production processes, including husbandry practices, harvesting, processing, storage and transport, and the marketing of organically produced food. In addition, these regulations encompass the inspection systems required to harmonize organic food production and the labeling of food to allow consumers to make their choice in retail outlets between organically produced foods and non-organic food.

10.1.3
Quality of Products

The risk of contamination with synthetic chemicals for organically produced food is low: however, currently, there is no total guarantee that this food is completely free of contaminants. Also, there is still no conclusive evidence that organically produced food is superior, in either nutritional value or in terms of flavor, to the same type of food produced by integrated farming.

10.1.4
Yield and Production Costs

The yields of organically produced crops can be 10 to 50% lower than those produced by conventional or integrated farming. This can be due to soil nutrient deficiencies, inefficient methods of weed control and the ravages of pests and diseases. All these limiting factors can be largely eliminated by the use of agrichemicals. The benefits that may be derived from the use of agrichemicals (i.e. increased yield × price) are generally much greater than the costs associated with their use. Thus, if organic production is to be as profitable as a production system using agrichemicals, the price paid for the final product must be higher.

10.1.5
Environmental Contamination

Due to the prohibition of synthetic chemicals and fertilizers, environmental pollution caused by organic farming is significantly minimized, but not totally eliminated. A low risk of contamination with organic materials (nitrates) and minerals (copper) still remains.

10.2
Organic Coffee Production

10.2.1
Introduction

With an international market that is ever more competitive, special coffees offer an excellent alternative for producers in the search of stable and innovative market niches. The more important types of special coffees presently offered on the international market can be classified as follows:

- Coffees of a particular geographical origin like Blue Mountain from Jamaica, Pluma Hidalgo from Mexico, café Cerrado from Brazil, Kona from Hawaii, Single Estate from Colombia, Mocha from Yemen, Sidamo from Ethiopia, etc.
- Pure coffees: coffees of unblended varieties like Arabica or Robusta
- Special blends: coffees to suit the taste of a particular market like Breakfast blend, continental blend, After Dinner blend, Viennese blend, etc.

- Flavored coffees: coffees with added aromatic substances like figs, vanilla, chocolate almonds, cardamom, etc.
- Decaffeinated coffees: coffees from which caffeine has been extracted
- Organically grown coffees: coffees free of contaminants that meet the IFOAM standards of organic agriculture.

Organic coffees are produced using environmentally friendly cultivation methods. A major objective of organic coffee production is to eliminate from the production system all contamination with synthetic inorganic chemicals and to develop a system of coffee plantation management that is sustainable in the long-term. Organic coffees have a higher labor requirement than would be required with the use of agrichemicals, particularly with regard to weed control, but their production costs are much lower. At the same time, organic coffee attracts significantly higher market prices. Given the aforementioned characteristics, organic coffee is especially attractive for small coffee farmers in underdeveloped areas where the cost of labor is relatively cheap and the cost of transport to market is relatively high.

10.2.2
Market Potential

Organic coffee, also called ecological or biological coffee, enjoys increasing popularity in Europe and in the US. Increasing numbers of consumers prefer to buy their food supply from ecologically sustainable agriculture where the use of agrichemicals (fertilizers, herbicides, insecticides, fungicides, nematicides, etc.) is avoided. The avoidance of the use of agrichemicals will eliminate both the risks posed to consumer's health by these chemicals and potential damage to the environment. At the same time, income levels of small coffee farmers will be improved through the higher prices being paid for the organic coffee.

While organic agriculture and the growing interest in "natural products" is a worldwide trend, the organic coffee market is basically made up of consumers that are sensitive to Third World problems and whose standard of living allows them to pay up to 50 % more for the product. In order to further expand the production of organic coffee, vigorous marketing campaigns will be essential to achieve selling prices that will compensate the farmer for the lower yields and the higher labor costs of organic coffee production (Table 10.1).

With prices peaking at over US$150 per quintal of green coffee, farmers could be tempted to produce organic coffee massively and saturate this special market. However, apart from a few exceptions, organic coffee production is not a likely alternative for large plantations because the size of this specialist market is limited and the stability of future market prices is precarious. In addition, the long-term maintenance of soil fertility on large estates cannot rely on an adequate supply of organic fertilizer (3–10 tons/ha/year required) and the availability of an increased labor supply at critical times is more difficult to guarantee on large estates.

Organic coffee production is much better suited to small producers who usually operate on family farms. The costs for hired labor are lower on the family farm than on large estates because of the low cost of labor for family members. Also, in many countries, land rental on the family farm and interest on capital investment are lower than on large estates. A further element in favor of organic production carried out by small producers is the laborious bookkeeping and the production controls needed for organic certification. The

Table 10.1 Return, variable costs and gross margins per hectare of one conventionally and two organically managed coffee plantations

Variables/management	Conventional intensive	Organic intensive	Organic extensive
Plants/ha	5000	3300	2500
Yield (kg/ha)	1265	870	500
Price/46 kg	52	130/62	130/62
Return/ha	1430	2459/1172	1413/674
Variable costs			
Materials			
compost 1.3 kg/tree/year	–	200	150
N fertilizer (300 kg)	60	–	–
NPK fertilizer + borax (300 kg)	85	–	–
foliar fertilizer	15	–	–
herbicides (3 kg)	35	–	–
fungicides, insecticides (3 kg)	30	–	–
food for pickers	105	70	60
miscellaneous	25	25	25
total cost, materials	*355*	*295*	*235*
Labor			
processing costs (US$10/46 kg)	275	190	109
manual weeding (6/40/40 days)	24	160	160
herbicide application (6 days)	24	–	–
foliar dressing (3 days)	10	–	–
fertilization (7 days)	24	–	–
compost preparation/application (25 days)	–	270	200
pruning (3/2/2/days)	12	8	8
shade tree pruning (0/4/4 days)	–	16	16
miscellaneous	11	11	11
harvest costs (US$15/46 kg)	412	284	163
total costs, labor	*792*	*939*	*667*
Administration/transport	200	200	200
Interest, 10%, 4 months	45	48	37
Total variable costs	1392	1482	1139
Gross Margin	+38	+977/–310	+274/–465

Coffee prices for conventional production based on world market prices, for organic production based on prices paid for organic TransFair-Coffee (130 US$) and on 20% surplus prices (US$62.).
Data sources: Compart, 1993; UNDP, 1992.

IFOAM guidelines for organic coffee production, for example, require annual protocols that record the "nutrient balance", among a host of other factors.

10.2.3
Interplanting Coffee as a Parallel Source of Income

Coffee plantations present an excellent opportunity to cultivate seasonal annual and perennial crops between the rows, in a polyculture system. In some circumstances these interplanted crops can generate a relatively lucrative source of supplementary income, whilst at the same time providing some protection to the soil during the rainy season. Successful experiments have been carried out with commercial intercropping. Furthermore, most of the small coffee growers own small plots of land, where it is essential to use the coffee-planting area to grow other crops to feed the family. In new coffee plantations, for example, farmers can produce annual crops like beans, corn, tomatoes, pepper, etc., and perennial crops like banana (fruits and leaves), orange, macadamia nut, zapote mamey, black pepper, avocado, custard apple, vanilla, etc.

10.2.4
Certification, Production and Consumption

As from 1 January 1993, imported products from "organic agriculture" may only be marketed within the EU if they come from a Third World country that has been accepted on the community list. A pre-requisite for acceptance of these products on EU markets is proof that the production methods meet the criteria laid down by EU regulations Coffee can only be labeled as organic when certified by an IFOAM-accredited certification organization. Thus,

an efficient inspection system must be in place in the countries of origin of the organic coffee.

Norms for inspection and certification are enforced by agricultural associations like Demeter, Naturland and the American OCIA (Organic Crop Improvement Association). These organizations are members of the IFOAM and their certification guidelines are based on IFOAM's basic standards of organic agriculture. According to Bio-Foundation CH, the worldwide organic coffee production was estimated at about 25 000 tons in 1991, but only 2 240 tons complied with EC regulation 2092/91. This represents 0.04 % of the world production estimated at about 6 000 000 tons of green coffee. The main organic coffee producing countries are Mexico, Guatemala, Kenya, Nicaragua, Tanzania, Brazil and Ethiopia (Table 10.2).

The main certified organic coffee importer countries are Holland, Germany and the US (Table 10.3).

Table 10.2 Main organic coffee-exporting countries

Country	Volume (60 kg/sacks)	Share (%)
Mexico	86250	20.5
Guatemala	40500	9.6
Kenya	40000	9.5
Nicaragua	29250	6.9
Tanzania	28250	6.7
Brazil	23750	5.7
Ethiopia	22750	5.5
Other 13 countries	149000	35,6
Total recorded	419750 (25185 MT)	100

Source: IFOAM, 1992 (only the exports recorded by IFOAM are included). Quoted by Santoyo *et al.* [6].

Table 10.3 Main certified organic coffee importers

Country	Volume (60 kg/sacks)	Share (%)	Main Importers
Holland	135000	28.8	Max Havelaar, Duque Import-Export, SOS Wereldhandel
Germany	175000	37.3	Gepa Aktion, D. Welt, Demeter DritteWelt Haus
Rest of Europe	30000	6.4	Max Havelaar, Artisans du Monde, Sandali Company
USA	118000	25.2	Frontier Herbs, Coffee Tierra, Café de Altura
Other countries	11500	2.5	
Total	469000 (28170 MT)	100	

Source: IFOAM, 1992. Quoted by Santoyo *et al.* [6].

In order to obtain the organic coffee label, coffee farmers must comply with the standards of ecological agriculture that include the following cultivation practices:

- Active measures to control soil erosion
- Mulching and cover cropping
- Composting
- Use of trees for shading and nitrogen fixation
- Renovation and pruning (including the recycling of pruned branches)
- Use of different leguminous plants
- Recording and limitation of external inputs
- Self-reliance for coffee seedlings and clones, as well as planting materials for shade trees, leguminous cover crops and compost ingredients
- Biological pest and disease control (exceptional spraying with accepted products).

In principle, all the above-mentioned practices should be implemented in fully organic production units. During the transition period, a very well-defined and approved conversion plan should be followed. The transition period is of 2–3 years de-pending on the type of plantation to be converted to a certified "biological" farm.

10.2.5
Certification of Organic Coffee in Mexico

10.2.5.1 General Procedures Required
As is customary with all organic agricultural products, organic coffee must be certified before being labeled and sold on the retail market. In order to obtain an organic coffee production certificate, the coffee must have been produced in accordance with the written standards for organic production. A certifying entity has to be established to issue organic quality certificates and supervise the correct implementation of the procedure. The certifying entity may be either sponsored by the Government of the producing country or carried out by private organizations involved in importing and marketing in the consuming country.

The staff of the certifying agency inspect the plantations, the warehouses, the processing areas, the means of transportation, and the inventory and marketing documents of the participating farmers and ensure compliance with organic production and processing standards. A Certifying Committee reviews the inspectors' reports and issues a certification statement. If the statement is favorable, the certifying agency issues a general certificate that is followed by transaction certificates of conformity for each sale.

Certification confirms that the producer, the processor and the trader have complied with the required organic production standards, and guarantee the organic quality of the product to the importing country consumers.

10.2.5.2 Mexican Certification System

Certification of organic coffee in Mexico began in 1962 when an inspector of the German certifier, Demeter, conducted an inspection of the plantations of a coffee farm located in the Soconusco area (Chiapas) for the first time. Demeter remains in Mexico today and, as well as certifying the coffee produced on a number of old established organic farms, the organization is also supervising farms which are in the process of converting to organic farming systems.

In 1983, two decades after the initial certification of organic coffee in Mexico, a number of small coffee-producer organizations started to produce organic coffee. The increasing supply of organic coffee producers increased the need for an expansion in certification capacity. This need was met by the arrival of certifiers from Imo-Control in Switzerland and from Naturland in Germany.

Since 1989, organic coffee production in Mexico has grown significantly. Cultivation areas have increased and markets have diversified. The number of foreign certifying companies has increased accordingly and recognized certification companies now include the OCIA from the US and, more

recently, QAI (Quality Assurance International).

10.2.5.3 Problems with Foreign Certifiers

Working with foreign certifiers enables the local producers of organic coffees to access various international markets and certifier's advice to producers generates constant improvement in organic production which will be ultimately beneficial to the reputation of locally produced organic coffee abroad. However, the involvement of international certification agencies also generates a few problems which are mainly cost-related. Foreign inspectors charge fees equivalent to those that prevail in the developed countries from which they come and these fees are, as a general rule, very high, particularly when compared to the socioeconomic conditions of organic producers in lesser-developed countries.

The standards required to match the rules and regulations of the developed world is also a source for some disquiet in lesser-developed countries. The standards that are imposed are generally drawn up in the countries of the foreign certifying agencies where environmental and cultivation conditions, as well as the types of producers, are very different from those that prevail in the majority of Third World countries.

Another factor responsible for the high costs of foreign involvement in the certification of organic produce are the heavy expenses for communication and transport. Measures to minimize the cost of communication and transport inevitably result in curtailing the time available for inspection. The staff of the foreign certifying agencies are not permanently on the spot to help overcome the problems posed by local organic agriculture but are only available for short, intermittent periods.

10.2.5.4 Domestic Inspection and Certification

In order to solve the shortcomings of foreign certifiers, proposals to localize the system of certification have been prepared by the associations of organic coffee producers representing the largest and the most experienced group of organic producers. These proposals have been adopted and in the early 1990s the Committee for Organic Product Certification (CUCEPRO) of the University of Colima, was created and has been conducting inspection and certification activities ever since.

In 1993, at the request of organic producers, CUCEPRO and the certifying agency OCIA implemented the first course for organic inspectors in Mexico. Mexican professionals, who had been performing inspection activities at lower costs, were certified as organic inspectors after following this course. Consequently, producers and advisors of various organizations, with the support of some inspectors, formed the certifying agency that is now known as the Mexican chapter of OCIA.

In 1997, members of organic producer organizations, mostly coffee growers who had been working with IMO-Control and Naturland certifying agencies, promoted the creation of CERTIMEX (Mexican Certifier of Ecological Products and Processes), which started its own inspection and certification activities in 1998.

10.2.5.5 Advantages of Domestic inspection and Certification

A system of local inspection and certification may bring about the following improvements:

- A significant reduction inspection costs which will encourage new groups of small coffee farmers to initiate organic production.
- The introduction of standards for certification which are adapted to the conditions of local producers.
- Domestic certifying agencies, located in the country which are easily accessible, and able to provide necessary assistance to producers, analyzing the various problems and answering enquiries on certification information.
- Locally trained inspectors and certifiers may participate in various events and contribute to the advance of organic agriculture in their country.
- The organization of periodical events where certifiers, technicians and promoters of producer organizations meet to identify deficiencies in compliance with standards, to analyze the problems, and to provide recommendations and guidance for the improvement of organic projects.

10.2.5.6 Important Items to be Included in Organic Coffee Projects

- Establish a technical team, which may include one or several professionals. The task of this team is to support organic producers, to provide advice and training and to prepare the necessary documents for the implementation of an organic project.
- Producer organizations should prepare their own internal production regulations for organic coffee.
- The producer or the producer's organization should prepare the blueprint for conversion to organic agriculture.
- Producers should undergo a conversion period. Work carried out during this period should be fully recorded and documented.
- Producers must comply with organic coffee production standards, this includes

aspects such as avoiding the use of agro-chemicals, maintaining shade trees, preventing erosion, improving soil fertility, preventing water pollution during wet processing, etc. The CERTIMEX agency outlines the specific requirements for organic coffee production.

- During storage and transportation, coffee contamination by any chemical product must be avoided. Mixing organic coffee with non-organic coffee must also be prevented.
- To avoid parallel production, producers must have completed the conversion of their entire coffee growing areas to an organic system within a period of 5 years at the most.
- In the case of producers' organizations, it is necessary to have set up an internal control system, which controls all the farmers who have subscribed to the organic coffee growing project.
- During the inspection of producer organizations, the inspector must visit between 15 and 30 % of the individual producers.
- In the case of a dry processing system, documents should state the quantities of coffee received, the quantities of stored coffee, the quantities that underwent processing, the processing yield and the quantities of sold or released coffee. Separate storage areas should exist and before processing organic coffee, all equipment should be thoroughly cleaned to avoid mixing organic coffee with non-organic coffee.
- Only coffee produced on certified coffee plantations should be marketed as organically grown coffee.

10.2.6
Agronomic Aspects

The first organic production was developed in 1960 by W. Peters Greather at "Finca Irlanda", located in the state of Chiapas, Mexico. This plantation uses an organic system based on three principles: maintenance of soil fertility, natural plant selection and adaptation to the environment. The production is characterized by agro-ecological criteria and techniques, such as:

- Conservation and development of biodiversity associated with the coffee tree
- Cultivation of several varieties, mainly Typica
- Low planting density (1200–1500 plants/ha)
- Production of compost and vermicompost from coffee pulp, cattle manure, dry leaves and green fertilizer
- Application of organic fertilizer to the coffee trees
- Soil conservation practices
- Biological control of pests and diseases
- Maintaining an acceptable level of social welfare for the farm workers
- An average production of 700–800 kg/ha.

The Irlanda farm is a mandatory reference when dealing with the subject of organic coffee in Mexico – a 40-year productive experiment that has served as the basic model for all the organizations that produce and export this type of coffee.

10.2.6.1 Soil Conservation Practices
In organic cultivation, it is essential to sustain, recover and increase the natural fertility of the soil on a long-term basis. When the plantation is laid out, soil conservation measures like contour planting facilitate the implementation of other conservation devices such as:

- Individual or bench terraces
- Anti-erosive live barriers
- Mulching
- Inter-row ridges

Picture 10.1 Organic agriculture could avoid such calamitous sights.

Picture 10.2 Putting in an anti-erosive live barrier in an organic coffee plantation (Lachiguiri-Oaxaca, Mexico). From: E. Escamilla P. – CRUO-UACH, Mexico.

- Hillside waterways
- Triangular planting. etc. (Pict. 10.1, 10.2)

10.2.6.2 Planting Materials

For organic coffee growing the following criteria should be respected for the propagation of coffee plants:

- Choice of vigorous and productive varieties which offer as much resistance as possible to coffee parasites and which are adapted to the local agroclimatic conditions; the use of genetically modified plants is prohibited, but hybrids like Catimor are allowed.

- Adequate preparation of the seeds (manual depulping and drying under shade).
- Use of physical means to disinfect the soil, like heat or hot water, for instance.
- Use of natural compounds like compost, vermicompost or other organic materials for fertilization.
- Ensure adequate shading by using well-adapted native trees.
- Ensure pest and disease control by using organic techniques such as biological control, biopesticides and cultural practices like, for example, grafting against nematodes.

10.2.6.3 Pest and Disease Control

Under prevailing regulations, the application of chemical insecticides, fungicides, acaricides, nematicides, rodenticides, etc., is strictly forbidden. Consequently pest and disease control in organic coffee must depend largely on the use of resistant varieties and on the active management of the ecosystem. Management or husbandry practices may include mulching and/or the application of organic fertilizer to maintain favorable soil physical characteristics and nutrient levels to strengthen plant vigor and resistance to pests and diseases. The use of shade to control growth combined with sound pruning techniques will also favor the development of strong healthy trees and provide optimum conditions so that the natural biological control of pests and diseases can take effect.

If serious damage due to pests and diseases occurs in a coffee plantation, other interventions may be allowed. The use of naturally occurring insecticides is permitted for "hot spot" spraying – such insecticides include the pyrethrins and extracts of neem and nettle. Some work has been done on the use of biological control methods. For example, Coffee Berry Borer (*Hy-*

Picture 10.3 Biological control of the Coffee Berry Borer by means of a trap impregnated with products to attract the adult beetles (Xicotepec, Puebla, Mexico). From: E. Escamilla P. – CRUO-UACH, Mexico.

Picture 10.4 Biological control of the Coffee Berry Borer by means of the *Beauveria bassiana* fungus supplied in bags (Lachiguiri, Oaxaca, Mexico). From: E. Escamilla P. – CRUO-UACH, Mexico.

pothenemus hampei) damage may be limited by using of the *Beauveria bassiana* fungus and also by parasitoid wasps, especially those of the *Cephalonomia stephanoderis*

species. Traps impregnated with substances to attract the Coffee Berry Borer are also successful (Pict. 10.3, 10.4).

For disease control, copper and other non-organic fungicides can be used if approved by the certification organization.

10.2.6.4 The Use of Fertilizers

The use of inorganic fertilizers is not permitted under the regulations of organic coffee certification. Thus the only source of additional plant nutrients is through the application of organic fertilizers. Recommended rates of application for maintaining soil fertility are 3–10 kg of organic matter per plant each year or every other year. As the use of inorganic fertilizers is not allowed, it is essential that organic coffee be restricted to deep soils with high inherent levels of soil fertility.

Organic fertilizer may be generated by compost or vermicompost. Residues from the primary processing of coffee, such as pulp and husks, provide a useful source of organic matter. Other non-polluted organic products like cow dung, pruned branches of leguminous shade trees and cut grass are also important sources of organic matter for mulching and composting (Pict. 10.5).

Picture 10.5 Composting box in an organic coffee plantation (Mexico). From: E. Escamilla P. – CRUO-UACH, Mexico.

Vermicomposting uses earthworms to accelerate the degradation of organic waste. The most widely used earthworms are *Eisenia* spp. Chicken or cattle manure issued from large-scale commercial livestock farming is not in line with the basic principles. Natural rock phosphate and dolomite are authorized as fertilizers and amenders.

Coffee trees respond very well to organic fertilizers and yield is highly increased. However, it is a cumbersome activity, as to fertilize only 1 ha, i.e. about 1000 trees, entails the preparation, the manual transportation and application of at least 3 tons of organic matter.

It has been reported that the total nitrogen requirements of coffee plantations shaded by leguminous trees, such as *Inga* spp. and *Erythrina* spp., can be largely met by the mineralization of the leaves of these shade trees and by nitrogen derived from the breakdown of nitrogen-fixing root nodules.

The use of young leaf material from leguminous trees for mulching has been a part of agroforestry systems that have existed in a number of tropical countries for centuries. However this practice does not comply with the regulations for organic coffee. The production of compost and mulching material by means of such agroforestry systems is so close to nature that is simply described as "unfertilized coffee" by certification associations and is not certified as organic coffee.

Due to low coffee prices many planters no longer use fertilizers and pesticides, and, therefore, believe that they fulfill the requirements for organic coffee. However, the withdrawal of agrichemicals alone is insufficient to qualify for certification as organic coffee. The most important aspect in organic coffee classification is the sustainability of the production cycle. This means that soil nutrients removed by the harvest must be replaced by allowing the natural weathering of the soil to release nutrients, the recycling of nutrients in applied organic matter which also maintains desirable soil physical characteristics.

10.2.6.5 Shading

Organic coffee is grown under shade. The shade trees are commonly a mix of various native and secondary vegetation and include a variety of fruit trees and leguminous trees *Inga* spp. are widely used as a shade tree in Mexico. These mixed coffee plantations, which may be described as a system of agroforestry, offer numerous ecological and economical benefits which include the following:

- Protection and conservation of biodiversity. Recently, the protection of migratory birds has become an increasingly important factor in coffee plantations – this has led to the concept of "shade" or "bird-friendly" green coffee [the Smithsonian Institute for Migratory Birds (SMBC) has initiated training programs for inspectors in Latin America and in 2002 the first certification inspections for "shade" coffee were initiated in several localities].
- Conservation of soil resources (shade leaves meet the nitrogen and mineralization needs of coffee plants)
- Protection against erosion
- Carbon pick-up
- Rainfall retention
- Regulation of environmental elements like rain, frosts, wind and hail
- Diversification of production.

Other shade benefits have been identified relating to coffee quality. According to new findings, shade contributes to the elaboration of aromatic compounds that

Picture 10.6 Conventional sun-exposed small grower's coffee plantation.

Picture 10.7 Organic intensive coffee plantation under natural forest shade (Sierra Norte de Puebla, Mexico). From: E. Escamilla P. – CRUO-UACH, Mexico.

Picture 10.8 Organic extensive coffee plantation under natural forest shade and fruit trees (Istmo de Tehuantepec, Oaxaca, Mexico). From: E. Escamilla P. – CRUO-UACH, Mexico.

Picture 10.9 Aerial view of organic coffee plantations. This type of vegetation can be almost assimilated to a forested cover. (Lachigiri, Oaxaca, Mexico). From: E. Escamilla P. – CRUO-UACH, Mexico.

largely determine coffee quality. Consequently, many important coffee-trading companies are promoting the use of shade trees in coffee plantations. Starbucks, for example, is aware of the potential improvements in quality which research findings have indicated and is currently investing US$ 150 000 over 3 years to provide technical and financial assistance to growers in the Chiapas region of Mexico.

Shade is controlled by lopping the lower branches of the shade trees in order to encourage high growth. The loppings are used for the production of mulch (Pict. 10.6, 10.7, 10.8, 10.9).

10.2.6.6 Weeding

Weeds compete with the coffee crop for vital space, water, light and nutrients. In organic production certain plants coexist favorably with the coffee trees. The frequency of weeding must be sufficient to prevent weed competition reaching the point at which coffee yields are adversely affected.

In some situations beneficial weed populations may develop naturally or their development will be encouraged by the farmer. Species of the *Commelina* genus are common and recognized as beneficial. Where beneficial species occur, weeding must be selective, eliminating only harmful plants. Beneficial plants can be controlled by slashing with a machete to a height of 5–7 cm so as to avoid exposing the soil to erosion and baking.

The use of a hoe is not allowed as it destroys the vegetation and exposes the soil. Cover crops have not yet been introduced. Synthetic herbicides are prohibited, but certain organic herbicides prepared with plants like nettles are authorized.

10.2.7
Positive Aspects of Organic Coffee Growing

Organic coffee production aims to be an ecologically stable farming system, a sustainable form of land use that has no harmful effects on the environment. Although its production volume is small in comparison to that of conventional coffee, organic coffee has a high economic and social im-portance for the farming communities and production units where it is implemented. Most of the organic coffee growers are small farm holders.

Considering the propensity to erosion in coffee growing areas, organic coffee production technology can revert or reduce negative environmental impacts, improving both biodiversity and the living conditions of the producers.

Due to its cost and its reliance on external agents, certification is a fundamental problem for organic coffee growers. The consolidation of domestic certifiers to help with the promotion of organically grown coffee and other organic products is therefore most important.

Finally, the potential of organically grown products, including coffee, will amplify as broader rural development projects are initiated and developed.

10.2.8
Problems Associated with Organic Coffee

Excessive numbers of varieties of coffee in the organic sector complicate the marketing and quality control aspects.

- Difficulties of controlling the marketing of organic coffee and making sure that all coffee sold as organic has a valid certificate
- Lack of agreement on unified standards to define internationally certified organic coffee
- Failure to market a high quality of organic coffee due to a lack of control at the plantation level where there is sometimes a mixing of different qualities
- The tendency of some extremists to exaggerate the ecological, ethical and social aspects of organic coffee is hardly acceptable to the majority of consumers – the excessive price of organic coffee can

only be justified by a truly improved quality and purity
- The difficulties inherent to the production and the marketing of organic coffee suggests that it is illusory to contemplate that enough volume will be available for large industrial production needs.

10.2.9
Solidarity Market

The "solidarity market" works on a parallel with the organic coffee market. Its main promoters are organizations like Max Havelaar in Holland, and GEPA in Germany, France and Great Britain. It also enjoys the support of the European Parliament.

The idea behind the notion of the "solidarity market" is that the coffee producer is paid a basic price guaranteed by the roaster and, as a result, bypasses intermediaries. This way the price offered is higher than that offered for conventional coffee. Hence, the end price for the consumer is also higher, but the consumer accepts it as the price for "solidarity" towards the producer.

10.3
Conclusions

Coffee is a major raw material, and its quality and wholesomeness must remain a constant concern. An ongoing effort must be maintained to ensure that it meets legal and corporate standards as well as consumer requirements.

Pesticides, herbicides, fungicides and fertilizers, in other words agrochemicals, play a key role in coffee production. Wide use of these products has enabled very significant increases in productivity. The practice of integrated management, which aims at minimizing agrochemical inputs while main-

taining yields and quality, is pertinent. Agrochemicals used only as needed and replaced, wherever possible, by biological methods of fertilization, weeding, pest and disease control, etc., is highly recommended.

As a general rule, when agrochemical residues are present in green coffee, the levels are very low. In any case, these residues are supposedly destroyed or strongly altered by the roasting process so that no, or only slight traces, are detectable in the final cup of coffee.

Up to now there is no tangible evidence that organically grown coffee has any organoleptic quality advantage over the coffee produced using conservative shaded farming methods inasmuch as optimal integrated agricultural practices have been applied throughout the entire production cycle.

Even though there is no absolute proof that organically grown coffee offers improved technological qualities or purity, organic farming remains highly justified because it seriously slows down the pace of deforestation and ambiental contamination (soil, atmosphere, water).

Over the last 3 years, organically grown coffee has led to the development of the concept of "sustainable coffee". This is now further enhanced by the notion of a "solidarity market" which takes the social aspects of coffee farming into consideration. The very recent recognition of "shade" coffee completes the new trend which encourages biodiversity in coffee plantations.

The beneficial combination of these three concepts will have to ultimately lead to the production of an excellent quality coffee so that sustainable coffee-growing practices can develop on a larger scale and become worthwhile to all concerned – producers and consumers alike.

Bibliography

[1] AMAE–IFOAM–UACH. *Memorias Conferencia Internacional sobre Café Orgánico.* Federación Internacional de Movimientos de Agricultura Orgánica (IFOAM), Asociación Mexicana de Agicultura Ecológica, Universidad Autónoma Chapingo, 1985.

[2] CERTIMEX. *Normas para la Producción y Procesamiento de Productos Ecológicos. Certificadora Mexicana de Productos y Procesos Ecológicos.* Universidad Autónoma Chapingo, 1988.

[3] Escarmilla, P. E. *El Café Cereza en México: Tecnología de la Producción.* CIEESTAAM-DCRU, Universidad Autónoma Chapingo, 1993.

[4] Figueroa, Z. R. B., Fischersworring, H. and Rosskamp, R. *Guía para la Caficultura Ecólogica Café Orgánico.* Lima, Peru, 1996.

[5] Marlin, Ch. Les stratégies des grands torréfacteurs et importateurs sur le marché international du café: quel espace pour les organisations de producteurs? *Collection Max Havelaar.* Max Havelaar, Montpellier, 1993.

[6] Santoyo, C. V. H., Díaz, C. S. and Rodríguez, B. P. *Sistema Agroindustrial Café en México. Diagnóstico, Problemática y Alternativas.* Universidad Autónoma Chapingo, 1994.

[7] Sosa, M. L. and González, J. V. *El Cultivo de Café Orgánico en México Universidad Autónoma Chapingo. Dirección de Centros Regionales.* Editorial Futura. Texcoco, 1995.

[8] Zapata, A. R. J. and Calderon, R. A. *Primer Foro Nacional sobre Agricultura Orgánica.* CONARAO, SAGAR, Universidad Autónoma Metropolitana. 1996.

11

Frost in Coffee Crops: Frost Characteristics, Damaging Effects on Coffee and Alleviation Options

A. Paes de Camargo and M. B. Paes de Camargo

11.1
Introduction

Frost in coffee-growing areas is a microclimatic phenomenon defined when the temperature falls below the freezing point of water. In agriculture, the damaging effects of frost result from the freezing of water in the cellular tissues. The formation of ice within the cell ruptures the cell walls causing irreparable damage.

Deposits of ice, called white frost, appear when the air is humid, but greater damage is caused by radiation frost, sometimes known as "true frost". Frost damage in tropical vegetation occurs when the temperature of the plant tissues drops well below freezing point, usually to between −3 and −4 °C. The lower the temperature and the longer it remains below freezing point, the more extensive the damage will be.

The simple presence of ice on the coffee plant does not have serious consequences as water freezes at 0 °C and internal liquids freeze even at a lower temperature. In many coffee-growing areas, ice deposits are commonly observed on the leaves of plants during the dry cool season, but do not cause any undue damage.

Coffee beans affected by frost turn black and their quality deteriorates in direct ratio to the severity of the frost damage.

This chapter attempts to describe various ways of controlling radiation frost.

11.2
Radiation Frost Formation

11.2.1
The Heat Balance and Radiation Frost

During the day, the soil receives heat from the sun in the form of short-wave radiation, whilst at the same time there is a continuing loss of heat from the soil to space in the form of long-wave radiation. During the day the sun dispenses more than enough heat to compensate for the losses by long-wave radiation to space, which causes the Earth to warm up. During the night, however, only the outgoing radiation takes place and, therefore, the Earth cools down. If the night is cloudy, the clouds intercept the radiation from the soil and reflect it back to the Earth. This means that on cloudy nights the soil and the atmosphere between the soil and clouds cools down more slowly (Fig. 11.1).

Coffee: Growing, Processing, Sustainable Production, Second Edition. Edited by J. N. Wintgens.
© 2012 WILEY-VCH Verlag GmbH & Co. KGaA. Published 2012 by Wiley-VCH Verlag GmbH & Co. KGaA.

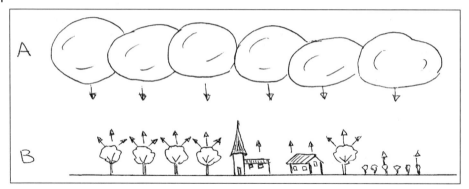

Figure 11.1 Exchange of soil and cloud radiation. (A) Clouds. (B) Soil, vegetation, civil works.

On cloudless nights, however, the cooling process of the soil is accelerated because all of the long-wave radiation is lost into space. As the soil cools down, it inevitably cools the air, which is in direct contact with it. Consequently, a layer of cold air commonly forms at ground level which is overlaid by warmer air a few meters above. When temperatures in the layer of cold air fall below freezing the condition is referred to as a radiation or a ground frost. It may be concluded that a radiation frost is most likely under clear, cloudless conditions.

11.2.2
Effect of Wind

Under calm windless conditions, the stable ground level layer of air will cool due to loss of heat by radiation and frost will occur. If, however, it is windy, the air movement will result in a continual mixing of the lower cool layers of air at ground level with the upper warm layers. The fall in air temperature at ground level will be reduced by the mixing and the incidence of frost will be reduced. It is therefore logical to conclude that the second condition where frost will be most likely to occur is on calm nights.

11.2.3
Short-term Prediction of Frost in Calm Conditions

The occurrence of frost may be forecast at or around sunset for the pre-dawn hours of the following day using various parameters related to the humidity of the air. The saturation water vapor content of the air is strongly dependent on temperature, e.g. at a temperature of 20 °C at saturation the air contains 14 g of water/kg of air, but at 5 °C saturated air contains only 3.5 g of water/kg of air. After sunset in most coffee-growing areas the air cools rapidly, in a few hours becoming fully saturated, and this leads to the formation of dew (condensation) and mist. The point at which the air is fully saturated and has a relative humidity of 100 % is known as the dewpoint. The time taken to reach dewpoint depends on the relative humidity of the air at the outset and the rate at which the temperature falls. In dry conditions a lower temperature is required to reach dewpoint than in moist conditions.

The relationship between temperature and dewpoint provides an understanding of the incidence of frost and forms the basis for the prediction of when frost is likely to occur. The relationship is illus-

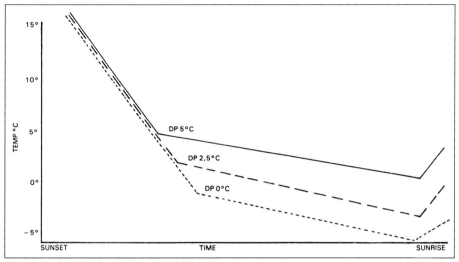

Figure 11.2 Relationship between temperature (DP = dewpoint) and time of night sunrise. From: *Coffee Handbook* [1].

trated by the three examples given in Fig. 11.2.

All the three cases illustrated in Fig. 11.2 begin with pre-sunset temperatures of above 15 °C. The continuous line in the figure relates to a high-humidity situation in which the dewpoint is reached at 5 °C. At this temperature condensation will occur and, with it, heat will be released which will slow down the decline in the air temperature so that absolute minimum temperatures may reach about 1 or 2 °C. In a medium-humidity situation (see dashed line), the dewpoint is reached at a temperature of 2.5 °C when the fall in temperature is slowed by the heat release from condensation, but a slight frost is still the likely consequence. In the driest situation, illustrated by the dotted line, the dewpoint is not reached until a temperature of 0 °C. In this situation the heat generated by condensation would not be enough to avoid a significant further fall in temperature and a quite severe frost.

One can, therefore, conclude that frost forecasts may be carried out with an acceptable level of accuracy to allow timely action to be taken to avoid frost damage in coffee. Measurements of temperature and relative humidity in the late afternoon or evening are used to calculate the dewpoint, the critical factor influencing the occurrence of a subsequent frost. An example of the use of dewpoint to forecast the severity of frosts in a coffee growing area, taken from Rhodesia (now Zimbabwe), is shown in Table 11.1.

It must be emphasized that Tab. 11.1 above only gives an indication of the severity of frost that may be expected under average conditions when a calm, cloudless night is expected. If a farmer wishes to have a better understanding of the frost risk, it may be helpful to implement the following procedure:

1) The dewpoint should be determined early in the afternoon with the aid of wet- and dry-bulb thermometers. Self-

Table 11.1 The relationship between dewpoint in the late afternoon and severity of frost in Zimbabwe

Dewpoint in the late afternoon	Approximate degree of frost to be expected at ground level
7 °C or above	no frost likely
4 °C	0 to –1 °C slight
2 °C	–1 to –2 °C slight to moderate
0 °C	–2 to –3 °C moderate
–1 °C	–3 to –4 °C moderate to severe
–2 °C	less than –4 °C severe

From: A. B. Law [19].

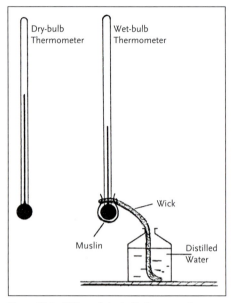

Figure 11.3 Wet- and dry-bulb thermometer.

explanatory tables obtained from the Meteorological Office provide simple conversion of the readings of the wet- and dry-bulb thermometers to dewpoint. The dewpoint then gives an indication of the severity of the frost to be expected the following morning under average conditions. It is advisable to make second readings of thermometers and calculation of dewpoint later in the afternoon to confirm original observations (Fig. 11.3).

2) In addition to checks on the wet- and dry-bulb thermometer, the farmer should make temperature observations directly on the threatened plots in order to correlate them with other observations, e.g. the relationship of afternoon dewpoint with the ground minimum temperature the following morning can be established.

This knowledge is essential for developing frost-prevention measures on an individual coffee plantation, with its own microclimatic circumstances.

11.3
Types of Frost

There are two types of frost formation: radiation frost and wind frost.

1) Radiation frost, or true frost, as noted earlier, occurs on cold, calm and cloudless nights when the air is dry and the dewpoint low. Even if it does not leave ice deposits on the leaves, it causes severe internal damage to the tissues of the coffee plants.

2) Wind frost is brought about by strong, sustained winds originating in cooler regions at higher latitudes or from cooler mountainous regions. Wind frost is more likely to occur in temperate climates or higher altitudes, when strong winds blow and the air temperature is below 3 °C. Wind frosts are common in the subtropical coffee-growing areas of Brazil and Paraguay, but the damage and prejudice are not severe (Pict. 11.1).

Picture 11.1 The effect of frost on coffee. (A) Radiation frost: the soft tissues and the upper leaves are more affected by the frost. (B) Wind frost: only the windside of the plant has been affected by the frost. From: J. N. Wintgens – Zimbabwe 1994.

11.4
Types of Radiation Frost

Radiation frost can occur in both subtropical climates and in temperate climates. In subtropical climates, i.e. at latitudes that are lower than 20 °C, radiation frost is rare, sudden and difficult to predict. Over the last century, severe frosts have been recorded at intervals that can vary from 5 to 15 years. The unpredictability of frost occurrence in subtropical areas makes decisions on protection measures difficult. Severe frosts in subtropical climates rarely affect the crop of the year in which they occur, but can damage the coffee trees themselves. On the other hand, in temperate climates the situation is different since frost does not unduly affect the plants themselves, but does cause damage to the crop of the year in which they occur. Radiation frosts are classified according to severity as outlined below:

1) Moderate frosts are relatively frequent in many coffee-growing areas. The occurrence of moderate frosts is often localized in low areas, or frost pockets, where there is restricted air circulation and where there is a continuous accumulation of cold air during calm, cloudless nights. Such frosts do not affect adjacent coffee plants located in better air-drained and elevated terrain.

2) Severe frosts are rare in most coffee-growing areas, but when they do occur they are more widespread affecting coffee plants even on well-drained terrain. These severe frosts usually kill the young top growth and the exposed branches of the coffee trees, and depress the yield of coffee at the succeeding harvest. Preventive protection methods of frost fighting, geared to the short, medium and long term, are usually economically worthwhile.

3) Very severe or catastrophic frosts are extremely rare. They occur only when the air temperature drops continually during the entire night at a rate of about 1 °C/h. Catastrophic frosts usually kill the entire foliage and young shoots of

Picture 11.2 Two examples of radiation frost.
(A) Very severe frost. Almost all the aerial parts of the coffee tree have been killed by frost. New shoots grow from the base of the plant but the new production cycle will be delayed for 2–3 years.
(B) Severe frost. Only the upper branches of the coffee tree have been killed by frost, therefore only the coffee crops for the following year will be affected. From: J. N. Wintgens.

the coffee plant, and result in massive yield reductions for the following 2–3 years. Normal preventive protection methods are then to no avail. Only direct protection by fogging during the night of the frost, or arborization, can ensure satisfactory protection (Pict. 11.2).

11.5
Factors Conducive to the Formation of Radiation Frost

11.5.1
Macroclimatic Factors

Macroclimatic factors are determined by overall geographical conditions such as latitude, altitude, inland location, relief, atmospheric pressure, etc.

The occurrence of frosts is an important factor influencing the global distribution of the coffee crop, and consequently the crop is largely restricted to the subtropical and tropical areas of the world, i.e. within the equatorial belt between latitudes 25° N and 25° S. Damaging radiation frosts are reported at the higher latitudes in this belt, but are usually associated with higher altitudes. However, coffee grown at sea level in Brazil on latitude 25° S is frequently affected by frost, and provides an exception to the association of latitude and altitude.

In general, the vast majority of coffee-producing countries are located in the latitudinal belt between 18° N and 18° S where the only frost problems that occur are associated with higher altitudes. Potential frost problems can be foreseen with an examination of average temperatures – the average temperatures decline by approximately 0.65 °C/100 m increase in altitude. In the higher areas of southern Brazil in the state of Paraná where plantations were located at an altitude of over 500 m, frosts became so frequent and severe in the 1970s and 1980s that commercial coffee farming was abandoned on a significant scale.

In Paraguay, the altitude limit set for coffee is 300 m above sea level at or about latitude 25° S; above this altitude the frost damage is so severe as to render coffee production uneconomic. In general, an inland continental location increases potential frost risks, as the oceans are known to be powerful thermal moderators. The further inland the plantations are located, the greater the thermal variations and, therefore, the higher the risk of frost occurrence.

Orographic factors have some influence on the occurrence of frost as well as on the formation of surface fog. In normal conditions, sloping lands have better air circulation and are less exposed to an accumulation of cold air than flat lands or lands with rugged topography. By the same token, flat or gently rolling lands are less exposed to the formation of fog and frost. In contrast, in "talwegs" or fluvial basins, low-lying layers of cold air accumulate, and bring about condensation and the formation of fog banks at ground level. When the atmosphere is dry and cold the fog banks can convert to radiation frost.

Changes in atmospheric pressure, chiefly brought about by conflicting winds, can also be conducive to frost occurrence. This is most noticeable when the jet stream winds meet a strong, dry polar anticyclone blowing from the opposite direction. When this occurs, wind movement is stilled and the ensuing rise in atmospheric pressure accompanied by a significant temperature drop is a sure indication of potential frosts. Such effects are likely to occur in those coffee-producing countries under continental influences as in South America and South and South East Asia.

11.5.2
Microclimatic Factors

Factors influencing the microclimate in coffee include a range of factors such as, vegetal cover (vegetation, mulch, organic material residues, etc.), soil moisture status, soil management practices, presence or absence of shade trees, etc.

During calm nights with no winds, the bare surface of the soil receives approximately 25 % of its heat from the air and 75 % of its heat from the lower layers of the soil. On a frosty night, the cold only affects the superficial layer of the soil to a depth of 5–10 cm. Below this level, the soil retains a certain amount of heat. Therefore, natural soil heat must be maintained at all costs and direct contact between the soil, and the surface air must be encouraged. This means that any screen between the soil surface and lower air layers should be avoided. Air is known to be a poor conductor of heat so screening practices that lead to the build-up of air pockets should be avoided.

As vegetation, mulch and organic residues are also poor heat conductors, they should be avoided in areas exposed to frost. In contrast, moist soil is a better heat conductor than dry soil. It is therefore wise to make sure that during potentially frosty months, the soil of frost-exposed areas is kept:

- Compact with no soil disturbance during the few weeks that precede the frosty season,
- Moist, as far as possible, and
- Free of weeds and mulch.

The presence of shade trees that form a dominant canopy over the coffee plants offers good protection against frost. The canopy effectively absorbs terrestrial radiation and reflects it back to the coffee plants, thus reducing the loss of atmospheric heat (Pict. 11.3, 11.4).

Picture 11.3 (A) Shading of coffee trees. Dense shading by *Erytrina* spp. This offers effective protection against frost. (B) Congusta hybrid under banana shade.

Picture 11.4 Intercropping with rows of maize between rows of coffee plants. The maize protects the coffee plants from strong winds and frost. From: D. Fernandes.

11.5.3

Topographic Influences on the Microclimate

The topography of the land has an important influence on air circulation and air movements. Cold air has a higher density than warm air. Thus, on sloping land the cold air, which accumulates at the soil surface as heat loss takes place during the night, tends to flow downwards and accumulate at the bottom of the slope in depressions or valleys. It is therefore to be expected that frosts will be far more frequent in depressions and in the bottom of valleys.

The downward movement of cold air, known as "katabatic breeze", can be compared to that of a viscous liquid flowing slowly down a slope. The lower the air drifts, the colder it gets. The downward flow of cold air can be effectively intercepted or dammed by a physical barrier such as a wall, a hedge, rows of trees or other barriers (Fig. 11.4).

11.5.4

Incidence of Topography and Land Cover on Frost

- Flat lands are subject to the stagnation of cold air at ground level which can lead to severe frosts in the whole area.
- Concave land is even more subject to the accumulation and stagnation of cold air at ground level; the highest frost risk areas are on concave land.
- Convex land and sloping hills have good air drainage; cold air is diverted and they are usually free from frost.
- An enclosure below the coffee plantation will lock cold air in and increase the frost hazard.
- A grass vegetation cover above coffee plots creates cold air pockets at ground level that drain towards the plants and increase the frost hazard.

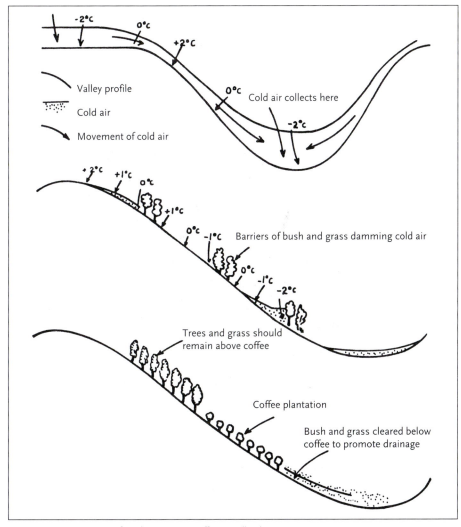

Figure 11.4   Movement of cool air. From: *Coffee Handbook* [1].

- A pond above the planted areas will absorb the katabatic breeze that will then gather heat and moisture from the water – this generates the formation of fog over the plantation, which prevents undue cooling of the soil and offers effective protection against frost (Pict. 11.5, 11.6, 11.7, 11.8).

Picture 11.5   Adult coffee trees affected by frost. The lower part of the foliage of the trees has suffered greater frost damage. The frost damage is more intense at lower levels.
From: J. N. Wintgens – Zimbabwe.

Picture 11.6 Young coffee trees damaged by frost due to the presence of a "frost dam" below. From: J. N. Wintgens – Zimbabwe.

Picture 11.7 Young coffee plantation hit by frost. The trees located in concave areas of the slope suffered the most. From: J. N. Wintgens – Mexico.

Picture 11.8 An adult coffee tree hit by frost. The lower part of the foliage skirt suffered more than the upper part because more exposed to the colder air pocket. From: J. N. Wintgens – Zimbabwe.

11.6
Destruction of the Plant Tissues

- Extracellular liquids are poorer in soluble solids than the intracellular liquids and therefore tend to freeze more easily.
- Intracellular liquids may seep out and form extracellular ice crystals.
- The cytoplasm may shrink away from the wall of the cell – this phenomenon, known as plasmolysis, is irreversible and leads to the death of the cell.
- In severe cases, the cells may be crushed by the pressure of the extracellular ice crystals (Fig. 11.5).

Tissues that have a high concentration of soluble solids, mainly potassium salts, in

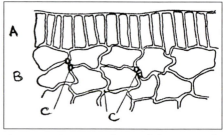

Figure 11.5 Formation of extracellular ice crystals. Tranversal cut through a leaf. (A) Palisade tissue. (B) Parenchymal tissue. (C) Sites where extracellular ice crystals start to form.

their liquids offer a higher resistance to frost as the freezing point is lower. Well-fertilized, vigorous plants or unproductive plants that have not yet consumed their potassium nutrients have the highest concentration of soluble nutrients in their sap – they are therefore less exposed to frost hazards.

11.7
Heat-stroke and Frost Damage

Frost damage is frequently attributed to "morning heat stroke", but the various explanations given to justify this theory are not supported by any scientific or experimental evidence. Based on a large number of observations on coffee trees which have suffered frost damage, no greater damage has been observed on the side exposed to the first hot rays of the early morning sun than on the shaded side. On the other hand, it has been noted that coffee plants that are underneath large trees have suffered less frost damage. This is due to the fact that the protective canopy of higher trees intercepts terrestrial radiation and reflects it back to the coffee plants.

11.8
Protection against Frost

Frost-protection methods can be divided into two categories: preventive protection and direct protection.

11.8.1
Preventive Protection

Preventive protection uses topoclimatic measures such as tree canopies, air-draining passages or pastures, ponds, etc., requires long-term planning. It is a simple and natural form of frost fighting which can give the best results if correctly managed.

11.8.1.1   Choice of Site and Frost-prevention Planning Criteria

Where there is a serious frost hazard it is essential that all available measures are taken to minimize the risk of frost damage. These measures should begin at the outset with the careful choice of the site. Having chosen the site, detailed consideration should be given to the factors that will reduce the frost hazard in the particular circumstances in which the new coffee plantation is being planned. The plantation should be established according to the frost-prevention planning criteria listed below:

- In highlands, the ideal location for a coffee plantation is near the top of a hillside.
- Plantations on a slope near low-lying terrain should be avoided; the same applies to plantations in valleys with a narrow exit where cold air accumulates.
- If there is a forest with high and dense vegetation below the plantation, a strip following the contour of the slope, of approximately 100 m width, should be opened through it to provide adequate drainage for cold air.
- In plantations located on a hillside, a high hedge or a row of high trees will provide good protection against the invasion of cold air from above.
- The establishment of leguminous shade trees in the coffee plantation is an effective long-term preventive measure against frost in addition to the other benefits.
- The choice of appropriate planting material should take account of the tolerance level to low temperatures – Robusta varieties are more susceptible to low temperatures than Arabicas with a minimum level of tolerance in the range 1–3 °C; on the other hand, Arabica varieties exhibit markedly different levels of tolerance to frost.

11.8.1.2 Crop Husbandry Practices and Frost Protection

Some crop husbandry practices contribute to frost protection when carried out in a timely manner in relation to the seasonal occurrence of frosts and they consist of the following:

- Maintaining a bare soil in the plantation during the cold months – this entails clearing all brushwood, straw, vegetation and mulch because soil covering of this type prevents the heat of the soil from radiating to the surface and warming the atmosphere during the night.
- Clearing the vegetation from the slopes below the plantation to provide for adequate drainage of air through the plantation and prevent the accumulation of cold air around the coffee trees.
- Providing sufficient potassium by correct fertilization of the coffee plants in order to increase their resistance to frost.
- Intercropping with rows of pigeon pea (*Cajanus cajan*) or other erect or tall plants – the additional vegetative cover of these tall-growing plants protects the coffee plants from frost by restricting outgoing radiation.
- Earthing-up the stems of young coffee plants before the frosts set in – the earth should be heaped up in a cone as high as possible around the stems this method provides excellent protection during the actual frosts, but as soon as the dangerous period is over, the earth should be flattened to its original level; this method should be avoided in areas where trunk fungus diseases are endemic.

11.8.2 Direct Protection

11.8.2.1 Fogging

This form of protection can only be used on the actual night when frost is expected. It is the most efficient method used to protect coffee from frost, being both practical and economic. The artificially created fog consists of droplets of moisture of 10–30 μm in diameter which are capable of absorbing terrestrial radiation and radiating it back to the coffee plants. Dense fogs protect the surface of the soil from drastic drops in temperature and the impact of frost.

Three major guidelines should be observed in order to obtain effective protection by atmospheric fogging:

1) The quantity of fog application should be calculated to absorb, reflect and disperse all terrestrial radiation.
2) The fog should be distributed from the top of the valley to protect the coffee plants in the depression that is threatened by frost.
3) The fog application needs to be started when the temperature at 50 cm above ground level among the coffee trees in the lower part of the plantation is at 2 °C or slightly below.

11.8.2.2 Overhead Irrigation

The overhead irrigation of coffee using sprinklers may be used as a dual-purpose system as in addition to applying irrigation water, the system may also be used for frost protection. For frost protection, the water is applied through the sprinklers just when the frost is about to occur or as it is actually occurring. The temperature of water used for sprinkling (overhead irrigation) will normally be above freezing point on leaving the nozzle and consequently will have an immediate warming effect. Even if the severity of the frost freezes the water on the leaves, the temperature within the leaf will not drop below freezing point as long as the sprinklers remain active and the layer of ice on the leaves is kept wet. This effect is due to the fact that the action of converting water into ice releases a certain quantity of heat (1 g of water when turning into ice release about 80 cal). Thus, in order to avoid frost damage, sufficient quantities of water must be used to ensure that the leaves are kept wet at all time during the period of frost particularly when they are covered with ice. For frost control sprinkler rotation should give each coffee tree a spray at least once per minute at a rate of approximately 2.5–3 mm/h (25–30 m³/h/ha).

In order to avoid the risk of soil-generated diseases, nitrogen leaching, denitrification, etc., special care must be taken to prevent soil saturation (water logging) after several nights of sprinkling. Sprinkling for frost control is expensive, and requires special techniques and equipment.

11.8.2.3 Frost Prevention by Measures Influencing Air Circulation

During radiation frost, temperatures at ground level are below freezing point, whereas just above there is a layer of warmer air. The occurrence of frost is often quite rare at levels higher than 2–3 m above ground so that any action that mixes the cool lower level air with the warmer layers above will generate an increased average temperature of the air mass. To achieve a mixing of air and an increase in the air temperature at the lower levels in coffee plantations, large fans are used. The best results have been obtained using fans tilted at an angle of about 60° from the surface to blow air along the planted rows. This type of air circulation can only

be used on flat lands and it is a generally accepted fact that they are not very cost-effective.

Any depression in the land may lead to an accumulation of a stagnant layer of cold air at, or just above, ground level and thus present a frost hazard to the coffee trees. The construction of irrigation basins or contour bunds for erosion control simulates the effect of a natural depression in the land in so far as the bunds provide an obstacle to the free movement of air. It is therefore advisable to make temporary openings in the retaining walls on the lower side of irrigation basins or in the contour bunds to improve cold air drainage during the frosty season.

11.8.2.4 Frost Prevention by Providing Heat Sources

Heat can be supplied by burning any available fuel, such as brushwood, paraffin, diesel oil, etc., in fires scattered amongst the coffee trees in the plantation. The convection currents generated by the fires mix the warmer air of the upper layers with the cold air at ground level. Many small fires scattered throughout the plantation are far more effective than a few large fires. The heating and circulation of air can be combined as a frost-control measure. However, such a measure can only be applied on flat lands, it is expensive and will not be effective in the case of severe frost and so, inevitably, is rarely used.

11.8.2.5 Smoke as a Frost-prevention Measure

Trials have been conducted using smoke generated by solid fibrils of coal as a frost-prevention measure in coffee. The theory was that the carbon particles in the smoke would absorb and reflect long-wave radiation. However, the smoke was ineffective in providing any frost protection because of the small size of the carbon particles and a failure to maintain a stable, smoke-filled environment around the coffee.

11.8.2.6 Coffee Husbandry Practices Designed to Minimize Frost Damage

Young coffee plants are more susceptible to frost because they are not very high and therefore more exposed to the layers of cold air just above ground level where freezing temperatures occur most frequently. Also, young coffee plants have a much higher proportion of soft fragile tissues which are highly susceptible to frost damage. A simple measure that is taken to reduce the exposure of coffee planting material to frost is to plant on raised mounds, 10–15 cm above ground level. This measure sets the plants above the coldest layer of air.

In the case of contour-planted plots, the mounds should be located in such a way as to allow for air drainage in all directions, across the rows and down the slope. Young coffee plants can also be protected from moderate frosts by wrapping their stems in sleeves of protective material or by covering the stems with earth.

Coffee growing in high-fertility conditions has been shown to be much more resistant to frost damage than coffee growing under conditions of inadequate nutrition. Also, strong coffee trees recover from frost damage at a much faster rate. Thus, an important part of an overall strategy to limit the damage due to frost must be to follow a well-balanced programme of fertilizer use. Fertilization should include high levels of potassium, whereas levels of nitrogen should be limited.

11.9 Prediction of Frost Occurrence

Experience and various studies have revealed that there are simple meteorological observations which can be used to predict an imminent frost. Understanding these observations and reading them correctly will enable the farmer to take necessary protective measures to limit or avoid potential frost damage in the coffee crop.

In coffee areas of the southern hemisphere of America, frost is very likely to occur if:

- Wind direction has been from the northwest, either continuously or sporadically, over the previous 2 days.
- The previous days have been either cloudy or rainy.
- After 6pm, the night skies are very clear, with bright stars, no clouds and no wind.
- The air temperature is below 7 °C at 7 p.m. or below 4 °C at 10 p.m.
- The relative humidity of the air is lower than 70 % at 10 p.m.

Frost is unlikely to occur if:

- The winds of the polar anticyclone have been blowing continuously from the south-east over the previous 2 days.
- The previous day has been very sunny.
- The weather is cloudy or windy at sunset.
- The air temperature is above 12 °C at 7 p.m. or above 8 °C at 10 p.m.
- The relative humidity of the air is higher than 90 % at 10 p.m.

11.10
Treatment of the Coffee Trees after Frost Damage

- If the lateral branches are still alive, do not intervene.

- If the lateral branches are damaged by frost, cut them back to 40 cm soon after harvesting.
- Fertilize with nitrogen by soil application, and apply Zn, Bo and Cu by foliage spraying.
- Remove affected branches.
- Reduce excessive suckers.

Bibliography

[1] Bisco, E. J., Logan, W. J. C. *Coffee Handbook*. Coffee Growers Association, Harare, 1987.

[2] Bouchet, R. C. M. Lutte contre les gelées de printemps. *La Météorologie* **1954**, *36*, 403–439.

[3] Brooks, F. A. Frost protection. In: *An Introduction to Physical Micro-meteorology*. Syllabus 397. University of California, Davis, CA, 1960.

[4] Brooks, F. A. Climatic environment. A thermal system. In: *Syllabus for Agricultural Engineering*. Syllabus 106. University of California, Davis, CA, 1951.

[5] Brooks, F. A., Kelly, C. F., Roades, D. G and Schultz, H. B. Heat transfers in citrus orchards using wind machines for frost protection. *Agricultural Engineering (University of California)* **1952**, *33* (2), 74–78 and 143–154.

[6] Camargo, A. P. Instruções para combate à geada em cafezais. *O Agonômico*, **1960**, *12*, 20–35.

[7] Camargo, A. P. *Instruções para Combate à Geada em Cafezais*. Insituto Agonômico, Campinas (SP), 1966.

[8] Camargo, A. P. Frequência de geadas exepcionais, como a de julho de 1975. In: *International Seminar on Climatology of the Southern Hemisphere*. Instituto Agronômico, Campinas (SP), 1997, 1–3.

[9] Camargo, A. P. O clima e a cafeicultura no Brasil. *Informe Agropecuàrio, Belo Horizonte (MG)* **1985**, *11* (126), 13–26.

[10] Camargo, A. P. Geada. O remedio e prevenir. *Boletim Tècnico 227*. CATI, Campinas (SP), 1997.

[11] Camargo, A. P. and Salti, E. *Pesquisas sore Combate à Geada Realizadas No Paraná, Brasil, em 1959*. Instituto Agronômico, Campinas (SP), 1960.

[12] Camargo, A. P, and Salati, E. Determinação da temperatura média letal da folhagem do cafeeiro em noite de geada. *Bragantia* **1966**, *25*, nota 14, LIXAI.

[13] Camargo, A. P. and Guizzi, S. M. Estimativa de temperaturas médas mensais com base em cartas de temperatura potencial normal ao nivel do mar para a Região Sudeste do Brasil. *Boletim Técnico 141*. CATI, Campinas (SP), 1991.

[14] Camargo, A. P. and Fernandes, D. R. Projects of frost and cold wind prevention on coffee crop at Southeast Paraguay. In: *Proceedings of the Symposium on Agronometeorology and Plant Protection*. Assunción, Paraguay, 1992, 25–33.

[15] Caramori, P. H., Correta, R. H., Chaves, J. C. D., et al. Proteçã de cafezais contra geada através do planteo intercalar de guandú (*Cajanus cajan*). In: *24° CBPC*, Poços de Caldas (MG), 1998, 146–147.

[16] Carneiro Filho, F., Viana, A. S., Kaizer, A. A., Matielo, J. B. and Camargo, A. P. Efeito da cobertura com terra em plantas jovens de café, como proteção contra geada. *Série Experimentação Cafeeira* **1978**, *1* (6), 51–56.

[17] Fournier, D. E. M. The modification of microclimate. In: *Climatology. Review of Research – X*. UNESCO, 1958, 126–146.

[18] Geiger, R. *The Climate near the Ground*. Harvard University Press, Cambridge, MA, 1950.

[19] Law, A. B. The use of wet- and dry-bulb thermometers to forecast frost at night in Rhodesia and frost protection in Rhodesia. *Notes Agricultural Meteorol.* **1968**, *16*.

[20] Levitt, J. *Responses of Plants to Environmental Stresses*, 2nd edn. *Vol. 1: Chilling, Freezing and High Temperature Stresses.* Academic Press, New York, 1980.

[21] Meyer, B. S. and Anderson, D. B. *Plant Physiology.* van Nostrand, New York, 1985.

[22] Serra, A. Monografia sobre mecanismo del "Tempo" em el Brasil. *Rev. Meteorologica* **1949**, *8*, 357–381.

[23] Serra, A. Previsão de geada. *Rev. Bras. Geo.* **1957**, 4.

[24] Volpe, C. A. and Andre, R. G. B. G. Prevenção e combate UNESP. *Boletim Técnico 2.* Facultade de Ciências Agrárias e Veterinária, Jaboticabal (SP), 1984.

12
Importance of Organic Matter and Biological Fertility in Coffee Soils

D. Snoeck and P. Vaast

12.1
Introduction

Soil fertility is the most important factor that fosters the growth and productivity of coffee trees. The topsoil, i.e. the 30 cm upper layer, is particularly rich in organic matter (OM) and it is in this layer that a large majority of the feeder roots are located. Consequently, the conservation of soil fertility and particularly its OM content is essential to ensure sustainable coffee cultivation [1]. The issue of soil OM conservation is crucial in mountain areas, which are exposed to erosion (Central America, East and Central Africa, Indonesia), but also in all areas where farmers use very little inputs and cultivate annual food crops between the coffee rows. This type of management practice, which is highly dependent on OM mineralization for its nutrients, produces poorly if soil management is inadequate.

In intensive farming systems, nutrients are provided by fertilizers and weeds are controlled by the frequent use of herbicides. Over time, the OM content of the soil is depleted and the mineral elements become unbalanced, which leads to leaching, erosion and soil compacting. Root growth cannot develop correctly and, ulti-mately, production is severely reduced due to the depletion of the nutritional reserves. At the same time, the coffee trees become more susceptible to pests such as nematodes or root diseases like *Fusarium* or *Rosellinia* [2].

The productivity of coffee trees is also very dependent on soil nitrogen (N) availability. In intensive farming systems, N is provided by frequent and intensive N fertilization (100–500 kg N/ha) to compensate for depletion through yields (50–70 kg N/ton of dry coffee beans) and to ensure the vegetative growth of the coffee trees. In contrast, little or no synthetic fertilizers are added in extensive farming systems. Consequently, the N availability is low due to OM-impoverished soil, which generates poor coffee growth and low yields [2].

Alternative practices to decrease the need for synthetic fertilizers and maintain yields consist of:

1) Reducing losses of OM due to run-off (contour planting, terracing)
2) Reducing losses of mineral nutrients due to leaching by improving ground cover
3) Adjusting fertilizer applications to synchronize better with plant needs

Coffee: Growing, Processing, Sustainable Production, Second Edition. Edited by J. N. Wintgens.
© 2012 WILEY-VCH Verlag GmbH & Co. KGaA. Published 2012 by Wiley-VCH Verlag GmbH & Co. KGaA.

4) Maintaining soil OM content by incorporating plant residues provided by organic amendment or by the pruning of associated trees and the cutting of living mulch
5) Creating soil conditions which encourage the efficiency of natural biological processes; such as symbiotic N fixation, which increases the N availability, and mycorrhizal symbiosis, which increases the mineral uptake capacity of roots.

In optimal ecological conditions, coffee trees grown under leguminous shade trees or associated with leguminous cover crops are generally less productive than in intensive farming systems under full sun regime. Nevertheless, these agroforestry (AF) systems, which encourage a higher level of mineral recycling, provided by the decomposition of plant residues such as leaves and branches, are certainly more sustainable due to a better conservation of OM. Such practices favor a more regular productivity over the years and are key factors to ensure the sustainability of coffee plantations under suboptimal ecological conditions [3]. Recent studies, mainly conducted in Central America and Central Africa, on N mineralization and losses have enriched our knowledge of N cycles in these AF coffee systems. These studies have shown that, every year, large amounts of N (100–150 kg/ha) are mineralized, of which 10–50 kg N/ha is lost by leaching in the form of nitrates. Additional studies have shown that associated leguminous plants can also improve the N cycle by increasing N mineralization and reducing the losses through leaching by an improvement of the soil properties, particularly its organic N content [4, 5].

More recently, environmental concerns have been raised regarding intensive farming coffee practices and the economic viability of these systems has been questioned, especially during periods when coffee prices are low. This reinforces the interest in promoting alternative coffee systems based on associations with either trees or cover crops, and in developing strategies to stimulate soil microbial activity and biological processes. Such approaches improve the availability of nutrients, and reduce the losses and environmental degradation. This leads to the improvement of the long-term social and economic well-being of coffee farmers.

12.2
The importance of the Soil OM

12.2.1
Description

The soil OM is composed of plant and animal residues at various stages of decomposition. It includes chemical and biological compounds synthesized from plant residues. Finally, OM also consists of residues of microorganisms, small animals and insects, which live in the soil and promote the process of decomposition [6].

12.2.2
Physical Properties

OM is an important factor for the soil structure, especially for its aeration and water-retention capacity. All of these factors are important for root growth and plant nutrition. The OM also encourages the activity of the microfauna (earthworms, termites, etc.), which contributes to a better soil porosity and facilitates the transformation of organic elements into nutrients [7].

Humus is the result of the decomposition of OM in the soil. It can be combined with the clay particles of the soil to increase its aggregation and to improve its struc-ture, especially micropores. This leads to better water-retention capacity, and a reduc-tion of water run-off and soil erosion. When the humus is leached away, the soil becomes compacted and its aeration is re-duced. These conditions result in a reduc-tion of nutrient uptake by the coffee roots. In 1998, Vaast *et al.* [8] observed that root absorption could be reduced by 30–50 % in unfavorable conditions.

12.2.3
Chemical Properties

In tropical regions where coffee is grown, soils are often acid, and rich in iron and aluminum oxides (e.g. andisol, inceptisol, oxisol). These types of soils are also charac-terized by a low cation-exchange capacity (CEC). As a result, it is quite common that the OM compounds represent up to 80 % of the topsoil CEC where most of the feeder roots are growing. Therefore, cultural practices that reduce the losses of OM must be a priority for coffee growers. Conserving or improving the quantity of main cations (K, Ca, Mg) is also essential to improving plant nutrition and ensuring the sustainability of the coffee production [1, 9].

The OM also improves the soil capacity to buffer pH variations. This is particularly important in intensive farming systems where high inputs of N fertilizers acidify the soil. Ten to 15 years of repetitive applica-tions of N can increase the soil acidity from pH 6 down to pH 4 [10], leading to Al toxi-city that impedes root growth and reduces nutrient assimilation [11]. This deteriora-tion of soil conditions can, in turn, increase plant sensitivity to nematodes and diseases [12]. Soil acidity also reduces the microbial activity, which is essential for the minerali-zation of soil OM compounds [13].

12.2.4
Biological Properties

The soil supports a large community of mi-croorganisms capable of transforming soil compounds into mineral forms readily available for root uptake. The OM also im-proves the activity of microorganisms not directly associated with its decomposition. It harbors fungi and bacteria that act against pests and root diseases, as well as mycorrhizae and fungi capable of forming a symbiotic association with coffee roots. Under adequate soil and climate condi-tions, leguminous cover crops or trees associated to coffee, can form a symbiotic association with rhizobial bacteria able to fix the atmospheric N_2. Nitrogen incorpora-tion to the soil via root, leaf and branch de-composition can reduce the dependence of coffee plantations on synthetic N inputs.

12.2.5
Cultural Practices to Improve OM in Coffee Soils

The addition of mulch made up of crop re-sidues, coffee pulp or pruning debris from coffee trees, associated shade trees or inter-crops is advised by many authors [1, 2]. These management practices maintain an adequate soil OM content and improve the nutritional cycle. Fermentation of cof-fee pulp prior to its use as mulch is sug-gested to eliminate the propagation of pests and diseases and avoid N stress. Other organic matter sources, like rice husks, are mentioned, but are not fre-quently used except in organic coffee farm-ing.

Mulching with materials from annual cash crops or grass from fallow fields and pastures is widely practiced in East and Central Africa to protect the soil from ero-sion and preserve its OM content. This

Table 12.1 Evolution of chemical properties of the topsoil (0–30 cm) in a coffee plantation after 4 years of regular application of various types of mulch in Burundi [14]

Organic matter	Ca^{++} (cmole/kg)	Mg^{++} (cmole/kg)	K$^+$ (cmole/kg)	Mg/K	Saturation (%)
Uncovered soil	1.07	1.24	0.19	6.5	20
Eragrostis	1.80	1.34	0.38	3.5	28
Pennisetum	2.49	1.70	1.26	1.3	43
Banana leaves	3.36	2.09	0.93	2.2	51
Coffee pulp	3.19	1.79	1.44	1.2	51
Leucaena K28 mulch	1.50	1.24	1.93	0.6	37

practice has the advantage of regularly introducing fresh organic material and increasing the soil nutrient content. In Burundi [14], important increases in main cations (K, Ca, Mg) have been noted after 4 years of mulch applications from various sources (Tab. 12.1). In 1993, Fassbender [15] also registered a high transfer of P, K, Ca and Mg in the soil of coffee grown under the *Erythrina poeppigiana* shade tree (Pict. 12.1). On the other hand, small farmers avoid the regular practice of mulching because it is labor intensive. Furthermore, mulch production requires sizeable acreage, as 1 ha of grassland is needed to produce enough mulch for 1 ha of coffee trees. Moreover, this practice often leads to an

export of nutrients from the fallow fields or pastures to the coffee plots. As a result, the former then have to be fertilized for long-term sustainability.

The importance of the association of leguminous species with coffee has long been recognized in many coffee-producing countries of Central and South America as well as in Central and East Africa. In 1993, Fassbender [15] registered a high increase in nutrient content (P, K, Ca and Mg) of coffee soils in Costa Rica where coffee was cultivated under the shade of *Erythrina poeppigiana*. This practice has many of the advantages of mulching without its inconveniences. The main disadvantage of this association is that, depending on the type of soil and the climatic conditions, leguminous species may compete with coffee trees for space, water, light and nutrients.

In areas where rainfall is not limited, or where irrigation is available annual leguminous cover crops (*Mucuna, Pueraria, Arachis, Mimosa*) and perennial ones such as *Flemingia* and trees like *Leucaena, Inga, Erythrina*, etc.) are highly recommended. They produce a large quantity of mulch through pruning and slashing. Under favorable climatic and soil conditions, leguminous cover crops can produce up to 20 tons of dry matter/ha/year.

Picture 12.1 Coffee–*Erythrina* association in Costa Rica.

In Cameroon, coffee trees associated with cover crops (*Mimosa, Flemingia*) produce 30% more than by traditional farming methods. These two leguminous crops have the advantage of not competing with coffee for light and water [16]. In Burundi, similar results have been recorded with *Desmodium intortum* or *Leucaena leucocephala* under favorable soil conditions. It has also been observed that coffee associated with *Desmodium* in acidic soils produced 50% less than traditional farming due to an inadequate N_2 fixation by the rhizobium. This caused N deficiency and soil nutrient exhaustion when the vegetative development of the cover crop was too important. In Cameroon, either *Flemingia* or *Mimosa* is recommended, as these leguminous cover crops help increase the production of Robusta coffee [16]. In Burundi, surveys have also noted large differences of dry matter produced by *Desmodium* (14 tons/ha/year), by *Leucaena* K28 (7 tons/ha/year) and by a native *Leucaena* (2 tons/ha/year) [14].

In lowland areas with high solar radiation and air temperature, leguminous trees (*Leucaena, Albizia, Gliricidia, Erythrina*, etc.) are highly recommended as shade trees. Adequate pruning regulates light availability for the coffee plants and produces mulch. High inputs of N as well as P, K, Ca and Mg to coffee soils under the shade of *Erythrina poeppigiana* have also been recorded [15]. Still, these AF systems produce less dry matter than cover crops. In Central America, several authors have reported that leguminous tree species such as *Inga* or *Erythrina* can produce from 2 to 14 tons/ha/year [3, 17, 18].

In these suboptimal zones, shade trees regulate adverse environmental effects and create a more favorable microclimate for coffee growth and production. Temperature variations are moderated by trees: temperatures are 3–5 °C lower during the hot period and 2 °C higher than the minima, according to measurements taken in Mexico [3] and Costa Rica [19]. The topsoil temperature is maintained at an optimal level for root development and nutrient uptake [8]. Shade trees also help to regulate coffee tree productivity by decreasing alternate bearing, thus providing farmers with a more regular income. The tree canopy also lengthens the productive lifespan of the coffee trees, since shade limits coffee tree flowering and production, thus reducing the early die-back of either branches or of the whole tree that is often observed in sun-exposed plantations when the cherry loads are heavy [20]. In Central America, recent studies have shown that shade trees guarantee better coffee quality due to the reduced fruit load and a longer ripening period [21, 22].

12.3
Biological Nitrogen Fixation (BNF) in Coffee Soils

12.3.1
Importance of BNF

Soil N availability is very important for coffee growth and productivity [1]. Among the plant species able to transform atmospheric N_2 into mineral N through BNF, only leguminous–rhizobium associations are used in coffee plantations. Depending on farming conditions, the dry matter produced by the leguminous crop provides 50 to 200 kg N/ha/year, of which 10–60 kg are actually derived from N_2 fixation [3, 17, 18]. The large differences in N quantity derived from BNF can be explained by:

1) The climatic conditions,
2) The quantities of dry matter produced by the leguminous species,

3) The N-fixing efficiency of the nodules; this N-fixing activity of the rhizobium also depends on agricultural practices, and, consequently, to ensure a sustainable and effective symbiosis, it is important to select the appropriate leguminous species and to implement adequate farming practices.

To ensure a healthy BNF, the rhizobium species must be adapted to the soil. Trials have shown that rhizobiae are not able to fix N if the soil is too acid (pH < 5) and if the soil CEC is too low. In such cases, the legume takes its N from the soil and competes with the coffee trees. Trials in nursery plots have shown that the addition of lime and phosphate to decrease the soil acidity and to increase its calcium and phosphorus content stimulates the nodule biomass from 5 to 50 g/plant and proportionally increases the N content of the plant from 15 to 150 g/plant [23].

When intercropped between coffee rows, the leguminous plants share part of the N fixed from BNF with the associated coffee tree. By using the natural isotopic abundance of the ^{15}N isotope techniques in Burundi, Snoeck *et al.* [24] highlighted the process of N transfer from the atmosphere to coffee trees by the activity of certain leguminous crops cultivated either as intercrops or as shade trees (Tab. 12.2)

If the amount of N fixed by the leguminous plant is low, the latter does not share its N with the coffee trees (see Tab. 12.2). The lowest level of fixation required depends on the type of farming: it can be lower for leguminous plants grown as an intercrop (Pict. 12.2) than for those grown as shade tree (Pict. 12.3) because the number of trees is considerably higher.

In the case of an efficient association, it has been shown [23] that in coffee–*Leucaena* associations, the coffee tree can benefit from about 25 % of the N coming from the associated leguminous plant through 17 % from the litter decomposition and 8 % from direct root contact and nodule decay (Fig. 12.1).

In Central America, several authors have reported that leguminous tree species such

Table 12.2 Percentage of N derived from atmosphere (NDFA) in leguminous crops and relative percentage of N transferred to coffee trees in two types of soils in Burundi [24]

Type of soil	Legume crop	Type of association	NDFA (%)	Percent N transferred (in % from NDFA)
High acidity (pH < 5)	*Flemingia macrophylla*	intercrop	42 ± 5	14.3 ± 2
	Leucaena diversifolia	intercrop	39 ± 19	12.8 ± 4
Light acidity (pH 5–6)	*Flemingia macrophylla*	intercrop	20 ± 9	30.0 ± 4
	Desmodium intortum	intercrop	50 ± 5	32.0 ± 3
	Leucaena leucocephala	intercrop	52 ± 4	42.3 ± 1
	Leucaena leucocephala	intercrop	0	
	Leucaena diversifolia	shade	48 ± 5	31.3 ± 3
	Calliandra calothyrsus	shade	20 ± 13	
	Erythrina abyssinica	shade	21 ± 4	

Picture 12.2 *Leucaena* as intercrop.

Picture 12.3 *Leucaena* as shade.

as *Inga* or *Erythrina* can restitute to the soil 60 to 350 kg N/ha/year via slashing and canopy pruning [3, 17, 18], whereas the direct contribution of BNF is an estimated 30–60 kg N/ha/year. The leguminous plants need to be pruned to limit their growth or to prevent excessive shading. Frequent and/or excessive prunings can, however, decrease the N_2 contribution to the agricultural system and ultimately lead to soil N competition between the associated plants.

In 1994, for instance, Snoeck *et al.* [14] observed that each time *Leucaena* was cut back, this brought about a severe decrease in the number of active nodules, which lasted up to 3 weeks and resulted in the absence of BNF activity for 2–3 months (Fig. 12.2).

Other researchers [25] found similar results in a coffee–*Erythrina poeppigiana* association in Costa Rica. Furthermore, the

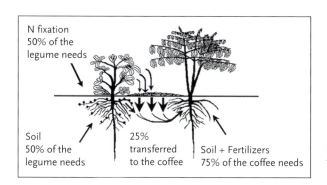

N fixation 50% of the legume needs

Soil 50% of the legume needs

25% transferred to the coffee

Soil + Fertilizers 75% of the coffee needs

Figure 12.1 Mechanism of N transfer from legume to coffee. From: D. Snoeck [23].

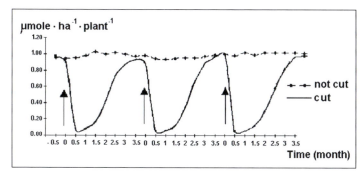

μmole $\cdot$ ha^{-1} $\cdot$ plant^{-1}

not cut
cut

Time (month)

Figure 12.2 Evolution of the BNF activity of *Leucaena* nodules in a coffee plantation due to pruning compared to uncut *Leucaena*. From: D. Snoeck [23].

authors reported that 67 % of the N was restored to the system by pruning, 24 % by litterfall and 9 % by nodule decay. This information leads to the conclusion that it is important to use adequate management practices for associate leguminous species by respecting the proper period for pruning, as well as its frequency and intensity, to ensure efficient and constant BNF.

12.3.2
The Effect of Leguminous Plants on Pest and Disease Tolerance

The association of coffee trees and leguminous plants can either increase or reduce the tolerance of coffee trees to pest and diseases by modifying environmental conditions like shade, air humidity and temperature. By improving the nutrient status of the coffee plants, shade trees can also reinforce their vigor and decrease their susceptibility to pathogens. Some leguminous plants can act as traps and others can harbor beneficial agents in their foliage.

Shade can increase the incidence of Coffee Leaf Rust (*Hemileia vastatrix*) and of other pathogenic fungi [26]. Excessive shade can also be responsible for an increase in Coffee Berry Borer attacks [2]; however, in their canopy or their root system, shade trees can also harbor fungi and insects which are not harmful to coffee trees, but are antagonists to coffee pests and diseases [27]. *Pueraria* or *Crotalaria*, for instance, are known to reduce the development and reproductive rate of root nematodes [28]. It would be worthwile to conduct further studies on these phenomena.

12.4
Mycorrhizal Symbiosis

12.4.1
Description and Importance

Most plants are symbiotically associated with either ectomycorrhizal or vesicular arbuscular mycorrhizal (VAM) root fungi. In coffee plantations, the degree of colonization by VAM varies according to the coffee root development, the type of soil, the farming system and the mycorrhizal species [29, 30].

Coffee roots can be colonized by germinated spores or by hyphae extending from adjacent roots. Once in contact with the root, the fungus penetrates the epidermal root cells, generally around the apex. After this penetration, the internal hyphae invade the root cells without generating any significant anatomic modifications. Arbuscules are produced; these are the sites of exchanges of carbohydrates for nutrients between the root and the fungus. The nutrients are absorbed by external hyphae in parts of the soil not explored by the roots (Fig. 12.3).

External hyphae can be several centimeters long and account for more than 1 m/g of soil [31]. As a result, they increase

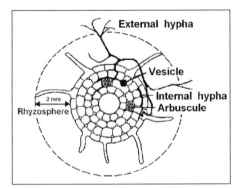

Figure 12.3 Schematic representation of the colonization of a root by VAM. From: Ph. Vaast [10].

the volume of the explored soil by 15–80%. Due to this enhanced soil exploration, the coffee plant can have access to nutrients of low mobility, especially P and micronutrients (B, Zn, Cu,). External hyphae also stimulate the development of soil microflora, particularly rhizobacteriae, which encourage a more rapid mineralization as well as a better use of organic compounds [32].

In certain circumstances, VAM can be inoculated to increase the symbiosis with coffee plant. This practice can be implemented when:

1) The soil has been partly sterilized by heavy applications of chemicals,
2) The natural density of propagules is very low, i.e. below 1 spore/g of soil,
3) Exogenous species are more efficient than native ones [33, 34].

Where inoculation is necessary, coffee plantlets are inoculated in nursery bags, prior to transplanting. Trials in nurseries have demonstrated that, depending on the type of soil, the growth of VAM-inoculated coffee plantlets could be 50–300% higher than non-inoculated ones and the growth period needed in the nursery to achieve

an adequate coffee plantlet vegetative development for field planting may be reduced by up to 50% [33, 34]. The beneficial effect remains visible after planting in the field as both a lower rate of mortality due to transplanting stress [35] and precocious yields [36] have been observed.

High inputs of N fertilizers significantly reduce the VAM colonization of coffee roots, mycorrhizal diversity and other remaining species [10].

12.4.2
The Effects of Mycorrhizae on Coffee Tolerance to Root Pests and Diseases

Mycorrhizal symbiosis increases the tolerance of coffee trees to the two main genera of nematodes that destroy their root systems – root-knot nematodes of the *Meloidogyne* genus [10, 37] and the root-lesion nematodes of the *Pratylenchus* genus [10] (Tab. 12.3).

The tolerance is generated by the increased vigor of the coffee plant due to its improved nutritional status in the presence of the VAM symbiosis and by the protecting effect of the hyphae around the coffee roots, which decrease and delay the pene-

Table 12.3 The effect of VAM inoculation with *Acaulospora mellea* (Mellea), 7.5 months after addition of nematode *Pratylenchus coffeae*, on various characteristics of 11.5-month-old coffee plantlets [10]

Mycorrhiza inoculation	P. coffeae	Foliar surface (cm²)	Root weight (g)	Percent colonized roots	Foliar P (%)	Nematode Density (g⁻¹)
Control	–	118 e	2.18 c d	–	0.058 c	–
	+	23 g	1.30 d	–	0.053 c	147 c
Mellea *ante*[a]	–	909 a	12.88 a	72 ab	0.108 a	–
	+	671 b	8.72 a b	56 b	0.102 a	1689 a
Mellea simultaneous[b]	–	215 d	2.36 c	53 b	0.085 a b	–
	+	65 f	1.30 d	23 d	0.059 c	209 c

[a]Inoculation of Mellea 4 months prior the addition of nematodes
[b]Simultaneous addition of Mellea and nematodes, 4 months after transplantation.

tration of the nematodes inside the roots [38].

Similarly, perennial plants like avocado and citrus are more tolerant to root soil-borne diseases when inoculated with VAM in the nursery [38]. However, this enhanced tolerance to root diseases has not yet been studied on coffee.

12.5
Other Microorganisms that have a Beneficial Effect on Coffee Roots

Over the last decade, several microorganisms which negatively impact pathogenic root fungi have been identified. Among these are fungi (*Trichoderma*, *Gliocladium*) and rhizobacteriae (*Bacillus*, *Pseudomona*, *Pasteuria*) [27]. Preliminary investigations have shown the positive effects of these microorganisms on the reduction of nematode damage [39, 40]. Synergies between mycorrhizae and the microorganisms which deter soil borne plant pathogens have also been identified [38].

12.6
Conclusions

In intensive coffee farming, large inputs of fertilizers and herbicides increase water contamination by nitrates, and lead to losses of OM and mineral elements by leaching and erosion. These farming practices also result in a reduction of the microflora and fauna biodiversity in coffee soils and probably increase soil parasitism.

In extensive farming with low or no N inputs, the N turnover is not rapid enough to ensure adequate N availability and acceptable yields. In both instances, coffee growers are researching alternative techniques to ensure a more sustainable coffee cultivation.

Depending on soil and climatic conditions, the use of leguminous plants is highly advisable, either as cover crops or shade trees. Furthermore, a well-planned management of OM will contribute to an increase of beneficial physical and chemical properties in the soil. It is also evident that efficient strategies to maintain or improve soil OM content result in conditions that encourage the higher biological activity that is greatly needed to achieve sustainable coffee cultivation.

As consumers are increasingly aware of environmental issues and demand healthier products, management schemes developed by research and extension institutions to enhance the role of soil OM and reduce the use of chemical inputs would appear to be ever more necessary and relevant.

Bibliography

[1] Carvajal, J. F. *Cafeto – Cultivo y Fertilización*, 2nd edn. Instituto Internacional de la Potasa, Berna, 1984.

[2] Wrigley, G. *Coffee. Tropical Agriculture Series*. Longman, London, 1988.

[3] Beer, J., Muschler, R., Kass, D. and Somarriba, E. Shade management in coffee and cacao plantations. *Agrofor. Syst.* **1998**, *38*, 139–164.

[4] Babbar, L. I. and Zak, D. R. Nitrogen cycling in coffee agroecosystems: net N mineralization and nitrification in the presence and absence of shade trees. *Agri. Ecosys. Environ.* **1994**, 48, 107–113.

[5] Reynolds-Vargas, J. S., Richter, D. D. and Bornemisza, E. Environmental impacts of nitrification and nitrate adsorption in fertilized andisols in the Valle Central of Costa Rica. *Soil Sci.* **1994**, *157*, 289–299.

[6] Schnitzer, M. Organic matter characterization. In: *Methods of Soil Analysis 2*. Page, A. L., Miller, R. H. and Keeney, D. R. (eds). American Society of Agronomists, Madison, WI, 1982, 581–594.

[7] Woomer, P. L., Martin, A., Albrecht, A., Resck, D. V. S. and Scharpenseel, H. W. The importance of management of soil organic matter in the tropics. In: *The Biological Management of Tropical Soil Fertility*. Woomer, P. L. and Swift, M. J. (eds). Wiley, Chichester, 1994, 47–80.

[8] Vaast, P., Zasoski, R. J. and Bledsoe, C. S. Effects of solution pH, temperature, NO_3^-/NH_4^+ ratios, and inhibitors on ammonium and nitrate uptake by arabica coffee in short-term solution culture. *J. Plant Nutr.* **1998**, *21* **(7)**, 1551–1564.

[9] Snoeck, J. and Jadin, P. Mode de calcul pour l'étude de la fertilisation minérale des caféiers basée sur l'analyse du sol. *Café Cacao Thé* **1990**, *34* (1), 3–16.

[10] Vaast, P. The effects of vesicular arbuscular mycorrhizae and nematodes on the growth and nutrition of coffee. *PhD Thesis*. University of California, Davis, 1995.

[11] Pavan, M. A. Efeito toxico de aluminio em cafeeiros. In: *Actes du Dixième Colloque Scientifique International du Café à Salvador de Bahia*, Jouve, Paris, 1983, 477–482.

[12] Lidell, C. M. Abiotic factors and soilborne diseases. In: *Soilborne Diseases of Tropical Crops*. Hillocks, R. J. and Waller, J. M. (eds). CAB International, Wallingford, 1997, 365–376.

[13] Robson, A. D. and Abbott, L. K. The effect of soil acidity on microbial activity in soils. In: *Soil Acidity and Plant Growth*. Robson A. D. (ed.). Academic Press, New York, 1988, 139–166.

[14] Snoeck, D., Bitoga, J. P. and Barantwaririje, C. Avantages et inconvénients des divers modes de couverture dans les caféières au Burundi. *Café Cacao Thé* **1994**, *38* (1), 41–48.

[15] Fassbender, H. W. *Modelos Edafologicos de Sistemas Agroforestales*. 2nd edn. CATIE, Turrialba, 1993.

[16] Bouharmont, P. L'utilisation des plantes de couverture et du paillage dans la culture du caféier Arabica au Cameroun. *Café Cacao Thé* **1979**, *23* (2), 75–102.

[17] Alpizar, L. A. Interacción de café y otras plantas, con especial referencia a la sombra de tipo permanente. In: *Curso Regional sobre Nutrición del Café*. IICA, San José, 1988, 55–82.

[18] Roskoski, J. P. Nodulation and N_2 fixation by *Inga jinicuil*, a woody legume in coffee plantations. I. Measurements of nodule biomass and field C_2H_2 reduction rates. *Plant and Soil* 1981, *59*, 201–206.

[19] Siles, P. and Vaast, P. Comportamiento fisiológico del café asociado con *Eucalyptus deglupta, Terminalia ivorensis, Erythrina poepigiana* y sin sombra. *Revista Agroforestal en las Americas* 2002, *9* (35), 44–49.

[20] Cannell, M. G. R. Physiology of the coffee crop. In: *Coffee: Botany, Biochemistry and Production of Bean and Beverage.* Clifford, M. N. and Willson, K. C. (eds). Croom Helm, London, 1985, 108–134.

[21] Vaast, P., Bertrand, B., Guyot, B. and Génard, M. Fruit thinning and shade influence bean characteristics and beverage quality of coffee (*Coffea arabica* L.) under optimal conditions. *Journal of Science of Food and Agriculture.* 2006, 86, 197–204.

[22] Guyot, B., Gueule, D., Manez, J. C., Perriot, J. J., Giron, J. and Villain, L. Influence de l'altitude et de l'ombrage sur la qualité des cafés Arabica. *Plantations, Recherche, Développement.* 1996, *3* (4), 272–283.

[23] Snoeck, D. Interaction entre végétaux fixateurs d'azote et non fixateurs en culture mixte: cas des *Leucaena* spp. associés à *Coffea arabica* L. au Burundi. *Thèse de Doctorat.* Université de Lyon 1, 1995.

[24] Snoeck, D., Zapata, F. and Domenach, A. M. Isotpic evidence of the transfer of nitrogen fixed by legumes to coffee trees. *Biotechnol. Agron. Soc. Environ.* 2000, *4* (2), 95–100.

[25] Nygren, P. and Ramirez, C. Production and turnover of N_2 fixing nodules in relation to foliage development in periodically pruned *Erythrina poeppigiana* (leguminosae) tress. *For. Ecol. Man.* 1995, *73*, 59–70.

[26] Avelino, J., Seibt, R., Zelaya, H., Ordonez, M. and Merlo, A. Encuesta-diagnostico sobre la roya anaranjada del cafeto en Honduras. In: *Memorias del XVIII Simposio Latinoamericano de Caficultura.* Editorama, San José, 1997, 379–385.

[27] Rodriguez-Kabana, R. and Kokalis-Burelle, N. *Chemical and biological control.* In: *Soilborne Diseases of Tropical Crops.* Hillocks, R. J. and Waller, J. M. (eds).

CAB International, Wallingford, 1997, 397–418.

[28] Bertrand, B., Anzueto, F., Pena, X. M., Anthony, F. and Eskes, A. Genetic improvment of coffee for resistance to root-knot nematodes (*Meloidogyne* spp.) in Central America. In: *Proceedings of the 16th International Conference on Coffee Research in Kyoto.* Jouve, Paris, 1995, 630–636.

[29] Lopes, E. S., Oliviera, E., Neptune, A. M. L. and Moraes, F. R. P. Efeito da inoculaçao do cafeeiro com differentes espécies de fungos micorrizicos vesicular-arbusculares. *Rev. Bras. Cienc. Solo.* 1983, *7*, 137–141.

[30] Riess, S. and Sanvito, A. Investigations on vesicular arbuscular mycorrhizae in different conditions of coffee cultivations in Mexico. *Micol. Ital.* 1985, *14*, 57–62.

[31] Brundrett, M., Melville, L. and Peterson, L. *Practical Methods in Mycorrhiza Research.* Mycologue Publications, Waterloo, 1994.

[32] Smith, S. E., Gianinazzi-Pearson, V., Koide, R. and Cairney, J. G. W. Nutrient transport in mycorrhizas: structure, physiology and consequences for efficiency of the symbiosis. *Plant and Soil* 1994, *159*, 103–113.

[33] Saggin, Jr, O. J., Siqueira, J. O., Colozzi-Filho, A. and Oliveira, E. A. Infestaçao do solo com fungos micorrizicos no crescimento post-transplante de mudas de cafeiro nao micorrizadas. *Rev. Bras. Cienc. Solo.* 1992, *16*, 39–46.

[34] Vaast, P. and Zasoski, R. J. Effects of nitrogen sources and mycorrhizal inoculation with different species on growth and nutrient composition of young coffee seedlings. *Café Cacao Thé* 1991, *35*, 121–129.

[35] Sieverding, E. and Toro, S. Efecto de la inoculación de hongos micorrízicos va en plantulas de café (*Coffea arabica* L.) y te (*Camellia sinensis* (L) O. Kuntze). In: *Mem. Sem. Sobre Micorrizas,* Medellín, 1986, 100–109.

[36] Siqueira, J. O., Colozzi-Filho, A., Saggin, Jr, O. J., Guimaraes, P. T. G. and Oliveira, E. Crescimento de mudas e producao do cafeeiro sob influencia de fungos micorrizicos e superfosfato. *Rev. Bras. Cienc. Solo.* 1993, *17*, 53–60.

[37] Dardon, J. E. M. Evaluación del efecto benéfico de las micorrizas en almácigos de café *Coffea arabica*, finca Buena Vista, San Sebastián Retalhuleu. *Tesis de Ingeniero Agrónomo*. Universidad Rafael Landivar, Guatemala, 1996.

[38] Azcon-Aguilar, C. and Barea, J. M. Arbuscular mycorrhizas and biological control of soil-borne plant pathogens – an overview of the mechanisms involved. *Mycorrhiza* **1996**, *6*, 457–464.

[39] Gowen, S. R. Some promising results from field application of *Pasteuria penetrans* for control of root-knot nematodes. In: *Proceedings of the 6th International Colloquium on Invertebrate Pathology and Microbiological Control*. FASEB/CIP, Bethesda, MD, 1997.

[40] Spiegel, Y., Chet, I., Mor, M., Kleifeld, O., Sharon, E. and Bar-Eyal, M. *Trichoderma harzianum* preparation as a biocontrol agent against the root-knot nematode, *Meloidogyne incognita*. In: *Proceedings of the 6th International Colloquium on Invertebrate Pathology and Microbiological Control*. FASEB/CIP, Bethesda, MD, 1997.

13
Sustainable Coffee Production

Moeko Saito

13.1
Background

Today, green coffee beans are the second largest commodity traded on the global market, second only to oil. Recently, however the coffee industry has experienced serious economic, environmental and social crises globaly.

13.1.1
Economic Crisis

The economic crisis occurred between 1997 and 2002 when the world commodity coffee price plummeted from US$1.27 to 0.45 per pound [6]. This crisis stemmed from an oversupply of coffee. According to the International Coffee Organization, the world coffee reserve in 2002 was 155 million bags (production of 115 million bags plus the previous year's stock of 40 million), in bags of 60 kg, against the total consumption of 108 million [6].

The most serious consequences of the drop in coffee prices are felt at the lower levels of the production system, where coffee growers are losing money on their activity.

The main causes of the crisis are:

- The rapid development of production in Vietnam – during the 1990s, Vietnam achieved a rapid and significant (1400 %) expansion in its coffee production [5].
- Little or no increase in the overall world consumption of coffee – while the total volume of the world production of coffee has been increasing significantly (15.9 % from 1997 to 2001), the world consumption has stayed relatively constant (1.05 % increase) [1].
- Brazil's frost management – Brazil, the largest coffee exporter in the world, has had a significant impact on both the world supply and the price of coffee because historically frost and drought problems have influenced the supply. Recently, Brazil's coffee-growing regions have been shifted to a frost-free zone, and this policy has enabled this country to ensure a higher and more constant volume of supply [5].

13.1.2
Environmental Crisis

The environmental crisis refers to the threat to natural ecosystems. This threat is brought about by the ecological stress caused by intensive farming associated with mass production by large coffee corporations. One of these intensive farming practices is the clear cutting of forests to provide full sun-exposure to the coffee plants as a stimulant for rapid growth and high productivity. This practice, associated with the intensive use of chemical fertilizers, causes soil erosion and depletion as well as the destruction of natural wild-life habitats.

13.1.3
Social Crisis

The combined effects of the economic and environmental issues have led to a social crisis in coffee-growing regions. The economic crisis has driven many coffee farmers out of the coffee business and forced them to seek work elsewhere. At the same time, the mass production system is detrimental to the lifestyle of local people because it often prevents the development of staple crops for everyday needs and, as a result, generates a dependency on food imports.

13.2
Sustainable Coffee

In an attempt to solve these problems, the notion of "sustainable coffee production" has become an important alternative for coffee producers [13]. Sustainable coffee production aims to achieve environmental, economic and social sustainability for the long-term development of coffee-growing regions. Its goals are to preserve water, soil and biodiversity resources, while maintaining the well-being of the people in their communities [13].

As a general rule, the concept of sustainable coffee production includes three major principles: "organic coffee growing", "shade-grown coffee" and "fair-trade" [4].

1) Organic coffee is produced by implementing methods that maintain soil health and fertility by encouraging natural biological cycles and prohibiting the use of synthetic chemicals (refer to Section I.10).
2) Shaded coffee is grown under a canopy of trees designed to maintain the biodiversity of the coffee-growing region [9].
3) Fair-trade refers to measures taken to improve the livelihood and well-being of coffee farmers by promoting equitable trade with a minimum guaranteed price for their crop – fair-trade coffee practices include a pre-financing credit to cover harvest costs, and the use of extra revenue to develop basic health care services, education and housing improvements within the coffee community [14].

13.3
A Comparison: Sustainable Coffee Growing Practices versus Intensified "Modern" Practices

Coffee grown by "modern" intensive technical practices is also known as sun coffee. Coffee trees have little or no shade cover and a high density of plants, approximately 3000–10 000/ha, or about 0.3–1 trees/m^2 [11]. Through this practice large companies produce a much higher volume of coffee cherries by using full sun-exposure and a significant amount of agrochemicals (fertilizers, herbicides, fungicides and nematocides) for rapid growth. While this system is highly productive, it not only reduces biodiversity, but also leads to soil erosion and contamination which, inevitably, will diminish yields in the long-term [2].

In contrast, sustainable coffee is characterized as an agro-ecosystem with moderate to heavy mixed shade cover where fruits trees, leguminous plants and hardwood species coexist alongside the coffee trees [12]. The density of coffee plants is lower, 1000–2000/ha, or about 0.1–0.2 trees/m^2, which results in lower immediate yields [11]. However this type of agroforestry structure guarantees long-term environmental sustainability.

To begin with, the coffee trees grow more slowly, which means that they do not need to be replaced as often, and, furthermore, the use of decaying leaf litter from shade trees and organic fertilizer for growth generally generates a higher quality aroma and taste.

Advantages of the shade-giving canopy combined with organic production methods (refer to Section I.10) are as follows:

- It enhances the recycling of nutrients by promoting the activity of nitrogen fixers and decomposers in the soil [10].
- The structural diversity of the coffee canopy helps to preserve the wildlife habitat, and, as a result, maintains the biodiversity of insects, animals and plants in the coffee region [3].
- The shade canopy often acts as a buffer against wind, rain and extreme temperatures thus reducing crop damage [10].
- The presence of a deep-rooted system preserves water and fertility, and stabilizes the soil against erosion [10].

The sustainable management of coffee farms requires a considerable amount of manual labor for fertilization, weeding, pest and disease control, and pruning to maintain the quality of beans (refer to Section I.10). Composted material, such as leaves or coffee cherry pulp, animal manure and green manure are used as fertilizers. The general weed management includes manual weeding and mulching to create soil cover for weed prevention. In addition, sustainable coffee requires hand-picking to select the best quality of beans.

In spite of the fact that sustainable coffee-growing practices are more labor-intensive, labor costs tend to be absorbed since sustainable coffee growers are mostly small-scale family farmers. As a result, the cost of producing coffee by "modern" intensive technical practices is actually higher because of the cost of the necessary chemical inputs. Even though the production volume is lower in sustainable coffee, the premium price benefits the small farmer.

13.4
Importance of Sustainable Coffee

The importance of sustainable coffee includes environmental, economic and social benefits to the entire coffee-growing community.

13.4.1
Environmental Benefits

While modern coffee production contaminates the environment with agrochemicals, sustainable coffee maintains natural biological cycles and helps to protect human health by restricting the use of agrochemicals. Shade-grown coffee, also known as "bird-friendly" coffee, creates a habitat for a greater variety of wild-life species: animals, insects and plants [3].

13.4.2
Economic Benefits

Table 13.1 compares the prices of sustainable and commodity coffee, and indicates that the premium prices of organic, shade-grown and fair-trade coffee generate significant economic benefits for producers. In 2002, for instance, both fair-trade

Table 13.1 Organic, shade-grown and fair-trade coffee prices (US$/lb) from 1999 to 2002

Production system	1999	2000	2001	2002
Shade-grown prices				
Chiapas, Mexico (including organic and fair-trade)				1.38
Fair-trade prices				
Washed Arabica	1.26	1.26	1.26	1.26
Organic washed Arabica	1.41	1.41	1.41	1.41
Washed Robusta	1.10	1.10	1.10	1.10
Organic washed Robusta	1.25	1.25	1.25	1.25
Average global price				
Arabica	1.02	0.71	0.60	0.45
Robusta	0.68	0.35	0.25	0.26

Source: World Bank 2002 [4].

and shade-grown prices were up to US$1.12 higher per pound than their equivalent commodity counterparts. These guaranteed premium prices help producers to maintain economic stability.

It is also important to note that shadgrown coffee is often grown with other cash crops such as cocoa, bananas, macadamia nuts and citrus fruits, amongst others. This diversifies the producer's sources of income and reduces the economic risks generated by the fluctuation of coffee prices.

13.4.3
Social Benefits

The major social benefit generated by implementing sustainable coffee-growing practices is a less precarious standard of living for local producers. This includes stable employment, improved conditions of health for workers and support for local community development. Community projects include basic health care, education and housing improvements [13].

A good example of this is the case of one Mexican cooperative, UCIRI (The Union of Indigenous Communities from the Isthmus Region), where benefits brought about by higher coffee prices enabled the community to build a mill, a hospital and a school, and to buy a truck for transporting the coffee [8].

13.5
Current Issues which Influence the Development of Sustainable Coffee-growing Practices

Currently several issues and obstacles exist in implementing the growth of sustainable coffee.

13.5.1
The Market for Sustainable Coffee

While sales of coffee grown by sustainable coffee-growing practices are increasing rapidly (10–20% in the US market), a rough estimate of the current share of sustainable coffee (including non-certified) is US$565 million, which accounts for less than 1% of the global market [4]. Consequently, although there is a developing interest in sustainable coffee for export, at present the current market outlets are still too small to support the potential supply.

13.5.2
Quality Requirements

Stringent quality requirements for exporting coffee as a premium quality product exist both in terms of size and quality of coffee beans [7]. First it is important to select suitable regions. The highest quality Arabica beans, for instance, are produced at an altitude of over 1000 m. Furthermore, careful picking, grading, processing and transportation are also important factors. Many of these quality criteria exclude a number of coffee growing regions because of their altitude, remote location, lack of knowledge of quality issues or inadequate processing facilities.

13.5.3
Certification

In order to sell coffee labeled as "organic", "shade-grown" and/or "fair-trade", it is necessary to obtain separate or multiple certifications.

It is frequently difficult for producers to comply with this for the following reasons:

- *The need for an organized cooperative.* In order to be certified as organic, shade-grown or fair-trade, producers need to form a co-operative with more than 15 members. This limits possibilities only to regions where the setting up of such

an organization can be achieved and prevents more isolated farmers from joining.

- *Period of transition from intensified cultivation practices to sustainable coffee growing.* A transition period of 2–3 years is required, depending on whether agrochemicals were used on the land or not. This means that generation of income as a sustainable certified coffee grower can only start after 2–3 years of implementation. If no financial aid during the transitional period is granted, it is extremely economically difficult for producers to decrease their incomes by reducing production to sustainable levels.

- *Certification costs.* An organized cooperative bears all the sustainable coffee certification costs. These costs vary according to the certification agency chosen and remain a huge obstacle for producers with no access to a secure market. Moreover, there is no uniform certification standard. Consequently, producers must acquire several certifications to export their coffee to different foreign buyers.

- *Fair-trade certification.* Products certified by the Fair-Trade Labeling Organization (FLO) can be sold as "fair-trade" and fair-trade organizations only focus on a clientele of small, family-run farms. This does not include "organic" or "shade-grown" certification. This can be a serious source of controversy when buyers are prepared to pay a fair price or more and the products cannot be certified as "fair-trade" unless the coffee was produced by a client of the FLO. Fair-trade certification requirements eliminate large-scale farms from participating if these farms are not involved in community development.

13.6
Potential Risks

The transition from intensified technical coffee-growing practices to sustainable coffee includes the costs of transition plus those of certification. Where there is no guaranteed access to a higher-price, sustainable coffee market, farmers who are interested in reducing their production volumes to meet sustainability criteria are faced with a serious challenge.

13.7
Suggestions and Recommendations

In conclusion it must be emphasized that the development of sustainable coffee-growing practices is a highly recommended way of ensuring the well-being of coffee growers as well as substancial environmental benefits.

In order to overcome the obstacles associated with the entry into sustainable coffee production, the following suggestions may be useful.

1) Awareness of the benefits of sustainable coffee-growing practices should be increased in developed countries. This should, in turn, help to increase the demand for coffee sold on the sustainable coffee market.
2) A sustainable coffee production model should be created. This model should include the logistics as well as the financial and technical aspects of the production process, quality management of picking, processing and transportation, and market access. These features would make the transition process easier.
3) An internationally recognized uniform certification system should be estab-lished. This should include a pre-financing program with low to zero interest in order to make certification more affordable to all producers.
4) An inventory of potential producers and buyers should be compiled. The goal of this inventory would be to create closer links between the various producers and interested consumers. It would also help producers to recognize the existence of a dependable, accessible and transparent market. A service of this type would encourage farmers to take on significant risks to implement sustainable coffee-growing production methods and, at the same time, it would ensure a constant supply of premium coffee.
5) Fair-trade opportunities for large-scale farms should be created. It is important for large-scale farmers to become involved because they have the potential for the conversion of sizeable areas to sustainable production. Large-scale farms could become increasingly committed to the well being of coffee producing communities. In this way, the creation of a well-managed, equitable operation on a large scale would be highly beneficial to both owners and employees, and the community as a whole.
6) The establishment of a network among stakeholders. A network consisting of banks, cooperatives, producers, technical advisors, local researchers, processing plants, transporters and exporters is critical for the success of sustainable coffee growing practices. It would strengthen cooperative implementation through the exchange of strategic, financial and technical information. A cooperative of producers could, for example, consult technical advisors and local researchers for farm management

and quality maintenance. Banks could assist in financial planning for implementation. Local experts, buyers, exporters and plants could cooperate to improve quality control. Finally, information and advice could be spread by means of localized workshops set up to teach producers about the methods and advantages of sustainable coffee production.

Bibliography

[1] Dicum, G. and Luttinger, N. *The Coffee Book: Anatomy of an Industry from Crop to the Last Drop.* New Press, New York, 1999.

[2] Faminow, M. D. and Rodriguez, E. A. *Biodiversity of Flora and Fauna in Shaded Coffee Systems.* International Center for Research in Agroforestry, Latin American Regional Office, 2001.

[3] Fernandez. C. and Muschler, R. *Aspectos de la Sostenibilidad de los Sistemas de Cultivo de Café en America Central, Desafíos de la Cafecultura en Centroamerica.* In B. Bertrand and B. Rapidel (ed.) Desafios de la cafecultura en centroamerica. CIRAD, IRD and IICA, San José, Costa Rica, 69–97, 1999.

[4] Giovannucci, D. *Sustainable Coffee Survey of the North American Specialty Coffee Industry.* Specialty Coffee Association of America, Long Beach, CA, 2001.

[5] Giovannucci, D., Varangis, P. and Lewin, B. Who shall we blame today: the international politics of coffee. *Tea and Coffee Trade J.* **2002**, *174* (1).

[6] International Coffee Organization. *Coffee Market Situation: Coffee Year 2001/02.* International Coffee Council Economic Report. ICO, London, 2002.

[7] Interview with Counter Culture Coffee. *Specialty Coffee Importer/ Roaster,* U.S.A, 2002, conducted by the author.

[8] Mace, B. *Alternative Trade & Small-Scale Coffee Production in Oaxaca.* Miami University, Oxford, OH, 1998.

[9] Moguel, P. and Toledo, V. M. Biodiversity conservation in traditional coffee systems. *Conservation Biol.* **1999**, *13*, 1–12.

[10] Nestel, D. Coffee and Mexico: International market, agricultural landscape and ecology. *Ecological Economics* **1995**, *15*, 165–179.

[11] Perfecto, I., Rice, R. A., Greenberg, R. and Van der Voort, M. E. Shade coffee: a disappearing refuge for biodiversity. *Bioscience* **1996**, *46* (83), 598–608.

[12] Rice, R.A. and Ward, J. R. *Coffee, Conservation, and Commerce in the Western Hemisphere.* Smithsonian Migratory Bird Center/Natural Resources Defence Council, Washington, DC, 1996.

[13] The World Bank. *Toward More Sustainable Coffee. Agriculture Technology Notes 30.* The World Bank, Geneva, 2002.

[14] TransFair USA. *Fair Trade: Now More Than Ever.* Annual Report. TransFair USA, Oakland, CA, 2001.

14
Shade Management and its Effect on Coffee Growth and Quality

Reinhold G. Muschler

14.1
Introduction

The wide range of climatic and soil conditions under which coffee is grown around the world have led to a high diversity of coffee production systems. Besides as environmental differences, the systems differ primarily in coffee species and varieties (cf. Sections I.2 and I.3), agronomic management (cf. Sections I.7 and I.8), and, most importantly, in the structure and function of their different plant components which range from small ground-covering herbs to dominant emergent trees that may rise more than 20 m above the coffee bushes. It is the dimensions and attributes of these associated plants which determine to a large extent the ecology and economy of coffee fields.

Around the world, coffee fields produce more than just coffee: they also produce food and medicinal plants, firewood and timber, they provide habitat for fauna and flora, and they supply essential ecological services such as biological nitrogen fixation and soil and water protection. Among the plant associates in coffee fields, trees dominate due to their longevity and size. Their positive effects on microclimate and soil fertility are increasingly recognized as key to the long-term ecological sustainability of coffee systems. The recent coffee price crisis has raised awareness about the economic contributions of trees through their products, and their biologically mediated weed control and nutrient contributions. This chapter explores the salient attributes of coffee systems differing in the number, species and management of their associated trees. (The expression "associated tree" is preferable to the widely used "shade tree" to eliminate undesirable ambiguities about the use of the tree; all trees cast shade, but only some are planted with the objective to provide shade. Consequently, "shade tree" should be used only for those species, e.g. intensely pruned legumes, which are planted with the primary objective to provide shade rather than other products.) The recognition of these attributes provides the basic guidelines for future work geared towards the development of coffee systems which are ecologically sustainable, economically attractive and socially viable.

The discussion on the sun-shade issue for coffee is as old as coffee production itself [6, 16, 18, 21]. In Latin America, most coffee (*Coffea arabica* L.) plantations include leguminous tree species such as *Inga*, *Erythrina* and *Gliricidia* to provide mulch and shade. Often, these trees are pruned up to 3 times

Coffee: Growing, Processing, Sustainable Production, Second Edition. Edited by J. N. Wintgens.
© 2012 WILEY-VCH Verlag GmbH & Co. KGaA. Published 2012 by Wiley-VCH Verlag GmbH & Co. KGaA.

per year in order to reduce the disease incidence on the coffee plants, and to stimulate flowering and maturing of the coffee fruits. Depending on the pruning intensity, the shade pattern can vary widely from light and dispersed shade of less than 30 % shade intensity to heavy and homogeneous shade exceeding 70 % intensity [3]. However, despite many discussions on the best levels of shade [3, 9, 21], there is surprisingly little quantitative information derived from field-level research on the effects of shade in different environments.

Most of the vast literature on shade reports data from lower levels of hierarchy, i.e. from studies at the level of individual cells, leaves, or plants and much of this information is derived from laboratory work (for a selection of references, see [3, 5, 37, 38]). Consequently, while these studies address important issues about coffee physiology and nutrition, their validity for deriving practical recommendations for plantation management by farmers is often restricted.

14.2
Early Shade Management: Traditional Coffee Farms

The two economically most important coffee species, *C. arabica* and *C. canephora*, evolved in the understorey of African forests [16]. Consequently, both species have a natural adaptation to shade. Since *C. arabica* is by far the most widely cultivated species and the ecological information on this species is vastly greater to that on *C. canephora*, this chapter focuses only on *C. arabica*. However, given the similar natural history under shade, the derived recommendations may also apply to most, if not all, systems with *C. canephora*.

C. arabica evolved in the understorey of mid-elevation forests in Ethiopia, i.e.

under high levels of shade [16]. From the beginning of commercial coffee cultivation until the mid-1900s the typical coffee field consisted primarily of tall-growing varieties of *Coffea arabica* L. (*cf.* Sections I.2 and I.3) planted at wide spacing of up to 3 m between plants. A common feature of all systems was the intense shading below a

Picture 14.1 High tree diversity, here in an organic and bird-friendly coffee field in Peru, generates multiple products and a diversified habitat for canopy-dwelling plants and animals. Such systems are of increasing importance for biodiversity conservation. Photo: R. G. Muschler.

Picture 14.2   Above: traditional coffee field under a dense overstorey of many tree species and bananas (upper right quadrant). Shade levels are often between 40 and 70 %. Pruning of all tree branches at the same time ("pollarding") leads to temporarily unshaded coffee until trees resprout (foreground and lower photo). The stems among the coffee bushes belong to the tree species *Erythrina poeppigiana*, a leguminous species widely used in Costa Rica. Below: "Intensified" coffee plantation without shade (foreground) after pollarding of the trees. The trees will resprout within 4–8 weeks. The flowering trees in the background are unpruned *E. poeppigiana* trees. Photo: R. G. Muschler.

An excellent example are today's coffee systems in Chiapas, Mexico, which often contain more than 15 tree species/ha and more than 80 tree species in individual coffee zones [39]. Similar systems can be found in most coffee-producing countries throughout the world (*cf.* Pict. 14.1).

Although the first recommendations to substitute the "natural shade" of the slightly thinned tree upper storey date back to the 1800s [18], large-scale substitution with carefully selected shade species, particularly bananas, fruit trees and nitrogen-fixing leguminous trees, did not occur until the middle of the 20th century. According to early recommendations, bananas were to be planted every 7 m and trees, preferably legumes, at the ratio of one for every three coffee plants. In this way, recommended tree densities were about 200 trees/ha.

Early research and development of coffee plantation technologies was primarily carried out on private estates, and, as a result, was dispersed and little divulged. The end of World War II marked the beginning of a profound change due to two principal factors; (1) the increasing availability of chemical inputs from the fledgling chemical industry and (2) the rapidly rising demand for coffee by an ever-increasing number of consumers. These factors provided the grounds for a period that has become to be known as the "intensification" of coffee production.

dense tree canopy which often had been intervened on only slightly, if at all, by man [18]. On most farms, coffee densities varied from 500 to 800 plants/ha and the per-area productivity was very low, often between 300 and 500 kg of green coffee/ha. Examples of such systems, typically requiring minimal or no external inputs, can be found up to today, predominantly in remote areas where the coffee intensification could not take hold [3].

14.3
"Intensification" of Coffee Production: Reduced Shade Levels

In Latin America, the 1940s and 1950s saw the foundation of most of the national coffee institutes responsible for developing and disseminating new technologies primarily geared to increase productivity. At the heart of the activities was work on coffee nutrition and physiology [19, 20, 24], on hormonal and photoperiodic control of flowering and fruiting, on pruning regimes [19], and on the effects of sun versus shade [14].

Studies from different countries demonstrated significant yield increases per hectare as a result of:

- Higher plant densities with up to 5000 or even more plants/ha of "new" compact and short-statured varieties
- The use of increasingly available external inputs, particularly fertilizers
- The intensive pruning, pollarding or even elimination of shade trees in the coffee plantations [14].

The practice of pollarding, i.e. the removal of all of the branches of a tree during a pruning (*cf.* Pict. 14.2 and 14.3), has become widespread in many parts of Costa Rica and has been hailed as an important factor which contributed to the spectacular yield increases in that country. However, productivity can only be raised substantially with this system as long as high levels of external inputs are available. As a result, the costs of implementing this system increase steadily as time goes by. Today, negative long-term consequences generated by excessive dependency on external inputs, and from the degradation of natural resources in the form of soil erosion, water contamination and biodiversity loss, are receiving increasing attention [26].

As a result of greater consumer awareness since the 1990s, ecological niche markets like "organic", "bird-friendly" or "Eco-OK" coffees have been created. Today, these markets are growing faster than most traditional segments in the coffee sector.

To return to the intensification in coffee production, it should be noted that depending on the environment and the condition of the coffee plants, the substitution of

Picture 14.3 Intensified coffee production without shade and complete pruning of coffee plants on slopes can have negative consequences: the soils are exposed to sun and rain for up to 1 year, leading to massive soil erosion (above). Also, the costs for weed control are greatly increased, often resulting in higher applications of herbicides (below). Photo: R. G. Muschler.

shade by external inputs can increase yields from 10 to 30 % and, occasionally, on a short-term basis, much higher. These changes transformed coffee production dramatically. In Costa Rica, for example, three-fold yield increases per hectare were reported between 1950 and 1974, a tendency which continued until a national average of approximately 1500 kg of green coffee/ha was reached at the end of the 20th century. Today, the country is recognized for having one of the highest per-hectare productivities in the world. To interpret these figures realistically, however, one should take the following aspects into account:

- The increased production costs, due to the intensive use of external inputs, particularly fertilizers to meet increased nutritional demands, herbicides to reduce weed competition, and pesticides to reduce disease and pest problems in these more homogeneous and less biodiverse systems [4].
- The damage associated with environmental and health externalities chiefly generated by contaminated water and soils [4].
- Occasionally reduced coffee quality when cultivated without shade [23].

Undoubtedly, the "intensification" of coffee production using a "technological package" was one of the most dramatic transformations of a tropical crop. However, its success or failure hinges on appropriate environmental conditions, particularly soil and climate, and on the use of the full "package", i.e. the adoption of all agronomic recommendations at the same time. In the past, many farmers who, in order to reduce costs, only made partial use of the recommendations, failed terribly. This was the case, for example, where farmers had eliminated shade or adopted high-yielding varieties without increasing their fertilization regime to meet the increased nutritional demands. Similar failures occurred on farms where the stress of coffee plants caused by inadequate soils or extreme climatic conditions was not corrected by soil amendments or shade. Typical examples are to be found in many farms exposed to pronounced dry seasons of up to 5–6 months. This is the case for most parts of the Pacific watersheds of Central America where water stress has repeatedly caused the collapse of unshaded plots.

In order to better understand the two principal production paradigms, the following section reviews the effects of shade on the ecosystem, coffee growth and production, and on coffee quality.

14.4
The Effects of Shade: How much Shade is Best?

Both shaded and unshaded systems have their merits and costs. The survival of the traditional high-biodiversity system even in countries with very intensive coffee production such as Colombia or Costa Rica illustrates its viability under specific conditions. The challenge is to understand the local environmental factors that determine which system is best and to understand the evaluation criteria used by farmers to judge the success of different systems. Understanding the farmers' criteria is essential to understand why a system which produces, for example, 1000 kg of green coffee/ha may be assessed differently by different farmers. From a pure production perspective, calculated in yield per hectare, this is a relatively low level of productivity since higher yields can be achieved, at least in the short run. However, from the point of view of a small organic coffee grower with limited resources, this productivity is quite attractive because it would most likely generate an acceptable income while requiring only few, if any, external inputs. The different factors which influence the decision of growing coffee with or without shade can be presented in three groups in the form of a decision tree as illustrated in Fig. 14.1.

The three groups of objectives of the grower which may include (1) the protection goals, (2) the environmental conditions which prevail on the particular site and (3) the availability of inputs to correct environmental limitations. The consideration of each factor suggests a decision

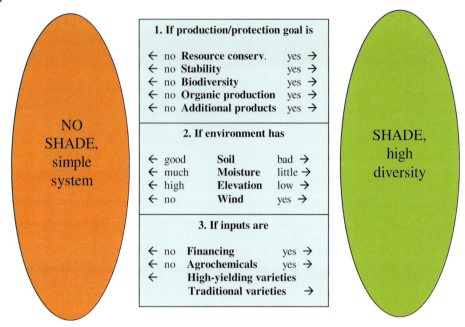

Figure 14.1 The three principal groups of factors which determine the decision about growing coffee under shade or not. The combination of the site-specific factors determines the number and diversity of trees associated with the coffee.

either towards a typically simple unshaded system (following the arrow towards the left) or towards a shaded system with, generally, high biodiversity (following the arrows to the right). The number and weight of the individual factors favoring one or the other system will determine the overall decision as well as the management details involved. The next sections explore the main advantages and disadvantages of shaded versus unshaded coffee systems in greater detail.

14.4.1
Effects of Trees and Shade on the Ecosystem

The main ecological arguments that favor high-diversity systems over species-poor systems are their capacity, like forests, to:

• Conserve a great diversity of potentially valuable plants and animals [10, 27]
• Moderate the microclimate [1, 21]
• Control weeds [12, 21]
• Retain water, reduce erosion and maintain soil fertility including soil microbial activity.

High diversity systems are ecologically more sustainable since they make better use of the local resources soil, water and sunlight [7, 11]. In contrast, this often does not apply to high-input and species-poor coffee plantations which frequently give rise to increased surface water run-off and ensuing soil erosion, as well as reduced light interception which can also lead to high topsoil temperatures which reduce microbial and root activity.

14.4.1.1 **Biodiversity**

In typical "intensified" coffee systems, most, if not all, organisms besides the coffee plants are considered undesirable because they offer potential competition or damage to the coffee plants. Consequently, management seeks to reduce or eliminate them, often by means of external inputs which include herbicides, fungicides, insecticides and nematicides. As a result, intensely managed coffee fields typically have fewer soil organisms, little or no vegetation under the coffee bushes and only a few trees above them (see Pict. 14.2 and 14.3).

This impoverishment in species has many negative effects which include a strong reduction of the biodiversity of birds [13], ants and other insects including predators such as spiders and parasitoids [27], mammals [10], and many soil and foliar microorganisms with different functions. This simplification of structure and biodiversity probably increases the susceptibility to outbreaks of epidemics, pests and diseases which did not occur before [36]. In contrast traditional shaded coffee fields can have levels of biodiversity similar to natural forests [27] (*cf.* Pict. 14.1).

Although discussions about the relationship between the biodiversity of an ecosystem and its stability and ecological functioning are still ongoing, the larger diversity of organisms in species-rich ecosystems is likely to provide important benefits particularly for the biological control of pests and diseases [35, 36] and by enabling a faster nutrient mineralisation, particularly of nitrogen.

Some problems like the outbreaks in intensified systems of certain pests and diseases such as Coffee Leaf Miners, antracnosis (*Colletotrichum* spp.) and Brown Eye Spot (*Cercospora coffeicola*) have already been partly attributed to the loss of biological diversity due to the "intensification" of the system [36]. Similar effects have been observed with outbreaks of nematodes and root diseases caused by a reduction of microbial activity and diversity in the soil brought about by the high application rates of soluble fertilizers, nematicides and residues of herbicides. In species-rich traditional systems these pests are usually held in check by the activity of antagonistic organisms which thrive in soils that are high in organic water. Far more research on this subject still needs to be completed, but there is mounting evidence pointing to the importance of high plant and microbial diversity for healthy and sustainable agroecosystems [7, 35, 36].

In the context of the Central American Biological Corridor, increasing attention will have to be given to the role of tree-dominated agroforestry systems, many of these with coffee, for linking fragmented conservation areas for the benefit of wildlife conservation. It is already commonly recognized that the role of shade-grown coffee is essential for many migratory birds in the Americas. Also, the benefits of organic farm management for the conservation of biodiversity is receiving increasing attention [8, 33]. The links between the management and the structure of coffee systems with the conservation of natural resources provide the foundation for certification criteria for niche products such as "bird-friendly" or "Eco-OK" coffee (Table 14.1).

Both certifications share the preference for native tree species and the need for the presence of different species of shade trees in any particular plot to provide as many habitat niches as possible for local wildlife and migrating birds.

Table 14.1 Tree-related criteria for the certification of coffee fields as "Bird-friendly" or "Eco-OK"

Criteria for inspection	"Bird-friendly coffee"[a]	"Eco-OK"[b]
Shade composition	at least 11 tree species, preferably native and non-deciduous species	at least 12 species, preferably native and non-deciduous species; each species must be represented by at least 1 tree/0.7 ha; trees should be well distributed
Shade cover	maintain shade cover throughout the year with a minimum of 40 %	at least 70 trees/ha; average shade of 40 % throughout the year
Structure of coffee system	tree crowns need to form three strata; emergent trees must be at least 15 m above the ground	tree crowns must form at least two strata; at least 20 % of the trees must be emergent; height of the tree crowns at least 15 m
Management	minimal pruning of trees; pruning should be avoided before and during the dry season; dead wood to be left in plantations; allow epiphytes to grow on shade trees	pruning should leave at least 50 % of flowers and fruits on the trees; pruning should be avoided before and during the dry season; epiphytes are to be conserved

[a]Norms according the Smithsonian Migratory Bird Center (www.natzoo.si.edu/smbc/coffee/criteria.html).
[b]Norms according to Rainforest Alliance, Washington DC.
These criteria are complemented by norms about authorized inputs, soil management, etc.

14.4.1.2 Soil and Water

Shade management also affects hydrological and pedological parameters. Soils in coffee monocultures often suffer strong erosion. An illustrative example is the loss of up to 3.4 cm of soil in just 2 years in coffee fields on volcanic soils in Colombia although the slope was only 12 % [30]. This example illustrates the potential dangers associated with certain "modern" technologies such as the total pruning of all coffee bushes per plot rather than by row or by individual plants (*cf.* Pict. 14.3, top). Consequently, the labor-saving benefits from large-scale mechanized pruning with chainsaws should be weighed against the degradation of the fertility of soils that are exposed to sun and rain for more than a year and against the social costs caused by the loss of employment opportunities in rural communities. Accurate assessment should look at more than one factor and take into account all costs and effects.

As to the effects on hydrology, coffee plants in unshaded plantations are often more water-stressed than shaded plants. In Central America, most experiments with shaded plantations in areas where dry seasons are severe have shown that the stress-alleviating shade effect on the coffee plants was more beneficial than the water consumption of the shade trees was detrimental. Although this contrasts with early observations made by Franco in the 1950s discouraging the planting of shade trees in coffee plantations exposed to drought stress on the grounds of water competition, the differences can probably be attributed to differing water consumption rates among tree species, to variable soil water storage capacities and, possibly, also to different evaluation criteria.

Consequently, generalizations without taking the specific species and environmental parameters into account are bound to be inadequate. For example, while the high water consumption of some species of *Eucalyptus* and *Grevillea robusta* [16] may prohibit their use in situations with limited water availability, their use might be of no consequence or even beneficial in areas with excessive moisture. This remains, however, another subject where further research is required to evaluate the water use efficiencies of tropical tree species and their effects on the hydrology of land use systems.

When examining the water budget of coffee plantations, Rice [29] reported that plantations with few or no trees had 72% less moisture in the soil compared to traditional shaded plantations with more trees. Apperently, condensation on the foliage of shade trees (estimated at about 210 mm/ year in the region of Carazo, Nicaragua) contributed significantly to this condition. Furthermore, the amount and persistence of litter on the ground affects the conservation of soil moisture and fertility. Rice [29] reported litter-fall of 7.8 tons/ha for an unshaded coffee field compared to 12 tons/ha in a traditional field. Shade trees can contribute typically between 5 and 10 tons of organic matter/ha and year [3]. Hence, they considerably increase the cation-exchange capacity and soil fertility in general (*cf.* Section I.2). The importance of soil organic matter is being increasingly recognized not only for nutrient release and retention, but also for the maintenance of an intact microbial community [32] within the soil.

14.4.2
Effects of Trees and Shade on Coffee

The common interest to intensify production in the second half of the 20th century rekindled the discussion about the utility of shade and trees and about the shade requirements of Arabica coffee [14, 16, 17, 37]. The dissemination of short-statured highly productive varieties that had been primarily selected for unshaded plantations led to a disregard of the true value of trees and shade. Based on studies that had demonstrated higher productivity without shade, at least on a short-term basis, the large-scale elimination of trees and shade became widespread [19]. However, these studies represented mostly the situation of coffee research sites located on fertile soils and at high altitudes, characterized by ideal temperatures for coffee. Unfortunately, in many cases, particularly for the low- and mid-altitude coffee zones of Central America, the biophysical constraints for coffee growing under suboptimal conditions were not considered. In contrast to the optimal conditions for Arabica coffee (mean annual temperatures around 20 °C, 3–5 month dry season, deep volcanic soils, moderate slopes) that can be found in the Central Valley of Costa Rica and similar areas in Colombia, Guatemala, Kenya, etc., the temperatures are higher, the dry season less pronounced or much stronger, and the soils less fertile due to high acidity and stoniness. Under such suboptimal conditions, which are typical for most Central American coffee areas, shade trees are far more beneficial than under the optimal conditions of many research centres (see Fig. 14.4).

14.4.2.1 Coffee Productivity
While an impressive amount of literature exists on the effects of shade on coffee, there is surprisingly little information derived from long-term field experiments. Strong evidence for the benefits of shading comes from a long-term experimental

study carried out for over 10 years in which the plots with the leguminous shade tree *Erythrina poeppigiana* (pollarded twice per year) produced the same amount of coffee as unshaded plots with approximately half of the fertilizer applications [28].

Further support comes from another field study on the effects of four different levels of shade on coffee in Turrialba, Costa Rica [21] (700 m.a.s.l., 2600 mm rain, slightly pronounced dry season, annual average temperature 23 °C). This study allowed to separate the effects of shade from the often confusing effects of free species and coffee varieties or management. It will be described in more detail here because of its relevance to many coffee zones with similar biophysical conditions. The four shade levels studied were generated by different pruning regimes applied to 13-year-old shade trees of the leguminous species *Erythrina poeppigiana* planted at about 123 trees/ha. The treatments were:

- "Sun": all branches removed every 2 months.
- "Pollarding": all branches removed every 4 months.
- "Open shade": half of the branches removed once per year while the others were allowed to grow freely to form an open homogeneous canopy of intermediate shade (40 to 60 % shade).
- "Dense shade": no pruning at all, resulting in a closed canopy casting more than 70 % shade [21].

"Pollarded shade" and, to a lesser extent, "open shade" exposed the coffee to abrupt microclimatic changes at the times of pruning or pollarding.

For both varieties studied, Caturra and Catimor 5175, fruit maturation occurred earlier when exposed to higher light levels than under shade. However, the total harvest period did not differ from one practice to another because the delay of maturation of shade-grown berries (30–40 days when compared to the berries from sun-exposed plots) was accompanied by a similar delay to complete the harvest under open and dense shade. For both coffee varieties, the yields of "quality berries", i.e. excluding all berries that had fallen to the ground prior to harvest or "rejects" that were deformed, sunburnt, mummified or diseased, over 4 consecutive years differed surprisingly little from one practice to another considering the large differences in shading.

Only the production of Caturra showed any significant differences. While the sun and pollarding treatments produced similar amounts of quality fruit, open and dense tree shade reduced relative production by up to 37 % for Caturra and up to 12 % for Catimor (Fig. 14.2). An unreplicated shade-cloth plot (50 % shade) produced the same amounts of quality fruit as coffee under sun and pollarding. This suggests that the light reduction of 50 % by itself was not detrimental to coffee production in this environment.

Shade also significantly reduces the percentage of fallen fruit, whereas unshaded coffee bushes often drop over 20 % of their fruit to the ground. This frequently observed effect is due to the fact that the tree canopy physically protects the ripe berries from the impact of raindrops and, possibly, also to an increased capacity of shaded plants to retain ripe berries on the plant.

Furthermore, shade tends to reduce the portion of reject fruits that are diseased, mummified, sunburnt or dried. In the study from Costa Rica rejects accounted for up to 10 % in the unshaded samples, whereas they were less than 1 % under shade.

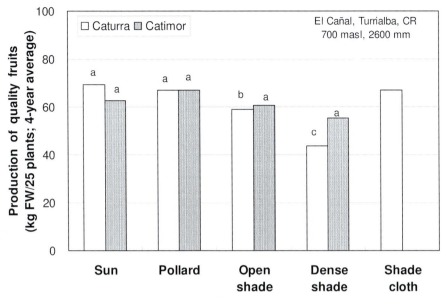

Figure 14.2 Production of "quality fruits" of *Coffea arabica* vars Caturra and Catimor 5175 under five levels of shade (3-year average for unreplicated shade-cloth; 4-year averages for others; FW = fresh weight). Within varieties, treatments with the same letter did not differ significantly (Duncan @ $\alpha = 0.05$). From: Muschler [21].

Picture 14.4 Random samples of coffee berries matured under more than 50% shade (left) and without shade (right) at a low-elevation site at 700 m.a.s.l. Notice that the shaded berries are larger, and show less defects from sunburn and fungal diseases than the unshaded berries. Photo: R. G. Muschler.

The higher number of rejects among harvested fruits of the unshaded and pollarded plots was probably due to the more intense heat and to a higher incidence of diseases combined with the reduced vigor of the heat-stressed plants carrying the fruits [21]. Physiological imbalances, which are also responsible for the higher incidence of chlorotic foliage in unshaded plantations, could also have increased the incidence of deformed fruit.

How can these findings, and similar ones from Mexico [34], which show maximum productivity under intermediate shade be reconciled with work that shows increasing production under increasing light levels? In addition to site-specific soil and climatic conditions, also the criteria chosen for evaluation can influence the conclusions. This is further illustrated with data assembled during the long-term study. The interpretation of the coffee response to increasing light levels changes according to whether one considers the number of berries produced (the potential production), the number of berries harvested (actual production) or the number of berries harvested which meet the quality standards of the processors and buyers (quality production).

While potential production, i.e. the total fruitset on the plants, increased linearly with increasing light availability (Fig. 14.3A), the quantity of fruits harvested (Fig. 14.3B) and of quality fruits harvested (Fig. 14.3C) showed quadratic relationships with the light levels. The curvature increased from Fig. 14.3A to 14.3C, because of the higher number of fallen fruits and the increasing percentage of rejects, i.e. sunburnt, mummified or diseased berries, in the pollarded and sun-exposed plots with less than 20% shade [21]. In these treatments, the percentage of rejects reached up to 50% of the production in some years. In contrast, under the open and dense shade, practically all the fruits harvested were well-formed and healthy. The regression in Fig. 14.3(C) suggests a 40–80% "window" of optimum light levels of photosynthetically active radiation (PAR), equivalent to 20–60% shade. A similar range of ideal shade was reported from Mexico [34] and is most likely valid for most of the low- or mid-altitude coffee areas where the plants are exposed to above-optimum temperatures.

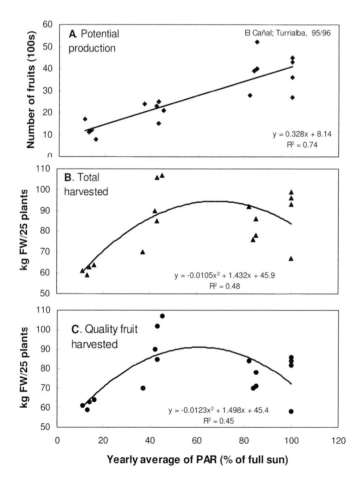

Figure 14.3 Regressions of estimated number of fruits (= potential production; A), total yield harvested (B) and total yield of quality fruits (C) as functions of the average annual PAR levels. Each dot corresponds to the estimated or harvested yield of a plot (FW = fresh weight). From: Muschler [21].

14.4.2.2 Microclimate and Coffee Physiology

In unshaded plantations, the temperatures of air, leaves and soil can be substantially higher than in shaded plantations, sometimes by more than 10 °C. High temperatures can reduce plant performance and coffee quality. It has been noted, for example, that photosynthetic rates were strongly reduced when the air temperature was above 26 °C [17]. Other studies have shown that coffee plants exposed to air temperatures above 25 °C suffer a 10% reduction of photosynthetic rate for every degree Centigrade increase [24]. Consequently, unshaded coffee plants which often experience temperatures above 25 °C for many hours per day suffer substantial losses. In contrast, in most environments, plants under at least 50% shade experience optimum air temperatures during the whole day, barely exceeding 25 °C at noon. Other important environmental parameters such as the vapor pressure deficit and leaf and soil temperatures are also more likely to exceed optimum levels during large parts of the day where no shade is available. The further the coffee fields are distant from the ideal altitude for coffee growing (either higher or lower), the more severely these microclimatic constraints will affect the coffee trees and, as a result, the need for shade will be greater.

In addition to the benefits to photosynthesis of moderate temperatures under "dappled" shade, coffee photosynthesis under shade may also benefit from the short direct sun as sunflecks pass over coffee leaves throughout the day. During short sunflecks, post-illumination carbon fixation can increase the actual carbon gain of plants to a level substantially above the predicted values. This effect may also have contributed to the unexpectedly high production under the high shade levels presented above. The shade benefits to coffee production at low altitudes documented in the shade study in Turrialba complement a 10-year fertilization × shade study under similar biophysical conditions. This study revealed that the nutrient inputs from pollarded *E. poeppigiana* (twice per year) led to a 65% yield increase of unfertilized coffee plants when compared to unfertilized plants without shade trees [28]. However, when the coffee plants were fertilized with 250 kg N/ha/year, the pollarded trees, similar to the data in Fig. 14.2, did not increase production when compared to the sun plots. In this case, fertilization probably masked the effects of the nutrient inputs from the shade trees, and the shade intensity under the pollarded trees was too low to reduce the microclimatic stress for the coffee plants. Both studies support the hypothesis that the beneficial effects of shade trees are more pronounced in areas with marginal environmental conditions.

The notion of the "window of optimum shade" mentioned earlier is also supported by many physiological studies at the leaf and plant level which have shown that the rates of photosynthesis or growth of *C. arabica* are highest at intermediate levels of shade, typically ranging from 30 to 50% [17, 25]. Since the light saturation of coffee occurs at relatively low levels, around 300 micromol/m^2/s for shade-adapted and 600 micromol/m^2/s for sun-adapted leaves [17], full sunlight (exceeding 2000 micromol/m^2/s) can cause strong photoinhibition [25]. Although nitrogen fertilization can reduce the extent of the protein damage in photosystem II, even fertilized plants can suffer long-term damage from sudden exposure to full sun [25]. Such effects become particularly important when the shade trees for coffee are severely pruned or pollarded periodically. This type of tree management, the tradition in

Picture 14.5 Sunburn of coffee leaves as a result of sudden removal of shade. Two days before taking the photo, the crown of the shade tree had been pollarded. Such sunburn is the extreme manifestation of the negative effects of photo-inhibition due to sudden shade removal.
Photo: R. G. Muschler.

many parts of Costa Rica, exposes the coffee repeatedly to abrupt changes from shade to sun and may result in extreme cases in severe foliar sunburn (Pict. 14.5).

Self-shading of the coffee plants can only partly buffer the effect of sudden exposure of coffee leaves to full sun. Overall, the acceptably high yields reported from densely shaded fields in Costa Rica and Mexico suggest that even "modern" varieties such as Caturra and Catimor maintain a relatively strong adaptation or, at least, tolerance to shade. However, more work is needed to characterize the specific shade requirements and tolerances of different varieties along environmental gradients of altitude, moisture and soil fertility.

Figure 14.4 attempts to reconcile the seemingly contradictory shade responses of coffee by taking environmental conditions into account. This graph is hypothetical, but it corresponds correctly to data assembled as a result of long-term experiments and observations made in different environments. A comparison of the long-term productivity of an imaginary unshaded coffee field with that of a field under 50 % shade at different altitudes (Fig. 14.4), and for soils without and with constraints ("good" and "bad" soil), would most likely lead to the following conclusions. Irrespective of soil condition, the highest production of unshaded coffee typically occurs at intermediate altitudes since they provide the ideal climate for coffee. In Central America this is often the case between altitudes of about 900 and 1300 m.a.s.l. The exact values for a particular region depend on site-specific conditions, including the effects of latitude on the average and extreme temperatures. At lower altitudes, unshaded coffee production decreases in response to increasing heat stress, while at higher altitudes, it decreases due to low temperatures and possible wind damage. Under such suboptimal conditions, trees can reduce the microclimatic stress to the coffee plants by shading at low altitudes and by providing wind protection at high altitudes. As a result trees tend to increase coffee production when compared to unshaded plots. This benefit is marked as the dappled area in Fig. 14.4 and referred to as "shade contribution". In contrast, for coffee grown within the optimum altitude range, trees do not exert the same beneficial effect because the microclimate is already ideal for coffee. Therefore, shading tends to limit coffee production. This is marked as the hatched area

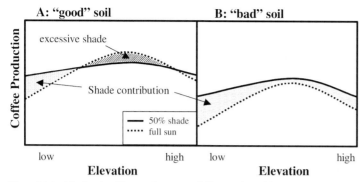

Figure 14.4 Idealized coffee production in full sun (dotted lines) and under trees giving 50% shade (thick lines) as a function of elevation for soils without (A) and with (B) limitations of rooting depth, nutrients or moisture. Notice that the benefit from shade ("shade contribution") is largest at elevations below or above the optimum elevation for coffee. For further explanation, see text.

Picture 14.6 Coffee fields at high elevations often do not require shading, but need protection from winds. Here, a shelterbelt of *Casuarina equisetifolia* around a coffee field at 1200 m.a.s.l. in Costa Rica. Photo: R. G. Muschler

labeled as "excessive shade" in the "good soil" scenario of Fig. 14.4. On "bad" soils, the productivity of both coffee systems is relatively lower as a result of nutrient and/or moisture limitations. However, the productivity of unshaded coffee is reduced even more under such conditions due to the absence of the beneficial effects of trees on nutrient cycling and water retention. Consequently, the "shade contribution" increases and probably extends across the whole elevational gradient. Considering these different environmental fac-

tors, the apparent contradiction between studies which report benefits of shade and others reporting shade-induced yield reductions no longer exist. Both theories are most likely to be correct within the limitations generated by specific local environmental conditions.

14.4.2.3 Plant Vigor and Nutrition

Provided similar nutrition, shaded coffee plants tend to be taller, more vigorous and have a higher leaf retention throughout the year compared to plants with little or no shade. This is usually most pronounced under suboptimal conditions for coffee and at the end of the harvest period, i.e. under conditions of maximum stress for the plants [20]. Under such conditions, the often higher leaf loss of unshaded plants is generally caused by an increased plant exhaustion due to a higher fruit load combined with heat stress, on the one hand, and higher susceptibility to diseases, on the other [20]. However, when high fertilizer inputs and chemical pest and disease protection can be provided, the recuperation of defoliated unshaded plants can be very rapid after the harvest. Also, in environments with ideal soil and climatic conditions for Arabica coffee, these differences are far less significant.

Nutritional stress and, hence, foliar deficiencies tend to be more pronounced in unshaded plants [37]. Sun-induced reduction of the levels of potassium and calcium in coffee plants was already reported in the 1940s. Reduced levels of potassium and zinc in unshaded coffee plants were also reported from the shade study in Costa Rica despite similarly high fertilization regimes for all shade treatments (Table 14.2). In addition to a higher demand for potassium of sun-exposed plants, potassium leaching from unshaded coffee foliage may contribute to reduce foliar concentrations, while plants under shade may experience less leaf leaching and, possibly, even some potassium inputs from the tree canopy.

During fruit filling, coffee plants have higher nitrogen requirements and, quite possibly, the same would also apply to other nutrients. If this higher demand is accompanied by microclimatic stress from a lack of shade, the plants may not be able to provide enough nutrients to the various sinks, resulting in higher fruit drop and in decreased plant vigor, leaf loss, and, ultimately, die-back of whole branches or plants [5, 37]. All these phenomena tend to be more pronounced in unshaded plantations, but as long as external inputs in the form of fertilizers and pesticides are available in abundance, the plants can be maintained and a large crop can be obtained. However, the current coffee price crisis coupled with growing environmental and health concerns impose increasing restrictions on this.

As a result, the development of sustainable coffee systems must increasingly focus on the design and management of agroforestry systems in which the beneficial functions of shade trees for improving the nutrient status and the microclimate of the systems are maximized. In other words external inputs need to be increasingly substituted by biological services provided by trees and other associated plants in coffee fields. The main mechanisms for improving the nutrient status result from the presence of the generally larger and more per-

Table 14.2 Nutrient concentrations (means, $n = 2$) in sun-exposed and shaded leaves[a] of *Coffea arabica* L, and significance levels for *t*-tests within years

Nutrient	December 1995			November 1997		
	Sun	Shade	P > t	Sun	Shade	P > t
N (mg/g)	25.6	25.5	NS	24.50	26.5	NS
p (mg/g)	1.2	1.1	NS	1.20	1.3	NS
K (mg/g)	11.2	13.9	0.16	12.15	19.7	0.01
Ca (mg/g)	16.2	16.8	NS	19.15	19.5	NS
Mg (mg/g)	6.3	5.0	NS	7.45	6.0	0.16
Zn (mg/kg)	14.0	18.0	NS	8.9	16.4	0.00
Mn (mg/kg)	479	396	NS	462	473	NS
Cu (mg/kg)	12	12	NS	9.9	19.4	0.06
B (mg/kg)	ND	ND		73.1	77.4	NS

[a]Fourth pair of fully developed leaves from the apical meristem, taken from branches in the upper half of the coffee bushes.
From: R. Muschler [21].

manent root systems of the trees which tend to increase nutrient cycling from the top and subsoil, reduce leaching losses, and provide more permanent nutrient uptake throughout the year as well as additional inputs of nutrient-rich organic matter and of biologically fixed nitrogen [31]. These processes are of key importance for organic production systems in which no synthetic fertilizers can be used. For these systems, the challenge consists in closing the nutrient cycle. Considering the inputs of up to 106 kg/ha/year of recycled and biologically fixed nitrogen from selectively pruned *E. poeppigiana* trees measured in the Costa Rican shade study [21], it seems possible to cover the extraction of about 27 kg of N/ha/year in 1500 kg green coffee harvested from 1 ha (Table 14.3).

As things stand now, it is difficult to prove whether this is actually feasible, but future studies on the nitrogen uptake efficiency and the percentage of nitrogen derived from biological fixation will clarify the issue. In the meantime, the amounts of nitrogen involved and empirical observations on successful pioneer organic farms suggest that it should be feasible. Even on farms with up to 10 years of productive organic coffee management, no signs of nitrogen deficiencies have appeared. Detailed reviews on nutrient cycling and evaluation methods have recently been presented [31, 32].

14.4.2.4 Pests, Diseases and Weeds

This issue is included here because of its dependence on shade and its relevance for coffee production and quality. In Central America, losses due to pests and diseases have increased strongly over the past decades [36]. The three main reasons are:

- The emergence of new pests, particularly the Coffee Berry Borer (*Hypothenemus hampei*) and Coffee Leaf Rust (*Hemileia vastatrix*).
- The greatly simplified and species-impoverished systems as a result of intensified production systems.
- The emergence of secondary pest problems such as the Mealybug (*Planococcus citri*) which began to be a problem in Nicaragua only after its natural enemies were decimated as a result of applying insecticides to control the Coffee Leaf Miner (*Leucoptera coffeella*) in unshaded coffee fields [36]. The second and third

Table 14.3 Nutrient extraction (kg/ha/year) in coffee harvest and inputs from pruned trees and from fertilizer inputs

Flux	N	P	K	Ca	Mg
Extraction (in 1500 kg green coffee)	−27	−1.5	−27	−3.6	−1.5
Input from total pruning of *Erythrina*	58	3.6	18	18	4
Input from partial pruning of *Erythrina*	106	6.6	33	33	7.3
Balance*	31 to 79	1.1 to 5.1	−9 to 6	14.4 to 29.4	2.5 to 5.8
Input from synthetic fertilizers	219	31.5	95	100	37.8

*The two values result from subtracting the extracted amount from one or the other input values from pruning of *Erythrina*.
From: R. Muschler [21].

reasons are primarily the result of reduced shade levels.

The combination of species and the intensity of tree shade modifies the environment for pests and diseases in three principal ways:

- It changes the microclimate by reducing average and extreme temperatures of air and soil and by increasing moisture.
- It increases soil organic matter inputs and soil cover.
- It increases the diversity of habitats for other organisms including pests, diseases and their natural antagonists.

The complex interactions among the different organisms (predators, herbivores, parasites, hyperparasites, etc.) in a multistrata coffee system and how to design a pest-suppressive system were recently reviewed [36]. The most important tool for designing a pest-suppressive system is to choose an appropriate shade level and species composition which minimizes the pest complex while maximizing the effects and persistence of beneficial microflora and fauna acting against it.

Of course, the best shade level for this purpose also varies with environmental conditions and production objectives. For low-altitude coffee zones (typically below 800 m.a.s.l.) with a pronounced dry season of 4–6 months, which is the case for many areas under Pacific climatic influence in Central America, the authors of the study suggest 35–65 % shade. This would favor leaf retention in the dry season, reduce the incidence of the Brown Eye Spot disease (*Cercospora coffeicola*), weeds and *Planococcus citri* while, at the same time, it would increase the efficiency of antagonists for diseases and pests. High shade levels should be avoided because of the potential increase of the American Leaf Spot (*My-*

cena citricolor) which is encouraged by higher humidity, a condition accentuated by shade. Conversely, lower shade levels should be avoided because of the increased incidence of Brown Eye Spot disease (*Cercospora coffeicola*) and of the Coffee Leaf Miner (*Leucoptera coffeella*) under unshaded conditions.

Similar results were also established by the mid-altitude coffee shade study in Turrialba [21]. For the two coffee varieties, Caturra and Catimor 5175, leaf infections were clearly more pronounced in the sun and pollarded plots, while the open and dense shade plots showed only minimal levels of infection (Fig. 14.5).

The most important disease was *C. coffeicola*, followed by *Colletotrichum* spp. and *H. vastatrix* in Caturra. This higher disease incidence in the sun and pollarding plots was probably linked to the extreme microclimatic conditions which prevailed there combined with the higher nutrient requirements of the sun-exposed coffee plants. Nutrient stress can predispose the plants to disease attack. Considering the key role of potassium in disease resistance, the often reduced levels of potassium (unless fertilized intensely) in sun-exposed plants (*cf.* Table 14.2) can reduce disease resistance and contribute to higher disease levels for unshaded plants. For *Cercospora*, an increased incidence at low levels of potassium is well documented. In contrast, the more moderate microclimate under shade may reduce the susceptibility of the plants to diseases. This is particularly marked for *Cercospora*, which does not develop under shade. Studies of the ecology of *C. arabica* in its native Ethiopia reported also a strong reduction or virtual absence of diseases (*Cercospora* and *Hemileia*) as long as coffee plants were under shade. It should also be noted that in several countries of the Asia–Pacific region, the activity

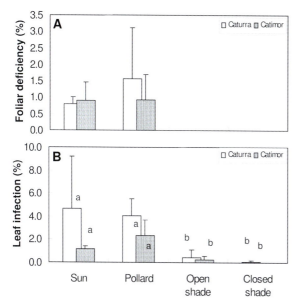

Figure 14.5 Foliar nutrient deficiencies (visual scoring) and infection of *Coffea arabica* vars Caturra and Catimor under four shading treatments in El Cañal, Turrialba, Costa Rica (means ± SE). The data was compiled during the fourth quarter of the harvest period 1995/96. Within varieties, means with the same letter are not significantly different.

of the White Stem Borer (*Xylotrechus quadripes*) is considerably inhibited under shade because egg laying is affected.

Another important effect of shade which benefits coffee vigor and helps to reduce management costs is the reduction or suppression of weeds. While even low levels of shade can modify the species composition of weeds, a permanent shade cover of 50 % and above often can control weeds completely. The main mechanisms for this are the reduction of light and the deposition of foliage on the soil surface acting as a barrier to the germination and emergence of weeds (Pict. 14.7). Weed inhibition induced by shade can be of particular importance in regions where environmental conditions encourage year-round weed growth. Under such conditions, as for example the ever-wet Atlantic-slope coffee fields in Central America, the costs of controlling weeds in unshaded coffee plots can amount to as much as 10 % of total management costs. These studies and examples demonstrate the importance of managing

Picture 14.7 Trees and shade for weed control: tree shade suppresses weeds (foreground) by reducing the light levels and by forming a mulch cover which acts as a barrier to weed germination and growth. Notice the aggressive growth of weeds in the unshaded center of the photo. Photo: R. G. Muschler.

an intermediate level of shade for minimizing the incidence of pests and diseases and for keeping weeds under control.

14.4.3
Effects of Trees and Shade on Coffee Quality

Although the effects of many factors on the development of coffee berries have been widely discussed [37], there is surprisingly little information on the influence of shade on coffee quality [15, 23]. In general, shade seems to benefit coffee quality as well, particularly under suboptimal conditions where unshaded plants suffer great environmental stress. Therefore, shade may be required for improving the quality of coffee grown at altitudes which are below the ideal level, while this may not be the case for certain high altitude zones

where coffee is grown under ideal climatic conditions. Again, the shade study from Costa Rica shall be used to discuss central aspects.

14.4.3.1 Fruit Weight
It is widely accepted that fruit weight increases under shade [37]. The shade study in Turrialba showed that plants of both Caturra and Catimor 5175 bore consistently heavier cherries (11–14 % higher fresh-fruit weight) under "open shade" and "dense shade" of *E. poeppigiana* than with no or little shade in the "unshaded" and "pollarded" treatments (Fig. 14.6).

Homogeneous intermediate shade provided by shade-cloth yielded intermediate fruit weights demonstrating the effect of shade *per se*, i.e. without the other effects

Picture 14.8 Harvest of high-quality coffee from a high-elevation field at 1800 m.a.s.l. with no shade (left) and from a low-elevation site at 700 m.a.s.l. under 50 % shade (right). Photo: R. G. Muschler.

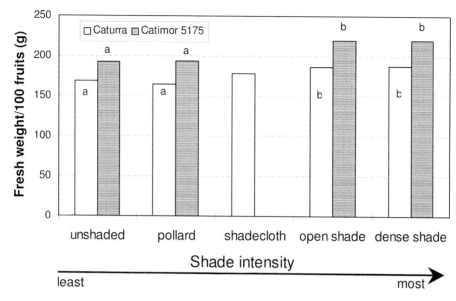

Figure 14.6 Fresh weight of 100 ripe fruits of two coffee varieties grown under increasing shade levels (El Cañal, Turrialba, Costa Rica; 1994/95/96; 700 m.a.s.l., 2600 mm). Within varieties, means with the same letter do not differ significantly ($n = 26$, $\alpha = 0.05$). From: Muschler [23].

of the trees (e.g. organic matter/nutrient cycling). The larger and heavier cherries produced under increased shade agree with results from other authors who also documented the beneficial effects of shading due to lower overall temperatures and an extension of the ripening period [15].

14.4.3.2 Fruit to Bean Conversion Factors

The ratios between fruit fresh weight (FW_{fruit}) and the corresponding dry weight of parchment coffee (DW_{parch}) and dry weight of green coffee (DW_{green}) did not differ among shade levels, although there was a slight tendency towards lower conversion rates for the unshaded treatment [23]. The ranges of the individual plot data were 15–19 % for DW_{parch}/FW_{fruit}, 12–16 % for DW_{green}/FW_{fruit} and 80–85 % for DW_{green}/DW_{parch}.

14.4.3.3 Bean Size

Bean size is also well known to increase under shade. Data from the Turrialba shade study illustrate the extent of this increase. The size distribution of green beans ("café oro") showed, for both coffee varieties, a significant and consistent increase in bean size with increasing shade levels (Fig. 14.7) [23]. The proportions of large beans (larger than 17/64 in) in the open and dense shade plots were significantly higher than those cultivated under little or no shade. While the increase in the percentage of large beans from "unshaded" to "dense shade" amounted to 20 % (from 49 to 69 %) for Caturra, it was 29 % (from 43 to 72 %) for Catimor (Fig. 14.7). Similarly, the reduction of small beans (smaller than 15/64 in) was larger for Catimor (14 %; from 18 to 4 %) than for Caturra (9 %; from 14 to 5 %).

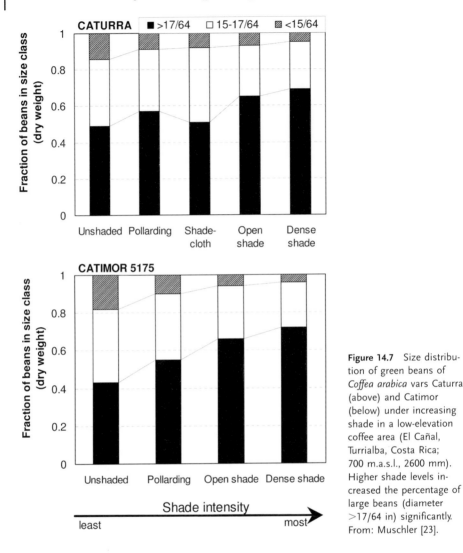

Figure 14.7 Size distribution of green beans of *Coffea arabica* vars Caturra (above) and Catimor (below) under increasing shade in a low-elevation coffee area (El Cañal, Turrialba, Costa Rica; 700 m.a.s.l., 2600 mm). Higher shade levels increased the percentage of large beans (diameter >17/64 in) significantly. From: Muschler [23].

Increased bean sizes under shade confirm studies carried out in other coffee areas in Puerto Rico, Cuba and Guatemala [15]. The relatively larger shade benefits for the bean size of Catimor versus Caturra could indicate a higher adaptation of Catimor to shade. This would concur with the slightly smaller yield reduction of 12% under dense shade versus no shade for Catimor compared to Caturra where dense shade reduced production by 37% [22]. However, the validity of this interpretation needs to be further verified in different environments.

14.4.3.4 Visual Appearance and Organoleptic Attributes

As with the effect of shade on production, the influence of shade on organoleptic attributes is also controversial. While shade may have little effect on the morphology and chemical composition of the cherries [37], studies carried out in Guatemala have demonstrated that shade and altitude had similar significant positive effects on coffee bean size and chemical composition, presumably because lower temperatures slowed the ripening process [15]. The shade study in Costa Rica also included visual evaluations and professional cupping of coffee from the different shade levels which revealed marked benefits for shaded coffee. With the exception of aroma, both varieties benefited consistently from shade (Fig. 14.8).

For Catimor, shade improved the appearance of green and roasted beans and also improved both taste parameters (body and acidity). In contrast, the index of aroma was lower for shaded Catimor. For Caturra, a similar pattern was found for all parameters, except for aroma and acidity, with the differences between treatments being slightly smaller. Again, the observation of larger differences for Catimor than for Caturra may indicate the higher shade requirements or tolerance of Catimor.

Given the improved taste and visual appearance of coffee under shade, it was concluded that organoleptic coffee quality was greatly improved by shade. The improvements were greater for Catimor than for Caturra, mainly due to the much lower marks for the unshaded Catimor compared to Caturra. Relatively low quality ratings of

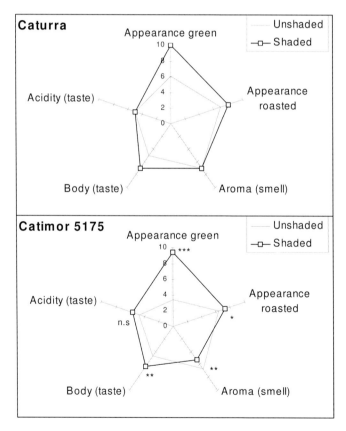

Figure 14.8 Quality attributes of green and roasted beans of two varieties of *Coffea arabica* grown without shade and with intermediate shade (above 50% shade) of *Erythrina poeppigiana* at a low-elevation site (El Cañal, Turrialba, Costa Rica; 700 m.a.s.l., 2600 mm). Beans were harvested at full maturity in September/October 1997. The scales use a relative index with a maximum of 10. Significance between sun and shade given as ***$\alpha = 0.01$, **$\alpha = 0.05$ and *$\alpha = 0.1$. From: Muschler [23].

unshaded Catimor 5175 compared to other varieties have been reported by different workers. However, it is interesting to note that the organoleptic quality of shaded Catimor was almost the same as that of shaded Caturra. This might indicate a way of improving the often deficient quality of certain lines of Catimor which have received much attention in recent years due to the presence of a Coffee Leaf Rust-resistance gene.

Improved organoleptic properties granted by shading were also reported from coffee zones in Honduras and for *C. arabica* var. Catuai from high-elevation zones in Guatemala [15]. In the latter study, shading also increased the acidity and the sucrose content of Catuai, both important ingredients of organoleptic evaluations. The main reason was reported to be delayed ripening due to the shade and its microclimatic effects. It is likely that the same factors were responsible for the higher quality of shaded coffee also in the Costa Rican study. At that mid-altitude coffee site with high temperatures, the reduction .of temperature extremes through shading may have played the dominant role for the uniform growth and ripening of berries [1, 21]. This might be the main reason for improved quality for coffee grown in sub-optimal environmental conditions.

14.5
How to Design the Ideal Shade

Based on the considerations given so far, how should the grower go about creating the ideal shade environment for his farm? The basic point of departure which cannot be ignored is the assessment of the determining factors for the sun-shade decision as discussed in Figs 14.1 and 14.4. The more coffee is stressed by biophysical con-

straints, the higher the required shade level should be. Where coffee is grown without such constraints, shade may not be required. However, even under ideal conditions it might be necessary to associate other trees with coffee though, possibly, for different reasons. For example, it could be useful to plant or retain trees in coffee fields on slopes to help prevent or reduce soil erosion or to act as wind-breaks. Trees may also be necessary as a habitat for wildlife and antagonists of pests and diseases.

Once the required level of shade has been determined, the appropriate tree species, and their spacing and management need to be defined. Information to help with this decision can come from three sources [22]. First and foremost, local experiences with site-adapted tree species should be considered. Secondly, research results on the compatibility between various tree species and coffee should be consulted. Thirdly, certification guidelines like the ones given in Table 14.1 should be taken into consideration. Based on the information provided from these sources, the grower should be able to assemble a list of appropriate species, spacing and management interventions to generate the desired amount of shade. Although more information is starting to emerge about growth rates and the ecological attributes of tree species, there still are large gaps in our understanding of their multiple roles in the ecosystems.

Among the most widely used tree species are the legumes of the genus *Inga* with more than 20 commonly planted species, *Erythrina* spp., *Gliricidia sepium*, and several species of *Albizia* and *Acacia*. Among timber and fuel-wood species, fast-growing trees belonging to the genera *Cedrela*, *Cordia*, *Eucalyptus*, *Grevillea*, *Tabebuia* and many others can be found throughout coffee fields of the world. The third group of widely used trees are fruit trees, which are particularly used in smallholder coffee fields. The most currently used species include bananas, citrus, avocados and many others. While there is a vast store of experiential information, much needs to be done to establish consistent evaluations of species-specific effects on coffee systems and to fine-tune planting arrangements and pruning regimes in such a way that the associated coffee receives the right amount of light and nutrients throughout its phenological cycle. For more information on species selection, the reader may want to consult reviews [2, 22, 38, 39].

14.6
Conclusions and Research Recommendations

Shade can play an important role for maintaining long-term coffee productivity, for conserving soils, water and biodiversity, and even for improving coffee quality. These effects tend to be more pronounced under suboptimal conditions where coffee plants are stressed. Since most coffee areas in the world are subject to at least one biophysical constraint, the benefits of shade are likely to be relevant for improving the majority of coffee systems.

As to coffee productivity, the main benefits of shade are to reduce plant exhaustion, extend crop life, reduce fruit loss, lower input requirements, encourage a more stable production, and improve the protection of natural resources like soil, water and biodiversity. For most suboptimal coffee-producing regions, intermediate levels of shade (Pict. 14.9) and a variety of tree species in coffee fields (Pict. 14.1 and 14.10) are likely to be the best options.

For coffee quality, the main benefits from shading tend to be a reduced number of deformed, sunburnt or diseased coffee berries, heavier cherries, larger beans, higher ratings for visual appearance of green and roasted beans, and may even include higher ratings for acidity and body. The extent of these improvements is largely determined by the environmental conditions of each particular site as well as by the attributes of coffee varieties and associated tree species. The overall validity of these assumptions, however, needs to be established by rigorous research.

Picture 14.9 Tree canopies generating about 50% shade. This amount of shade is ideal for many low- or mid-elevation coffee areas exposed to temperatures above the ideal for coffee. Without shade, the air, soil and leaf temperatures would rise above the optimum for coffee. Above: *Erythrina poeppigiana*. Below: *Inga* spp. Notice how the clustering of foliage allows direct sunlight to penetrate through the gaps thus generating "sunflecks" that are beneficial to coffee. Photo: R. G. Muschler.

Picture 14.10 High-value timber trees such as *Cordia alliodora* (left) or fast-growing species such as *Eucalyptus deglupta* (right) in Central America generate long-term additional income and provide shade to the coffee. Notice the smaller leguminous shade trees *Erythrina poeppigiana* in the left photo which form an intermediate canopy between the timber trees and the coffee. These trees are regu-larly pruned to adjust the shade intensity for the coffee, and to provide biomass and mulch material. *E. deglupta* has been widely planted as a fast-growing tree in coffee plantations for firewood and timber. However, its ecological value is limited due to suspected allelopathic effects and its limited value as habitat for birds and other animals. Photo: R. G. Muschler.

The observation that intermediate shade ("open shade") may produce larger beans of better organoleptic attributes while sustaining harvest levels of quality cherries similar to those in unshaded plots [22] suggests that intermediate shade may allow the grower to get the "best of both worlds", at least in environments which resemble the one studied.

Future research should:

- Evaluate the apparent differences in shade response and input requirements among coffee varieties – this work should be carried out along environmental gradients of altitude, soil fertility and moisture availability.

- Substantiate the effects of shading on coffee quality according to environment and coffee varieties – improved quality should be considered as essential to meet the growing requirements of specialty markets including gourmet and organic coffee.

- Quantify the ecological and economic value of managing species-rich agroforestry systems with coffee to generate multi-

ple benefits, including the conservation of biodiversity, soils and water.
- Characterize the ecological compatibility of different tree species with coffee pro-

duction and other objectives, such as wildlife conservation, nutrient cycling and the maintenance of soil fertility among other issues.

Bibliography

[1] Barradas, V. L. and Fanjul, L. Microclimatic characterization of shaded and open-grown coffee (*Coffea arabica* L.) plantations in Mexico. *Agric. For. Meteorol.* **1986**, *38*, 101–112.

[2] Beer, J. W. Advantages, disadvantages and desirable characteristics of shade trees for coffee, cacao and tea. *Agrofor. Syst.* **1987**, *5*, 3–13.

[3] Beer, J. W., Muschler, R. G., Kass, D. and Somarriba E. Shade management in coffee and cocoa plantations. *Agrofor. Syst.* **1998**, *38*, 139–164.

[4] Boyce, J. K., Fernández, A., Fürst, E. and Segura, O. *Café y Desarrollo Sostenible: del Cultivo Agroquímico a la Producción Orgánica en Costa Rica.* EFUNA, Heredia, Costa Rica, 1994.

[5] Carvajal, J. F. *Cafeto: Cultivo y Fertilización*, 2nd edn. Instituto Internacional de la Potassa, Bern, 1984.

[6] Cook, O. F. *Shade in Coffee Culture.* USDA Bulletin 25. USDA, Washington, DC, 1901.

[7] Ewel, J. J. Designing agricultural ecosystems for the humid tropics. *Ann. Rev. Ecol. Syst.* **1986**, *17*, 245–71.

[8] FAO. *Biodiversity and the Ecosystem Approach in Agriculture, Forestry and Fisheries.* FAO Interdepartmental Working Group on Biological Diversity for Food and Agriculture. FAO, Rome, 2003.

[9] Fernandez, C. E. and Muschler, R. G. Aspectos de la sostenibilidad de los sistemas de cultivo de café en América Central. In: *Desafíos de la Caficultura en Centroamérica.* Bertrand, B. and Rapidel, B. (eds). IICA–PROMECAFE–CIRAD, 1999, 69–96.

[10] Gallina, S., Mandujano, S. and Gonzalez-Romero, A. Conservation of mammalian biodiversity in coffee plantations of Central Veracruz, Mexico. *Agrofor. Syst.* **1996**, *33*, 13–27.

[11] Gliessman, S. R. *Agroecology: Ecological Processes in Sustainable Agriculture.* Ann Arbor Press, Ann Arbor, MI, 1998.

[12] Goldberg, A. D. and Kigel, J. Dynamics of the weed community in coffee plantations grown under shade trees: effect of clearing. *Israel J. Bot.* **1986**, *35*, 121–131.

[13] Greenberg, R., Bichier, P. and Sterling, J. Bird populations in rustic and planted shade coffee plantations of eastern Chiapas, Mexico. *Biotropica* **1997**, *29*, 501–514.

[14] Guiscafre-Arrillaga, J. Sombra o sol para el cafeto? *El Café de El Salvador* **1957**, *308* (9), 320–364.

[15] Guyot, B., Gueule, D., Manez, J. C., Perriot, J. J., Giron, J. and Villain, L. Influence de l'altitude et de l'ombrage sur la qualité des cafés Arabica. *Plantations, Recherche, Développement* **1996**, *3* (4), 272–283.

[16] Haarer, A. E. *Modern Coffee Production.* Leonard Hill, London, 1962.

[17] Kumar, D. and Tieszen, L. L. Photosynthesis in *Coffea arabica*. I. Effects of light and temperature. *Expl. Agric.* **1980**, *16*, 13–19.

[18] Lock, C. G. W. *Coffee: Its Culture and Commerce.* Spon, London, 1888, 264 pp.

[19] Machado, A. Transformación de plantaciones de café. *Cenicafé* **1959**, *10*, 217–264.

[20] Müller, L. E. Algunas deficiencias minerales comunes en el cafeto (*Coffea arabica* L.). *Boletín Técnico 4.* IICA, Turrialba, Costa Rica, 1959.

[21] Muschler, R. G. Tree-crop compatibility in agroforestry: production and quality of

coffee grown under managed tree shade in Costa Rica. *PhD Dissertation*. University of Florida, Gainesville, FL, 1998.

[22] Muschler, R. G. *Árboles en Cafetales*. Módulo de Enseñanza Agroforestal No. 5. Proyecto Agroforestal CATIE/GTZ. CATIE, Turrialba, Costa Rica, 2000.

[23] Muschler, R. G. Shade improves quality of *Coffea arabica* L. in a sub-optimal coffee zone of Costa Rica. *Agrofor. Syst.* **2001**, *51*, 131–139.

[24] Nunes, M. A., Bierhuizen, J. F. and Ploegman C. Studies on productivity of coffee. I. Effect of light, temperature and CO_2 concentration on photosynthesis of *Coffea arabica*. *Acta Bot. Neerl.* **1968**, *17*, 93–102.

[25] Nunes, M. A., Ramalho, J. D. and Dias, M. A. Effect of nitrogen supply on the photosynthetic performance of leaves from coffee plants exposed to bright light. *J. Exp. Bot.* **1993**, *44*, 893–899.

[26] Pendergrast, M. *Uncommon Grounds. The History of Coffee and how it transformed our World*. Basic Books, New York, 1999.

[27] Perfecto, I., Rice, R. A., Greenberg, R. and Van der Voort, M. E. Shade coffee: a disappearing refuge for biodiversity. Shade coffee plantations can contain as much biodiversity as forest habitats. *BioScience* **1996**, *46*, 598–608.

[28] Ramirez, L. G. Producción de café (*Coffea arabica*) bajo diferentes niveles de fertilización con y sin sombra de *Erythrina poeppigiana* (Walpers) O. F. Cook. In: *Erythrina in the New and Old Worlds*. Westley, S. B. and Powell, M. H. (eds). Nitrogen Fixing Tree Association, Paia, HI, 1993, 121–24.

[29] Rice, R. A. Observaciones sobre la transición en el sector cafetalero en Centroamérica. *Agroecología Neotropical* **1991**, *2*, 1–6.

[30] Ruppentahl, M., Leihner, D. E., Hilger, T. H. and Castillo, J. A. Rainfall erosivity and erodibility of inceptisols in the southwest Colombian Andes. *Expl. Agric.* **1996**, *32*, 91–101.

[31] Schroth, G., Lehmann, J., Rodrigues, M. R. L., Barros, E. and Macedo, J. L. V. Plant–soil interactions in multistrata agroforestry in the humid tropics. *Agrofor. Syst.* **2001**, *53*, 85–102.

[32] Schroth, G. and Sinclair, F. L. (eds). *Trees, Crops and Soil Fertility. Concepts and Research Methods*. CAB International, Wallingford, 2003.

[33] Scialabba, N. and Hattam, C. (eds). *Organic Agriculture, Environment and Food Security*. FAO, Rome, 2002.

[34] Soto-Pinto, L., Perfecto, I., Castillo-Hernandez, J. and Caballero-Nieto, J. Shade effect on coffee production in the northern Tzeltal zone of Chiapas, Mexico. *Agriculture, Ecosystems and Environment* **2000**, *80*, 61–69.

[35] Soto-Pinto, L., Perfecto, I. and Caballero-Nieto, J. Shade over coffee: its effects on berry borer, leaf rust and spontaneous herbs in Chiapas, Mexico. *Agrofor. Syst.* **2002**, *55*, 37–45.

[36] Staver, C., Guharay, F., Monterroso, D. and Muschler, R. G. Designing pest-suppressive multistrata perennial crop systems: shade-grown coffee in Central America. *Agrofor. Syst.* **2001**, *53*, 151–170.

[37] Willey, R. W. The use of shade in coffee, cocoa and tea. *Horticult. Abstr.* **1975**, *45* (12), 791–98.

[38] Wrigley, G. *Coffee*. Longman. London, 1988.

[39] Yépez, C., Muschler, R., Benjamin, T. and Musalem M. Selección de especies para sombra en cafetales diversificados de Chiapas, México. *Agroforestería en las Américas* **2002**, *9*, 55–61.

Part II
Pests & Diseases

1
Coffee Pests in Africa

T. J. Crowe

1.1
Introduction

Coffee, particularly Arabica coffee, is very susceptible to attack by insect pests. Also, being a perennial plant, crop loss, due to ill-timed or injudicious sprays, may not only occur in the current season but continue into the next season. It is essential, therefore, to evolve a proper pest-management program.

The details of the program can only be worked out in the context of the pest complex and farming system prevailing in a particular area. However, some general principles can be recognized:

1) A scouting system must be introduced. The coffee bushes should be checked at least once a week. In some cases only the "presence" or "absence" of pests needs be noted. For others, regular counts need to be made.
2) For major pests, threshold levels for spraying must be established. In a few cases some guidelines are given below under specific pests, but in all cases the level will need to be adjusted, in the light of experience, to suit local needs.
3) Cultural and mechanical measures may give adequate control without the use of chemicals, e.g. mulching, pruning and the collection of fallen berries can all affect the pest population. Some examples are given below under specific pests.
   On large estates, where coffee is grown in monoculture, some degree of weed growth may be desirable. It will provide essential food (nectar and pollen) for adult hymenopterous parasites of coffee pests. Certain useful general predators such as Green Lacewings (Neuroptera: Chrysopidae) also need nectar and pollen.
4) When spraying foliage avoid the use of persistent insecticides if at all possible. They kill natural enemies of coffee pests and may lead to outbreaks of another pest.
5) Use as selective an insecticide as possible, i.e. one that kills the pest but has little or no effect on natural enemies. Examples are given below under specific pests.
6) Avoid overall spraying if possible. Spot-spray heavily infested areas leaving lightly infested areas as reservoirs of parasites and predators.

Coffee: Growing, Processing, Sustainable Production, Second Edition. Edited by J. N. Wintgens.
© 2012 WILEY-VCH Verlag GmbH & Co. KGaA. Published 2012 by Wiley-VCH Verlag GmbH & Co. KGaA.

Further reading: Dent, D. *Insect Pest Management.* CAB International, Wallingford, 1991.

1.2
Grasshoppers and Crickets

Soil-burrowing crickets (Orthoptera: Gryllidae) including the Tobacco Cricket *Brachytrypes membranaceus* (Drury) and the Two Spotted Cricket *Gryllus bimaculatus* DeGeer were once serious nursery pests. Soil treatment with persistent insecticides was often necessary. Nowadays, however, the practice of raising seedlings in polythene bags has largely solved the problem. The same considerations apply to the Mole Cricket *Gryllotalpa africana* Palisot de Beauvois (Orthoptera: Gryllotalpidae) (Pict. 1.1, Fig. 1.1).

Short-horned grasshoppers and locusts (Orthoptera: Acrididae) sometimes nibble near the tips of suckers, and cause breakage and unwanted branching. Locusts do not normally eat coffee, but the roosting of a large swarm can cause breakage of branches.

Green Coffee Locust
Phymateus viridipes Stå
(Orthoptera: Pyrgomorphidae)

This is a large green grasshopper up to 10 cm long. It sometimes forms small swarms that settle on coffee. Diazinon sprays have been recommended for its control (Fig. 1.2).

Figure 1.2 Green Coffee Locust – *Phymateus viridipes*. Adult.

Picture 1.1 Tobacco Cricket – *Brachytrypes membranaceus*. Adult insect. From: A. Autrique.

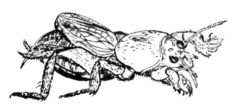

Figure 1.1 African Mole Cricket – *Gryllotalpa africana*.

Elegant Grasshopper
Zonocerus elegans Thunberg
(Orthoptera: Pyrgomorphidae)

Stinking Grasshopper
Zonocerus variegatus (Linnaeus)
(Orthoptera: Pyrgomorphidae)

These are large, brightly colored grasshoppers, green with red and black markings. Adults are about 5 cm long. They both occur throughout East and West Africa, but they are more damaging in the west. All species of coffee and a wide range of other crops are attacked. There is only one generation a year.

They are mostly pests of smallholder plantations because the young stages feed on wild plants and weeds. The most effective control is to observe the sites where the

Figure 1.3 Stinking Grasshopper – *Zonocerus variegatus*. Adult male.

Picture 1.2 Stinking Grasshopper – *Zonocerus variegatus*. Adult insect and damage on coffee leaves. From: A. Autrique.

egg packets are laid in the soil and then organize a campaign to dig them up before hatching. Chickens will eat the egg masses but the active stages are distasteful.

Another solution consists of spraying the larvae with a contact insecticide (e.g. diazinon) 4–6 weeks after hatching, when they are still concentrated. Poison baits, as used in locust control are also effective (Fig. 1.3, Pict. 1.2).

Further reading: COPR. *The Locust and Grasshopper Agricultural Manual*. Centre for Overseas Pest Research, London, 1982.

1.3
Termites

Many species of termites are common in coffee farms. The following are the commonest genera:

Macrotermes
Microtermes
Nasutitermes
Odontotermes
(Isoptera: Termitidae)

Termites are both beneficial and detrimental in their effects on coffee. Beneficial effects include:

1) The soil burrows improve rainfall infiltration.
2) Large burrows allow topsoil to penetrate to subsoil zones providing favorable paths for root growth.
3) The breakdown of mulch is speeded up (this is important especially with hard or woody material).
4) The mating flight of winged reproductives provides, for a few days, a highly nutritious source of food, popular in many parts of Africa.

Detrimental effects include:

1) Many species produce mounds that inhibit cultivation work and/or reduce cultivated areas – these mounds have to be leveled and poisoned to prevent recurrence.
2) Termite holes may be dug out by Ant Bears *Orycteropus afar* leaving large holes hazardous to wheeled vehicles – coffee bushes may even be uprooted.
3) Termites eat the dead bark on healthy trees – this is normally harmless; however, if the tree is weak, wounded or droughted, the termites may then penetrate into the living wood and kill the tree.

For nest destruction, aldrin is the most efficient product. Pour a solution of about 10 litres of 0.3 % aldrin down natural burrows or augur holes into each nest. If aldrin is not available, synthetic pyrethroids (e.g. permethrin) or chlorpyrifos may be used instead (Pict. 1.3).

Picture 1.3 White termites. Damage to coffee trunk (basal bark has been gnawed). From: J. N. Wintgens.

Nests can also be fumigated using a commercial dichloropropane/dichloropropene mixture (e.g. "D-D") injected into the nest.

Further reading: Robinson, J. B. D. Some chemical characteristics of termite soils in Kenya coffee fields. *J. Soil. Sci.* **1958**, *9*, 58–65.

1.4
Aphids

Despite their great importance on other crops, aphids are not usually a pest on coffee. One species is, however, very occasionally damaging:

Black Citrus Aphid
Toxoptera aurantii (Boyer de Fonscolombe)
(Hemiptera: Homoptera: Aphididae)

Colonies of these black or dark brown, soft-bodied insects with black siphunculi and banded antennae are common, clustering on sucker tips and growing shoots (Pict. 1.4).

Picture 1.4 Black Citrus Aphid – *Toxoptera aurantii*. A colony of aphids on coffee leaves. From: A. Autrique.

Hover flies (Diptera: Syrphidae) are extremely efficient predators and large white syrphid eggs and/or leech-like larvae are usually to be found, even in the smallest aphid colony. No control is usually needed. If it is, then a spray of any systemic organophosphate (e.g. dimethoate) will be effective.

The honeydew produced by this aphid may be important as a source of food for leaf miner moths (Pict. 1.5).

Picture 1.5 Black Citrus Aphid – *Toxoptera aurantii*. Adult female. From: A. Autrique.

Further reading: Carver, M. The black citrus aphids *Toxoptera citricidus* (Kirkaldy) and *T. aurantii* (Boyer de Fonscolombe). *J. Aust. Ent. Soc.* **1976**, *17*, 263–270.

1.5
Scale Insects and Mealybugs

Scale insects are the wingless forms of certain families of Homoptera. They are flattened and immobile, and so-called because they resemble the scales of a fish or reptile. The males may be scale-like in the young stages, but the adult male is a frail two-winged insect not often seen. In some species males are rare or non-existent.

The adult female produces eggs or living young, usually under her body. When the eggs hatch the first instar (called a "crawler") walks away from the mother, finds a suitable feeding place, inserts its long tube-like mouth parts and begins to feed.

Some species never move from the initial feeding place; other species may withdraw their mouth tube and move once or twice during their life span to a new feeding site.

Mealybugs have a similar life cycle but the body differs in being soft and covered with a secretion of white wax, often fringed round the edges.

Many scales and Mealybugs take in excess plant sugar during their feeding and excrete it as drops of a sweet liquid called "honeydew". This is highly attractive to ants that, while collecting it, chase away or attack parasites and predators and thus aggravate the scale infestation.

The damage caused by the scale is:

- Loss of plant sap causing poor growth and reduced yields.
- Development of "sooty mold" – this is a fungus that grows on the honeydew and covers the leaves with an opaque black crust, thus reducing their photosynthetic ability. Note that it is quite superficial and can be rubbed off with the thumb to reveal a healthy leaf beneath (Pict. 1.6).
- Prevention of satisfactory fungicide deposits, due to the presence of honeydew and sooty mold, leading to secondary infection by Coffee Leaf Rust *Hemileia*.

Important scale and Mealybug species are described on the following pages.

Picture 1.6   Sooty mold on upper coffee leaf surface. From: T. J. Crowe.

Soft Green Scales
Coccus spp.
(Hemiptera: Homoptera: Coccidae)

Coccus spp. are worldwide pests of both Arabica and Robusta coffee. The commonest species in Africa are *C. alpinus* De Lotto, mostly at high attitudes, and *C. celatus* De Lotto and *C. viridis* (Green) at lower altitudes.

Apart from coffee, many tree crops (e.g. citrus) and wild trees are attacked (Pict. 1.7).

The female scale is roughly oval in outline, about 4 mm by 2 mm in size and pale green in color. The skin is soft and elastic. Scales tend to be arranged in lines along the main veins of the leaf. Reproduction is rapid – one female in favorable circumstances being able to produce over

Picture 1.7 Soft Green Scales – *Coccus viridis*. Scales together with black sooty mold. From: A. Autrique.

Picture 1.8 Soft Green Scales – *Coccus* sp. being eaten by the Ladybird beetle. *Hyperaspis senegalensis*. A larva on the beetle is also shown covered with white down. From: T. J. Crowe.

500 eggs. Honeydew production is copious and the presence of sooty mold is often the first indication that the coffee bushes are attacked (Pict. 1.8).

Parasitized scales turn brown or black. Ladybird beetles (Coccinellidae) are common and efficient predators. *Hyperaspis senegalensis* Mulsant is a particularly common predator in Africa. The adult is shiny black with two anterior and two posterior red spots. The larvae are covered with white wax and, unfortunately, is often mistaken for a Mealybug. Entomophagous fungi are very important in the natural control of soft scales in hot humid areas but rare in the drier zones.

Many other species of soft scale occur on coffee, especially the "black" scales *Saissetia* spp. and those covered with copious soft wax, *Ceroplastes* spp. However, serious infestations are very rare (Pict. 1.9).

Picture 1.9 Soft Scales. Dark brown scale – *Saissetia hemisphaerica* syn. *S. Coffeae*. (5) Male adult and larva greatly enlarged. (6) Infested twig. From: Bayer.

Kenya Mealybug
Planococcus kenyae (Le Pelley)
(Hemiptera: Homoptera: Pseudococcidae)

This Mealybug was a major pest of Kenya Arabica coffee in the areas east of the Rift Valley from its accidental introduction around 1923 to its successful biological control in 1938. It proved to be a species from Uganda, where it is common on both Arabica and Robusta coffee. It is controlled in Kenya by parasites introduced from Uganda (Pict. 1.10).

Picture 1.10   Kenya Mealybug – *Planococcus
kenyae*. Coffee tree heavily infested.
From: J. N. Wintgens.

Picture 1.11   Kenya Mealybug – *Planococcus
kenyae*. Colony of Kenya Mealybug near a young
berry cluster. From: T. J. Crowe.

The female Mealybug is oval in outline
and about 2.5 mm by 1.5 mm. The body
color is yellowish or pinkish, but the color
is obscured by a waxy secretion. This secretion is absent or less dense in the longitudinal mid-dorsal line and in cross lines
marking the segments.

The female lays between 50 and 200 eggs.
The preferred feeding sites are around flowers and young, fruits but all green parts of
the plant can be attacked. Honeydew production is copious (Pict. 1.11).

Further reading: Le Pelley, R. H. *Pests of
Coffee*. Longmans, London, 1968.

Coffee Root Mealybug
Planococcus fungicola Watson and Cox
(Hemiptera: Homoptera: Pseudococcidae)

This Mealybug feeds below ground on the
roots of both Arabica and Robusta coffee.
The infestation is covered by a loose sheath
of fungal growth that encloses the roots
like a sock. The fungus *Diacanthodes* sp.
feeds on Mealybug honeydew – not on
the coffee. Affected plants, which are
usually found singly or in small groups,
turn yellow, wilt and often die. Certain
soil types appear to favor the infestation
of this pest.

Further reading: Watson, G. W. and Cox,
J. M. Identity of the coffee root Mealybug.
Bull. Ent. Res. **1990**, *80*, 99–105.

Many other Mealybugs live on the aerial
parts of the coffee bush, notably the Citrus
Mealybug *Planococcus citri* (Risso). They
can usually be identified by the pattern of
wax distribution. The control strategy for
all is essentially the same.

Fried-egg Scale
Aspidiotus (ruandensis? Balachowsky)
(Hemiptera: Homoptera: Diaspididae)

Unlike most perennial crops, coffee is not
very susceptible to attack by armored scales
(Diaspididae). The Fried-egg Scale appeared suddenly in Kenya in 1977 and
caused severe damage for some years, but

Picture 1.12   Fried-egg Scales – *Aspidiotus* sp. Scales on the upper side of coffee leaf. The smaller specimens with a dark center are immature males. From: T. J. Crowe.

is now well controlled by local ladybird beetles (Coccinellidae) and especially by the exotic *Chilocorus nigrita* (Fabricius) that is established in many parts of the country.

The very small, shiny, black, indigenous beetle *Cybocephalus* sp. (Coleoptera: Nitidulidae) has also become a useful predator (Pict. 1.12).

Star Scale; Yellow Fringed Scale

Asterolecanium coffeae Newstead
(Hemiptera: Homoptera: Asterolecaniidae)

Most individuals of this pest are found in bark crevices on the main stem of the coffee bush. Some are found on fruiting branches. It is potentially a very serious pest that can lead to the death of the tree. It is found throughout tropical Africa from Kenya to the Democratic Republic of Congo, usually on Arabica coffee.

The mature female is pear-shaped, the head being at the wider end. It is redbrown in color, about 1.5 mm long and hairy with a distinct fringe. When mature the scale body shrinks and is slowly replaced by about 50 yellow eggs, visible through the transparent skin.

Further reading: Crowe, T. J. The star scale. *Kenya Coffee* **1962**, *27*, 9–11.

Control measures for scale insects and Mealybugs include:

1) Heavily infested branches should be cut off and thrown on the ground. Parasites and predators can then breed through to adults and fly back to the tree. The wingless scales will die.

2) Control of attendant ants is essential for scale control. The tree trunk is painted with an insecticidal band about 15 cm wide and any drooping branches or pieces of mulch should be removed that may allow the ground-dwelling ants to bypass the band. The band should be 10 cm or more above ground level to avoid mud splash. For many years the favored pesticide was 0.5 % dieldrin. This toxicant is now banned in many countries and none of the substitutes are quite as effective. However, ethion and chlorpyrifos are fairly satisfactory. For root Mealybugs, ant control by soil treatment is the only effective control measure.

3) Suppression of road dust. Inert dusts are fatal to many small parasitic wasps. Therefore trees covered with road dust are more susceptible to scale attack than clean trees. Old sump oil is useful for treating dirt roads or it may be necessary to grow a living screen between the road and the coffee.

4) Organophosphate insecticides are effective against most scales. Those that act mainly by contact (e.g. diazinon) are usually more effective than those that act systemically. As always, spraying should be restricted to the more heavily infested trees. Admixture with white oil often enhances the effectiveness of the spray.

5) For Star Scale, the trunk should be painted with tar oil or a white oil/insecticide mixture. Mixtures used for treating apple trees in the European winter are recommended.

Weak trees are particularly susceptible to Star Scale attack and should receive extra applications of mulch and fertilizer.

1.6
Sucking Bugs

Antestia Bugs
Antestiopsis spp.
(Hemiptera: Heteroptera: Pentatomidae)

There are several species of *Antestiopis* in various parts of Africa. They normally feed on Arabica coffee, although sometimes they are seen on Robusta. In eastern Africa *Antestiopsis orbitalis bechuana* (Kirkaldy) is the most important: in West Africa *A. o. ghesquierei* Carayon and *A. intricata* (Ghesquière and Carayon) are major pests. From the standpoint of a practical farmer, however, they may be treated as a single species.

All active stages of the bug suck green berries, flower buds and growing tips injecting toxic saliva, and, in most areas, the spores of the fungus *Ashbya* sp. (formerly called *Nematospora*).

The results of this feeding are the following types of damage:

1) Blackening of flower buds
2) Fall of immature berries
3) Rotting of the beans within the berry or conversion of the substance of the bean into a soft paste; on the drying tables the parchment shows longitudinal brown streaks ("zebra beans")
4) Multiple branching and shortening of internodes.

Eggs are laid on the underside of leaves in batches of about 12. They are barrel-shaped. white and laid touching. The size is about 1.2 mm tall and 1 mm across. There are five nymphal stages; the last

two showing rudimentary wing pads (Pict. 1.13 and 1.14).

The adult bug is roughly triangular in shape, colored black, orange and white and, in most species, is about 8 mm long. The adult female can live over 1 year and lay up to 500 eggs, although 150

Picture 1.13 Antestia Bug – *Antestiopsis facetoides*. An adult and two late-instar nymphs. This is the species found on low-altitude coffee in Kenya and Tanzania. From: T. J. Crowe.

Picture 1.14 Antestia Bug – *Antestiopsis orbitalis ghesquierei*. Damaged berries and eggs. From: A. Autrique.

Picture 1.15 Antestia Bug – *Antestiopsis orbitalis ghesquierei*. Adult insect. From: A. Autrique.

is nearer the average. The egg stage lasts 1–2 weeks in the field and the nymphal stages 7–14 weeks (Pict. 1.15).

Parasitized eggs turn a grey color. Adults and nymphs are attacked by hymenopterous parasites (*Braconidae*) and by parasitic flies (Tachinidae). Common predators include mantids (Mantidae) and assassin bugs (Reduviidae). In Tanzania, the bug is also attacked by *Corioxenos antestiae* Blair, a species of Strepsiptera (Pict. 1.16 and 1.17).

Control measures include the following:

1) Pruning to open the bush. Antestia Bugs prefer dense foliage.
2) Hand collection can be successful on small plots of coffee but not on large plantations. It is, however, essential to

Picture 1.17 Antestia Bug – *Antestiopsis orbitalis ghesquierei*. Above: healthy coffee beans. Below: damaged coffee beans. From: A. Autrique.

collect the eggs as well as the active stages. Collection should take place once a week. Smoldering plant material is carried in a metal tray and smoke is allowed to drift through the foliage. This causes the bugs to run to the center of the tree where they can be caught and dropped into a tin containing a little kerosene. Leaves with egg masses are also collected in, for example, a small basket. This is then hung in a tree 1–2 m upwind of coffee. Most egg parasites that breed through can then fly or be blown back to the coffee.

3) Spraying is necessary when the number of bugs per tree exceeds two (approximately 2500/ha). A sample of trees is sprayed with pyrethrum and dead or parasitized bugs are collected from

Picture 1.16 Antestia Bug – *Antestiopsis orbitalis ghesquierei*. Eggs hatching on leaf.
From: A. Autrique.

cloth sheets under the tree. Effective insecticides include fenitrothion, trichlorfon and diflubenzuron. Synthetic pyrethroids are also highly effective but tend to cause outbreaks of scale insects.

4) There are possibilities for classical biological control but much research would be needed, e.g. the parasite complex of *Antestiopsis* in South Africa is little known. Also there are nymphal parasites of closely related *Antestia* spp. in Asia and Madagascar that might be useful.

Further reading: Le Pelley, R. H. *Pests of Coffee*. Longmans, London, 1968.

Coffee Capsid Bug
Lamprocapsidea coffeae (China)
(Hemiptera: Heteroptera: Miridae)

This bug was for long considered a serious pest of Arabica coffee, especially in East Africa, but also in the Democratic Republic of Congo. Nowadays, however, apparently due to more intensive cultivation, it is only considered to be a very minor pest.

The bug sucks the flower bud causing death of petals and stamens that blacken. The style however continues to elongate causing the typical symptoms of a black head on a pale green shaft (Pict. 1.18).

The adult bug is greenish-brown in color and about 6 mm long. The posterior halves of the wings are folded sharply downwards covering the abdomen. The eggs are laid embedded in the flowers. There are five, pale-green nymphal instars; the nymphal period lasting 2–3 weeks.

Other mirids causing similar damage are *Taylorilygus* spp. in Uganda and the Democratic Republic of Congo, and *Volumnus obscurus* Poppius, especially in West Africa. If flower buds are few in number then a population of four bugs per tree (5000/ ha) will merit spraying. However, if the

Picture 1.18 Coffee Capsid Bug – *Lamprocapsidea coffeae*. Damage to flower buds. From: T. J. Crowe.

flowering is excessive and liable to lead to overbearing, a much higher number of bugs can be tolerated.

All the chemicals recommended for Antestia Bug are effective, mostly at a somewhat lower dosage (Pict. 1.19).

Picture 1.19 Coffee Capsid Bug. Adult bugs. Left: *Lamprocapsidea coffeae*. Right: *Volumnus obscurus*. From: A. Autrique.

Further reading: Le Pelley, R. H. Lygus, a pest of coffee in Kenya Colony. *Bull. Ent. Res.* **1932**, *23*, 85–99.

Coffee Lace Bugs
Habrochila spp.
(Hemiptera: Heteroptera: Tingidae)

The taxonomy of *Habrochila* spp. is in a confused state. There are at least two species, *H. placida* Horvath and *H. ghesquierei*. Schouteden, which are coffee pests, and possibly one or two more species. Both Arabica and Robusta coffee are attacked (Pict. 1.20).

Picture 1.21 Lace Bug – *Habrochila* sp. Damage (yellowing and spots of shiny black excreta) and the gregarious nymphs. From: T. J. Crowe.

Picture 1.20 Coffee Lace Bug – *Habrochila ghesquierei*. Typical damage on leaves. From: A. Autrique.

The damage is seen on the underside of leaves. The feeding area turns yellow and is dotted with shiny, black spots of excreta. In West Africa, Lace Bug attacks can be severe but are sporadic and local (Pict. 1.21).

In the main coffee area of Kenya east of the Rift Valley the pest was accidentally introduced in 1956 and for some years was very damaging. Recently it has been well controlled by natural enemies.

The adult bug is brown in color and 4 mm long. The wings have a lace-like pattern. Eggs are laid embedded in the leaves. There are five nymphal instars lasting 16–36 days. The nymphs feed gregariously on the underside of leaves (Pict. 1.22).

Picture 1.22 Lace Bug – *Habrochila ghesquierei*. Adult bug. From: A. Autrique.

The main controlling agents are voracious mirid predators *Stethoconus* spp. If *Stethoconus* nymphs are present in good numbers chemical control will not be needed.

The predator nymph has a whitish body tinged with red and dark antennae with a broad white band.

Organophosphate insecticides (e.g. fenitrothion) give good control. Synthetic pyrethroids are also effective but should only be spot-sprayed on badly infested trees.

Malagasy Lace Bug
Dulinius unicolor (Signoret)

This lace bug, sometimes called "tigre du caféier", is a serious pest of coffee.

Stethoconus frappae Carayon is an important predator of *Dulinius unicolor*. Use the same chemical control as for *Habrochila* sp.

Further reading: McCrae, D. J. The control of coffee lace bug. *Coffee Bd. Kenya Mon. Bull.* **1958**, *23*, 153–154.

1.7
Thrips

Only a few species of Thrips have been recorded as serious pests.

Coffee Thrips
Diarthrothrips coffeae Williams
(Thysanoptera: Thripidae)

For many years Arabica coffee in Kenya and northern Tanzania was regularly attacked in the hot dry season before the main rains. Nowadays, however, it is only a minor pest.

The adult Thrips is a very small (1–1.5 mm), slender, grey–brown insect. It feeds on the underside of leaves in company with the yellow cigar-shaped nymphs. The fresh damage is characteristic. There are silvery patches on the leaf dotted with very small, shiny black spots of excreta. These spots are much smaller than those produced by the Lace Bug.

It appears that only trees suffering from water stress are badly attacked. Since the general use of dried grass mulch (starting around 1955), outbreaks have become very rare. This, supplemented if possible with irrigation, should be the main control measure. In a hot, dry season, if the population exceeds two per leaf (adults plus nymphs), then spraying may be necessary.

Dicrotophos has given the best results, but, if a less toxic product is indicated, fenitrothion, fenthion and chlorpyrifos have proved satisfactory.

Panchaetothrips noxius Priesner has been recorded doing similar damage in the Democratic Republic of Congo and Uganda but little is known of its biology. The eggs are heavily parasitized by a *Plegaphragma* sp. (Hymenoptera: Trichogrammatidae) that may prove to be a useful biocontrol agent in other countries.

Further reading: Evans, D. E. Insecticide field trials against the coffee thrips, (*Diarthrothrips coffeae* Williams) in Kenya. *Turrialba* **1967**, *17*, 376–380.

Leaf Rolling Thrips
Hoplandothrips spp.
(Thysanoptera: Phlaeothripidae)

Three species of *Hoplandothrips* are recorded from coffee – *H. bredoi* Priesner as a minor pest in the Democratic Republic of Congo, *H. marshalli* Karny as a more serious pest in Uganda and the Democratic Republic of Congo, and *H. coffeae* Bagnall as a minor pest in western Tanzania (Pict. 1.23).

The adults are blackish and about 2 mm long. The larvae are yellowish. Feeding on young leaves causes them to develop tightly rolled with loss of photosynthetic area, but crop loss is usually negligible. Very severe attacks have been controlled with sprays of dimethoate (Pict. 1.24).

Picture 1.23 Leaf Rolling Thrips – *Hoplandothrips marshalli*. Damage on leaves. From: A. Autrique.

1.8
Wood-boring Beetles

Generally speaking the species of Coleoptera that bore into coffee trees prefer weak or droughted ones. Good standards of cultivation, the use of mulches, planting of multiple stem coffee trees with hard wood and irrigation will do much to reduce the potential for damage (Pict. 1.25).

Picture 1.24 Leaf Rolling Thrips – *Hoplandothrips marshalli*. Adult Thrips. From: A. Autrique.

Picture 1.25 Wood-boring beetles. Old Robusta single stem tree infested by borers. Observe abundant warts and nodules along the trunk as reaction to the wounds caused by the boring beetles. From: J. N. Wintgens.

Taeniothrips xanthocerus (Hood) and *T. antennatus* Bagnall are found in pustules of Coffee Leaf Rust, apparently sucking out the contents of the spores. (Their gut contents are orange). They may be of some importance in spreading the disease.

Black Borers
Apate monachus Fabricius and other *Apate* spp.
(Coleoptera: Bostrychidae)

Picture 1.26 Black Borer –
Apate monachus. Adult insect.

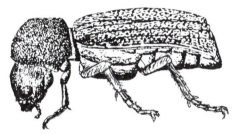

Figure 1.4 Black Borer – *Apate monachus*.
Adult insect. Observe the head concealed below
the thorax.

like frass stops coming out of the tunnel
entrance after a few days, the beetle has
been killed.

Coffee Twig Borer
Xylosandrus compactus
(Synonym: *Xyleborus morstatti*)
(Coleoptera: Scolytidae)

This beetle is a very serious pest, especially
of Robusta coffee. It occurs in many Asian
countries as well as tropical Africa south to
Zimbabwe. Apart from coffee it is a pest of
avocado pear, cocoa, mango and tea. It tunnels out small galleries in twigs, choosing
those 1–2 cm in diameter. Wilting and
death of the twig soon follow (Pict. 1.27).

A whole brood of the beetle lives in the
gallery, feeding, not on the wood itself,
but on fungi growing on the gallery walls.

These beetles live on forest trees where
they complete their life cycle. Sometimes
adults fly into coffee plantations and bore
tunnels about 20 cm long upwards in a
main stem. Both Arabica and Robusta coffee are attacked. The black beetles are
about 2 cm long, rather square at the anterior end and with the head concealed below
the thorax (Pict. 1.26 and Fig. 1.4).

Control is by pushing a springy wire (e.g.
a bicycle spoke) into the hole. If sawdust-

Picture 1.27 Coffee Twig Borer – *Xylosandrus compactus*. Infested twig. Observe hole communicating
with the gallery bored by the beetle.
From: Nestlé.

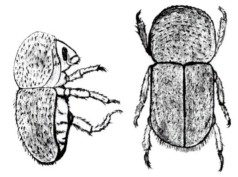

Figure 1.5 Coffee Twig Borer – *Xylosandrus compactus.*

The dark brown males are only about 2 mm long and are flightless. The females are 4–5 mm long. Eggs are laid in the gallery. There are three larva instars and the pupal stage, all found in the gallery. One generation lasts 20–40 days (Fig. 1.5).

No effective control methods have been developed for the pest. Pruning of infested twigs and other cultural methods have proved useless. Some control has been achieved by spraying a mixture of dieldrin and a fungicide.

A certain varietal resistance has been observed in Robusta according to B. Decazy.

Further reading: Entwistle, P. F. In: *Diseases, Pests and Weeds in Tropical Crops,* J. Kranz, H. Schmutterer and W. Koch (eds). Paul Parey, Berlin, 1977.

White Coffee Borer
Monochamus leuconotus (Pascoe)
(Synonym*: Anthores leuconotus*)
(Coleoptera: Cerambycidae)

Until about 1966 this was an extremely serious pest of coffee, especially Arabica coffee, in eastern Africa from Kenya to South Africa. Apart from coffee, it attacks many wild shrubs in the coffee family Rubiaceae.

The adult is a typical "longhorn" beetle up to 30 mm long. It is grey in color, but the elytra are mainly white. The antennae of the female are slightly longer than the body. Male antennae are much longer (Pict. 1.28).

Picture 1.28 White Coffee Borer – *Monochamus Leuconotus.* Adult insect. From: A. Autrique.

Eggs are laid in the bark. Young larvae burrow in the inner layers of the bark. Older larvae bore into the heart wood. Larval feeding can cause ring-barking leading to yellowing and death in young trees (Pict. 1.29).

Picture 1.29 White Coffee Borer – *Monochamus Leuconotus.* Ring-barking damage by larvae. From: A. Autrique.

Picture 1.30 White Coffee Borer – *Monochamus Leuconotus*. Tunnels excavated by larvae in coffee trunk. From: A. Autrique.

Picture 1.31 White Coffee Borer – *Monochamus Leuconotus*. Bundles of wood tissue (frass strands) produced by larva. From: A. Autrique.

Yields in mature trees can be reduced to totally uneconomic levels. Pupation takes place in a prepared chamber just under the bark. The total life cycle takes 1–2 years (Pict. 1.30 and 1.31).

Highly effective control was achieved in the 1960s by painting a 0.5 % dieldrin emulsion on the trunk from ground level up to 45 cm. Nowadays, even without chemical treatment, the pest is rare due probably to increased use of mulch and irrigation in recent years.

Further reading: Tapley, R. G. The white coffee borer *Anthores leuconotus* Pascoe and its control. *Bull. Ent. Res.* **1966**, *51*, 279–301.

West African Coffee Borer
Bixadus sierricola (White)
(Coleoptera: Cerambycidae)

This beetle that attacks both Arabica and Robusta coffee is a severe pest in West and Central Africa as far east as Uganda. Its biology is very similar to White Coffee Borer and the same dieldrin treatment gives effective control (Pict. 1.32).

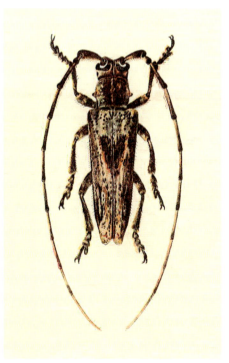

Picture 1.32 West African Coffee Borer – *Bixadus sierricola*. Adult male.

Yellow-headed Borer
Dirphya nigricornis (Olivier)
(Coleoptera: Cerambycidae)

This beetle is a minor pest, mostly of Arabica coffee in eastern Africa.

Eggs are laid at the tips of fruiting branches that then wilt and die. The larva bores inwards to the main stem and then downwards towards the root. As it bores, it makes a series of flute-like holes along the branch and down the side of the trunk. These are used to eject the sawdust-like frass. Pupation takes place in the trunk. The whole life cycle takes 1 year.

Control is by cutting of affected primary branches before the larvae reach the main stem. If the main stem has already been reached, then the lowest frass ejection hole should be enlarged and a little persistent insecticide squirted into the tunnel with an oil can or other suitable applicator.

The closely related *Neonitocris princeps* (Jordan) does similar damage in West Africa.

Recurrent attacks of Yellow-headed Borer only occur in areas where there are numerous shrubs of the coffee family Rubiaceae growing near the coffee. A campaign to eradicate these shrubs may be necessary if the pest is particularly troublesome (Fig. 1.6, 1.7 and Pict. 1.33).

Further reading: Crowe, T. J. The biology and control of *Dirphya nigricornis* (Olivier), a pest of coffee in Kenya. *J. Ent. Soc. Afr.* **1962**, *25*, 304–312.

Coffee Stem Borer
Ancylonotus tribulus (Fabricius)
(Coleoptera: Cerambycidae)

This borer, widespread in West Africa and the Democratic Republic of Congo bores in coffee stems causing serious damage to the plant (Fig. 1.8).

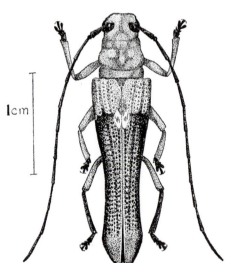

Figure 1.6 Yellow-headed borer –
Dirphya nigricornis. Adult beetle. From: T. J. Crowe.

Picture 1.33 Yellow Headed Borer –
Neonitocris princeps. Adult beetle.

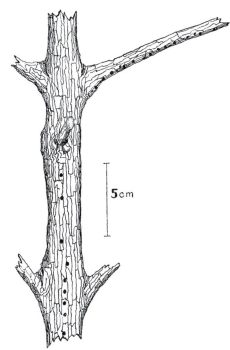

Figure 1.7 Yellow-headed Borer – *Dirphya nigricornis*. Coffee stem and primary branch showing frass-ejection hole made by the borer.
From: T. J. Crowe.

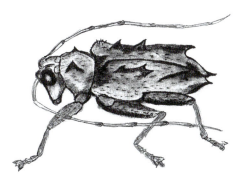

Figure 1.8 Coffee Stem Borer. *Ancylonotus tribulus* (Cast). Adult insect. Observe spiny protuberances on thorax and body.

1.9
Coffee Berry Borer

Coffee Berry Borer
Hypothenemus hampei (Ferrari)
(Coleoptera: Scolytidae)

This is an indigenous African beetle, often a serious pest of Robusta and Arabica coffee grown at low altitudes.

The female beetle is dark brown and about 2.5 mm long. The males are only about 1.5 mm and are outnumbered by females by a ratio of 10 : 1 (Pict. 1.34).

The female cuts a characteristic clean circular hole at the blossom end of large green berry and begins to tunnel inside (Pict. 1.35).

Picture 1.34 Coffee Berry Borer – *Hypothenemus hampei*. Adult beetle. From: Nestlé.

Picture 1.35 Coffee Berry Borer – *Hypothenemus hampei*. Holes in the tip of the berries opened by female borers. From: A. Autrique.

Picture 1.36 Coffee Berry Borer – Hypothenemus hampei. Coffee berries cut open to show larvae, adult insects and damage to the berries. From: A. Autrique.

Picture 1.37 Coffee Berry Borer – *Hypothenemus hampei*. Damaged beans. From: Nestlé.

Eggs are laid in the berry. The white, legless larvae eat out the bean tissue. There may be up to 20 larvae in a single berry. Many berries fall prematurely. The naked white pupae are also found in the tunnel (Pict. 1.36).

The total life cycle takes 25–35 days. Between peaks of fruiting, the infestation is carried over by breeding in overripe berries either on the tree or fallen to the ground. Females can also survive by eating young berries.

The following cultural measures, if conscientiously applied, do much to reduce the infestation:

1) Reduce heavy shade.
2) Prune the coffee to keep the bush as open as possible.
3) Picking should take place at least once a week in the main harvest season and once a month at other times.
4) Berries should be left on the ground as little as possible; if the coffee is mulched, it may be necessary to spread sheets under the trees during picking.
5) All infested berries should be destroyed by burning, deep burying or, if possible, rapid sun-drying.
   The berry disposal unit described below under Coffee Berry Moth could probably be modified to deal with beetle-in-

fested berries. The mesh size would, however, have to be reduced so that female beetles cannot pass through. The (smaller) males can be ignored.
6) Before a main flowering the crop should be stripped completely.

Chemical control should only be used as a supplement to cultural measures. Among the chemicals recommended are carbosulfan, chlorpyriphos and endosulfan. Regular spraying is not recommended due to the danger of causing outbreaks of other pests (Pict. 1.37).

Further reading: Le Pelley, R. H. In: *Diseases, Pests and Weeds in Tropical Crops*, J. Kranz, H. Schmutterer and W. Koch (eds). Paul Parey, Berlin, 1977.

1.10
Leaf-eating Beetles

Systates Weevils
Systates spp.
(Coleoptera: Curculionidae)

The adult weevil lives in the soil or mulch and climbs the tree at night causing characteristic damage to the edges of the leaves (Fig. 1.9).

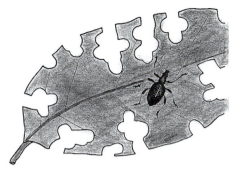

Figure 1.9 Systates weevil – *Systates* sp. Adult weevil on damaged leaf. From: T. J. Crowe.

Coffee Chafers
Pseudotrochalus spp.
(Coleoptera: Scarabaeidae)

At least two species of *Pseudotrochalus* are recorded in West Africa. The larvae (Chafer

Picture 1.38 Coffee Chafer Beetle – *Pseudotrochalus schubotzi*. Adult beetle and damage on coffee leaves. From: A. Autrique.

Picture 1.39 Heavy attack of Coffee Chafer Beetle *Pseudotrochalus* sp. on coffee tree. From: A. Autrique.

Grubs) feed on coffee roots and other vegetable matter. The adults climb the tree at night and eat leaves. No effective control for these pests is known (Pict. 1.38 and 1.39).

1.11
Fruit Flies

Larvae of several species of fruit fly are to be found feeding in the pulp of ripe coffee cherries. The commonest are:

Medfly, Mediterranean Fruit Fly
Ceratitis capitata (Wiedemann)
(Diptera: Tephritidae) (Fig. 1.10)

Coffee Fruit Fly
Trirhithrum coffeae Bezzi
(Diptera: Tephritidae)

The white maggot-like larvae of *C. capitata* are particularly common in Arabica coffee. There is no direct damage to the coffee cherry, but the fly has been suspected of introducing organisms into the pulp that cause an undesirable flavor when the coffee is brewed. After much investigation, this is now considered highly unlikely. Very occasionally the female flies lay their eggs in young green berries. No larvae develop, but the stung berries fall prematurely. This damage is not predictable and no control measures are possible.

T. coffeae is the commonest species on Robusta coffee. The closely related *T. inscriptum* (Graham) attacks Arabica coffee in the Democratic Republic of Congo. It is said to introduce a bacterium into the pulp that causes the undesirable "potato flavor" in the liquored coffee (Pict. 1.40 and 1.41).

Further reading: White, I. M. and Elson-Harris, M. M. *Fruit Flies of Economic Significance*. CAB International, Wallingford, 1992.

Figure 1.10 Medfly – *Ceratitis capitata* (1–3). Mexican Fruit Fly – *Anastrepha ludens* (4–7). From: Bayer.

Picture 1.40 Coffee Fruit Fly – *Trirhithrum inscriptum*. Adult fly. From: A. Autrique.

Picture 1.41 Coffee Fruit Fly – *Trirhithrum coffeae*. Larva localized in the mucilage. From: A. Autrique.

1.12
Lepidopterous Leaf Miners

Coffee Leaf Miners; Coffee Blotch Miners
Leucoptera caffeina Washbourn
Leucoptera meyricki (Ghesquière)
(Lepidoptera: Lyonetiidae)

These moths are major pests of Arabica coffee especially in eastern Africa from Ethiopia to South Africa.

L. meyricki and *L. caffeina* have both been recorded attacking the indigenous wild coffee *Coffea eugenioides* and other shrubs of the coffee family *Rubiaceae*, and this was presumably the source of the present infestation on cultivated coffee.

The larvae (caterpillars) produce brown, irregular, blotch-like brown areas on the upper surface of fully hardened leaves (Pict. 1.42).

This is in contrast to dipterous miners which produce a silvery serpentine mine.

When the fruiting branches of the coffee bush are bearing a heavy crop it is essential that these branches carry an adequate number of leaves to feed that crop through to harvest.

If the leaves fall prematurely, the tips of the branches will die causing the condition called "overbearing die-back". *This die-back causes corresponding death in the root system* and the tree will not only produce a poor crop in the current season, but also in one or more subsequent seasons. Leaf Miner damage (like Leaf Rust) tends to induce premature leaf fall and that is the reason for its major economic importance. *L. meyricki* is a serious pest particularly in Kenya, Tanzania and Malawi. At the ends of its range in Ethiopia and South Africa it is only a minor pest.

L. caffeina is particularly associated with shaded coffee. Probably for this reason it is the more important species in Ethiopia, but is less important than *L. meyricki* now in other countries following the removal of many shade trees (Pict. 1.43).

The eggs are shaped like inverted oval pie-dishes. Fresh eggs are silver in color when first laid and just visible to the naked eye being about 0.3 mm long. They are laid on the upper surface of fully hardened leaves. *L. caffeina* eggs are laid touching in a line along a main vein in groups averaging six.

L. meyricki eggs are laid in small groups usually around four or five, not touching and orientated at random. For *L. caffeina* and *L. meyricki* hatching takes place after 5–15 days in the temperature range 24–17 °C.

Picture 1.42 Coffee Leaf Miner – *Leucoptera meyricki*. Typical damage to the leaves. From: A. Autrique.

Picture 1.43 Coffee Leaf Miner – *Leucoptera caffeina*. Typical damage on coffee leaf. Green areas occur where the palisade tissue has been recently eaten. In the brown area, the epidermis has died. The mine originated from eggs laid on the main vein of the leaf. From: T. J. Crowe.

The larva bores through the base of the egg and the leaf epidermis, and feeds in the palisade tissue. Some excreta is left in the egg shell causing hatched eggs to appear brown. If a fresh, active mine is examined, three zones can be distinguished:

1) A central dark brown area where the epidermis has died and black excreta has been pasted on its lower side.
2) A narrow lighter brown zone where the epidermis is dying but no excreta has been deposited.
3) A narrow green zone on the outside of the blotch where the palisade tissue is still alive.

The mature larva is a flattened, whitish caterpillar with a dark head and distinct segmentation (Fig. 1.11).

When healthy, the gut contents are green and can be clearly seen through the translucent body.

L. meyricki larvae average about 5.5 mm long; *L. caffeina* larva about 6 mm.

After a larval period ranging from about 20 days at 20 °C to 34 days at 17 °C, the full grown larva cuts a semi-circular slit in the upper surface of the mine and, around mid-day, swings down on a silk thread seeking pupation quarters on the underside of one of the leaves or on dead leaves on the ground (Pict. 1.44).

Here the caterpillar spins an H-shaped white silk cocoon in which it pupates.

Picture 1.44 Coffee Leaf Miner – *Leucoptera meyricki*. Full-grown larvae leaving the mine. From: A. Autrique.

Picture 1.45 Coffee Leaf Miner – *Leucoptera meyricki*. Cocoon on the underside of a coffee leaf. From: T. J. Crowe.

The pupal period varies from 7 days at 20 °C to 14 days at 17 °C (Pict. 1.45 and 1.46).

The adult of both species is a tiny white moth *L. meyricki* moths are about 2 mm

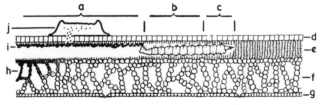

Figure 1.11 Coffee Leaf Miner – *Leucoptera meyricki*. Diagram of a section through a coffee leaf showing a Leaf Miner caterpillar and parts of its mine. (a) Dark central zone of mine. (b) Light-brown zone. (c) Green zone. (d) Upper epidermis. (e) Palisade tissue. (f) Spongy mesophyll. (g) Lower epidermis with stomata. (h) Dying mesophyll. (i) Larval excreta. (j) Empty eggshell. From: T. J. Crowe.

Picture 1.46 Coffee Leaf Miner – *Leucoptera meyricki*. Enlarged view of cocoon on the underside of a coffee leaf. Observe the larvae inside the cocoon. From: A. Autrique.

Picture 1.47 Coffee Leaf Miner – *Leucoptera caffeina*. Adult moth. Caterpillar. From: Bayer.

days after emergence. The number of eggs laid is variable depending on the quality of the food. Leaves rich in starch appear to give rise to particularly fecund females. Also, adult moths, which feed on honeydew, produce many more eggs than those that feed only on water. A typical moth may lay about 75 eggs.

Each year there is a Leaf Miner season lasting some seven to nine months. During this season there may be six peaks in the moth population. Spraying should take place 1 week after a moth peak. This is because the spray will kill eggs, larvae and moths but not the pupae in their silk cocoons. The moth population can be checked by shaking a sample number of trees and estimating how many moths flutter out. If the moths average somewhere between 20 and 40 per tree (30 000–60 000/ha), then spraying will probably be needed. The exact threshold level will depend on the number of leaves present and the level of crop so has to be worked out locally.

Samples of cocoons should also be squeezed out regularly with the thumb. If they produce a drop of yellow fluid, they still contain a fresh pupa and spraying on that day will not be effective.

On a large coffee plantation many of the small moths may be carried in air currents and dropped upwind of lines of trees or bushes at the edge of the farm. (The same can happen with Coffee Thrips and other very small flying insects.) Thus, next to windbreaks there may be so-called "hot spots" where the infestation comes earlier and is more severe than in other parts of the plantation. The farmer should try to identify these "hot spots" and give them special attention when scouting for Leaf Miner moths. Sometimes they can be spot-sprayed without treating the rest of the farm.

long and have dark wing tips. *L. caffeina* moths are a little larger and have a dark spot near the wing tip (Pict. 1.47).

The female moth lives about two weeks laying most of her eggs on the first four

Organophosphate insecticides (e.g. fenitrothion) give the best control. Synthetic pyrethroids kill the pest, but are liable to *lead to infestations of other pests.*

Before application a sample of fully grown coffee leaves should be dipped in the prepared spray mixture. When the leaves are removed, a film of spray liquid should remain covering most of the leaf surface.

If, on the other hand when they are removed, the liquid film breaks up immediately into discrete drops, then addition of more wetting agent (surfactant) to the spray mixture will increase its efficiency.

Insecticide granules applied to the soil and watered in, offer a convenient alternative to spraying. Aldicarb, carbofuran and disulfoton are all effective, but their very high mammalian toxicity makes them unsuitable for use except where very close supervision is possible.

There are possibilities for classical biological control using parasites of the closely related South American Leaf Miner *Perileucoptera coffeella*, but this would need extensive research (Pict. 1.48).

Picture 1.48 Coffee Leaf Miner – *Leucoptera caffeina*. Biological control. An old empty Leaf Miner mine. The Leaf Miner caterpillars were attacked by natural parasites and killed before emergence. Two circular holes can be seen where adult natural parasites have emerged. From: T. J. Crowe.

Further reading: Crowe, T. J. Coffee leaf miners in Kenya. Species and life histories. *Kenya Coffee* **1964**, *29*, 173–178; Causes of outbreaks. *Kenya Coffee* **1964**, *29*, 222–227; Control measures. *Kenya Coffee* **1964**, *29*, 261–265.

Leucoptera coma Ghesquière
(Lepidoptera: Lyonetiidae)

This Leaf Miner is a common pest of Robusta coffee in the Democratic Republic of Congo. The life cycle is very similar to that of the *Leucoptera* spp. described above, except that the eggs are laid in groups of over 20. Similar control measures are effective.

1.13
Berry-boring Lepidoptera

Coffee Berry Moth
Prophantis smaragdina (Butler)
(Lepidoptera: Pyralidae)

This pest was for many years regarded as a minor pest of Arabica coffee in eastern Africa (from Ethiopia to South Africa). In recent years, however, damage has become much more serious, especially on intensively cultivated coffee which is receiving high rates of fertilizer and mulch. Economic loss is not only due to the destruction of berries and the labor costs of stripping, but also because larval damage allows penetration of the Coffee Berry Disease fungus and the silk webbing prevents the efficient deposition of fungicide.

No wild hosts have been recorded.

Eggs are laid singly usually on green berries. They are scale-like and translucent and difficult to see in the field. They hatch after about 1 week.

The reddish larvae graze flowers and pinhead berries. On older berry clusters, the larvae eats out first one berry and then moves to another. Meanwhile, the whole cluster is webbed together with silk. The berries that were eaten first turn brown. The symptoms of attack are, therefore, berry clusters that are webbed together with silk, one or more of the berries being brown and hollow (Pict. 1.49).

After about 2 weeks the caterpillar is 1.3 cm long. At night, it emerges from the berry cluster and drops to the ground where it pupates between two leaves tied together with silk. The pupal stage is vari-

Picture 1.49 Coffee Berry Moth – *Prophantis smaragdina*. A berry cluster attacked by Berry Moth. Observe the silk. From: A. Autrique.

Picture 1.50 Coffee Berry Moth – *Prophantis smaragdina*. Mature caterpillar seeking pupation quarter. From: T. J. Crowe.

able from 1 to 6 weeks or more, apparently depending on soil moisture (Pict. 1.50).

The adult is a golden-brown, night-flying moth.

The control of Berry Moth is very difficult. The larvae are susceptible to insecticides but, to be effective, sprays have to be applied just before or just after hatching. Unfortunately, both eggs and young larvae are difficult to detect in the field. Once the berry cluster is webbed, the spray will not penetrate inside it.

Early warning of an attack can be achieved by scouting for moths at, or soon after, flowering. For example two Field Assistants, starting in the cool of the morning, can search for adult moths for a fixed period (e.g. for 1 h). Moths are mostly found on the lower branches on the underside of leaves. It is essential to fix the time of counting since moths in the heat of midday are more difficult to find. If two technicians can find more than 50 moths in 1 h further counts should be made. Berries on random branches should be checked for eggs and pin-head berries checked for signs of grazing. Many insecticides have proved effective including chlorpyrifos, deltamethrin, fenitrothion and trichlorfon.

According to B. Decazy, endosulfan is also efficient against the Coffee Berry Moth, even when the larvae are inside the berries.

There are indications that certain *Bacillus thuringiensis* formulations also kill the young larvae and, if this is confirmed, this would be the best product. The development of a sex pheromone for this pest would be a great help because it would make it easier to time the sprays. Unfortunately nothing has been done in this field up to now (B. Decazy).

If crop damage has already occurred, the infested berry clusters should be stripped off by hand. The most efficient way of disposing of the strippings is as follows. Dig a hole like a pit latrine. Cover it, as usual, with a concrete slab. In the concrete slab, set two structures, a manhole cover or similar "door" and a small "window". The window should be closed with 2-mm stainless steel mesh. The strippings are put in the pit through the manhole cover that is otherwise kept closed. Moths and parasite wasps breed through in the pit, but the moths are trapped in the pit and die. Adult parasites, being smaller, pass through the 2-mm mesh and return to the coffee bushes.

Coffee Berry Butterfly, Coffee Playboy
Deudorix lorisona Hewitson
(Synonym: *Virachola bimaculata*)
(Lepidoptera: Lycaenidae)

This pest has caused sporadic locally, but serious damage to all species of coffee throughout tropical Africa and down to Mozambique in the east.

Eggs are like minute sea urchins stuck to the side of green berries. The larva is a green and brown caterpillar that feeds inside green berries but drops its excreta outside. It grows to a length of 20 mm. Single holes are bored in large green berries and both beans are usually eaten. As the berry dies and turns brown, the edge of the hole turns up like a rim.

It appears that there are only serious attacks when there are numerous shrubs of the coffee family Rubiaceae growing near the coffee. *Heinsia pulchella* is an important host plant recorded from Sierra Leone (Fig. 1.12).

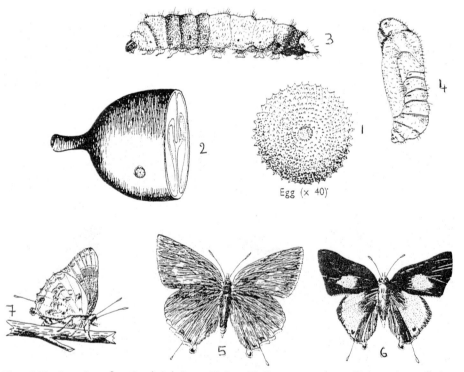

Figure 1.12 Berry Butterfly – *Deudorix lorisona*. (1) Egg. (2) Egg on green berry. (3) Larva (caterpillar). (4). Pupa. (5) Female butterfly. (6) Male butterfly. (7) Butterfly in natural position. From: T. J. Crowe.

If these are eradicated, then attacks usually subside.

Like Berry Moth, once the caterpillar is in a large berry, chemical control is not effective. If an attack is expected, a scouting system based on egg counts is possible. Chemicals that control Berry Moth are also effective against this pest.

Many other species of Lepidoptera have berry-boring caterpillars, including the Coffee Tortrix mentioned below but serious attacks are very rare.

1.14
Leaf-eating Caterpillars

The larvae of many species of Lepidoptera can cause serious damage to coffee by eating the leaves. The attacks, however, are usually localized and sporadic. Among the commoner species are the following:

Jelly Grub, Gelatine Caterpillar
Niphadolepis alianta Karsch
(Lepidoptera: Limacodidae)

This species is recorded from Kenya, Malawi and Tanzania as a minor pest of tea and Arabica coffee. The adult moth lays scale-like eggs on the leaves. The caterpillars are oval, smooth and slug-like. The color is very pale blue or green. Young caterpillars make small shot holes in the leaf; older ones feed at the leaf edge. After some 6–8 weeks, when they are about 1.3 cm long, they pupate between two leaves or a folded leaf (Fig. 1.13).

Further reading: Smee, C. Gelatine grub on tea in Nyasaland. *E. Afr. Agric. J.* **1939**, *5*, 134–142.

Stinging Caterpillar, Nettle Grub
Parasa vivida (Walker)
(Synonym: *Latoia vivida*)
(Lepidoptera: Limacodidae)

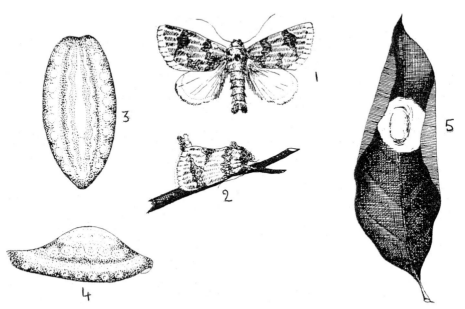

Figure 1.13 Jelly Grub – *Niphadolepis alianta*. (1) Moth. (2) Moth in natural position. (3) Larva (caterpillar), top view. (4) Larva (caterpillar), side view. (5) Cocoon in folded leaf. From: C. Smee.

This species is recorded as a sporadic pest of Arabica coffee in most East African countries and also the Ivory Coast.

The adult moth lays groups of scale-like eggs (overlapping like tiles) on the underside of leaves. Young caterpillars are white and feed gregariously making small "windows" in the leaf. Mature caterpillars are green and feed singly at the leaf edge. All stages bear short, finger-like projections covered with stinging hairs. Pupation takes place in a very tough, white cocoon often found on the tree trunks.

In addition to chemical control (discussed below), on small plots of coffee, it is useful to destroy pupae with nail or knife (Fig. 1.14).

Apart from the leaf damage, the stinging hairs on this caterpillar can cause pain and distress to coffee pickers.

Further reading: Smee, C. Leaf-eating caterpillars on tea. *Nyasaland Tea Ass. Q. J.* **1939**, *3*, 1–8.

Green Coffee Tortrix
Archips occidentalis (Walsingham)
(Lepidoptera: Tortricidae)

Red Coffee Tortrix
Tortrix dinota Meyrick
(Lepidoptera: Tortricidae)

Sporadic attacks of these two caterpillars are common on Arabica coffee in Kenya. Both species are polyphagous and are distributed right across Africa.

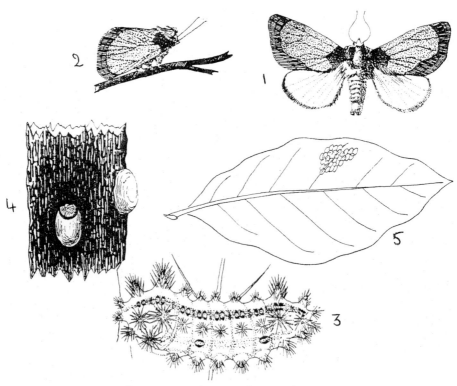

Figure 1.14 Stinging Caterpillar – *Parasa vivida*. (1) Moth. (2) Moth in natural position. (3) Larva (caterpillar). (4) Cocoon on stem of coffee bush. (5) Egg mass on leaf. From: T. J. Crowe.

The eggs are laid overlapping like fish scales in masses of more than 200 eggs on the upper side of leaves.

The larvae that grow to a length of 2.5 cm, feed on soft leaves, webbing them together with silk and also sometimes bore into berries. Damage to sucker tips can be particularly troublesome.

The caterpillars are difficult to control chemically but trichlorfon is effective.

Further reading: Evans, D. E. Coffee tortrix caterpillars. *Kenya Coffee* **1968**, *33*, 195–197.

Coffee Feeding Pyralid
Dichocrosis crocodora (Meyrick)
(Lepidoptera: Pyralidae)

This pest is a pest of Robusta coffee, especially in the Democratic Republic of Congo, but also in Ivory Coast.

Eggs are laid in large masses usually on the underside of the leaf. Young caterpillars feed communally in partly folded leaves. Later they live singly in the shelter of rolled leaves moving to new ones when one is skeletonized (Pict. 1.51).

Further reading: Schmitz, G. La pyrale du caféier Robusta. *Dichocrosis crocodora*. *Bull. Agric. Congo Belge* **1948**, *39*, 571–580.

Tailed Caterpillars
Epicampoptera andersoni Tams
Epicampoptera pallida Tams
Epicampoptera strandi Bryk
Epicampoptera marantica Tams
(Lepidoptera: Drepanidae)

These caterpillars are sporadically serious defoliators of coffee throughout tropical Africa, especially of plantings next to rain forest. *E. andersoni* is recorded from Ivory Coast, western Kenya and the Democratic Republic of Congo; *C. marantica* from Tanzania, Uganda and the Democratic Republic of Congo (Pict. 1.52).

Robusta coffee is the most adversely affected, but severe attacks on Arabica are also possible.

The brown caterpillars have a swollen, humped thorax and a thin tail-like extension of the abdomen. Young larvae make "shot holes" in the leaves; older ones feed at the leaf edge. Pupation takes place in a rolled leaf.

Picture 1.51 Coffee Feeding Pyralid – *Dichocrosis crocodora*. (1) Adult insect. (2) Mass of eggs. (3) Larvae of different stages. (4) Caterpillars in rolled and spun leaf. From: Bayer.

Picture 1.52 Tailed Caterpillar – *Epicampoptera andersoni*. Larvae feeding on leaves. (5) Butterfly. (6) Eggs, larvae of different stages. Above: adult caterpillar with "rat's tail". From: Bayer.

Further reading: Notley, F. B. Leaf eating caterpillar of coffee. *E. Afr. Agric. J.* **1935**, *1*, 119–126.

Giant Coffee Looper

Ascotis selenaria (Denis and Schiffermüller) (Lepidoptera: Geometridae)

This pest came into prominence in Kenya in 1961 following excessive spraying with parathion. It is highly polyphagous, but a strain that prefers Arabica coffee seems to have been inadvertently selected and it continues to cause sporadic loss despite general reduction in the use of sprays (Pict. 1.53).

Eggs are laid in bark crevices. The larva moves with the typical looping motion of a geometric. It is brown and twig-like and grows to a length of 5 cm. Pupation takes place in the soil.

Picture 1.53 Giant Looper – *Ascotis selenaria.* Caterpillar and moth. From: T. J. Crowe.

Further reading: Wheatley, P. E. The giant coffee looper *Ascotis selenaria reciprocaria. E. Afr. Agric. For. J.* **1963**, *29*, 143–146.

Coffee Leaf Skeletonizer

Leucoplema dohertyi Warren (Lepidoptera: Epiplemidae)

This pest is recorded as a minor pest of Arabica and Robusta coffee in Kenya, Uganda and the Democratic Republic of Congo.

Eggs are laid on the underside of leaves, often on patches of old skeletonizer damage. The caterpillars eat everything except the veins and upper epidermis leaving irregular, lace-like patches of damage on the leaves (Pict. 1.54).

The caterpillars, which grow to about 8 mm in length, turn from grayish white to red before pupating in the soil.

Picture 1.54 Coffee Leaf Skeletonizer – *Leucoplema dohertyi.* (1) Moth. (2) Larva. (3) Injury on leaf. From: Bayer.

Further reading: Crowe, T. J. The leaf skeletonizer. *Kenya Coffee* **1960**, *5*, 256–257.

Bee Hawk Moth

Cephonodes hylas (Linnaeus) (Lepidoptera: Sphingidae)

Picture 1.55 Bee Hawk Moth – *Cephonodes hylas*. Above: caterpillar (6) and injury on leaves (7). Below: moth (5). From: Bayer.

This species is a sporadically serious defoliator of coffee in West Africa, Uganda, Tanzania and Central Africa. Its range extends through Malaysia to Japan and Australia. All species of coffee are liable to attack.

The caterpillar grows to a length of 4 cm. Like all Sphingidae, it has a horn-like mid-dorsal projection on the eighth (final) abdominal segment (Pict. 1.55).

Further reading: Corbett, G. H. and Yusope, M. The coffee clear wing hawk moth (*Cephonodes hylas L*). *Malay. Agric. J.*, *20*, 508–517.

Control measures for leaf-eating caterpillars include:

- When assessing the need to spray it is important to remember that, unlike *Leucoptera* leaf miners, the caterpillars do not tend to cause premature leaf fall. The damage is simply the loss of functional leaf area.
- If possible, the caterpillars should be sprayed when still young. This is because not only are they easier to kill than full grown ones, but also damage increases logarithmically as they grow.
- Natural control is normally by hymenopterous parasitoids and/or tachinid flies. Insecticide application should, therefore, be as selective as possible to preserve some of the parasite population. The insecticides of choice are either a *Bacillus thuringiensis* formulation – preferably one known to be effective against Lepidoptera – or a feeding repellent such as fentin hydroxide. Otherwise a broad-spectrum, non-persistent insecticide of low mammalian toxicity should be used.
- Only heavily infested trees should be sprayed. Lightly infested ones will act as a parasite reservoir. Similarly, if the infestation is largely confined to sucker growth, only the suckers should be spot-sprayed.

1.15
Ants

Ants are important in coffee plantations:

1) As attendants to scale insects and Mealybugs as described above;
2) As direct predators of pests, e.g. mature *Leucoptera* larvae may be eaten while seeking pupation quarters on the ground;
3) As nuisance to human beings, e.g. the biting ants *Macromischoides aculeatus* (Mayr) and, to a lesser extent, *Oecophylla longinoda* Latreille make their nests in Robusta coffee trees and attack pickers and pruners.

If practicable, the nests should be treated individually. An oil can with a long, sharp-pointed spout is charged with a persistent insecticide and a quick squirt is directed into each nest (Pict. 1.56).

Control of biting ants can be performed by dusting or spraying the nests with lindane (0.6–1.0%), dieldrin (0.1%) or parathion (0.5–1.0%) (Pict. 1.57).

Further reading: McNutt, D. N. The control of biting ants *Macromischoides aculeatus* Mayr (Formicidae) on Robusta coffee. *E. Afr. Agric. For. J.* **1963**, *29*, 122–124.

Picture 1.56   Red Fire Ant – *Oecophylla longinoda*.
(3) Adult and (4) nest. From: Bayer.

Picture 1.57   Biting Ants – *Macromischoides aculeatus*. View of opened nests. From: J. N. Wintgens.

Picture 1.58   Adult Leaf-cutting Ant. Affected branch. From: Bayer.

Carpenter Ant

Atopomyrmex mocquerysi André
(Hymenoptera: Formicidae)

Common in the Democratic Republic of Congo and Tanzania. These ants bore galleries in the top of the coffee trunks to establish their nest. These ants are directly damaging by nibbling the terminal buds and sucking the sap. They can be controlled with the same pesticides as those recommended for biting ants injecting them into the galleries to reach the nests.

Leaf-cutting Ant

Atta spp (Pict. 1.58)

Other Ants

Crematogaster sp. and *Pheidole* sp.
(Hymenoptera: Formicidae)

They are common in East Africa. These ants attend aphids, coccids and some other *Homoptera*, and rely on their excreta (honeydew) for part of their food, the rest being supplied by predation.

They can be controlled by poisoned baits or with the same specific insecticides as for biting ants.

1.16
Mites

These occur on most perennial crops. The excessive use of insecticides is liable to lead to attacks by mites. This is not, however, true for coffee. A few species have been recorded doing slight damage but none is a major pest.

Picture 1.59 Coffee Mites. Damage on leaves. From: J. N. Wintgens.

Picture 1.60 Coffee Red Spider Mites. Adults, larva, eggs and damaged leaves. From: Bayer.

Coffee Red Spider Mite

Oligonychus coffeae (Nietner)
(Prostigmata: Tetranychidae)

This mite is widespread in the tropics on a very wide range of crops including cotton and tea. It causes a bronze discoloration on the upper surface of fully hardened leaves (Pict. 1.59).

The red mites, white cast skins and bright red spherical eggs can all be seen with a hand lens (Pict. 1.60).

Red Crevice Mite, False Spider Mite

Brevipalpus phoenicis (Geijskes)
(Prostigmata: Tenuipalpidae)

This mite is also widespread in the tropics and is recorded from tea, citrus and many other crops.

It is found commonly where corky areas have developed on leaves or green berries and in the groove at the blossom end of developing berries. The red mites are just visible to the naked eye (0.25 mm).

B. phoenicis is a considerable pest of citrus but on coffee the damage is very rarely important.

Yellow Tea Mite

Polyphagotarsonemus latus (Banks)
(Prostigmata: Tarsonemidae)

Once again, this is a highly polyphagous mite with a pantropical distribution. It is recorded from cotton, sesame, tea and many other crops. On coffee it causes leaf distortion, but there is no clear evidence that this leads to crop loss.

The females are 0.2 mm long; the males somewhat smaller. They feed on soft, unexpanded leaves. Damaged leaves have a rippled surface and, on the underside, the main veins are outlined by corky areas.

Further reading: Crowe, T. J. Mites as coffee pests. *Kenya Coffee* **1960**, *25*, 504–505.

1.17
Storage Pests

Washed coffee of low moisture content is virtually immune from attack by insect pests. If, however, the coffee is produced in a humid area or is stored in a humid tropical port, it can be attacked seriously.

Coffee Weevil
Araecerus fasciculatus (DeGeer)
(Coleoptera: Anthribidae)

This beetle, which now has a pantropical distribution, is usually known as the Nutmeg Weevil or Coffee Weevil, although it is not, in fact, a weevil. It is the only storage pest that has been reported doing damage in various parts of the world. Contamination of the coffee by the presence of dead beetles is often more important than the actual damage. It is more serious as a pest of dried cassava, spices and nuts.

The adult is 3–5 mm in length and mottled brown or grey in color. The wings are slightly shorter than the abdomen and the three terminal segments of the antennae are thickened (Fig. 1.15).

Eggs are laid singly in dried pulp (if present). Feeding by the white, legless larva is initially in the pulp but later the bean is penetrated. Pupation is in the bean. Duration of stages: 5–7 days; larva 46–66 days; pupa 6–9 days; pre-oviposition 4–5 days (Pict. 1.61).

To control the pest every effort should be made to avoid storage in humid areas. Sundrying of the coffee may kill the pest because it cannot survive temperatures above 37 °C. Cold also kills the beetle and it cannot survive a European winter.

Fumigation gives good control but does not, of course, eliminate the contamination. Methyl bromide and ethylene dibromide are effective, but are now being

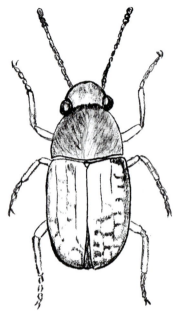

Figure 1.15 Coffee Weevil – *Araecerus fasciculatus*. Adult weevil.

Picture 1.61 Coffee Weevil Araecerus fasciculatus Damage on coffee beans. From: Nestlé S.A.

phased out. Phosphine gas, generated from aluminum phosphide tablets, is also effective and there appear to be no reports of resistance to this product. As with all storage pests, cleanliness and good warehouse management are basic to good pest control.

Further reading: Mphuru, A. N. *Araecerus fasciculatus* DeGeer (Coleoptera: Anthribidae): a review. *Tropical Stored Products Information* **1974**, *26*, 7–15.

1.18
Vertebrates

There are a few scattered records of vertebrate activity in coffee but there are no reports of serious attacks.

1.18.1
Amphibia

Tree frogs *Leptopelis* spp. are fairly common in coffee. They are insectivorous and may be beneficial, but they appear to have a preference for predaceous ladybird beetles (Coccinellidae) and might therefore be harmful.

1.18.2
Reptilia

Insectivorous lizards are also found on coffee, but, once again, they appear to prefer predators to pests. Snakes, especially puff adders *Bitis* spp., cobras *Nuja* spp. and mambas *Dendroaspis* spp., are quite common in coffee plantations, and represent a potential hazard for pickers and pruners. They are, however, very rarely aggressive and records of snake bite are extremely few.

1.18.3
Aves

Birds sometimes eat the pulp of ripe berries, but this type of damage is very rare. Insect-eating birds are often abundant in coffee plantations, but their beneficial effect, if any, has not been investigated. Woodpeckers are recorded eating the young larvae of White Borer *Monochamus leuconotus* in Kenya and may well contribute significantly to the natural control of this beetle.

1.18.4
Mammalia

Mole rats *Tachoryctes* spp. eat coffee roots and very occasionally kill a few newly-planted coffee seedlings. Local traps are usually effective.

Fruit-eating bats have been recorded eating coffee in the Democratic Republic of Congo, Cameroon and Brazil, but coffee is only eaten in the absence of more attractive fruit. Control was by shooting, presumably while they were roosting on trees during the day.

The hippopotamus has been recorded in Ugandan and Tanzanian plantations near Lake Victoria.

Duikers *Cephalophus* spp. are recorded eating coffee berries and young shoots in West Africa, and sometimes causing significant damage.

Baboons *Papio* spp. and colobus monkeys *Colobus* spp. sometimes invade coffee plantations and strip ripe berries and break branches, often destroying much more than they consume.

Elephants do not feed on coffee, but occasionally cause damage by trampling or playfully uprooting seedlings. For the control of large mammals the local Game Department should be consulted.

1.19
Future Trends in Pest Control

The most important objective of future research should be a reduction in the use of pesticides. These can cause problems of residues in the crop, off flavors in the brewed coffee, pollution of the environment and toxic hazards to spray operators. Biological and cultural control measures avoid these problems and are cheaper in the long run than pesticide application.

There are a few cases where classical biological control (the introduction to a country of exotic natural enemies) is a possible option. The conservation of *indigenous* parasites and predators, however, offers far more scope and is a badly neglected subject. Basically one determines if the population of a beneficial species is held down by the shortage of some resource (e.g. nesting sites). That resource is then supplied artificially. The population of the beneficial species will then rise and pest control becomes more effective. Some possibilities that might be worth a trial are:

1) Provision of nest-boxes for insect-eating birds (the Chinese already use this method for forest insect control);
2) Provision of structures for colonies of insect-eating bats (some bats prefer beetles, some moths so the correct species is encouraged if possible);
3) Provision of breeding places for spiders (spiders like to lay eggs in crevices. Even pairs leaves of stapled together may increase the population);
4) Provision of plants with accessible pollen and nectar for adult parasites (Clean weeding on large plantation means that very little pollen and nectar is available for feeding adult parasitoids, thus reducing their fecundity and fertility).

The advantages of this type of pest control are:

1) Only simple, cheap, locally available equipment is needed.
2) If a new method proves to have adverse rather than beneficial effects it can be immediately abandoned without long-term damage.

Good standards of cultivation will do much to reduce the potential for damage of coffee pests.

2
Major Pests of Coffee in the Asia–Pacific Region

C. C. Lan and J. N. Wintgens

2.1
Introduction

Worldwide, coffee is attacked by numerous pests. In the Asia–Pacific region, the stem borers *Xylotrechus quadripes* Chevrolat and *Acalolepta cervina* Hope are severe pests of Arabica coffee in Thailand and China, while the berry borer *Hypothenemus hampei* Ferr. is a serious problem of Robusta coffee in the Philippines, Indonesia and India. In this chapter, the major pests of coffee in this region, with special reference to the above countries, are elaborated. The management of these pests using an integrated approach is discussed, based on practical experience in the respective countries.

2.2
Integrated Management of Insect Pests

The integrated pest management (IPM) approach is necessary for sustained control of pests in perennial crops under tropical environments. It attempts to bring the pest population to an acceptable level by combining all possible control measures (mechanical, physical, cultural, chemical and biological) in a complementary and comprehensive manner. There are four basic principles underlying the IPM approach:

1) Potentially harmful pest species will continue to exist at tolerable levels. Low-level infestation of some pests may sometimes be desirable, as it provides important sources of food or shelter for natural enemies. It is, therefore, important to know the relationship between population levels of a pest and its natural enemies and crop loss. A reliable system of pest population monitoring is necessary to determine the optimum timing for application of appropriate control measures.

2) The ecosystem is recognized as a management unit and any measures taken to control pests may upset its initial natural balance and, therefore, create other, often worse, problems. For example, the indiscriminate use of broad-spectrum, long-residual-contact insecticides for the control of key pests may result in the outbreak of a less common pest due to the destruction of its natural enemies. Usually, the activities of natural enemies in their natural state keep in check a large number of pests that otherwise could increase to levels damaging to crops.

Coffee: Growing, Processing, Sustainable Production, Second Edition. Edited by J. N. Wintgens.
© 2012 WILEY-VCH Verlag GmbH & Co. KGaA. Published 2012 by Wiley-VCH Verlag GmbH & Co. KGaA.

3) The use of natural control agents is emphasized. In nature, insect mortality is regulated by abiotic factors, such as weather parameters, and biotic factors, such as natural enemies (predators, parasites, phytophagous organisms and pathogens) and the host plant itself.
4) Pesticides still form an important component of IPM; however, they are used more intelligently and with care. Chemicals with the least harmful effect on natural enemies are always preferred. Correct application technique is stressed – emphasizing the use of appropriate applicators, right dosage, good coverage and correct timing.

Instead of using chemicals on a calendar basis or indiscriminately, pest monitoring is implemented periodically to determine the status of pest attack. Chemicals are used only when the number of insects or level of attack exceeds the economic threshold (ET). The ET is defined as the level of pest attack at which it is profitable to carry out a controlled action. The ET is a concept to help farmers to decide when a control measure is required so that chemicals will not be used unnecessarily, killing the natural enemies along with the pest, which may be present at a very low level.

2.3
Major Coffee Pests

2.3.1
Coffee White Stem Borer

Xylotrechus quadripes Chevrolat
(Coleoptera: Cerambycidae)
French	Scolyte blanc du tronc
Spanish	Broca del tallo
Portuguese	Broca do tronco
German	Weißer Kaffeebohrer
Dutch	Witte Koffieboorder

This is a severe pest of Arabica coffee. The larvae bore into stem, resulting in yellowing, defoliation and wilting. Affected plants are easily broken by the wind and death occurs. Older plants are able to withstand the infestation for a few seasons with reduced yield.

It has been reported in Kwangtung, Kwangxi and Yunnan Provinces of China, Thailand, Vietnam, India, and Java.

The insect is a slender elongated Cerambycid, 13–16 mm long by 3.4–4.5 mm wide, black with grey pubescence on the head, thorax and abdomen. The inverted "V" markings on the forewing are characteristic of the species (Pict. 2.1A).

The female beetles are very active, flying freely and laying eggs on stems, mainly in the afternoon in bright sunshine. Mated female lay eggs in small groups of one to 10 in crevices in the bark of the main stem and thick primaries. The eggs are small, milk white at first and slightly yellowish later; 1.25 mm by 0.5 mm, elongated, oval in shape, with one end more pointed than the other. The eggs are hatched in 6 days. On hatching, the larva is a whitish grub, broad at the head end and gradually tapered towards the tail end. The young grub bores and tunnels into the bark to about 3 mm depth, where it may spend about 20 days. These tunnels appear as ridges in the bark, particularly around 50 cm from ground level, and are clear symptoms of attack. It begins to tunnel the wood in all directions for 1–2 months, and then forms an exit tunnel and pupates. The pupal stage lasts 10–12 days. Adults live for 8–14 days (Pict. 2.1B, 2.1C, 2.1D, 2.2, 2.3).

There are three generations (flight period/season) per year in Sechwanbana and Yunnan; two generations in Poseh and Lungzhou; rarely three generations in Kwangxi; and more than three generations in Kwangtung and Hainan. The beetle

Picture 2.1 (A) Adult White Stem Borer, *Xylotrechus quadripes.* (B) Coffee stem showing formation of ridges. (C) Scrapping revealing tunnels beneath the ridges. (D) Longitudinal section of stem showing the haphazard tunnels made by *X. quadripes.*

Picture 2.2 Larva *of X. quadripes,* with powerful mandible, broad at the shoulder, gradually tapering towards the tail end.

Picture 2.3 Pupa of *X. quadripes* in the pupation chamber.

completed its lifecycle in about 90 days when reared under captivity in Lungzhou and Kwangxi.

Control

An integrated control consisting of physical and mechanical, cultural, chemical, and biological measures is described.

Physical or mechanical control

Scrubbing the bark with a brush or a piece of metal is commonly practiced by farmers in India and has been proven to be very effective. In China, scrubbing is done by wearing a pair of gloves made from thick fabric (Pict. 2.4).

The smooth stem surfaces discourage the beetle from laying eggs. Scrubbing also dislodges the eggs and grubs, and this reduces the chances of survival of the

Picture 2.4 Scrubbing the stem of plant 2–3 years or older using hand gloves.

insect. There are no adverse effects on the coffee stem as a result of scrubbing. It is usually done on young plants aged 2–3 years old and, once completed, the bark need not be scrubbed until 6–8 years later. However, this method has not received much attention in Thailand and is not popularly practiced by farmers as it is too laborious.

Cultural control

- *Sanitation*. The coffee farm should be inspected for plants damaged by the pest, especially from June to July and from November to December. They are easily traced by the yellowing of leaves and confirmed by further examining the stem for the presence of ridges. Infested plants should be uprooted and burned. This is important as the plant can be a serious source of infestation. The larvae can complete their development in uprooted plants lying in the farm. Transporting affected trees outside infested areas should be avoided. Alternate hosts should also be destroyed.
- *Shade*. Shading has been recommended as a control measure in India and Vietnam. The use of shade has a direct relationship with the behavior of the beetle. The adults are very active during bright, warm and sunny days, flying, mating and laying eggs. If sufficient shade is maintained, egg laying will be affected. In Mysore, India, the shade provided by the coffee leaves themselves is considered important. The shade tree *Alleviates Montana* was recommended as the most suitable in Vietnam, as its root system is mainly at a lower level than that of coffee, which prevents water and nutrient competition.

Pheromones

Sex pheromones are natural compounds produced by either male or female insects to increase the probability of successful mating. Early studies in India have shown the presence of a sex pheromone released by the male beetles. The pheromone could be exploited for mass trapping of the female beetles. It could also be used for population monitoring. Control measures such as chemical swabbing could then be applied more timely to improve the efficiency.

Researchers have identified the major pheromone component and tested it in the field. The results suggested that trapping the female beetles with pheromone-baited traps in the field is possible, although catches were low. This is because the pest is present and causes significant damage even at very low densities. Nestlé R & D/Singapore and the department of Biological Sciences, Simon Fraser University, Canada had collaborated in developing the pheromone as a monitoring tool and mass-trapping device for the management the Coffee Stem Borer (Pict. 2.5).

Picture 2.5 Field testing of pheromone in Yunnan, China, April 1998.

Chemical control

Swabbing or painting the stem with chemicals is a popular way of controlling the pest. The main stems and primaries of each plant are swabbed with insecticide such as fenitrothion or chlopyriphos. Swabbing has to coincide with the flight period of the beetle, so that the chemical can come in contact with the adult during egg laying and after hatching, before boring into the stem, to the improve efficiency of the control. Systemic insecticides such as monocrotophos and oxydemeton-methyl were not effective. In addition to swabbing or painting, plugging the exit hole with 10 % solution of Didiwei 80 EC (dichlorvos) in March and July, i.e. before the emergence and egg laying period, has been recommended in China.

2.3.2
Coffee Brown Beetle

Acalolepta cervina Hope
(Coleoptera: Cerambycidae)

This is a serious pest of Arabica coffee. The larva bores and tunnels into the stem, causing yellowing and drooping of leaves. In severe cases, the affected plants die.

It is distributed in the Province of Yunnan, Sechwan in China and in the Doi Tung area in Northern Thailand (first reported by Nestlé).

The adult is a robust, brown beetle of 15–27 mm long by 5–8 mm wide (Pict. 2.6).

Antennae are cylindrical with 10 segments tapering gradually towards the last segment and extend beyond the length of the body. The prothorax appears squarish with two pronounced lateral spines. The upper two-thirds of the elytra is covered with minute punctures. There is a tiny blotch of light brown fur on either corner of the elytra below the central portion of the prothorax, which appears as a dorsal white spot.

Eggs are 3.5–4 mm long, narrow at both ends and slightly curved. They are white when first laid, turning to yellowish-white

Picture 2.6 (A and B) Adult Brown Stem Borer, *Acalolepta cervina*.

Picture 2.7 Larva of Brown Stem Borer, note the coarse frass as a result of feeding and tunneling in the stem.

later and yellowish-brown before hatching. The larva has a small dark brown head (Pict. 2.7).

Its antennae extend from the head to thorax, and are folded neatly, curled up by the side of the abdomen. The body color of the larva is milky white to light brown. Only the prothorax (first segment of the thorax) has small light brown patches. The mesothorax (second segment) has a pair of spiracles. The larva is without legs. There are eight or nine segments in the abdomen. The full-grown larva is 38 mm in length.

There is only one generation per year in Sechwanbana. Eggs are hatched in 6–8 days and the larva goes through six instars in 288 days. Pupation takes 10–15 days. The adult lives for 60 days. The full-grown larva winters in the affected stem and pupates in March the following year. Pupation peaks from early April to mid-May and the adult emerges in June (Pict. 2.8).

The next generation of eggs begins hatching at the end of June, and the larva emerges and bores into the stem. Damage is most serious from August to September (Pict. 2.9).

On emerging, the beetles feed on the young bark, petiole and main veins of the leaves for 5–10 days before mating. They hide in shady places such as the basal portion of the trunk and leaf litters in the day, actively mating and laying eggs from 8.00 p.m. After mating, the female makes a 1–2-mm slit in the bark before inserting an egg into it. The females prefer to lay eggs on the tree trunk facing the sun. On hatching, the larvae bore into the tree bark, feeding and tunneling into the woody portion, spiraling upwards and severely damaging the stem. The larvae leave the tunnel and pupate in a chamber made 2–9 mm below the bark.

Control

1) Swabbing the stem with an insecticide from the basal portion up to 1 m above ground to prevent egg laying.
2) Swabbing with 0.1 % solution of Didiwei 80 EC in late June to early July to prevent larvae from boring into the stem.

Picture 2.8 Pupa of Brown Stem Borer with curly antennae neatly folded by the side of the thorax.

Picture 2.9 Stem with tunnels on the bark – a symptom of Brown Stem Borer attack.

3) Cultural practices as for the White Stem Borer, i.e. sanitation: removal of infested trees in August and September; avoid movement of infested plants to non-infested areas; planting of shade trees; proper pruning and fertilizer application to improve the general health of plants.

2.3.3
Red Branch Borer

Zeuzera coffeae
(Lepidoptera: Cossidae)
French   Ver rouge
German   Roter Kaffeebohrer
Dutch   Rode Koffieboorder

This is found in India, Papua New Guinea, Taiwan and South-East Asia. In China, it is reported in Kwangtung, Fukien and Hainan Provinces. It has a very wide range of hosts, including tea, coffee, cocoa, loquat, lychee, cotton, orange, teak and many other forest trees. It has not yet gone beyond the Asia–Pacific region.

The moth is grayish-white in color with several dark spots on the forewings. A bigger dark spot is present in the middle of the forewing. The wing expanse varies from 26 to 52 mm depending on sex (Pict. 2.10).

Eggs are laid in masses on the new shoots or bud. These are yellowish-brown and oval in shape. The larvae are red to violet–brown, often with some yellowish rings. The pupa is reddish-brown with short spines. Pupation takes place within the stem.

The larva bores into the woody stem and branches (up to several centimeters in

Picture 2.10 Adult of the Red Branch Borer, *Zeuzera coffeae*, a grayish-white moth with several dark spots on the forewings.

Picture 2.11 Longitudinal section through the middle of a stem showing a full-grown larva of *Zeuzera coffeae* feeding on the heartwood.

Picture 2.13 Woody stem and branches up to several centimeters in diameter attacked by Red Branch Borer and snapped under strong wind.

Picture 2.12 Pellet-like excrement of the Red Stem Borer – a clear sign of its attack.

diameter) of shrubs and small trees (Pict. 2.11).

At the attack site, a circular tunnel is formed under the bark, causing the death of the distal part of the branch. The larva bores a longitudinal tunnel in the drying wood. Also typical are the cylindrical reddish-brown or yellow fecal pellets, which are ejected from the tunnels and accumulated at the base of the stem (Pict. 2.12).

Control

1) All attacked branches must be pruned and burnt (Pict. 2.13).
2) Spray Samingsong EC 50 % (fenitrothion) and Leguo EC 40 % (dimethoate) at 0.1–0.15 % solution (product) before

the eggs hatch and larvae bore into the stem. However, chemical application to prevent the pest is seldom justified as the larvae are subjected to natural control agents such as the woodpecker, a braconid (*Bracon zeuzerae*) and several tachinids [*Isosturmia chatterjeeana* (Bar.) and *Carcelia kockiana* (Towns)]. Natural mortality is high during hatching.

3) Plug the tunnel with Leguo EC 40 % (dimethoate) at 2.5 % solution (product) and Didiwei 50 EC at 10 % solution or inject the chemical into the tunnel to kill the larvae in the bored stem.

2.3.4
Shot-Hole Borer, Twig Borer

Xylosandrus compactus Eichhoff
(Coleoptera: Scolytidae)

French	Scolyte des rameaux
German	Kaffeezweigbohrer
Dutch	Bruine Takkenboeboek

This can be a serious pest of Robusta coffee in many countries. It is found in India, Indonesia, Malaysia, Vietnam, Sri Lanka, Madagascar, Hawaii, Mauritius, Seychelles, Florida, Fiji and tropical Africa. The insect attacks both suckers as well as branches and twigs 1–2 cm in diameter. Affected stems or branches wither or dry up with shot holes between the nodes.

The adult is a small brown to black beetle with a short, subcylindrical body. The female is darker and larger (1.5–1.8 mm), and the male is dull and small (0.8–1.0 mm). The female enters the twig by making a shot hole. It makes a tunnel in the twig and lines the wall with ambrosia fungus for 11–21 days. Pupation takes 7–11 days. On average, from the egg to the adult takes 4–5 weeks.

The pest population builds up very fast in the wet season and declines during the dry period. It is most serious in plantations under heavy shade, as humidity helps in the establishment of the ambrosia fungus, which the young stages of the borer feed on.

Picture 2.14 Drooping of leaves – a clear sign of Twig Borer attack.

Control

The nature of the infestation makes control by insecticide difficult and uneconomical. Affected twigs may be pruned 5–8 cm behind the shot hole and burnt. This is done as soon as the first symptom of attack, like drooping of leaves, is noticed (Pict. 2.14).

As the pest prefers to breed in the suckers during dry periods, unwanted suckers should be removed. Thin shade should be maintained in the plantation, as the infestation is more serious under heavy shade.

2.3.5
Coffee Berry Borer

Hypothenemus hampei
(Coleoptera: Scolytidae)

German	Kaffeebeerenbohrer
Dutch	Koffiebessenboeboek
French	Scolyte des baies du caféier
Spanish	Broca del café
Portuguese	Broca do café

Coffee berry borer (CBB) is a severe pest of coffee affecting all *Coffea* species, of which *C. canephora* is the most preferred. It is the most important pest of *C. canephora*. Damage is caused by the females feeding on young berries, resulting in significant losses through premature falling. The bulk of the damage is on the endosperm of the mature bean, which can be extensively damaged or completely destroyed by the adult or larvae feeding and tunneling in it.

It was first found in Java in 1909 and in the Philippines in 1963, and, only recently, in India. It is also present in Malaysia, Thailand, Vietnam and New Caledonia, but not yet in Yunnan, China.

The adult is a small black beetle about 2 mm long by 1 mm wide. The females bore a hole at the tip of the berry and tunnel into the bean (Pict. 2.15).

The female lays eggs in batches of eight to 12 over 3–7 weeks, producing about 30–70 eggs in her life time. The larvae are legless, white with brownish heads and feed on the beans (Pict. 2.16).

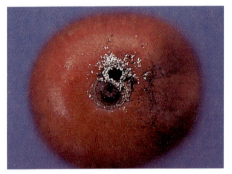

Picture 2.15 Female Coffee Berry Borer enters the coffee berry through a circular hole it makes, usually at the tip of berry.

Picture 2.16 Larva and pupas of Coffee Berry Borer.

Picture 2.17 Pupa of Coffee Berry Borer.

The larval stage lasts 10 to 26 days. The pupal stage lasts 4–9 days in the larval galleries (Pict. 2.17).

The adult males emerge from the pupa earlier than the females. They fertilize the female siblings as they emerge. From egg laying to adult takes 25–35 days.

Only the females fly off the berries in search of ripe berries for new colonization. The males cannot fly and do not leave the berries. The beetles also breed in dry berries that are left on the tree after harvest or in fallen berries. Females outnumber males by at least 10 : 1. They live longer – 156 days on average. The larvae can feed or breed in stored coffee if it is not fully dried (moisture above 13.5 %).

Control

Cultural control

- *Sanitation and harvesting.* The best method of control is crop hygiene, both in the field and in the stores. The crop should be picked regularly, every fortnight in the heavy fruiting season and monthly at other times. No ripe or dried berries should be left on the ground or on trees and all infested berries should be burnt. These infested berries have been shown to generate significant numbers of beetles affecting the next crop. In Columbia, green berries left on the ground are treated with 10 % urea to hasten their decomposition and improve soil fertility.
- *Rampassen.* Rampassen, which means complete removal of all developing berries in the trough period, is an old method employed to break the development cycle of the beetle. A period of 3 months is recommended. Young berries that are formed within the period are continuously stripped as these serve as a source of food for the adult beetles. In areas where there is a dry spell of

2–4 months, there will be no berries – a condition resembling that created by rampassen.

- *Pruning*. Pruning plays an important role in the management of CBB. There are basically two types of pruning in coffee: formative and routine. In formative pruning, one could select either a single stem or multiple stems (three or four) in a bush. In either case, the coffee bush is maintained at a comfortable height of 2.0–2.2 m to facilitate complete and clean harvesting. Plants which are too high will result in berries being left on the branches beyond the reach of the harvester. If infested by CBB, these berries will eventually encourage more infestation. In maintenance pruning, water shoots, pest- and disease-infected branches, and interlocking, upright, downwards and inwards branches are removed periodically to reduce the competition for nutrients and open up the canopy to encourage natural enemies. Heavy shading brought about by inefficient pruning favors the survival of CBB, and is unfavorable to natural enemies. Proper pruning is, therefore, necessary for direct or indirect control of CBB.

Biological control using parasitic wasps

There are three important natural enemies which are indigenous to central Africa, i.e. two Bethylids (*Prorops nasuta* and *Cephalonomia stephanoderis*) and an Eulophid (*Phymasticus coffea*). Other natural enemies of the CBB are birds and ants.

Cephalonomia stephanoderis was introduced to Indonesia in 1989, but not much progress has been made since. Only the Indonesia Coffee and Cocoa Research Institute (ICCRI) managed to keep a small population in the laboratory and possibly a few insects could still be found in the fields where small-scale release has

been conducted in the past. No other countries in this region have experimented with this wasp.

In Columbia, however, millions of *Cephalonomia* and *Prorops* had been reared and released since 1990. The objective is to mass release and achieve an equilibrium after a number of years. The two wasps were aimed at reducing the population during the trough period where berries were left on the trees or on the ground. They were specific on the young stages of the CBB.

The Eulophid *Phymastichus coffea*, however, has only recently been studied in Columbia. Apparently, it could be used during the early stages of berry development, i.e. the first 4 months after flowering. This wasp parasitizes the adult female CBB when she begins to tunnel into the berry. The wasp also attacks the male in the bean. However, the adult *Phymastichus* lives only a few days. This makes the rearing program much more intensive and more care has to be given to ensure a continuous laboratory population. The present program on the development of *Phymastichus* as a component of integrated management of CBB involves the rearing and handling of the wasp, and evaluating its range of alternate host.

Biological control using entomogenous fungus

The parasitic fungus *Beauveria bassiana* can infect the beetles and larvae under warm and humid conditions.

It has been found in many countries including the Philippines, Brazil, Jamaica, Cameroon, Congo, Ivory Coast, Malaysia and Indonesia. Recently, it has been marketed commercially in Latin America under the name "Brocarial" by Laverlam and "Conidia" by AgrEvo.

In Indonesia, the fungus is produced at the farm level for CBB control. It is tar-

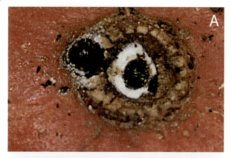

Picture 2.18 (A) *Hypothenemus hampei* adult beetle infested by *Beauveria bassiana* with extensive growth of the fungus at the entrance hole on the distal end of the berry. (B) *Beauveria bassiana* mummified larva found in a gallery of an infested bean.

geted at reducing the CBB population during the first 4 months after flowering. The fungus has been used like an insecticide. Thus, control would be difficult once the beetles have tunneled into the berry. However, the use of this fungus was readily accepted by the farmers because they could see the effect easily. The fungus appeared as a white cottony ball covering the CBB, usually found at the beetle's entrance hole (Pict. 2.18).

Chemical control
Chemical control is recommended only when the pest is out of control and cultural practices have failed. The biology of the CBB makes it difficult to control by chemical means. Only females in flight searching for breeding sites are vulnerable to contact poisons. They are not susceptible to

stomach poisons since they reject the portions of the skin of the berry that they remove in boring the berries. A systemic insecticide with sufficient translocation into the berries, but that does not taint the coffee, would be ideal. Chemicals with minimum effect on the natural enemies are emphasized.

2.3.6
Mealybugs

Planococcus spp.
(Homoptera: Pseudococcidae)
French   Cochenille farineuse
Spanish   Projo blanco aereo
Portuguese Escama branca
German   Kaffeelaus
Dutch   Koffielvis

Mealybugs are one of the most important sucking pests of coffee. They are soft-bodied insects covered with a white waxy coat and capable of movement. Both the nymph and adult infest the tender branches, nodes, leaves, flower clusters, berries and roots, feeding on plant sap. Leaves turn chlorotic, flower buds abort and berries fail to develop properly as a result. A superficial sooty black, saprophytic mold is usually associated with Mealybug infestation. The fungus thrives and develops on the sugary excretion of the Mealybugs, and may cover the leaves and affect photosynthesis. It may also affect the flowers and berries, and interfere with their development.

Another common feature is the association of Mealybugs with ants which attend them on the host plant. The main importance of the ant is that it kills and carries off any predators of Mealybugs. It feeds on the honeydew secreted by the Mealybugs, gaining some vital nutrients, and removes any dead Mealybugs, cast skins and other debris. Hence, controlling the ants will have a direct effect on reducing the population of the Mealybugs.

Mealybugs are found in almost all coffee growing areas in the Philippines, Indonesia, Thailand, Malaysia, India and China. Their infestation is, however, sporadic and localized. It is most serious during the dry season with intermittent showers, and in plantations with non-uniform and limited shade.

Two species, i.e. *Planococcus citri* Risso and *Planococcus lilacinus* Cockerell, are the most common species attacking coffee (Pict. 2.19).

The adult of *P. citri* is oval, wingless and is covered with mealy secretion. The female lays 10–1000 eggs in an ovisac, which hatch in about 10 days. *P. lilacinus* does not have an ovisac and the eggs hatch within a few minutes after they are laid. The nymphs migrate to different parts of the plant, settling on tender branches, nodes, and leaves, flower clusters, berries, and roots in large numbers. The female passes through three nymphal instars, while the male forms a white cocoon and transforms into a winged adult after the second instar. The development period from egg to adult is approximately 30 days.

Planococcus minor (Maskell), also known as *Planococcus pacificus* Cox, is found in Malaysia (Pict. 2.20).

A species that attacks the root of coffee in Yunnan, China is identified as *Planococcoides robustus* (Pict. 2.21).

This species is common in the Indian subcontinent, especially on fruit trees such as mango. Its effect on the Arabica coffee plant is being monitored in Yunnan, China.

Picture 2.19 *Planococcus lilacinus*, another common species of Mealybug in coffee.

Picture 2.20 *Planococcus minor*, a species of Mealybug attacking fruit and branches of Robusta. Malaysia.

Picture 2.21 A species of Mealybug, *Planococcoides robustus*, attacking the root of coffee in Yunnan, China.

Control

Mealybugs may be effectively managed through the following methods:

1) Maintain adequate shade of 30% for Arabica cultivation and 20–25% for Robusta.
2) Control ants by spraying soil around the base of coffee trees or shade trees with insecticides such as quinalphos 1.5% or methyl parathion 2%; or Malathion 5%. Ant nests should be destroyed. Plants with their canopies touching each other are pruned to prevent bridging.
3) Affected patches can be sprayed with the following insecticides: quinalphos, fenitrothion, fenthion and kerosene.

Biological control may be achieved through introduction of two exotic natural enemies, i.e. the predatory ladybird beetle, *Cryptolaemus montrouzieri* (Coleoptera: Coccinellidae) and the parasitoid *Leptomastix dactylopii* (Hymenoptera: Encyrtidae) (Pict. 2.22, 2.23).

In India, several laboratories were set up to rear the two natural enemies for mass release (Pict. 2.24).

If *P. citri* is predominant, 17 500–25 000 parasitoids would have to be released initially and 5000–7500 in subsequent years until control is achieved. If other species of Mealybug dominate, 10 000–15 000 predators would need to be released. Both the parasitoids and predators could be reared on pumpkin. At about 6 weeks after the pumpkin is infested with Mealybugs, the progeny of the parasitoid start emerging. In the case of the predator, this takes about 8 weeks in total. The release of biological control agents is preferably done in the evening on infested plants.

Picture 2.22 Adult *Cryptolaemus montrouzieri*, a 3–4-mm long predatory ladybird beetle of Mealybug.

Picture 2.23 *Leptomastix dactylopii*, a parasitoid of *Planococcus citri*, introduced into India in 1983 from Trinidad.

Picture 2.24 *Cryptolaemus* and *Leptomastix* being reared for mass release to control Mealybug in India.

2.3.7
Green Scale

Coccus viridis (Green)
(Homoptera: Coccidae)
French   Cochenille verte
Spanish   Cochenilla verde
Portuguese   Cochenilha verde

Green Scale is a serious pest of coffee, especially Arabica. The female is oval, about 4 mm by 2 mm, fairly flat and pale green in color with a few black spots in the middle of the back (Pict. 2.25).

Each female may lay 50–600 eggs which hatch within a few hours. There are three nymphal instars, each is larger than the previous one, over duration of 4–6 weeks. Adult live for 2–5 months. This pest has a wide range of host, plants including many important tree crops such as cassava, citrus, guava, mango and tea.

The insect sucks the plant sap and debilitates the plant, particularly the young bushes. The honeydew that is excreted by the scale forms a film on the leaves and acts as a medium for the growth of sooty mold. Like Mealybugs, Green Scale is attended by various species of ants.

Control

The strategy for management is similar to that for Mealybugs. Ants could be controlled in a similar way by spraying affected patches with the insecticides malathion, quinalphos, fenitrothion or fenthion.

In nature, Green Scale has a number of parasitoids and predators; and a few fungal pathogens. Of these, *Coccophagus bogoriensis*, *Coccophagus cowperi*, and the pathogenic fungi *Verticillium lecanii* and *Empusa lecanni* are more important (Pict. 2.26).

As the pest prefers to breed in the suckers during dry periods, unwanted suckers should be removed. Thin shade should be maintained in the plantation, as infestation more serious under heavy shade.

Picture 2.25 Adult Green Scale, a flat, oval, light green with an irregular, distinct intestinal loop of blackish spot.

Picture 2.26 A Green Scale parasitized by *Coccophagus bogoriensis*.

3
Nematodes in Coffee

G. Castillo P., J. N. Wintgens and J. W. Kimenju

3.1
The Parasites

Nematodes are multicellular microorganisms 0.1–5 mm long and shaped like cylindrical worms. They are found in the soil and inside the roots of coffee trees where they feed on the sap through their fine stylets. As they feed and develop, they weaken the host plant and cause open wounds which disrupt root functions and allow access to infections of bacteria and fungi.

The importance of actual damage caused by nematodes varies according to the tolerance threshold and the overall condition of the coffee trees, as well as the type of pathogen and environmental factors. Nematodes reproduce sexually and parthogenetically. They are very prolific – one female can produce over 2000 eggs. Their lifespan lasts from 25 to 70 days and develops in three stages: eggs, larvae and adult.

Symptoms of nematode infestation are variable and can often be mistaken for nutritional disorders or the results of vascular problems. Multiplication takes place through the roots via soil and water, and, above all, through young plants contaminated in the nursery.

There are a number of species of nematodes which attack coffee trees but the following are the two most important groups.

3.1.1
Root Lesion Nematodes (*Pratylenchus* spp.)

Pratylenchus spp. are migratory endoparasites which feed on the root cortex, turning it first yellow, then brown and finally killing it. The lesions made by *Pratylenchus* spp. can be very large. They can even encircle the root and detach the dead cortex, hence their name of "root lesion nematodes".

3.1.2
Root-knot Nematodes (*Meloidogyne* spp.)

French	Anguillule des racines
German	Wurzelgallenälchen
Dutch	Wortelknobbelaaltje
Spanish	Nematodo de agallas
Portuguese	Nematóides de galhas

Meloidogyne spp. are sedentary nematodes. The females settle into the rootlets of the coffee trees causing distorted knots known as galls which involve the entire section of the root. In normal growing con-

Coffee: Growing, Processing, Sustainable Production, Second Edition. Edited by J. N. Wintgens.
© 2012 WILEY-VCH Verlag GmbH & Co. KGaA. Published 2012 by Wiley-VCH Verlag GmbH & Co. KGaA.

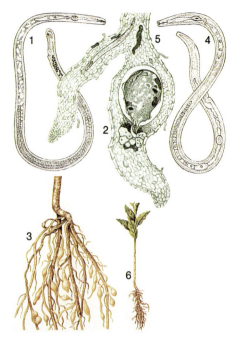

Picture 3.1 *Meloidogyne javanica* and *Pratylenchus coffea* nematodes. From: Bayer.

Picture 3.2 Infestation of root-knot nematodes showing swollen roots and small pyroid galls. From: Nestlé.

ditions, infected coffee trees are debilitated, but they do not necessarily die (Pict. 3.1 and 3.2).

Different species of nematodes can cause many other forms of damage to roots depending on their respective interactions and their association with fungi.

3.2
Geographical Distribution

In the coffee-growing countries of America, Asia and Africa, nematologists have found a wide range of plant parasitic nematodes which affect coffee. They have been known for over a century and, since they were first recognized, a considerable list of species which attack coffee plant roots has been recorded. Table 3.1 shows the most important coffee-parasitic nematodes and their geographical distribution.

Undoubtedly, the most widespread genus in coffee-growing areas is *Meloidogyne*. New species have been added to this genus such as: *M. arabicida*, described in Costa Rica [8], or *M. africana* in Kenya, as a root parasite of *C. arabica* [18].

So far, *Meloidogyne* spp. have been identified as one of the most aggressive species whose characteristics differ from one population to another.

In Central America, a study of this species has led to a more precise definition of the genus phenotypes. The investigation conducted by Hernández *et al.* [15] by means of advanced electrophoresis techniques has identified at least four distinct species, among which the most widespread are *M. exigua* (Costa Rica, Honduras and Nicaragua), *M. incognita*, esterase phenotype F_1 (Guatemala), *M. arenaria* (El Salvador) and *M. arabicida* (Costa Rica).

The list of other nematodes less frequently associated with coffee is long.

Table 3.1 Geographical distribution of the main coffee plant parasitic nematodes

Nematode	Latin American countries	Other continents/countries
Meloidogyne incognita[a]	Brazil, Colombia, Costa Rica, Cuba, El Salvador, Guatemala, Mexico, Nicaragua, Venezuela	Africa
M. arenaria	El Salvador	
M. exigua	Brazil, Colombia, Costa Rica, Dominican Republic, Guatemala, Honduras, Nicaragua	Africa, India
M. arabicida	Costa Rica	
M. coffeicola	Brazil	
M. javanica	Colombia, Cuba, El Salvador	
M. africana		Kenya
M. spp.	Brazil, Guatemala, Honduras,	Kenya
Pratylenchus coffeae	Brazil, Costa Rica, Dominican Republic, El Salvador, Guatemala, Venezuela	Madagascar, India, Africa, Asia
P. brachirus	Brazil	Africa
P. spp.	Colombia, Costa Rica, Cuba, El Salvador, Guatemala, Honduras, Mexico, Nicaragua	Kenya, India
P. vulnus		India
Helicotylenchus sp.	Brazil, Colombia, Costa Rica, Cuba, Dominican Republic, El Salvador, Honduras, Guatemala, Mexico, Nicaragua, Venezuela	Africa
Xiphinema sp.	Brazil, Costa Rica, Honduras, Mexico	
Radopholus sp.	Mexico	India
Others	Costa Rica, Cuba, El Salvador, Honduras, Nicaragua, Venezuela	India

[a]Esterase phenotypes linked to this species have been found in electrophoresis tests, mainly in central American countries.

The most outstanding genera being *Xiphinema, Helicotylenchus, Radopholus, Hemicriconemoides, Macrophostonia, Rotylenchulus, Criconema, Criconemella, Tylenchus, Trichodorus, Longidorus, Aphelenchus* and *Aphelenchoides*.

In India, over the last few years, new species such as *Rotylenchoides desouzai, Hemicriconemoides coffeae, Nothocriconema indicum* and *Macropostonia degrissei* have been identified [1, 4, 6, 12, 13, 17–19, 22, 24, 28].

3.3
Importance and Economic Damage

In Latin America, from an economic point of view, nematodes are of significant importance because they limit coffee production. One can even say that, after the coffee berry borer, *Hypothenemus hampei*, nematodes have become a major focus of interest to phytosanitary professionals since the losses they cause are higher than those caused by Coffee Leaf Rust, and also because adequate treatment is both expensive and difficult to implement.

Sasser [27] estimated the losses in Central America at 10% of the total production and Alvarado [2] has calculated the losses in Guatemala at 40%. In Brazil, particulary in the main coffee areas like Sao Paulo, Rio de Janeiro, Paraná and Minas Gerais, where at least three species of the *Meloidogyne* genus are coffee parasites, the losses result in a production decrease of 15–35%.

In the central area of the State of Veracruz in Mexico, the affected area covers approximately 20 000 ha, whereas in Costa Rica the damage to coffee trees has only been attributed to *M. exigua* and *M. arabicida* in very localized producing areas like San José Viñas [24].

Overall, in Latin America, the amounts invested to control infestation by nematodes have considerably increased production costs. In Guatemala, for instance, the cost of two applications of nematicide per year ranges from US$228 to 678/ha, depending on the chemical used [26]. In Brazil, over the last 10 years, attempts to control the *Meloidogyne* genus in states like Rio de Janeiro, Paraná and São Paulo have added thousands of dollars to investments in coffee culture [6].

Research in the field of nematode control in Latin America is developing rapidly and considerable sums of money are allocated to train specialists to further investigate this problem.

3.4
Symptoms and Damage

3.4.1
Nematodes in Seedbeds and Nurseries

At this stage in the development of the coffee plantlet the losses caused by nematode attacks can be total, since damage to roots may well extend to all the plants. Between 3 and 6 months after germination, the knots or galls caused by the *Meloidogyne* genus are well defined, but when they associate with fungi, in particular, a much more destructive process is initiated. The gall loses its shape and the root cortex is transformed into a cork-like sleeve with a spongy humid appearance (Pict. 3.3).

This is particularly noticeable with the *M. incognita* and *M. arabicida* species where the symptoms become well defined especially when the plant is approximately 1 year old (Pict. 3.4).

This is the stage when the aerial symptoms begin to appear. A generalized chlorosis caused by the clogging of the vascular tissues can be observed as well as the fall of older leaves (Pict. 3.5).

In Guatemala, where *Pratylenchus* spp. predominate, lesion attacks by this nematodes have reportedly damaged secondary

Picture 3.3 Young coffee plant roots showing nematode attack symptoms of "corchosis". The root cortex has been transformed into a soft cork-like sleeve. From: G. Castillo Ponce.

Picture 3.4 Well-defined "corchosis" on a coffee root.

Picture 3.6 Coffee plantlet affected by *Meloidogyne* in the seed bed. From: G. Castillo Ponce.

Picture 3.5 Foliage of a young coffee tree showing symptoms of nematode attack. *Pratylenchus* and *Meloidogyne*. From: Bayer.

roots and caused cortical necrosis, cortex detachment and the destruction of feeder roots [27].

In Mexico, mixed populations of the *Pratylenchus* and *Meloidogyne* (Pict. 3.6) have been found in coffee fields.

3.4.2
Nematodes in Plantations

C. arabica plantations of all ages are frequently affected by nematodes. In heavily infested fields, plant mortality may appear between 3 and 4 months after planting. Even if this does not occur, growth remains slow and weak, in spite of adequate fertilization. The foliage begins to thin out during maturation, production drops and the lifespan of the coffee tree can be shortened to only 3 or 4 years.

Symptoms are even more severe when the plant starts to produce. The plant becomes stressed and defoliation is evident, mainly at lower levels. No marked flacidity can be observed on the foliage, unless the soil is very poor.

The symptoms caused by the lack of nutrients due to the reduced activity of the feeder roots are: chlorosis, necrosis on the leaf edges and tips, thin, brittle stems, and, above all, a gradual decrease in the yield (Pict. 3.7 and 3.8).

In Guatemala, it has been reported that where large populations of *Pratylenchus* are present, the coffee plant only produces once [3].

In Mexico, where mixed populations of nematodes are present, three to four harvests have been recorded, but yield decrease every year and, finally, the coffee plant dies [26].

In Southern India, early attacks by *Hemicriconemoides* have made coffee culture economically unviable [18].

In each of the cases mentioned above, the symptoms appear in the plantation in the form of "large spots".

Amongst varieties of *C. arabica*, there is a range of susceptibility to *M. incognita*. Caturra, Catuaí and Garnica tend to decline faster when attacked by this nematode. This is not the case for Typica, Mundo Novo and Bourbon, as these varieties appear to have a stronger defense capacity due to the development of a large quantity of small roots just below the collar. This phenomenon, known as the "wig", confers a certain anchorage and secures nutrition to the plant, even though the remaining roots may be affected by "corchosis" (Pict. 3.9).

Intensive production can also alter the resistance of the coffee plants. In Brazil and Costa Rica, for instance, where intensive production systems are practiced (agricultural mechanization, high-density plantation, homogeneous varieties, etc.) nematodes spread in a more uniform manner. As a result, compact areas of affected plants are easily identified (Pict. 3.10).

Conversely, where diversified production systems are used on small cultivated plots, the symptoms may be masked by a num-

Picture 3.7 Spot of adult coffee trees affected by nematodes (Guatemala). From: A. B. Eskes.

Picture 3.8 Patch of coffee trees suffering from "corchosis" (Mexico). From: G. Castillo Ponce.

Picture 3.9 One-year-old "Tipica" coffee plant affected by "corchosis" showing the superficial root proliferation under the collar (wig). From: G. Castillo Ponce.

Picture 3.10   Coffee trees killed by nematodes.

ber of different factors. In this case, a wide experience in the identification of the nematode attack is required in order to make a reliable diagnosis.

In Robusta plantations, aerial symptoms are difficult to define, since this species has a certain tolerance to nematodes. Even where there is evidence of root parasitism, symptoms rarely appear in the foliage. Moreover, Robusta production is not affected by nematodes [10].

As a result, the use of Robusta rootstock for grafting more susceptible varieties like *C. arabica* is presently being researched and developed.

3.5
Association with Fungi

In Latin American coffee-growing countries, the nematode problem in coffee is amplified due to a complex in the pathogenesis, where at least two genera of fungi are present. The nematodes, in this case *Meloidogyne* and *Pratylenchus*, predispose the plants to fungal infection and cause physiological alterations to the tissues. *Pratylenchus* spp., for example, can cause protoplasm lysis in the cells. The fungus takes advantage of this to invade

the plants later on. The most commonly reported fungus is *Fusarium* spp., which appears to be present in most countries. Another relatively common fungus is *Rhizoctonia solani*. Both are pathogenic in coffee, mainly during early stages of planting.

Meloidogyne and *Fusarium* are frequently associated in coffee cultivation, and their combined effect accelerates the process of withering and the death of the plant. Furthermore, in Costa Rica, Brazil and Mexico, the presence of *Trichoderma* sp. together with *Fusarium* spp., in the "corchosis" has been confirmed. In Costa Rica *Trichoderma* sp. has also been isolated inside the "corchosis" in association with *M. arabicida*, *M. cylindrocladium* spp. and *Phylaphora* spp. [6, 8, 25].

The specific role most of these fungi play in this complex is unknown, although Taylor and Sasser [25] mention that the normally non-pathogenic fungi become pathogenic after the root tissues have been invaded by *Meloidogyne*. Gaumann, quoted by these authors, points out that the nematode "breaks down" the host's resistance to the penetration of other secondary agents present in the rhizosphere, when the local medium changes. Even if these secondary agents live most of their time as saprophytes, their characteristics may change once the cell change is completed and this could cause them to invade the host's tissues. This may be the case for *Trichoderma*, normally a saprophytic fungus in the soil fungi and which has been consistently found inside the "corchosis" outgrowth in Mexican plantations. It should also be mentioned that *Trichoderma* spp. probably plays an antagonistic role with the other organisms involved.

This interaction may turn out to be very interesting for biological control purposes, but more research in this field is needed.

3.6
Loss Estimation

The damage caused by nematodes affects different regions in different ways, and, in each case, the outcome can only be determined locally. This involves taking into account the various factors that directly affect the losses and the damage thresholds. The major factors that should be considered are the cultivation system (intensive, extensive, multi-cropping, etc.), the varieties used, the density of the plantation and, above all, soil characteristics and the genera of the nematodes present.

As a result, a damage assessment scale can only be established on the basis of specific conditions in different geographical areas.

The following scale is used in Mexico to make an on site assessment of the damages caused by the "corchosis" problem. It has been suggested by Hernández [16] and is a modified version of the scale proposed by Krener and Unterstenhofer.

Value	Condition
1	Healthy coffee plant
2	Coffee plant with yellowing and scant production
3	Coffee plant with marked malnutrition, scant foliage and no production
4	Totally defoliated non-productive coffee plant with no probability of recovery
5	Dead or non-existent coffee plant

These values are converted in the following equation:

$$PDP = \frac{(nv2)+(nv3)+(nv4)+(nv5)-(nv1)}{ZN} \times 100$$

where:

PDP	= loss percentage
n	= number of plants evaluated in each category of the scale
v	= numerical values of the infestation categories (1, 2, 3, ...)
Z	= numerical value of the maximum category (in this case 5)
N	= total number of plants evaluated.

This simple equation is useful for making a rapid assessment of damage on site by observation of the aerial symptoms and, indirectly, assessing the loss in production.

The formula should be applied to batches of 20 coffee plants in the field and the number of batches assessed will depend on the size of the field. As a general rule, five batches per hectare are taken for assessment and the sites are selected in an "X" shape across the field.

20		20
	20	
20		20

Five batches of 20 plants = 100 plants on 1 hectare

Each value in the scale is added up and included in the equation as follows:

Category	1	2	3	4	5	total
No. of Coffee plants	44	21	15	13	7	100

and then converted into the formula:

$$PDP = \frac{(21\times2)+(15\times3)+(13\times4)+(7\times5)-(44\times1)}{5\times100} = 26\%$$

These calculations reflect the number of coffee trees that are out of production and, consequently, the estimated loss for the coffee grower.

3.7
Estimation of Root Infestation

Sampling the soil and the roots to determine the importance of the nematode population and the species involved is far more sophisticated than the rapid assessment of damage established by on site observation of the aerial damage. Sampling is also far more expensive since it entails costly extraction techniques. This means that, when faced with nematode damage to the plantation, the local coffee grower needs to choose the most cost-efficient way of assessing damage and potential loss in production.

Assessment scales for estimating root damage have, however, been established for the *Meliodogyne* species as shown in Tab. 3.2 and Fig. 3.1.

3.7.1
Monitoring

If the presence of nematodes is suspected, sampling is the first step which needs to be taken to detect them. It should be carried out every year, in the middle of the rainy season.

Table 3.2 Assessment scales of reaction to the *Meliodogyne* genus

Galling scales and indexes used				% of galled roots
0–4	*0–5*	*1–6*	*0–10*	
0	0	1	0	0
	1	2	1	10
	2	3	2	20
1			3	30
			4	40
2		4	5	50
	3		6	60
			7	70
3				75
	4	5	8	80
			9	90
4	5	6	10	100

Source: APS (USA) 1978: Special Publication.

A root sample of at least 10 g is necessary for a reliable analysis. This sample should be taken from several different representative plants out of a maximum of 100.

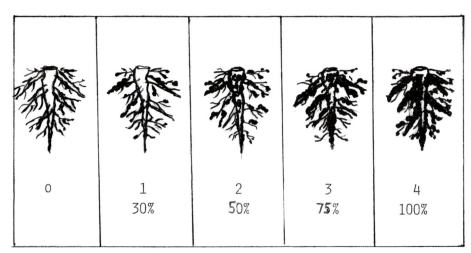

Figure 3.1 Galling scale graduated from 0 to 4 with corresponding percent of galled roots.

Table 3.3 Some herbaceous plants that are hosts of root knot nematodes

Scientific name	Family	Nematode	Other characteristics
Amaranthus dubius Mart	Amarantacea	*M. incognita*	allelopathy
Amaranthus spinosus L.	Amarantacea	*M. incognita, Pratylenchus* spp.	
Bidens pilosa L.	Compositae		allelopathy
Borreria leavis (Lam)	Rubiaceae	*M. javanica*	
Cuphea micrantha H.B.K.	Lythraceae	*M. incognita,M. exigua*	melliferous
Cuphea racemosa L.	Lythraceae	*M. javanica, M. hapla*	melliferous
Cynodon dactylon	Gramineae	*M. incognita*	allelopathy
Cyperus rotundus L.	Cyperaceae	*M. incognita, M. exigua*	allelopathy, melliferous
Echinochloa colonum L.	Gramineae	*M. incognita*	
Eleusine indica L.	Gramineae	*M. incognita*	toxic for bovines and horses
Emilia sonchiflora L.	Compositae	*M. incognita*	
Euphorbia hirta L.	Euphorbiaceae	*M. incognita*	
Euphorbia hirta L.	Euphorbiaceae	*Rotylenchus reniformis*	
Euphorbia hypericifolia	Euphorbiaceae	*Radopholus* sp., *Pratylenchus* spp., *Helicotylenchus* sp.	
Gallinsoga caracasana (DC)	Compositae	*M. incognita, M. exigua*	
Gallinsoga ciliata (Raf)	Compositae	*M. incognita, M. exigua*	
Gallinsoga parviflora Cav	Compositae	*M. incognita*	
Heliopsis buphthalmoides (Jacq)	Compositae	*M. incognita, M. javanica*	melliferous
Impatiens balsamica L.	Balsaminaceae	*Meloidogyne* spp.	melliferous
Impatiens wallerana Hooker F.	Balsaminaceae	*Meloidogyne* spp.	melliferous
Mormordica charantia L.		*M. incognita*	toxic, melliferous
Oxalis corniculata	Oxalidaceae	*Meloidogyne* spp.	melliferous
Oxalis latifolia	Oxalidaceae	*M. exigua*	melliferous
Phyllanthus niruri L.	Euphorbiaceae	*M. incognita*	
Physalis nicandroides Schl.	Solanaceae	*M. incognita, M. exigua, M. javanica*	melliferous
Sida acuta Burm f.	Lytraceae	*M. incognita, M. exigua, Radopholus* sp., *Pratylenchus* spp., *Helicotylenchus* sp.	melliferous
Solanum nigrum Sendt	Solanaceae	*M. incognita, M. exigua, Rotylenchus* sp.	
Spananthe paniculata (Jacq)	Umbelleferae	*M. incognita*	
Synedrella nodiflora L.	Compositae	*Radopholus* sp., *Helicotylenchus* spp., *Pratylenchus* spp.	
Talinum paniculatum L.	Portulacaceae	*M. incognita, M. exigua, M. javanica*	melliferous

Reliability will be improved where root sampling is complemented by soil sampling.

In traditional cultivating systems, plants generally grow in a natural way under tall coffee trees. These plants can be excellent indicators of the existence of nodulating nematodes, at least, since from the beginning of their development (two to three pairs of leaves), they may show nodules on their roots (see Tab. 3.3).

Table 3.4 Strategies for the direct and the indirect management of nematodes in coffee plants

Practices	Strategy and effect	Percent of surface treated annually
Severe pruning of coffee trees over 10 years old	indirect; stimulates functional roots and new shoots	25
Stumping	indirect; stimulates new foliage and functional roots	15
Density increase, replacement of coffee trees over 20 years old	indirect; prevention of yield decrease due to plant reduction	10
Fertilization with NPK and micro-elements	indirect; complete nutrition strengthens plants with a response capacity	100
Mixed weed control with machete and glyphosate	direct and indirect; eliminates alternative hosts and avoids competition	100
Shade regulation	direct and indirect; controls the soil moisture and improves the nutrition capacity	100
Grafting	direct; encourages "corchosis" tolerance and production sustainability for a longer period	33
Nematicide applications	direct; nematode populations decrease at the beginning of the season and during sowing	30
Application of organic fertilizers and liming	direct and indirect; it has an antagonistic and repressive effect on "corchosis" and improves the nutrition and conditioning of the soil	30
Soil erosion control	direct and indirect; reduces spread of plant parasite nematodes	100
Planting nematodes free seedlings	direct; reduces the nematode load at the early stages of crop growth	100
Regular soil sampling	indirect; serves as the basis of the course of action	100
Cover cropping	direct and indirect; increases organic matter in the soil and reduces soil erosion	100
spot treatment with nematicides	direct; reduces the build up and spread of nematodes	10

3.8
Practices for Nematode Control

Traditional cultural practices in coffee plantations may well need to be changed, and new practices implemented in order to either prevent nematode attacks or, at least, limit their development and spreading. Table 3.4 provides a review of practical strategies.

3.8.1
Soil Disinfecting

Soil disinfecting is a preventive measure intended to curb the primary infection. Its goal is not only to eliminate nematodes, but also other undesirable pathogens and small weed seeds. If this is carried out correctly, in the first stages of the development of the coffee tree, the infection will be considerably reduced later on.

Soil disinfection can be achieved by using appropriate chemicals, applied according to the rates prescribed by the manufacturer and in respect of prevailing legislation.

In hot, sunny climates soil solarization is recommended as an "environmentally friendly" and efficient method for disinfecting soil in seedbeds. To obtain best results, the beds should be covered with a thick, transparent sheet of plastic for 2–3 days to raise the temperature beneath the sheet to 50 °C for at least 2 h each day.

Prior to transplanting in the nursery, soil and roots should be checked once more against any latent sources of infection.

3.8.2
Weed Control

A large number of weeds are hosts to nematodes which is one of the reasons why they should be regularly slashed. Table 3.3 provides a list of the best-known nematode host weeds.

Efficient programs for weed control will take into account the period during which they compete with coffee plants, the need for soil conservation and, of course, the costs involved.

To effectively eliminate nematodes, weeds should be controlled. The use of systemic herbicides combined with manual weeding with tools like the machete is a good practice since the herbicide eliminates the weed roots and reduces the nematode populations.

Sublethal doses of glyphosate may be used after weeding. This is an inexpensive method which delays the weeds growth and helps to maintain the balance in the agro-ecosystem.

3.8.3
Pruning

Plants with low infection levels respond favorably to pruning as this practice helps to prolong their productive life. In addition systematic elimination of unproductive aerial tissues strengthens the root system. The development of superficial roots also enables the coffee tree to resist the nematode attack for a longer period.

3.8.4
Removal of Dead Plants

Dead plants should be uprooted during the dry season and exposed to the sun. This eliminates the major sources of infection. When the dead coffee plant is uprooted and removed, the root pit should be disinfected with a nematicide.

3.8.5
Organic Fertilization

The input of any organic fertilizer (coffee pulp, bovine or horse manure, sugar cane bagasse, hen droppings, guano, etc.), either in a compost or separately, efficiently stimulates the root systems of coffee plants and improves nutrition. At the same time, this practice increases the populations of soil organisms that are antagonistic to nematodes.

The use of organic fertilizers should, however, be based on knowledge of the chemical composition of the soil since excessive rates of application or inadequate mineralization may cause pH or nutrient imbalance.

In nurseries, nematode control combined with organic fertilization encourages the development of the plants.

3.8.6
Genetic Resistance

This is one of the safest and most economical practices for nematode control in coffee plants. Basically, research has focused on the resistance levels of *C. canephora* and *C. arabica*, as well as on slight tolerance differences which have been observed in Catimor hybrids and Caturra (Pict. 3.11).

The high level of tolerance which has been recorded in *C. canephora*, specifically

in the Robusta variety, is extremely important and has led to its incorporation into grafting schemes. This practice is already being applied in many countries.

The chief goals of regional improvement programs in Latin American countries are the selection, hybridization and fast multiplication of materials from *C. canephora*, to be used in commercial grafting programs.

Anzueto *et al.* [3] worked for several years with materials of this species, selected from various areas of Central America, until they found that descendents of the Robusta breed T3561 × T3751 showed a marked resistance to *Meloidogyne arenaria* in El Salvador, to the *Meloidogyne* sp. in Guatemala, to *M. exigua* and *M. arabicida* in Costa Rica, and to *M. incognita* in Nicaragua. This new variety is now known as Nemaya and its seeds will be available to coffee growers for rootstock production.

In Costa Rica, some selected Robustas have shown a satisfactory response, mainly to the *M. exigua*. T3757, T3751 and T3561 materials are outstanding – the last two being the ancestors of the Nemaya variety [20].

In Brazil, great progress is being made in the field of genetic resistance research and one of the main objectives is to identify rootstock that is resistant to *Meloidogyne*

Picture 3.11 Differences in tolerance to nematodes amongst several Catimor hybrids and Caturra. From: A. B. Eskes.

Picture 3.12 Romex (R34) highly tolerant to "corchosis" (Mexico). From: G. Castillo Ponce.

spp. So far, the Robusta hybrids Apoata and Nemaya has been found to have considerable resistance to many *M. incognita* pathotypes [14].

Research for materials that are highly tolerant to "corchosis" has been ongoing since 1990. In Mexico "Romex", Mexican Robusta, is being used, and clones R34, R37 and R48 have shown a high tolerance to "corchosis" (Pict. 3.12).

Currently, the National Institute of Forestry, Agricultural and Cattle Research (INIFAP) is extending this Robusta collection by the introduction of new materials from Africa and from other countries.

3.8.7
Grafting

Grafting coffee plants has been a recognized practice for four decades. Initiated by Reyna [21] in Guatemala, its major aim was to strengthen the coffee plant. Later on, this was further developed by the need to provide a higher tolerance to nematodes.

The practice consists of a graft, conducted shortly after the emergence of the plantlet (soldier stage), which joins the rootstock with the aerial part of the graft. The joining point should be fastened with self-destructive material (parafilm "M"), which does not need to be removed (Pict. 3.13).

Picture 3.13 Young coffee plantlets grafted by the Reyna method and tied with the self-destructive parafilm "M". From: G. Castillo Ponce.

This technique has been adopted in most Central American countries, Brazil and Mexico. Many improvements have been made to the original method and the current success rate averages at about 95 %.

Grafted trees have a more stable yield, and they produce 30 % more than nongrafted plants. Their tolerance to nematodes makes them the best option in the short term.

In Mexico, the author of this paper has conducted cup tests with coffee from all the grafts developed since 1992 and has come to the conclusion that coffee quality is not affected by grafting.

3.8.8
Chemical Control

The majority of the nematicides currently on the market are effective against nematodes in coffee plants. They are, however, not extensively used since they are costly. As a general rule, only approximately 5–10 % of the larger producers use this form of control. Furthermore, since the active life of most nematicides is short, applications have to be made twice a year and this could lead to an accumulation of residues in the beans coupled with health and environmental risks.

In cases where a new plantation is being established on plots with a known history of nematode infection or in the case of the rejuvenation of a plantation in similar circumstances, nematicides may be incorporated with an integrated control system. Granulated nematicides are best suited for the treatment of seedbeds and nurseries, and they should always be used, especially as a preventive measure.

Nematicides are highly toxic pesticides. Consequently, every precaution should be taken to ensure adequate protection for both the workers and the crop. The manufacturers directions for use should be strictly followed as well as local legislation requirements.

3.8.9
Biological Control

This tactic has not been widely studied in the case of nematodes in coffee cultivation. A few isolated experiments at a regional level are known but they have not been validated in plantations at a commercial level.

Certain microorganisms, such as bacteria of the *Bacillus* genus and nematode destroying fungi have been detected, by sweeping microscopy, as parasites of nematode eggs and larvae. This may turn out to be the choice strategy for the future, since these organisms can easily be reproduced [5]. Certain fungi and other nematodes have also been detected as having a potentially destructive action on nematodes but, again, further research is needed.

3.8.10
Antagonistic Plants

There are many plants which are antagonists to nematodes because they contain toxins which are lethal to these parasites. These plants include the genus *Mucuna crotallaria*, *Tagetes*, the Neem tree, and several graminacca.

The most widely recognized plant which is known to be an effective antagonist of nematodes is the common marigold (*Tagetes erecta*) and the *Crotalaria juncea*.

They can either be sown on the whole plot prior to planting the coffee or between the rows of the planted trees. This method of control is practiced in many countries and has also had good results in association with other cultures like bananas, citrus fruits and certain annual crops.

Picture 3.14 *Tagetes patula.*

Again, however, though examples of this type may seem simple and efficient, further research is needed to consolidate them before they can be advised as "foolproof" strategies.

3.8.11
Integrated Management

Table 3.4 shows an example of integrated management in Mexico on a plantation severely afflicted by "corchosis", where annual losses due to nematodes, were as high as 22%.

The combined effect of the various cultivation practices implemented over a period of 4 years turned the situation round. After that period, the plantation was showing a 150% increase in yield.

3.9
Conclusions

Any cultivation practice which encourages and maintains healthy growing conditions for coffee trees will, automatically, strengthen their resistance or tolerance to nematodes.

The following are, however, essentially important practices which should be part and parcel of the management program in all plantations.

- Correct maintenance of soil fertility through inputs of both organic and chemical fertilizers, or even *mycorrhiza.*
- Avoiding high soil acidity by the adequate dosing of fertilizers and, particularly, avoiding an excessive input of nitrogen and acidifying fertilizers.
- Regular liming to maintain the appropriate soil pH.
- Abandoning the notion of forcing higher yields by reducing or discarding shading.
- Maintaining a sustained organic matter content in the soil either by using cover plants and shade trees or by an extra input of compost – this will encourage water retention, secure nutrient adsorption and favor the development of micro-organisms which antagonize nematodes.

It should be noted that perennial monoculture crops, like coffee, inevitably generate conditions which encourage the development of nematodes and other parasites.

Strategies to combat nematode damage should be developed on the basis of regular assessments of the degree of infestation and the type of nematode involved. When starting up a plantation, special care should be taken to assess and treat infected areas.

Treatments applied to the soil in seedbeds and nurseries are imperative as a preventive measure against nematodes. In established plantations, soil treatments only provide a temporary solution to the problem since they cannot reach the deeper roots and soil layers infected by nematodes.

At present, the most efficient strategy to reduce nematode infestation is grafting with resistant rootstock. Biological control and the development of resistant varieties of coffee are potential solutions for the future but, so far, research and investigations have, unfortunately, scarcely begun.

Bibliography

[1] Abrego, L. and Holdeman, Q. L. Informe de progresos en el estudio del problema de los nemátodos del café en El Salvador. *Bol. Inf. ISIC* **1961**, *Suppl. 8*, 16.

[2] Alvarado, J. Diagnóstico sobre el parasitismo de los nemátodos y cochinillas de la raíz en la zona cafetalera del Suroccidente del Guatemala. *Tesis de Ingeniero Agrónomo.* Centro Universitario de Occidente, Quetzaltenango, Universidad de San Carlos, Guatemala, 1997.

[3] Anzueto, F., Bertrand, B., Peña, M., Marban, N. and Villain, L. Desarrollo de una variedad porta injerto resistente a los principales nemátodos de América Central. *XVII Simposium sobre Cafeticultura Latinoamericana.* PROCAFE-PROMECAFE, Consejo Salvadoreño de Café, El Salvador, 1995.

[4] Baeza, A. C. A. Nemátodos asociados al cultivo del cafeto en Colombia. *Fitopatología (Alf)* **1975**, *10*, 4.

[5] Baeza, A. C. A. Parasitismo de *Bacillus penetrans* en *Meloidogyne exigua* establecido en *Coffea arabica. Cenicafé* **1978**, *29(3)*, 94–97.

[6] Brasil. *Cultura de Café no Brasil.* Manual de Recomendaciones. IBC-MAC-GERCA, Río de Janeiro, 1981.

[7] Bruno, B. G. Metodología para evaluar la reacción del cafeto al nemátodo *Meloidogyne exigua. Tesis MSc.* CATIE, Turrialba Costa Rica, 1984.

[8] Calderón, V. M. Reacción de diferentes genotipos de café a *Meloidogyne arabicida h. y s.,* gama de hospedantes y hongos fitopatógenos asociados. *Tesis MSc.* CATIE, Turrialba, Costa Rica, 1989.

[9] Castillo, P. G. Estudio y evaluación del daño causado por el nemátodo nodulador de las raíces en semillero de café (*Coffea arabica* L.) en condiciones de inverna-dero. *Tesis Prof. Lic. en Biól.* Xalapa, Veracruz, Mexico, 1977.

[10] Castillo, P. G. *Informe de Actividades Realizadas en el Recorrido Técnico a las Zonas Afectadas por Nemátodos en Guate-mala.* IICA-PROMOCAGF. Mexico, 1988.

[11] Castillo, P. G., Castillo, H. J. and Tejeda, O. I. La injertación como una opción para mejorar la problemática fitosanitaria del cafeto. VII reunión científica de Veracruz, INIFAP. *Serie Mem. Cient.* **1994**, *7*, 77–86.

[12] Di Pietro, C. D., Guerra, N. E. G., Freitas, V. L. V. de and Passos, A. Levantamiento preliminar de ocurrencia de nematoides do género Meloidogyne, no Estado de Sao Paulo. *Inst. Bras. do Café 9th Congreso de Pesquisas Caffeeiras,* 1981.

[13] Ferraz, S. Reconocimiento das especies de fitonematoides presentes nos solos do Estado de Minas Gerais. *Experientia* **1980**, *26*, 255–328.

[14] Goncalves, W. and Ferraz, L. C. C. B. Resistencia do cafeeiro. *Informe Agrope-cuario* **1987**, *16* (172), 66–72.

[15] Hernández, A., Farguette, M., Sarah, J. L., Decazy, B., Eskes, A., Molinier, V. and Boisseau, M. Caractérisation biochimi-que, biologique et morphologique de différentes populations de *Meloidogyne* sp. parasites du café en Amerique Cen-trale. *Resumés du 10e Colloque de l'ASIC,* Kyoto, 1995.

[16] Hernández, V. E. E. Determinación y cuantificación de los nemátodos asocia-dos a las raíces del cafeto (*C. arabica* L.) en la cabecera municipal de Tlaltetela, Ver., México. *Tesis Lic. Biología.* Veracruz, Mexico, 1992.

[17] Kermarec, A. and Beliard, L. Étude pré-liminaire sur les némlatodes des plantes cultivées de Saint Domingue. *Turrialba* **1977**, *27*, 17–22.

[18] Luc, M. Nematological problems in the former French African tropical territories and Madagascar. In: *Tropical Nematology.* G. C. Smart Jr and V. G. Perry (eds). University of Florida Press. Gainesville, FL, 1968, 107.

[19] Macias, T. N. Situación nematológica de la cafeticultura en Honduras. *II Semi-nario Regional de Nematología en el Cultivo de Café.* PROMECAFE-ANACAFE, Anti-gua, Guatemala, 1993.

[20] Muñoz P. J. and Morera, G. N. Resis-tencia genética e injertación como alter-nativa de combate de nemátodos en el cultivo del cafeto. *Propuesta de un Pro-grama Nac. de Investigación.* Programa Cooperativo ICAFE-MAG-PROMECAFE, San José, Costa Rica, 1989.

[21] Reyna, E. H. Un nuevo método de injer-tación en café. *Bol. Téc. 21.* Ministerio de Agricultura, Guatemala, 1966.

[22] Roario, D. Situación nematológica del cultivo del café en la República Domini-cana. *II Seminario Regional de Nematolo-gía en el Cultivo de Café.* PROMECAFE-ANACAFE, Antigua, Guatemala, 1993.

[23] Sasser, J. L. Economic importance of *Meloidogyne* in tropical countries. In: *Rootknot Nematodes (Meloidogyne Species), Systematic Biology and Control.* F. Lam-berti and C. E. Taylor (eds). Academic Press, London, 1979, 257–268.

[24] Schieber, E. Nematode problems of cof-fee. In: *Tropical Nematology.* G. C. Smart Jr and V. G. Perry (eds). University of Florida Press. Gainesville, FL, 1968, 81–92.

[25] Taylor, A. L. and Sasser, J. N. Biología, *Identificación y Control de los Nemátodos de Nódulo en la Raíz [Trans.].* University of North Carolina, NC, 1985, 13–25.

[26] Teliz, O. D., Castillo, P. G. and Nieto, A. D. La Corchosis del cafeto en México. *Resúmenes XVI Simposio de Cafeticultura Latinoamericana.* PROMECAFE-CONCAFE-CATIE, Managua, Nicaragua, 1993.

[27] Villain, L. Dynamique de populations de *Pratylenchus* spp. sur café dans le sud ouest du Guatemala. *Proc. XVI Colloque Scientifique International sur le Café.* San Francisco, CA 1991.

[28] Villain, L., Licardo, D., Toledo, J. C. and Molina, A. Evaluación de tres nematici-das y la práctica de injerto hipocotiledo-nal en el control de *Pratylenchus* spp. *II Seminario Regional de Nematología en el Cultivo de Café.* PROMECAFE-ANACAFE, Antigua, Guatemala, 1993.

4
Coffee Diseases

R. A. Muller, D. Berry, J. Avelino, and D. Bieysse

4.1
Introduction

This chapter will only discuss coffee diseases due to fungal and bacterial infections. Diseases induced by nematodes and viruses will be discussed in the following chapters in Part II.

4.1.1
General Comments

Due to their different reactions to specific environmental conditions, the impact of the vectors of major coffee diseases varies greatly according to the environment. A typical example of this phenomenon is the case of Orange Leaf Rust (*Hemileia vastatrix*), a disease which is far more serious in hot climates and, therefore, less active at higher altitudes. Another typical case is that of the Coffee Berry Disease (CBD) found mainly in plantations at over 1500 m in Equatorial regions. Here, lower temperatures are not only favorable for the development of the pathogen, they also slow down the ripening process of the fruit which means that the berries are vulnerable for a longer period [35]. It is, therefore, important to bear in mind that the fact that a given pathogen is prevalent

does not necessarily mean that it will cause a serious disease with major economic consequences.

In certain cases, even if the plant shows symptoms of the disease, these only reveal a physiological disorder which, by weakening the plant, makes it more receptive to the disease itself. This is typically the case for "die-back", due to *Colletotrichum* sp., or for Cercospora Blotch, due to *Cercospora coffeicola*. In both cases, as the parasites have only a secondary influence, the problem can be better solved by improved agricultural care.

Recent research, conducted in Honduras, could lead us to conclude that a better knowledge of plant physiology and the influence of certain environmental factors on the plant will enable us, in due course, to determine how nutrition in the coffee plant can influence certain diseases. Such knowledge could then be included in the measures taken to prevent the disease. Improved plant nutrition or appropriate agricultural practices may not eradicate the disease, but could certainly reduce its impact and lessen the possibilities for it to develop. Knowledge of this can help to reduce the use of fungicides for disease control.

In any case, determining the factors which favor the development of diseases

Coffee: Growing, Processing, Sustainable Production, Second Edition. Edited by J. N. Wintgens.

should be of great help when recommending measures to be taken for fighting diseases in specific environmental conditions and areas [4].

Some diseases only attack the plant where it has been accidentally injured by agricultural tools, especially when the cut is at the base of the trunk, which often happens when cleaning weeds with a machete. Other injuries caused by insects or animals, as well as cultural practices like pruning or trimming, can also favor pathogen development. Obviously, care must be taken to avoid damaging the trees; however, where this is unavoidable, wounds should be treated with an antiseptic healing paste. This simple, but efficient, preventive measure should be a regular practice on all plantations as it also offers real protection against serious diseases like *Fusarium* wilt and canker.

Certain diseases like Coffee Wilt Disease due to *Fusarium xylariodes* which were a major preoccupation at a given time, seem to have disappeared over a period of 20 years, but have now resurfaced in certain African countries like the Democratic Republic of Congo and Uganda. From the 1930s to the 1960s, this disease, which attacked *Coffea excelsa* as well as *C. canephora*, was the main subject of a number of surveys and research in West and Central Africa. Then, for no discernible reason, it almost disappeared. It is generally thought that this can be explained by the introduction of a new, naturally resistant, variety of *C. canephora*, i.e. Robusta, coupled with the associated rejuvenation of plantations as well as improved agricultural practices including appropriate fertilization and better sanitary control. The momentary disappearance of this threat should not, however, lead to the belief that the disease is definitely eradicated. In fact, it reappeared in 1995 in the north-east region of the Democratic Repub-

lic of Congo and in 1993 in Uganda, where it has become a lethal threat to Robusta. At present, it has reached the stage of an epidemic disease and, so far, it has not been possible to determine whether this is due to mutations in the fungus or to a lack of care in agricultural practices and sanitary control, or, more likely, a combination of both factors. Whatever the reasons, the situation goes to show that vigilant supervision should never let up. In the south of Ethiopia, Coffee Wilt Disease on *C. arabica* now has an impact on the economy of the region.

Certain afflictions have been reported which cause the overall death of plantations for no identifiable reason. Research has not yet been able to pinpoint a common denominator for these disasters. This is the case for "sudden death" in Angola or the "mal de Viñas" in Guatemala.

4.1.2
Chemical Control

4.1.2.1 Pesticides

For a very long time, the available pesticides for disease control were limited to products based on copper sulfates, notably the famous and widely used "Bordeaux Mixture", which is still frequently considered to be the reference fungicide. At present, however, the specialized chemical industry offers a large spectrum of different products which are not only preventive, like the copper sulfate compounds, but which, given their systemic or penetrating properties, also heal and act as an antisporulant. The fact that the systemics penetrate the tissues of the plant not only reduces the risk of them being washed away by rains, but also, if they are not degraded too rapidly, allows for enduring long-term action. An additional plus for these fungicides is the fact that they pollute the soil far less as they deposit no copper, lead or other metallic residues.

This means that today we are well equipped to efficiently control the major leaf and branch diseases; however, pesticides are still not much help when it comes to fighting soil-borne diseases.

In spite of their many drawbacks, however, copper compounds are still popular and widely used, sometimes alternating with more recently developed products, for fighting leaf diseases. They may lack in penetrating action and require frequent, repetitive application due to rain, but they remain cheaper and, more importantly, they are less selective. Copper compounds act globally by coagulating the cytoplasm of the fungus, whereas more selective products have, in certain cases, generated new, resistant breeds of pathogens. The use of copper compounds alternating with new products has proved to be efficient [29, 49]. Copper compounds allow the new products to prolong their action and prevent the development of new mutant pathogens. The combination of the two has, over recent years, reduced the number of applications by about 30 %, or even 50 % in certain cases.

4.1.2.2 Productivity

In spite of the introduction of new pesticides with penetrating and systemic actions requiring fewer applications, chemical treatment remains exacting and expensive. The number of applications required involves labor costs, and the purchase and maintenance of special equipment, over and above the actual cost of the products. This, inevitably, increases production costs and, frequently, entails the need for available labor at a time when work has to be done elsewhere on the farm. This is especially true for smallholders.

Thus, the benefits of chemical control will be weighed against costs and productivity. Lack of treatment at a crucial time, too long an interval between treatments or a reduced concentration of the product in the mixture are essential factors which may well counteract all benefits.

Even if the cost of correct treatment is higher than the cost of defective treatment, the end result will be much better and profits will increase accordingly.

4.1.2.3 Epidemiology

Although the choice of the pesticide to be used is important, it is even more important to be well versed in basic epidemiology. It is necessary to know when to intervene, particularly when the first treatment should be made, and the sequence and the number of applications required. This can only be ascertained by careful observation and comparison of results carried out over a number of successive campaigns, taking into account the annual evolution of the disease under specific growing conditions. As all diseases caused by parasites develop according to climatic conditions and the vulnerability of the plants themselves, it is essential to determine the exact state of the plant combined with the climatic factors (rainfall and temperature) which spark off the development of the disease and the time it takes to spread.

For chemical treatment to be efficient, three key factors need to be respected:

- *When to begin*. Treatment must begin just when the infection sets in and at the time when climatic factors are most favorable to its spreading. This generally means that the effects of the disease are not yet clearly visible, but action is necessary, particularly when using preventive chemical products like copper compounds.

- *How long treatment is required.* Treatment must be sustained for the entire period during which the disease is likely to develop and spread. Treatment can be stopped just before the infection period peaks – a time which generally corresponds to a slow-down in plant growth and, occasionally, the end of the vulnerable period. If treatment has been sustained during the critical stages, the effects of the disease are reduced, and it is time to strike a balance between the costs of later treatments and potential benefits. One needs to judge whether the disease can still have an impact on either the year's production or that of the following year.
- *How often treatment is required.* Treatments need to be made sequentially in such a way that total protection of the plant is ensured during the critical period, this depends on the type and remanence of the pesticide used as well as on climatic conditions, particularly rainfall. For optimal results, the sequence of the treatments should be geared to the development of the disease and the specific rainfall conditions of the year, whereas, only too often, treatments are applied according to an empirically accepted pattern.

It is therefore of paramount importance that epidemiology be recognized as an applied science providing the necessary guidelines to ensure that chemical treatment is economically viable, efficient and cost-productive.

4.1.2.4 Treatment Methods

Application methods also need to be carefully chosen taking into account the areas which require treatment, the topography and the layout of the plantation. The equipment can be manual or motorized, it can be carried or drawn, in large or small tanks, but in all cases it must be adapted to the layout and the needs of the plantation. The width between lines is a decisive factor as, for example, in larger coffee plantations, narrow-gauge tractors, similar to those used in vineyards, are quite suitable even for plantations where the lines are relatively close.

Another aspect that needs to be taken into account is the fact that fighting diseases requires high-volume spraying in order to cover, as far as possible, all the vulnerable parts of the plant. In this way, disease control differs considerably from pest control where very low-volume spraying or even fumigation is used. High-volume spraying, on the other hand, entails using large quantities of water. Even if part of the product sprayed, diluted by rains, frequently filters through to the lower leaves when dripping off the upper leaves, this cannot be relied upon to ensure total coverage. Therefore, a liberal use of spray mixture is necessary and, consequently, expensive. A number of experiments, particularly for fighting CBD, have proved this fact beyond doubt. Recently, however, good results have been reported on certain methods of low volume spraying used in controlling fungi, but this still needs to be checked out and confirmed.

Further experiments using mineral oils in conjunction with fungicides have been carried out in Cameroon as this type of "plantation oil" is used successfully in banana plantations for controlling "Sigatoka disease" i.e. cercosporiosis. Treating banana plantations with "plantation oil" is carried out by spraying the mixture in a fine, regular spray from airplanes and, as the banana leaves which need covering are all at the top of the tree, this method is suitable. In coffee plantations, however, the quantities of the mixture required to penetrate the thick foliage down to the lower leaves, which also need protection, are such that experiments resulted in burning the young leaves and the tips of the upper branches.

The need for significant quantities of water for high-volume spraying may create a problem since sufficient water storage is not always available on the spot when required. It is therefore recommended to build "impluviums" or water tanks on site, inside the plantation itself. For larger estates, permanent tanks or water towers serve the purpose; in smaller plantations, more primitive water collectors like a simple pit lined with a sheet of plastic and topped with corrugated metal sheets set at an angle on each side to funnel the water are quite suitable (Pict. 4.1).

Picture 4.1 Primitive rainwater collector "impluvium" on a small plantation. From: J. N. Wintgens.

Another, very similar and simple solution to the problem would be a 200 l drum with similar corrugated metal or plastic sheets to funnel the water into the container.

4.1.2.5 Application Rates

When calculating the quantities of fungicide required, considerations like "kilograms of product (either active or commercial) per hectare" or "hectoliters of mixture per hectare" should be discarded. Specific calculations should be made taking into account factors like the actual size of the trees and their density of plantation. As these factors vary considerably from one plantation to another, commercial indications regarding the quantities required are of little use. Therefore, when advising producers, one should think in terms of the concentration of the mixture (kilograms of product per hectoliter). The concentration should be calculated to ensure efficient maximum protection with a product input, allowing for the "dripping effect" from the upper to the lower leaves. In this way costs can be geared to real needs and the producer can learn to estimate quantities with greater precision.

4.1.2.6 Assessment of Damage related to the Efficiency of the Treatment

When setting up the most efficient method for disease control, making the right decisions is no easy task. On the one hand, it is necessary to assess the potential efficiency of the treatment and, on the other, to assess the potential damage risks if no treatment is implemented.

If, for example, the disease attacks the foliage, damaged leaves should be periodically counted in order to observe the evolution of the disease and to evaluate its spread [1, 43]. However, this type of assessment does not always reveal the true impact of the disease. This aspect needs to be examined in terms of its effect on the potential annual yield. It is necessary to ascertain just how far the disease can be allowed to spread before chemical spraying becomes imperative and, if chemical spraying is implemented, whether the costs involved will be worth the investment. The only way to answer these questions is by comparing crop results on treated and untreated control plots in a given region.

The evaluation of damage due to disease at a given time is often complicated by the fact that significant damage can also be caused by physiological weaknesses in the plant and it is sometimes difficult to distinguish this from the true damage. Specific observation methods need to be implemented in order to come to the right conclusions and decide on what type of action needs to be taken.

4.1.3
Genetic Control

4.1.3.1 The Benefits of Genetic Engineering
Although the use of chemicals for controlling cryptogamic diseases is efficient, even very efficient if carried out correctly, this method still has to be continually repeated, year after year. For this reason, producers are beginning to examine the advantages of genetic engineering. It may well be more costly and take far longer to implement, but the end result is durable protection and consequently represents a worthwhile investment.

4.1.3.2 Progress of Research on the Coffee Plant
The great diversity of planting material already existing within the various collections has been significantly increased over recent decades by surveys throughout the African continent, and now represents an impressive gene bank available for use in the more traditional methods of hybridization and breeding.

Recent biotechnological progress may improve selection methods and create more resistant varieties. Studies in cellular and molecular biology have not only helped to identify and select quality genitors to be used for breeding, but have also brought about a more precise identification of pathogens, and a better knowledge of their geographical distribution and their genetic variability. Research can, therefore, be focused with greater precision and is now developing methods for using gene transfers to control certain types of insects. The possibility of using similar strategies for controlling disease seems likely in the near future. *In vitro* propagation techniques, the use of micro-cuttings and somatic embryogenesis are techniques which now enable the early discernment of resistant genotypes issued from hybridization. These techniques bypass the need to wait for a number of generations to observe the fixed characteristic. When compared to traditional horticultural methods, they are, therefore, great time-savers in the reproduction process of the genotype.

Unfortunately, however, as reported in Section II.6, resistant characteristics do not necessarily become permanent acquisitions. The use of resistant varieties should be carefully monitored taking into account the fact that pathogens can also modify their behavior to overcome the plant's resistance and even amplify their virulence. Strategies exist to compensate for shortfalls in resistance and they must be implemented without fail.

4.1.4
Biological and Agronomical Control

Apart from the use of pesticides and genetically acquired resistance, it is also worthwhile examining biological and agronomic methods of disease control.

4.1.4.1   Prophylactic Measures

Among the simplest management methods, healthy growing conditions are always beneficial in fighting disease. The sick parts of the plant are the areas where pathogens develop and breed so that, wherever possible, it is better to remove and burn this material in order to avoid further contamination. This was common practice in the old days when little was known about more sophisticated controlling methods and, even now, this practice should not be neglected. "Removing and burning" should be systematically carried out between production cycles as this method delays the inception of infection and, consequently, limits the number of reproduction cycles for the pathogen. It is even possible to "remove and burn" during the cycle which would efficiently prevent the infection from developing into an epidemic. In certain cases, particularly for "Pink Disease" or "Thread Disease" or even for fighting CBD in Arabica, this method is highly recommended. It is also valid when preparing the soil for a new plantation as it eliminates plant debris which could be contaminated by root rot, for instance.

Although prophylactic measures alone will not solve all problems, they will certainly reduce the impact of the disease and diminish the importance of the attacks as well as help to improve the efficiency of chemical treatment.

Prophylactic management methods like removing and burning or protecting accidental injuries with an antiseptic healing paste do not require undue effort, and should always be included in general maintenance practices.

4.1.4.2   Agricultural Practices

Apart from prophylactic management methods, cultivation practices should provide the most favorable environmental conditions for the crop. Here, it is important to distinguish between the different types of infection:

- In cases where the presence of pathogens reveals a physiological deficiency like, for example, "die-back", it is obviously necessary to improve conditions by an adapted input of nutrients, suitable drainage, appropriate shade, etc.
- In cases where environmental conditions like excessive humidity or too much shade may not be detrimental to the coffee plant, but are conducive to the development of pathogens, pruning, trimming and suitable drainage should be implemented.

One must be aware, however, that these methods need to be practiced with caution as they could be more harmful to the plantation than the pathogen itself. In the case of Coffee Leaf Rust, for example, shade tree pruning will increase the effect of the sun on the plant and thus hinder the development of the pathogen, but, at the same time, it will increase productivity. This means that the plant will need an extra input of nutrients, particularly nitrogen, to compensate for its increased production. Furthermore, it has been proved that a higher production of fruit provides more scope to the spread of diseases like CBD and Coffee Leaf Rust itself. It is wise, therefore, to weigh the advantages against the disadvantages when implementing certain cultivation practices to fight coffee diseases.

The above comments go to show that present-day agricultural practices are somewhat limited when it comes to controlling coffee diseases. In future, however, research may well improve our knowledge of specific physiological conditions of the complex host–parasite interaction where corrective and preventive cultivation methods could be implemented to efficiently control it. For the time being, however, we can only recommend that the cultivation of coffee plants takes place in ideal environmental conditions for the plant itself. This will avoid a number of setbacks and ensure a profitable crop to cover the costs of controlling disease.

4.1.4.3 Biological control

Biological disease control is already used to destroy certain insects which attack plantations and, similarly, the idea of using natural enemies to destroy coffee plant pathogens has been explored for quite some time. Hyperparasites of *Hemileia* sp., e.g. especially *Verticillium hemileiae* = *V. lecanii*, frequently accompany *H. vastatrix* whose spores they destroy. This also applies to *Paranectria* spp. which attacks the spores of *H. coffeicola*. To date, however, studies and research into these possibilities have not yet come to any practical conclusions which could be implemented.

For the control of soil-borne pathogens, the use of certain types of *Trichoderma*, reputed antagonists for a number of species, is being seriously considered. *T. harzianum*, which generates antibiotics that hinder the development of filamentous fungi and hydrolytics which destroy the mycelium of pathogens, could efficiently counteract Basidiomycetes (*Armillaria* sp.), Ascomycetes (*Rosellinia* sp.) and various

other fungi of different groups like *Pythium*. Similarly, the highly competitive strain of a non-pathogen, *Fusarium oxysporum*, could be used to prevent the development of other microorganisms in the soil and also because, after inoculation, it produces phytoalexines in the coffee plant which generate a certain resistance. Biological control of soil pathogens with natural enemies can thus be viewed as a hopeful development for the near future, which would be a real breakthrough as soil pathogens are the most difficult to fight.

Some research scientists also hope to use strains of these fungi to control foliage diseases as well, but here research is still at the hypothetical stage.

4.1.4.4 Escape Strategies

A typical example of "escape strategies" can be seen in the method used to reduce the propagation of CBD. This strategy is carried out by using early irrigation to bring forward berry ripening. The crucially vulnerable stage when the cherries develop is shifted to a time when natural climatic conditions are not favorable for the pathogen to become active, thus reducing infection [35]. This is, of course, a unique case; however, research could develop the idea and possibly discover similar escape strategies to be used in other cases.

4.1.5
Eradication Measures

When a new disease starts up in a given region, the first reaction is often to attempt to eradicate it immediately by destroying the first focus of infection. The only condition for successful eradication in this case is to be sure of catching the disease at its onset, before the pathogen has time to develop and spread. Once it has started to spread, eradication becomes illusory. In the early stages, eradication does not only mean the destruction of diseased plants, but also the elimination of neighboring plants which may appear to be healthy but could already be carriers. To be certain of having destroyed the disease at its onset, a careful observation, on a daily basis, of the plants on all the plantations is necessary over a long period. This is a major task which, initially, could well proscribe the practice altogether.

Orange Leaf Rust (*H. vastatrix* Berk. and Br.) is a typical example of the need for vigilance when implementing escape measures.

This disease existed worldwide with the exception of America until 1970; however, the threat was greatly feared by all groups involved in coffee production on this continent and rightly so, at least to a certain extent. As all the plantations had originated from the same material, imported to the Caribbean and Guyana in the 18th century, and this source was known to be receptive to the disease, it was feared that once the pathogen reached the continent, the entire coffee cultivation would be condemned. Nevertheless, as it turned out, the fear was somewhat overstressed as experience had already shown that the great diversity of environmental conditions in the tropical areas of America would lead to an equally varied development of the disease, running from a total disaster to a relatively minor onslaught. Inevitably, when the disease first appeared in the state of Bahia in Brazil in 1970, there was a reaction of panic. Initially, it was decreed that all the coffee trees in the state should be destroyed and a safety belt stretching over 385 km between Rio de Janeiro and Bello Horizonte was set up to prevent the spread of the disease. Fortunately, things calmed down fairly quickly when it was noted that there was no way the disease could be controlled simply by eradication as it had been caught too late and had already contaminated over 500 000 km^2, i.e. an area which was far too extensive to hope to control the spread by eradication [34].

In Nicaragua, Orange Leaf Rust was first detected in 1976 and eradication was immediately planned. In 1977, diseased coffee plants and those within a radius of 20–30 m were destroyed with chemical brush-killers or burnt. Plants on the fringe of the condemned zones were sprayed with a fungicide and the canopy of trees further away was treated preventively [47]. Although official reports claimed that the campaign was a success as the number of diseased trees had significantly decreased, the operation had to be repeated every year as new disease foci appeared. At the end of the campaign, as Orange Leaf Rust only persisted in one isolated zone, it was decided to destroy every plantation in the contaminated area (7 000 ha) and to reinstall them on the basis of intensive cultivation methods with dwarf plants of the Caturra variety which, it was thought, could not be infected as coffee trees in the entire zone had been destroyed. In actual fact, one or two plantations escaped destruction so the costly operation was unsuccessful and Orange Leaf Rust had come to stay. The final result of this stubborn undertaking had done far more harm than the disease could possibly have done over many years. Potential productivity had been ruined in the entire area for a number of years and a great deal of money had been spent to create new plantations. A 55 ha nursery was set up which used 30 tons of seed and employed 4 000 workers in an area which, although distant from the infected zone, turned out to be unsuitable for coffee growing. One way and another, the operation turned out to be a total and costly failure.

It was, actually, the success of the first eradication campaign in Papua New Guinea [48] which led to the conviction that an eradication campaign could also be successful in the tropical zones of America. Conditions, however, were radically different. The insular isolation of Papua New Guinea and the scarcity of coffee plantations at the time were the major keys to the success of the first campaign there, whereas the entire American continent, with very extensive areas of coffee cultivation could not hope to meet with similar success. It should also be noted that when the disease re-appeared in Papua New Guinea in 1986, eradication was not even attempted as, by that time, coffee cultivation had developed to an extent where eradication was no longer justified.

The lesson to be drawn from these various examples is that, as a general rule, eradication campaigns are counter-productive and costly. It is wiser to learn to "live with the disease" and to use chemical or other methods available to keep it under control. It should also be remembered that the presence of any given pathogen does not necessarily lead to total disaster, particularly as environmental conditions frequently limit its impact.

4.1.6
International Exchange

To round off this introduction, it should be recorded that certain serious diseases, e.g. CBD, are still contained on one continent. This means that necessary precautions should be taken when transferring plants. This does not only apply to inter-continental transfers, but also to transfers from one country to another on the same continent, particularly as, due to differences in the

impact of local strains of the pathogen, the same disease has different levels of virulence in each area.

4.2
Cryptogamic Diseases

A great number of various types of fungus can be found on the coffee plant but, as many of them are insignificant, only six or seven serious cryptogamic diseases need to be taken into account from an economic point of view. Others may also have their importance under specific climatic or agronomic conditions.

4.2.1
Nursery Diseases

4.2.1.1 Damping Off

French	Fonte des semis, pourriture du collet
Spanish	Volcamiento, mal del talluelo, Podredumbre del cuello, rhizoctoniasis
Portuguese	Tombamento, Estiolamento, Rhizoctoniose das mudas
German	Keimpflanzenkrankheit
Dutch	Kiemplantenziekte

This is the most frequent disease and the coffee plant is susceptible to attacks at a very early age. It occurs in seed beds and in nurseries alike. It is a type of rot brought on by *Rhizoctonia* – *Rhizoctonia bataticola* (Taub.) Butler, which is very widespread, or by *Rhizoctonia solani* Hühn, which is more often reported in the Americas. It is active in most tropical countries and attacks all species of coffee as well as tea, cocoa, rubber plants, etc (Pict. 4.2).

In seed beds *Rhizoctonia* sp. attacks the collar, rotting rings around the bark of the tiny stem, thus preventing the sap

Picture 4.2 "Damping off" due to Rhizoctoniasis – observe the damage to the collar of the seedlings. From: Nestlé.

from circulating correctly and causing the seedling to wither away. The disease develops in round spots which spread fast in humid soil and can destroy entire seed beds in only a few weeks. In nurseries, the new young roots are attacked first. From there, the main roots are attacked, and, again, the plant dries up and withers.

The most appropriate preventive measure is to disinfect the soil before sowing or transplanting and the best traditional method for disinfecting is to use a solution of formalin at a concentration of 2–3 % sprinkled over the soil with a watering can. Other products like cryptonol are available now, but should only be used in strict compliance with the manufacturer's instructions. In Central America, methyl bromide is used frequently with the added advantage of also acting on small seeds, other pathogens and the nematodes which are particularly aggressive in this area. Due to its toxicity, however, this product is likely to be forbidden in the near future, but it can be replaced by Basamid (Dazomet) which is similar but less dangerous and easier to apply.

Nurseries should also be carefully drained to avoid stagnant water and excess humidity.

Whenever the disease actually appears, the seedlings should be burnt and the soil, as well as the other seedlings in the vicinity, should be sprayed with an appropriate fungicide.

4.2.1.2 Seedling Blight

Causal organism
Fusarium stilboides Wollenw.

Occurrence/distribution
The disease is common in germination beds and nurseries. It is found in most countries where coffee is grown.

Symptoms/disease development
On infected seedlings, the cotyledons often fail to unfold, the stem has necrotic lesions and the seedling wilts. In some cases, necrotic lesions appear on the stalk below the unfolded cotyledons. These lesions enlarge and eventually result in the death of the seedling (Pict. 4.3).

Picture 4.3 "Seedling Blight" – observe the necrotic lesions on the stalk.

The fungus is seed-borne, and can also be spread by insects and rain. Infection is severe during periods of high and frequent rainfall. The optimum temperature for infection is around 25 °C.

Control

The control methods used are hygiene and chemical treatment. Disease-free seeds should be used. Seeds may be treated with Benomyl (1 g of Benomyl 50% WP/kg seeds) before sowing.

4.2.2
Root Diseases

4.2.2.1 Root Rot Diseases

French	Pourridiés
Spanish	Podredumbre de las raices
Portuguese	Pedridas da raiz
German	Keimkrankheiten des Wurzel-systems und der Stammbasis

Causes and symptoms

Major root afflictions, in coffee trees and in most other ligneous species, can be ascribed to very polyphagous fungi, chiefly Basidiomycetes and certain Ascomycetes, more popularly known as "root rotters".

Root rotters are active in a number of very varied conditions and their polyphagous nature indicates that environmental conditions like vegetation (forests) and cultivation practices (the presence of shade trees) are determining factors for their development.

Injuries to roots brought about either by insects or nematodes or, more importantly, by mechanical tools used for cultivation open the way for the fungi to penetrate the roots and to spread from plant to plant. Generating from specific localized sites, the fungi progress via rhyzomorphous mycelium, web-like growths that stretch from root to root. The natural evolution of the mycelium is generally a relatively slow process; however, where mechanical tools spread fragments of mycelium, evolution accelerates and the fungi propagate rapidly through the rows.

This phenomenon was noticed and reported in a high-altitude plantation of Arabica in Cameroon (1650 m) which was afflicted by *Clitocybe elegans*. The first signs of the disease were pinpointed in 1957 and after 9 years only 9% of the trees were diseased. In the following 4 years (1966–1970), however, with more extensive use of mechanical tools, a further 10% of the trees were infected. This represents a fairly typical exponential curve.

Root rot is fairly easy to detect as the leaves wither visibly and the branches decay into dry rot. The plant weakens and finally dies. These diseases are serious as they directly affect the crop and the productivity of the plantation often involving a significant loss of investment capital. The death of the afflicted trees can take place within a few months only, but depends on the hardness of the wood. Diseased trees may live on for a number of years, although the speed at which the disease brings about the death of the tree is directly related to the length of time it has taken to mature. Thus, an Arabica grown at high altitude will resist longer as it has taken longer to develop and its wood is denser, whereas the same species of Arabica grown at a lower altitude will have matured faster and its wood will, therefore, be more vulnerable to rapid decay by root rot.

The causes can be precisely identified by examining the roots and the base of the tree trunks. The more common pathogens can be described as follows:

• Two Ascomycetes, *Rosellinia bunodes* (Berk. and Br.) and *Rosellinia pepo* Pat., generally develop best where the collar of the tree has been injured. Known as black root rot, (French = Pourridié noir, Spanish = Podredumbre negra, Hongos negros de las raizes, Portuguese = Mal de quatro años, Podridão negra, roseli-

Picture 4.4 *Rosellinia* sp. – black root rot. From: G. Castillo.

guese = Podridão branca) The whitish mycelium webs cling to the diseased tissues and along the roots generating small ledges, built by carpophores, at the base of the trunks. This type of rot is prevalent in Africa.

Three types of agaric fungi, often confused by non-specialists, *Armillaria mellea* (Vahl) Quel., or *Armillariella mellea* (Vahl) Pat., which is very widespread, *A. fuscipes* Petch. and *Clitcybe tabescens* (Pers. ex-Fr.) Bres or *Armillaria (Clitocybe) elegans* R. Heim, Bres. These fungi, generally known as agaric rot (French = Pourridiés agarics, Spanish = Podredumbres agaricos) generate an easily recognizable pathology called "split disease" where the diseased organs like roots, collars or stems, split lengthwise (Pict. 4.5 and 4.6).

niose) these fungi generate black rot on the bark and roots where a black mycelium develops groups of round, blue–black spots. This disease is particularly prevalent in El Salvador and although it is relatively unimportant in neighboring countries, who is to say that it could not prevail elsewhere under similar environmental conditions (Pict. 4.4)?

- *Phellinus lamoensis* (Murr.) Heim, causes brown rot of the root tissues (French = Pourriture brune des racines, Spanish = Mal pardo de las raïzes, Portuguese = Podridão parda). Its mycelium mixes with the soil to form scabs on the outer tissues of the root.
- *Leptoporus (ex-Fomes) lignosus* (Klot.) Heim ex-Pat., causes white root rot of all tissues. (French = Pourridié blanc, Spanish = Podredumbre blanca, Portu-

Picture 4.5 Agaric root rot (*Clitocybe elegans*): splitting the collar and the trunk of a coffee tree. From: G. Blaha – CIRAD.

Picture 4.6　Agaric root rot (*Clitocybe elegans*): carpophores of the fungus emerging from the split of the collar and the trunk of the coffee tree. From: G. Blaha – CIRAD.

Picture 4.7　Agaric root rot (*Clitocybe elegans*): cross-section of a coffee tree trunk showing the xylotromes of the fungus developed along the medullary rays. From: R. A. Muller.

"Root splitting", "collar crack" or "stem splitting" comes from the mycelium generating spikes or xylostromes of a few millimeters in diameter with carpophores along the medullary rays of the core of the diseased organ, causing it to split (Pict. 4.7).

C. elegans is prevalent worldwide. In the 1940s and 1950s it caused very serious problems in Madagascar where it attacked both Robusta and Kouillou coffee plants as well as the surrounding shade trees, particularly *Albizzia lebbeck*. Likewise, in Cameroon [10], where it attacked Arabica plantations, including the surrounding shade trees (*Albizzia malacocarpa, Leucaena glauca*).

Root rot control: curative measures

Overall, one can do little more than limiting the spread of the disease. As the pathogens develop via their rhizomorphous network from root to root, and thus from plant to plant, attempts to restrict the spread from an initial site where the disease is festering can be made by digging a circular ditch 60 cm deep around the site to cut the web. When digging the ditch, the contaminated soil should be thrown towards the inside of the circle, not the outside. In practice, the diseased trees and their surrounding shade trees are pulled out.

Special care should be taken to remove as much of the woody roots as possible using a levered hoist if available. Trunks, twigs and roots must be burnt on the spot. Before re-planting any coffee trees, grass, leguminous plants or any other crop impervious to fungi diseases should be sown over the dug-up area to prevent erosion and leaching. Leaving the land fallow has a cleansing effect, but it is slow as illustrated by Tab. 4.1 compiled after an experiment carried out in Cameroon to follow the development of *C. elegans* on a high-altitude Arabica plantation [10]. This method consisted of placing small sticks previously

Table 4.1 Initial results of the "trap-stick" experiment

Period during which the land was left fallow (months)	Percentage of contaminated sticks at a depth of	
	40 cm	80 cm
2	41	59
12	34	35

sterilized and cut from wood of a species receptive to cryptogamic diseases like Hevea or Leucaena in the contaminated soil. The sticks (30–40 cm long and 3–4 cm in diameter) should be placed at varying intervals and depth. If the fungus is still present, they will be rapidly contaminated, hence their nickname: "trap-sticks".

The experiment clearly indicates that "natural" cleansing of the land is a slow process and that it would be a wise precaution to avoid replanting coffee trees in contaminated areas for at least 3–4 years.

Chemical decontamination can, however, accelerate the process. The holes made when uprooting the diseased trees can be treated with Basamid or even methyl bromide, which is commonly used in Central America.

In-depth fumigation of contaminated soil with methyl bromide has also been carried out experimentally on high-altitude Arabica plantations in Cameroon and has shown positive results in the destruction of fragments of C. elegans mycelium which remain in the soil after the tree has been removed. Although the product does not seem to have a direct effect on the mycelium itself, it accelerates the generation of Trichoderma, a natural enemy, resistant to methyl bromide, which efficiently destroys the mycelium.

From an economic point of view, this method is only valid for small, localized sites, but it can certainly be recommended when the disease has been pinpointed in its early stages.

From a practical point of view, the best way to implement the method is to loosen the soil with a hoe and then cover the hoed area with a plastic film. Once this is securely in place with the edges tucked into the soil to make sure it is airtight, insert a pressurized cartridge containing the methyl bromide gas (1 400 g) under the sheet, open it and let the gas permeate the soil. This method is efficient on 9 m^2 (3 m × 3 m) areas centered on the hole where the tree was uprooted. The operation should be repeated once a week for 3 consecutive weeks. When the treatment is spread out over a period of 3 weeks, better results have been obtained than when a more concentrated dose is injected in one treatment only. Table 4.2 illustrates this point.

Even after treatment of a given site, contamination may still spread. This can be monitored by placing trap-sticks on the fringe of the treated area or even further outside the zone. Any residual contamination will be revealed by the white mycelium on the trap-sticks.

Table 4.2 Results after 2 months' treatment with methyl bromide (obtained by the "trap-stick" method)

Quantities of methyl bromide used on 9 m^2 (g)	Number of applications (with an interval of 1 week)	Percentage of contaminated trap-sticks at a depth of	
		40 cm	80 cm
1 400	1	17.70	19.38
	3	1.02	22.45
2 800	2	5.10	9.18
	3	0	4.08

Another method which can be used to limit the extension of root rot is to ring the bark of the trunks of healthy coffee trees in the immediate vicinity of the infected area. The plant can survive for a time on its reserves, but the root rot pathogens, which survive on starches and not on the lignin, starve. This practice entails sacrificing at least one row of healthy coffee trees around the infected area and would obviously be far too costly if many infected areas were scattered throughout the plantation.

In conclusion, there is no simple recipe for controlling root rot, but, possibly, future research should concentrate on encouraging the presence of natural enemies like *Trichoderma*, particularly *T. harzianum* or non-pathogenic strains like *Fusarium oxysporum*.

Once the plantation has been completed, control is still required to detect the development of any pocket of infection so as to eradicate it when still in its early stages. Residual stumps and/or dead wood and even certain shade trees may still expose the young plantation to contamination.

After destroying the stumps and dead wood, a further precaution would be to cultivate a leguminous crop like peanuts or beans, etc., for 3 years on the cleared land prior to planting the young coffee trees. This practice, although not generally very popular with producers, has the merit of cleansing the soil more thoroughly.

4.2.2.2 Mealybug Root Disease or Phtiriosis

French	Phtiriose
Spanish	Ftiriosis
Portuguese	Ftiriose
Dutch	Wortelluisschimmel

Known worldwide, this disease is caused by different species of Mealybugs (*Pseudococcus* sp., *Formicoccus* sp., etc.) spread by ants (*Paratrechyna* sp., *Pheidole* sp. and other similar species). The bugs colonize the collar and the roots of the coffee trees. The conidial phase of *Bornetina coryum* develops on their sweet excretions and, by mingling with the surrounding soil, generates a brown mycelium sleeve-like coating which encircles the vital parts of the plant.

The afflicted coffee trees are weakened and chlorotic. In extreme cases, where the stifling effect of the external mycelium sleeves combined with the internal action of the bug on the sap are not brought under control soon enough, the plant suffocates and dies.

Here, obviously, preventive measures should target both the bug and the ants. The most effective method of control is to sprinkle the base of the coffee trees with a liquid solution of Aldrin or Dieldrin (3–5 g per tree).

4.2.3
Diseases of Aerial Organs

4.2.3.1 Coffee Wilt Disease

French	Tracheomycose
Spanish	Tracheomicosis
Portuguese	Traqueomicose, Fusariose

Causal organism

Fusarium xylarioïdes Steyaert, *Gibberella xylarioïdes* (Steyaert) Heim and Saccas, or *Carbuncularia xylarioïdes* (Heim and Saccas).

General comments

This disease was first detected in 1927 in Oubangui-Chari (now the Central African Republic) and was initially thought to be caused by root rot. The true pathogen was only identified in 1939 by examining samples of dead coffee trees from the Democratic Republic of Congo, former Belgian

Congo. In the Central African Republic the disease spread and developed drastically over the following decade. Initially it chiefly affected Excelsa coffee trees, but by 1949 it also affected C. neo-Arnoldiana [46] and, to a lesser extent, C. canephora var. robusta. At the same time, an overall deterioration of coffee trees in the Ivory Coast was also attributed to this disease where it affected the two local species of C. canephora: the Kouillou Bandama and the Kouillou Touba as well as C. abeokutae, also called Indinié.

All in all, the disease had extremely serious effects in Central and West Africa from 1938 right up until the 1950s. In these areas, it practically wiped out the entire cultivation of C. excelsa and similar species like C. abeokutae. For C. canephora, the Kouillou variety was particularly vulnerable to the disease and, whereas the Robusta variety showed greater resistance to the disease in West Africa [33], it seemed more vulnerable in Central Africa. This leads us to suppose that the pathogen has considerable variations in its virulence.

Researchers and producers alike concentrated their efforts on investigating this disease during the crucial period, but by the mid-1960s interest waned and the disease was practically forgotten. The chief explanation for the decline of the disease can be attributed to the fact that the initial vulnerable plantations which had been wiped out by the epidemic were replaced by the more resistant Robusta variety and also to the regeneration of the plantations, coupled with the development of improved methods of control. In Central Africa, however, Robusta is still more vulnerable to the disease than in West Africa. This can only mean that the disease is still latent and therefore vigilant supervision is still necessary as there is no way to predict how the pathogen will evolve.

The disease was first reported in 1957 (Stewart) on C. arabica in Ethiopia and has spread very little to date. The disease re-surfaced in the Democratic Republic of Congo in the middle of the 1980s and had a significant impact in the East and in the province of Kivu. The disease was reported in Uganda in 1993 and nearly all the coffee areas of this country were attacked. The reason why the disease resurfaced is unknown. Various hypotheses have been suggested like bad sanitary conditions in the plantations or a mutation of the pathogen. Detailed investigation will be required to explain this phenomenon.

Symptoms and biology

The leaves turn yellow, dry up and fall, although sometimes this only happens on part of the canopy (hemiplegia), branches die, and, finally, within a few months or even a few weeks, the entire tree withers and dies. These symptoms, however, can be classed as "first-degree" symptoms. Common to a number of afflictions and potentially due to a number of different causes, they cannot lead to a precise diagnosis of the disease.

"Second-degree" symptoms, more specific to the disease itself, therefore need to be examined. They can be found on the trunks, lodged in small cracks of the bark where tiny blue–black pellets, the perithecium of the pathogen, develop. One way of checking for the disease is to examine the layer underneath the bark where there is a long black strip bordered by healthy areas (Fig. 4.8 and 4.9).

The decay of the coffee plant is due to colonization of the sap vessels by the mycelium, and the tyloses that block water and sap circulation. The infection can set in any time from the cotyledon stage to maturity. The death of the tree depends on the age, the vigor of the plant or the agro-

Picture 4.8 Coffee Wilt Disease on the trunk of *C. canephora* var. *robusta*.
From: R. A. Muller – CIRAD.

Picture 4.9 Coffee Wilt Disease on *C. canephora* var. *robusta*. Observe the black, brownish or purple color of the section. From: GENAGRO.

environmental conditions and can occur within 9–18 months. Both micronidia and macronidia abound shortly before death and when the tree dies. The infection spreads via the micronidia, the macronidia and the ascospores because the perithecia continue to develop for a long time after the death of the tree.

Control
Chemical treatment cannot efficiently protect the trees against the disease. Contamination can be limited by applying a suitable antiseptic healing paste on injuries caused either by pruning, insects or, more importantly, cultivation tools, as such injuries create wounds where the pathogen can penetrate the sap vessels beneath the bark.

In actual fact, little is known about the epidemiology of this disease. At the time when it ravaged the plantations of Central and West Africa, very few studies were made of the environmental conditions which were likely to affect its development. In the Ivory Coast, however, it was noted that the first areas to be affected and those that suffered the most had a dry climate which is not best suited to coffee cultivation.

Recently, in Uganda and in the Democratic Republic of Congo, it was also noted that old and badly maintained plantations were more exposed to the disease. This is, however, very piecemeal information and not much to go on. More serious surveys are needed to clearly define the importance of the effects of the pathogen itself on the one hand and environmental or cultivation conditions on the other.

In the past, attempts were made to prevent the spread of the disease by destroying the affected trees and those in their immediate vicinity either with flame torches or by spraying with carbonyleum to prevent the spores from spreading. It is doubtful that these measures were efficient, as most attempts to eliminate the disease by eradication have failed where cryptogamic diseases are concerned.

Only the careful selection of resistant varieties or clones seems to be the best answer. Important leads, generated by studies carried out in the 1950s on Canephora, Excelsa and, more recently, on Arabica in Ethiopia seem to point towards this solution, and one can but hope that it will provide the answer, particularly at a time when the threat of a serious, widespread outbreak of the disease is again to be feared.

Finally, as in the near future extensive replanting will have to be carried out in Uganda and in the Democratic Republic of Congo, it is to be hoped that this will provide the opportunity for a significant transfer of technology to rural areas so that replanting and maintenance can progress according to the new rules generated by research.

4.2.3.2 Canker or "Machete Disease"

French	Chancre du caféier
Spanish	Mal de machete,
	Llaga macana, cancer
Portuguese	Cancro

This affliction, common to coffee, cocoa and various citrus trees, is caused by a fungus, *Ceratocystis fimbriata* (Ellis and Halst.) which decays the woody tissues of the trunk from the collar to a height of about 1 m. The external symptoms are the withering of the leaves, with dead leaves remaining attached to the branches. In the beginning this may only affect part of the canopy, but shoots can also wither.

This disease owes its name to the machete, a tool which often causes injuries to the tree which lead to the development of canker. In actual fact, however, canker is more frequently caused by insects, particularly *Xyleborus*, which bore the holes where the fungi develop in the wood.

Again, apart from careful use of the machete to avoid injuries and the application of a suitable antiseptic healing paste, little can be undertaken to control the disease. Residual wood should be burnt to prevent the multiplication of the *Xyleborus*.

4.2.3.3 Phloem Necrosis

French	Déperissement à phytomonas du caféier, Nécrose du phloème, Nécrose du liber
Spanish	Necrosis del floéma
Portuguese	Nécrose do floema

As previously mentioned, this disease is not actually a cryptogamic disease.

The plant withers and dies because a flagelliform protozoan, *Phytomonas leptovaorum* (Stahel), develops in the vessels of the phloem causing necrosis of the tissues. This disease only prevails in the northern reaches of South America, Guyana and Surinam.

In this case, however, climatic and environmental conditions are very likely to be more significant factors for the development of the disease than the protozoan itself. In these areas, coffee trees are planted in very humid zones where water frequently floods the soil and drainage is not always sufficient. The submerged roots are asphyxiated by water, and consequently the plant withers and dies. Thus, the presence of the flagelliform protozoan parasite is more likely to be due to the fragility of the tree rather than a pathogen causing the disease. One can only recommend coffee cultivation in the most suitable conditions possible.

4.2.3.4 Leaf rust diseases

Two widespread foliage diseases caused by Uredinales of the Pucciniaceae family, Orange Leaf Rust, also known as Coffee Leaf Rust (CLR), and Grey Leaf Rust are discussed below.

4.2.3.4.1 Coffee Leaf Rust

French	Rouille orangée
Spanish	Roya Herrumbre
Portuguese	Ferrugem do cafeeiro
German	Kaffeerost

Causal organism

Hemileia vastatrix (Berk. and Br.).

Historical and geographic survey

Identified for the first time in 1861 in Ceylon (now Sri Lanka), this disease was long considered to prevail exclusively in Asia and

Africa. By a pure fluke, a Brazilian phyto-pathologist, Arnaldo Gomes Medeiros, identified its presence in a small, more or less abandoned, Arabica coffee plot, in the midst of a cocoa plantation in the State of Bahia in Brazil in 1970. Since then, the disease is known to be prevalent in most coffee-producing countries world-wide. It may seem surprising that the presence of *H. vastatrix* in Brazil was only identified by chance; however, it must be remembered that, in Bahia, at the time, the majority of plantations were given over to cocoa growing, not coffee. Coffee plantations were scattered and did not have any significant impact on the economy of the state. This means that producers and agronomists alike focused on cocoa plantations rather than coffee.

The end result was that once the disease was detected, it became apparent that an important area of coffee plantations were already contaminated. It would seem, therefore, that the pathogen had been present in Brazil for a long time before it was initially detected [34].

CLR is considered to be the most severe foliar coffee disease known to date. This may be due to the fact that it led to the abandoning of coffee growing in Sri Lanka only 6 years after the disease was first identified. Coffee cultivation was then replaced by tea plantations and this factor largely contributed to the widespread consumption of tea in the UK. At the time, of course, no treatment was available to stem the spread of cryptogamic diseases, not even the famous "Bordeaux Mixture". The fact remains, however, that this disease is severe and its impact on the economy is most significant, particularly as Arabica plantations are the most vulnerable and they represent four-fifths of the world's coffee production.

In South America, prior to its detection in Bahia, the merest threat of CLR was liable to generate real panic, as already described. This is, of course, quite understandable as Latin American countries are the major producers of Arabica coffee and, until recently, all their plants stemmed from one specific plant reputedly vulnerable to CLR. On the other hand, the panic reaction was also irrational as it has long been a recognized fact that climatic and environmental factors greatly influence both the impact of the disease on the plant and its potential evolution to epidemic proportions. On the American continent, like elsewhere, climatic diversity means that, although the pathogen is present wherever coffee is grown, the extent of its onslaught ranges from a major disaster to a relatively minor affliction.

Generally speaking, *Coffea arabica* L. is the most receptive of all *Coffea* species to CLR, but, again, receptivity can be more or less active depending on the different varieties (Pict. 4.10 and 4.11).

On the other hand, receptivity has also been noted even in coffee plant species reputedly resistant to CLR, like *C. liberica*, *C. excelsa* or *C. canephora* Pierre, as well as in different varieties of these species like Robusta or Kouillou. This can be explained

Picture 4.10 *Hemileia vastatrix*: leaves of *C. arabica* affected by CLR. View of upper and lower sides of the leaf. From: A. Autrique.

Picture 4.11 *Hemileia vastatrix*: close-up of CLR spots proliferating on a *C. arabica* leaf. From: Bieysse – CIRAD.

Picture 4.13 Spot of *Hemileia vastatrix* (yellow) surrounded by *Verticillium lecanii* parasiting the spores of *Hemileia vastatrix*. The black blotch in the center is composed of necrotic tissues of the coffee leaf. From: R. A. Muller – CIRAD.

Picture 4.12 *Hemileia vastatrix*: leaves of *C. canephora* var. *robusta* affected by CLR. From: J.N. Wintgens.

by the allogamy of these varieties issuing coffee types with different characteristics (Pict. 4.12).

Symptoms

The first symptoms of the disease are small, discolored spots which develop on the underside of the leaves. Within 10 days these small spots increase in size and are powdered with the spores of the pathogen ranging in color from yellowish-orange to bright orange. These cuneiform (wedge-shaped) uredospores have spines along half of their back, hence the name "*Hemileia*", which means "semi-smooth". To date, no other host of propagation has been identified for this parasite; only coffee

plants seem to be receptive to *H. vastatrix*. The spots on the leaves can grow to a diameter of approximately 2 cm and, as their nucleus on the coffee leaf decays, they eventually cover significant areas on the limb. The development of a white pulverulent fungus, *Verticillium lecanii*, which is parasitical to the spores of *Hemileia*, is frequently observed (Pict. 4.13).

The germ tubes generated by the uredospores of *H. vastatrix* penetrate the stoma of the leaves where they form a blister, shaped like an anchor, which develops an abundant, ramified intercellular mycelium (Pict. 4.14).

The mycelium colonizes the tissues of the leaf and, from there, can spread in any direction.

Damage caused

CLR brings about the loss of physiological activity in the affected parts of the limb and causes the leaves to fall. Potent onslaughts of the disease can cause branches to wither completely. This weakens the plant and hinders, or even stops, its development. Frequently, badly diseased and weakened coffee trees do not survive.

Studies made on the loss of physiological activity of the leaves by CIRAD in the

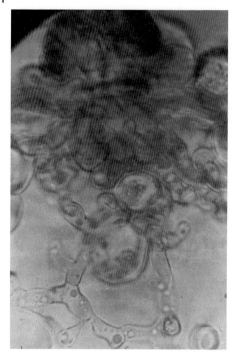

Picture 4.14 Longitudinal cut through a coffee leaf showing the abundant and ramified mycelium of *Hemileia vastatrix*.
From: R. A. Muller – CIRAD.

Centre d'Etudes Nucléaires de Cadarache have ascertained that at least 20% of the leaves have to be affected before any significant disruptions can be detected in the photosynthesis or the photo-respiratory system of the plant. The effects of the onslaught, however, are not limited to the current crop as they have also been seen to affect the following crop by reducing the production of young wood and diminishing stock-piling in the plants [6].

On the other hand, the real damage caused by the onslaught of the disease is assessed in terms of a loss of yield; nevertheless, the level of infected foliage should also be used as an indication of the damage inflicted. To date, however, no real correlation has been established between the extent of the foliar damage and its impact on yield. No precise data exist as yet which would enable a clear prediction of the extent of damage to potential yield. It is generally assumed that 1% of damaged foliage can lead to a yield loss of 1% in the following year. In Brazil, it is thought that, depending on the location and the year, the disease can cause losses which vary from 25 to 45%.

On the basis of general observation, one could come to the conclusion that, essentially, it is chiefly when young leaves are affected that the disease inflicts the most serious damage, as older leaves can be considered to have already lost most of their potential for physiological activity and are due to fall anyway.

Propagation

The spores only germinate in damp conditions, in the presence of water. Epidemics are, therefore, prevalent during the wet season. Rainy spells show an increase in the spread of the disease and periods of intense infection correspond to those of high rainfall.

Germination of *H. vastatrix* spores is also more active at temperatures between 20 and 25 °C, showing its highest activity at 22 °C [38]. At temperatures below or above this level, germination is either slow or nil. The temperature is, therefore, a further important factor to take into consideration.

Another element is the fact that the germination of spores has been observed to take place chiefly in the dark, at night [42].

These observations do not necessarily mean that germination never takes place during the day or in cooler temperatures as, hypothetically, the disease could propagate in the centre of the foliage, particularly under shade trees or on a cloudy day when daylight scarcely penetrates the canopy.

As temperatures are directly related to altitude, the risk of infection can be defined on the basis of this parameter. If, in equatorial zones, the onslaught of the disease is more intense at low altitudes, it will be less so higher up and, in fact, practically nil above the 1 300 m level. Nearer the tropics, the danger level has been drawn at the 800–900 m level [2].

The humidity generated by the presence of shade trees is generally thought to be favorable to the propagation of the disease with a ratio of intensity that increases with the density of the canopy. This "hands-on" observation has, however, been countered by research carried out by CIRAD in France which has revealed that sunlight can increase the plant's receptivity to the disease.

Furthermore, it has also been ascertained [20] that coffee trees, induced to high yield crops by exposure to the sun, are more receptive to *H. vastatrix* [6, 30, 50].

Significantly, it has been noted, too, in both Cameroon and Guatemala, that plants of the same genotype can be receptive to the disease when in the fructification stage and practically immune when in the purely vegetative stage like, for instance, just after stumping. This cannot only be explained by the lack of inoculum which is generated by the leaves of the preceding year.

Physiological observations of this type are worth following up with careful research as they may lead to interesting developments where corrective nutritional measures, like increasing macro- or oligo-elements during the yielding period, can be used as a means to increase resistance to the disease.

Surveys carried out in Honduras by Avelino between 1994 and 1997 have actually established a link between shading, yield and the intensity of CLR infection. Schematically, one can now say that a coffee plantation with a relatively low yield, with or without shade, will only suffer a slight onslaught of the disease, whereas for a plantation where the yield is high, the degree of infection will be more intense in proportion to the density of the shade. In other words, the denser the shade, the more intense the infection will be.

The same author has also noted that a low soil pH combined with a high yield can also predispose coffee trees to the disease. A similar relationship between soil pH and CLR intensity has also been found in New Caledonia [39], although these observations have yet to be substantiated.

H. vastatrix spores, like the spores of other rust diseases, can be carried by the wind over long distances. It has been assumed that spores have been carried by wind across the Atlantic from Africa to Latin America by high altitude currents [12].

Although *H. vastatrix* spores can be carried by wind or by any other factor that moves within the plantation, including people who are active vectors, basically, in any given plantation, the chief source of contamination at the beginning of the infection is to be found in the infected leaves of the preceding campaign. These leaves host the pathogen so that, when the first rains generate the new vegetation, the fungi are also activated to produce new uredospores [35]. Contamination then trails from leaf to leaf and accelerates as the infected spots multiply the production of spores. As residual intrinsic inoculum indisputably plays a major role in the start-up of the infection, the use of a selective defoliant should be considered. This selective defoliant should be geared to target old leaves or leaves that have been previously infected; consequently, their elimination would also destroy the pathogen which would delay and reduce the onslaught of the disease. Experiments along this line that were initiated in Cameroon 30 years ago should be re-activated and developed. Such measures, which aim to destroy the residual inoculum left over from the preceding year, would be all the more successful if they were extensively implemented. If all the plantations in a given zone were treated the same way, the early effects of a residual inoculum strong enough to launch the disease would be avoided, contained or, at least, delayed.

In some areas, like, for instance, the Bamoun in Cameroon, natural phenomena have this very effect. Coffee plantations in this area, situated at an altitude of approximately 1100 m, suffer from drought, aggravated by an ashy-porous soil. This causes the trees to lose practically all their leaves by the time the first rains set in with the result that, due to a lack of residual inoculum, the onslaught of the disease is significantly delayed and little or no damage affects either the ongoing crop or future crops.

An epidemiological survey carried out in areas with a very dry season in the South of the Chiapas in Mexico drew similar conclusions [2].

Another alternative to the use of a selective leaf-killer would be the use of systemic fungicides to destroy or reduce the residual inoculum on old leaves [3, 27]. This method would have the added advantage of preventing the propagation of the pathogen while maintaining the old leaves so that they can still play their original role by transferring nutritional elements to the developing leaf tissues.

Chemical control

As the significant damage caused by the disease has been universally recognized for the last 30 years, research has been actively attempting to develop methods for controlling it.

Chemical control with copper-based products has proved to be efficient if applied at regular intervals at least twice and up to 5 times a year, depending on the area. The first application should take place just after the first rains and subsequent applications should be made approximately every month.

The most widely recognized of the modern fungicides is Triadimefon. Its systemic action, combined with preventive, curative and anti-sporulant properties, generates good results with fewer applications. Other fungicides of the triazole group like, for instance, Cyproconazole or Hexaconazole, have also been successful. Fungicides of this type can complement copper based fungicides and be used alternately.

No rigid rules can be established for the application of fungicide treatments. Epidemiological surveys substantiated by precise experimentation should be established for each zone. They should indicate the sequence, length and number of applications recommended for the area. Treatment should start with the first rains, as soon as possible before the initial attacks appear, and should continue for as long as the pathogens are active.

Obviously, in areas where a very dry season has caused the fall of many infected old leaves, the lack of residual inoculum will cause the pathogen to develop far later. In this case, if chemical control is still necessary, it can be initiated at a later date and spaced out accordingly. It may not even be necessary at all as chemical treatment carried out when the peak of the infection comes about at the end of the berry-ripening period is of very little use and, from an economic point of view, scarcely to be recommended.

On the other hand, treatment should not be stopped too soon. A delayed onslaught of the disease may still infect the leaves and cause a physiological weakening which would affect the transfer of nutritional elements to the new shoots due to carry the following year's crop. This further highlights the need for specific surveys to monitor chemical control in each area.

Genetic control

Specific or "vertical" resistance

Solutions for genetic control have been, and are still being, actively researched.

Operational results have already been ascertained, but care needs to be taken to maintain them and research still has to develop replacement solutions to be implemented in cases where the pathogen has managed to break down the defenses introduced genetically.

Following initial research [32], which now dates back considerably, further research carried out at Oeiras in Portugal by the Centro de Investigaçao das Ferrugens do Cafeeiro (CIFC) along the lines of specific or vertical resistance has identified virulence factors in the pathogen [45]. To date, 40 distinct physiological races of *H. vastatrix* have been recognized as carriers of one or more virulence factors.

The specific resistance mechanism is illustrated in Tab. 4.3 where the virulence factors of the parasite (v) are portrayed by keys and the resistance factors in the host plant by padlocks (SH). Where there is compatibility between a key and a padlock, vulnerability has been established, and where there is no compatibility between a key and a padlock, the opposite resistant phenomenon is noted.

Resistance factors SH6 to SH9 (or SH10) issued from *C. canephora* and found in Timor Hybrids (spontaneous hybrids generated between *C. arabica* and *C. canephora*) are especially interesting.

Timor Hybrid 1 (832/1) represents resistance group "A", and resists all known vari-

Table 4.3 *Hemileia vastatrix*: chart illustrating compatibility (susceptibility) and incompatibility (resistance) between CLR and the coffee tree

		v5		SH5		
Race II		🔑	→	🔓		Group E
	v2	v5		SH2	SH5	
Race I	🔑	🔑	→	🔓	🔓	Group D
	v2	v5		SH5		
Race I	🔑	🔑	→	🔓		Group E
	v1	v5		SH5		
Race III	🔑	🔑	→	🔓		Group E
	v1	v5		SH2	SH5	
Race III	🔑	🔑	⤄	🔒	🔓	Group D
		v5		SH2	SH5	
Race II		🔑	⤄	🔒	🔓	Group D

→ = compatibility (susceptibility).
⤄ = incompatibility (resistance).
From: J. Avelino – CIRAD.

eties of the pathogen and carries at least four resistance factors not to be found in SH1 to SH5.

Timor Hybrid 2 (1343/269) represents resistance group "R" and carries at least two resistance factors not to be found in SH1 to SH5. It is resistant to all known types of the pathogen except eight.

Other types of the Timor Hybrid with different genetic components represent resistance groups 1, 2 and 3, and resist all known breeds of the pathogen with a few recognized exceptions.

To date, this specific research carried out by the Portuguese team of Oeiras and a Brazilian team based at the University of Viçósa has lead to the development of the "Catimor", a coffee hybrid issued from a cross of Caturra (a dwarf Arabica Bourbon very vulnerable to leaf rust) and the Timor Hybrid. Colombian research has developed the "Colombia" and breeders in Kenya have developed the "Ruiru 11". These varieties are now available for use in coffee plantations and offer significant resistance to leaf rust diseases. The same applies to the "Icatu" developed in Brazil, which is an Arabusta (an artificial hybrid from Arabica and Robusta) back-crossed to Arabica without using the Timor Hybrid. It carries the resistance genes found in Robusta and offers good genetic resistance to CLR.

The introduction of resistant varieties issued from the Timor Hybrid should be implemented with all due precaution, bearing in mind the fact that *H. vastatrix* can be easily mutated and that diversity in coffee genotypes stimulates genetic diversity in the pathogen. This is due to the fact that mutants of the pathogen may well find genotypes which suit them in the diversified varieties of coffee trees and will propagate all the more easily. Several indices have been noted which substantiate this theory.

As to the performance of Catimor plants in the field, producers are not unanimous. Many strains have been developed in an attempt to parry the rapid exhaustion of the plant (called "agotamiento" in Central America) which can appear after the first main crop. Small-scale trials are, therefore, recommended prior to committing the entire plantation to the Catimor variety.

To date, the best strains of Catimor are the T5175 and the T8667, but even the T5175 can show serious physiological failures under certain climatic conditions, particularly at low altitudes, in sites where a lengthy drought is followed by heavy rains. The major recognized affliction is known as "black bean" and can affect 80% of the production. In this case, although the berries appear to be healthy, the beans are black and withered.

Moreover, some of these resistant commercial varieties are reputed to have off-tastes, mainly when the yields are very high. In the current crisis period which has led prices to drop to an all-time low, quality is a means of obtaining higher prices. Thus, resistance and high yields are not enough. Today, the real challenge of research is to create resistant varieties whilst preserving cup quality.

General or "horizontal" resistance

To hold a full hand of aces, however, one cannot bank solely on specific resistance as described above. While, at present, the Catimor, Colombia and Icatu offer the most advanced and efficient material, it may not be the case on a long-term basis, particularly as, due to mismanagement, these breeds are already losing genetic resistance.

It is, therefore, advisable to examine the possibilities of a different method of genetic control, by associating "general" or "horizontal" resistance.

Where "specific" resistance reacts qualitatively, in other words the trees are either victims of an onslaught of the disease or they resist it, "general" resistance reacts quantitatively and will only be noticeable once the disease has actually taken hold. Consequently, "general" resistance aims at reducing the intensity of the onslaught and lengthening the period of latency, thus reducing sporulation. Observations made on the reactions of several genotypes to infection substantiate this theory. Pic-

Picture 4.16 Evidence of "general" resistance to *Hemileia vastatrix* in *C. arabica*. For inoculations made with the same pathogen strain under the same conditions, the intensity of the disease is variable depending on the genotype of the coffee. From: D. Bieysse – CIRAD.

Picture 4.15 Evidence of resistance to *Hemileia vastatrix* between Catimor and susceptible Arabicas in the field. From: J. N. Wintgens.

tures 4.15 and 4.16 clearly illustrate significant differences in the reactions of different genotypes to standard inoculations made on the same date.

As all the aspects of the resistance can be measured, the results can be determined globally under the heading "Intensity Index". This enables straightforward comparisons between the various genotypes observed (Fig. 4.1).

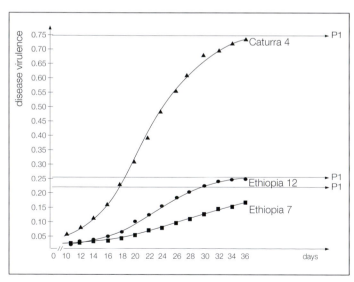

Figure 4.1 Expression of the "general" resistance to *Hemileia vastatrix*. Comparison of the evolution of different intensities (Intensity Index) of *Hemileia vastatrix*, race 2 on three distinct genotypes of *C. arabica*. From: J. Leguizamon.

Recent research carried out at CIRAD has revealed that this type of resistance exists naturally in Arabica, particularly in certain wild genotypes to be found in Ethiopia, in "vulnerable" varieties of Canephora and in Catimor genotypes which have lost their "specific" Timor Hybrid resistance factors [23, 26, 28]. Earlier, less in-depth, research carried out in Brazil had already discovered this type of resistance both in *C. arabica* and in varieties issued from the Timor Hybrid [13, 14, 17–19, 52].

To date, research carried out on this type of resistance has been promising, but has not yet been developed sufficiently to justify implementation. A few questions remain unanswered, particularly the fact that, although distinct evidence of "general" resistance can be noted in young coffee trees, it seems to dwindle as the tree grows older. Added to which, fructification increases the "vulnerability" of the coffee tree to the pathogen.

Research should be carried out to examine the real potential of "general" resistance and the methods to be set up for its implementation as it may well be the best answer in cases where "specific" resistance is either lost or greatly reduced. The two methods of genetic control could also be complementary as "general" genetic control could be used to back up "specific" genetic control and ensure its efficiency for a longer period.

It should also be remembered that the quality of the coffee produced from genotypes issued from the Timor Hybrid is not fully satisfactory; this criterion would justify the need to introduce pure Arabica leaf rust resistant varieties issued from the Ethiopian genotypes.

4.2.3.4.2 Grey Leaf Rust

French	Rouille farineuse, Rouille grise
Spanish	Roya harinosa

Causal organism

Hemileia coffeicola Maublanc and Roger.

Historical and geographic survey

Noticed for the first time on Arabica in Cameroon by Coste in 1929, Grey Rust was given its name by Maublanc and Roger. In 1953, Muller first noted its appearance on Robusta, again in Cameroon. Later on, Saccas noted that it was also present on Excelsa and *Coffea de la Nana* in the Central Africa Republic. Grey Rust spreads slowly and, to date, is only prevalent in Central African (Cameroon. Central African Republic, Nigeria, Sao Tome and Angola). In Togo, it has been noticed on a Robusta clone by Partiot and, in Ivory Coast, by Lourd on certain varieties of wild coffee. In the latter case, however, although this has yet to be checked, there is some doubt as to whether it might not be another, morphologically similar, type of the disease.

Picture 4.17 Comparison of the symptoms of attacks of *Hemileia vastatrix* (right) showing clearly defined spots and *Hemileia coffeicola* (left) showing unevenly spread blotches on the leaf surface. From: R. A. Muller – CIRAD.

Symptoms

The macroscopic and microscopic characteristics of Grey Leaf Rust make it easy to distinguish from CLR. This is chiefly due to the shape of the spots generated by *H. vastatrix* and *H. coffeicola*, respectively. Where *H. vastatrix* spots are clearly defined and contained, *H. coffeicola* shows uneven blotches spread on the underside of the leaf. The leaves do not lose their color, they are just covered by the yellow spread which is the first visible symptom of the disease (Pict. 4.17).

From a microscopic point of view, the mycelium inside *H. coffeicola* is not abundant with a number of tiny haustoria like *H. vastatrix*. On the contrary, it generates only two or three large intercellular filaments with no ramifications. Each filament is terminated by a large globular and complex sucker which invades the cells of the host. The spines on the back of the uredospores of *H. coffeicola* are also better defined than those of *H. vastatrix* (Fig. 4.2).

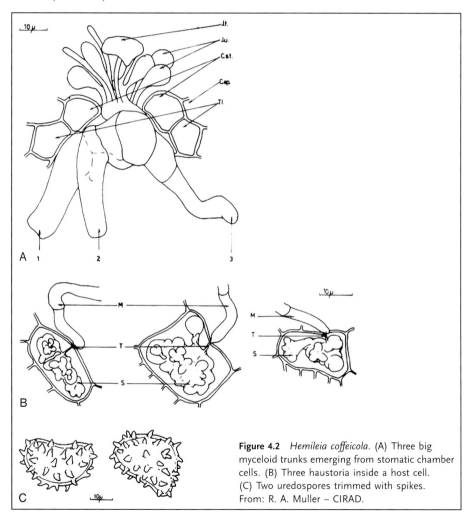

Figure 4.2 *Hemileia coffeicola*. (A) Three big myceloid trunks emerging from stomatic chamber cells. (B) Three haustoria inside a host cell. (C) Two uredospores trimmed with spikes. From: R. A. Muller – CIRAD.

Damage

From a macroscopic point of view, Grey Leaf Rust is less visible and, therefore, less easy to detect than CLR. Consequently, planters tend to minimize its importance. In Cameroon, however, surveys have shown that Grey Leaf Rust disrupts photosynthesis and the assimilation of phosphorus and that it causes considerable loss of leaves, although not quite as much as the loss of leaves caused by CLR. It can also be a factor which seriously curtails production as intense onslaughts of the disease can either significantly reduce or totally destroy the crop yield of the following year [35].

Propagation

As, geographically speaking, *H. coffeicola* is relatively localized and its discret symptoms are not taken too seriously, this disease has not been extensively studied except in Cameroon. Like *H. vastatrix*, *H. coffeicola* propagates fastest in wet weather and its annual development depends on rainfall. Surveys carried out in Cameroon at high altitudes reveal that the first symptoms appear approximately 1 month after the first rains and that the development of the epidemic is rapid. The peak of the attack on coffee leaves can be very high towards the middle or the end of the rainy season.

Apart from the fact that the rainy season favors the development of *H. coffeicola*, no other specific environmental conditions seem to be required for the disease to spread. While *H. vastatrix* does not proliferate at high altitudes, *H. coffeicola* can be as dangerous to high-altitude plantations as to those at lower altitudes. It has been abundantly detected in high-altitude plantations as well as on Arabica and on "susceptible" Robusta genotypes right down to sea level (Tab. 4.4). This clearly indicates that Grey Leaf Rust is a potential danger to coffee cultivation worldwide and coffee producers should be more aware of the threat of this pathogen.

Table 4.4 Chart showing the degree of virulence of *Hemileia coffeicola* according to altitude

Species	Locations	Cultivars	Years	Aggression rate (%)
C. arabica	High altitude, 1 650 m		1970	14.82
	Babadjou	Jamaica	1971	22.42
			1972	37.16
			1973	16.81
	Low altitude, 1 110 m	Jamaica		12.84
	Foumbot	Bourbon Mayaguez	1974	36.52
		Mulungu 5		39.65
		Ethiopia 37 (217/9)		38.11
C. canephora var. robusta	Hot zone, 175 m Barombi-Kang	B 49		38.87
		C 1	1974	36.52
		B 21		31.68
		N 3		28.15

From: Tarjot and Lotode [52].

Chemical control

Chemical control is possible. Copper-based products can be efficient and the most efficient is cuprous oxide (Red Copper) with a 50% Cu content and used as a brew at 0.5%. In Cameroon it has been shown that three treatments per year at monthly intervals are enough to control the onslaughts, even on irrigated plantations. At high altitudes, it has also been noted that cupric treatments used against CBD are also efficient against Grey Leaf Rust [35].

Triazoles, used to control *H. vastatrix*, can also be used to control *H. coffeicola*.

Genetic control

Little is known about genetic control of Grey Leaf Rust but, according to observations made in Cameroon by [52], it would seem to be possible. A collection of Arabica including approximately 100 of Ethiopian origin has been classified and compared according to their level of resistance in the field. It is also evident that the Canephora species holds genotypes which offer different levels of "susceptibility". On the other hand, apparently, genes that resist *H. vastatrix* would not necessarily be the same as those that resist *H. coffeicola*.

4.2.3.5 American Leaf Spot

French	Petite vérole, Maladie américaine des feuilles
Spanish	Ojo de Gallo, Gotera, Viruela, Mancha americana
German	Amerikanische blattkrankheit
Dutch	Stilbella-ziekte

Causal organism

Omphalia flavida (Maublanc and Rangel) = *Mycena citricolor* (Berk. and Curt.) Sacc.

Historical and geographic survey

This disease is essentially prevalent in Latin America; more specifically, in Costa Rica, where it has a significant impact, and in the Caribbean. It affects all the organs, stems, branches, fruit and leaves of all cultivated coffee trees. Its onslaught is not limited to coffee trees as it also attacks a number of plants of the family of the Leguminosae and Compositea as well as a few adventitious plants, which could well be fungus banks, representing a potentially active source of contamination for coffee trees.

Symptoms

American Leaf Spot is recognizable by its numerous round spots which appear on the limbs of the leaf. The tissues of the spots wither and fall out, leaving a hole in the leaf which, eventually, also falls. As a general rule, these round spots vary in color from chestnut brown to a deeper brown, but other types of spots, reddish in color and more jagged in shape, have also been detected (Pict. 4.18).

The Basidiomycete from the Agaricales order, named *Mycena citricolor* in 1950 by Dennis, is the responsible pathogen. It has two types of fructifications: vegetative and sexual. The vegetative fructification looks like a series of little headpins a few

Picture 4.18 *Mycena citricolor.* View of an attack on a coffee leaf. From: J.N. Wintgens.

Pictures 4.19 The two upper pictures: Asexual fructifications on a coffee leaf and fruit shaped like small headpins. Lower picture: Carpophores on a coffee leaf. From: Sandoz, Guatemala.

millimeters high, and the sexual fructifications are carpophores with a diameter of a few millimeters and 1 cm high. The latter are yellow and of a mossy consistence, but as they usually appear on fallen leaves, they are not believed to be active in spreading the disease (Pict. 4.19).

Propagation

A survey carried out in Guatemala by Avelino [6] revealed that, in that country, near to the tropics, American Leaf Spot is particularly serious at high altitudes, above 1200 m. Its development relies on very damp conditions and chiefly prevails in plantations where bad maintenance is combined with excessive shade. On the other hand, it also prevails in the absence of shade, but in specific environmental conditions, at high altitudes where the weather is often cloudy or foggy and persistent rains are frequent.

The disease can cause losses of up to 35 % of the expected yield over 2 years [7]. These losses are due to the infected fruits which drop and to the negative impact of the disease on the growth of the branches which, in turn, affects the following crop.

Chemical and agronomical control

For many years this disease was controlled by spraying lead arsenate, but this practice is no longer authorized so growers have resorted to more traditional, copper-based fungicides like "Bordeaux Mixture" which is also proving to be efficient [5]. It is worthwhile noting that the calcium phase of "Bordeaux Mixture" actively controls the disease by preventing the development of oxalic acid, the toxin generated by the fungus when penetrating the leaf of the plant [40]. Modern triazoles with systemic effect like Cyproconazole or Hexaconazole can also be used, either on their own or alternating with cupric fungicides, in order to reduce any residual inoculum after an onslaught which would otherwise generate sources of contamination for a further onslaught.

Correct maintenance methods like weeding, pruning and shade control are also necessary as they either prevent the onslaught of the disease or, at least, reduce its intensity. A well-ventilated plantation and stumping in alternate rows will offer fewer possibilities for the disease to develop and spread. Surveys carried out in Guatemala have shown that the percentage of diseased leaves on adult coffee trees in a row next to a stumped row could be as little as 15 %, whereas the same trees in a row next to another non-stumped row of adult trees could have up to 45 % of diseased leaves [5].

Genetic control

As American Leaf Spot originated in Latin America and coffee plants were initially imported from Africa, their native continent, there can be no parallel evolution between the host and the pathogen. There is, therefore, little chance of developing "specific" resistance. This means that possibilities for "general" resistance would need to be examined. To date, very little research has been carried out along these lines, although comparisons between Catimor and Catuai, both inoculated cleanly, have pointed to different levels of resistance to both the penetration and the fructification of the disease. Catimor appears to be more resistant to penetration and Catuai more resistant to its fructification [37].

Resistance of this type, however, tends to be ineffective in places where the environment is highly favorable to the fungus. For more conclusive results, the scope of investigation would need to be extended to more varied genotypes, including Ethiopian Arabica or interspecific hybrids. On the other hand, different levels of field resistance could be pinpointed by taking the structure of the plant into consideration – the more compact the plant, the higher the contagion and the greater the development of the disease within the plant itself.

4.2.3.6 Blister Spot

French   Maladie des taches huileuses
Spanish   Mancha mantecosa,
           Mancha aceitosa
Portuguese Mancha manteigosa

The economic impact of this disease is relatively low, but it is mentioned here because specialized literature highlighted it in years gone by, F. L. Wellman in 1961 and 1972. It remains, however, localized in America and is characterized by small light green, greasy-looking translucent spots which appear on the leaves. These spots, which are slightly bulbous in appearance, spread over the leaf and link; when this occurs, they decay and cause the leaves to fall which, in turn, causes the branches to whither. Young leaves are particularly vulnerable. Frequent attacks lead to the decay and the death of the entire tree. This disease is to be found mostly in nurseries. Initially, it was thought to be generated and propagated by aphids (*Toxoptera auratii*), but more recent studies seem to point to a cryptogamic disease generated by *Collectotrichum* sp.

In 1991, Matiello mentioned it as being especially virulent in Conillon of the *C. canephora* variety where, in Brazil, up to 15 % of the plants were exposed. Icatu would seem to be less vulnerable and Arabica very rarely attacked by this disease in Brazil. In the state of Vera Cruz, in Mexico, however, Blister Spot infected a high percentage of Arabica plants.

Chemical control is possible, but considered, in Brazil, to be too costly as several repeated treatments with systemic fungicides, like Benlate (no longer authorized), would be needed to ensure efficient control. Other methods would, therefore, need to be researched and implemented as the economic impact of the disease would justify the cost of the research.

4.2.3.7 Brown Eye Spot or *Cercospora* Blotch

French   La maladie des yeux bruns, Cercosporiose

Spanish   Mancha parda, Mancha de hierro

Portuguese   Mancha do ôlho pardo

German   Braunaugenkrankheit

The causal organism of the disease is *Cercospora coffeicola* (Berk and Cooke).

Causal organism

Cercospora coffeicola (Berk. and Cooke).

General comments

This disease attacks the leaves of coffee trees of all ages, but its impact is most serious on young coffee plants both in germination beds or in nurseries. The infected plants may lose the majority of their leaves, or even all of them, and the berries can also be diseased. Arabica trees seem to be the most vulnerable.

The symptoms which appear on the leaves are round, grey–brown spots approximately 1 cm in diameter and with a lighter-colored dot in the center (Pict. 4.20).

Concave pressure points, known as "thumb dimples", appear on the berries. These marks are brown and covered with a visible grayish fuzz which is the fungus

Picture 4.20 *Cercospora coffeicola* – Brown Eye Spot. Attack on a coffee leaf. From: F. Kohler IRD.

fructification. Frequently, these marks are rimmed in red and the surrounding fruit pulp tends to ripen prematurely.

As a general rule, the coffee trees which show these symptoms are more yellow than green because this disease is, typically, a disease due to a physiological deficiency within the plant – the fungi only develop as a result of the general condition of the plant. Studies carried out on Arabica in Cameroon and in Colombia and on Arabusta in the Ivory Coast have revealed that the principal causes of this disease are a lack of nitrogen and, possibly, potassium as well as an over-exposure to sunlight. The lack of essential minerals is aggravated by sunlight as the sun activates production regardless of nutritional deficiencies. It also activates the growth of grasses which are known to be greedy consumers of nitrogen [35].

Consequently, obvious methods for controlling this disease reside in correct maintenance of the plantation as a whole, with special care given to weeding out the seedling beds and the nurseries. Nitrogen fertilizers are also recommended. Fungicide treatments may reduce the presence of *Cercospora* but they do not solve the basic problem.

4.2.3.8 Other Smudgy Foliar Diseases

4.2.3.8.1 Coffee Canker

French   La lèpre du caféier

Spanish   Lepra del cafeto

This is caused by green algae, *Cephaleuros virescens*, which develop on the leaves. Initially reddish-brown and of a velvety texture, these smudges become grayish or greenish and are purely superficial. They are generally due to an excess of humidity and do not cause much damage. In the worst case, they can hamper the assimilation of chlorophyll and are only mentioned here because they are the object of widespread discussions. They can be compared to lichen which develop on old leaves in wet climates and are no more important.

4.2.3.8.2 Sooty Blotch Mold or Sooty Fungus

French	Fumagine
Spanish	Fumagina, Carbon
Portuguese	Fumagina
German	Russtaupilze
Dutch	Roetdauwschimmels

Sooty Blotch Mold is a fungus which develops on the sweet excretions of aphids and scale insects, most often *Capnodium coffeae* Pat. As it covers the leaves, the branches and even the berries with a thick, black crusty coating, this disease looks more virulent than it really is. Basically, it chiefly only hampers photosynthesis but, more significantly, it indicates the presence of an unwarranted number of aphids and scale insects. Treatment should, therefore, aim at eradicating these unnecessary hosts as well as the ants which transport them (Pict. 4.21).

Picture 4.21 Sooty Black Mold. Observe the whitish mealy bugs and black papery films of fungal growth on leaves. From: Nestlé.

4.2.3.9 Thread Blight Diseases

Traditionally, the different types of Thread Blights were considered to be of minor importance as they were chiefly found on neglected coffee trees or those under excessive shade in a damp environment. More recently, however, Thread Blights have been detected in non-shaded plantations, although at low altitudes and in damp climates in various Central American countries. It has also been noted that they prevail in high density plantations which use dwarf varieties of Arabica, especially Caturra, by maintaining confinement. They have become a real problem in western Guatemala and are also a serious concern in the Chiapas state of Mexico as well as in the Lake Yojoa region and in the Santa Barbara department of Honduras.

The increasing impact of Thread Blights may be due either to the evolution of agricultural practices or to the fact that coffee-growing areas have now extended into zones which favor the development of the pathogen.

The three major Thread Blights are Koleroga or Black Rot, Marasmoid Thread Blight and Horse Hair Blight. All three are caused by Basidiomycetes.

4.2.3.9.1 Koleroga

French	Maladie des fils blancs
Spanish	Mal de koleroga, Moho de hilachas, Infierno del cafeto, Mal de hilachas
Portuguese	Koleroga
German	Silberdrahtkrankheit

Caused by *Pellicularia koleroga* (Cke.) = *Corticium koleroga* (Cke.) Höhn, this disease prevails on every continent and on many varieties of woody plants like citrus, ficus, cocoa, etc. Fine whitish threads develop on the branches and the twigs, and reach the leaves which they cover in a loose web. The leaves wither and die. They then fall off and hang in the web of the fungus (Pict. 4.22 and 4.23).

Picture 4.22 *Corticium koleroga.* (1) Infected leaves and twigs. (2) Dead leaf still hanging on the twig but retained in the web developed by the hypha of the fungus. (3) Cut through the mycelium and the parasite. From: Bayer.

Picture 4.23 *Corticium koleroga.* Infected branch. Observe the whitish web developed on the twig and the leaf. The progress of the disease is centrifugal. From: J. N. Wintgens.

4.2.3.9.2 Marasmoid Thread Blight

French   Maladie de la toile d'araignée
Spanish   Enfermedad de la telaraña, arañera

Caused by *Marasmius scandens* Mas, this disease is similar in appearance to Koleroga. A fine film of grayish mycelium envelopes the leaves and the twigs, causing the leaves to wither and die. Again they remain hanging within the enveloping mycelium.

4.2.3.9.3 Horse Hair Blight

French   Maladie du crin de cheval
Spanish   Moho de hilacha

Caused, in this case, by *Marasmius equicrinis* (Muell.), long thin black threads develop on the leaves and the twigs. These threads are capped by tiny fungal carpophores. As in the two previous cases, the diseased leaves wither and die only to remain hanging by the filaments.

The best way to avoid attacks of this type of fungus is, obviously, correct maintenance of the plantation. This entails strict control of shade trees or their removal and careful pruning of the coffee trees themselves with a view to eliminating infected branches. All diseased branches should be cut off and burnt. Chemical control with cupric compounds or other fungicides can be used in extreme cases, but further investigation is needed to pinpoint the most efficient fungicides.

4.2.3.10 Die-back

French   Anthracnose des rameaux
Spanish   Antracnosis de las ramas,
          Muerta regresiva

When this disease attacks a coffee tree, the branches wither from their tips down towards the older limbs and the core of the plant, hence the name "die-back". Leaves and berries on the affected branches also wither and fall (Pict. 4.24).

Fungi of different varieties have been associated with this disease, but the most common cause is *Colletotrichum coffeanum* Noack which develops on mummifying tissues. This *Colletotrichum* is often associated with *Cercospora coffeicola* as pathological characteristics known in Central America by the all-inclusive name of "chasparria negra".

It should be clearly noted, however, that this *Colletotrichum* is not to be confused with *Colletotrichum kahawae* Waller and Bridge (ex. *Colletotrichum coffeanum* Noak *sensu* Hindorf) which causes CBD.

In fact, die-back is a physiological disease generated by unsuitable environmental conditions, over-production and inadequate cultural methods.

As a general rule, any circumstances that weaken the coffee tree can lead to die-back. More specifically, these can be one or a combination of the following:

- Over-exposure to sunlight with deficient nitrogen nutrition
- "Hot and cold" shocks due to sharp differences in temperature, during the day and the night
- Difficult acclimatization of certain genotypes
- Over-production with deficient compensation in fertilization
- Exhaustion after the first major yield in some Catimor varieties
- Intense attacks of CLR with loss of leaves and general weakening of the plant
- Inadequate soil maintenance
- Erosion which prevents the soil from retaining moisture and essential nutrients.

Pictures 4.24 *Colletotrichum coffeanum.*
The fungus causes the necrosis of the still green terminal tissues of the young twigs.
From: A. Autrique.

Picture 4.25 Coffee cherries affected by *Colletotrichum coffeanum* (Anthracnose) showing brown concave lesions on ripening cherries.
From: G. Castillo.

Consequently, the most efficient way of controlling die-back entails the practice of sustainable cultivation. In plantations where Arabica is farmed intensively, the entire range of technical support and input available should be correctly implemented. Dwarf varieties, like Caturra, which can be grown densely in full sunlight, need exacting treatment. Fertilization should be adapted to the native soil fertility and rates should adequately support the high yield expected of the coffee tree, particularly where the soil is not fertile enough.

It should also be noted that shade generally protects coffee trees from die-back but, at the same time, it is not conducive to high yields. The more modest yields on shaded plantations do not exhaust the coffee plant to the same extent as the high yields harvested on sun-exposed plantations.

On *C. canephora* var. *robusta*, the die-back *Colletotrichum* attacks the leaves where it generates brown spots rimmed in yellow and induces defoliation (Pict. 4.26).

When this serious affection occurs in propagation plots, it stints the vegetative growth of shoots assigned to the production of cuttings.

Picture 4.26 *Collectotrichum coffeanum*.
Attack on the leaves of *C. canephora* var. *robusta*.
From: R. A. Muller – CIRAD.

Only a limited number of leaf nodes remain on the afflicted plants which decreases their potential for cutting production. This is yet another aspect of physiological weakness, although it has not been possible to determine the exact cause. Some Robusta clones are susceptible when they are very young and seem to develop with no after-effects when they are cultivated at a later date provided they are treated with due care. These symptoms, however, should be taken as pointers to possible poor acclimatization. Susceptible clones should be discarded before they are distributed to farmers as they may never develop into productive trees.

Collectotrichum coffeanum can also cause the small, discolored spots found on Arabica known as "weak spots". This minor affliction has no economic impact.

4.2.3.11 Pink Disease
French   Maladie rose
Spanish   Mal rosado
Portuguese Mal rosado

Caused by *Corticium salmonicolor* Berk. and Br., this disease covers the bark of stems and branches with a powder-like salmon-pink coating which pales later on and thickens into a crust. The diseased parts wither and die (Pict. 4.27 and 4.28).

Little is known about this disease which, luckily, does not cause much damage, but which can sometimes have unexpected consequences. Reputedly linked to excessively damp conditions, it has also been known to develop in full sunlight.

Pink Disease can be treated with cupric fungicides, but the diseased branches should be removed and burnt.

Tephrosia candida is known to be particularly susceptible to this disease and should,

Pictures 4.27 *Corticium salmonicolor* – Pink Disease. The fungus produces a salmon-pink crusty coating on the surface of the bark which pales later on. From: Nestlé.

Picture 4.28 *Corticium salmonicolor* – Pink Disease. General aspect of a coffee tree severely affected by Pink Disease. From: J. Avelino – CIRAD.

therefore, not be used as temporary shade or as a cover plant in coffee plots.

As is the case in cocoa cultivation, different levels of vulnerability to Pink Disease have been detected between coffee varieties and clones.

4.2.3.12 Burn or Blight

French   La Brûlure des pointes des rameaux
Spanish   Requemo, Derrite, Quema
Portuguese Phoma

Causal organism
Phoma costarricensis (Pict. 4.29).

General comments
This disease prevails more specifically at high altitudes in Central America but it can be found throughout the American continent.

The initial symptoms of "Burn" appear in the form of small necrotic spots of an undefined shape on the edges of the young limbs of the first pair of leaves. When the attack is of minor importance, the leaves grow in a sickle shape as the limb does not develop at points where the fungus has taken hold. When the attack is more intense, the dark spots spread to cover the entire surface of the leaf. Tiny nubs then appear, these are the pycnids of the pathogen which will develop the

Picture 4.29 *Phoma costarricensis.* Affected leaves showing necrotic tissues with the typical non-symmetric development into the sickle shape. From: J. Avelino – CIRAD.

spores. The terminal buds are attacked as well as the young shoots which then appear to be totally necrotic. This happens when *Collectotrichum* sp. combine their activity with phoma. This form of the disease is more frequently to be found on young plants or stumped plants and can hinder, or even prevent, the growth of coffee plants. These attacks are often associated with insect damage.

The propagation of the disease is greatly increased by low temperatures, persistent fine rains and winds. Adequate shade and wind-breaks can efficiently prevent it. In the more exposed zones, fungicide treatments are also recommended. Previously, Captafol was most efficient but since it has been forbidden, cupric fungicides or organic fungicides like Rovral are advised.

4.2.4
Berry Diseases

4.2.4.1 **Coffee Berry Disease (CBD)**

French	Anthracnose des baies
Spanish	Antracnosis
Portuguese	Antracnose
German	Kaffeekirschenkrankheit

Causal organism

The fungus which causes this disease is a *Colletotrichum* known for many years as *Colletotrichum coffeanum* Noack as no distinct morphological characteristics could be attributed to it. Later on, to highlight its specific action on berries and to distinguish it from the *Colletotrichum* which attack leaves, this variety of *Colletotrichum* was called *"virulans"* [41] and then, in honor of the extensive studies on the action of *Colletotrichum* on coffee contributed by [25], its name was changed to *Colletotrichum coffeanum sensu* Hindorf. Finally, in order to avoid further confusion and to highlight the specific characteristics and impact of this pathogen, it is now called *Colletotrichum kahawae* Waller and Bridge [55].

General Comments

CBD was first detected and identified by McDonald in Kenya in 1922. It spread rapidly, first to the Kivu district of the Democratic Republic of Congo, and then on to Uganda, Burundi, Ruanda, Tanzania and Angola. In Cameroon, the disease was identified in 1958 by R. A. Muller and, far later, in 1970, in Ethiopia [36]. It only affects Arabica plants and, to date, is limited to the African continent.

The impact of the disease on productivity is stronger in high-altitude plantations, above 1 500 m in the tropics, but, occasionally, it takes hold in plantations at lower altitudes which have a specifically cold and damp micro-climate, similar to the prevalent climatic conditions found at higher altitudes.

From the point of view of Arabica cultivation worldwide, CBD would not appear to be a major problem. On the other hand, its impact has significant consequences on losses. Surveys carried out over a number of years in Cameroon by Muller

[35] and Bieysse [8], reveal losses which can be as high as 80%, or more, of the total production. The intensity of the attacks is closely related to climatic conditions as persistently cool, wet weather favors a rapid development and spread of the fungus. If, at the same time, the berries are plentiful and dense, the fungus propagates even faster and the impact of the disease is disastrous. CBD is a serious threat for all the main growing zones, particularly in Latin America.

Though CBD is still contained on the African continent, the threat of its spreading to other continents is real. South and Central America are particularly vulnerable as Arabica cultivation is widespread and climatic conditions in many areas are similar to those found in the high altitude plantations of Africa (Pict. 4.30).

Symptoms
CBD can be found on all berries at every stage of their development but only the symptoms detected on young green berries can lead to a clear diagnosis of the disease.

On young berries, symptoms appear in two forms:

- "Scabs" which are buff-colored, slightly concave spots of irregular shape with scattered black pinpoints. This form is of minor importance and does not greatly impact the yield.
- The "active" form of the disease, on the other hand, generates darker wounds which are more concave and carry a greater quantity of acervuli on the injured surface. The pinkish mass of spores develops rapidly. The fungus destroys the pulp and the bean in a few days, leaving little black, empty pouches instead of healthy young berries. The berries are dry and mummified and most of the afflicted berries fall off. The "active form" of the disease causes losses (Pict. 4.31–4.33).

It is interesting to note that the "active" form of the disease chiefly attacks young berries, whereas the "scab" form is found on more mature berries. This would seem to indicate that the "scabs" are, in reality, generated by resistance of the tissues to the disease.

CBD symptoms which appear on mature berries are not characteristic of the disease. Known as "Brown Blight", they closely re-

Picture 4.30 *Colletotrichum kahawae* – CBD. Berries affected by CBD. From: A. Autrique.

Picture 4.31 *Colletotrichum kahawae* – CBD. Young Arabica berry affected by the "scab" form of the disease. The concave fawn-colored spots are of irregular shape, slightly corky. A few black specks called acervuli arise on the surface of the spot. The lesion evolves slowly and without gravity. From: R. A. Muller – CIRAD.

Picture 4.32 *Colletotrichum kahawae* – CBD. Young cherries of *C. arabica* affected by the "active" form of the disease. The dark brown concave spots look slimy. They are colonized by many well-defined pinpoints, the acervuli of the pathogen which produce the spores. This damp rot evolves rapidly and affects the pulp of the fruit as well as the seeds. From: R. A. Muller – CIRAD.

Picture 4.33 *Colletotrichum kahawae* – CBD. Close-up of a young Arabica berry affected by the "active" form of the disease which has reached the final stage of decay. The fruit is completely destroyed. From: R. A. Muller – CIRAD.

semble the attacks of other saprophytes (*Colletotrichum*) and extend to the pulp of the berry to cause premature ripening or softening. They have very little impact on the crop itself as the coffee beans are rarely damaged and only occasionally touched by the blight.

To complete the diagnosis and in order to avoid confusion, other significant aspects should be highlighted:

- CBD attacks very green berries on coffee trees which are otherwise healthy and have well-developed vegetative growth. It should not be confused with attacks of *Cercospora coffeicola* which attacks the leaves and berries of trees suffering from a lack of nitrogen. These trees are already yellowed and the symptoms caused by the pathogen have a different appearance.
- CBD can destroy practically all the berries on the coffee tree without having any impact on the leaves or the branches. This point is very significant when it comes to establishing a correct diagnosis of the afflicted coffee trees and clearly distinguishes the pathogen which attacks berries from other pathogens which attack leaves, branches or the entire tree.
- Berries destroyed by CBD are in no way to be confused with the dried, blackened pouches left hanging on an equally withered and blackened branch after an attack of die-back.

Propagation

The great number of spores of the pathogen generated on diseased berries are dispersed by rains and transported from tree to tree by anything that moves through the rows of the plantation. Wind, however, is not a major propagation factor. The development of the disease to a level of epidemic importance depends chiefly on rains and fog as the spores of the pathogen need moisture to germinate and develop.

Studies carried out in 1969 by S. K. Mulinge in Kenya and, more importantly, by R. A. Muller in Cameroon show that the susceptibility of the berries varies during the different stages of their development.

R. A. Muller has drawn the following conclusions from the chart (see Fig. 4.3):

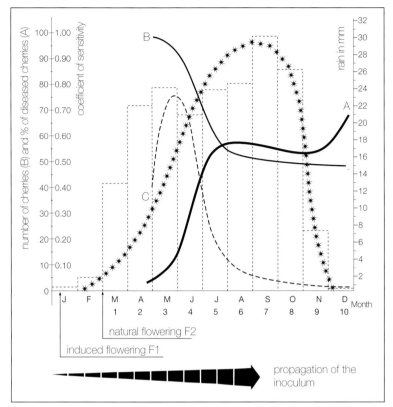

Figure 4.3 *Colletotrichum kahawae* – CBD.
Diagram showing the coefficient of intensity of
the infection (C) comparing the number of berries
lost (B) and the percentage of infected cherries

(A) against the rainfall (in mm) over the months
of the year. Blossoming is given to be at the end
of February. From: R. A. Muller – CIRAD.

1) The evolution of the disease is expressed by the percentage of infected berries in ratio to their total number. The curve (A), based on weekly observations, follows the stages of the development of the berries and indicates periods of expansion, stabilization, regression and a renewed outbreak in the evolution of the disease. Where the period of expansion can easily be explained by heavy rains which favor the development of the pathogen, the periods of stabilization and regression, which take place in the same climatic conditions, can only be attributed to different factors which do not depend on climatic conditions. The same would apply to the renewed outbreak of the disease, noted at the beginning of the dry season (October–November), which would not seem to be conducive to the germination and dissemination of the spores.

2) The curve (B) shows the losses due to CBD with a peak at the end of April, 2 months after blossoming, followed by

a regression and a period of little variation.

3) The curve (C) shows the evolution of the coefficient of "susceptibility" of the berries in both quality and quantity. It rises rapidly to a sharp peak when the berries are young and quickly stabilizes when the berries are in their fifth month of development. This coefficient is calculated on a weekly basis taking into account the number of berries lost over 2 weeks "$n + 1$" and "n", and the number of berries existing in week "n". This susceptibility coefficient is independent of climatic conditions. It depends on the physiological stage of the development of the berries. This could possibly explain the periods of stabilization and regression as well as the new outbreak of the disease shown in curve (A). The periods of stabilization and regression could be attributed to the fact that the development of the berries had reached a stage of "non-susceptibility" during their maturation (July), while the new outbreak of the disease (November) could be caused by the fact that once the pulp had ripened; its sweet, juicy content offers the favorable conditions needed for development of the fungus remaining in the berries after the initial attack of the pathogen. In this case, the new outbreak of the disease would not mean a new infection

of the pathogen but rather a regeneration of the initial infection. Table 4.5 sustains this hypothesis.

Further observation also shows that, during the period of "non-susceptibility" of the pulp, the parchment hardens and thickens, providing a strong barrier against the invasive pathogen and efficiently protecting the coffee bean.

Hence the above observations would lead to the hypothesis that the disastrous impact of the disease is due to the fact that heavy seasonal rains, which favor the development of the spores of the pathogen, coincide with the stage in the development of the berries where their susceptibility is at its peak.

As mentioned previously, the pathogen is common in coffee plantations at altitudes above 1 500 m close to the Equator. The pathogen has, however, been observed occasionally at lower altitudes under microclimates similar to the high altitude and especially in plantations which cultivate Caturra, varieties known to be especially exposed to the pathogen. In Cameroon, observations made on Caturra have revealed that vulnerability is marked in plantations situated at an altitude of 1 100 m, whereas *C. arabica* var. *Jamaica* is severely afflicted at an altitude of 1 500 m or more, but not at 1 100 m. Studies carried out by R. A. Muller in 1980 have led to the conclusion

Table 4.5 Moisture content of fresh pulp and beans, shown as % of their weight

Age of the berries (weeks)										
	8	11	13	15	17	20	22	26	28	33	35
Pulp (%)	80.2	83.3	86.5	78.4	77.7	74.8	72.9	72.3	72.5	79.8	77.4
Beans (%)	88.6	91.0	91.5	90.7	90.8	88.5	86.8	74.2	69.7	53.3	43.9

From: G. Blaha – CIRAD.

that this phenomenon is due to the fact that, at low altitudes, the berries develop much faster, and thus the pathogen has less time to take hold and penetrate them. At a higher altitude, however, growth is correspondingly slower and offers the pathogen more time to infect the berries. This theory is substantiated by the fact that the maturation cycle of the same variety of Arabica takes 32 weeks at 1100 m and 42 weeks at 1500 m. This, however, remains a theory which requires further research to justify changes in cultivation practices.

Chemical control

In geographical zones like Cameroon, Malawi or Zimbabwe which only have one rainy season and one flowering period, the spray program is based on the evolution of the disease given earlier in this chapter with curves highlighting the peaks and regression of the impact of the pathogen that clearly indicate the best periods for scheduling chemical treatments. The spray program should begin just after the blossoming of the coffee trees to protect the young berries before they reach their first stage of "susceptibility" and should be repeated at precise intervals calculated according to the remanence of the chemical used. The frequency of the treatments could, however, be increased in the case of important rainfall. Treatments should stop at week 20 after blossoming, at which point the berries are no longer vulnerable to the pathogen.

Thus the basic schedule for cupric fungicides would be as follows:

Treatment 1	just after flowering
Treatment 2	5 weeks after the first treatment
Treatment 3	4 weeks after the second treatment
Treatment 4	3 weeks after the third treatment
Treatment 5	2 weeks after the fourth treatment
Treatment 6	2 weeks after the fifth treatment
Treatment 7	2 weeks after the sixth treatment

In Cameroon, the recommended fungicides are NORDOX 50 WP (cuprous oxide 58%), NORDOX SUPER 75 WP (cuprous oxide 86%) and KOCIDE 101 (cupric hydroxide 56%).

Over recent years, several molecules have been developed and tested with a view to broadening the scope of chemical treatments available, and, more importantly, to generate efficient systemic, in-depth treatments which would reduce the number of applications required to protect the plantation against the pathogen.

It should be noted that in tropical climates, like in Cameroon, where there is only one flowering during the year, seven applications of cupric fungicides are required to efficiently protect coffee trees.

In equatorial bimodal climates like Kenya, however, where there are two annual flowering periods, berries generated by the April flowering contaminate those generated by the October flowering. Similarly, berries developing from October onwards would contaminate those developing in April. In this way, on plantations in bimodal climates, contamination is ongoing throughout the year so that 10 or 12 annual treatments are required. This, inevitably, means that high doses of cupric fungicides are used extensively. While copper, in small doses, has a tonic effect on coffee trees, the massive doses required to provide efficient protection against CBD have a toxic effect, sapping their vitality and impairing their vegetative growth. Nevertheless, and in spite of these drawbacks, cupric fungicides are still used extensively, although, more often, they are now mixed with efficient, but more expensive, organic pesticides. Mixing them with cupric fungicides reduces costs and offers the added advantage of immediately destroying any strains of the pathogen that develop resistance to organic pesticides. In Kenya, in certain areas where *Pseudomonas syringae* pv. *garcae* is active at the same time as CBD, the use of cupric fungicides remains essential (Tab. 4.6). In this case, cupric hydroxide or "Bordeaux Mixture" is recommended.

In countries where the climatic conditions only favor one major annual blossoming, the situation is somewhat different, but there are significant similarities. For instance, berries generated by small intermediary flowerings are not picked during the main harvest, but they are contami-

nated and could, therefore, transmit the pathogen to the next crop. The same applies to berries left on the trees by the pickers, those that have dropped to the ground as well as the peduncles which remain on the tree.

Based on the assumption that berry-to-berry contamination is a major factor in the propagation of the disease and that the denser the berries, the greater the contamination, it becomes obvious that the epidemic will be far more serious when the coffee trees produce a high yield. Consequently, it is recommended to specifically schedule chemical treatments on the basis of the importance of the crop and the density of the berries on the branches. Dwarf varieties with short internodes are more vulnerable than taller trees with longer internodes.

Table 4.6 Tested and recommended fungicides (active substance) for CBD control and tank mixture for CBD and CLR control in Kenya

CBD control	CBD and leaf control (tank mixture)	
Active substance	Organic fungicide	Copper 50% WP
Chlorothalonil	Daconil 2787 W75	cuprous oxide
Fluazinam		cupric chloride
Dithianon		cupric hydroxide
Azoxystrobin	Delan 75% WP	cuprous oxide
Anilazine		cupric chloride
Anilazine/copper		cupric hydroxide
Chlorothalonil/copper	Dyrene 75% WP	cuprous oxide
50% Copper formulation		cupric chloride
(a) Cuprous oxide	Chlortocaffaro 75% WP	cupric chloride
(b) Cupric chloride	Rova 75% WP	cupric chloride
(c) Cupric hydroxide		
Copper sulfate Lime	Quadris	cupric chloride
	Delan 500 SC	cupric chloride

From: Technical Circular 804-2003.

As to the methods used for applying chemical treatment, both past and recent research lead to the same conclusion: high volume spraying of 700–1 000 l of brew/ha are the most efficient and spraying should aim at the tops of the trees. In Kenya, special, mechanical sprayers have been designed for this purpose; they "shower" the trees from the top downwards. Pneumatic low-volume spraying is generally less efficient, but is widely used in Ethiopia where the height of the trees entails the use of powerful mechanical equipment to reach the canopy which is not accessible with current knapsack sprayers.

On the other hand, supposedly efficient results have been obtained by ultra-low-volume spraying techniques in Kenya. These techniques are tempting because they do not consume a lot of water, but it is not certain that they are really adapted to controlling fungus as the most efficient fungicides need to be sprayed in large volumes throughout the coffee tree. Ultra-low-volume techniques could, however, provide better results with new organic pesticides whose action is not geared to protective covering, but to penetration within the plant at specific points. Once experiments with this type of treatment have proved its real efficiency and implementation has become current, significant changes will take place in the fumigation practices used for chemical control. Meanwhile, as discussed earlier in this chapter, simple "impluviums" can be cheaply and easily installed on the plantation to provide the water necessary for high-volume applications on the coffee trees.

Picture 4.34 shows the differences between a crop that had been efficiently protected from the pathogen by cupric compounds applied at high volume as compared to that of a control tree which had not been treated.

Pictures 4.34 *Colletotrichum kahawae* – CBD. Appearance of a coffee crop treated against CBD with cupric compounds (plot 3) as compared to a non-treated control (plot 2).
From: R. A. Muller – CIRAD.

Agronomical control

Based on the evolution of the "susceptibility" cycle of coffee fruit during their development, R. A. Muller set up an "early irrigation strategy" in Cameroon. Investigated, experimented and implemented over a period of 10 years between 1973 and 1982, this strategy has been proved extremely valid and provides most satisfactory results.

The strategy consists in providing adequate irrigation to the coffee tree as early as the middle of January; in other words, artificially generating climatic conditions which are the same as those which prevail at the onset of the rainy season. This induces the dormant buds on the tree to flower during the dry season, approximately 1.5 months before their normal blossoming at a time which does not favor the activity of the pathogen.

Once the rainy season sets in and the pathogen becomes active, the berries will have already outgrown their "susceptibility" stage. Thus, the majority of them will not be infected by the pathogen.

Early irrigation strategies, combined with a distinctly reduced number of chemical treatments, have also been noted to increase the productivity of Arabica coffee trees. Consequently, such cultivation practices have been introduced in Cameroon with highly satisfactory results on plantations where sufficient water was available during the dry season to ensure adequate irrigation. Large-scale implementation would, however, require installing an adequate irrigation system on all the plantations threatened by CBD.

This would, undoubtedly, provide an efficient method for controlling a lethal disease which combines environmentally friendly cultivation practices with limited chemical control and offers the added advantage of increasing both the quality and the quantity of the coffee.

Integrated control

The convincing results obtained by implementing early irrigation strategies would appear to be an important factor when examining the different aspects of "integrated estate management" or the ways and means of combining all available strategies to improve disease control and the prosperity of the plantation as a whole. To date, however, little research has been carried out to further develop other practices and only a few useful strategies can be recommended.

The following points should, nevertheless, be considered:

1) In regions where only one major flowering takes place during the year, special care should be taken, after harvesting the main crop, to remove all remaining berries, even those that are not ripe. These berries are a source of inoculum and would, therefore, contaminate the following crop. Removing them delays the impact of the disease, a fact which has been substantiated experimentally by R. A. Muller in Cameroon. Furthermore, it is also a preventive measure against the Coffee Berry Borer *Stephanoderes hampei*.

2) It would also appear that an excess of potassium chloride generates a situation which is favorable to the CBD pathogen. The same applies to zinc sulfate or zinc chelate.

3) Similarly, certain fungicides, particularly cupric fungicides, also seem to favor the sporulation of the CBD pathogen as well an increase in the activity of other related pathogens. This theory, developed by Gibbs in 1972, has, however, never yet been fully researched and substantiated, and cannot, therefore, lead to any concrete recommendations. It is suspected that fungicides destroy the antagonist microflora and thus create conditions which favor epidemic development [31].

4) Recent observations in Cameroon show the incidence of the agro-environmental situation of the coffee plantation on the development of the epidemic. Shading, for instance, has a negative impact on

disease development. These studies also revealed the importance of "physiological berry fall" in relation to cultivation conditions. Precise knowledge of climatic conditions favoring disease development and the phenological cycle of the trees would help advisors to make recommendations adapted to different types of farms, in order to implement optimal management strategies and decrease phytosanitary treatments [9].

Genetic Control

Researching resistant varieties, another aspect of integrated farming practices, has been ongoing for a number of decades [21].

Initiated by Cook [15], and carried on by van der Vossen [54], research was carried out with a view to developing specific oligogenic resistance based on wild varieties like the Timor Hybrid and Rume Sudan, which both showed different levels of resistance. Experiments relied on systematic tests carried out by inoculating young seedlings at the "soldier" stage whose reactions were judged to be similar to those of the berries. As a result of this experimentation, the Catimor variety, which also shows valid resistance to CLR, was defined as the most interesting possibility. Variety Ruiru 11 appears to a worthwhile choice as it combines resistant strains to both C. kahawae and to H. vastatrix, but, again, suitable precautions need to be taken as certain Catimor descendants already show signs of susceptibility to the pathogen.

In Cameroon, experiments carried out by Bouharmont [11] led to selection of the Java variety as offering promising resistance to CBD and CLR as well as the added advantage of being vigorous and high-yielding with an excellent cup quality. The resistance level of the Java variety has, however, yet to be put to the test as, for the time being, this does not appear to be a constant characteristic. Further selection to better define it would appear to be necessary.

When CBD was first detected in Ethiopia in 1970, the center of the diversity of C. arabica, a broad scope of possibilities for research on a number of Arabica varieties was opened to researchers like Robinson [44] and van der Graaf [53]. It became possible to observe the reactions of a great number of varieties of C. arabica and to determine those which offer the most promising resistance to CBD.

Once this was achieved, it became possible to select specific varieties to be used for experimentation and research prior to introducing them into plantations as suitably resistant varieties on a long-term basis.

More recently, a project financed by the European Union has grouped the efforts and results obtained by CIRAD in France, CIFC in Portugal, IRAD in Cameroon and CRF in Kenya. This has greatly developed research into genetic resistance to the CBD pathogen.

In its first stages, this international cooperation program identified two distinct groups of strains: the East African strains and the Cameroon strains. This identification was based on research carried out on a great number of strains of the pathogen collected in various African countries where CBD is present. The geographic and genetic diversity of C. kahawae was then drawn up by Bella Manga et al. in 1997. Molecular markers, using randomly amplified polymorphic DNA (RAPD) analysis techniques, have led to the conclusion that there is not, however, a very significant genetic variability between the two groups.

At the present stage of this research program, it has not yet been possible to define the type of resistance coffee generates against C. kahawae, the CBD pathogen. This seems to lead to the conclusion that we are faced with a "general" as opposed to a "specific" type of resistance. It has also been possible to identify different strains which show different characteristics of the disease within the pathogen itself. Thus, CBD strains in Cameroon appear to show more aggressive characteristics than those in Eastern Africa. These findings point to the possible necessity that selection made in certain geographical zones will have to be repeated with local pathogen strains.

Apart from the study of the genetic structure of *C. kahawae* populations, the program also includes the study of the development processes of the infection itself. This involves, among other things, an evaluation of environmental influences, a histopathological study in the field of the reactions of the berries of different genotypes to the pathogen, and the setting up of precise and reliable methods of observation to assess losses caused by the disease.

The major target this program has been set is, however, to determine the best sources of resistance to the pathogen and the corresponding mechanisms of the transmission of this resistance in order to better assess the resistance of new descendants issued from the breeding of identified genotypes.

To achieve this result, a reliable system of testing is required. At face value, it would seem obvious that tests should be carried out on the berries as they are the vulnerable organs of the coffee trees. This, however, is a time-consuming procedure as it is necessary to test each annual fructification over a number of years before obtaining reliable results.

To save time, Kenyan research, as already mentioned, was carried out in research nurseries on young seedlings at the "soldier" stage, believed to replicate the reactions of the berries in the field. As no other testing method has yet been investigated, this method is still implemented today, but as it does not involve the actual berries in the field, it cannot be considered truly reliable until further proof is established. On the other hand, it is quite likely a useful tool for the first stages of selection as it would indicate potentially exploitable new genotypes, prior to their actual field test.

Thus, active research work is now ongoing in the hope that a genotype with true resistance to the CBD pathogen will be developed soon.

Assessment of losses caused by CBD

Global assessment of losses caused by CBD on any given plantation is relatively easy as the final yield of the annual crop provides an indisputable answer. While this may be adequate when it comes to an overall evaluation of the efficiency of chemical treatments, it is totally inadequate when precise studies of the epidemiology of the disease are required in a given area. Thus, a "physiological berry fall" can reach 50% of the potential crop and can be confused with CBD losses.

In Cameroon, weekly berry counting has provided a useful tool as it also distinguishes between CBD diseased berries and "physiological berry fall". This method has been improved by not only counting the berries on marked branches where the number of berries had previously been established, but also by delicately marking diseased berries, detected during a given week count, with a label attached to a fine thread. In this way, the next weekly observation can also distinguish the quantity of "physiological berry fall" from the number of berries lost due to the disease. This method is, obviously, extremely painstaking, but, given the rapid rate at which young berries drop at the vulnerable stage of their development, it seems, to date, to be the only way to clearly define the impact of the disease. These few comments on the considerable difficulties encountered by researchers when investigating the impact of the disease are only mentioned as an indication of the precision and care taken to either substantiate or invalidate the various theories and questions that arise in the course of a scientifically conducted research program.

Conclusion

The many different methods which can be implemented to fight CBD, chemical control linked to improved cultivation practices and genetic control, among others, indicate that the solution to the serious problem of controlling CBD in Arabica can be considered as an excellent illustration of how integrated cultivation practices based on scientifically substantiated counseling by qualified agronomists can produce the most satisfactory results.

4.3
Bacterial Diseases

4.3.1
Elgon Die-back or Bacterial Blight

Causal organism
Pseudomonas syringae pv. *garcae*.

General comments

This disease was first observed in Kenya, in Uganda and on the slopes of Mount Elgon, above an altitude of 1800 m. In Brazil, it was identified at Garça (Sao Paolo), hence its name "garcea". In 1945, Thorold, on the basis of the symptoms of the disease, deduced that it must exist in Zaire or in India, but according to J. Avelino, Elgon die-back is latent in most coffee producing countries, but only develops under very specific conditions.

Symptoms

Basically, Elgon die-back is a nursery disease which essentially affects leaves and young tissues.

When the leaves are damaged, they first appear to be water-soaked. This is the initial stage of the infection. Later on, the leaves curl, dry out, blacken and die, but they do not drop off the tree. The initial attack affects the terminal buds and works downwards, causing die-back of the twig. A small cluster of blackened leaves generally remains at the tip of the branch. This is the factor which differentiates Elgon die-back from die-back caused by overbearing when the twigs die after losing their leaves.

On young, succulent shoots, infection can be noticed on the stipules of the nodes just below the tips then, progressively, the nodes are totally invaded. Necrosis of the upper part of the shoot is caused by the destruction of the vascular tissues. A typical symptom is the blackening of the part which corresponds to the petiole at the base of the shoot (Pict. 4.35).

Where, occasionally, it affects established plantations, the symptoms appear on flowers and pin-head berries which blacken and fall.

Attacks can affect the entire plant or only parts of it. In severe cases, the coffee tree looks as if it has been scorched by fire (Fireblight).

Biology

The bacteria live in epiphytes on all parts of the plants and remain "expectantly inactive" until the ideal conditions prevail. When this happens they will become active pathogens. Ideal conditions for *P. syringae* are rain or dampness and cold winds, the

Picture 4.35 Elgon die-back on a young shoot. From: A. Autrique.

type of conditions which are frequent on the slopes of Mount Elgon, the Solai Valley, Paranà or Sur de Minas. According to Kairu, on the basis of observations recorded in 1983, the incubation period of the disease is approximately 4 weeks.

Though high altitudes generate favorable conditions for the development of both CBD and Bacterial Blight, these two diseases are rarely to be found in the same place at the same time.

Control

In 1945, Thorold noted that "French Mission" coffee trees resisted Bacterial Blight well, and that they were "bronze-tipped" trees with broad and rather coarse leaves. Further observations led him to record that nearly all the "green-tipped" trees were susceptible to the disease.

If nurseries are well protected against wind with efficient windbreaks, outbreaks of Bacterial Blight are less frequent. If there is potential danger of an outbreak, trimming and pruning should be reduced.

Copper oxychloride is the best chemical control to date. In nurseries it should be applied twice at an interval of 20 days in a solution of 50 % Cu at 0.2–0.3 % + 0.2 % of either Agrimicine or Distreptine 20.

On established plantations, in the field, one or two copper oxychloride treatments with 50 % Cu at 1 % are recommended.

4.4
Physiological Diseases

4.4.1
Hot and Cold Disease

This affection, also known as Black Tip, was first reported in Uganda. Initially, it was thought to be caused by a virus as its various symptoms can also be typical of viral diseases.

The tips of the young shoots blacken, and the adjoining leaves crinkle, distort and reduce in size. Terminal growth is excessively branched and bunched. The discoloration of the leaves starts from the outer edge and works inwards. In severe cases, the leaves may scorch and fall or, if they are subsequently infected by *Phoma* spp., die-back of the young shoots may follow (Pict. 4.36).

This physiological disease is caused by environmental stress, particularly high temperature variation between night and day. It can also be caused by insect stings, but it only prevails in high-altitude plantations. According to a survey carried out in Cameroon by Bruneau de Mire in the late 1960s and early 1970s, hymenopterous insects (*Ulopidea, Collborhis* sp.), whose sting could cause this syndrome, would only be present under specific climatic conditions.

Picture 4.36   Hot and Cold Disease.

Bibliography

[1] Almeida, S. R. *Doenças do cafeeiro.* In: *Cultura do Cafeeiro, Fatores que Afetam a Produtividade.* A. B. Rena, E. Malavolta, M. Rocha and T. Yamada (eds). Associação Brasileira para Pesquisa da Potassa e do Fosfato, Piracicaba, 1986, 391–399.

[2] Avelino, J., Muller, R. A., Cilas, C. and Velasco, H. Développement de la rouille orangée dans des plantations en cours de modernisation plantées de variétés naines dans le Sud-Est du Mexique. *Café Cacao Thé* **1991**, *XXXV* (1), 21–42.

[3] Avelino, J. and Savary, S. La lutte chimique raisonnée et optimisée contre la rouille orangée (*Hemileia vastatrix*). L'expérience de l'Amérique latine. In: *Recherche et Caféiculture.* CIRAD-CP, Montpellier, 2002, 134–140.

[4] Avelino, J., Seibt, R., Zelaya, H., Ordoñez, M. and Merlo, A. Enquête-diagnostic sur la rouille orangée du caféier Arabica au Honduras. In: *17th Colloque Scientifique International sur le Café.* ASIC, Paris, 1998, 613–620.

[5] Avelino, J., Toledo, J. C. and Medina, B. El caldo bordelés y la recepa en el control del ojo de gallo. In: *Memoria Técnica de Investigaciones en Café 1990-1991.* ANACAFE, Guatemala, 1992, 123–129.

[6] Avelino, J., Toledo, J. C. and Medina, B. Développement de la rouille orangée (*Hemileia vastatrix*) dans une plantation du Sud-Ouest du Guatemala et évaluation des dégâts qu'elle provoque. In: *15th Colloque Scientifique International sur le Café.* ASIC, Paris, 1993, 293–302.

[7] Avelino, J., Toledo, J. C. and Medina, B. Desarrollo del ojo de gallo (*Mycena citricolor*) en una finca del norte de Guatemala y evaluación de los daños provocados por esta enfermedad. In: *XVI Simposio sobre Caficultura Latinoamericana.* IICA, Honduras, 1995.

[8] Bieysse, D., Mouen Bedimo, N'Deumeni, J. P. and Berry, D. Effet de différentes conditions agro-écologiques sur le développement de l'anthracnose des baies du caféier arabica dans l'ouest du Cameroun. In: *18th Colloque Scientifique International sur le Café.* ASIC, Paris, 1999.

[9] Bieysse, D., Bella Manga, Mouen Bedimo, N'Deumeni, J. P., Roussel, V., Fabre, J. V. and Berry, D. L'anthracnose des baies du caféier. Une menace potentielle pour la caféiculture mondiale. *Plantation, Recherche, Développement* **2002**, *May*, 144–152.

[10] Blaha, G. Un grave pourridié du *C. arabica* au Cameroun: *Clytocybe elegans* Heim. *Café Cacao Thé* **1978**, *XXII* (3), 203–216.

[11] Bouharmont, P. Sélection de la variété Java et son utilisation pour la régénération de la caféière arabica au Cameroun. *Café Cacao Thé* **1992**, *XXXVI* (4), 247–262.

[12] Bowden, J., Gregory, P. H. and Johnson, C. G. Possible wind transport of coffee leaf rust across the Atlantic ocean. *Nature* **1971**, *229*, 500–501.

[13] Cadena-Gomez, G. and Buritica-Cespedes, P. Expresión de resistencia horizontal a la roya (*Hemileia vastatrix* Berk.

and Br.) en *Coffea canephora* variedad Conilón. *Cenicafé* **1980**, *31* (1), 3–27.

[14] Chaves, G. M. Melhoramento do cafeeiro visando à obtençao de cultivares resistentes à *Hemileia vastatrix* Berk. et Br. *Revista Ceres*. **1976**, *23* (128), 321–332.

[15] Cook, R. T. A. Screening coffee plants for CBD resistance. *Coffee Res. Found. Kenya Annu. Rep.* **1972/73**, 66–68.

[16] Dennis, R. W. G. An earlier name for *Omphalia flavida* Maubl. and Rangel. *Kew Bull.* **1950**, *3*, 434.

[17] Eskes, A. B. The effect of light intensity on incomplete resistance of coffee to *Hemileia vastatrix*. *Neth. J. Plant. Pathol.* **1982**, *88* (5), 191–202.

[18] Eskes, A. B. Incomplete resistance to coffee leaf rust (*Hemileia vastatrix*). *Thèse de Doctorat*. Agricultural University of Wageningen, The Netherlands, **1983**.

[19] Eskes, A. B. Resistance. In: *Coffee Rust: Epidemiology, Resistance and Management*. A. C. Kushalappa and A. B. Eskes (eds). CRC Press, Boca Raton, FL, 1989, 171–291.

[20] Eskes, A. B. and Souza, E. Z. Ataque da ferrugem em ramos com e sem produção, de plantas do cultivar catuaí. In: *9th Congresso Brasileiro de Pesquisas Cafeeiras*. IBC, Brasil, 1981, 186–188.

[21] Firman, I. D. Screening of coffee for resistance to coffee berry disease. *E. Afr. Agric. J.* **1964**, *29*, 192–194.

[22] Gibbs, J. N. Inoculum sources for coffee berry disease. *Ann. Appl. Biol.* **1969**, *64*, 515–522.

[23] Gil, S. L. Recherches sur la résistance à *Hemileia vastatrix* Berk. *et* Br. de génotypes de *Coffea arabica* L. d'origines éthiopiennes. *Thèse de Docteur-Ingénieur*. ENSA, Montpellier, France, 1988.

[24] Heim, R. and Saccas A. M. Observations préliminaires sur la Trachéomycose des *Coffea excelsa, neo-arnoldiana* et Robusta de l'Oubangui-Chari. *C. R. Acad. Sci., Paris* 1950, *231*, 536–538.

[25] Hindorf, H. *Colletotrichum* spp. isolated from *Coffea arabica* L. in Kenya. *Z. Pflkrankh. Pflschutz.* **1970**, *77*, 328–331.

[26] Holguin, F. Contribution à la recherche d'une résistance durable du caféier (*Coffea* spp.) À la rouille orangée *Hemileia vastatrix* Berk. et Br. Etude de la variabi-

lité génétique du pathogène. *Thèse de Doctorat*. USTL, Montpellier, 1993.

[27] Kushalappa, A. C. and Chaves, G. M. An analysis of the developement of coffee rust in the field. *Fitopatol. Bras.* **1980**, *5*, 95–103.

[28] Leguizamon, J. Contribution à la connaissance de la résistance incomplète du caféier à *Hemileia vastatrix* Berk. et Br. *Thèse de Doctorat*. ENSA, Montpellier, France, 1983.

[29] Mansk, Z., Matiello, J. B. and Almeida, S. R. Efeito da aplicaçao do fungicida Bayleton associado ao oxicloreto de cobre a diferentes no de aplicaçoes no controle da ferrugem do cafeeiro, (*H. vastatrix*, Berk et Br.). In: *6th Congresso Brasileiro de Pesquisas Cafeeiras*. IBC, Brasil, 1978, 339–341.

[30] Mariotto, P. R., Geraldo, C. Jr, Silveira, A. P. De, Arruda, H. V. De, Figueredo, P. and Braga, J. B. R. Efeito da produção sobre a incidencia da ferrugem do cafeeiro. In: *2nd Congresso Brasileiro de Pesquisas Cafeeiras*. IBC, Brasil, 1974, 144.

[31] Masaba, D. The role of saprophytic surface microflora in the development of coffee berry disease (*Colletotrichum coffeanum*) in Kenya. *Thesis*. University of Reading, 1991.

[32] Mayne, W. W. Physiological specialization of *Hemileia vastatrix* B. and Br. *Nature* **1932**, *129* (3257), 510.

[33] Meiffren, M. Contribution aux recherches sur la trachéomycose du caféier en Côte d'Ivoire. *Café Cacao Thé* **1961**, *X* (1), 28–37

[34] Muller, R. A. La rouille orangée du caféier (*Hemileia vastatrix*) sur le continent américain. *Café Cacao Thé* **1971**, *XV* (1), 24–30.

[35] Muller, R. A. Contribution à la connaissance de la phycomycocénose, *Coffea arabica* L., *Colletotrichum coffeanum* Noack *sensu* Hindorf, *Hemileia vastatrix* B. et Br., *Hemileia coffeicola* Maublanc et Roger. *Thèse de Doctorat d'Etat*. Université de Paris VI, 1978.

[36] Mulinge, S. K. Outbreaks and new records in Ethiopia, coffee berry disease. *FAO Plant Protect. Bull.* **1973**, *21*, 85–86.

[37] Nuñez, C., Bertrand, B., Vargas, L. and Avelino, J. Estudio preliminar sobre el

modo de inoculación del hongo *Mycena citricolor* (ojo de gallo) sin heridas, en la hoja del cafeto: importancia de diferentes factores que intervienen en la penetración. In: *XVI Simposio sobre Caficultura Latinoamericana*. IICA, Honduras, 1995.

[38] Nutman, F. J. and Roberts, F. M. Studies on the biology of *Hemileia vastatrix* Berk. and Br. Trans. *Br. Mycol. Soc.* **1963**, *46*, 27–48.

[39] Pellegrin, F., Nandris, D., Waestrelin, S. and Kohler, F. Situation pathologique des Arabica en Nouvelle-Calédonie; corrélations entre pathogenèse et environnement. In: *16th Colloque Scientifique International sur le Café*. ASIC, Paris, 1995, 690–698.

[40] Rao, D. V. and Tewari, J. P. Production of oxalic acid by *Mycena citricolor*, causal agent of the American leaf spot of coffee. *Phytopathology* **1987**, *77*, 780–785.

[41] Rayner, R. W. Coffee berry disease. A survey of investigations carried out up to 1950. *East. Afric. Agric. J.* **1952**, *17*, 130–158.

[42] Rayner, R. W. Germination and penetration studies on coffee rust (*Hemileia vastatrix* B. and Br.). *Ann. Appl. Biol.* **1961**, *49*, 497–505.

[43] Rivillas Osorio, C. A., Leguizamon Caycedo, J. E. and Gil Vallejo, L. F. Recomendaciones para el manejo de la roya del cafeto en Colombia. *Bol. Téc. Cenicafé* **1999**, *19*.

[44] Robinson, R. A. *Terminal Report of the FAO Coffee Pathologist to the Government of Ethiopia*. FAO, Rome, 1974.

[45] Rodrigues Jr, C. J., Bettencourt, A. J. and Rijo, L. Races of the pathogen and resistance to coffee rust. *Annu. Rev. Phytopathol.* **1975**, *13*, 49–70.

[46] Saccas, A. M. La Trachéomycose (Carbunculariose) des *Coffea excelsa, neo-ar-*noldiana et Robusta en l'Oubangui-Chari. *Agron. Trop.* **1951**, *VI* (9/10), 453–506.

[47] Schuppener, H., Harr, J., Sequeira, F. and Gonzalez, A. First occurrence of the coffee leaf rust *Hemileia vastatrix* in Nicaragua, 1976, and its control. *Café Cacao Thé* **1977**, *XXI* (3), 197–202.

[48] Shaw, D. E. Coffee rust outbreaks in Papua from 1892 to 1965 and the 1965 eradication campaign. *New Guinea Dept. Agric. Stock and Fisheries Res. Bull.* **1968**, 2, 20–52.

[49] Silva-Acuña, R. Control químico de la roya del cafeto (*Hemileia vastatrix*) con el uso de un fungicida sistémico y uno protector. *Fitopatología Venezolana*. **1990**, *3* (2), 22–27.

[50] Silva acuña, R. Intensidad de la roya (*Hemileia vastatrix* Berk. et Br.) en cafetos con diferentes niveles de producción controlada en Venezuela. *Café Cacao Thé* **1994**, *XXXVIII* (1), 19–24.

[51] Stewart, R. B. *Some Diseases Occurring in Kaffa Province, Ethiopia*. Imperial Ethiopian College of Agriculture and Mechanical Arts, Alemaya, 1957.

[52] Tarjot M., Lotode, R. Contribution à l'étude des rouilles orangée et farineuse du caféier au Cameroun. *Café Cacao Thé* **1979**, *XXIII* (2), 103–118.

[53] Van der Graaff, N. A. The principles of scaling and the inheritance of resistance to coffee berry disease in *Coffea arabica*. *Euphytica* **1982**, *31*, 735–740.

[54] Van der Vossen, H. A. M. and Walyaro, D. J. Breeding for resistance to coffee berry disease in *Coffea arabica* L. 2. Inheritance of the resistance. *Euphytica* **1980**, *29*, 777–791.

[55] Waller, J. M., Bridge, P. D., Black, R. and Hakiza, G. Characterization of the coffee berry disease pathogen, *Colletotrichum kahawae* sp. nov. *Mycol. Res.* **1993**, *97*, 989–994.

5

Viral Diseases in Coffee

E. W. Kitajima and C. M. Chagas

5.1
Introduction

Coffee Ringspot Virus (CoRSV) (Portuguese = mancha anular) is the only virus known to date that spontaneously infects coffee plants, although Tomato Spotted Wilt and Tobacco Mosaic Viruses have been known to infect the leaves of coffee seedlings under experimental conditions [7, 11].

5.2
Coffee Ringspot Virus

5.2.1
History

CoRSV is apparently restricted to Brazil. The symptoms were initially described by Bitancourt [1, 2]. His observations on coffee leaves and fruit in plantations at Caçapava region (state of São Paulo) led him to diagnose a viral infection.

5.2.2
Symptoms

Symptoms usually appear in:

- The lower leaves where they are characterized by chlorotic spots or rings which sometimes follow the veins (Pict. 5.1)
- The berries, which show ring-shaped depressions (Pict. 5.2)

Local chlorotic/necrotic lesions have also been induced in experimentally infected *Chenopodium quinoa* Willd and *C. amaranticolor* Coste and Reyn (Pict. 5.3).

Recently a distinct type of ringspot symptom has been observed in coffee plants growing in public parks of the city of São Paulo, SP. The leaf lesions have a more reddish, rusty appearance (Pict. 5.4), though in the berries the symptoms are the same. As described below these symptoms are induced by a distinct isolate of the CoRSV, referred to as CoRSV-SP [6].

5.2.3
Nature of the Causal Agent and Transmission

CoRSV particles vary from a short rod-like appearance to baciliform. They can occasionally be detected by electron microscopy in negatively stained extracts from infected plants (Pict. 5.5).

Electron microscopic examinations (Pict. 5.6) of the leaf tissues from the ringspots

Picture 5.1 Ringspot symptoms in *Coffea arabica* cv. "Mundo Novo".

Picture 5.4 Detail of the reddish ringspot symptoms caused in *C. arabica* by CoRSV-SP.

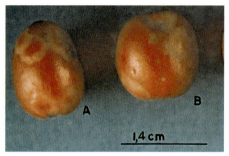

Picture 5.2 Symptoms caused by CoRSV in coffee berries. In spots A and B the rings are clearly visible.

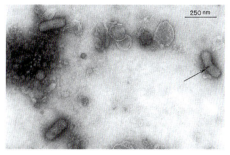

Picture 5.5 Bacilliform particles present in negatively stained fruit pulp extract of CoRSV-infected *C. arabica*. One of them is partially broken (arrow).

Picture 5.3 Local lesions induced by mechanical inoculation of CoRSV in *Chenopodium amaranticolor* (left) and *C. quinoa* (right): (a) inoculated leaf and (b) non-inoculated control leaf.

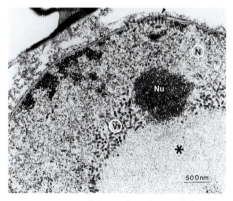

Picture 5.6 Part of the nuclei of mesophyl parenchyma in the ringspot area in CoRSV-infected coffee leaf. Note a group of rod-like particles (V) in the periphery of an electron-lucent viroplasm (*) near a nucleolus (Nu). Some of the rod-like particles appear arranged perpendicularly onto the nuclear envelope (arrowheads). The insert shows these rod-like particles in detail.

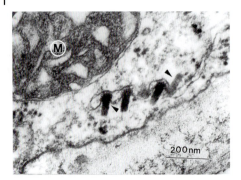

Picture 5.7 Rod-like particles in the cytoplasm, associated to the membrane of the endoplasmic reticulum (arrowheads). M: mitochondrion.

Picture 5.8 The Tenuipalpidae mite *Brevipalpus phoenicis*. Courtesy of: Dr C. C. Childers.

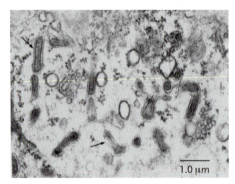

Picture 5.9 Some bacilliform particles within the lumen of the endoplasmic reticulum (arrows). CL: chloroplast; Cy: cytoplasm; M: mitochondrion; PC: cell wall; Vc: vacuole.

reveal short, rod-like particles (around 40 nm × 100–120 nm) mostly located in the nucleus commonly associated with an electron-lucent mass, referred to as viroplasm. These rods may appear associated with the nuclear envelope in a partial budding process. Rods may also be present in the cytoplasm associated to the endoplasmic reticulum membrane (Pict. 5.7).

This type of particle and associated cell changes are basically similar to those previously observed in tissues infected with Orchid Fleck Virus (OFV) [15]. Further evidence of the viral nature of the causal agent of the CoRSV has been obtained by Chagas [5] who succeeded in transmitting the disease by means of the Tenuipalpidae mite *Brevipalpus phoenicis* Geijskes (Pict. 5.8). He also transmitted it to *C. amaranticolor* and *C. quinoa* by mechanical means using a somewhat elaborate extraction buffer. These two species reacted by generating localized lesions (see Pict. 5.3). He also confirmed the previous cytopathology works carried out by Kitajima and Costa [12].

In the tissue infectec by CoRSV-SP the presumed viral particles are short bacilliform, membrane-bounded and occur longitudinally and individually arranged within the cisternae of the endoplasmic reticulum (Pict. 5.9). Nuclear viroplasms are rarely seen [6]. Primers have been designed to detect CoRSV by RT-PCr [17] which did not amplify vDNA from CoRSV-SP (Locali E.C., unpublished data). CoRSV-SP was also transmitted by *B. phoenicis* to coffee seedlings reproducing the original symptoms. *B. phoenicis* eggs collected on symptomless coffee trees gaave rise to colonies raised on orange fruits. Individuals from this colony in the larval, nymphal and adult stages were able to transmit CoRSV-SP after a previous feeding period between 18 to 24 h (Chagas, C.M. Unpublished data). However, so far mechanical transmission of CoRSV-SP to coffee or other host plants has been unsuccessful.

CoRSV has been purified from *C. quinoa*, which was systemically infected after treatment with higher temperature, and a specific antiserum was produced. Comparative ELISA revealed distant serological relationship between CoRSV and OFV [3].

This cumulative evidence indicates that coffee ringspot is indeed caused by a virus, officially designated CoRSV, which may well be related to OFV. OFV is transmitted by *B. californicus* (Banks) and infects certain herbaceous hosts, both locally and systemically [8, 15]. A purified, genome analysis indicated that it has a bipartite single-stranded RNA with genes similar to those known for Rhabdoviruses [16].

In experiments carried out by Chagas [4] and more recently by J. V. C. Rodrigues (personal communication), it would appear that *B. phoenicis* does not transmit CoRSV to citrus, *Ligustrum* and passion fruit, although these species can show viruses transmitted by this type of mite. On the other hand, citrus, *Ligustrum* and passion fruit viruses do not contaminate coffee plants. Evidence also exists to show that populations of *B. phoenicis* specialize in certain hosts. Hence, efforts are now ongoing to check whether *B. phoenicis* may possibly represent more than one species (C. Childers, personal communication).

Electron microscopy of thin sections of the mite vector *B. phoenicis* collected from CoRSV-infected coffee plants from two different localities (Lavras, MG and Campinas, SP) revealed the presence of the same cell alterations found in plant tissues (nuclear viroplasm and rodlike particles) in the cells of prosomal gland (= salivary gland) (Pict. 5.10 and 5.11) The viroplasm is though to be the site of the viral replication/assembly and its presence in mite

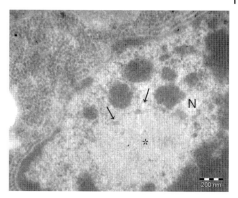

Picture 5.10 Part of the prosomal gland cell of the mite *B. phoenicis* viruliferous for CoRSV. Note the electron lucent viroplasm (*) and some presumed virions (arrows).

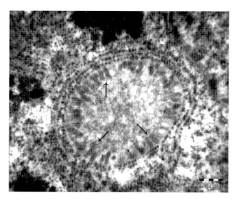

Picture 5.11 Similar to 5.10 showing an enlarged view of a group of presumed virions (arrows) arranged radially in the cytoplasm, in a configuration known as "spokewheel".

cells is strongly suggestive that CoRSV is multiplying in the mite vector, thus the virus/vector relationship would be of circulative/propagative type [14]. Similar conclusion was drawn to the Orchid fleck virus (OFV) based on the existence of a latent period of about 2 weeks for transmission after viral acquisition [13]. CoRSV shares many common characteristics with OFV and may be also a member of the recently proposed genus *Dichorhabdovirus*, family *Rhabdoviridae* for OFV [14].

Until recently *Coffee arabica* was the only known natural host for CoRSV. A survey made in a germplasm bank of the Centro APTA Café Alcides Carvalho, Instituto Agronomico, Campins, SP, Brazil, revealed that some other *Coffea* species and hybrids (*C. kapakata*, *C. devewrei cv. excelsa*, *C. canephora cv. robusta*, hybrid *C. arabica x C. racemosa*, *C. arabica x C. dewevrei*, hybrid Timor (a natural hybrid of *C. arabica* and *C. canephora*) as well as *Psylanthus ebracteolatus*, a rubiaceae plant close to the genus *Coffea* presented ringspot symptoms in the leaves and/or berreis essentially similar to those caused by CoRSV in *C. arabica*. Infection by CoRSV was confirmed by electron microscopy and RT-PCR [13].

5.2.4
Geographical Distribution

Besides the state of São Paulo, coffee ringspot has been reported in the Federal District [3] and in the state of Minas Gerais. It may well also be present in other coffee-growing states in Brazil like Paraná, Bahia, and Espírito Santo. Outside Brazil, occurrence of CoRSV is confirmed to be present in Costa Rica [20]. Reyes [18] reported a disease similar to coffee ringspot in several *Coffea* species in the Philippines, which was graft transmitted to *Coffea* plants as well as to several herbaceous hosts. Seed transmission in *C. excelsa* has also been observed [19]. These pathological properties suggest that the Philippine version of ringspot disease in coffee differs from the version described in Brazil. Another disease considered similar to the Brazilian coffee ringspot was described by Wellman [22] in Costa Rica and referred to as blister spot. At a later date, however,

Vargas and Gonzales [21] demonstrated that this was caused by a *Colletotrichum* fungus.

5.2.5
Economic Importance

Due to its very low incidence, this disease has, as a general rule, been dismissed as relatively unimportant from an economic point of view. It did, however, give rise to a certain amount of concern in 1995 when its incidence increased sharply in at least two regions of Minas Gerais State (Araguari and Alto Parnaiba), causing intense defoliation. As the damage at the time went no further, it was thought that specific, but unusual, conditions generated an explosion of the *Brevipalpus* mite population [9, 10]. In all events, this freak rise in CoRSV should not be considered lightly as it clearly indicates that, under favorable conditions, the disease may cause considerable yield losses.

5.2.6
Control

As is normally the case for viral infections, control should be preventive but, due to the marginal importance of the coffee ringspot, very few steps have been taken in this direction. No information regarding genetic resistance in *Coffea* to either the virus or the vector mite is available. Monitoring the level of infestation of coffee plants by the vector mite *B. phoenicis* [it is reddish, very active and small, around 0.3 mm (Pict. 5.8)] would be advisable and the use of chemical controls recommended if the population were to increase significantly.

Bibliography

[1] Bitancourt, A. A. A mancha anular, uma nova doença do cafeeiro. *O Biológico* **1938**, *4*, 404–405.

[2] Bitancourt, A. A. Lesões nas frutas da mancha anular do cafeeiro. *O Biológico* **1939**, *5*, 33–34.

[3] Boari, A. J., Freitas-Astua, J., Ferreira, P. T. O., Nader D. J., Nogueira, N. L., Rossi, M. L. and Kitajima, E. W. Purification and serology of the Coffee ringspot virus. *Summa Phytopathologica* **2004**, 30 in press.

[4] Branquinho, W. G., Cupertino, F. P., Takatsu, A. Kitajima, E. W. and Boiteux, L. S. Doenças que afetam plantios de café e de mandioca no Distrito Federal. *Fitopatol. Bras.* **1988**, *12*, 140.

[5] Chagas, C. M. Mancha anular do cafeeiro: transmissibilidade, identificação do vetor e aspectos anatomo-patológicos da espécie *Coffea arabica* L. afetada pela moléstia. *Tese de Doutorado.* Instituto de Biociências, Universidade de São Paulo, 1978.

[6] Chagas, C. M., Kitajima, E. W. & Locali-Fabris, E. C. Isolado distinto do vírus da mancha anular do cafeeiro (CoRSV). *Fitopatol. Bras.* **2007**, *32* (supl.), S135.

[7] Costa, A. S., Silva, D. M. and Carvalho, A. M. B. Infecção de cafeeiros com o vírus de vira-cabeça. *Bragantia* **1960**, *19*, XLVII–LII.

[8] Doi, Y., Chang, M. U. and Yora, K. Orchid fleck virus. *CMI/AAB Description of Plant Viruses* **1977**, *183*.

[9] Figueira, A. R., Reis, P. R. and Pinto, A. C. S. Vírus da mancha anular do cafeeiro tem causado prejuízos relevantes aos cafeicultores da região do alto Parnaiba. *Fitopatol. Bras.* **1995**, *20*, 299.

[10] Juliatti, F. C., Báo, S. N., Araújo, A. C. G., Kitajima, E. W., Neves, J. B. and Peixoto, J. R. Mancha anular do cafeeiro: etiologia e danos em lavoura da região de Araguari, MG. *Fitopatol. Bras.* **1995**, *20*, 337.

[11] Kitajima, E. W. and Costa, A. S. Evidência electrono-microscópica de multiplicação dos vírus do mosaico do fumo e de vira-cabeça, em tecido foliar de cafeeiro. *Bragantia* **1970**, *29*, XXXV–XL.

[12] Kitajima, E. W. and Costa, A. S. Partículas baciliformes associadas à mancha anular do cafeeiro. *Ciência e Cultura* **1972**, *24*, 542–545.

[13] Kitajima, E. W., Chagas, C. M., Braghini, M. T., Fazuoli, L. C., Locali-Fabris, E. C. Infecção natural de várias espécies e hibridos de Coffea e uma outra espécie de rubiaceae pelo virus da mancha anular do cafeeiro (CoRSV). *Fitopatol. Bras.* **2007**, *32* (supl.), S134.

[14] Kitajima, E. W., Boari, A. J. & Chagas, C. M. Detecção do vírus da mancha anular do cafeeiro nos tecidos do ácaro vetor Brevipalpus phoenicis (Acari: Tenuipidae). *Fitopatol. Bras.* **2007**, *32* (supl.), S135.

[15] Kondo, H., Maeda, T. & Tamada, T. Orchid fleck virus: *Brevipalpus califoricus* mite transmission, biological properties and genome structure. *Exp. Appl. Acarology* **2003**, *30*, 215–223.

[16] Kondo, H., Maeda, T., Shirako, Y. & Tamada, T. Orchid fleck virus is a rhabdovirus with an unusual bipartite genome. *J. gen. Virol.* **2006**, *87*, 2413–2421.

[17] Locali, E. C., Freitas-Astua, J., Antonioni-Luizon, R., Boari, A. J. & Machado, M. A. Diagnose da mancha anualr do cafeeiro

através do RT-PCR. *Fitopatol. Bras.* **2005**, *30* (supl.), S185.

[18] Reyes, T. T. Ringspot of coffee in the Philippines. *FAO Plant Protect. Bull.* **1959**, *8*, 11–12.

[19] Reyes, T. T. Seed transmission of coffee ringspot by Excelsa coffee (*Coffea excelsa*). *Plant Dis. Reptr* **1961**, *45*, 185.

[20] Rodrigues, J. C. V., Rodriguez, C. M., Moreira, L., Villalobos, W., Rivera, C. and Childers, C. C. Occurence of Coffea ringspot virus, a *Brevipalpus* mite borne virus in coffee in Costa Rica. *Plant Disease* **2002**, 564.

[21] Vargas, G. E. and Gonzales, U. L. C. La mancha mantecosa del café causada por *Colletotrichum. Turrialba* **1972**, *22*, 129–135.

[22] Wellman, F. L. Blister spot of Arabica coffee from virus in Costa Rica. *Turrialba* **1959**, *7*, 13–15.

6

Resistance to Coffee Leaf Rust and Coffee Berry Disease

C. J. Rodrigues Jr and A. B. Eskes

6.1
Introduction

Among the various fungal diseases affecting coffee, Coffee Leaf Rust (CLR) or Orange Rust (*Hemileia vastatrix* Berk. and
Br.) and the Coffee Berry Disease (CBD)
(*Colletotrichum kahawae* Waller and Bridge)
have the most significant economical impact. Although CLR is present wherever
coffee is grown, CBD has so far only
been reported in African Arabica-growing
countries. If no chemical control is used,
the former may cause annual losses of
about 30 % and the latter up to 80 %. The
number of spray applications necessary to
contain these diseases is usually high (up
to six to eight sprays in some areas), and
a review of the various disadvantages of
the fungicides (cost, inapplicability in sloping areas, possibility of inducing off-flavors, environmental pollution, etc.) has
led to increased attention being given to
the development of resistant coffee varieties by coffee pathologists and breeders
worldwide [9, 14, 22]. In the present chapter, the two diseases will be treated separately with emphasis on the available
knowledge on resistant varieties and on
their future prospects.

6.2
Coffee Leaf Rust

6.2.1
History

Between 1850 and 1870 coffee cultivation
was a flourishing industry in Ceylon (now
Sri Lanka). With as much as 50 000 metric
tons exported during the peak period, Ceylon was, at the time, the greatest coffee producer in the world. In 1869, the first coffee
leaves affected by CLR were detected in the
country and the heavy attacks caused by
the disease, linked to a fall in coffee prices,
caused coffee cultivation to be hastily replaced by tea [20]. That same year the disease spread to India as well, and, a few
years later, to Sumatra (1876), Java (1878)
and to other countries of the same continent. Although the disease is believed to
have originated in Africa, it was only detected in Natal in 1878, and, during the following decade, in Tanzania and Madagascar. It was not detected in West Africa (Angola) until 1956. It appeared in Brazil in
1970 and from then on spread rapidly to
the other coffee-growing countries of the
Americas. Wherever the disease appeared,
economic losses where high due to the
susceptibility of the uniform traditional

Coffee: Growing, Processing, Sustainable Production, Second Edition. Edited by J. N. Wintgens.
© 2012 WILEY-VCH Verlag GmbH & Co. KGaA. Published 2012 by Wiley-VCH Verlag GmbH & Co. KGaA.

Arabica varieties. This led to research into resistant planting material.

Following the ravages caused by CLR on Arabica coffee in Ceylon and neighboring countries, other coffee species like *Coffea liberica* and *C. canephora* were introduced as alternatives to fight the disease. To begin with, *C. liberica* was quite popular in Ceylon and Java, but by the end of the century this species also became more susceptible to CLR. In Indonesia and India, spontaneous hybrids also appeared between Arabica and Liberica coffees. The Kalimas and Kawisari hybrids discovered in Java were quite resistant to begin with, but were not used in further breeding. Liberica and spontaneous hybrids between Arabica and Liberica were, however, used more extensively in India. Following the foundation of the Central Coffee Research Institute in 1925–26, outstanding mother plants derived from crosses between S.26, S.31, S.71, etc. were selected and used. Progenies of these plants were backcrossed to Kent's and gave rise to some interesting selections such as S.288, S.333, S.353 and S.795. The latter selection was released in 1946 and was widely used afterwards, although unfortunately it, in turn, also became susceptible in the field after 1960.

The introduction of *C. canephora* to Java from the ex-Belgian Congo at the beginning of the 20th century led to a revival in coffee production and may be considered as a success, except at high altitudes. This species showed high levels of resistance to CLR, and soon replaced Arabica and Liberica at the low and medium altitudes. Spontaneous hybrids between Arabica and Canephora as well as artificial ones were studied. Some backcross populations were very poor croppers, whereas others generated plants of an Arabica type and yielded a product similar to Arabica coffee. Perhaps the most significant of these interspecific hybrids is the tetraploid Arabica-like Timor Hybrid, a natural cross found in East Timor that has been the basis of the coffee breeding program for rust resistance carried out at the Centro de Investigação das Ferrugens do Cafeeiro (CIFC) [12] and in other research centers over the last 30 years (see also Section I.3). In India, the Devamachy hybrid, discovered in the 1930s, was backcrossed to Arabica on several occasions. Artificial crosses between Arabica and Canephora have been made mainly by previous duplication of the Canephora parent. This approach has been followed by the French in Africa for the production of "Arabusta" varieties (tetraploid F_1 hybrids) and, in Brazil, for the creation of the Icatu population, by backcrossing the tetraploid F_1 hybrids to local Arabica cultivars [8]. Triploid hybrids have been used more recently in Colombia and Brazil.

In terms of Arabica, it appears that the first selection for resistance to CLR was carried out in India where, in the early 1900s, the highly susceptible "Old Chiks" variety was replaced by the more resistant "Coorg". In 1911, a plant showing resistance to CLR was discovered in Mr Kent's plantation at Doddengoda and descendants of this plant gave rise to the well-known Kent's cultivar. Derivatives of Kent's were further selected in Tanzania, giving rise to the KP, F, H and X series, and in Kenya resulting in selections like K7 and SL6. Improvement of CLR resistance in Arabica was attempted in these two countries by the introduction of coffee varieties from Ethiopia, Sudan and India.

In the 1950s experts of the FAO, USDA (USA) and ICO (USA) in collaboration with the Jimma Agricultural Technical School of Ethiopia began to organize a germplasm collection of Arabica in its diversity center in Ethiopia. In 1952–53, the

American scientists F. Wellman of the USDA and W. Cowgill of USAID made a worldwide trip to collect CLR-resistant coffee varieties from African and Asiatic countries. A coffee mission to Ethiopia organized by FAO took place in 1965, and seed collections were sent to research stations in India, Tanzania, Portugal, Peru and Costa Rica for germplasm collections and other studies, among which the search for rust resistance was one of the main objectives [14]. In 1955, the CIFC was founded in Oeiras, Portugal. As the name indicates, this center was dedicated to do research on CLR and, particularly, on the search for CLR-resistant coffee varieties. A great deal of the above-mentioned collected material was screened for rust resistance to all known variants (races) of CLR at CIFC. This center has provided very valuable support for the synthesis of modern coffee varieties, which offer CLR resistance to a wide spectrum of rust races (e.g. Catimors) available in several coffee-producing countries since 1985 [4].

6.2.2
Races of CLR and Resistance Genes

6.2.2.1 Races of CLR
After the outbreak of the CLR in Ceylon and in other Asian countries, and during the process of searching for CLR-resistant coffee types, it became evident that some varieties of Liberica and, to a lesser extent, of Canephora, introduced to replace the susceptible Arabica, also became susceptible after a few years of cultivation. The same happened in relation to some Arabica cultivars, particularly in India, where the highly susceptible original coffee strain of Arabica, the "Old Chick", was replaced by the apparently more resistant variety "Coorg" in the 1880s. A few decades later, the Coorg coffee became intensely contaminated by CLR and

was in turn replaced around 1920 by the then resistant Kent's selection. Kent's selection, however, also lost its resistance about 10 years later. The phytopathologist Wilson Mayne, working in India at that time, was aware of these facts and by experimental inoculations he demonstrated that one sample of rust could attack two varieties of coffee, whereas another rust sample could only attack one of these two varieties. For the first time ever, this proved the presence of rust types (hereafter called "races" or "physiological races") with different capacities to infect coffee varieties (called a "virulence" spectrum). Mayne duly numbered these as race 1 and race 2. Later on he discovered two more different races (named 3 and 4). Mayne concluded from his discoveries that what appeared to be apparent successive losses in resistance were, in reality, the development of new races of CLR able to attack different varieties of coffee.

Following the creation of CIFC at Oeiras, D'Oliveira and collaborators received many samples of leaf rust and coffee genotypes from different regions of the world. Thanks to this work, the existence of Mayne's races 1, 2 and 4 was confirmed, and many other races were discovered. At present a collection of 40 races, named I to XXXX, exists at CIFC. These are identified on a range of coffee "differential" varieties according to the virulence or avirulence reactions they induce when inoculated on these "differentials" [21].

The most widespread race in the world is the race II (Mayne's race 1), one of the races with a narrow virulence spectrum. This is the race that infects the highly susceptible traditional cultivars such as Bourbon, Caturra, Catuai, Mundo Novo, SL series, etc., and that very probably destroyed the Arabica cultivation in Ceylon, Java and other countries during the 19th century. The presence of race II is still frequently identified in rust samples collected on these varieties in different countries. Other races, with broader infection spectra, are usually collected in coffee varieties with resistance genes different from the above-mentioned commercial varieties. This is the case for race I (Mayne's race 2) which was responsible for the loss of resistance of the Kent's selection in India. It is the second most widespread race, and is frequently found in India and in Kenya where varieties with the resistance from the Kent's variety still constitute a large proportion of the local Arabica varieties. The existence of races contained within certain regions may therefore be explained by the presence of resistant varieties, which are more or less confined to those regions [22].

6.2.2.2 Differential Resistance Groups of Coffee and Resistance Genes

Screening for leaf rust resistance has been carried out at CIFC since its foundation. It led to the characterization of differential resistance groups distinguished from each other essentially by reactions of resistance and susceptibility to specific pathogen races. Twenty-four groups have been identified by screening CLR races on existing coffee germplasm and 16 more have been synthesized through crosses between coffee varieties. Plants of group A, characterized by resistance to all known CLR races, have been detected in interspecific hybrid populations (*C. arabica* × *C. canephora*) and in varieties derived from such hybrids (Catimor, Sarchimor, Icatu). The resistance of the interspecific hybrids is due to the parent *C. canephora*. Plants of group A have also been found in other diploid coffees. Group F, in contrast to group A, presents susceptibility to all the known races and includes the large majority of the *C. ra-*

cemosa accessions received at CIFC. Group E, including most traditional commercial varieties (Bourbon, Mundo Novo, Caturra, Catuai, etc.), is susceptible to 31 of the 40 known races. Group D (including Kent's and its derivatives) is susceptible to 16 races. Groups X, C, L, J and W are found in spontaneous Arabica accessions collected in Ethiopia and include a variable range of susceptibility. Groups G and H are found in interspecific hybrids derived from *C. arabica* × *C. liberica* selected in India. They present a large spectrum of resistance, and are susceptible only to races 4 and 5, respectively.

The resistance traits identified in *C. arabica* and *C. canephora* have been linked to the presence of specific resistance genes that have been denominated as "SH" genes. In *C. arabica*, four genes (SH1, SH2, SH4 and SH5), present either alone or in association, have been identified [2, 17]. These genes have a "dominant" action, which means that the resistance governed by these genes, when present in homozygous varieties (or pure lines), is transferred to all its seed progeny. The Arabica varieties that contain the same resistance genes are part of the same "resistance group". For example, traditional varieties such as Bourbon and Typica belong to group E (SH5), and Kent's derivatives are part of group D, that possesses the genes SH2 and SH5 . Ethiopian coffee varieties like Dilla and Alghe, Geisha, S.4 Agaro, S.12 Kaffa, etc., possess the four SH genes, in a different gene association. These four resistance genes from *C. arabica* may be defeated by rust races bearing different combinations of "matching" virulence genes denominated v1, v2, v4 and v5. The existence of these genes is inferred according to the gene-for-gene concept. The resistance gene SH3 comes from *C. liberica*. It is present in groups G and H, in which several Indian selections, like S795, are included. The resistance of these selections may be defeated by rust races with different combinations of the virulence genes v2, v3 and v5. The Timor Hybrid populations derived from *C. arabica* × *C. canephora* possess at least five dominant or partially dominant resistance genes conditioning, either singly or in association, the resistance groups R, 1, 2 and 3 [3, 5]. These genes may be defeated by CLR races with different combinations of the virulent genes v5, v6, v7, v8 and v9.

At CIFC, previously identified group A plants of Catimor still offer resistance to new arrivals of CLR races from countries like India where Catimor lines show various degrees of susceptibility in field plots. This would suggest the existence of resistance genes not previously pinpointed.

Apart from the SH genes identified at CIFC, it is quite likely that other major and minor genes, which offer resistance to CLR, are present in other germplasm of *Coffea* [4].

6.2.3
Evaluation of Resistance

6.2.3.1 Resistance Tests

Resistance tests have been developed for field, nursery or greenhouse plants as well as for detached leaves or leaf disks in the laboratory. Factors involved in the expression of resistance and which are to be taken into account for reliable resistance evaluation have been described by Eskes [14]. Favorable growing conditions including high light intensity (direct sunshine or light shade) are necessary for a correct expression of resistance symptoms. The fungal spores used may be collected fresh from the field or stored for some time either in refrigeration (4 °C) or in liquid nitrogen. Before inoculation, the germina-

tion capacity of the spores should be verified (minimum 10–20%). Spores may be applied either dry, to be wetted immediately afterwards, or in an aqueous suspension on the lower leaf surface. In suspension, spores should be adequately dispersed by manual shaking or by using a stirring device, keeping the suspension in motion during the inoculation experiment to avoid drowning the spores. The spore suspension should not be exposed to sunlight as this inhibits germination. Due to the different levels of resistance caused by the age of the leaf (partial resistance), leaves of different ages should be inoculated so as to correctly identify the resistance level of the entire plant. The inoculated plants or leaves should be incubated in the dark for 24–48 h at 100% relative humidity. During incubation, environmental conditions should be favorable to fungal development with temperatures set between 18 and 28 °C and shady conditions with a fairly high level of humidity. Symptoms will develop after about 20 days and, ideally, resistance should be recorded once the susceptible control variety is fully sporulating (25–50 days after inoculation).

6.2.3.2 Recording of Resistance Reactions

The traditional recording system of CLR lesions developed by D'Oliveira at CIFC [11, 12] is based on artificial inoculation of recently expanded young coffee leaves of greenhouse plants. The identification of individual "lesion types" should be done once the lesion is fully developed. Depending on the climate (mainly temperature), this may take place between 3 and 4 weeks after inoculation. The types of lesions may be small chlorotic "flecks", small groups of dark green swollen cells ("tumefactions"), larger chlorotic areas without sporulation (lesion type 0) or lesions with little chlorosis but showing an increasing intensity of spore production (lesion types 1–4). Plants presenting predominantly resistant lesions up to type 1 are considered as resistant (R) and those with predominantly "susceptible" lesions are considered as susceptible (S) [22]. Resistance due to major genes in Arabica coffee is generally characterized by the presence of predominantly "resistant" lesion types. However, many other coffee genotypes show mixtures of susceptible and resistant lesion types on the same leaf or on leaves of different ages. This is frequently observed on plants which exhibit intermediate levels of resistance, as is the case with derivatives of interspecific crosses (Catimor, Icatu) and also in certain populations of *C. canephora*. A resistance scoring using "reaction types" rather than lesion types has therefore been suggested to assess the intermediate resistance of entire plants. For detailed resistance observations, a 10-point scale (0–9) for reaction type has been successfully applied [14]. For more routine observations in the field and in the nursery, this scale can be transformed into a five-point scale (0–4), with the following description of reaction types:

0 Absence of visible symptoms (highly resistant, HR),

1 Presence of resistant types of lesions only: flecks and/or small chlorotic points (resistant, R),

2 Presence of resistant types of lesions and chlorotic lesions with sparse sporulation, often the case for older leaves only (moderately resistant, MR),

3 Presence of predominantly susceptible types of lesions (two to four), but also of resistant lesions, often on younger leaves (moderately susceptible, MS),

4 Presence of highly susceptible types of lesions only, with intensive sporulation, often accompanied by leaf fall (susceptible, S).

6.2.4
Resistant Varieties

Selection for resistance to CLR has mainly concerned *C. arabica*, because all traditionally grown varieties of this species (Bourbon, Typica, Mundo Novo, Catuai, Caturra) are highly susceptible to common races of the fungus. Three types of resistance factors have been selected in commercial varieties: those originally present in *C. arabica* and those introduced from *C. liberica* or from *C. canephora* [22, 14]. The selection programs carried out to develop resistant varieties in Brazil, Colombia, India and Central America have been described respectively by Carvalho *et al.* [8, 9], Castillo [10], Sreenivasan [24] and Echeverri and Fernandez [13].

6.2.4.1 Resistance from *C. arabica*

Commercially used resistance from *C. arabica* is mainly limited to the Kent variety, selected in India, and its derivatives, like the K7 variety. The K7 variety is a tall-growing variety with reasonable yield potential and is still used in the lower coffee belt of Kenya. The Kent and K7 varieties carry the SH2 resistance gene, which has generally shown not to be very effective, due to the quick appearance of new races of the fungus that overcome this type of resistance. The same has occurred with the other resistance factors present in certain Arabica accessions from Ethiopia, as SH1 in "Geisha" and "Dille and Alghe", and SH4 in "Agaro". This was clearly demonstrated in Brazil, were CLR was first detected in 1970 and these varieties were first attacked 2–4 years later [14]. The selection of the IARANA variety, already initiated in the 1960s at the Instituto Agronômico in Sao Paulo, was based on combinations of accessions with different resistance genes from *C. arabica*. The distribution of this variety was interrupted early in the 1970s because a majority of the plants became susceptible, and also because its yield potential was lower than that of Mundo Novo and Catuai.

6.2.4.2 Resistance from *C. liberica*

The resistance factor derived from *C. liberica* is called SH3. It is present in many Indian selections, like S795 that was released in the 1940s and maintained its resistance in the field for about two decades. This variety is not completely fixed because it was released as the progeny of an F_1 cross between S26 and Kent [24]. It therefore shows a variation for resistance as well as other characteristics, like elephant beans, which, under certain conditions, can be unduly prevalent. Generally speaking, it is a very hardy variety, adapted to rather marginal conditions, but can also be quite susceptible to the *Fusarium* disease. An advanced selection of S795, the S1934, shows more uniform characteristics and has been introduced into a few countries like Indonesia. The resistance capacity of these "S" varieties has been partially overcome by the CLR fungus in most Arabica coffee-growing areas in Indonesia, but in certain newer areas it apparently still remains effective. In Brazil, and also in other parts of America, the field resistance due to the SH3 factor is still largely operative [14].

6.2.4.3 Resistance from *C. canephora*

The most important source of resistance to leaf rust comes from *C. canephora* and was introduced into *C. arabica* through natural or artificial crosses between these species. The Timor Hybrid appeared early this century in Timor [15]. This hybrid, as it was selected by the farmers, maintained a large variation for all traits, including yield, seed defects (peaberries, empty locules) and resistance. Consequently, progenies of the Timor Hybrid have generally no direct commercial value. However, in the 1960s, the CIFC identified plants of the Timor Hybrid with resistance to all known races of the fungus and subsequently crossed two of these, CIFC 832/1 and 832/2, with several commercial varieties. Derivatives of crosses with "Caturra" and with "Vila Sarchi", called "Catimor" and "Sarchimor", respectively, are the more widely known [1]. In Colombia, coffee breeders used another population of the Timor Hybrid, i.e. CIFC 1343, which was crossed with Yellow Caturra [10]. CIFC collaborated with countries like Brazil and Colombia to verify the resistance of advanced selections of Catimor progenies. In Brazil, artificial crosses between Canephora and Arabica were created in the 1950s, and backcrosses of these hybrids were selected intensively in the 1970s and 1980s under the name "Icatu" [8]. Varieties with resistance from Canephora that were released to farmers in different countries from 1985 onward have been described in Section I.3.

6.2.5
Prospects

The release of the Catimor, Sarchimor and Icatu varieties has been followed by the widespread cultivation of these varieties in regions where these varieties show promising adaptation and long-lasting resistance to CLR (like in Colombia). The breeding populations that gave rise to these varieties show other interesting traits, and are being further selected for resistance to CBD and nematodes, as well as for agronomic and even technological traits. The cup quality of advanced selections derived from canephora, which offer resistance to CLR, seems to be roughly similar to traditional Arabica varieties. However, variations in quality components and caffeine content have been observed, and these will be selected for countries that traditionally produce high-quality Arabica coffees.

With regard to long-lasting resistance to CLR in these varieties, the recently observed diminished resistance of Catimor lines in Asian countries is a warning that resistance may gradually fail in African and American countries as well. Breeders and pathologists have been actively researching long-lasting types of resistance, particularly partial resistance. Some of the partial resistance encountered in interspecific cross populations has not been lasting, but other types of partial resistance which last do seem to exist [14]. Meanwhile, different strategies for major gene resistance management have been suggested so that this resistance may become sufficiently durable. One of the strategies examined is the use of composite varieties, like the Colombia variety [16], where each line of the variety mixture may carry different resistance factors, which should help to avoid a sudden breakdown of resistance. Another strategy is the accumulation of several resistance genes into one line, thus increasing the barrier to be overcome by the fungus. In fact, the presence of several genes in the Catimor and Icatu varieties appear be a significant factor for maintaining resistance in many coffee-growing areas.

Further breeding of Arabica coffee will make ample use of the resistant varieties developed so far. These varieties will be improved for other desirable traits, like resis-

tance to nematodes or insects as well as for quality. In order to avoid the breakdown of resistance, combinations of already identified effective resistance factors should be considered [14], along the same lines as the combination of resistance factors derived from Canephora with that of Liberica (SH3).

6.3
Coffee Berry Disease

6.3.1
History

In 1922, in Kenya, CBD was reported for the first time by MacDonald as attacking the green fruit of coffee. The identification of the causal agent of a coffee disease as *Colletotrichum coffeanum* had been determined by Noack in 1901, in Brazil, but as it is known that the pathogen inducing CBD is not present in this country, the causal agent described by Noack would not be the same as the one described by MacDonald. This ambiguity was maintained for many years both for the pathogenic and for the saprophytic strains of *Colletotrichum* isolated from coffee. For some mycologists, the original description of *C. coffeanum* by Noack corresponds to *C. gloeosporioides*, a fungus that is frequently isolated from coffee fruits and branches, but does not cause CBD. In the 1940s, Rayner designated the causal agent of CBD as *C. coffeanum* var. *virulans* on the basis of the grayish color of the mycelium growing in culture. Gibbs proposed four different forms of *C. coffeanum* according to the mycelium color and production of acervuli in malt extract agar. In 1970, Hindorf made an intensive revision of the genus *Colletotrichum* and suggested identification of the following species:

1) *C. coffeanum* Noack (*sensu stricto*) for the CBD strain;
2) *C. acutatum* and *C. gloeosporioides* (white mycelium form; greenish mycelium form).

After this revision by Hindorf, the CBD strain was designated as *C. coffeanum* Noack *sensu* Hindorf.

The fact that both *C. acutatum* and *C. gloeosporioides* may be also isolated from fruit infected with CBD as well as the great variability of *C. coffeanum* Noack *sensu* Hindorf when grown in agar medium induced further review of the issue. In 1993, Waller *et al.* worked on pathogenic, metabolic and cultural characteristics of fresh isolates of *C. colletotrichum*, and suggested the alteration of the designation of *C. coffeanum* Noack *sensu* Hindorf to *C. kahawae* Waller and Bridge. sp. nv. Consequently, from now on, the causal agent of CBD will be referred to here as *C. kahawae*.

The fungus is mainly conspicuous on green fruit at any stage of development. This includes the so-called "active" lesions, which first appear as small dark concave patches spread over the entire surface of the berry. In wet weather, pinkish masses of spores are produced on the berry surface. As a result of the infection the berries may be shed prematurely or destroyed inside once the fungus has penetrated the beans. The berries may then dry out and become mummified. The so-called "scab" lesions are usually superficial and pale tan in color, sometimes with concentric rings of black incipient acervuli. This type of lesion is usually associated with unfavorable conditions for disease development or with defense mechanisms of tolerant or resistant varieties. Generally speaking, the scab lesion does not interfere with berry development, although it may become "active" if environmental conditions become more favorable or when the berries ripen. The fungus may also infect the flowers and, occasionally, the leaves.

Following the report of CBD in Kenya in 1922, the disease was detected in the following countries: Democratic Republic of Congo in 1938, Angola in 1940, Cameroon in 1955, Tanzania in 1964, Ethiopia in 1971, Malawi and Zimbabwe in 1985, and Zambia in 1986. Hence, so far, it is restricted to Africa. CBD spores are released and dispersed by rain, but they may travel long distances on latently infected parts of the plant or infected seeds. It can also be carried by birds, insects, man, etc. Diseased berries are, however, the main source of CBD inoculum in coffee fields.

6.3.2
Resistance Tests

The use of resistance tests for CBD has been described by Van der Vossen [26] and by Van der Graaff [25]. Selection of coffee plants for resistance to CBD can be made in the field by observing the natural behavior of the plants in the presence of the fungus. The percentage of infected berries can be used to evaluate field resistance; however, care should be taken as the microclimate in each individual canopy may affect the fungus. Trees with a low yield tend to be less affected than trees that have a high production of fruit, due to the fact that the fungus can more easily infect berries growing nearby than berries that are far from the source of infection. Consequently, many trees of the same variety must be repeatedly screened at suitable intervals.

Artificial inoculations are needed for rapid assessment of resistance of young plants and may also be used for pre-screening individual seedlings. Careful checks should be made of the level of pathogenicity of the isolated subject, as variations may occur after subculturing the fungus in the laboratory. Inoculation tests have been developed using conidial suspensions of the fungus applied to different plant parts, mainly developing attached or detached berries, young stems and hypocotyls of young seedlings [25, 26]. The hypocotyl test was developed in Kenya by Cook 30 years ago and this test proved to give quite good correlations with the level of field resistance. It has been largely used in many laboratories as a pre-selection test for CBD resistance. The traditional method of inoculation is to spray young seedlings in the pinhead stage with an aqueous suspension of conidia.

Some laboratories also use uprooted and washed seedlings placed horizontally in laboratory trays. This method offers the added advantage of easier control of inoculation and incubation conditions, but appears to increase the vulnerability of the seedlings when compared to the traditional hypocotyl test (D. Bieysse, unpublished data). Attached berry tests on field plants have been successfully used to confirm the resistance of field plants [25], but, in this case, success depends greatly on weather conditions. The use of detached berries, inoculated in the laboratory, has apparently given good results in Ethiopia, but in other countries the results are less consistent.

6.3.3
Breeding for Resistance

Selection and breeding for resistance to CBD started in the 1970s in Kenya, Ethiopia and Cameroon [7, 26, 25]. Evaluation of collections in the field showed that the majority of cultivated varieties were highly susceptible to the disease. However, some lines of the Typica variety as well as the

K7 variety showed some degree of field resistance in Kenya. High levels of resistance were found in the Rume Sudan accession and in the Hybrid of Timor. Resistance of these varieties appeared to be related to specific genes, the recessive "k" gene is present in K7 and in Rume Sudan, the "T" gene in Typica and in the Hybrid of Timor, and the "R" gene in Rume Sudan. Also variable degrees of resistance were identified in Ethiopian accessions, indicating a quantitative nature of this trait, although some accessions show very little attack [7, 25]. It is not yet known if the K, R and T genes are present in Ethiopian material.

In Kenya, crosses of resistant varieties with local commercial varieties (mainly SL28) were made in the 1960s. In 1971, a formal breeding program was initiated, based on backcrossing such hybrids in order to introduce resistance from different sources into the SL28 variety. The hypocotyl inoculation method was used efficiently to screen large numbers of progenies.

After a few cycles of backcrossing and selfing, resistant plants were crossed to Catimor lines, introduced mainly from Colombia, which had inherited resistance from Timor Hybrid. Since 1985, a mixture of selected dwarf hybrids, resulting from these crosses, were released to the farmers under the name "Ruiru 11" [26]. This variety is reproduced by hand pollination, and is presently further selected for adaptation to different environments and for certain quality traits. The resistance of Ruiru 11 appears so far quite stable, although slight levels of disease may be found where climatic conditions are especially favorable to the disease.

In Ethiopia, a program was initiated in 1973 to exploit the large variation in field resistance found in spontaneous or locally cultivated varieties [25]. Mother trees were selected for field resistance to CBD, as well as for other traits, in areas of high natural infection. Resistance of these trees was confirmed by inoculations carried out on attached berries in the field. Progenies of selected mother trees were planted in a large collection created in the Jimma Research Station, as well as in other areas in the country. Progenies have been assessed for field resistance and detached berry tests were carried out. Further breeding was done by inter-crossing resistant lines and out-crossing these to other selected lines. Some of the crosses between resistant varieties combine hybrid vigor with good resistance to CBD.

In Cameroon, resistance to CBD was identified in field collections, containing cultivated varieties as well as introductions from Ethiopia. Some highly resistant accessions were not only identified within the Ethiopian collection, but also in some other introductions. Variety trials were carried out in the 1960s and 1970s. Local cultivars as well as most introduced cultivars, like Caturra and Mundo Novo, were poor yielders, probably because they were highly susceptible to rust and CBD. Among the varieties assessed, the most productive one appeared to be the "Java" variety, which is vigorous and only moderately susceptible to rust and field resistant to CBD [6]. This variety slightly resembles "Abyssinia", a variety introduced by Cramer into Java in the beginning of the 19th century and is now known as "AB" in Indonesia.

The Java variety was released to the farmers in Cameroon in the early 1980s. Selected plants of this variety were used to establish seed production plots. Artificial inoculations carried out recently in Cameroon and by CIRAD in Montpellier indicate that some variation for resistance between individual plant progenies of this variety may exist.

6.3.4
Fungus Races

A large number of fungal cultures isolated from green coffee berries from different African countries (Angola, Burundi, Cameroon, Ethiopia, Kenya, Malawi, Rwanda, Zimbabwe) have been gathered and maintained at the CIFC laboratory as well as at CIRAD. After proving their pathogenicity on coffee berries, these isolates were inoculated, over recent years by the hypocotyl test, on a large number of coffee introductions including Arabica cultivars and tetraploid interspecific hybrids.

The results of this screening at CIFC showed that some of the isolates offer a different virulence towards the same progeny and differential interactions between the host genotype and the fungus isolate have been observed. This fact suggests the existence of physiologic races or "pathotypes" in *C. kahawae* [23, 27]. It has also been observed that isolates display different degrees of aggressiveness, expressed by a more or less rapid killing of the hypocotyls of the same susceptible coffee accession. These aspects continue to be studied in the context of an international project financed by the EC, in which collaborative research is carried out between Portugal, France, Cameroon and Kenya, and include a "ring-test" for resistance and pathogenicity.

6.3.5
Prospects

Most of the knowledge currently available on CBD is the result of investigations carried out in Kenya, Ethiopia and Cameroon concerning the biology and epidemiology of the fungus, chemical control and selection and breeding for resistance. Although chemical control of the disease is to some extent successful, all of its negative implications makes the synthesis of resistant varieties the main objective of the coffee scientists. The varieties developed so far, such as the Ruiru 11 in Kenya and Java in Cameroon, need to be followed in the field and laboratory for evaluating the durability and stability of their resistance. Screening for resistance can be more informative when done by using several CBD isolates from different geographic regions of Africa. This should at the same time allow us to verify the existence of physiologic races in *C. kahawae*. If the presence of races is confirmed, then future breeding work must take this fact into consideration. In other words, screening for CBD resistance should not be limited to the fungal isolates of a single region, but include isolates from other regions as well. In addition, the breeders should use different sources of resistance trying to accumulate them into single varieties, as has been done in the selection of Ruiru 11 [26]. The possibility of introducing resistance from the large number of Ethiopian accessions should be considered.

6.4
International Cooperation

Over several decades a great deal of research has been carried out on CLR. This has been made possible by international cooperation established between the Coffee Rusts Research Center and other Institutes and Experimental Stations of coffee-growing countries. An important breakthrough for the pragmatic achievements of this study was the identification of the Timor Hybrid as a rust-resistant progenitor, and the creation of the Catimor and Sarchimor lines that were freely distributed and used with success in many countries. The subsequent successful development of the Colombia variety [26] is an example of the benefits of this collaboration.

Since the resistance of the new varieties may not be long-lasting, complementary work must follow, with a view to integrating the major genes like, for instance, the SH3 gene, genes from Canephora and from other coffees species into selected material. Alternatively, lasting partial resistance may be sought in breeding populations. This is a goal which would be beneficial for all coffee growers worldwide; hence, International Institutes and Research Stations in tropical countries should collaborate.

With regard to CBD, coffee breeders have also resorted to the Timor Hybrid as well as to other resistance sources with a view to generating resistant commercial varieties. However, local varieties or breeding populations synthesized in three African countries, which show resistance or tolerance to the disease, are not easily available to other countries. Consequently, its behavior in relation to fungal strains from other countries is mostly unknown. For safety reasons, a broad screening for resistance with isolates from different African regions cannot be carried out in any of the coffee-growing countries. Therefore, a well-established system of centralized resistance testing carried out in a non-coffee growing country would benefit coffee growers both in Africa and elsewhere. An important step towards increased collaboration between European and African countries is the E-funded STD project on resistance to CBD which was initiated in 1995.

Bibliography

[1] Bettencourt, A. J. Características agronómicas de seleções derivados de cruzamentos entre Híbrido de Timor e as variedades Caturra, Villa Sarchi e Catuaí. In:
*Comunicações Simposio sobre Ferrugens do
Cafeeiro*. CIFC, Oeiras, 1983, 351–374.

[2] Bettencourt, A. J. and Noronha-Wagner,
M. Genetic factors conditioning resistance of *Coffea arabica* L. to *Hemileia
vastatrix* B. and Br. *Agronomia Lusitana*
1971, *31*, 285–292.

[3] Bettencourt, A. J., Noronha-Wagner, M.
and Lopes, J. Factor genético que condi-
ciona a résistencia do clone 1343/269
(Hibrido de Timor) á *Hemileia vastatrix*
Berk. and Br. *Broteria Genética* **1980**, *I*
(LXXXVI), 53–58.

[4] Bettencourt, A. J. and Rodrigues, Jr, C. J.
Principles and practice of coffee breeding
for resistance to rust and other diseases.
In: *Coffee. Vol. 4: Agronomy*. Clark, R. J.
and Macrae, R. (eds). Elsevier, London,
1988, 199–234.

[5] Bettencourt, A. J., Lopes, J. and Palma, S.
Factor genético que condiciona a resis-
tencia às raças de *Hemileia vastatrix* Berk.
and Br. dos clones-tipo dos grupos 1, 2 e
3 de derivados de Hibrido de Timor.
Broteria Genética **1992**, *XIII* (LXXX),
185–194.

[6] Bouharmont, P. Sélection de la variété
Java et son utilisation pour la régénéra-
tion de la caféière Arabica au Cameroun.
Café Cacao Thé **1992**, *36*, 247–262.

[7] Bouharmont, P. *La Sélection du Caféier
Arabica au Cameroun (1964–1991)*.
Document 1-95, CIRAD, Montpellier,
1995.

[8] Carvalho, A. Principles and practice of
coffee plant breeding for productivity and
quality factors: *Coffea arabica*. In: *Coffee.
Vol. 4: Agronomy*. Clark, R. J. and Macrae,
R. (eds). Elsevier, London, 1988, 129–166.

[9] Carvalho, A., Ferwerda, F. P., Frahm-Leliveld, J. A., Medina, D. M., Mendes, A. J.
T. and Monaco, L. C. Coffee. In: *Outlines
of Perennial Crop Breeding in the Tropics*.
Ferwerda, F. P. and Wit, F. (eds). Veenman, Wageningen, 1969, 189–241.

[10] Castillo-Z, J. Breeding for rust resistance
in Colombia. In: *Coffee Rust: Epidemiology, Resistance and Management*. Kushalappa, A. C. and Eskes, A. B. (eds). CRC
Press, Boca Raton, FL, 1989, 307–316.

[11] D'Oliveira, B. As ferrugens do cafeeiro.
Revista do Café Português **1954–57**, *1* (4),
5–13; *2* (5), 5–12; *2* (6), 5–13; *2* (7), 9–17;
2 (8), 5–22; *4* (16), 5–15.

[12] D'Oliveira B. and Rodrigues, Jr, C. J. O
problema das ferrugens do cafeeiro.
Revista do café Português **1961**, *8* (29),
5–50.

[13] Echeverri, J. H. and Fernandez, C. E. The
PROMECAFE programme for Central
America. In: *Coffee Rust: Epidemiology,
Resistance and Management*. Kushalappa,
A. C. and Eskes, A. B. (eds). CRC Press,
Boca Raton, FL, 1989, 323–329.

[14] Eskes, A. B. Resistance. In: *Coffee Rust:
Epidemiology, Resistance and Management*.
Kushalappa, A. C. and Eskes, A. B. (eds).
CRC Press, Boca Raton, FL, 1989,
171–292.

[15] Gonçalves, M. M. and Daenhardt, F. A
Hemileia vastatrix em Timor. Nota sobre a
sua importância económica e melhora-

mento do cafeeiro face à doença. *Missào de Estudos Agronómicos do Ultramar.* Publ. 666. Lisboa, 1971.

[16] Moreno-R, G. and Castillo-Z, J. The variety Colombia: a variety of coffee with resistance to rust (*Hemileia vastatrix*). Berk. and Br. *Technical Bulletin 9.* CENICAFE, Chinchina, Caldas, Colombia, 1990.

[17] Noronha-Wagner, M. and Bettencourt, A. J. Genetic study of the resistance of *Coffee* spp. to leaf rust. I. Identification and behaviour of four factors conditioning disease reaction in *Coffea arabica* to physiologic races of *Hemileia vastatrix*. *Can. J. Bot.* **1967**, *45*, 2021–2031.

[18] Osorto J. J. El programa de selección y evaluación de variedades de café en America Central. IICA. *Bol. Promecafé* **1990**, *49*, 13–16.

[19] *PNG Coffee*, **1993**, 9 (2), April.

[20] Rodrigues, Jr, C. J. Coffee rusts: history, taxonomy, morphology, distribution and host resistance. *Fitopatol. Brasileira* **1990**, *15* (1), 5–9.

[21] Rodrigues, Jr, C. J., Varzea, V. M. P., Godinho, I. L., Palma, S. and Rato, R. C. New physiologic races of *Hemileia vastatrix*. In: *Proc. 15th Int. Sci. Coll. on Coffee.* ASIC, Paris, 1993, 318–321.

[22] Rodrigues, Jr, C. J., Bettencourt, A. J. and Rijo, L. Races of the pathogen and resistance to coffee rust. *Annu. Rev. Phytopathol.* **1975**, *13*, 49–70.

[23] Rodrigues, Jr, C. J., Varzea, V. M. P and Medeiros, E. F. Evidence for the existence of physiological races of *Colletotrichum coffeanum* Noack *sensu* Hindorf. *Kenya Coffee* **1992**, *57* (672), 1417–1420.

[24] Sreenivasan, M. S., Breeding coffee for leaf rust resistance in India. In: *Coffee Rust: Epidemiology, Resistance and Management.* Kushalappa, A. C. and Eskes, A. B. (eds). CRC Press, Boca Raton, FL, 1989, 316–323.

[25] Van Der Graaff, N. A. Coffee berry disease. In: *Plant Diseases of International Importance. Vol. IV: Diseases of Sugar, Forest and Plantation Crops.* Mukhopadhyay, A. N., Kumar, T., Singh, U. S. and Chaube, H. S. (eds). Prentice-Hall, Englewood Cliffs, NJ, 1992, 202–230.

[26] Van Der Vossen, H. A. M. Coffee selection and breeding. In: *Coffee, Botany, Biochemistry and Production of Beans and Beverage.* Clifford, M. N. and Willson, K. C. (eds). Croom Helm, London, **1985**, 48–96.

[27] Varzea, V. M. P. and Rodrigues, Jr, C. J. Different pathogenicity of CBD isolates on coffee genotypes. In: *Proc. 15th Int. Sci. Coll. on Coffee.* ASIC, Paris, 1993, 318–321.

7
Spraying Equipment for Coffee

H. Pfalzer

7.1
Introduction

A variety of spraying equipment is available on the market for the application of crop-protection products (CPP) to coffee trees. This equipment ranges from hand-held or backpack applicators to tractor-mounted or -drawn machinery.

A number of factors need to be considered when deciding which equipment to chose, e.g.:

- The size and spacing of the coffee trees
- Topography of the area to be treated
- The spray volume considered to be necessary
- The spray coverage required on leaves, berries and branches
- The toxicity hazard of the pesticide to the operator
- The degree of sophistication of the equipment in view of handling and potential repairs
- The cost of the equipment

Whether the equipment is a simple manually operated sprayer or sophisticated machinery, four key points need to be taken into consideration:

- The timing of the treatment

- The coverage of the biological target with spray droplets
- The dosage of the product
- The safety of both the operator and the environment.

7.1.1
Timing of the Treatment

This is the most important aspect of application. Treatment can only be fully effective if it is implemented at the most vulnerable stage of the development of the insect, the weed or the disease (the latter is often a preventive measure).

Table 7.1 shows the dosage required during the stages of development of *Spodoptera littoralis*.

This simple experiment proves that any intervention must be aimed at the first and second instar larvae. Older larvae are more difficult to control because of the progressively higher quantities of product which must be picked up by the larval body.

In the case of insecticides, decisions as to the timing of the treatment should include considerations regarding the development of natural enemies, e.g. the risk of damage to beneficial insects like bees. The timing of fungicide applications depends on the epidemiology of the pathogen and the per-

Coffee: Growing, Processing, Sustainable Production, Second Edition. Edited by J. N. Wintgens.
© 2012 WILEY-VCH Verlag GmbH & Co. KGaA. Published 2012 by WILEY-VCH Verlag GmbH & Co. KGaA.

Table 7.1 Dosage required during the stages of development of *Spodoptera littoralis*

Larval instar	Dose (ng/larva)	Droplet diameter[a]	Relative volume
1	1.7	19	1.0
2	5	27	2.93
3	50	58	29.3
4	500	126	293.8

[a]Formulation containing 50 % a.i. monocrotophos; average LD_{50} = 500 ng for larval instar 4, with contact and stomach action.
From: Novartis Crop Protection, Application Services.

sistence of the chemical element used. Weather conditions, particularly rainfall, are important in determining when pathogens may spread. In addition, the plant may only be susceptible to infection at certain stages in its development, so that fungicide application should be limited to the times when these susceptible stages occur. Contact fungicides act by preventing infection. Timely applications are therefore essential.

7.1.2
Spray Coverage

The second priority for best biological results is the degree of spray coverage required. In general, efficiency increases with the number of spray droplets per unit target area. This basic principle is applicable to all pesticides, although the effect of coverage may be concealed, for instance, by the systemic properties of fungicides or insecticides, or in the case of pre-emergent herbicides. When using products with contact effect a complete coverage where the pest attacks is essential. The degree of coverage required is also determined by the mobility of the pest at a given stage in its development Thus, mobile pests may require a lower density of droplets than scale insects, for example (Pict. 7.1).

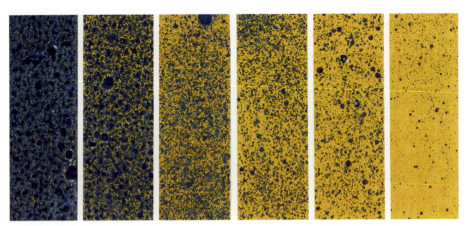

Picture 7.1 Spray coverage visualized on sensitive paper (see also Pict. 7.10). A quick glance reveals positions of over-dosing (left), under-dosing (right) and nozzle dripping (second from left).
From: Novartis CP.

7.1.3
Product Dosage

The dosage recommended on the product label is the amount which controls the pest under most circumstances. Over-dosing increases the persistence of the deposit and may cause a selection for resistance. Under-dosing can also lead to the selection of resistant species of weeds, fungi and insects, as well as to the lack of efficiency of the treatment. It is therefore important to adhere strictly to the instructions on the label.

7.1.4
Safety

All pesticides must be considered as toxic. The best way to minimize the hazards potentially involved in the use of a plant-protection product is to know as much as possible about the precautions to be taken while using it. As the given toxicity of a plant-protection product cannot be altered, the only way of reducing the risk of exposure to harmful amounts is to be aware of it, to have the right attitude, and to translate awareness and required attitude into practice. Hands and arms run the highest risk of exposure when measuring and mixing the concentrate.

7.2
Application Equipment

A variety of spraying machines are used in coffee – the type to be used depends largely on the terrain and suitability of the plantation for mechanization. Tractor-drawn or -mounted sprayers are suitable on flat estates with well-spaced coffee plants. However, a large majority of smallholders with coffee trees on steep land need manually operated equipment such as knapsack

Picture 7.2 Coffee grown on steep slopes in Central America where only hand- and shoulder-carried spray equipment can be used, and, in some cases, spray guns combined with stationary pumps or long hoses.

sprayers or backpack motorized mistblowers (Pict. 7.2).

7.2.1
The Lever-operated Knapsack Sprayer

This type of sprayer is the best choice for the small coffee farmer. In addition, it is the only one which can be used in extremely hilly areas, where even the backpack-carried motorized mistblower is too cumbersome and even dangerous. The equipment employs a metal or plastic tank holding about 10–20 l that is carried on the operator's back with straps like a knapsack. It also consists of a hand-operated pump, a pressure vessel and a lance with one or more nozzles. The tank itself is not pressurized and is not airtight. These sprayers may be built for left (or right)-handed use or adjustable for either. They are fitted with an over- or under-arm lever. The over-arm lever type is easier to operate when walking between tall coffee trees, but is more tiring than the under-arm lever types. The pressure vessel, either in-built or fixed on the outside, serves to maintain pressure on the spray liquid between the pump strokes. This is necessary for the delivery of a continuous flow of liquid at the nozzle with sufficient velocity to cause the liquid to break up into droplets. Fitting a pressure-regulating valve to the outlet of the tank or lance maintains constant pressure (Pict. 7.3).

Lever-operated sprayers, as well as the compression sprayer, are equipped with hydraulic nozzles, preferably with hollow cone nozzles for insecticide and fungicide treatments. Spray volumes per hectare vary according to the density of the coffee trees per hectare, the topography (which may limit the forward speed), the height and the density of the foliage. Productivity also varies according to the above-mentioned factors. In hilly areas like the Andean hills of Colombia, 0.2–0.5 ha of coffee trees can be covered in a day, whereas in flat areas, productivity may rise to 1–1.5 ha per day.

Bearing in mind the safety of the operator as well as increased spray coverage, industry has designed vertical tailbooms which are fixed to the back of the sprayer with a nozzle arrangement pointing horizontally to the coffee tree. This nozzle positioning enhances spray penetration and coverage of underleaf surfaces. A second benefit is the remarkable increase of safety as the operator does not have to walk through the treated crop (Fig. 7.1).

Features

- Medium-volume sprayer
- Tank capacity 10–20 l
- To be used for insecticides, fungicides and herbicides
- Spray volume in coffee: 300–700 l/ha according to crop density and size
- Recommended for small coffee plantations, especially in hilly areas.

Picture 7.3 Lever-operated knapsack sprayer used in weed control. From: Novartis CP.

7.2.2
Compression Sprayer

This type of sprayer has a cylindrical tank in which air is pressurized with a plunger pump that passes through the top of the sprayer and is screwed into the filler opening as part of the lid. The entire tank is airtight and acts as a pressure vessel. This equipment is usually fitted with one or two nozzles on the lance. The sprayer should only be filled with liquid to about 75 % of its capacity, leaving a space above the liquid level to pressurize the air. In theory no pumping is required during spraying. In practice, however, output decreases as the spray pressure decreases.

Variations in output due to pumping rate can be eliminated by fitting a pressure regulating valve [also called a spray manage-

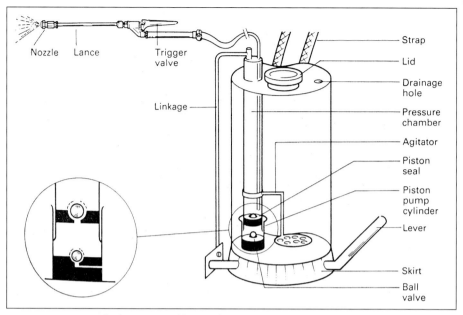

Figure 7.1 Diagram of a lever operated knapsack sprayer with piston pump.

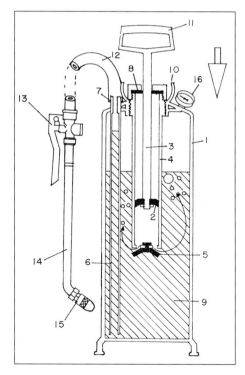

Figure 7.2 Cross-section of a compression sprayer with piston pump: (1) tank, (2) piston washer, (3) pump piston, (4) pump cylinder, (5) flap valve, (6) dip tube, (7) hose attachment, (8) sealing washer, (9) spray solution, (10) filler hole, (11) handle, (12) delivery hose, (13) trigger cut-off valve, (14) lance, (15) nozzle and (16) pressure gauge.

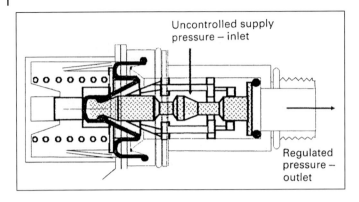

Uncontrolled supply
pressure – inlet

Regulated
pressure –
outlet

Figure 7.3 Spray
Management Valve.
From: Matthews and
Hislop [9].

ment valve (SMV)] on the spray lance. This valve opens at a set pressure and prevents the spray being applied at a pressure exceeding that set at the valve. According to Matthews and Hislop [9], output on knapsack compression sprayers was higher, in practice, and can vary from 6 to 65 %, resulting in wastage of agrochemicals.

Features

- Medium volume sprayer
- Tank capacity 5–15 l
- To be used for insecticides, fungicides and herbicides
- Spray volume in coffee: 300–700 l/ha according to crop density and size
- Recommended for small coffee plantations, especially in hilly areas.

7.2.3
Power-operated Spray Gun

The equipment consists of an integrated or external spray tank, a petrol-run engine connected to a high-pressure pump, a pressure-regulating valve, and a spray hose and gun; the length of the hose varies according to needs. The flow rate and spray pressure are relatively high, they vary between 10 and 30 bars. This power-operated spray

gun is basically developed for treating bushes and tree crops (Fig. 7.3).

The described equipment increases productivity and guarantees good spray coverage with a certain risk of passing the point of runoff (spray not fully retained by the target with consequent loss of spray liquid to the ground). It also requires additional manpower to manipulate long hoses (Pict. 7.4).

Features

- High volume, high pressure sprayer
- To be used for insecticides, fungicides
- Spray volume in coffee above 500 l/ha
- To be used only for bush and tree crops.

7.2.4
Motorized Knapsack Mistblower

Motorized mistblowers have the advantage of carrying the spray droplets into the canopy in a flow of air; this disturbs the canopy and results in good coverage of most of the plant surfaces. A further advantage is in that no pumping is required. The weight and the quite high noise level are, however, negative points.

Generally the spray tank, located above the engine/fan, holds about 10 l. The fan produces a high-velocity air stream which is directed through a flexible hose to a re-

Picture 7.4 High-pressure spray gun used for treating tall and dense coffee trees.
From: Novartis CP.

Features

- Low volume sprayer
- Tank capacity 10–15 l
- To be used for insecticides, fungicides
- Spray volume in coffee: 100–300 l/ha according to crop density and size
- Recommended for small to medium-sized coffee plantations.

Picture 7.5 Fungicide treatment of coffee with a motorized knapsack mistblower.
From: Novartis CP.

7.2.5
Tractor-mounted or -drawn Mistblowers

strictor nozzle mounted on its end. Droplet production is brought about by the combination of air stream and liquid fed through a tube, called a shear nozzle, into the air stream. Normally the spray tank is slightly pressurized by the introduction of air produced by the fan. Pressurization of the tank improves the flow of the spray mixture to the shear nozzle and allows the operator to raise the nozzle above the liquid level in the tank. This is necessary for treating higher coffee trees. The tank must be airtight. The machine must always run at full throttle to guarantee optimal droplet production.

Once the high-speed air stream has left the air tube, the air blast very rapidly looses momentum due to the very low volume of air output. At a distance of 5 m from the outlet the speed will have dropped from its initial 250 to less than 20 km/h. This loss of momentum results in limited swath and limited penetration into the large masses of vegetation which can sometimes be found in certain orchards (Pict. 7.5 and Fig. 7.4).

The basic type of blower deflects the air stream produced by an axial fan through 90°. Hydraulic-type nozzles are mounted close to the outlet. The air stream carries the droplets into the coffee tree canopy. In this context it is worthwhile mentioning that it is mainly the volume of air produced by the equipment which is decisive for the quality of the treatment and not the air speed. Modern equipment is fitted with baffle plates in order to better direct the air stream. A directed air stream minimizes losses into the soil or into the air. The pressure of the liquid varies between

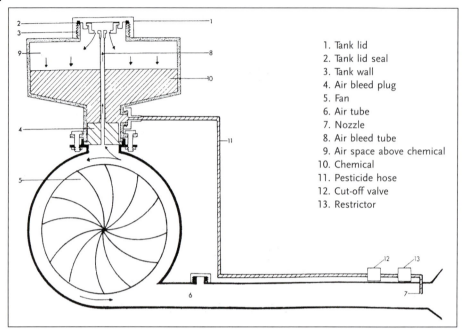

Figure 7.4   Cross-section of a motorized knapsack mistblower. From: Novartis CP.

1. Tank lid
2. Tank lid seal
3. Tank wall
4. Air bleed plug
5. Fan
6. Air tube
7. Nozzle
8. Air bleed tube
9. Air space above chemical
10. Chemical
11. Pesticide hose
12. Cut-off valve
13. Restrictor

Picture 7.6   Tractor-mounted mistblower on a well-spaced and leveled plantation. From: Novartis CP.

5–15 bar. The power or volume of the air stream and the speed of the tractor should be adapted to the density of foliage and the distance from the sprayer outlet to the tar-

get. All the leaves of a coffee tree should be moved, but not harmed.

The spray volumes applied are normally rates well below run-off. When adapting the sprayer to relatively low spray volumes, the concentration of the spray mixtures should be increased (Pict. 7.6 and 7.7).

Features

- Low volume sprayer
- Spray tank capacity 600–4000 l
- Spray volume in coffee: 200 to more than 1000 l/ha according to crop density and size.

7.2.6
Canon-type Mistblower

This type of application equipment offers high productivity covering greater areas in less time, but it only pays off when used

Picture 7.7 Vertical booms fitted with hydraulic nozzles without air assistance. From: Novartis CP.

Picture 7.8 A truck-mounted cannon-type mistblower (SWISSATOM) operating in a coffee plantation. From: Fischer.

can drastically influence the behavior of the smaller droplet fraction. This is why sprayers of this type should preferably be used during windless nights. They are less suitable for disease control because of the high volume and the full coverage of the plants required (Pict. 7.8).

Features

- Low- to high-volume mistblower
- To be used for insecticides and fungicides
- Spray volume in coffee: 10–500 l/ha
- Recommended for tree and bush-type crops
- The effective spray blast should be carefully checked
- High application productivity

(Based on the technical leaflet of the manufacturer: Fischer, 1800 Vevey, Switzerland.)

in extended coffee plantations with well-leveled terrain and non-shaded crops of uniform size. Its air volume production (i.e. 120 000 m^3/h) is higher compared with the common tractor mounted mistblowers. The blast of the spray can reach up to 65 m when moving forward at a speed of 1.2 km/h. The effective cloud may, however, be somewhat less. The spray jet is drastically reduced when moving at higher speeds, e.g. 4 km/h. The simultaneous coverage over a number of rows of coffee trees (i.e. 20 rows) is not as even as that achieved with a mistblower passing through each row, so care must be taken to overlap the various swaths efficiently in order to achieve the required biological effect.

The spray volume can be varied from 10 to 500 l/ha according to the technical information provided by the manufacturer.

Correct timing of the treatment is of vital importance as wind and thermal currents

7.2.7
Other Application Equipment

7.2.7.1 Reduced-volume Sprayers, also known as Controlled Droplet Application (CDA)

Sprayers of this type have become quite popular for field crops like cotton and in areas where water is in short supply. CDA is generally achieved by centrifugal-energy nozzles, i.e. spinning disks. CDA uses quite low volumes of spray, i.e. down to 1 l/ha in certain field crops. In extremely hilly coffee plantations, tests have been made with backpack carried spinning disk equipment provided with a fan to enhance penetration into the canopy and improve operator safety. Coffee Leaf Rust was effectively controlled at 50 l/ha with a 5-fold improvement in workrate compared to the traditional high volume spraying techniques, i.e. with knapsack sprayers (Fig. 7.5).

7.2.7.2 Electrostatic Spraying

The Electrodyne™ was invented in the 1980s with a view to improving the control of droplet distribution and reducing the reliance on wind dispersal for spray droplets generated by CDA sprayers.

The Electrodyne™ is similar to a CDA sprayer, but it is equipped with a high-voltage generator combined with a special nozzle and product container which is filled with specially formulated CPPs based on mineral oil solvents. The high-voltage generator charges the spray liquid and causes it to form a series of ligaments as it emerges from the nozzle.

According to Matthews [9], the droplets produced are very uniform in size. The electric charge ensures that they are equally distributed to both the upper and the under surfaces of the leaves. This is most important for mite control.

The charged spray collects on the object nearest to the nozzle. As a result, to avoid contamination, the nozzle should always be closer to the crop than to the operator. This is sometimes difficult to achieve, particularly in hilly areas like coffee plots in Columbia. In such cases, the Electrodyne™ nozzle should always be aimed downwind from the operator.

The distance of the spray drift is generally lower than that of conventional CDA sprayers so penetration to the lower leaves in closely planted coffee trees can be poor.

It should also be noted that the use of the Electrodyne™ is limited to a selected range of CPPs which are adapted to this type of spray technique which, in turn, means that it has not been adopted on a wide scale.

Another type of sprayer called air assisted electrostatic spraying (ESS) is based on following principle:

– Air and liquid enter the rear of the nozzle separately and then interact at the tip of the nozzle to form spray droplets. The diameter of the droplet is between 30 and 60 microns. Air pressure

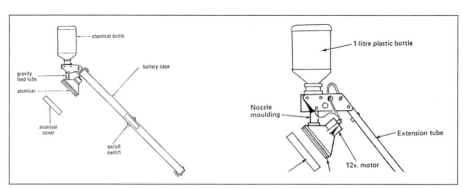

Figure 7.5 Reduced-volume sprayer, also known as CDA sprayer. From: Micron Sprayers.

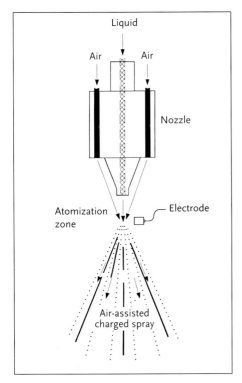

Figure 7.5a Air atomizing induction charging nozzle. From: Electrostatic spraying systems, Inc. Watkinsville, Georgia, USA.

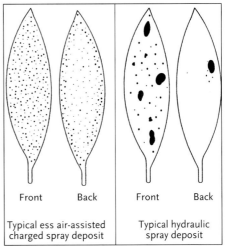

Front	Back	Front	Back
Typical ess air-assisted charged spray deposit		Typical hydraulic spray deposit	

Figure 7.5b These sketches were made from microscope evaluations of spray deposits on plant leaves. The coverage with the air-assisted electrostatic sprayer is a fine powder coat. The spray is well distributed on both sides of the leaves. Hydraulic sprayers produce droplets which vary widely in size and often "puddle" on the leaf. From: Electrostatic spraying systems, Inc. Watkinsville, Georgia, USA.

is between 30 and 40 psi and liquid pressure below 15 psi.

- As the spray is atomized, the droplets pass an electrode which induces a negative electric charge on each droplet.
- The force of the turbulent air-stream then propels the charged droplets into the plant cover (see fig. 7.5a).
- The positive charge of the plants attracts the droplets which coat the leaves, fruit and stems (see fig. 7.5b).
- The charged droplets have a force of attraction which is 7.5 times that of gravity which means that droplets will also move upwards to coat hidden surface as well (Pict. 7.9).

Comments

Air assisted electrostatic sprayers provide 4 to 10 times improved coverage with a low volume mix (100 to 250 l/ha, when compared to traditional electrostatic sprayers, added to which chemical inputs are halved.

The turbulent air assistance generated by the mist-blower combined with the low volume electrostatic system provides optimal crop penetration and enhances the biological efficiency of the CPPs.

Tractor-drawn equipment of this type is now used in a number of large African coffee estates with, on the whole, satisfactory results.

Weak points are, inevitably, the price of the equipment and the fact that high level technology is involved which means

Picture 7.9 Ari assisted electrostatic sprayer (Egg). Detail of layout. From: J.N. Wintgens.

pends on prevailing horizontal and vertical air currents which can cause the width of the swath itself to vary drastically, added to which it is difficult to obtain a correct distribution on leaves like coffee at a range of up to 10 µm.

7.2.7.4 Granule Applicators

This is, typically, the equipment to choose when controlling rust or insects, like the Leaf miner and nematodes with systemic pesticides. Herbicides are also applied as granules. According to Matthews [8], the production of dust formulations has dropped over the years, whereas granule application has increased.

When selecting a granule applicator, four important points should be taken into consideration:

- Accurate delivery of the required amount
- Even spread of particles
- Avoidance of damage to the granules by grinding or impaction
- Adequate mixing and feeding of the granules to the metering device

Special attention must be paid to the metering system of a granule applicator in order to avoid the risk of blockage and/or

that skilled operators need to be in charge of the spraying operations.

7.2.7.3 Fogging Equipment using Shear Force for Droplet Production

The information available on the use of cold- and hot-fogging equipment in outdoor crop protection is very limited and usually targets mosquitoes and flies. This type of equipment is more often used under indoor conditions, e.g. insect control in stored product protection and insect and/or fungi control in greenhouses. As the droplet spectrum ranges from 0.1 to about 20 µm, the effectiveness of this "aerosol" or "fog" protection greatly de-

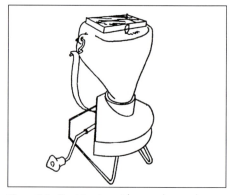

Figure 7.6 Spot treatment knapsack granule applicator. From: Horstine Farmery.

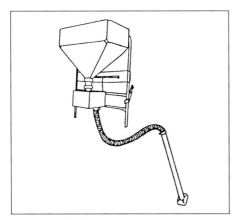

Figure 7.7 Shoulder-carried granule applicator.
From: American Spring & Pressing Works, India.

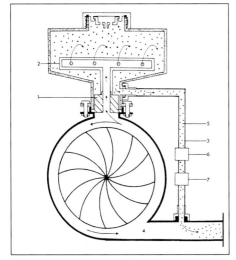

Figure 7.8 Diagram of a knapsack-carried
mistblower fitted for solid product formulations:
(1) air bleed plug, (2) agitator bar, (3) pesticide
discharge tube, (4) air tube, (5) chemical,
(6) cut-off valve and (7) restrictor.

damage of the granules (grinding). There
are two types of metering system – either
a device where the cross-sectional area of
a chute can be altered by means of a
lever or screw, or, alternatively, the granules
are dropped through one or more holes,
the size and number of which can be regulated (Fig. 7.6 and 7.7).

For coffee, mainly manually operated
hand-, shoulder- or backpack-carried applicators are in use for band or spot application of granules. In addition, and for broadcast treatments, most motorized backpack-
carried mistblowers can be converted for
granule application by removing the spray
hose and inserting a wider tube, feeding
the granules through a metering device
into the air stream (Fig. 7.8).

Note that granule formulations have two
potentially environmentally friendly aspects
in their favor. A good quality dust-free granule is unlikely to cause problems associated
with spray drift, and, at the end of treatment, surplus granules can be retrieved
from the spreader and replaced in their original container, eliminating the problem of
disposing of leftover spray mix.

7.3
Calibrating the Application Equipment

Calibrating a sprayer is a simple, but
important, procedure. Under-dosing and
over-dosing are both wasteful of chemicals,
and can result in crop damage and yield loss.

There are different calibration methods.
The most easy method is to:

- Determine the volume sprayed in a defined area or on a defined number of
  coffee trees
- To calculate the volume sprayed for 1 ha
  or a given number of coffee trees

Example
Step 1: Determine and mark a representative crop area of, for example, 200 m^2
Step 2: Spray the coffee trees and measure
the spray volume used, e.g. 12 l
Step 3: Calculate the spray volume per unit
area

$$(12 \times 10\,000) / 200 = 600 \text{ l/ha}$$

To calculate the average spray volume per coffee tree, the volume/ha should be divided by the number of trees (e.g. 6500/ha).

600 l / 6500 = 0.09 l per tree

If the spray volume is considered to low or to high, the following points should be taken into account:

Method to increase the spray volume	Method to decrease the spray volume
Bigger nozzle orifice	Smaller nozzle orifice
Higher spray pressure	Lower spray pressure
Lower forward speed	Higher forward speed

7.4
Preparing for Spraying

7.4.1
Product Measurement and Mixing

7.4.1.1 Portable Equipment
Where no mixing tank is used, fill half of the sprayer with clean water, add the calculated and measured quantity of product, close the knapsack lid, shake well, and continue to fill the sprayer with water and shake again.

Another method is to premix in a bucket, transfer the mixture into the knapsack, preferably using a funnel, and filling up the sprayer with water. This mixing method must be used with all solid preparations like "wettable" powders or water-soluble granules.

7.4.1.2 Tractor-drawn Equipment
Fill a quarter of the spray tank with clean water, start agitation, add the calculated and measured quantity of product (solid preparations are best pre-mixed in a bucket), and continue to refill the spray tank while agitating. To maintain a uniform spray mixture continue agitating during the treatment. When filling the spray tank, the filling or loading hose should always be above water level in order to prevent reflux. Water quality is important – extreme alkaline or acid pH can hydrolyze the active ingredient of a plant protection product, thus reducing efficiency. Product compatibility should first be checked out on a small patch with the partner formulation.

7.5
Spraying

7.5.1
General Points to be observed when spraying Coffee Trees

- High temperatures at noon and early in the afternoon can cause drift by thermal currents and quick evaporation of small water droplets.
- Relative air humidity influences the evaporation rate of water-based sprays – under tropical conditions, the speed at which the droplet size decreases is faster when the temperature is high and humidity low.
- Wet foliage after dew or rain can cause an easy run-off of the spray.
- Coffee trees are best treated from both sides.
- Adjust the spray volume to the size and volume of the coffee trees in order to obtain thorough coverage.
- Stop spraying in winds higher than 5 m/s (Pict. 7.10).
- With motorized knapsack mistblowers, the engine should always run at full throttle (speed) in order to obtain the best droplet formation.

Picture 7.10 Wind meter to determine wind speed and wind direction. From: Spraying Systems, Germany.

- Adjust the air stream of tractor-drawn mistblowers to the foliage height in order to minimize product losses by drift and on soil.

7.5.2
Spray Volumes

One must bear in mind that high spray volumes, leading to run-off from leaves and branches, induce product losses and unnecessary contamination of the soil. The same applies to the use of extremely high spray pressures which can cause extreme spray drift causing contamination to nearby crops and the environment in general.

The spray volume up to the run-off point depends on the height and width of the coffee tree; in other words, small trees will receive less, bigger trees will receive higher volumes in order to ensure the coverage needed for satisfactory biological control. The concentration of the product in the carrier, mainly water, is the one indicated on the product label, e.g. 40 g/ml in 100 l of water. As soon as the spray volume is in-

creased, the product dose rate per hectare is also increased. This method is called crop-adapted spraying (CAS) and was developed by the Application Services of Novartis Crop Protection.

Example:

- Product use rate given on the label: 40 g per 100 l of water.
- Spray volume adapted to crop height: 100 up to 750 l/ha.
- Product dose rate adapted to tree height: 40–300 g (g/ha = g/100 l x spray volume/100).
- This ensures a constant product quantity per leaf area.

Results of above example:

Height of coffee tree (cm)	25	40	70	90
Spay volume just before run-off (l)	100	230	540	750
Product dose rate (g/ha)	40	92	216	300
Percentage of full dose (%)	13	31	72	100

From: Novartis Crop Protection, Application Services.

Reducing the spray volume – when using low volume spray equipment and due to reasons of productivity, farmers can significantly increase the number of hectares sprayed per day. However, one should bear in mind that the lower the spray volume, the more important it is to accurately calibrate the equipment. When lowering the spray volume the concentration of the product/water mixture has to be increased proportionally.

Example:

- Spray volume reduced by 50 % and compared to the spray volumes mentioned in above example (100–230–540–750 l/ha).

Results of above example:

Height of coffee tree (cm)	25	40	70	90
Spray volume[a] (just before run-off) (l/ha)	50	115	270	375
Product dose (g/ha)		40	92	216	300
Percentage of full dose (%)		13	31	72	100

[a]If the spray volume is reduced by 50 %, the product/water concentration is increased by a factor of two, i.e. 80 ml/100 l.
From: Novartis Crop Protection, Application Services.

7.5.3
Spray Deposit Monitoring with Water-sensitive Papers

Water-sensitive papers containing a layer sensitive to water makes the spray visible. Spray droplets falling onto the layer cause blue spots. The paper is ideal to check spray droplet density and spray penetration

Picture 7.11 Water sensitive paper covered with spray droplets. From: Spraying Systems, Germany.

into the coffee tree. It shows where the spray has reached the target and where not. It can be stuck to leaves or branches or laid out on the ground. The material was developed by CIBA. Today, it is sold by Application Services of Novartis Crop Protection exclusively to Spraying Systems, one of the world's largest spray nozzle manufacturers (Pict. 7.11).

7.5.4
The Spray Nozzle

Nozzles have two important functions:

- To break the spray mixture or the concentrate (in the case of ultra-low-volume formulations) into droplets;
- To disperse the droplets by means of hydraulic pressure, air shear or the rotation of a disk.

There are three main methods by which liquids are transformed into droplets:

- By hydraulic pressure, whereby the liquid is forced through different types and sizes of a hydraulic nozzle either fixed to knapsack sprayers or to tractor-drawn mistblowers;
- By centrifugal force, where a rotating disk is fed with the spray mixture – this nozzle is called the spinning disk nozzle and is used on spinning disk sprayers;
- By air blast or shear force, involving the use of a high-speed air stream and the shear nozzle, used on backpack-carried motorized mistblowers.

7.5.4.1   Nozzle Types and Parts

7.5.4.1.1   The Hydraulic Nozzle
Different types of hydraulic nozzles are shown in Pictures 7.12a and 7.12b and Figures 7.9 and 7.10.

Picture 7.12a Parts of a flat nozzle, from left to right: nozzle body, cap, strainer, nozzle tip or orifice. From: Novartis CP.

Picture 7.12b Parts of a flood jet nozzle for herbicide application. From: Novartis CP.

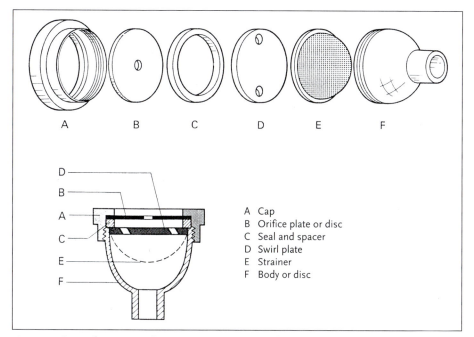

A Cap
B Orifice plate or disc
C Seal and spacer
D Swirl plate
E Strainer
F Body or disc

Figure 7.9 Parts of a cone nozzle. From: WHO Equipment for Vector Control, 1974.

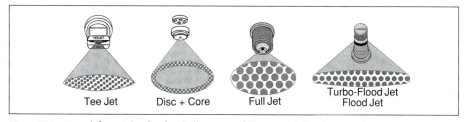

Tee Jet   Disc + Core   Full Jet   Turbo-Flood Jet / Flood Jet

Figure 7.10 From left to right: flat fan, hollow cone, full cone and flood jet nozzle. From: Spraying Systems.

Features (to be used with):

- Manually operated knapsack and pre-pressurized knapsack sprayers
- High-pressure spray guns
- Tractor-mounted or -drawn mistblowers
- Flat fan and floodjet recommended for herbicide treatments
- Hollow and full cone recommended for insecticide and fungicide treatments

7.5.4.1.2 The Spinning Disc Nozzle (Centrifugal Energy Nozzle)

Details of spinning disc nozzles are illustrated in Figure 7.11.

Features (to be used with):

- Controlled droplet application equipment – either manually operated spinning disk sprayers or integrated on tractor mounted or drawn mistblowers

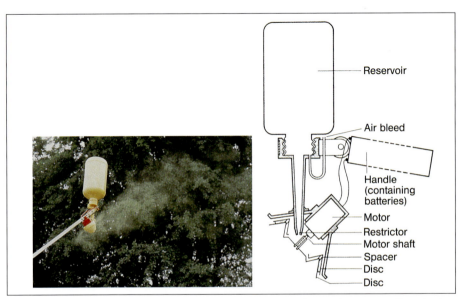

Figure 7.11 Cross-section showing details of spinning disc. Observe the very fine droplet cloud produced by the spinning disc. By adjusting the rotational speed of the disc, smaller or larger droplets are produced. This nozzle operates efficiently when the volume of spray applied is restricted, i.e. 1–10 l/ha (ultra-low-volume to very-low-volume spraying). From: Novartis CP.

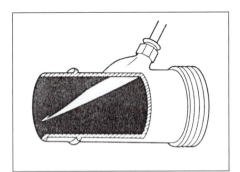

Figure 7.12 Cross-section of a shear nozzle. From: Matthews [8].

Picture 7.13 Shear nozzle used on a motorized knapsack mistblower. From: Novartis CP.

- For insecticide and fungicide treatments
- Adapted equipment for herbicide treatments can also be found on the market

7.5.4.1.3 The Shear Nozzle
(Gaseous Energy or Twin-fluid Nozzle)
Details of Shear Nozzles are shown in Fig. 7.12 and Pict. 7.13.

Features (to be used with):

- Motorized knapsack mistblower
- For insecticide and fungicide treatments

7.5.5
Nozzle Material

Orifices of hydraulic nozzles are made from different materials (Tab. 7.2). The choice of material depends on the abrasiveness of the product formulation in the spray mixture. "Wettable" powders and flowables are more abrasive than emulsions and solutions.

The nozzle orifice is the heart of every hydraulic nozzle – its quality is essential for spray quality. Wear or damage of the orifice, which can be caused by imperfect cleaning, can change the characteristics of the spray pattern as shown in Pict. 7.14.

7.6
The Spray Droplets

The spray mixture, applied through a nozzle with either a hydraulic, a shear or a spinning disc, will result in a spray cloud. This spray cloud consists of numerous droplets of different sizes. In order to obtain the right perception of droplet sizes, Tab. 7.3 illustrates the size range of different natural and artificial particles.

The diameter of the droplets of the mist and spray spectrum produced by the different nozzles varies from smallest up to about 1 mm. The smallest droplets may invacle the lungs. Particles of mist and all types of spray larger than 10 microns can be trapped in the nose.

Table 7.2 Materials for hydraulic nozzle orifices

Material	Rate of wear	Cost
Brass	rapid wear	low
Stainless steel	slow wear	high
Nylon	slow wear	medium
Ceramic	very slow wear	high

From: Novartis Crop Protection, Application Services.

Table 7.3 Droplet spectrum produced by different types of spray nozzles

Type of nozzle	Definition of spray
Hydraulic nozzles	fine to coarse spray
Shear nozzle	mist to fine/medium spray
Spinning disk nozzle	aerosol to mist

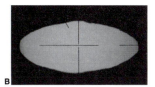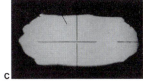

Picture 7.14 Spray pattern of hydraulic nozzles:
(A) new orifice, (B) worn orifice and (C) damaged orifice, due to imperfect cleaning.

The optimal droplet size spectrum for a given application will depend on the pesticide used (systemic or contact), the target configuration and the tolerable loss to non-target areas.

7.7
Operator Safety

7.7.1
Exposure

The risk of exposure to CPPs during the various handling operations like measuring, mixing and applying is closely related to the operator's awareness of the risk, his standard of hygiene, the application technique used and the quality of the spray equipment. However, hazard exposure depends largely on the toxicity of a given formulation, the quantity of product which is finally absorbed and the physical condition of the exposed person.

The human body can absorb CPPs in three ways:

- Exposure through the skin (dermal absorption)
- Exposure through inhalation (nasal absorption)
- Exposure through the mouth (oral absorption).

Dermal absorption represents the most frequent form of contamination in plant-protection activities. It is assumed that 90 % of all exposure cases worldwide are accounted for by skin contact. Penetration through the skin, i.e. unprotected hands, arms, feet and face, can happen rather fast. High temperatures and skin perspiration increase absorption. EC and ultra-low-volume preparations generally penetrate the skin much more rapidly than EC-water mixtures. The handling of liquid

Picture 7.15 In the particular case of treating young coffee trees, inhalation protection is not absolutely necessary. From: Novartis CP.

concentrates presents a high risk of exposure.

Inhalation is the second form of exposure to CPPs in agriculture. Inhalation hazard is mainly relevant in the case of vapors, particularly when CPPs are applied in an enclosed space and in the case of fine sprays from ULV equipment or dust. The critical drop size to be absorbed effectively through the lungs is below 10 μm, whereas droplets of up to 100 μm are inhaled and impact the mucous membrane of the nasal canal (Pict. 7.15).

Ingestion of CPPs can occur through the consumption of contaminated food or the use of contaminated eating utensils. Contaminated hands may also lead to intake of CPPs, particularly when smoking. Contamination is also frequent when blowing through the spray nozzle to clean it or when using the mouth to siphon products out of a container.

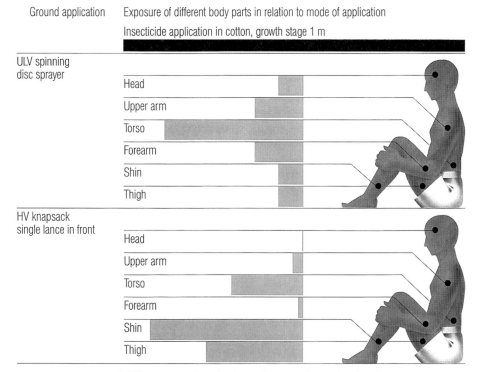

Ground application   Exposure of different body parts in relation to mode of application

Insecticide application in cotton, growth stage 1 m

Figure 7.13 Exposure of different parts of the body in relation to the mode of application.
From: Novartis CP.

Application equipment and the type of crop treated lead to risk exposure of different parts of the body. Figure 7.13 illustrates different dermal exposure patterns. This is useful to increase user awareness of potential risks.

7.7.2
Classification of Pesticides by Hazard

The toxicity values of formulated products are a guide to the toxic effects of pesticides on humans. The active ingredient (a.i.) toxicity does not reflect the hazard associated with exposure to the formulated pesticide.

For instance, the a.i. in a product may be categorized as "extremely toxic" to humans, but only present a low to moderate hazard to the user because of the low con-centration of a.i. in the formulation or because it is marketed as a solid formulation of low volatility (i.e. granules).

On the other hand, a product containing an a.i. with low toxicity may present an increased hazard to humans because it is marketed as a liquid formulation containing solvents which accelerate skin penetration.

Table 7.4 provides a more practice-oriented recommendation with regard to personal protection.

- Pesticides of class Ia and Ib are considered as "high hazard products" and those of II and III are considered as "low hazard products"

- The judgment of hazard is more rigorous for liquid pesticides due to the higher risk of penetration through protective

Table 7.4 Classification by hazard on the basis of LD_{50} for the rat (mg/kg of body weight)

Class	Oral hazard		Dermal hazard	
	Solids	Liquids	Solids	Liquids
Ia: Extremely hazardous	5 or less	20 or less	10 or less	40 or less
Ib: Highly hazardous	5–50	20–200	10–100	40–400
II: Moderately hazardous	50–500	200–2000	100–1000	400–4000
III: Slightly hazardous	500–2000	2000–3000	over 100	over 400

The terms solids and liquids refer to the physical state of the product or formulation being classified.
Source: WHO Recommended Classification of Pesticides by Hazard.

clothing and the possibility of ready absorption through the skin

- Pesticides with very high LD_{50} values (above 2000/solids and above 3000/liquids) are considered as "unlikely to present an acute hazard in normal use"

7.7.3
Necessary Precautions when Handling CPPs

- The operator should be aware of the toxicity of the product
- The operator should make sure that the application equipment, including the spray nozzles is in good working condition
- The operator should wear appropriate skin protection
- Basic hygiene rules should be strictly followed: no eating or smoking during preparation and treatment, wear clean work clothes (avoid putting on clothes which have already been in contact with CPPs and which have not been washed), wash and bathe or shower immediately after work, and wash off spills on unprotected skin immediately

Extreme care must be taken to avoid spills into the eyes.

Wind velocity and wind direction have to be taken into consideration before starting the application. Table 7.5 gives some examples of potential contamination sources for both humans and the environment.

7.7.4
Factors which Influence the Dermal Absorption of CPPs

The main factors which influence the dermal absorption of CPPs are listed in Table 7.6.

7.7.5
Prevention of Adverse Health Effects

7.7.5.1 Education
Misuse of CPPs is often the result of ignorance, which can only be dealt with by information, education and training. Among other training programmes and, in the long run, formal school education can be positively associated with the adoption of protective gear and improved hygiene practices.

7.7.5.2 Protective Practices
Practices that influence exposure and health risks include hygiene, careful handling procedure, safe storage and proper disposal of packaging, etc. Personal hygiene requires the greatest attention. Soap and

Table 7.5 Examples of potential contamination sources for both humans and the environment

Potential sources of contamination	Contamination risk and main route of exposure
Transport of crop protection products together with food	Cross-contamination of foods; farmer and his family (oral)
Road accidents, leakage of product containers	Soil, water, farmer, population
Storage on the farm close to food	Cross-contamination, farmer (oral)
Leaky product container	Farmer and his family (dermal)
Application close to inhabited areas (drift)	Farmer and his family, house animals, population (dermal, inhalation)
Disposing rinse water of spray equipment into waterways	Farmer, population, wildlife (oral)
Insufficient dermal protection	Farmer, spray operator (dermal)
Leaky spray equipment	Farmer, spray operator (dermal)
Cleaning the nozzle by "blowing through"	Farmer, spray operator (oral)
No awareness of personal hygiene, i.e. spills during handling of the concentrate are not washed away immediately, not taking a bath/shower after field work, etc.	Farmer, spray operator (dermal)
Using unrinsed product containers for home use, i.e. for water or grain storage	Farmer and his family (oral)
Leaving "empty" product containers in the field	Children, wildlife, water, soil (oral)
Non-observation of the re-entry interval after spraying	Farmer, scouting team (dermal)
Non-observation of the pre-harvest interval	Population (oral)
Using crop protection products (not approved for this purpose) for hygiene pests like cockroaches, flies, mosquitoes, etc.	Farmer family (inhalation, oral, dermal)
Using crop protection products as "home cures", i.e. control of fleas	Farmer family (dermal, oral)
Re-use of working clothes which have not been properly washed, i.e. gloves, boots, trousers	Farmer, spray operator (dermal)
Handling of product packages contaminated on the outer surfaces	Farmer, spray operator, children (dermal)

water should be available where mixing, loading and spraying take place, and should be used to wash skin and clothing. In order to reduce absorption, washing of spills on the skin should take place as soon as possible after exposure. Careful handling is crucial to prevention, as spills of the concentrate entail very high levels of exposure (Pict. 7.16 and 7.17).

Table 7.6 Factors which influence the dermal absorption of CPPs

Skin characteristics	sores and abrasions wetness of skin location on the body (absorption occurs readily through eyes and lips)
Environmental factors	temperature humidity
Product characteristics	acidity physical state concentration of active ingredient

Source: Angehrn [1].

Picture 7.17 Good skin protection is already a positive step forward for the protection of the health of the operator. A plastic cape worn over normal clothing avoids soaking and is also relatively comfortable to wear even in hot climates. From: Novartis CP.

7.7.5.3 Protective Devices

Product labels generally state which protective devices should be worn, but one should bear in mind that the actual protective effect of these devices, i.e. trousers, boots or gloves, may differ considerably according to the quality of the materials.

Protective devices are only as good as the maintenance they receive, therefore washing and appropriate storage are of paramount importance. Rubber or plastic gloves are particularly important as hands are subject to a large proportion of exposure especially while measuring and mixing the concentrate.

Heavy protective gear like waterproof trousers or jackets may not be comfortable to wear in warm climates. It is therefore recommended to select only devices and material which allow for minimum working comfort.

Picture 7.16 This is not the way agro-chemicals should be applied. From: Novartis CP.

7.7.5.4 Substitution

Methods for replacing exposure hazard products by relatively harmless products are presently being investigated. Advances have been made in developing safer CPPs, safer application methods, and safer preparations and packages. The use of seed dressing, spot treatments, solid preparations, micro-encapsulation, water-soluble product packages, etc., are examples of this research.

7.7.5.5 Specificity

Many newly developed CPPs are very specific to the target organism. The fact that they are selective products means that they are less hazardous to human health. Hence, the use of modern CPPs requires specific knowledge, particularly with regard to correct pest identification.

7.8 Conclusions

When growing coffee, there is a permanent need to protect the crop against damage caused by pests, diseases, and weed competition. Infestation of any type reduces yields and has a negative effect on quality.

In order to control crop damage, two requirements are of paramount importance:

a) the use of specifically targeted crop protection products

b) the availability of efficient spraying equipment.

During recent decades, crop protection products have become more selective and much less toxic to both humans and warm-blooded animals. The product formulations have been improved to simplify and facilitate the measuring and mixing process and reduce exposure risks to the operators.

It should be noted, however, that the best product is only as good as is its application: this calls for efficient and well maintained spraying equipment.

Environmental lobbies and consumers alike are ever more aware of the negative impact of incorrectly handled crop protection tasks. This can include excessive drift, overdosing and wrong timing.

Furthermore, the current economic difficulties encountered in the coffee trade call for safe and cost controlled crop protection management including necessarily the adoption of integrated pest management.

On the other hand, spraying equipment manufacturers would be well advised to pay particular attention to safety measures for both man and the environment.

Innovation is the key to the future!

New technologies should enable the spraying equipment industry to satisfy modern requirements and make plant protection treatment ever more reliable and safe.

Bibliography

[1] Angehrn, B. *Plant Protection Agents in Developing Country Agriculture.* ETH 11938. Swiss Federal Institute of Technology, Zurich, 1996.

[2] Felber H. U. Novartis Crop Protection, Application Services, 4002 Basle, Switzerland. Personal communication.

[3] GTZ/GATE. *Tools for Agriculture, A Buyer's Guide to Appropriate Equipment.* Intermediate Technology Publications, London, 1985.

[4] Hastings, J. J. and Quick, G. R. *Small Sprayer Standards, Safety and Future directions for Asia.* IRRI, Manila, 1992.

[5] Hiller, M. Novartis de Colombia SA, Bogotá, Colombia. Personal communication.

[6] Hoffmann, H. Investigación y Asistencia Agrícola SA, San José, Costa Rica. Personal communication.

[7] Landesanstalt für Pflanzenschutz/Pflanzenschutzdienst. *Pflanzenschutzmaschinen und deren Einsatz.* Landesanstalt für Pflanzenschutz/Pflanzenschutzdienst. Stuttgart, Germany, 1979.

[8] Matthews, G. A. *Pesticide Application Methods.* Blackwell, Oxford, 1979.

[9] Matthews, G. A. and Hislop, E. C. *Application Technology for Crop Protection.* Oxford University Press, New York, 1993.

[10] Matthews G. A. and Clayphon J. E. Safety precautions for pesticide applications in the tropics. *Pest Articles & News Summaries* **1973**, *19*, 1–12.

[11] Novartis Crop Protection, Application Services. *Safety depends on You/Operator Safety: A Key Element in Successful Pesticide Application.* Novartis Crop Protection, Application Services, 1988.

[12] Novartis Crop Protection, Application Services. *Application Techniques for Plant Protection in Field Crops,* 2nd edn. Novartis Crop Protection, Application Services, 1985.

[13] Raisigl, U. Novartis Crop Protection, Application Services, Basel, Switzerland. Personal communication.

8
Quarantine for Coffee

D. Bieysse

8.1
The Reasons behind Quarantine Measures

Quarantine measures are implemented when any plant materials are transported from one area of cultivation to another. When such transfers are planned, special care should be taken to avoid contaminating the destination area. The plants must therefore undergo a period of quarantine in a nursery located well outside cultivation areas. The plants should be carefully monitored in the quarantine nursery in order to ensure that they are healthy and not likely to carry pathogens to their new location. Where necessary, the reproduction cycle of the plants can be completed in these nurseries. The length of the quarantine period is determined by the normal incubation and development period of the suspected or detected pathogen or parasite. Climatic conditions in the quarantine nursery should be similar to those where the pathogen would develop naturally and the quarantine period should be longer than its normal development cycle.

For coffee plants, supervision is essentially geared to the detection of fungi, insects, nematodes and bacteria. To date, no lethal virus has yet been noted in the *Coffea* species, thus a 6-month quarantine period is generally considered to be an adequate guarantee when transporting coffee plants from one area to another.

Since the Coffee Ringspot Virus has become active, several countries are establishing compulsory protective protocoles (see chapter 2.5).

8.2
General Recommendations for the Transfer of Plant Material

In order to reduce the risk of introducing known or unknown diseases when plant material is to be transferred, the following precautions must be taken.

Preparing the transfer:

- Avoid collecting plants in an area where pathogens have been detected.
- If this is not possible, only select plants which appear to be healthy.
- Treat the plants with either a cupric-based product or a product which is specific to the suspected or detected pathogen.
- Transfer the plants to a quarantine nursery well outside cultivation areas.

Coffee: Growing, Processing, Sustainable Production, Second Edition. Edited by J. N. Wintgens.
© 2012 WILEY-VCH Verlag GmbH & Co. KGaA. Published 2012 by Wiley-VCH Verlag GmbH & Co. KGaA.

Reception of the plants in the quarantine nursery:

- Isolate the plants on reception in specific premises.
- Restrict access to the isolation premises to specialized staff only.
- Examine the plants on arrival and destroy any suspect material.
- Destroy all excess or suspect material either by incineration or by autoclaving.
- Incinerate all packaging and other refuse.

As no lethal virus has yet been detected in coffee plants, the transfer of plant material for reproduction by *in vitro* methods is considered to be an efficient way of avoiding contamination by pathogens.

8.3
Type of Plant Material

8.3.1
Ripe or Ripening Fruit

- Select fruits that appear healthy.
- Where possible, remove the pulp, rinse and dry the seeds.
- The seeds should be dusted with a fungicide and an insecticide, and packed in sealed plastic envelopes.

8.3.2
Grafts and Shoots

- Grafts, with or without leaves, should be taken from plants that appear healthy.
- They should be dipped in a fungicide and an insecticide solution, and wrapped in newspaper that has been soaked in the solution.

8.3.3
Shoots and Plants with Roots

- Special attention should be given to the roots when checking that the plant appears to be healthy.
- Clean the roots; remove, as far as possible, plant debris earth and decayed roots.
- The plants and their roots should be either dusted with a fungicide and an insecticide or dipped in a corresponding solution.
- Wrap the plants and their roots in newspaper which has been soaked in the fungicide/insecticide solution.

8.4
Transfer

Once the adequate quarantine period has been completed and the plants have been checked as healthy specimens, they can be transferred to the cultivation area.

In tropical conditions, the plants should be placed under supervision for a period prior to transplanting in the plantation nursery or the field.

In quarantine nurseries, sterile frames should be set up to ensure strict separation between reception premises and premises used to prepare plants for transferral.

Part III
Harvesting & Processing

1
Yield Estimation and Harvest Period

Ch. Cilas and F. Descroix

1.1
Introduction

The early estimation of coffee yields is becoming more and more important for the different parties involved in the production and commercialization of coffee. This estimation can also be useful both to forecast production and for agricultural experimentation.

At the plantation plot level, these estimates are useful for the farmer who wishes to know, as early as possible, the return he can expect from his crop. For the experimenter, whose objective is often to determine the most profitable techniques in coffee farming.

Harvest estimates are also of interest to other participants in coffee production.

In an industrial plantation, of a cooperative or of coffee-growers associations, this can be used to determine more accurately the logistical necessities for transport and post-harvest treatment for both immediate and overall requirements.

On a national scale, early coffee harvest assessments allow the people in charge to fix and negotiate the prices of green coffee.

Buyers and coffee brokers have equal needs for national and international forecasts in order to fix the rates and regulate the markets. Fluctuations in world production lead to the evolution of international prices, which are sometimes based merely on partial information. For example, knowledge that frosts have recently affected the coffee trees in certain regions of Brazil has led to an inflation of prices on the international markets despite the fact that no precise quantification of the losses had been undertaken. More reliable world forecasts enable a better regulation of market prices and a more equitable remuneration for different parts of the network.

1.2
Different Approaches to Harvest Estimates

The estimate of yields depends on the following objectives: forecast of yields at different amplitude (by plantation plot, regional or national level) or labour economy in experimentation. It also depends on the desired accuracy and the means that are available.

Two approaches are generally used to predict harvests of vegetal crops:

- Methods based on a model of variations of yield from exogenous variables, most often climatic [3].

Coffee: Growing, Processing, Sustainable Production, Second Edition. Edited by J. N. Wintgens.
© 2012 WILEY-VCH Verlag GmbH & Co. KGaA. Published 2012 by Wiley-VCH Verlag GmbH & Co. KGaA.

- Methods using models of endogenous variables, as different components of the yield [4].

The first approach requires access to reliable meteorological readings from observation points distributed throughout the production zones in which the estimates are early desired. The second approach requires the counting on samples of plots on established plants.

The method shown in this chapter calls upon the second approach, which is known to be more efficient for estimating coffee yields. In addition, this approach can be used both for the forecast of yield and for agronomic experimentation.

1.3
Presentation of the Method

1.3.1
Principles

The method presented here was established with the collaboration of the biometrics research unit of CIRAD-CP and the Office of Coffee in Burundi (OCIBU).

The initial objective of OCIBU was to predict, approximately 6 months in advance, the coffee production of two producing regions in Burundi: Cibitoké and Bubanza. The forecast estimation method developed in this case relies on an early evaluation of fruit numbers. It concerns estimating the average number of fruit per tree, by a means of arranging different yield components defined in an equation (1):

The average number of fruit =

number of stems per coffee tree $\times$
number of fruit-bearing branches perstem $\times$
number of glomerules per fruit-bearing branch $\times$
number of cherries per glomerule   (1)

This estimate of the average number of fruits per coffee tree is, nevertheless, not sufficient alone to predict the yield of a production unit (plot, village or region). Between this early count and the harvest of ripe fruit, various phenomena will intervene, e.g. the physiological shedding of the fruits. Otherwise, the weight of the coffee grains, the other component of the yield, can only be known at the harvest, on the basis of representative samples of the production unit. Nevertheless, it has often been stated that the physiological shedding or the weight of the grains was generally constant over time on the production plots. The hypothesis has therefore been made that the principal source of yield variation between successive years was due to variations in the number of set fruit. The relation of true production between successive years can thus be considered proportional to the relation of the number of estimated fruit by equation (1) for the years concerned. This reference is expressed in the following equation (2):

E_n = estimate of the average number of fruit in year n

E_{n+1} = estimate of the average number of fruit in year $n+1$

P_n = true production in year n

P_{n+1} = estimated production in year $n+1$

Therefore:

$$E_{n+1}/E_n = P_{n+1}/P_n \qquad (2)$$

where:

$$P_{n+1} \times E_n = E_{n+1} \times P_n$$

$$P_{n+1} = (E_{n+1} \times P_n)/E_n \qquad (3)$$

Knowing the number of estimated fruit E_n and E_{n+1} by equation (1), and the true production P_n of year n, it will be possible to estimate the true production of year $n + 1$ by equation (3). Thus it appears possible to estimate the yield of a production unit as of the second year of counting.

To implement this work it is also necessary to take into account the structure of the production unit, notably to define which sampling procedure to use for estimating the average number of fruits (E). In effect, a projected yield estimate can be envisaged for different levels: plot, village, region or national. At the plot level, it will merely be necessary to define the number of coffee trees to sample in order to have good estimates of the average number of fruits. The type of survey used can be risky if there is no structure at the plot level. On a higher level, involving a greater number of plots, it is necessary to study the distribution of the production according to various factors, such as altitude or soil type, in order to structure the productivity at different levels.

1.3.2
Example of Counting and Sampling Methods

These methods have been established in order to estimate the production on the scale of two large coffee farming regions in Burundi.

For all components of the concerned yield, a specific plot effect was shown. For this reason, it has been thought preferable to estimate the components of yield on a significant number of plots but on a limited number of coffee trees per plot.

To estimate the average number of fruit (E_i) produced per plot of approximately 100 coffee trees, the sampling protocol has been determined as follows:

- Estimate of the average number of stems per coffee tree: 5 coffee trees
- Estimate of the average number of fruit-bearing branches: 1 stem
- Estimate of the average number of glomerules: 4 fruit-bearing branches
- Estimate of the average number of cherries: 2 glomerules

The estimate of the number of fruit is carried out by plot. The number of plots to take into account obviously depends on the variable components of the yield in the concerned area. In the case of Burundi, a survey rate of approximately 0.2 % was applied. This survey rate should increase on a smaller production unit (e.g. a large plantation) and decrease for a larger production unit (e.g. a national estimate).

1.3.3
Improving the Sampling Technique

Previous studies have shown that components of the yield essentially vary according to different factors such as altitude or soil type [1, 2]. An improvement of the type of sampling could then be envisaged by splitting the production units into several strata.

For example, in the case of Burundi, the production unit (the region) was split into different strata, corresponding to the different classified altitudes. The number of coffee trees sampled at each strata is proportional to the magnitude of the plots at each strata, meaning this is a type of sampling "stratified to proportional allocation". This survey plan was adopted for estimates of the number of fruit per tree from 1994: the number of coffee trees sampled at each stratum is proportional to the number of coffee trees on each stratum concerned, with a fixed survey rate. This method should allow a better understanding of the disparities between successive years, taking into account the importance of the plot in the different identified strata.

Under these conditions, a survey rate of approximately 0.1 % seems to be sufficient in the case of regional estimates in Burundi, but this rate must be adapted according to the desired precision and the structure of the plantation for which the estimate must be forecast.

1.4
Application of the Method in Burundi

The aim was to predict the production of two coffee growing provinces in Burundi: Bubanza and Cibitoké. The different methods proposed use various levels of stratification to estimate the average number of fruits (village, altitude, type of soil), following the previously described method.

At the end of the 1993 season, it was possible to compare forecasts (P_2^*) carried out with the recorded harvests (P_2) for two surveyed provinces as shown in Table 1.1.

The forecasts obtained are satisfactory and allow an estimate of production with

sufficient precision (deviation <5 %). The stratification methods by altitude or by soil can therefore be retained for the following years, by making improvements to the survey plan.

A stratified survey plan with proportional allocation has therefore been adopted as of 1994. Several production estimates P_3^* for 1994 are possible following the given stratification:

Based on the 1992 harvest $P_3^* = (P_1 \times E_3)/E_1$
Based on the 1993 harvest $P_3^* = (P_2 \times E_3)/E_2$

Details of these two estimates are shown in Tables 1.2 and 1.3.

Other estimates intervening during these 2 years could be proposed, but the modification of the survey plan between the successive years can lead to significant bias.

Two estimates are proposed for the stratification according to the areas (first method of Table 1.2). The values for the average number of fruit (E_1, E_2 and E_3) are calculated by area and production values (P_1, P_2 and P_3) are expressed in kg.

A production increase in the range of 15 % (+ 567 tons) is expected for the two provinces compared with 1992.

Table 1.1 Real and estimated production for 1993

P_2: real 1993 production				
Stratification method	Bubanza 1774290 P_2^* (Bubanza)	Cibitoké 1868840 P_2^* (Cibitoké)	global 3643130 P_2^* (global)	$(P_2^* - P_2)/P_2$
Area	2038812	1545155	3583967	−1.62 %
Altitude	1995263	1635765	3631028	−0.33 %
Soil	2180411	1442789	3623200	−0.55 %
Area/altitude	2044603	1482404	3527007	−3.19 %
Area/soil	2140224	1585278	3725502	+2.26 %

Table 1.2   Estimated production P_3^* for 1994 based on the 1992 harvests

Village	E_1	P_1	E_3	P_3^* (based on the 1992 harvest)
Cibitoké				
Buganda	4125.72	273371	7060.27	467815
Bukina		11898	8155.88	11898
Mabayi	4461.08	77516	11405.35	198180
Mugina	9752.99	824361	8932.10	754976
Murwi	13549.02	286191	11166.69	235870
Rugombo	5958.01	184201	3932.46	121578
Total		1657538		1790317
Bubanza				
Bubanza	6574.30	848000	6302.50	812941
Mpanda	5410.43	350053	6108.00	395186
Musigati	7132.05	867878	10460.00	1272846
Rugazi	6345.24	63602	8258.00	82775
Total		2129533		2563748
Total		3787071		4354065

Table 1.3   Estimated production P_3^* for 1994 based on the 1993 harvests

Village	E_2	P_2	E_3	P_3^* (based on the 1993 harvest)
Cibitoké				
Buganda	5016.86	159480	7060.27	224438
Bukina	7975.66	10530	8155.88	10768
Mabayi	6894.30	35010	11405.35	57918
Mugina	8648.94	943240	8932.10	974121
Murwi	9093.27	418950	11166.69	514478
Rugombo	5108.16	301630	3932.46	232207
Total		1868840		2013930
Bubanza				
Bubanza	5370.38	527360	6302.50	618892
Mpanda	4428.76	282110	6108.00	389077
Musigati	8389.26	891590	10460.00	111663
Rugazi	5074.82	73230	8258.00	119164
Total		1774290		2238796
Total		3643130		4252726

A production increase in the range of 17% (+ 610 tons) is expected for the two provinces.

This second estimate should be more reliable than the former, as the survey basis was already improved between 1992 and 1993.

1.5
Conclusion

A provisional estimation method of coffee yields has been shown above. It is a method based on an early estimate of the average number of fruit using different components of the yield. This early estimate enables prediction 6 months ahead of the production of a coffee-growing region for a given year based on the relation between the early estimation and true production of the previous year.

This method can be used to predict the production of a plot, a village, a province or a country. Results obtained on coffee yield forecasts from two provinces in Burundi were satisfactory. This method could now be applied on a national scale in several countries.

Improvements can still be made, notably in the area of sampling. Nevertheless, this simple and reliable method can be applied by industrial planters or agricultural cooperatives. It can also be adapted for national forecasts of coffee production.

Following Table 1.4 presents the harvesting periods in coffee-producing countries.

Bibliography

[1] Cilas, Ch. Mumirwa project support and the coffee programme of ISABU. Methods of yield forecasts in coffee. *Mission Report in Burundi.* DOC CP 97, 1993, 28.

[2] Descroix, F. Résultats de l'enquête: Identification des exploitations des régions du projet Mumirwa – Burundi, Ministère de l'Agriculture et de l'Elevage Direction générale de la vulgarisation, 1992.

[3] In, K. Yield forecasts of cultivated plants from meteorological facts by multiple throw-back. *Bull. Rech. Agron. Gembloux* 1977, *12*, 37–54.

[4] Upreti, G., Bittenbender, H. C. and Ingamells, J. L. Rapid estimation of coffee yield. In *Proc. ASIC 14th Colloq.,* San Francisco, CA, 1991, 585–593.

Table 1.4 Harvesting Periods in Coffee-Producing Countries

	Sept	Oct	Nov	Dec	Jan	Feb	Mar	Apr	May	June	July	Aug
Angola	R	r						A	A	A	R	R
Australia	A	r						A	A	A	A	A
Brazil	A	a							a	a	A	A
Burundi	a	a					a	a	a			a
Cameroon	a / R	a / r	A / R	A / R	a / R	a / r						
China			a / r	A / R	A / R	A / r	a					
Colombia												
Antioquia, Boyaca, Caldas, Risaralda	A	A	A	A	a			a	a	a		
Cundinamarca, Cauca, Huila	a	a	a	a	a			a	a	a		
Cesar, Guajira, Magdalena	A	A	A	A	A				A	A		
Santander	A	A	A	a	a		A	A	A	A		A
Nariño, N. Santander, Quindio, Tolima, Valle	A	a	a	a	a			A	A	A		
Costa Rica	a	a	A	A	A	a	a	a				
Cuba	A	A	A	A	a	a	a	A	A			
Dominican Republic	a	a	A	A	A	a	a					
Democratic Republic of Congo[1]	a / R	a / r	A / r	A / R	a / R	R	r	r				
Ecuador	a / R	a / r	A / r	A / R	A / R	A / R	A / r	A	A	A / r	A / R	A / R
El Salvador[2]		a	a	A	A	A	a					A
Ethiopia[3]	a / R	a	A	A	A	A	a			r	R	a / R

[1] In the South of the Country, Robusta is harvested from May to July; [2] Includes Standard, High Grown and Strictly High grown Coffee
[3] Harvesting period for Natural Arabica coffee only

Table 1.4 Continued

	Sept	Oct	Nov	Dec	Jan	Feb	Mar	Apr	May	June	July	Aug
Guatemala	a	A	A	A	A	A	A	a				a
Haiti	a	a	A	A	A	a	a					
Hawaii												
Kona	A	A	a	a								a
other islands	A	A	A	A	a							
Honduras	a	a	a	A	A	A	a					
India			r	r	r	R	R	r				
Indonesia												
E. and W. Nusa Tenggara, N., Central and S. Sulawesi, E. Timor, E Java, Bali	a		a				a	A	a	a	A	A
W. and Central Java			r				r		r	r	A	A
ACEH				A	A	A	a	A	A	A	R	R
N. Sumatra	R		a	R	R	R	R	R	R	R	R	R
Lampung (Highland), S. Sumatra						a		r	r	r	r	a
Lampung (Lowland)				r		r				r	r	r
Jambi, Riau Bengkulu										R	R	
Ivory Coast	A	r	R	R	R	r	r	r	R	R		r
Jamaica	a	A	A	A	A	A	A	a	a	a	a	r
Kenya	a	a	A	A	a				a	r		a
Malaysia		R	R	R	R	A	R					
Mexico	a	a	A	A	A	A	a	r	r	r	r	a

[4] Includes low and high areas for Arabica

Table 1.4 Continued

	Sept	Oct	Nov	Dec	Jan	Feb	Mar	Apr	May	June	July	Aug
New Caledonia	a							a	a	A	A	A
	R	r								r	R	R
Nicaragua[5]	A	a	A	A	A	a	a			r	r	a
Panama		A	A	A	A	a	a					
Peru	A	a	A	A	A	A	A	a	a	a	A	A
Philippines												
Surigao, Agusan				A	A	A	A					
N., W. and S. Mindanao, Visayas			R	R	R	R	R	r				
Luzon					r	r	r	r				
Puerto Rico	A	A	A	a	a	r						
		r		R	R	R	r	r				
Tanzania	A	A	A	a	a	a	a	a	r	R	R	R
	r	r	r	r					r			
Thailand		r	r	R	R	R	r	r	r	R	R	R
			a	a	A	A	a					
Trinidad and Tobago		r	r	r	r	r	r	r				
Uganda, Central, E. and W. areas	a	a	A	A	A	a	a					
	r	R	R	R	R	R	R					
Venezuela	a	A	A	A	A	a	a	a				
Vietnam	r	R	R	R	r	r	r	r				
Yemen (Arabica)[6]		a	a	A	A	a	a					
Zimbabwe	A	A	A	A	A	A	a		r	a	A	A
						R	a	r	r	R	R	R

[5] Includes low and high areas; [6] Includes low and high areas

A = Arabica main picking period, a = Arabica minor picking period

R = Robusta (Conillon) main picking period, r = Robusta (Conillon) minor picking period

2
Harvesting and Green Coffee Processing

Carlos H. J. Brando

2.1
General Principles of Coffee Processing

Commercial coffee beans belong mainly to two species: *Coffea arabica* and *Coffea canephora* var. *robusta*, whose intrinsic characteristics are very distinct.

The different parts of a coffee cherry are:

- An external skin (exocarp), red or yellow when the fruit is ripe
- A mucilaginous flesh (mesocarp), known as pulp and mucilage
- Generally two grains or beans (each one called an endosperm), which contain a germ (embryo).

Each grain, or bean, is covered by a spermoderm called silverskin and surrounded by parchment (endocarp). If one grain aborts, its place remains empty and the other one grows into a more rounded shape (a peaberry).

The water content of the whole ripe fresh cherry is about 65 %.

Arabica, generally grown at higher altitudes, is a weak-bodied, acidic and aromatic coffee. Its caffeine content is low (about 1.5 % of the dry material).

Canephora, mainly represented by the Robusta variety, is a lower-altitude coffee. It is full-bodied, but more bitter, less aromatic and less acidic than Arabica. The Canephora beans are smaller than the Arabica ones and their caffeine content may exceed 2.5 %.

These quality criteria, which are linked to genetic origin, can be also influenced by environmental factors such as cultivation method, soil or climate. Technological operations which are carried out on the beans such as post-harvest processing, special wet milling and drying as well as hulling and sorting can also affect quality.

Coffee processing aims to lower the water content of fresh cherries to a level which allows the preservation of beans (about 11–12 %), removing all the covering which surround the beans and preparing the beans according to market requirements.

After harvesting, three different systems are used for processing: natural coffees are dry-processed and washed coffees are wet-processed; more recently a third, intermediate semi-dry (pulped natural) process has been introduced.

Dry processing implies that the whole cherry is dried together (exocarp, mesocarp and endosperm, i.e. pulp, parchment and bean). The whole hull (dried pulp and parchment) is then removed mechanically to obtain green coffee. The dry process is

Coffee: Growing, Processing, Sustainable Production, Second Edition. Edited by J. N. Wintgens.
© 2012 WILEY-VCH Verlag GmbH & Co. KGaA. Published 2012 by Wiley-VCH Verlag GmbH & Co. KGaA.

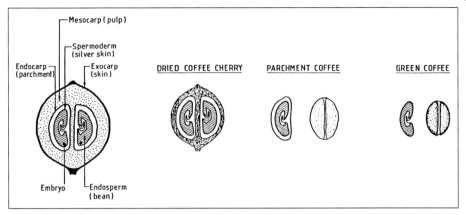

Figure 2.1   Coffee cherry: its modification through processing. From: CIRAD-CP.

used for more than 80 % of Brazilian, Ethiopian and Yemen Arabicas, and for almost all Robusta coffees in the world.

During wet processing, the pulp (i.e. the exocarp and a part of the mesocarp) is removed mechanically. The remaining mesocarp, called mucilage, sticks to the parchment and is also removed before drying. Hulling of dry parchment coffee leads to green coffee. The wet process is used for all Arabicas other than those mentioned above and only for a small percentage of Robustas, although the trend to wet process Robustas is increasing.

In the semi-dry process, that Brazilians call the pulped natural process, the mucilage is not fully removed after pulping, and parchment is dried together with most or all its mucilage. Green coffee is obtained by hulling dry parchment with dried mucilage adhering to it. The semi-dry process, originally used in Brazil, is now being introduced in other countries too.

Green coffee, obtained by any of the three methods mentioned above, is submitted to three successive types of sorting according to size, density and color. Blending of different coffee types or bulking of equal qualities may be required for export and or roasting.

Because coffee harvesting is seasonal, the dry-, semi-dry- or wet-processing centers run only during the coffee picking months of the year. Dry cherry or dry parchment coffee is then stored and the further processing operations may be staggered over a longer period.

2.2
Harvesting

2.2.1
Objective

Coffee harvesting may have different objectives depending on the method of processing as well as the availability and cost of labor.

Where the wet or semi-dry method is to be used, traditionally the main objective is to maximize the percentage of ripe cherries harvested. On the other hand, if the dry method is to be used, the usual objective is to harvest all cherries simultaneously with the least percentage of unripe ones. What had, for many years, been a clear-cut decision has been complicated by recent increases in harvesting costs in most coffee-producing areas of the world. As a

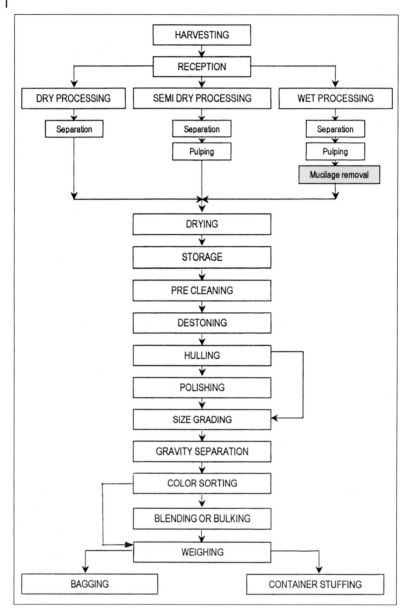

Figure 2.2 Complete green coffee processing.

result, coffee harvesting now has to be analyzed from the conflicting standpoints of product quality and harvesting costs.

The ideal situation is to harvest all fresh, ripe cherries with the least possible damage to the tree, irrespective of the processing system to be used. With the harvesting technology available today, 100 % ripe cherry harvesting may be only achieved by selective hand-picking, which generally corresponds to the most expensive option available. In situations where labor is

Dry Processing						Wet Processing			
Products		Arabica		Robusta		Products		Arabica	
(at each stage of processing)	Moisture Content (%)	Weight (in kg)	Bulk Density (kg/m³)	Weight (in kg)	Bulk Density (kg/m³)	(at each stage of processing)	Moisture Content (%)	Weight (in kg)	Bulk Density (kg/m³)
Products						**Products**			
Fresh cherries	65	100.00	616	100.00	645	Fresh cherries ↓	65	100.00 ↓	616
						Pulped coffee ↓	n.d.	54.00 ↓	846
↓						Wet parchment coffee ↓	55	45.00 ↓	665
Dry coffee cherries ↓	12	37.20	422	42.20	440	Dry parchment coffee ↓	12	23.30 ↓	352
Green coffee	12	19.00	650	22.00	750	Green coffee	12	19.00	650
Byproducts						**Byproducts**			
						Pulp	84	46.00	420
Husk	12	18.20	230	20.00	230	Parchment	12	4.30	230

Figure 2.3 Technical data on post-harvest processing. From: CIRAD-CP.

scarce or expensive in relation to coffee prices, selection may have to be overlooked so unripe and over-ripe cherries must then be picked. This is mostly the case today, with 100 % ripe cherry harvesting nearly impossible to achieve with or without selective picking. The extreme case corresponds to the situation where all cherries (unripe, ripe and over-ripe) have to be picked at once. High quality coffee may still be produced in any case from the fresh, ripe cherries alone, but the total volume of high quality coffee available is then smaller. Markets are progressively being developed for coffees of different qualities that reach consumers at more affordable prices.

The choice of harvesting system is, by and large, not made rationally at present. It results from long-established practices and traditions. Whereas countries producing washed coffees still insist on selective

Figure 2.4 Harvesting coffee. From: Nestlé.

hand-picking without questioning the choice, producers of naturals favor the simultaneous stripping of all cherries in the tree. The situation is, however, changing quickly.

Whether to achieve coffee quality by harvesting ripe cherries or harvesting a mixed product and complementing with proper post-harvest treatment is a cost–benefit decision that coffee growers are already having to face. If only ripe cherries are picked, the volume of quality coffee is higher, but harvesting costs are higher, too. If a mixed product is picked, the volumes of quality coffee are smaller, but harvesting costs fall. The decision facing the grower is whether the savings in harvesting cost offset the loss of income from less quality coffee. If they do, the grower should move away from selective hand-picking and into partial or full stripping and modern mechanical harvesting systems to maximize his profits.

Selective coffee picking is not the only way to ensure that quality in the tree is transferred to the cup. The fact is that selective picking is no more than an indicator that only sound, fresh, ripe coffee cherries should be used as raw materials to produce the finest beans from which a perfect cup is brewed. Sound, fresh, ripe cherries may be obtained from a variety of picking practices combined with processing techniques. Top-quality coffee may be produced regardless of the harvesting techniques employed, as it has been shown in several countries.

2.2.2
Principle

In the case of selective hand-picking, the choice of ripe cherries has been traditionally made by visual means only using color as the criterion. Ripe cherries are either red or yellow depending on the cof-

fee variety. Unripe cherries are several shades of green, whereas over-ripe cherries go from grayish-red or yellow to full black.

In the case of mechanical harvesting, the acting harvesting principle is vibration. Selectivity may be obtained because unripe cherries are more firmly attached to the branches than ripe cherries. As a result, by changing the intensity of mechanical vibration, the detachment effort applied may vary thus enabling some measure of selectivity according to the degree of ripeness.

The growing use of high-pressure espresso preparation techniques, that emphasize the astringency of unripe cherries, means that color and visual evaluation alone are insufficient to separate 100 % ripe cherries, because partially ripe cherries may also be red or yellow. Post-harvesting separation of cherries by softness, which is an excellent indicator of maturation, is becoming a good alternative to selection by color or degree of attachement to the branch.

2.2.3
Techniques and Equipment

2.2.3.1 Manual Harvesting
The two most commonly used manual harvesting techniques are selective harvesting and stripping.

Figure 2.5 Ripe cherries. From: Nestlé.

Selective Harvesting

Selective harvesting consists of the hand-picking of ripe cherries only. In the course of a harvesting season every tree is visited several times, up to 10 times in some countries, and only the ripe cherries should be selectively picked each time. The cherries are hand placed into bags or baskets usually held at waist level by the pickers. At the end of the harvesting season one final collection of all cherries, irrespective of their degree of maturation, usually takes place.

No matter how good the pickers are and how uniform maturation is, a small percentage of unripe and over-ripe cherries is to be expected. If the time which elapses between harvesting rounds is too long, a percentage of over-ripe cherries is also to be expected. Scarcity of skilled labor and the cost pressure to harvest more in a

Figure 2.6 Selective harvesting.

shorter time are rendering the percentages of unripe and over-ripe cherries greater year after year.

A worker can selectively harvest from 50 to 120 kg of fresh cherry per working day. This large variation is not only due to individual skills and training but also to the uniformity of maturation, the yield per hectare, tree density, the gradient of the slope, etc.

Most countries that produce washed coffee resort to selective picking. The notable exceptions are Hawaii and Australia where labor costs render selective manual harvesting unfeasible. Harvesting costs may force the more developed producers of washed coffees to re-evaluate their option for selective manual picking in the very near future.

Stripping

Stripping, also known as "milking", consists of removing all the cherries present on a branch irrespective of their degree of ripeness. There is usually only one picking round, although a few Arabica growers may favor two or three rounds, concentrating each round on the branches where maturation is more uniform. The cherries at all maturation stages are dropped either directly to the ground, that has been previously cleared, or onto plastic sheets, canvas or cloth spread under the trees. If the weather is dry and coffee is collected soon after harvesting, there is no major damage to quality when the cherries are dropped to the ground. In all other conditions it is preferable to have the soil covered for the sake of quality.

The choice of the moment when strip picking takes place is critical in terms of the composition of the harvested product. Ideally harvesting should be delayed until the percentage of unripe cherries is under 5 %. However, in most conditions this

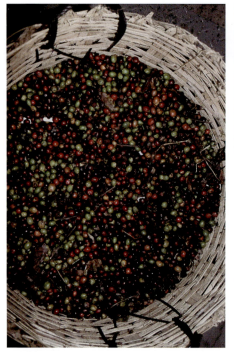

Figure 2.7 Cherries at different maturation stages (together and separate).

means that the cherries already dried on the tree will start falling to the ground and the quality losses from these cherries may offset the gains from the maturation of the unripe cherries.

A worker can strip harvest between 120 and 250 kg of cherries per working day. This large variation is explained by the same factors mentioned before for selective harvesting.

Stripping is practiced by Arabica growers that produce natural coffees, notably in Brazil, and by most Robusta producers.

2.2.3.2 Mechanical Harvesting

There are two alternative technologies for mechanical coffee harvesting today: large mechanical harvesters on wheels and light hand-held harvesters.

Large Harvesters on Wheels
These large machines with their own engines, made originally in Brazil, then in the U.S. and more recently in Australia, have been in the market for over three decades. Lighter, tractor-driven versions are more recent.

Basically all large harvesters work on the same principle: one or two vertical or forward-leaning shaking heads that travel along the coffee row causing the cherries

Figure 2.8 Stripping.

Figure 2.9 Self-propelled harvester.

areas, like the large Hawaiian plantations and the flat areas of the Brazilian Cerrado (Minas Gerais and Bahia) these machines are being increasingly considered in other countries. Although they are an alternative for all flat to moderately sloped coffee areas, they require planting along rows with a minimum space between rows to allow the passage of the harvester.

Tractor-driven machines usually have a single shaking head which drops the cher-

to drop. The shaking head is composed of a center shaft and spoke-like rods made of synthetic materials (e.g. fiberglass, nylon, graphite, etc.).

In the self-propelled machines the usual arrangement is two shaking heads, one on each side of the coffee row, dropping coffee onto a "fish-plate" collection system which unloads the cherries into a conveyance system that lifts them to a pneumatic separator to discharge leaves and other light impurities and to load the clean cherries into big-bags or boxes attached to the machine or into tractor carts moving alongside the harvester.

Large self-propelled harvesters are still restricted to coffee plantations with moderate slopes. Recent developments have made these machines less wide and compatible with narrow spacing between rows of trees. Originallay an alternative to hand-picking in only a few coffee growing

Figure 2.10 (A) Inside view of large harvester showing shaking heads and fish-plates. (B) Tractor-driven harvester. From: Jacto.

ries directly onto the ground or onto a manually or mechanically spread ground cover which is also used for collection. Research on mechanical collection is under way and new prototypes are being tested.

Tractor-driven harvesters on wheels can negotiate steeper slopes and closer-spaced planting than self-propelled machines. They also cost a lot less. Their use is increasing rapidly in many areas of Brazil, including South Minas that alone produces more coffee than Colombia or Vietnam.

The high speed of picking of the mechanical harvesters makes them cost-effective in spite of their high price. The large initial capital investment is rapidly offset by the stream of cost savings in consecutive crop years. Machine rental and shared ownership are other ways to cope with the high investment cost.

Three basic parameters are used to evaluate the performance of the large coffee harvesters: selectivity, crop removal and recovery.

Though selectivity has been the focus of much research and development, these machines are still far from being selective. The shaking heads can either harvest the full height of the tree at the same time or concentrate on the branches were the cherries are ripe if some rods are partially removed. Adjustments in the speed of vibration also help, but do not solve the problem.

Crop removal is measured by the percentage of cherries that should have been harvested which are left on the tree. Full removal is nearly impossible unless a large percentage of unripe cherries is removed and the tree is damaged. In spite of recent advances in crop removal, a second round is still required to avoid losing part of the crop.

Recovery is measured by the percentage of the cherries effectively removed by the harvester, but which end up lost on the ground. This has to do with the efficiency of the "fish-plate" collection system. It is in this area, rather than in the other two, that the more recent machines are improving efficiency.

Damage to the coffee trees depends not only on the speed of vibration and travel, but also on the driver's skills. Although some degree of leaf removal is unavoidable, it can be minimized by correct operation. The breakage of branches should definitely be avoided.

The damage sustained by plantations which have been mechanically harvested for many years has not been reported to affect yields.

Light Hand-held Harvesters

This is a new system which was tested in Brazil in 1994 and commercially sold from 1995 onwards. It is beeing increasingly used in Brazil, especially in areas where the slope and tree spacing do not allow the use of the larger machines on wheels. Labor scarcity and costs are forcing trials and the introduction of these harvesters in other countries, specially in North and Central America.

The hand-held harvesting tool is composed of one or two sets of rods driven by a pneumatic piston. As the tips of these finger-like rods move, they transmit vibrations to the branches of the tree and cause the ripe cherries to fall on the previously covered ground.

This small shaking head is assembled at the end of a pole that is available in several lengths in order to harvest trees of different heights. This individual shaking head is light and manually operated so it can be directed to the branches or parts of branches where maturation is more uniform. This improves selectivity which may also be enhaced by keeping the intensity of vibration at a level that is enough to detach ripe cherries, but does not drop the unripe ones.

The hand-held pneumatic harvester is connected by a long plastic hose to a compressor that may be carried and driven by a tractor or have its own diesel or gasoline engine. The pressure valve and gauge which adjust the speed of vibration are assembled on the compressor itself. Depending on its size, the compressor can power one to six harvesting tools.

The pneumatic harvesting sets, with two to six hand-held harvesting tools, are being replaced by less costly single harvesting tools driven by a two-stroke gas engine that is carried by the operator. These single rod machines with an engine at one end of the pole and a harvesting tool at the other use a variety of new designs for the shaking head. Different suppliers are offering vibrating, reciprocating or rotary heads, with marked gains in harvesting efficiency and product durability.

Hand-held harvesters may be used in all conditions, i.e. on all types of slopes as well as in flat areas, with coffee trees of all heights and any spacing between trees. Selectivity, which is not fully achieved by using the machine alone, may be possible if the pneumatic system is combined with a limited amount of hand-picking.

Hand held harvesters cause less damage to the trees than harvesters on wheels.

2.2.3.3 Equipment to Pick Coffee from the Ground

When cherries are manually or mechanically dropped to the ground, they have to be collected and conveyed to the processing facilities.

In cases where mats are used to cover the ground, they are also used to collect the cherries manually or mechanically by raising the corners of the mat to create a make-shift "bag" that is unloaded into trucks, carts, big-bags or bags.

If the cherries are dropped directly on the ground or left behind by harvesters, they

Figure 2.11 (A) Pneumatic hand-held harvester. From: Agromatica. (B) Single rod hand-held harvester. From: Brudden.

have to be either raked up and collected manually or collected by machines.

Several types of machines are available to collect coffee from the ground They are mostly based on a suction or aspiration principle, similar to an oversized vacuum cleaner. The majority of these machines use built-in separation devices to eliminate the impurities (leaves, soil, etc.) from the coffee cherries. Smaller machines using conveyor sweeping and other systems to collect coffee are also available in Brazil.

2.2.3.4 Winnowers

The self-propelled harvesters and some types of equipment to collect coffee from the ground have their own winnowers, i.e. their own systems to clean the freshly harvested cherries by blowing off light impurities. Other harvesting systems may require separate winnowers, especially when the volumes harvested are large or the content of foreign matter is substantial.

Winnowers may be mobile, for use in the plantation, or stationary, for use at the mill.

Although a winnower is a device to remove light impurities from the product, most machines used today for coffee also include screening systems to remove impurities that are larger or smaller than the cherries.

Separation of light impurities (dust, light sand, leaves, sticks, etc.) is performed by either blowing or sucking out foreign materials with the help of an air current created by a fan. The separation of other impurities, larger or smaller than the cherries, is obtain-

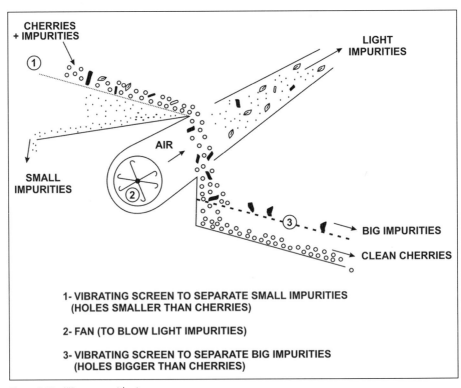

Figure 2.12 Winnower with sieves.

ed by two vibrating screens, one with holes bigger than the cherries, which retains the larger impurities (stones, big sticks, etc.), and one with holes smaller than the cherries, which retains the cherries themselves and lets small impurities (sand, etc.) pass through. The two cleaning devices, fan and vibrating screens, are arranged in different sequences by various suppliers.

The separation of heavy impurities the same size as the cherries cannot be made by winnowers with screens but rather by flotation in water, which is described elsewhere.

2.2.3.5 Post-harvest Quality Corrections

If the preparation of quality coffee requires the use of ripe cherries only, coffee-picking systems should strive to harvest 100 % ripe cherries. This is the case for selective hand-picking, although scarcity of labor and rising costs are becoming a major problem.

One answer to the lack of expert labor and its rising costs is the use of simplified hand-picking techniques like, for instance, a modified approach to the traditional stripping or milking systems used in Brazil or a switch to mechanical harvesting which has already taken place in the larger plantations of Brazil and Hawaii. These alternative techniques, which generally entail fair to poor selective picking – hand-held harvesters used without complementary labor or large mechanical harvesters – frequently lead to harvests which include unripe or over-ripe cherries as well as impurities like leaves or twigs. Careful post-harvest separation of these undesirable materials can be carried out so that quality coffee may still be produced from the remaining ripe cherries.

Coffee growers must therefore be prepared to cope with harvesting out-turns that are mixed. Their ability to separate sound, fresh, ripe cherries from immature,

partially ripe, over-ripe, semi-dry and dry cherries is becoming a crucial factor in the production of top-quality coffees.

Modern post-harvest processing equipment now available, to be described later in this chapter, can handle the mixed out-turns efficiently while using less water and power and causing less pollution. Winnowers remove light impurities; mechanical siphons remove heavy impurities smaller or larger than the cherries and the "floaters" (over-ripe, semi-dry and dry cherries). The latest generation of unripe cherry separators sort out these immature products. Only 100 % ripe cherries remain in the main stream to be further processed, while the materials separated are either discarded (impurities) or submitted to different types of processing (unripe, over-ripe and dry cherries) in order to retain as much quality and value as possible.

2.2.4 Problems and Impact on Quality

Unintentional or intentional harvesting of cherries at several stages of maturation may have adverse impacts on coffee quality if these materials are processed together. Cherries at different stages of maturation should be separated and processed using different techniques in order to retain as much quality as possible.

Figure 2.13 Manual separation of cherries at different maturation stages. From: Nestlé.

When the wet process is used to produce washed coffees one should not pulp fully ripe cherries together with unripe, over-ripe or dried out cherries nor fruit that has fallen onto the ground and then gathered with the crop. If these materials are pulped and further processed along with the ripe cherries they will jeopardize the quality of the best coffees.

If, on the other hand, undesirable cherries are separated, it will be possible to produce top-quality washed coffees from the fresh ripe cherries only, to pulp some cherries at other stages of maturation separately, and to process further cherries as naturals in order to maximize the quality of all the product fractions harvested.

When the dry process is used to produce natural coffees, the main quality problem related to harvesting is the presence of immature cherries in the harvested product. Whereas the over-ripe cherries as well as those partially or fully dried on the tree may be separated by flotation and dried separately from the ripe cherries, flotation cannot separate unripe cherries from ripe cherries because their densities are similar.

When unripe cherries are dried along with ripe cherries, the unripe beans will produce an unpleasant harsh taste in the cup. In addition, many unripe cherries will become black beans. The higher the temperature in natural or artificial drying, the more intense will be the blackening of unripe beans. To solve the unripe cherry problem harvesting must either be selective, which is very seldom the case with natural coffees, or color sorters should be used at the end of the process to sort the unripe and black beans. However, color sorting unripe cherries is a specially difficult and costly operation.

One intermediate solution is the semi-dry (pulped natural) system developed in the 1990s. This system separates the unripe cherries and produces a cup that is more often closer to naturals than washed coffees. Brazilians call the method "pulped natural" and the coffee it produces "pulped natural coffee" (or CD for Cereja Descascado) because of the natural coffee taste produced.

A final warning about unripe cherries in any process relates to weight loss as these cherries would eventually develop to full maturity and full weight. However, we insist here that the quality and weight losses in unripe cherries may well be countered by the savings introduced by cost-effective labor-saving harvesting systems, provided that proper post-harvest separation and processing takes place.

The remarkable growth of the specialty coffee sector and its emphasis on quality should not be misinterpreted as signals that quality coffee is bought at any price. More to the point, the world trend in all markets is to seek quality at affordable prices. The coffee market is no exception, so growers should be aware of rising labor costs and try to shift to more cost-effective harvesting systems in order to produce quality coffees at affordable prices.

2.2.5
Comments

Cherries must undergo either dry, semi-dry or wet processing as soon as possible after harvesting.

The storage of fresh cherries in bags, heaps, hoppers or silos should not last more than 8 h. Otherwise unwanted fermentation will generate the dreaded "stinker" beans, which appear when the temperature before pulping exceeds 40–42 °C. Even if the temperature does not rise to the point of creating stinkers, uncontrolled fermentation may have other negative impacts on the cup.

In emergency situations fresh cherries may be stored for a longer period if they are kept under water.

Coffee cherries should not be kept in plastic bags. The damp air-tight atmosphere that develops inside plastic bags favors unwanted fermentation that damages coffee quality.

2.3
Dry, Semi-dry and Wet Processing

2.3.1
Dry Processing

2.3.1.1　Objective

Most often the dry process is used after non-selective harvesting, i.e. after stripping or mechanical harvesting. In this case, when the coffee reaches the processing line it is a mixture of unripe, ripe, over-ripe and partially dry cherries along with leaves and sticks as well as earth and stones when coffee is harvested directly on the ground.

The objective of dry processing is to clean the coffee cherries and to separate them, to the extent possible, according to their moisture content. Cherries at different moisture levels are then dried separately to obtain an evenly dried product.

Unfortunately many growers who produce natural coffees simply ignore these steps and go directly from harvesting to drying without any cleaning or separation. The result is a product of lower quality.

Natural coffees that have been properly dry-processed can be a quality product with their own market. Natural coffees are a basic ingredient of espresso blends and, as such, they enjoy a growing demand, especially for high-quality products.

2.3.1.2　Principle

The cleaning of coffee cherries is performed by winnowing and sifting.

Separation according to moisture content is based on density and made by flotation in water. Flotation is also used to separate stones from coffee.

2.3.1.3　Techniques and Equipment

The diagram in Fig. 2.14 shows the two ways used to obtain natural coffees.

Drying the mixed cherries without any cleaning should be avoided. In fact, great benefits at little cost may be obtained if the simple cleaning and separation procedures described below are implemented.

Winnowing

Light impurities such as leaves, stems and dust are separated from the coffee cherries by winnowing, i.e. with the help of an air flow.

The traditional approach to winnowing is the one still used by small traditional farmers in Brazil and which is often portrayed in postcards. The coffee cherries are gathered on a round sieve. The leaves that move to the top as coffee is gathered are removed manually. The cherries are then thrown up and caught again on the sieve several times. Dust, leaves and other light impurities are removed as the coffee moves up and down in response to the special motion transmitted to the cherries by trained hands. Sand and small impurities pass through the holes of the sieve and fall onto the ground.

The traditional manual system is being gradually replaced by mechanical winnowers that use a fan to either blow or suck out light impurities. There are several types of winnowers on the market. Some combine a fan with vibrating sieves to separate impurities which are smaller and larger than the cherries (see "sifting" below).

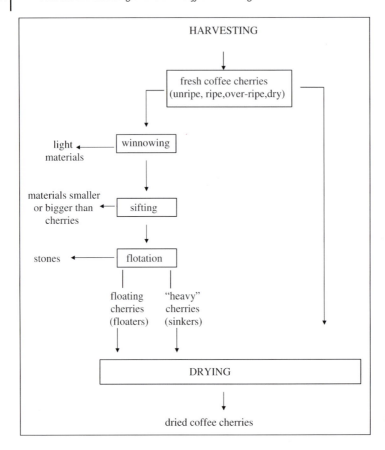

HARVESTING

fresh coffee cherries
(unripe, ripe, over-ripe, dry)

light
materials ← winnowing

materials smaller
or bigger than ← sifting
cherries

stones ← flotation

floating   "heavy"
cherries   cherries
(floaters)   (sinkers)

DRYING

dried coffee cherries

Figure 2.14
Dry processing.

Some models are mobile, powered by the PTO of a tractor. These offer the advantage of leaving the organic rejects (leaves and stems) and soil on the plantation itself.

The large self-propelled mechanical harvesters have their own built-in winnowing system, whereas tractor-driven harvesters may have a winnower or simply drop the cherries down to be collected and winnowed separately. Equipment that collects coffee from the ground usually has built-in winnowers.

Most mechanical winnowing systems have one single adjustment that controls the intensity of sucking or blowing. This must be set to ensure that only impurities are separated and that no coffee is dis-

Figure 2.15 Person winnowing coffee.
From: Nestlé.

carded with the impurities. Although, in theory, the speed of vibration of the sieves could be adjusted and a choice of hole sizes in the sieves could be available, in practice these are defined and fixed by the manufacturer.

Sifting

Heavy impurities smaller or bigger than the cherries like sand, stones, mud balls, etc., are separated from the coffee cherries by sifting, i.e. by using two vibrating screens with holes which are smaller or bigger than the cherries. The screens may be arranged one after the other or one above the other.

In some machines sifting is combined with winnowing in one single structure that performs the two functions.

Flotation

Flotation is used to separate stones and cherries with different moisture contents.

The initial moisture content of the mixture of strip-picked or mechanically harvested cherries varies widely, from 65 % in unripe and ripe fruits to 25–30 % in partially dried ones. Unless cherries with very different moisture contents are separated and dried separately it will be nearly impossible to obtain a product with a uniform moisture content. Drying will be more efficient and less costly if the cherries are separated according to moisture content and dried separately.

Several systems are used for separation by flotation in water: static water tanks, siphon tanks, channels with traps and mechanical washer-separators that are also called mechanical siphons. All systems are based on the difference of density between the "sinkers" (ripe and unripe cherries which are denser and sink) and the "floaters" (partially dried cherries which are less dense and float).

Stones, which are denser than all cherries, also sink and can be separated from coffee by flotation. Even though stones that are bigger or smaller than the cherries are separated by sifting, stones of the same size as the cherries can only be separated by flotation.

Static Water Tanks

Static tanks are largely abandoned today because they have limited output and require a lot of labor for loading and the removal of the various outgoing products.

Siphon Tanks

These are masonry or concrete receiving tanks filled with running water. The denser ("heavier") cherries which sink are recovered by a siphon, whereas the "lighter" ones which float are removed manually or through a weir along with leaves and twigs. Stones sink to the bottom and are removed manually when the tank is emptied.

Siphon tanks are still in use although they have the disadvantage of consuming a lot of water to move the "sinkers" through the siphon. There is no separation of leaves, twigs and other light impurities from the floaters. Finally, labor is required to remove the stones from the tank.

Small continuous siphon tanks, usually metallic, with a volume of 1–2 m^3, have been introduced in recent years. Although stone removal may be made easier in these tanks, floaters and light impurities remain mixed.

Although siphon tanks are a means to separate floaters, sinkers and stones, their use is much more prevalent in the wet than in the dry method because the production of naturals begins with more impurities due to the picking method (stripping or mechanical) and a high percentage of the final product consists of floaters that must be separated from impurities.

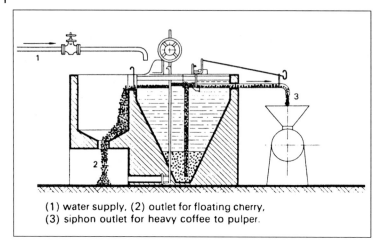

(1) water supply, (2) outlet for floating cherry,
(3) siphon outlet for heavy coffee to pulper.

Figure 2.16
Siphon tank.
From: R. Wilbaux.

Channels with Traps

As cherries are carried by water along a channel, floaters and sinkers are separated into two layers. A trap at the bottom of the channel enables the separation of the lower layer, i.e. the sinkers, which are deviated from the main flow. Stones may also be separated when a metallic trap is used which enables control of the water pressure to ensure that only sinkers are pushed up by water while the stones remain at the bottom of the trap.

This system uses large volumes of water which must run continuously along the channel to ensure the separation of floaters and sinkers.

Channels with traps were frequently used in areas of Brazil where water is plentiful. Their use is now decreasing in favor of water saving mechanical washer-separators.

Mechanical Washer-Separators (Siphon Tanks)

These separators use the same principle as the channels with traps with the major difference that all separation takes place inside a metallic tank where the same volume of water is used throughout the process. Hence, water consumption is lower than any of the systems described above.

Mechanical washer-separators have a separation channel with one or two traps. The older models use a single trap to separate sinkers and stones from floaters, whereas stones are separated from sinkers by adjusting water pressure. The newer models have two traps, one for stones and one for sinkers. In both models sinkers and stones first sink and are then pushed up by an upward moving water current created by a water propeller or a pump that links the two divisions of the metallic tank.

When the mixture of cherries reaches the separator, the floaters remain on the surface of the water and exit through one channel, while the stones and the cherries (ripe and unripe) sink into trap(s). Water moving upwards through the trap(s) pushes up both the stones and the sinker cherries. The stones are taken to a vibrating conveyor that discharges them behind the separator after draining off all the water. The sinkers are conveyed to a channel parallel to the floater channel. The separate floater and sinker channels end on

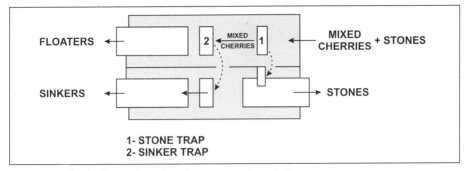

1- STONE TRAP
2- SINKER TRAP

Figure 2.17 Sketch of a mechanical washer separator (top view).

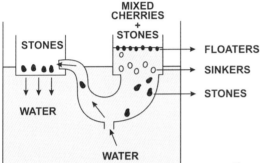

Figure 2.18 The stone trap (cross-section).

the top of a vibrating conveyor with a perforated drainage screen and two separate outlets, one for floaters and one for sinkers.

Invented in Brazil and progressively extending to other countries, the mechanical washer-separator greatly reduces water consumption and labor requirements.

Combined Winnower, Cleaner and Washer-Separator

The mechanical washer-separator can be fed by a two-sieve vibrating conveyor on top of which a winnower is assembled to create a machine that performs all three functions described in this section: winnowing, sifting and floating. This machine removes all light impurities by winnowing, then it removes impurities larger and smaller than the cherries by sifting, and, finally, it separates stones, sinkers and floaters by flotation thus performing all separation functions required in the dry process.

2.3.1.4 Problems and Impact on Quality

The overall quality of coffee is improved by the mere separation of floating and sinking cherries, which may then be dried, stored, processed and sold separately as they have different qualities.

While the advantage of this early separation is widely accepted in Brazil, the country with the largest production of Arabica naturals, it is still to be recognized and adopted by Robusta producers, who all chiefly use the dry system. Although many Robusta growers in Brazil already use this technique, it is still seldom used in other Robusta-producing countries.

Figure 2.19 Separation of floaters from ripe and unripe cherries.

Figure 2.20 Mechanical washer-separator with cleaning screens.

When sinkers and floaters are dried separately there are substantial gains in drying time, labor and space (sun drying) as well as fuel and energy consumption (mechanical drying). In addition, the final product has a uniform moisture content, which leads to an improved aspect and an even roasting. Finally, when substantial percentages of unripe cherries are found in the sinkers, drying may be slowed down in order to avoid that they become black beans. It is known that unripe cherries do not develop into black beans if dried at temperatures below 30 °C. This last advantage of cherry separation and separate drying is even more beneficial with Robusta, whose harvest often yields larger percentages of unripe cherries because less care is taken than in the case of Arabicas.

Even though the separation of sinkers and floaters has a very positive impact on quality, the problem of the mixture of unripe and ripe cherries in the sinkers remains. As stated in the section about harvesting, if unripe cherries are dried along with ripe cherries, the unripe beans will produce an unpleasant harsh taste in the cup. To solve this problem harvesting must either be selective, which is very seldom the case with natural coffees, or the unripe beans must be sorted out with the help of costly color sorters at the end of the grading and sorting process prior to export or roasting.

A recent solution to the problem of unripe cherries is discussed in the section about the semi-dry (pulped natural) system.

2.3.1.5 Comments

Given that static water tanks are no longer used and channels with traps are disappearing rapidly, the comparison should be made between siphon tanks and mechanical washer-separators.

The performance of small siphon tanks remains questionable. The full-sized masonry or concrete siphon tanks imply high costs of civil works and require labor to remove stones. Even though siphon tanks may still have a role in the wet process, they have no place in the dry system.

In dry processing there remains a choice between sending mixed cherries directly to drying, as picked, with the quality problem discussed above, or using mechanical washer-separators. It now seems clear that quality improvements are only possible by cleaning and separating the mixed product harvested by manual or mechanical stripping. Since the present trend is also towards saving water, the best option is mechanical washer-separators preceded by winnowing and cleaning.

Besides the new semi-dry system, to be discussed in the next section, another potential solution to the problem of unripe (green) cherries mixed with ripe (red or yellow) ones could be color sorting. Attempts have been and are still being made to separate the green cherries by sophisticated electronic color sorting, but the obstacles range from the high cost of the equipment to the need to handle a product with high moisture content, not to mention the high volumes of coffee that require processing. As the volume ratio between fresh cherries and coffee beans is about 5 : 1, it would be more sensible to implement color sorting at the end of the processing chain when the volume of coffee is 5 times smaller.

Sucessful trials are currently under way in Brazil to sort over-ripe cherries out from the floaters and to process them separately with quality and price gains.

2.3.2
Semi-dry (Pulped Natural) Processing

The semi-dry process was initiated in the 1990s as an intermediate system between the traditional dry and wet methods. There is a new tendency to call this process "pulped natural" in order to clearly differentiate it from the dry and wet processes.

The pulped natural coffee produced by this systems is known as "Cereja Descascado" or simply CD in Brazil.

2.3.2.1 Objective
The semi-dry process was developed to address the problem of the mixture of unripe and ripe cherries found in the dry process after the use of flotation to separate the over-ripe and partially dry cherries. The semi-dry process is an answer to the need to separate immature cherries from mature ones when non-selective harvesting is used.

The objective of the semi-dry process is to go one step beyond the dry process and to mechanically separate unripe cherries from ripe cherries in order to treat the ripe ones separately to obtain a better cup quality. Although the ripe cherries are pulped as they are separated, the semi-dry process falls short of the wet process because mucilage is not removed from parchment, which is dried with some or all of its mucilage. As a result, this new process creates a product that has its own organoleptic characteristics.

Semi-dry (pulped natural) processed coffee is increasingly popular on the market either as a high quality natural, free from unripe cherries, or as a substitute for washed coffee, specially if more body is desired.

Figure 2.21 Unripe green cherries separated from ripe cherries.

2.3.2.2 **Principle**

After the steps described earlier for dry processing are carried out, i.e. after coffee cherries are cleaned by winnowing and sifting, the stones are removed and the cherries are separated into floaters (over-ripe and partially dry cherries) and sinkers (unripe and ripe cherries), the mixture of unripe and ripe cherries is separated again, this time by pressure, into three products: unripe (green cherries), parchment (which results from depulping the ripe cherries) and pulp.

The separation of unripe and ripe cherries is possible because the ripe cherries are soft and can pass through a screen with long slotted holes, whereas the unripe cherries are hard and cannot go through. The ripe cherries are pulped as they pass through the screen. The resulting pulp and parchment are separated later on.

2.3.2.3 **Techniques and Equipment**

The diagram in Fig. 2.22 summarizes the processing steps required to produce semi-dry/pulped natural coffees. Winnowing, sifting and flotation have been described earlier in the dry processing section. Unripe cherry separation and pulping are described below.

Unripe and ripe cherries are separated by forcing the mixture to pass through a specially designed screen, the so-called immature (green) separator which also acts as a pulper and is known as a screen pulper. The separator, shown in Fig. 2.23, consists of a cylinder with long slotted holes which is fitted with an inside rotor. The rotor forces the cherries to move along the length of the cylinder to an outlet that is controlled by a gate with a counterweight to adjust the pressure on the cherries. As the rotor creates pressure the soft ripe cherries pass through the holes of the screen and lose their pulp as they pass. In other words, the soft ripe cherries are pulped in a very gentle manner as they pass through the screen. The hard, unripe cherries cannot "squeeze" themselves to pass through the holes. As a result, they are forced to travel along the length of the cylinder and to leave through the outlet controlled by the counterweight.

The mixture of parchment and pulp falls into a double-drum vertical pulper, shown in Fig. 2.23, that acts as a pulp separator and re-passer. This vertical drum pulper, to be described in the section about wet processing, separates pulp from parchment and depulps the small cherries that have passed through the holes of the unripe separator without being pulped. Figure 2.24 shows the screen pulper with a vertical drum re-passer and pulp separator.

The three products which come out of the unripe separator and re-passer – unripe cherries, parchment coffee and pulp – are processed separately. Unripe cherries, parchment and pulp will follow different processing steps.

Unripe cherries must be dried at temperatures below 30 °C to prevent the coffee beans from turning black. Even though coffee obtained from unripe cherries is of lower quality and produces a harsh cup,

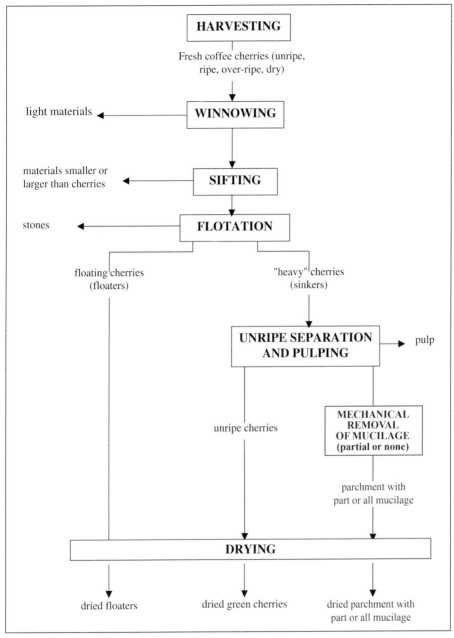

Figure 2.22 Semi-dry (pulped natural) processing.

slow drying at a low temperature helps to preserve quality and aspect and prevents further damage and losses.

Parchment does not have the mucilage removed but is dried with it. This requires special techniques. Parchment with muci-

Figure 2.23 Unripe cherry separator/screen pulper.

Figure 2.24 Pulper with unripe cherry separator (screen pulper) and vertical drum re-passer.

Figure 2.25 Parchment with mucilage.

lage can only be mechanically dried after the mucilage is dry enough not to adhere to the walls of the drier. Therefore both mechanical and conventional sun drying of parchment with mucilage must be preceded by sun drying with special techniques for at least one full day.

Parchment with mucilage must be spread on a concrete or tiled drying ground in thin layers no more than 2.5 cm thick. While being pre-dried, parchment must be turned over very often, at least every half-hour and more often if the day is overcast. This is done with a wooden rake with the special profile shown in Fig. 2.26. Pre-drying is

completed when parchment taken into a person's hand no longer sticks to the palm when it is facing downwards. Pre-drying is over when the mucilage is dried. Conventional sun or mechanical drying may then take place until parchment reaches the correct moisture level required for storage.

The semi-dry (pulped natural) method has been modified for use in areas where weather conditions do not allow pre-drying under the sun. In this case, parchment with mucilage is passed through a mechanical mucilage remover to remove some of the mucilage in order to accelerate or even eliminate pre-drying under the sun. The amount of mucilage removed must be just enough to facilitate the drying of the parchment beans leaving as much mucilage covering as possible. The mucilage remover is described in the section about wet processing.

Figure 2.26   Wooden rake.

Pulp will be disposed of by any of the methods listed in the section about wet processing.

New, more sophisticated techniques and equipment to separate unripe cherries and to pulp only the 100 % ripe ones now allow cherries that are not yet fully mature to be separated from fully ripe ones. As a result it is now possible to produce very high quality semi-washed (or fully washed) coffee that it was not possible to obtain before because conventional pulping systems do not select the cherries by degree of ripeness before pulping. Even manual separation before pulping using a visual criterion does not discriminate between cherries that are not yet fully mature and the fully mature ones because their color is practically the same. The screen pulpers, that use hardness as the selection criterion, can now discard cherries that are not fully ripe, with the important benefit that quality in the cup will not be more negatively affected by the astringency of unripe cherries.

These new machines and techniques now allow Brazilian growers to produce the highest quality pulped naturals, that they call CD1 and derive from only 100 % ripe cherries. They also process CD2, that derives from cherries that are not yet fully mature, and CD3, deriving from the pulp-

Figure 2.27   Sun drying of semi-dry/pulped natural coffee.

ing of unripe cherries using special procedures and techniques. The same approach can be applied in countries that produce washed coffee, specially in those experiencing problems with the selectivity of harvesting, to produce parchment 1, 2 and 3, with the benefit of obtaining a top quality coffee derived from coffee beans without any astringency.

2.3.2.4   Problems and Impact on Quality

The use of the semi-dry/pulped natural process has major impacts on the quality of coffee that is strip harvested manually or mechanically. In spite of the fact that cherries in all stages of maturity are harvested together, the semi-dry system enables each product to be processed separately; this includes floaters (over-ripe and partially dry cherries), immature (unripe, green) cherries and ripe cherries. More importantly, it enables the separate processing of the top-quality fresh ripe cherries that produce the best pulped natural coffees. From a quality standpoint, the semi-dry/pulped natural process is a major positive enhancement over the traditional dry process and an excellent way to produce top quality coffees in conditions where selective harvesting is not feasible.

The cup features of semi-dry/pulped natural processed coffee change according to altitude. Low-grown coffee has a cup closer to naturals, whereas high-grown pulped naturals come closer to washed coffees as altitude increases. In all cases the cup is free from the adverse tastes typical of immature cherries as these are eliminated by the new process. Growers are finding that one of the greatest advantages of semi-dry processed coffee is that it blends well with either naturals or washed coffees. This means additional flexibility at the time of selling. Pulped natural coffees have also become known as an excellent ingredient for high quality espressos, single origin or blends. Specific demand has already developed for this new type of coffee. Brazilian production of pulped natural coffee already exceeds 5 million bags in large crop years.

2.3.2.5 Comments

The semi-dry process improves quality in areas where selective harvesting is not possible, e.g. Brazil. The growing costs of labor and its scarcity are increasing the interest for the semi-dry process in other parts of the world. Picking that is actually fully selective is hardly the norm anywhere in the world today.

It is known that Burundi has used the semi-dry system for many years. Other countries resort involuntarily to this system in peak conditions because they do not have enough capacity to demucilage their parchment.

In Malaysia, Liberica coffee cherries are submitted to a special treatment because of the strong adherence of the skin to the beans. This is shown in Fig. 2.28. The cherries have their pulp cut and bruised by an aggressive type of pulper prior to sun-drying and hulling. The combined effect of the coffee variety and this special treatment produces a harsh, bitter beverage.

2.3.3
Wet Processing

Even though the disk pulper dates back to 1810 and the drum pulper to 1850, wet processing remained practically the same until the early 1980s. It is only during the last 30 years, and more actively over the last 20 years, that wet processing has started to change, first to accommodate the harvesting of cherries at different stages of maturity and then to control the damage caused to the environment by the waste products of the wet process. These changes caused the introduction of the unripe cherry separator (that is in fact a pulper) and the development of technologies that decrease water consumption and contamination like dry pulping, new types of mucilage removers, the dry transport of coffee and pulp, etc. Ecological wet milling is the combination of these new water saving and contamination control techniques.

The "re-conversion" of wet mills to shift to a new ecological approach that requires less water and decreases environmental damage started slowly, but has been gaining strength over the last 15 years. Some coffee quality experts are now wondering whether this re-conversion has not been too fast and too drastic. In the same way that the introduction of unripe cherry separators created

Figure 2.28 A different approach to the semi-dry system in Malaysia.
(A) Incoming cherries (B) Cherries after pulping (C) Dry cherries and pulp (D) Green coffee.

the fear that harvesting conditions would deteriorate, it is now feared that dry pulping and other techniques to control environmental damage may have a negative effect on coffee quality. Although it has already been proven that very substantial reductions in water consumption are possible without any quality loss, complete "dry wet processing" is yet to be achieved in a way that does not jeopardize cup quality. Ecological wet processing needs to balance environmental control and coffee quality objectives in a sensible way.

The objective of wet processing is to remove pulp and mucilage from ripe coffee cherries in an environmentally friendly way. In order to process only fresh ripe cherries, it is necessary to harvest only such cherries or to separate the undesirable products before pulp and mucilage are removed. The description of the wet process will be divided into cleaning and separa-

tion, pulping and mucilage removal, as shown in Fig. 2.29.

The diagram shows that the intensity of cleaning and separation required before pulping depends on the harvesting system used. Whether to separate unripe cherries mechanically or not also depends on the type of product harvested. The solid lines indicate the more frequently used processing sequences, whereas the dotted lines represent other common sequences.

Most producers of washed Arabica coffees have traditionally used selective harvesting. Wet processing after strip harvesting, manual or mechanical, is only the case for Arabicas in Brazil, Hawaii (large plantations outside Kona) and Australia, as well as for Robustas in a few countries. Labor shortages and high harvesting costs are, however, forcing more countries to consider partial stripping or even mechanical harvesting as a means of lowering production costs.

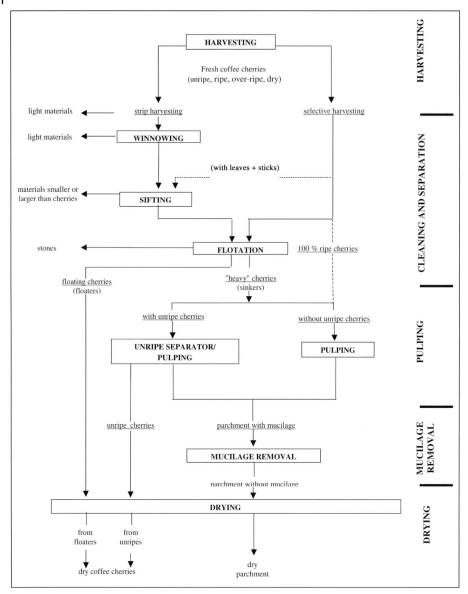

Figure 2.29 Wet processing.

Even if harvesting is selective, the types of products harvested must be considered when deciding which cleaning steps and pulping systems to use. When the harvested product contains only a few impurities, for instance, it may not be necessary to sift it. Flotation may be bypassed if there are no over-ripe cherries. Dry pulping should only be attempted in cases where there is almost 100% of ripe cherries.

In the following sections we do not cover cleaning and separation but only pulping and mucilage removal. Cleaning and separation have been covered under dry processing and all that has been stated there applies to wet processing as well, including the types of equipment recommended.

2.3.3.1 Pulping

2.3.3.1.1 Objective
The objective of pulping is to separate the pulp from the coffee bean.

2.3.3.1.2 Principle
The pulp, which consists of the outer skin and a major part of the mesocarp, is torn off by squeezing the cherries in one of the following ways:

- Between a pulping bar and a rotating disk (disk pulper),
- between a breast plate and a rotating drum (drum pulper),
- as they pass through the slots of a screen (unripe cherry separator and pulper, i.e. screen pulper), or

- between a rotary drum and a stationary screen case (Raoeng pulper).

In conventional devices coffee cherries are pulped in a water flow. As the water content of coffee cherries is high, most pulpers can operate with little or no water provided that alterations are made to ensure that the feeding of cherries and the removal of parchment and pulp can take place without water. Research along these lines has led to a substantial reduction of water consumption in the pulping operation.

2.3.3.1.3 Techniques and Equipment

Disk Pulpers
Disk pulpers consist of one or more disks with a diameter of approximately 45 cm assembled on a horizontal rotating shaft, pulping bars that squeeze the cherries against the disk and plates that separate the pulp from parchment beans. The disks are made of cast iron with bulbs on the surface. They can also have a replaceable punched copper plate or stainless steel coating.

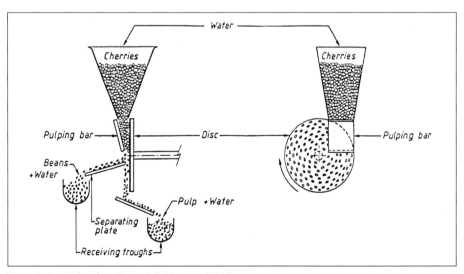

Figure 2.30 Disk pulper. From: J. C. Vincent, CIRAD.

Figure 2.31 Three-disk pulpers.

Coffee cherries are fed to either side of the disk, whose rough surface forces the cherries into the breast channel (the space between the disk and the inclined pulping bar). The pulp is removed as the cherries are squeezed between the pulping bar and the rough surface of the rotating disk. Pulp is separated from the parchment by a plate with a straight sharp edge which allows the pulp to go through, but retains parchment beans. To reduce the amount of unpulped cherries and damaged beans, both the pulping bar and the separation plate can be adjusted according to the diameter of the cherries.

Disk pulpers on the market have from one to four disks, some include re-pass disks. The average flow rate for one disk is about 1.0 ton of fresh cherries/h; the power required ranges from 1.0 hp for a single disk to 2.5 hp for four disks.

Horizontal Drum Pulpers

Horizontal drum pulpers consist of a rotating metallic cylinder, 20–30 cm in diameter, with bulbs (indentations), a

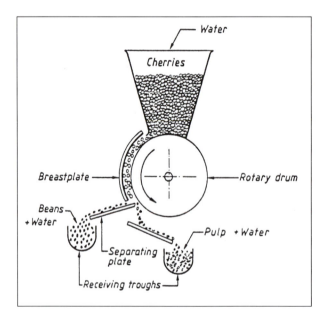

Figure 2.32 Horizontal drum pulper with metallic breastplate. From: J. C. Vincent, CIRAD.

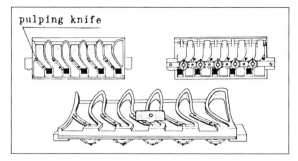

pulping knife

Figure 2.33 Breast-plate with channels. From: R. Wilbaux.

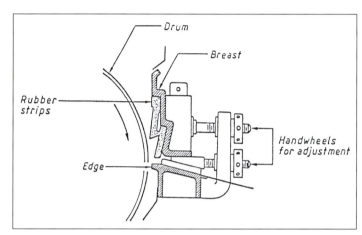

Drum

Breast

Rubber strips

Handwheels for adjustment

Edge

Figure 2.34
Horizontal drum pulper with rubber breast.
From: R. Wilbaux.

breastplate with or without channels and a plate that separates pulp from parchment.

The breastplate with channels is metallic and the ribs that operate inside the channels to force the cherries against the drum may be metallic or rubber. In either case the clearance between the ribs and the drum decreases from top to bottom so that the cherries are submitted to increasing pressure which forces the pulp to be removed. The variable clearance of the channels allows this machine to pulp cherries of all sizes simultaneously.

Channels are not needed where rubber breastplates span the full length of the cylinder and offer an increased working surface for the same size of cylinder.

Cherries are evenly fed over the whole length of the cylinder. The indented rotating drum carries the cherries into the pulp-

Figure 2.35 Small horizontal drum pulper.

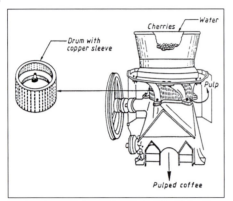
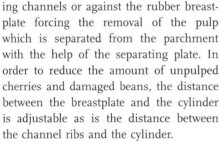

Figure 2.36 Single-drum vertical pulper.
From: CIRAD-CP.

Figure 2.37 Pulping screen and channels of vertical drum pulper.

ing channels or against the rubber breastplate forcing the removal of the pulp which is separated from the parchment with the help of the separating plate. In order to reduce the amount of unpulped cherries and damaged beans, the distance between the breastplate and the cylinder is adjustable as is the distance between the channel ribs and the cylinder.

Rubber channels or rubber breastplates introduce a certain degree of selectivity into the machine because they allow some unripe and partially dry cherries to pass through unpulped. However, these unripe and partially dry cherries are often partially pulped or damaged which makes their further separation difficult.

Horizontal drum pulpers are made in a large variety of sizes with capacities ranging from 0.25 (for a hand-driven machine) to 4 tons of fresh cherries/h. The required power is about 2 hp per ton of fresh cherry/h.

Vertical Drum Pulpers
Vertical drum pulpers consist of one or two narrow vertical rotary cylinders covered with a copper or metallic sleeve with bulbs, around which three to six cast iron channels are assembled.

Single-drum vertical pulpers are shown in Fig. 2.36 and 2.37 whereas a double-drum vertical pulper is found in Fig. 2.23 presented earlier.

The cherries are spread on the top of the cylinder by a stationary feeding disk. As the channels become progressively narrower from top to bottom, the cherries are submitted to increasing pressure that forces the removal of the pulp which is carried through the clearance between the channel and the cylinder. Parchment beans remain inside the channels and are collected beneath the rotating cylinder, whereas pulp is gathered outside.

As is the case for horizontal drum pulpers with channels, the progressively reduced clearance between the channel and the cylinder allows for the simultaneous pulping of cherries of different sizes. However, there is no quick way to adjust this clearance. It can only be changed by opening the machine and releasing the fastening screws.

Vertical drum pulpers have capacities ranging from 0.25 (for the manual versions) to 2 tons of fresh cherries/h. The required power is about 0.7 hp per ton of fresh cherry/h.

Screen Pulpers (Unripe Cherry Separator and Pulper)
Screen pulpers perform two functions: the separation of unripe cherries and the pulping of ripe ones. There are two distinct lines of screen pulpers: one of small to moderate capacity with slotted screens, which should be used with a re-pass pulper, and the other of large capacity with welded wire screens and its own electric motor for independent use. The operation of the screen pulper has already been described in the section about equipment for semi-dry processing. Where Fig. 2.23 shows the slotted screen pulper alone, Fig. 2.24 shows the slotted screen pulper coupled with a vertical drum re-pass pulper.

Slotted screen pulpers coupled with double-drum pulp separators/repassers can be used to produce different qualities of washed coffee that derive from cherries at different stages of maturity. Top quality washed coffee results from the pulping of only 100 % ripe cherries, free from any cherries that are only partially mature. The next quality derives from pulping partially mature cherries that are free from unripe cherries. A third quality washed coffee may result from the pulping of unripe cherries using special procedures and techniques. This three-fold pulping/repassing process in unripe separators/screen pulpers is not included in the diagram of Fig. 2.29 for the sake of simplicity. Coffee would go three times through the same box marked "unripe separator/pulper", this box being either a single pulper where coffee would be repassed or three different pulpers in the case of large mills.

The large wire screen pulpers were traditionally installed after the siphon tanks to remove the unripe cherries and to pulp the ripe ones. Nowadays, they are more likely to be installed after the main pulper(s) in order to function chiefly as an unripe cherry separator. The idea is to save water consumption by using pulper(s) that require little or no water, loosely adjusted to let the unripe cherries pass through to be separated by the wire screens afterwards. For this arrangement to work efficiently, the main pulper(s) should neither damage nor partially pulp the unripe cherries.

Screen pulpers have capacities ranging from 0.7 to 15 tons of fresh cherries/h. The required power is about 1.5–2.0 hp per ton of fresh cherry/h.

Raoeng Pulper
This machine consists of a horizontal cast iron rotating drum assembled inside a fixed screen cylinder. The inner drum has screw-shaped channels on the feed side, to force the cherries to move forward, with longitudinal and transverse cleats further on. A gate with a counterweight is fitted at the outlet of the screen cylinder.

Unlike the machines described earlier, the Raoeng pulper removes both pulp and mucilage in a single operation that consumes a large amount of water and power. The Raoeng machine may be used either as a pulper and mucilage remover, mostly for Robustas, or only as a mucilage remover.

Cherries are pulped and demucilaged by rubbing against each other and against the metallic parts as they pass between the cleated drum and the perforated outer cylinder. Water under pressure is fed from perforations in the inner drum in order to wash and carry the fine ground pulp and the mucilage through the holes of the screen which retains parchment. The counterweight at the outlet gate controls pressure and the time the product remains inside the machine.

Raoeng pulpers have capacities ranging from 0.75–3.0 tons of fresh cherry/h with power requirements of 7–10 hp per ton of fresh cherry/h.

Dry Pulping

As the water content of coffee cherries is high, most pulpers are able to operate without water inasmuch as the feeding of cherries and the outflow of parchment and pulp can take place without water. Recent development work to reduce and eliminate water consumption in pulping has concentrated on drum and disk pulpers. Work is presently under way on screen pulpers as well. The high water and power consumption of Raoeng machines has eliminated them as a feasible option for further development as a pulper.

To decrease contamination due to pulping is more and more important as coffee producing countries become aware of the environmental damage caused by the wet process. As it is more costly to treat a lot of water with little contamination than a little water that is highly contaminated, the efforts to make pulping environmentally friendly have concentrated on cutting down or eliminating water consumption, provided that coffee quality is not negatively affected.

In theory most pulpers can operate without water. However, a lot of research and development work has been required to cut down water consumption. Although all the pulping principles described above are compatible with dry operation, the challenge is to ensure the feeding of cherries, the separation of pulp from parchment, as well as the outflow of parchment beans and pulp without water and without damaging the coffee.

A ripe cherry contains a lot of water so the parchment beans will pop out easily when the cherry is squeezed. This is what happens when we press a ripe coffee cherry between our fingers. The problem is to ensure that the inflows and outflows and the separation of products can take place without water at the capacities required in pulping facilities.

Although each machine requires specific alterations, the basic practical changes introduced in drum and disk pulpers have been:

- To increase the slope of the cherry hoppers so that feeding can take place without water.
- To increase the speed of rotation and to change certain features of the drum or disk so that pulp can be separated without water.
- To improve systems for transporting parchment beans and pulp without water so that the pulper does not clog.

Practical examples of these changes are the conic drums and the double drums in vertical pulpers, the double rubber breasts in horizontal drum pulpers (Fig. 2.34) and mechanical devices to facilitate the outflow of parchment beans and pulp.

The changes to use less water in screen pulpers will apparently require a greater development effort because it will be necessary to redesign key components of the machine. The work of research institutes and private companies has enabled the manufacture of a number of vertical and horizontal drum pulpers and disk pulpers that operate with little or no water. Screen pulpers that operate without water will soon be available.

Dry pulping should not be achieved at the expense of coffee quality or at unduly high costs. It does not help to have dry pulpers that damage the beans or cannot properly separate the pulp which remains mixed with parchment beans and affects fermentation and mucilage removal. It does not help to have dry pulpers that allow parchment coffee to be lost with the pulp with obvious cost impacts. A little water consumption and excellent coffee quality may be better than no water consumption and lower quality coffee or no water consumption and high processing costs or both!

Most dry pulpers on the market today require incoming cherry to be 100 % ripe because they have little or no ability to cope with unripe, over-ripe and partially dry cherries. Dry pulpers tend to require more power than the equivalent wet models.

Pre-grading and Re-passing

Pulping is an industrial process subject to inordinately high variations caused by the changing characteristics of incoming cherries. The size of cherries and beans, the thickness of the pulp, the mucilage content, the degree of maturation, the impurity content, etc., tend to vary depending on coffee variety, climate, time and system of harvesting, degree of separation, etc.

In order to introduce some rationality into the system and to attempt to standardize the process, pre-grading systems have been created, whereby cherries are sorted by size before pulping. The objective is to avoid nipping large beans or to have smaller beans that are partially pulped or intact. Pre-grading systems can be cumbersome because after grading, different pulpers are required to handle different sizes. Since the size of the cherries may change from season to season on the same farm, one is likely to get different

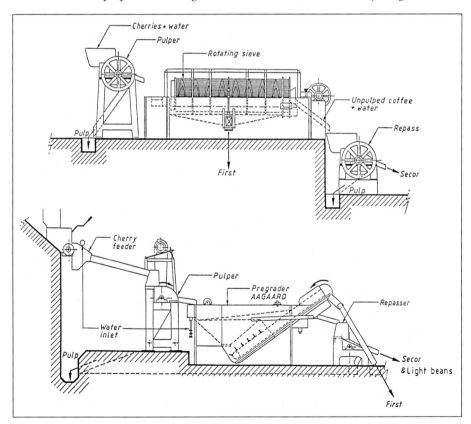

Figure 2.38 Two stage (re-pass) pulping with rotary sieve (top) and Aagaard pre-grader (bottom). From: R. Wilbaux, adapted by CIRAD-CP.

percentages for each size every year which, in turn, calls for different pulping capacities for each size.

The modern tendency of ecological wet-milling is to avoid water-intensive pre-grading of cherries by using pulpers that are able to handle different cherry sizes simultaneously like the horizontal and vertical drum pulpers with conic channels or horizontal drums with rubber breasts. The pulpers can be adjusted from season to season or even daily (horizontal drum pulpers only) to handle different cherry features.

Another alternative to pre-grading cherries is the use of a re-pass pulper. In this case the main pulper is less tightly adjusted so that the larger beans are not nipped. The small cherries which pass unpulped through the first pulper are separated by a rotary or flat sieve and directed to a re-pass pulper that is more tightly adjusted. The re-passer is a pulper of smaller capacity than the main pulper and easily adjustable, like the horizontal drum pulpers. Re-passing may also be performed in a main pulper provided it is adjusted more tightly. In this case un-pulped cherries to be re-passed are stored during regular processing hours and re-passed later in the same machine.

From a water consumption standpoint it is better to use a single pulper only, preferably one that is able to handle cherries of several sizes simultaneously. If that fails, a re-pass pulper may be a better alternative than pre-grading, which should only be used when absolutely necessary. In any case both the screen after the main pulper and the pre-grading screen(s) should operate with little or no water.

2.3.3.1.4 Problems and Impact on Quality

With so many changes and new developments in pulping one is often confused about which systems and machines to use

to carry out ecological wet milling. The starting point is the diagram in Fig. 2.29 where it is clear that the first variable to be considered in choosing a pulper is the harvesting system used. The second is the degree of separation required before pulping. If, for instance, unripe cherries are mixed with ripe cherries, which is increasingly the case today, the choice of pulpers has to be made accordingly. Other variables are water consumption, power requirements, physical damage to beans, farm size and coffee variety.

When comparing pulpers according to water consumption, one should always bear in mind that water re-circulation may be a good alternative to a fully dry pulper. Where unripe cherries are a problem, for instance, it is unlikely that a dry pulper will be an adequate solution. In this case the best solution may be a screen pulper that consumes a little water which can be re-circulated.

If availability of power is a constraint, the vertical drum pulpers appear to be the best solution because they are the ones that consume the least power per ton of fresh cherry processed.

When physical damage is a critical factor, screen pulpers coupled with drum re-passers become an obvious choice because they cause the least physical damage and lose the least possible parchment beans with pulp, although they may let somewhat more pulp pass through.

The size of the farm affects the choice of pulper because a small farmer is more likely to ensure that ripe cherries reach the pulper, either by careful selective picking or by hand sorting before pulping. However, even small farmers cannot visually separate 100 % ripe cherries from those that are only partially ripe. Harvesting quality is much more difficult to control in a mid-to-large farm setting where the number of people involved, labor availability and harvesting speed may all become critical factors. Farm size also affects the choice of pulping systems because small fully dry systems are easier to design than large facilities where coffee transport issues and the costs involved may also become critical factors.

Apart from the choice of pulping systems, the adjustment of the machines is another critical issue. The best possible system will provide poor results if maintenance and adjustments are not adequate. Damaged parchment, nipped beans, unpulped cherries and mixed products result more often from poor adjustment of machines than from the choice of systems or the quality of harvesting. Suppliers of pulpers are striving to simplify the adjustments or even to eliminate some in order to ensure better results. If faced with unskilled labor it may be better to settle for a pulper that is simple to adjust (e.g. a vertical drum pulper or a screen pulper) than to try to get the best results with a horizontal drum pulper with individual adjustments for each channel!

2.3.3.1.5 Comments

Arabica cherries are easier to pulp than Robusta cherries. As a result the capacity of pulpers falls when processing Robusta coffees.

Raoeng pulpers should not be used in new wet mills because of their high water and energy consumption.

Disk pulpers have a tendency to lose the capacity and efficiency of separation of pulp from parchment rather quickly as the bulbs wear out. Frequent monitoring and maintenance (replacement and refurbishing of disks) are therefore critical for optimum performance.

Stones and sticks can be very damaging to all pulping systems and even more so in the case of the copper sleeves of drum pulpers. If it cannot be guaranteed that the harvested cherries are free from foreign materials, pulping should be preceded by a flotation device to remove stones and other impurities.

Steel sleeves are more resistant to wear and tear and damage than copper sleeves.

The generation of pulp fiber is a problem for many pulping systems. Fiber is very difficult to separate from water and high cost special equipment is required. The pulping system that produces the least fiber is the slotted screen one. A pulping system that produces whole pulp rather than pulp pieces and fiber is environmentally superior.

2.3.3.2 Mucilage Removal

2.3.3.2.1 Objective

This operation aims to remove the residual part of the mesocarp, called mucilage, which remains stuck to the endocarp (parchment envelope) after pulping.

2.3.3.2.2 Principle

Mucilage is insoluble in water and clings to parchment too strongly to be removed by simple washing. Mucilage can be removed by fermentation followed by washing or by strong friction in machines called mucilage removers.

Fermentation may be natural or accelerated by chemicals or enzymes.

Mechanical mucilage removers operate by rubbing parchment beans against each other and against the mobile and static parts of the machines.

2.3.3.2.3 Techniques and Equipment

Natural Fermentation

Natural fermentation is usually carried out in masonry or concrete tanks which may vary considerably in size and shape. A practical system is rectangular tanks with a sloped bottom, used both as a tank and a means of conveyance for coffee. Sluice gates with screens allow the discharge of either water alone, as the tank fills, or water and coffee together at the end of fermentation.

The size of the tank depends on the capacity of the pulping equipment because the tank must be filled rapidly e.g. in 1 hour, to avoid substantially different fermentation times for the first and last beans admitted into the tank. The shorter the period of fermentation, the greater the need to fill the tank quickly.

Natural fermentation may be dry (without water) or under water. Dry fermentation is faster, but more difficult to control. Fermentation tanks must be protected by a roof or covered to avoid drying the top layers and to protect the coffee from rain. Fermentation under water is often more homogeneous and does not require a roof over the tanks. In both methods, the water conveying the coffee into the tanks must be completely drained. In the case of under-water fermentation the tank is then refilled with clean water.

It should be noted that the term "fermentation" is not appropriate in the case of coffee because, unlike cocoa, no biochemical reaction takes place inside the coffee bean. It would be more correct to refer to mucilage removal by means of a biochemical reaction or hydrolysis of the mucilage which covers the parchment beans. This reaction is caused by enzymes (pectinases and pectase) which are naturally present in coffee cherries.

Fermentation times can vary substantially, from 6 to 72 h, depending on the temperature, the amount of mucilage, and the concentration of peptic enzymes. Higher temperatures and thicker mucilage

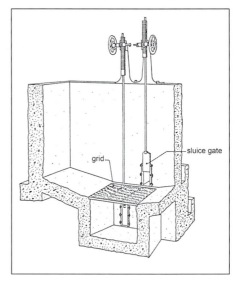

Figure 2.39 Outlets of fermentation tanks. From: R. Wilbaux, adapted by CIRAD-CP.

Figure 2.40 Dry fermentation.

layers accelerate fermentation. Higher concentrations of peptic enzymes are found in ripe cherries and in parchment that has been dry pulped. Dry fermentation is faster because the peptic enzymes are not diluted and oxygenation is more intensive.

Coffee is held in fermentation tanks till the mucilage is completely digested and ready for washing. This is verified by rubbing and washing a few beans by hand. A grating noise should be clearly heard and the clean beans should feel gritty, like pebbles, when fermentation is complete. To

Figure 2.41 Tanks for wet fermentation.

stop fermentation at the right time is critical for coffee quality as this avoids over-fermentation and the formation of stinkers. On the other hand, when fermentation is deficient or too slow, butyric acid or propionic acid can develop, both of which have an undesirable impact on coffee quality.

Fermented parchment should be washed immediately after fermentation is completed.

Figure 2.42 Washing channels.

Shortened Fermentation

Natural fermentation can be accelerated either by adding certain types of enzymes or yeast (*Saccharomyces*) or with hot water. These methods were more common before efficient mechanical demucilagers became available and pressures to decrease water consumption and to protect the environment increased. At present they are rarely used commercially, although interesting development work and trials are currently under way with enzymes.

Washing of Fermented Coffee

Washing of fermented coffee is one of the main sources of water consumption and contamination in wet processing. Fermented coffee can be washed manually in the tank itself or in channels, by centrifugal pumps or by several types of specific machines. Any of the mechanical mucilage re-

movers described below can also be used to wash fermented coffee.

Manual washing in the fermentation tanks is water and labor intensive, especially in large tanks. Water is added to parchment which is then stirred by hand with wooden paddles. The process is not always thorough when tanks are large or particularly deep.

Washing channels is another water intensive washing system. However, the same water may be reused for grading. This method is likely to be progressively abandoned as concerns over high water consumption grow stronger.

Centrifugal pumps vigorously stir coffee and water as they pass through the pump. This washing method has the added advantage of transporting coffee to the drying areas. Any centrifugal pump compatible with solids the size of parchment coffee may be used provided that the recommendded mixture of solids (coffee) and water is respected. These pumps usually have an open rotor.

Many types of mechanical stirrers and agitators have been made in several countries. However, the latest generation of low-cost, high-capacity mucilage removers is curtailing the use of these machines. Mucilage removers wash fermented coffee and can also complement the task of the tank in case of incomplete fermentation. They offer spare demucilaging capacity at peak periods and grant greater control over the process. Whereas in the past a mucilage remover cost substantially more and required a lot more power than a mechanical washer, the new mucilage removers are far more cost effective and efficient.

Mucilage Removers

As natural fermentation is water, labor and time consuming, the quest for mechanical mucilage removers started a long time ago. The first successful machine was the Raoeng pulper which was invented over 50 years ago to pulp and demucilage coffee. Later on it was only used as a mucilage remover.

The environmental concerns of recent years, which brought about the concept of ecological wet milling, pressed for new machines that would consume less water and power. The first response was the ELMU type vertical machine created 30 years ago. It has now been replaced by the more efficient upward flow machines developed in the early 1990s.

(1) *Aquapulpa*. It was soon clear that the Raoeng pulper could have a much larger capacity if it was only used to remove the mucilage of coffee. As a result the Raoeng machines used initially to pulp, demucilage and wash Robusta coffee became known as the Aquapulpa, the only mucilage remover in widespread use for nearly three decades. Many versions of the Aquapulpa have been produced worldwide.

The Raoeng pulper/Aquapulpa has already been described in the section about pulping. When used exclusively as a mucilage remover the machine has more capacity. Although it is very efficient for removing mucilage, it requires careful adjustment to avoid hulling parchment and physical damage to the beans. The main drawbacks are associated with the high power requirements (8–10 hp per ton of pulped coffee/h) and the huge water consumption of conventional models (over 4 m^3 of water/ton of pulped coffee). Although lower water consumption versions of this machine have been developed, their high power requirement remans a problem. Usual capacities range from 0.75 to 3.0 tons of pulped coffee/h.

(2) *ELMU*. Invented in the early 1980s the ELMU has a shaft with metallic fingers placed at right angles and pointing out radially. The shaft and fingers revolve inside a vertical jacket which has another set of similar fingers pointing to its center and located between the other fingers. The cylinder is divided horizontally into three or four compartments that are connected by openings in alternate positions to force coffee to move sideways inside the machine.

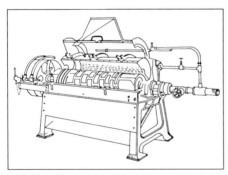

Figure 2.43 Aquapulpa. From: R. Wilbaux.

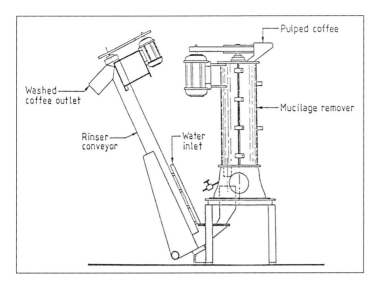

Figure 2.44
Mucilage remover
ELMU with washing
conveyor.
From: CIRAD-CP.

Drained parchment is fed into the top of the machine. Mucilage is removed by rubbing the beans against each other and against the fixed and mobile fingers as parchment moves downwards by gravity. A counterweight is located at the outlet at the bottom of the machine to regulate the degree of washing. Water is injected in the upper and lower thirds of the cylinder to facilitate washing and to help parchment flow out of the machine.

As the ELMU discharges parchment mixed with mucilage, a washing device like a screw conveyor should be installed after the machine.

Total removal of mucilage may not be achieved by the ELMU but the little that remains is not likely to cause quality or drying problems as is sometimes feared.

The ELMU mucilage remover has capacities ranging from 1 to 3 tons of pulped coffee/h. Power requirements are from 3 to 5 hp/ton of pulped coffee/h.

(3) *Upward flow mucilage removers.* New additions to the family of mucilage removers were developed first in the early 1990's and then in the mid 2000's. Several versions are currently offered on the market and hundreds of units are already operating successfully in many countries.

These machines consist of a vertical rotor composed of an endless screw in its lower section and a set of cogwheel-shaped or vertical bar agitators in its upper section. The rotor is assembled inside a cylindrical screen cage made of a perforated steel plate or welded wire. The screen cage and the rotor are set in a case that retains water and mucilage and directs their flow downwards.

Drained pulped coffee is fed into the base of the machine. The endless screw moves parchment up to the section with agitators where mucilage is removed as the beans rub against each other and against the rotor and the screen. Mucilage flows through the holes of the screen in the form of a very viscous liquid that is discharged at the bottom of the machine. Parchment without mucilage exits at the top of the machine. Water is injected in small quantities at points along the cylinder to wash the mucilage and ease the flow.

The capacity of the upward flow mucilage removers ranges from 0.75 to

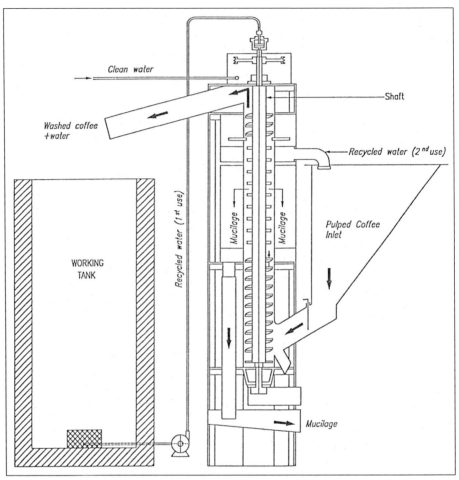

Figure 2.45 Upward flow mucilage remover. From: CIRAD-CP according to Penagos.

10.0 tons of pulped coffee/h. The power requirement is from 1.5 to 3.0 hp per ton of pulped coffee/h. The innovative principle of this machine reduces power and water consumption significantly. Water consumption, claimed to be under 1 m³/ton of dry parchment, is the lowest of any mucilage removal system known to date.

2.3.3.2.4 Problems and Impact on Quality

Even though mucilage removers have become far more efficient in terms of water and power consumption, a heated debate is still taking place over the actual impact on coffee quality caused by the elimination of natural fermentation. The critical consideration behind all discussions appears to be that several experiments associate natural fermentation with less bitterness and more pronounced acidity and aroma in coffee. The fundamental question raised is: can the same quality improvement be obtained when mucilage is removed mechanically? Recent research has consistently replied "yes" by demonstrating that mechanically demucilaged coffees have

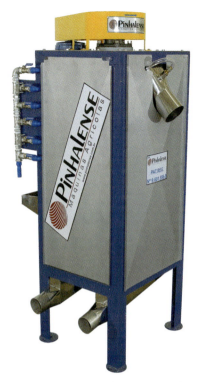 

Figure 2.46 Outside and inside view of upward flow mucilage removers.

the same organoleptic features as those that are naturally fermented.

Costa Rica's CICAFE was perhaps the first research institute to undertake blind liquoring of samples of the same coffee processed by the two methods. At the time the ELMU mucilage remover was probably used. The results pointed out that it was not possible to find significant statistical differences between samples naturally fermented and those mechanically demucilaged. Colombia's CENICAFE came to the same conclusions when using the upward flow mucilage remover. Kenya's Coffee Research Foundation also reported that following trials with the upward flow mucilage remover, it was not possible to state, after blind cupping, that naturally fermented and mechanically de-

mucilaged samples were significantly different. More recently, series of trials with the same coffees emphasized the quality consistency of coffee demucilaged mechanically when compared to those naturally fermented.

Further evidence in favor of mechanical demucilaging is the fact that in all the trials mentioned above natural fermentation was closely monitored and precisely performed in laboratory-like conditions. This is very seldom the case in real life wet processing, where controls are not strict and over fermentation is frequent. In other words, mechanical mucilage removal, which does not require any special control, was compared to natural fermentation in ideal control conditions seldom to be found in actual processing. If no quality differences were found under these specific conditions, one is tempted to say that machines are superior to tanks in everyday processing because machines do not require the close supervision that natural fermentation does!

To these results one should add the frequent tests carried out by machine manufacturers in several countries. Their views may be biased, but they do have moral and ethical obligations with respect to their clients, not to mention the need to retain these clients by stating the truth about their products.

It is not necessary to elaborate here on the problems linked to the control of natural fermentation. The adverse results of poor control are well known. They range from onion flavor to stinker beans, among other defects. Practical processing problems caused by natural fermentation are also well known: costly fermentation tanks, high labor requirements, batch rather than continuous processsing, and, worst of all, the consumption of huge amounts of water that becomes highly contaminated and is difficult to treat. In traditional wet processing, fermentation accounts for nearly 50 % of the total water consumption and the wastewater generated is not suitable for recycling. When low water consumption or dry pulping is used, natural fermentation accounts for practically all the water consumption.

A final argument against natural fermentation relates to weight loss. It is widely accepted that pulped and washed coffee loses solids as long as it remains moist and alive, i.e. with a moisture content above 12 %. These losses are caused by normal metabolism, respiration and exosmosis. Natural fermentation increases such losses by delaying processing and raising the temperature. Losses are therefore larger in hotter areas. Weight losses associated with fermentation have been found to range from 0.5 to 6.0 % (9.0 % in extreme cases) depending on local temperatures and the number of hours required for fermentation.

Mechanical removal of mucilage avoids weight loss and the risk of further quality deterioration during the long fermentation process. The value of the increased weight of coffee should more than offset any quality losses due to the absence of fermentation, even though all recent evidence indicates that there are no quality losses when fermentation is not used.

Purists may still insist that unfermented coffee does not develop a "white center cut" when roasted. Even though this point has never been properly demonstrated by formal research, it is accepted as true by many people in the industry. However, this discussion is pointless because only a very small percentage of coffee is sold in whole bean and even the most refined consumer is not aware of the information about the white center cut.

The trend towards eliminating natural fermentation in favor of the mechanical removal of mucilage seems clear. If the cost and quality arguments were not strong enough in the past, new scientific evidence and new machines combined with environmental pressures will cause the progressive replacement of fermentation tanks by mucilage removers as per with the modern concept of ecological wet milling.

For the conservative miller, hesitating before a drastic change, a more gradual approach exists. As mentioned above, all mechanical mucilage removers can function as washers after fermentation. Overly cautious millers can always start by adding a mucilage remover to work as a washer after their fermentation tanks. They may progressively ferment fewer hours and complete the work with the help of the mucilage remover until they are fully convinced that natural fermentation can be eliminated.

In spite of this approach, some specialists still assert that natural fermentation cannot be bypassed. Their arguments are based on the results of research carried out in Kenya by Wootton in 1965 and 1967. This school of thought maintains that certain bitter and soluble components of the coffee bean (polyphenols, diterpenes, etc.) permeate the liquid medium by osmosis during fermentation. This phenomenon, compounded by loss of bean weight, is claimed to considerably improve coffee quality by reducing the bitterness and the harshness of the liquor. Can the phenomenon above take place when mechanically demucilaged coffee beans are soaked in clean water for 12 or 24 h as described in the following item?

One can but hope that further research, conducted on the basis of more up-to-date scientific knowledge, will finally supersede these various schools of thought and end these ever-recurring long-standing arguments. Or, perhaps, there is no room for this debate any longer considering that today's market calls for diversity and there is space for a multitude of coffees with different nuances in quality

Recent trends either to increase the body of fully washed coffees for espresso blends or to facilitate the drying of semi-washed coffee require that mucilage be only partially removed. This cannot be achieved with natural fermentation but only with the help of mucilage removers.

2.3.3.2.5 Comments

The upward flow mucilage remover is widely accepted today as a better machine than the ELMU and the Aquapulpa.

In all three types of mucilage removers described above pulped coffee must be drained before entering the machine to en-sure the friction required to rub the mucilage off.

Some researchers claim that soaking, i.e. keeping parchment coffee under clean water for a number of hours after it has been demucilaged, lessens bitterness and favors quality. This could be an interesting option for quality conscious millers because the parchment could be soaked after the mechanical removal of mucilage. Although this procedure requires tanks and batch processing and causes weight losses, the water will not be contaminated and can easily be recycled. Weight losses will be lower than in fermentation provided that coffee is soaked for a shorter period than for the fermentation process.

The weight gains due to mechanical mucilage removal may be lost if drying does not start shortly after demucilaging. The interval between mucilage removal and drying should not exceed 6 hours.

Furthermore, the mechanical removal of mucilage allows the design of more compact mills and the construction of compact modules for wet processing.

2.3.3.3 Design of Wet Mills

2.3.3.3.1 Objective

Wet-processing facilities are designed so that their layout provides rational systems for cleaning and separation, pulping and demucilaging.

The modern water saving machinery used in each phase of the process is not much help if water is consumed and contaminated as coffee is fed from one machine to the other. A key design objective is to minimize water consumption, to recycle the water that is used, to ensure the safe disposal of contaminated wastewater and solid byproducts while, above all, preserving coffee quality.

Some concepts presented here are also developed in Section III.3 "Ecological Processing of Coffee and Use of Byproducts".

2.3.3.3.2 Principle

The traditional approach to the design of wet mills needs to change. Reception of coffee with water, conveying coffee and by-products in channels with water, the use of machines that consume high quantities of water, natural fermentation, etc., defy the purpose to wet-process coffee in an ecologically friendly way.

Conventional wet-processing may consume from 20 to 100 m^3 of water per ton of green coffee, with the lower end of the range achieved only by recycling. Ecological wet milling should require under 10 m^3, ideally under 5 m^3 of water per ton of green coffee. Wastewater and all other by-products must be safely disposed of.

The following basic principles must be applied to the design of ecological wet mills:

- All machines should operate with as little water as possible or, better still, no water at all.
- The reception of cherries must be without water, i.e., dry.
- All conveyance, by gravity or by mechanical means, must be carried out without water.
- The contact of pulp and parchment with water must be minimized or avoided.
- Water that has to be used must be recycled as far as possible.

Water contact with parchment and pulp should be avoided so that soluble contaminats do not transfer to the water. It is easier and cheaper to deal with solid waste products. Water recycling reduces water consumption and concentrates the organic loads. It is easier and cheaper to treat small volumes of water that are heavily contaminated than large volumes which are less contaminated.

The design of any facility must start with the analysis of the wet processing diagram in Fig. 2.29 which indicates the processing steps required according to the type of harvesting system and products harvested.

2.3.3.3.3 Techniques and Equipment

Dry Transport

The machines commonly used for the dry transport of coffee are screw conveyors, belt conveyors with cleats (inclined or horizontal) and bucket elevators. All machines may require changes or adaptation when used for transporting wet products.

Screw conveyors are used to transport products and to drain and transport wet products. Screw conveyors are excellent for draining pulp and parchment after pulping.

Belts with cleats (Fig. 2.47) are easier to clean, although they should only receive pre-drained products. Parchment with mucilage should not be transported by these conveyors because it sticks to the belt and the cleats.

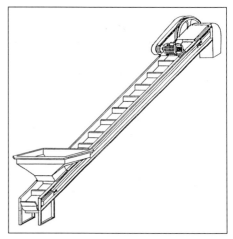

Figure 2.47 Conveyor belt with cleats.

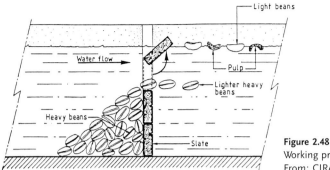

Figure 2.48
Working principle of a grading channel.
From: CIRAD-CP.

Bucket elevators must have perforated buckets to transport products with free water.

Separation Screens and Grading Channels

Screens to separate coffee from unwanted products may be either rotary or flat. In either case they must preferably be used without water.

Grading channels are masonry, concrete or metallic structures used to separate de-mucilaged parchment by flotation. Heavier beans sink, while the lighter ones float in running water that is required in large volumes. Although this water can be recycled, grading channels are being discontinued along with fermentation tanks as wet mills become more compact and water efficient.

Layouts

It is useful to compare the layouts of conventional and ecological wet-processing units.

Figures 2.49 shows the conventional wet mill still common in many coffee areas of the world. A large siphon tank which consumes a lot of water feeds a pulper with an Aargard separator (vibrating screen with water). Channels and pipes carry parchment and water into large concrete fermentation tanks. A water intensive washing and grading channel completes the process. The space required, the cost of the structures and the large amounts of water needed are all too obvious from the drawing.

Figures 2.50 and 2.51 show the design of wet mills with dry reception and transport, dry pulpers for 100% ripe cherries only, fermentation tanks and grading systems.

Figures 2.52 and 2.53 show ecological wet mills – dry reception and transport, mechanical removal of mucilage and water recycling – that can process 100% ripe cherries or ripe cherries received mixed with unripe and over-ripe cherries. This mixture happens at the beginning and end of the picking season; in some areas mixed harvesting may result from lack of labor or stripping. These mills are prepared to receive any type of incoming product and to pulp and demucilage only the ripe cherries using only a minimum amount of water which can be recycled.

The main difference between Figs 2.52 and 2.53 is the availability of a slope. This highlights another important point. Traditionally pulpers were installed in sloped areas so as to use gravity for feeding from one terrace to the other (see Fig. 2.49). As transport was carried out along channels with water, not much slope was required. However, with dry gravity transport (Fig. 2.52), slopes need to be steeper to ensure that products run freely from one machine to the other. Increasing the height of the terraces inevitably increases the costs of terracing.

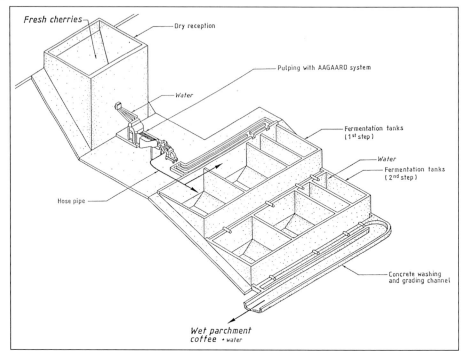

Figure 2.49 Conventional wet mill (sloped area). From: CIRAD-CP.

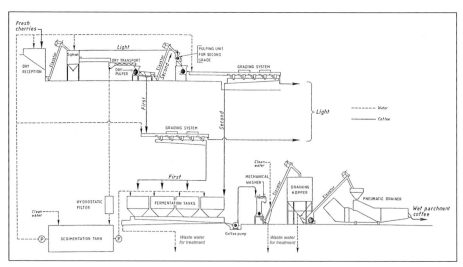

Figure 2.50 Wet mill with dry pulper for 100% ripe cherries (flat area). From: CIRAD-CP.

In designing modern water efficient ecological wet mills one must compare the costs of terracing (Fig. 2.52) with the costs of equipment required for dry mechanical transport (Fig. 2.53). In many cases it will be found that the costs of

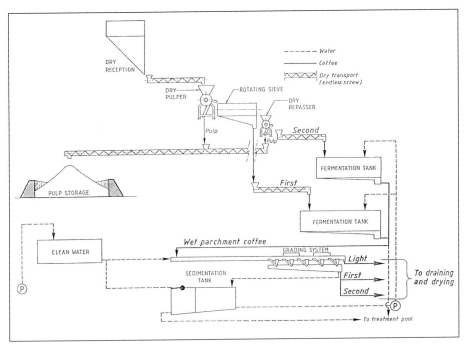

Figure 2.51 Wet mill with dry pulpers for 100% ripe cherries (sloped area). From: CIRAD-CP.

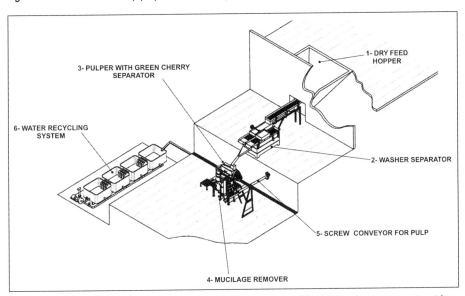

Figure 2.52 Ecological wet mill with dry gravity transport (compatible with any harvesting system/slope available). From: Pinhalense.

civil works are higher than the costs of mechanical conveyors. Another alternative is to use gravity feeding combined with dry mechanical transport.

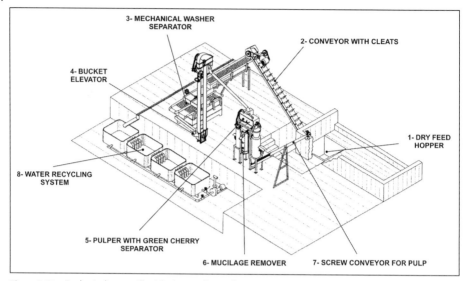

Figure 2.53 Ecological wet mill with dry mechanical transport (compatible with any harvesting system/ slope not available). From: Pinhalense.

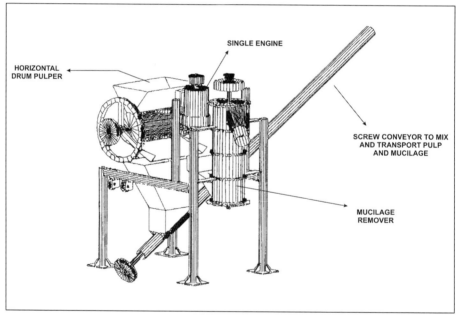

Figure 2.54 Compact ecological wet mill with horizontal drum pulper. From: Federación Nacional de Cafeteros de Colombia (FEDECAFE).

Compact Ecological Wet Mills

A quick analysis of Figs 2.52 and 2.53 brings to mind a new idea: now that these machines are so close together, practically on top of one another, why not assemble them in one single structure? The concept of the compact ecological wet mill is the next logical step.

A typical compact ecological wet mill comprises a vertical or horizontal drum pulper, an upward flow mucilage remover and a screw conveyor for pulp, all assembled in one single structure.

A dry hopper is assembled on a drum pulper that discharges parchment into the hopper of the mucilage remover and pulp into the hopper of the screw conveyor. The upward flow mucilage remover delivers washed parchment, whereas the screw conveyor piles the pulp or loads it into a cart. In the compact mills of small capacity, mucilage may be incorporated into the pulp in the screw conveyor. In the larger units mucilage is drained at the base of the mucilage remover for subsequent disposal.

The compact mills with drum pulpers are essentially designed to handle ripe cherries.

Compact mills with screen pulpers (unripe cherry separators) are also available to handle a mixture of ripe and unripe cherries and to pulp ripe cherries only.

Compact ecological wet mills are available in capacities ranging from 0.5 to 10.0 tons of cherries/h. The power required is between 2.5 and 8.0 hp per ton of cherry/h. The lower end of the range corresponds to machines without unripe cherry separators.

Water Recycling and Disposal of Waste Products

Wastewater is drained and separated from coffee or pulp by means of screw conveyors and screens. Solid separation is performed by filters like the one shown in Fig. 2.56A. Once this water is separated from coffee and larger solid particles it can be recycled or disposed of.

Recycling wastewater from mucilage removers may not be possible if it is too con-

Figure 2.55 Compact ecological wet mill with vertical drum pulper.

Figure 2.56 Compact ecological wet mill with screen pulper (unripe cherry separator) and vertical drum re-passer.

Figure 2.56A Solid particles separator.

centrated and there is not enough additional wastewater to dilute it.

Wastewater to be recycled must go through settling tanks where small solids and other impurities sink and form a deposit. Depending on the volume of water to be processed and the amount of impurities it contains, more than one tank may be necessary. The same pump that injects recycled water into the processing flow is used to wash settling tanks periodically. Depending on the degree of impurities it still contains, recycled water can be either injected into specific machines or mixed with incoming clean water.

Several studies have shown that recycled water does not affect coffee quality even after several days of reuse. However, there seems to be consensus among researchers and suppliers of equipment that recycling of the same water should not exceed two days.

Wastewater due to be returned to rivers, streams or lakes must be treated beforehand. If treatment is not possible this water must be retained in lagoons, seeped

into the ground or used for irrigation. The treatment and disposal of wastewater and pulp is covered in another chapter of this book.

Pulp may be used for composting (with or without lombrices), animal feed or fertilizer. The easiest way to dispose of pulp is to drain it and then stack it along the coffee fields where it is left to decompose in heaps prior to distribution to the plantation as fertilizer. Coffee pulp is rich in potassium.

2.3.3.3.4 Problems and Impact on Quality

The layout of the wet mill is critical to ensure processing efficiency and to save water. A terraced or flat design depends on the cost of equipment and civil works.

As compact mills do not require specific layouts and design work, one could assume that individual machines will eventually be replaced by combined sets. However, the compact sets do not always allow for re-passing and may be cumbersome to combine in mills that require large capacity or flexibility of process.

The compact mills do, indeed, eliminate design work and customization in cases where their capacity and features meet the farmer's need. They represent a practical "off-the-shelf" solution not only for small and mid-size growers but also for small central wet milling facilities to be used by groups of micro-growers. Nevertheless, large flows may entail the grouping of machines by function (e.g. a battery of pulpers), specific separation and transport equipment between machines, re-passing, water collection and re-circulation, etc. The quality of harvesting may also elude compact mills and require separation equipment that they do not incorporate.

It is important to remember that standard compact mills are designed to process clean cherries. Impurities, over-ripe cher-

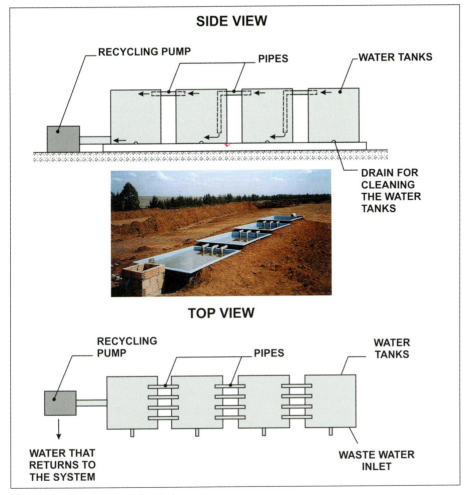

Figure 2.57 Wastewater recycling tanks.

Figure 2.58 Drained coffee pulp.

ries and partially dry cherries require separation by floating and perhaps sifting before the cherries can go through the compact mill. Unripe cherries require a compact mill with a screen pulper.

2.3.3.3.5 **Comments**

This section on the design of wet mills is complemented by the sections on ecological wet milling and waste treatment and disposal (Part III, Section 3).

Although the bulk of Robusta production is dry processed, washed Robustas have been offered by some countries (e.g. India, Democratic Republic of Congo and Indonesia) for many years. Recent tendencies point to a slow shift to Robusta washing in several areas of the world. Some experts claim that washed Robustas may be a good substitute for low-quality Arabicas, especially in espresso blends.

Some Robusta growers, particularly in Uganda, Vietnam and Brazil, are shifting to the pulped natural system.

2.4
Drying

2.4.1
Objective

The objective of drying is to lower the moisture content of parchment or cherry coffee to about 12% so as to preserve the beans safely in storage.

Coffee moisture remains at 12% when air moisture at storage is between 60 and 65%. In humid areas it is usual to dry coffee to 10 or 11% to increase storage time before coffee picks up moisture from the air. In other areas drying stops at 13 or 14% to account for moisture losses in hulling.

Moisture levels below 10 or 11% increase the breakage of beans at hulling, specially with Robustas that are more brittle. Moisture contents above 12% increase the risks of quality and weight loss in storage.

2.4.2
Principle

Coffee is dried by increasing the temperature of the bean to evaporate water.

In sun drying beans are heated by direct exposure to the sun and by radiation from a heated surface (in the case of drying grounds). Convection and wind move the saturated air away. In mechanical drying beans are heated by the passage of hot air which also carries the moisture away.

Temperatures must be monitored during natural and artificial drying. Coffee temperatures should not exceed 40 °C for parchment and 45 °C for cherries. It is often thought that overheating can only occur in mechanical dryers. Reality points to the opposite. Maximum tolerated temperatures may well be exceeded in sun drying if the beans are not revolved frequently or, in the case of fine Arabica beans, protected by plastic sheets, a tarpaulin or a roof during the hottest hours of the day.

Temperature control becomes more critical in the later stages of the drying process when moisture levels are low. In the early stages, there is a lot of water to be removed and relatively high air temperatures are not likely to induce the beans to overheat. At this stage, air flow to remove the surface moisture is more important than the temperature itself. As coffee moisture decreases water must move from the center to the periphery of the beans at a speed that depends on the outer air temperature and on the intrinsic physical characteristics of the bean itself. Since there is no more free moisture near or at the surface, the beans heat quickly. At this point the difference between air and coffee temperatures are considerably reduced and the drying temperature must be carefully monitored to avoid damaging the beans.

In sun drying, temperature is controlled by revolving the coffee frequently. Parchment should be covered before it becomes too hot. In machine drying, the temperature is controlled by fuel feeding, air flows, etc.

Figure 2.59 shows schematically how air and coffee temperature behave in a dryer.

Coffee moisture is high, 50–55 % or even more, at the beginning of the drying period so high air temperatures may be used without risk of overheating the coffee. In most types of dryers the coffee will not heat beyond 30 °C while the moisture is high even though the air temperature may reach 90 °C. It is widely accepted that the color and quality of Robusta coffees benefit from a high air temperature at the initial stages of drying.

Air temperature must be lowered as drying progresses. Control of the air temperature is critical towards the end of the drying process because the coffee starts to gain temperature rapidly. In order to preserve quality, temperatures for parchment coffee should not go above 40 °C, whereas for cherry coffee they should not exceed 45 °C.

The speed at which coffee picks up temperature and, consequently, the pace at which the air temperature must be lowered to avoid damage to the coffee will depend on the features of each dryer and the moisture levels of incoming coffee. In Fig. 2.59 it is assumed that coffee enters the drier "fully wet" (recently picked cherries or parchment that has just been washed) and leaves the dryer with a moisture content of 12 %.

The damage caused by overheating depends on coffee moisture content, the temperature and the time coffee is exposed to overheating. For example, a sudden high increase in the coffee temperature for a few minutes may be less harmful than 60 °C for several hours.

The damage to beans may take several forms: parchment cracking, bending, bleaching, undue widening of the center cut, toasted beans, etc., all of which lead to quality losses. Parchment cracking exposes the beans to attacks by microorganisms during storage. Bent (instead of normally flat) beans indicate severe overheat-

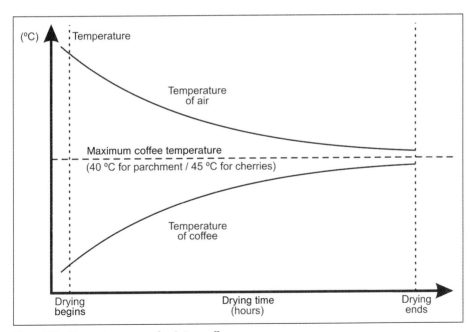

Figure 2.59 Temperature curves for drying coffee.

ing that some claim causes sour flavors and a "flat" cup. Overheating at the early stages of drying without a sufficient air flow to remove moisture may cause the "stewing" of coffee that can give it a cooked flavor. Beans that are overheated may have their surface "crystallized", i.e. the surface becomes impermeable so that moisture can no longer leave the interior of the bean. This impermeable surface will crack in storage and the bean will bleach as moisture gradually moves out.

The speed of drying cannot be increased above the speed at which moisture migrates from the center to the periphery of the bean. The speed at which water migrates in response to the temperature gradient artificially created is limited. This is a physical characteristic of the coffee beans and it cannot be changed. Attempts to accelerate drying beyond accepted limits will always leave moisture in the center of the bean. This moisture will move out later, which will have a negative effect on the aspect and the quality of the coffee.

Slow drying of coffee is a good solution to homogenize coffees received with uneven moisture contents. As a rule, coffee that is dried more slowly will have a more homogeneous moisture content and a more uniform color because longer drying grants the beans more time to exchange moisture and to equalize their moisture content.

In recent years some experts have associated metallic tastes in brewed coffee with unduly short drying times.

Evenly dried coffees fetch better prices because they have a more uniform color, a better aspect and roast better. It is highly desirable to have coffees with uniform moisture content at the end of the drying process.

The fact that the moisture meter indicates 12% has nothing to do with homoge-

neous drying because meters measure *average* moisture. There may be beans with moisture levels ranging from 9 to 15% although the average moisture may still read the desirable 12%!

Cherry drying often poses the problem of beans entering the drier with an uneven moisture content. This is specially the case for floaters whose moisture content can range from 25 to 40%. A wide disparity of this type can only be leveled out by excellent dryers and special drying techniques. It is more unfortunate, however, when coffee enters the dryer with a uniform moisture content, which is the case for washed coffees, but leaves the machine with an uneven moisture content. This can only be attributed to poor heat distribution in the drier.

It is of paramount importance that heat is transmitted uniformly to the drying beans. The majority of dryers revolve coffee as it dries because this process grants the possibility of heating each and every bean equally even if the distribution of hot air in the dryer is not perfect.

An efficient drying system must ensure a good temperature control, homogeneous air distribution, and frequent revolving of coffee. The art of sun-drying coffee and the techniques of mechanical drying converge to the same point. In the same way that years of experience have taught the farmer where to locate the drying ground to get the best exposure to sunlight and even ventilation, when to revolve the beans to ensure even drying, and when to shelter coffee to control moisture and temperature, manufacturers must equip the modern coffee dryer with features that enable even better control and independence from adverse weather conditions.

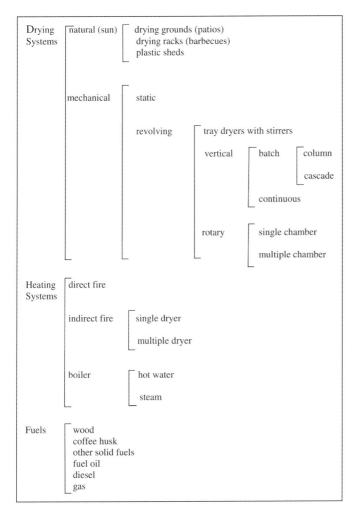

Figure 2.60
Drying systems.

Figure 2.61 Drying coffee cherries on a patio.
From: Nestlé.

2.4.3
Techniques and Equipment

Coffee drying may be performed in so many different ways that it is useful to summarize the possibilities before we discuss natural and artificial drying techniques as well as the machines commonly used today.

Figure 2.60 lists all the choices that have to be made to define the drying system required.

Not all the possible combinations in Fig. 2.60 are common. Some of the possibilities have historical interest only because they are no longer used. We will concentrate on the systems and machines that are widely used today.

2.4.3.1 Natural or Sun Drying

2.4.3.1.1 Drying Grounds (Patios)

Drying grounds are flat surfaces built with a small slope (0.5–1 %) to drain rainwater. Screens with holes smaller than the coffee beans are placed at the low points to drain rainwater. Drying grounds are usually made of concrete, tiles or asphalt. Compacted soil patios, still found in some areas, should not be used if coffee quality is a concern.

The length of the drying grounds should be laid out at an east-west orientation to maximize the reception of sunlight and

Figure 2.62 Drying parchment coffee on a patio. From: Nestlé.

Figure 2.63 Drying coffee on sliding trays.

to avoid shading by buildings and trees located alongside.

Parchment and cherries are handled essentially in the same way. Wet coffee is spread in thin layers and revolved 8–10 times per day with a flat rake. During the initial stages of drying coffee should not be covered at night or when it rains. Once coffee is partially dried, the thickness of the layer is increased and the coffee is piled and covered with plastic or canvas to shelter it from dew and rain. The raking frequency remains the same or is decreased depending on local weather. Raking may be mechanized when the drying grounds are large.

Figure 2.64 Drying racks with parchment coffee. From: Ch. Lambot.

In some areas sliding roofs or sliding trays are used to shelter coffee. The high cost of roofs restricts this solution to small patios and areas with high labor costs.

Parchment requires more careful handling than cherry to avoid cracking and physical damage to the beans. Raking must be more gentle. In tropical areas parchment is often covered during the hottest hours of the day to avoid cracking caused by overheating.

Depending on climatic conditions, sun drying of coffee in patios takes from 7 to 15 days for parchment and from 12 to 21 days for cherries.

Figure 2.65 Parchment on racks covered during the hottest hours of the day. From: J. De Smet.

2.4.3.1.2 Drying Racks

Drying racks are wire or plastic mesh trays assembled on table legs. They are mostly used to dry parchment coffee. Parchment is revolved by hand and the drying procedures are similar to those used in drying grounds.

Racks keep coffee cleaner and protect it from contamination from the ground. Aeration from above and below helps to accelerate the drying process. Racks are also more exposed to wind; this helps remove saturated air and shorten drying times. Racks require more labor than patios and are difficult to mechanize.

When drying parchment coffee on racks it is advisable to cover it during the hottest hours of the day to protect from excess heat that may damage the beans.

In areas subject to an adverse combination of high moisture above the surface of the ground and lack of wind, coffee may ferment on the racks during the earlier stages of drying.

Depending on climatic conditions, parchment takes 5–10 days to dry on raised racks.

2.4.3.1.3 Plastic Sheds

Plastic sheds are light wooden or metallic structures with a plastic roof and walls. The flat floor is made of concrete or tiles

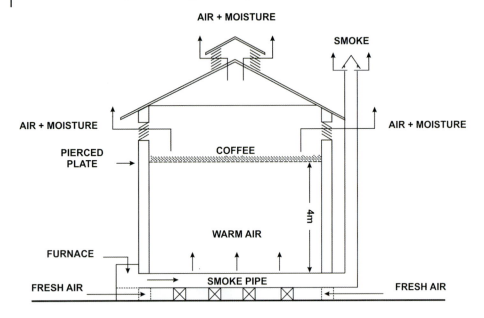

Figure 2.66 Simple static dryer (sketch and building).

like a patio. Transparent or translucid plastic creates a greenhouse effect that may raise the temperature 10–15 °C. Fans may be used to help remove the saturated air.

Drying procedures are similar to those in a patio except that coffee is permanently sheltered from dew and rain.

High costs restrict plastic sheds to small areas and, sometimes, to dry premium coffees.

2.4.3.2 Artificial or Mechanical Dryers

2.4.3.2.1 Static Dryers

Static dryers are the most primitive type of dryers. Easy and cheap to build anywhere, they only require a tray made of a perforated metal plate, a fan (optional) and a source of heat.

Vertical walls hold a round, rectangular or square horizontal tray and create an enclosed chamber underneath the tray and an open-top chamber above it. Coffee is loaded by hand onto the perforated screen tray to fill the upper chamber. A fan installed on the wall of the lower chamber blows hot air through the screen and the layer of coffee to be dried. Dry coffee is unloaded by hand.

If the layer of coffee is not stirred, the beans closer to the screen will dry faster than the beans at the top of the layer. In this case, the coffee layer must be very thin. The thicker the layer of coffee the less homogeneous the drying.

Thick layers of coffee need to be used for a static dryer to be economical. Unfortunately, this causes drying to be uneven. As their top is open, these dryers can be severely overloaded at peak times which causes drying to be even more heterogeneous.

In order to even out the drying process and produce a more uniform result, stir-rers may be used to revolve coffee or air may be blown in alternatively from below and from above. Colombia, which relies heavily on static dryers, has developed several types of two-tray static dryers that enable alternate air flows with a view to improving homogeneity.

Figure 2.67 shows a two-tray drier with a set of gates that grants an air flow at the bottom or at the top of each tray. Loading and unloading is manual. The same air flow passes through the two-trays to maximize moisture removal with hot air passing through the drier coffee first. In spite of all improvements, however, the main drawback of static dryers still prevails: coffee is not revolved so drying remains faster at the top and bottom of the layers and slower in the center. The final product will have an uneven moisture content.

Tray dryers take from 25 to 30 h to dry wet parchment coffee and more to dry coffee cherries.

In static dryers, homogeneity can only be improved by making the coffee layers thinner. Slowing the drying process does not make drying more homogeneous.

2.4.3.2.2 Dryers that Revolve Coffee

Dryers with Stirrers

These dryers are similar in construction to static dryers but they have stirrers added to revolve coffee.

Dryers with stirrers are usually round with rotating stirring structures or rectangular with stirring structures that travel lengthwise. Stirrers are either rake-like structures or paddles or variations of these two.

Although stirring may improve uniformity of drying, the structure of the stirrers is sometimes cumbersome and difficult to keep in constant operation, especially

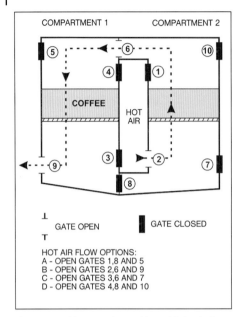

COMPARTMENT 1   COMPARTMENT 2

COFFEE

HOT AIR

GATE OPEN   GATE CLOSED

HOT AIR FLOW OPTIONS:
A - OPEN GATES 1,8 AND 5
B - OPEN GATES 2,6 AND 9
C - OPEN GATES 3,6 AND 7
D - OPEN GATES 4,8 AND 10

Figure 2.67 Two-tray dryer with multiple air flow options.

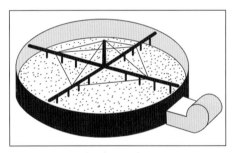

Figure 2.68 Circular dryer with stirrers.

when the coffee layer is thick. Worse still, they may even crack parchment in the later stages of the drying process. Dryers with stirrers are only common in a few countries. During the initial stages of drying there is less risk of damaging parchment which is damper and more flexible. For this reason, existing dryers with stirrers could still be used as pre-dryers only, to be combined with more modern machines.

Vertical Dryers

Vertical dryers are available in many models and sizes. Although they are used extensively to dry grain, one should not assume that every vertical grain drier is also a coffee drier.

Drying coffee is very different from drying grain. Grain is harvested with a moisture content under 20 % and this only needs to be lowered by a few percentage points while coffee moisture content must be lowered from 55 to 12 %. Grain can withstand much higher temperatures than coffee without being damaged.

The large volumes of water to be removed, the lower temperatures required, the specific features of parchment and cherry beans and the length of the drying process require the development of specific dryers for coffee, some of which are vertical and may resemble, or share features with, grain dryers. The closest resemblance to grain drying is when green (hulled) coffee only needs to be dried a few percentage points to reach 12 %. Even in this case a standard grain drier has to be modified to achieve the desired results.

There are many types of vertical coffee dryers, so we will group them into three families according to their system of operation.

Column (Tower) Batch Dryers

The column or tower vertical drier consists of a round, rectangular or square tower, several meters high. Inside this, there is a smaller perforated structure of similar shape which receives the hot air and distributes it through the coffee layer. The space between the two structures is completely filled with coffee which moves downwards slowly. After passing through the layer of coffee beans, the moisture-saturated air leaves the drier either through its open

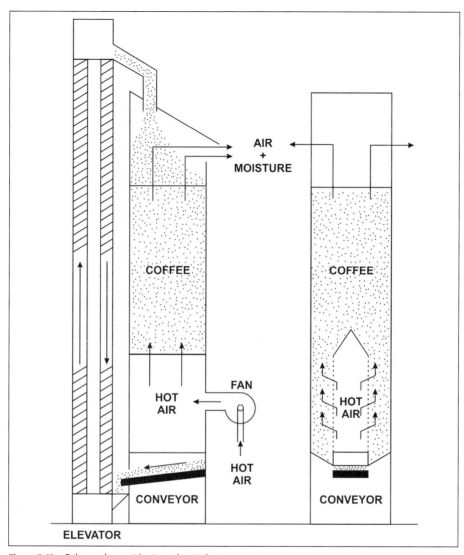

Figure 2.69 Column dryer with air outlet at the top.

top or through perforations in the walls of the outside tower.

Figure 2.69 shows a tower drier with an air outlet at the top. In this version hot air must go through a thick layer of coffee that moves downwards slowly. Saturated air leaves the dryer through its top.

Once the drier is full, coffee is discharged at an adjustable rate and fed by a bucket elevator back into the top of the drier. This circuit continues until the coffee reaches the desired moisture content.

The deep layer of the product causes the air to be fully saturated at the lower level of the tower. The upper level of the tower works as an equalizing bin where moisture is transferred from damper to drier beans.

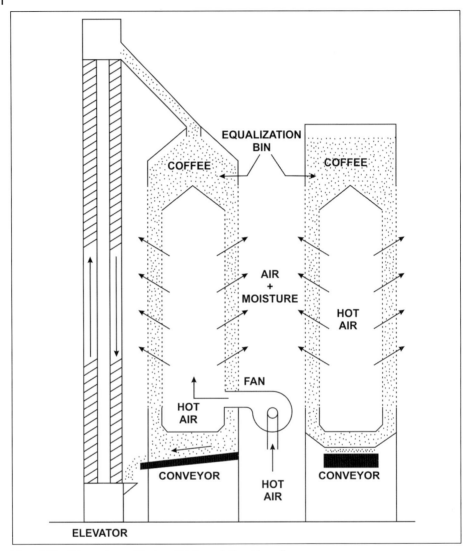

Figure 2.70 Column dryer with air outlet through the side walls.

Though this drier uses the full drying capacity of air and releases it fully saturated, drying is slow. Depending on the extent of pre-drying, drying may take 50–60 h.

Figure 2.70 shows a tower dryer with air outlets in the sidewalls. Hot air passes through the thin layer of coffee and leaves through the perforations on the side walls.

The coffee circuit is the same as in tower dryers with an outlet at the top.

There is an interaction between the thickness of the coffee layer, uniform drying, and drying efficiency. A thinner coffee layer allows more uniform drying but lowers drying efficiency because the drying capacity of the air is poorly utilized. As coffee layers become thicker, the drying capacity of air is used more efficiently but the beans dry at different rates depending on how close they are to the heat chamber. If the homogenization bin is designed correctly, it may be better to use thicker layers to improve drying efficiency.

Although the drier with perforated walls dries coffee faster, heat and fuel efficiency are lower in the column dryers with perforated walls than in those with an air outlet at the top.

Neither of these column dryers can receive freshly washed parchment or freshly harvested cherries. Both these products must be pre-dried before they are loaded into the drier. Although column dryers are suitable for cherry and parchment coffee they are not recommended for the latter because the repeated circulation of the beans can crack parchment and cause physical damage, especially at the end of the process when parchment is more brittle.

Cascade Dryers

These dryers remove moisture as large volumes of hot air pass through thin layers of coffee which move downward by gravity.

(1) *Cascade skin dryers.* The cascade skin dryers, mostly used in Central America, remove surface water from parchment immediately after washing. They can only be used for parchment that is processed continuously and then sent to other dryers for actual drying.

Cascade skin dryers have a steeply inclined metallic screen on which parchment slides to reach a second screen with slope of less than 1%. Hot air is blown in beneath the screens and through the parchment coffee to remove its surface moisture. Coffee takes between 3 and 5 min to travel through the machine. Layers should not be thicker than 10 cm.

Cascade skin dryers require big fans (20–40 hp) and considerable sources of heat to ensure an adequate supply of hot air. Skin drying in these machines reduces the overall drying time by 3–4 h.

(2) *Louver batch dryers.* This type of vertical drier has columns with louvers through which coffee flows by gravity. Hot air is blown into the columns from below and forced to pass through the cascading layers of coffee inside the columns. At the bottom of the columns there are gates to adjust the coffee flow and a mechanical conveyance system to convey the coffee to an elevator so that it repeats the circuit. An equalization bin may be placed at the top of the columns so that beans can exchange moisture before they enter the columns again for another circuit. Damp air can be recycled to increase drying efficiency.

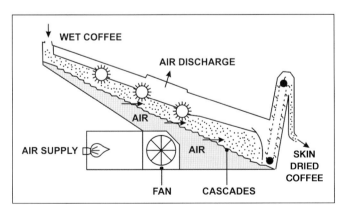

Figure 2.71
Cascade skin dryer.

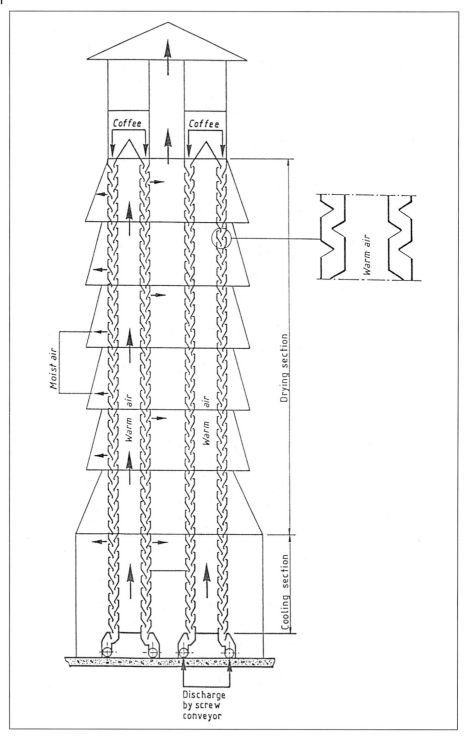

Figure 2.72 Louver dryer.

These dryers are mostly used for parchment coffee received directly after washing with a moisture content of about 55 %. Although they can be used for the full drying cycle, the tendency is to use these machines chiefly at the pre-drying stage and to use other types of dryers to complete the drying process.

A series of louver dryers may be installed next to each other, with elevators feeding coffee from one to the other, so that coffee received wet in the first drier leaves the last one dry. However, the need to dry coffee slowly may require many machines which entails undue investment costs and power requirements.

Louver dryers require big fans, sometimes 100 hp or larger, and heat sources of compatible capability. Although large motors may be acceptable in a coffee mill they can be a nuisance in a farm setting.

(3) *Continuous dryers.* These are grain dryers adapted to dry green (hulled) Robusta coffee.

In several Robusta producing countries, coffee arrives at the dry mills already hulled but with moisture contents that may be as high as 20 %. This green coffee must be dried to 12 % before it is stored or processed for exporting.

Many of the coffee dryers described in this section (e.g., column, louver and rotary) can be used to lower the moisture of green coffee from 20 to 12 %. However, due to the high volumes to be dried, the short drying period, the capacity of Robusta beans to withstand higher temperatures without as much quality damage as Arabicas, Robusta coffee millers generally prefer adapted continuous dryers. This tendency is further supported by the fact that quality is often a secondary concern in Robusta processing.

Continuous grain dryers adapted for green Robusta coffee will not be described because there are many types of continuous grain dryers and the modifications are specific to each type of drier. The changes required involve the following specific features of coffee drying:

- It is more difficult to remove moisture from coffee than from grain in the moisture range of 20–12 %.
- Robusta coffee is more brittle than most grain.
- Physical damage is not acceptable in coffee whereas, to a certain extent, it may be acceptable in grain.
- Silverskin detaches from coffee during drying.
- Heat exchangers may be required by some processors, etc.

Continuous grain dryers adapted for coffee have high power requirements and need a lot of heat especially if clean air, exempt from smoke, must be used. A 10–20 ton/h drier may require 50–100 hp.

Rotary dryers

The rotary drier, also called Guardiola after its Guatemalan inventor, is a horizontal rotating drum whose cylindrical walls are made of a perforated metal sheet. Heated air is blown into the hollow shaft of the drum from one or both ends. The shaft may be perforated and have different shapes or attachments like radial perforated pipes to distribute warm air evenly along the drum and to force it to pass through the tumbling coffee. Moist air leaves the drum through perforations in the outer walls.

The rotary drum may have one single chamber or it may be divided into radial or cylindrical (ring) compartments. Stirrers, paddles and wing plates are fixed to the shaft or to the inner side of the drum to help move the coffee. The drum revolves slowly at 2–4 rpm depending on the model.

Coffee is loaded by an elevator, conveyor or overhead loading silo through doors fitted on the outer walls of the drier. The drum must be fully filled to avoid heat losses as the volume of coffee shrinks through drying. Coffee is discharged through the loading doors.

Coffee and air temperatures are measured by thermometers located in the drum wall and hot air entrances. Coffee samplers can be installed on the perforated outer wall to enable sampling without stopping the drum.

Rotary dryers are the most widely used dryers today due to their flexibility, uniformity of drying, simple operation and durability. They may be used for parchment, cherry and green coffee. They receive parchment directly from washing or cherries directly from harvesting and dry them all the way to 12%. They may also be used at different stages of drying, e.g. to complete drying, to homogenize lots of green coffee received with high moisture at dry mills, etc. Rotary dryers are said to be the best machines to homogenize coffee batches with an uneven moisture content. Many users believe that rotary dryers are the best machines for the final stages of drying irrespective of the machines used before. The design of the air distribution system and the layout of stirrers and wings are important to ensure good homogenization.

Rotary dryers revolve coffee continuously and gently without the need of elevators and conveyors for recycling. Gentle handling is excellent for dry parchment which may crack or be peeled in machines that handle coffee more vigorously.

The rotary drier is one of the few machines that is able to receive wet cherries, specially those with thick pulp, i.e. a lot of mucilage. These cherries cannot be received by most dryers unless they are pre-dried. Special precautions must be taken in rotary dryers to make sure these wet cherries are not "steamed" during the initial stages of drying.

Rotary dryers are offered with capacities of 1.5–15 m^3 per batch which are equivalent to 1–10 tons of wet parchment and slightly less for cherry. Power requirements are 1–5 hp to drive the drum and 3–15 hp to drive the fan.

Rotary dryers have been in use for over 100 years. As a result, a number of different models and types are available today although not every machine offers all the features described above. Over the last 20 years many improvements have been made to rotary dryers with a view to increasing drying efficiency, improving heat distribution, and shortening drying time. Substantial performance differences can be noted between a traditional Guardiola and a modern rotary drier.

2.4.3.2.3 Combination of Dryers

As no drier has been found to be perfect under all conditions, a combination of dryers is used in some areas. On the one hand, the combination of different types of dryers allows heat efficiency to be maximized and the best features of each drier to be used in a given stage of the process. On the other hand, handling increases as coffee is moved from one machine to the other, and operation and control become more cumbersome. Additional handling may have an undesirable impact on quality when parchment is dried.

As a rule, a combination of dryers should only be attempted in large installations that require many dryers and a lot of handling whatever the type(s) of dryer used. To combine dryers requires skilled personnel for operation and good understanding of the process, otherwise the final results may not be worth the trouble.

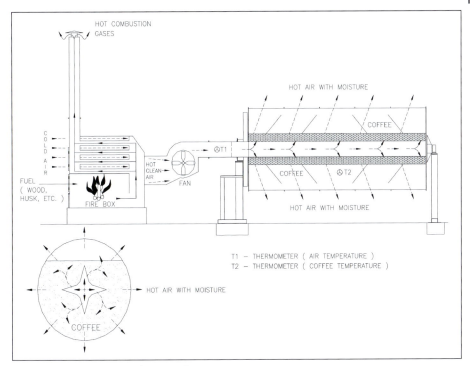

Figure 2.73   Cross-sections of a rotary dryer.

When only one or just a few dryers are required the best solution is not to combine machines but to choose the one type of drier that will perform better under the average conditions faced. The same is true when personnel is unskilled, control is difficult or the energy supply is not reliable.

Figure 2.74   Rotary dryer.

A combination of dryers is more common in Central America, especially in Costa Rica, where large drying facilities are used. One typical combination found in this area is cascade skin dryers followed by vertical louver dryers and then rotary dryers. Other mills go directly from cascade skin dryers to rotary dryers and some, that do not have skin dryers, go from louver dryers to rotary dryers.

In most other areas of the world the combination is restricted to a pre-dryer followed by a dryer, e.g. a static dryer with stirrers followed by a rotary or column dryer.

Most drying facilities around the world have one type of dryer only as the concept of a combination of dryers is more attractive in theory than in practice.

2.4.3.3 **Sun Drying versus Artificial Drying**

Ever since the first installation of coffee dryers, the debate between the merits of sun drying as opposed to artificial drying has been ongoing. Modern improvements in drying, combined with the lack of skilled laborers and escalating labor costs are currently tilting the balance in favor of mechanical drying. Nowadays coffee is increasingly mechanically dried as revealed by the sales figures of drier manufacturers.

Literature on coffee drying often refers to the merits of sun drying. Some sources mention the beneficial effect of sunlight on the quality of fine washed Arabicas. They claim that the desirable yellowish color of dry parchment is due to UV rays and that the visible frequencies cause the bright bluish-green color. Most of this literature mentions results of one single study made over 40 years ago! Very little research work has been given over to assessing the actual benefits of sun drying or the positive and negative aspects of either system.

From a purely pragmatic point of view, assuming that sun dried coffee has somewhat better quality than mechanically dried coffee, it should be noted that ideal sun-drying conditions can never be constant because they depend on climatic conditions. Hence the quality of sun-dried coffee may well be superior if the weather is nearly perfect, but this does not happen all the time. On the other hand, the quality of mechanically dried coffee is constant irrespective of weather conditions. As a result, the average quality of coffee over the whole season will be better for mechanically dried coffee. The obvious conclusion is that, over the full season, it is safer to have mechanical rather than sun drying in order to maximize quality!

The grower or the miller has to make decisions which take the full processing season into consideration. Therefore laboratory results that show that sun drying is better than artificial drying are only of interest where laboratory conditions can be duplicated in real scale over the full length of the processing season. This is definitely not the case.

This logical line of thought is causing an increasing number of farmers and processors to switch from sun to mechanical drying. Mechanical drying first entered areas where sun drying was difficult because of adverse weather, then it entered areas with high labor costs. Now it is becoming widespread as it offers a reliable means of producing quality coffee throughout the season and from year to year.

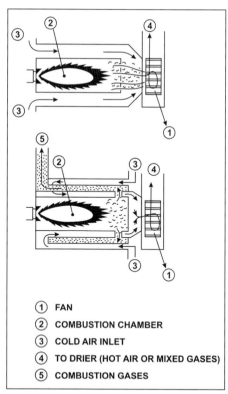

① FAN
② COMBUSTION CHAMBER
③ COLD AIR INLET
④ TO DRIER (HOT AIR OR MIXED GASES)
⑤ COMBUSTION GASES

Figure 2.75 Sketch of a direct fire heater (top) and heat exchanger (bottom). From: J. C. Vincent.

If we take the specialty coffee market as a reference for coffee quality we will soon realize that several high-priced specialty coffees are mechanically dried.

A word of warning – in the same way that the poor operation of a patio can destroy quality, a poorly operated coffee drier can be disastrous. Coffee drying must be carefully performed whether under the sun or in machines.

Finally, the same rule about weight loss discussed when we compared fermentation and mechanical removal of mucilage, applies here. When coffee takes 7–10 days to dry on a patio it will lose weight due to metabolism and respiration until it reaches a moisture content of 12 %. When coffee is mechanically dried in 40 h the process of weight loss is interrupted much earlier. This is yet another argument in favor of mechanical drying. The sooner coffee is safely dried to 12 %, the less the weight loss and the lower the risk of quality loss.

2.4.3.4 Heating Systems

Air blown into coffee dryers is heated by direct or indirect fire systems. Direct fire heating systems force hot combustion gases to pass through coffee. Indirect systems use a heat exchanger system to heat clean air that passes through coffee. Direct fire systems transmit smoke odors to coffee whereas indirect systems do not.

As coffee is consumed for its organoleptic (taste) characteristics, a product with smoke odors is not acceptable. However, most Robusta coffees are still dried with direct fire systems.

Direct fire systems use a fan to blow the combustion gases directly into the dryer. Depending on the fuel used, filtering systems are required to prevent sparks coming into contact with the coffee. When wood, coffee husks or other solid fuels are used, it is usual to have a cyclone between the heat source and the dryer to protect the coffee from fire. Unfortunately, fire accidents in direct-fired dryers still occur rather often.

The only direct fire heating system that does not transmit odors to coffee is the one that burns gas. Natural gas, propane, etc. produce odorless combustion gases that may be safely used to dry quality coffees.

Indirect fire systems like the one in Figs 2.75 (bottom) and 2.76 force the hot combustion gases to pass through heat exchangers which heat clean air. A heat exchanger is a cast iron or thick steel sheet structure composed of a fire box and a series of pipes or passageways. The high temperature combustion gases flow inside the pipes and radiated heat warms up clean air. The dryer fan sucks in the air from outside and forces it to pass around the tubes so that it is heated before it is blown into the dryer. Alternatively, combustion gases may be outside the pipes and clean air inside. There is no contact between combustion gases that are discharged through a chimney and the clean air that is heated and blown through the coffee.

Heat exchangers may supply hot air to one or several dryers.

2.4.3.4.1 Single-dryer Heat Exchanger

Single-dryer heat exchangers are much more common. They are usually supplied with the dryer and connected to it by a fan. It is usual in the market for dryers to be sold with their respective heat exchangers, fans, thermometers and other accessories.

2.4.3.4.2 Multiple-dryer Heat Exchanger

Heat exchangers that supply several dryers simultaneously are used in Central America. The idea behind multiple-dryer heat exchangers is to save on investment and op-

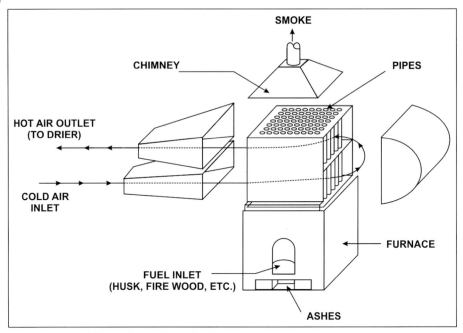

Figure 2.76 Typical heat exchanger for coffee drying.

eration costs. Although cost savings are, in fact, achieved by these large heat exchangers, it is often difficult to distribute hot air evenly among dryers. Pipes with large cross-sections and a complex system of dampers are required for efficient operation. When different sizes of dryers are used or some machines are not in operation, inefficient hot air distribution may counteract fuel savings. Coffee quality may also be impacted by uneven temperatures and their variations.

New computer controlled, modern technology, vertical multiple-dryer heat exchangers developed in the 1990s were designed to cope with the shortcomings above. Though not many units are currently being used, their performance seems very promising.

2.4.3.4.3 Boilers and Radiators
Savings in the production of heat can also be made by using boilers that generate hot

water or steam, carried to radiators by small-diameter pipes. Each drier has its own radiator and fan. The fan blows air through the radiator whose high temperature heats the air instantly before it enters the drier.

Central boilers with individual radiators enable excellent control of temperature in the dryers but as their initial costs are high, central boilers are only recommended for plants with significant energy needs, e.g. six to eight rotary dryers or several lines of skin and column dryers. In spite of their high investment costs, boilers with radiators have the lowest operational costs of all systems.

2.4.3.5 Fuels
The most common fuels used in coffee dryers are coffee husks, wood and oil. Gas, coal and other solid fuels are only used in a few areas. Most direct fire systems

and heat exchangers described above may burn any solid fuel with only minor adaptations. Compact direct fire oil and gas systems are offered by several makers. Oil and gas burners may be coupled with any heat exchanger or boiler. The use of wood as fuel to dry coffee is already forbidden in some countries. Environmental protection agencies are actively defending their remaining forests and restricting wood burning to the wood produced in reforestation projects. Increasing costs and unavailability are forcing the use of other fuels, especially coffee husks, instead of wood.

Figure 2.77 Feeding parchment into a burner.

2.4.3.5.1 Coffee Husks

Coffee husks are the ideal fuel for coffee dryers provided that the coffee industry is organized in a way that coffee is dried and hulled in the same area or close enough to justify the cost of transport.

The husks obtained from hulling 1 ton of parchment are enough to dry 0.7 tons of parchment coffee. Cherry husks obtained by hulling 1 ton of dry cherry are enough to dry 1.3 tons of cherry coffee.

Parchment husks can be used for little else than burning. Cherry husks can be used as a fertilizer. Consequently, the value of cherry husk as fuel should be analyzed in view of that competitive use.

Husks are usually blown by a fan into air heating systems. Otherwise they are fed by gravity, using a ladder device to facilitate aspiration by the draft of the chimney. The choice of husk feeding system depends on the type of huller used because the husks may be more or less pulverized.

2.4.3.5.2 Gas

The use of gas should be favored wherever it is economically feasible to burn it.

Gas burning does not require a heat exchanger because there are no offending odors in the combustion gases. The elimi-

nation of the heat exchanger makes drying much more heat efficient. The heat exchanging process has energy losses of 50–60 % depending on the design of the system. In other words, only 40–50 % of the energy supplied to a heat exchanger is effectively used to dry coffee. Gas burning without a heat exchanger allows most of the energy to be used for drying coffee.

2.4.4
Problems and Impact on Quality

Quality in coffee drying entails:

- Keeping coffee temperatures under 40 °C for parchment and 45 °C for cherry coffee.
- Distributing hot air evenly to each and every bean and revolving the beans.
- Air flows which are compatible with the quantity of moisture to be removed.
- Allowing enough time for moisture to migrate from the inside to the outside of the bean.

The perfect drier is the machine that fulfils all the above requirements. These requirements are easier to fulfill at the beginning of the drying process when it is easy to remove moisture. They become progressively more difficult as the moisture content of the beans drops.

A few comments regarding drying and quality.

As dryers are expensive machines, coffee drying is often a bottleneck. This gets worse at peak harvesting periods when there is a tendency to increase coffee temperatures to shorten drying time. Quality may be severely affected in the process.

There are some interesting alternatives to cope with drying bottlenecks. Parchment coffee may be dried to 20–25 % and then stored in aerated silos to free drying capacity for more humid lots. The chances are that damage to coffee will be distinctly lower at 20–25 % than at 50–55 %. Depending on local climate conditions, it may be better to have several lots at 20 % (low risk) than a few at 55 % (high risk) and a few at 12 % (no risk).

A bottleneck in cherry coffee drying may be solved by pulping coffee. A pulper is far less expensive than a dryer. After pulping, the volumes to be dried will be reduced by approximately 50 % and drying time will be reduced accordingly. Surprisingly few people realize that they may double the capacity of their dryers if they pulp their cherry coffee first!

The best way to homogenize coffee moisture is to slow down the drying process so that the beans can exchange moisture. Slow drying, however, cannot be achieved when the volume of coffee to be dried creates a bottleneck!

One way of avoiding this situation is to interrupt drying and to place hot coffee in a silo for a few hours while the dryers are used to dry different batches. The first batch can then be removed from the silo and returned to the dryers. This procedure can be repeated more than once provided coffee moisture is not very high. The period the partially dry coffee spends in the resting silos has the same effect as the equalizing bins in column dryers.

Insufficient drying and poor drying techniques favor the development of ochratoxin A, a mycotoxin in coffee. This toxin thrives in the warm, damp environments that may develop during the drying process, especially when sun dried coffee is not revolved often enough or when it is piled and covered too early in the process. Efficient coffee dryers reduce the risk of ochratoxin A which is more likely to occur in areas that are hot and/or damp during the harvesting season.

2.4.5
Comments

Though many techniques of drying and machines have been described above there are still many others that are chiefly of local interest or limited use only. Practically all the other approaches to drying fall within the major categories shown in Fig. 2.60.

This is true for large two-story "houses" used to dry Robusta in the Democratic Republic of Congo (Fig. 2.66) which are static dryers with coffee revolved by manual raking and the "patio quindiano" found in Colombia, which is a drier with stirrers. The masonry furnaces with metal pipes used in Brazil and several other countries are low-cost versions of the metallic heat exchangers.

Dry processing separates "floaters" and "sinkers" which have different moisture contents and should be dried separately. The drying techniques remain the same: floaters and sinkers should be dried in different areas of the drying ground, in different dryers or one after the other in the same dryer. Separations are also made in the semi-dry (pulped natural) process:

parchment, floaters and unripe cherries should be dried separately.

Research and development in coffee drying is intense at the institutional and private levels. In the same way that the 100-year-old Guardiola has been completely redeveloped over the last 25 years to become the rotary dryers of today, other machines will be modernized and totally new dryers will develop. Microwaves and fluidized beds are some of the technologies that may be incorporated into tomorrow's dryers.

In recent years, there has been a lot of talk about building solar dryers. Proposals include collecting solar energy to heat air that is then used to dry coffee. As a certain amount of energy is lost in the process of generating heat and conveying it to the coffee, it is argued that it would be more logical to use solar energy in the same simple direct way it has been used for centuries, in other words, sun drying!

A sun-drying patio with a plastic roof and walls helps to increase the effects of solar energy. Other possible improvements are to use solar energy to pre-heat the air as it is conducted into the dryers and/or to re-use non-saturated air as it comes out of the drier.

2.5
Cleaning

2.5.1
Objective

The cleaning or pre-cleaning of coffee aims to remove dust and light impurities as well as other impurities that are larger or smaller than the coffee beans.

Cleaning aims to provide the market with a product that is free from impurities and to protect the processing equipment from any damage that may be caused by them.

2.5.2
Principle

The three principles used in pre-cleaning are suction, sifting and vibration:

- Suction is used to remove dust and light impurities like small leaves, husks, etc.
- Sifting is used to remove impurities that are larger (heavy leaves, sticks, stones, etc.) or smaller (sand, stones, etc.) than coffee.
- Vibration is used to move coffee along the screens for sifting.

2.5.3
Techniques and Equipment

Most pre-cleaners (Fig. 2.78) consist of at least two inclined screens, one with holes that are larger than coffee beans and one with holes that are smaller than coffee beans. These two screens are assembled either vertically, one above the other, or horizontally, one after the other. They are used for sifting the impurities out of the coffee. When the screens are one above the other, the screen with large holes is placed on the top. If they are assembled one after the other, the screen with small holes is placed first.

Light impurities are sucked up or blown away as coffee crosses an upward moving air current on entering the machine. The screen with large holes retains large impurities and lets coffee pass through. The screen with small holes holds the beans back and lets the small impurities pass through. Clean coffee is discharged at the front of the machine, whereas impurities are discharged either at the front or at the side of the machine. As it leaves the machine, the coffee may go through a second upward moving air current for the further removal of light impurities.

As pre-cleaners are used for dry cherries, dry parchment and green coffee, the screens

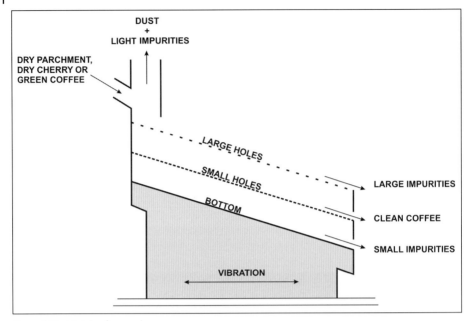

Figure 2.78 Sketch of a pre-cleaner.

should be inter-changeable. Some pre-cleaners have a third screen for the separation of undesirable coffee beans, i.e. cherry beans mixed with parchment or cherry and parchment mixed with green coffee.

Air with light impurities must be conveyed to a cyclone or to a silo with filters.

The capacity of pre-cleaners range from a few tons to 20 or even 30 tons/h. Power requirements range from 0.3 to 1.0 hp per ton of coffee/h.

2.5.4
Problems and Impact on Quality

Many coffee mills do not have pre-cleaners. This is particularly true for those that receive parchment coffee because it is frequently argued that parchment has gone through a pulper so it cannot contain impurities.

This may seem correct in theory but, in practice, impurities are known to find their way into parchment during drying and handling. Another aspect of impurities in parchment is caused by "unfair trading" practices whereby the volume and the weight of coffee is sometimes increased with "additions".

In the absence of a pre-cleaner, impurities can be removed at a later processing stage. This means, however, that dirty coffee is present in the hullers and other machines, which reduces their useful life by abrasion and damage caused by stones, pieces of metal and other undesirable elements. Dust and light impurities which have not been removed early in the process can cause problems throughout the mill because they are released into the air at other processing stages.

In coffee processing it should always be assumed that light and small impurities will be present during the entire process no matter how much cleaning is carried out. Parchment may crack at any time and release fragments, cherries may release husk fragments and adherent sand, green coffee releases silverskin, and there is always the chance of breakage of light and hollow beans, elephant beans, ears, etc. Cleaning should start at coffee reception and dust should be sucked out as often as necessary throughout the process.

2.5.5
Comments

Large capacity pre-cleaners are used for the reception of coffee even in small to medium size mills. To avoid congestion at peak coffee delivery periods, it is advisable to have large cleaning capacities at reception.

2.5.5.1 Magnets

In some areas ferrous materials (nails, screws, bolts, etc.) are found mixed with parchment, cherry and green coffee. They must be separated by magnets.

Magnets range from simple plate models that require periodic hand cleaning to sophisticated self-cleaning rotary models that are machines in themselves with their own hoppers, motors, etc.

Plate magnets may be installed in many places at the cleaning area, e.g. at the outlet of the pre-cleaner, at the inlet or outlet of the destoner, etc. Rotary magnets are usually installed between the pre-cleaner and the destoner or immediately after the destoner.

2.6
Destoning

2.6.1
Objective

Destoning is necessary to separate by density and remove stones that are the same size as coffee and cannot be removed during the pre-cleaning process.

2.6.2
Principle

Destoning is achieved by product flotation (density separation) and vibration.

- Product flotation (or stratification) with the help of an air current separates the stones from the coffee by density.
- Vibration moves the stones to the upper part of the deck after separation.

2.6.3
Techniques and Equipment

Destoners consist of an inclined vibrating metal screen with scales, bulbs or cleats. One or more fans installed below or above the screen create a strong upward air current that passes through the screen and the product to be separated.

Coffee is fed through the top of the sloped screen and distributed to cover it completely. As the beans move down the screen they meet with the upward air current that forces the "lighter" coffee to float, whereas the "heavier" stones remain in contact with the screen. The vibration of the screen is transmitted to the stones by the scales, bulbs or cleats. This moves the stones upwards for discharge behind the machine. Floating coffee (without stones) flows by gravity and is discharged at the front of the machine.

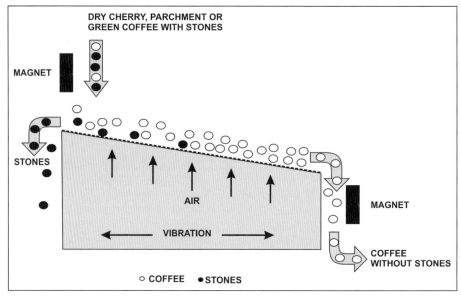

DRY CHERRY, PARCHMENT OR
GREEN COFFEE WITH STONES

MAGNET

STONES

AIR

MAGNET

VIBRATION

COFFEE
WITHOUT STONES

○ COFFEE   ● STONES

Figure 2.79 Sketch of a destoner (magnet at inlet or outlet).

Positive-pressure destoners (fan below the screen) have a dust suction column where the coffee enters, before it reaches the screen, or a dust hood that completely covers the screen. Negative-pressure destoners (fan above the screen) suck up dust as the coffee is floated.

Some types of destoners use different screen profiles for different products. As parchment, cherry and green coffee may interact differently with scales and bulbs, separation is improved by using different screen designs for each product.

The capacity of destoners range from 2 to 10 tons/h. The power required by destoners is 1–2 hp per ton of coffee/h for positive-pressure destoners and 3–4 hp per ton of coffee/h for negative-pressure destoners.

2.6.4
Problems and Impact on Quality

Many coffee mills do not have destoners. This is particularly true for those that re-

ceive parchment coffee because it is frequently argued that parchment has gone through a pulper so it cannot contain stones. However, stones are heavier and more efficient than other impurities for generating an unseemly increase in coffee weight. One way or another, stones always

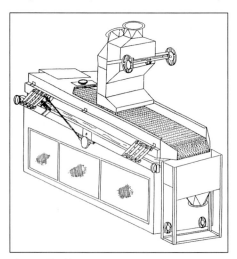

Figure 2.80 Destoner.

find their way into coffee, be it parchment, dry cherry or green coffee.

The myth which maintains that parchment and green coffee do not require pre-cleaning and destoning when they enter a mill is totally untrue. Stones, impurities and dust found in finished coffee affect quality and taint the image of the supplier. Most coffee classification (defect count) systems used in coffee trading penalize stones heavily.

Destoners are required for incoming cherry coffee as well as for parchment and green coffee. Stones, like pieces of metal, are very damaging to machinery.

2.6.5
Comments

In some areas, mud balls, small nuts or other seeds create very difficult separation challenges to destoners. The same size as coffee, they often have nearly the same density, especially the mud balls. When different adjustments of the destoner fail to separate these undesirable elements it is sometimes necessary to call in the manufacturer to investigate the problem and suggest specific solutions.

Mud balls are particularly damaging to quality because they break up during hulling. The resulting earth affects the color of green coffee and causes an earthy taste in brewed coffee.

2.7
Hulling and Polishing

2.7.1
Objective

The objective of hulling is to remove the husk layer(s) that cover the dry coffee beans. Coffee is hulled to remove the

parchment husks from dry parchment coffee or the cherry husks from dry cherry coffee.

Coffee is polished to remove the thin layer of silverskin (spermoderm) that is closely attached to give the beans an extra gloss.

Some hulling machines are also polishers and perform both tasks in one single operation.

2.7.2
Principle

Hulling is performed by:

- Rubbing the beans against each other and against the metal parts of the machines or
- Tearing the husk as the coffee is pushed against a sharp edge of a blade, of a screen hole or both.

Polishing is carried out by friction as the beans rub against each other and against the machine.

2.7.3
Techniques and Equipment

Table 2.1 summarizes the types of hullers and polishers available, and their features and applications.

Some machines only hull parchment, whereas others hull both parchment and cherries. Some hullers are considered hot hullers because they increase the temperature of the product while hulling. Other machines, cold hullers, hull coffee without increasing its temperature. The friction created by hot hullers polishes the beans. Cold hullers do not polish the coffee beans although they partially remove the silverskin.

Table 2.1 Types of hullers, their applications and their feature

Type of huller	Product	Temperature	Degree of polishing
Screw rotor and ribs (Smout/Squire)	parchment only	hot	polished
Double compartment (Apolo/Okrassa)	parchment only	hot	polished
Cylinder with cleats and knife (Engelberg/Africa)	parchment and cherry	hot	partially polished
Cross-beater (Brazilian)	parchment and cherry	cold	unpolished

2.7.3.1 Hullers/Polishers with Screw Rotors and Ribs

These machines are commonly known as Smout- or Squire-type hullers. They can be used as hullers and polishers or as polishers only. They only hull parchment coffee.

The machine consists of two metallic horizontal half-cylinders fastened together by hinges on one side and bolts with wing nuts on the other side so that one of the halves swings open. The upper half-cylinder is lined with ribs which are in a spiral at the feeding end and parallel to the axis at the unloading end. The lower half-cylinder can be similar to the upper one or else it can be equipped with a screen. It can also combine both systems, starting with the ribs and ending with the screen. The rotor is a helix that starts closely spiraled at the feeding end and finishes parallel to the axis at the opposite end.

Parchment coffee passes between rotor and ribs which curve in opposite directions. Pressure is created by the screw shape of the rotor and by the gate with counterweight that controls the coffee flow at the outlet. Parchment is removed by the friction of the beans against each other and against ribs and rotor. Parchment husks and silverskin are sucked out by a fan connected to the outlet column and to the screen (if there is one) in the lower half-cylinder.

The degree of hulling and/or polishing is controlled by the rate of feeding and the pressure exerted by the counterweight. Excessive pressure will unduly heat coffee and affect quality by causing fading, a "flat" cup, etc. Overheating may also cause a loss in weight.

This type of huller is usually made of cast iron. The contact parts may be made of phosphor bronze to enhance the color of the beans.

Screw type hullers have capacities ranging from 0.15 to 2.0 tons of green coffee/h when hulling and polishing. Capacities are larger when polishing only. The power required is 20–25 hp per ton of green coffee hulled.

Figure 2.81 Screw huller/polisher opened to show rotor and ribs.

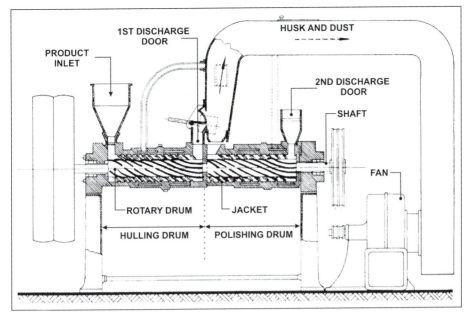

Figure 2.82 Okrassa huller/polisher. From: R. Wilbaux adapted by CIRAD-CP.

2.7.3.2 Double-compartment Hullers/Polishers

There are two types of double-compart-
ment hullers: the Okrassa and the Apolo.
Both machines have two compartments
assembled on the same shaft and powered
by the same motor.

The Okrassa huller has two cylinders
which resemble the screw huller with
ribs. Figure 2.82 shows how the Okrassa
huller operates. These machines are seldom used today.

The Apolo huller has two octagonal ro-
tors with replaceable cast iron plates
whose ribs are inclined at the feeding
ends and axial at the discharging ends.
Each rotor operates inside its own separate
eight-sided case that opens in two halves
which are linked by hinges and bolts with
nuts. The inner part of the case also has
replaceable cast iron plates with ribs.

Parchment coffee is hulled mainly in the
first compartment as a result of the pres-

Figure 2.83 Apolo-type huller.

sure created by the inclined ribs and the outlet gate with counterweight. A fan sucks out the husks as the coffee leaves the first compartment and falls into a bucket elevator that feeds the second compartment. The coffee then enters the second compartment where the remaining unhulled beans are re-passed (i.e. hulled), the silverskin is removed and the beans are polished. The same fan sucks out husks, silverskin and dust as the polished coffee leaves the second compartment.

An Apolo huller is able to operate at a wide range of capacities depending on the degree of silverskin removal and polishing desired. High degrees of silverskin removal and polishing require high pressures with resulting overheating, which can lead to loss in quality and weight. To avoid overheating, it is advisable to adjust the capacity to below maximum and to aim at moderate polishing only.

Apolo hullers are not designed to hull dry cherry coffee. Their usual capacities range from 2 to 7 tons/h of green coffee with a power requirement of 15 to 25 hp per ton of green coffee/h, depending on the degree of polishing.

2.7.3.3 Hullers/Polishers with Cylinders with Cleats and Knives

These machines, known as Engelberg or Africa hullers, can process both cherries and parchment coffee.

This huller, shown in Fig. 2.84, has a cylinder with short, spiraled (heavily inclined) cleats at the feeding end and long, lightly inclined cleats along the rest of its length. The rotor case has a smooth cylindrical upper half and a lower half equipped with either a perforated sheet screen (for parchment) or a thick wire mesh (for cherries). An adjustable steel knife passes through the length of the cylindrical case

and constricts the space between rotor and case. The knife not only increases pressure on the coffee but its edge also introduces a tearing effect that facilitates hulling, especially for dry cherry coffee.

Parchment or cherry coffee is hulled by friction and tearing as it moves between the rotor and the case and hits the blade. Pressure is created by the spiraled cleats, the counterweight controlled coffee outlet gate and the knife. The knife is adjustable both inwards and outwards to change pressure and to maximize the tearing action of its sharp edge without causing undue damage to the beans. Husks pass through the screen in the lower half of the case and are sucked out by a fan that also creates an upward air current in the hulled coffee outlet in order to complete dust and husk removal.

Cherry coffee is harder to hull than parchment and requires both active knife action and an adequate wire mesh to obtain good results. Machine capacity falls by 30–40 % when cherry coffee is hulled.

Hullers with cleats and knives lose efficiency quickly as coffee moisture increases above 12 %. Critical moisture content levels can even cause the machine to jam. High moisture cherry coffee may be severely overheated while hulling with this system.

Hullers of this type exist in small hand powered versions and large power-driven models. The processing capacity of the power-driven hullers ranges from 50 kg to 2 tons of green coffee/h and their power requirements are between 20 and 50 hp per ton. This range is wide because hulling cherry coffee requires more power than hulling parchment coffee.

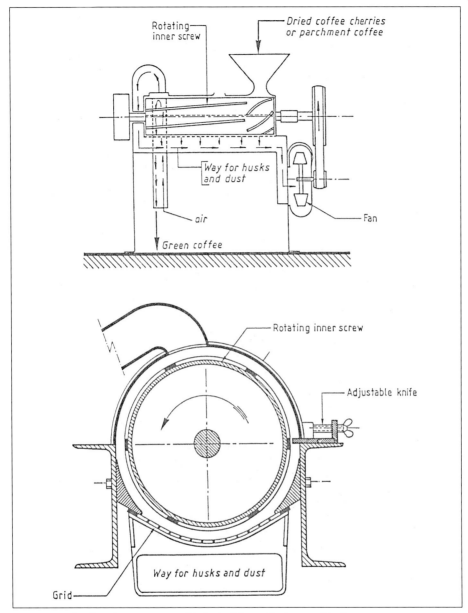

Figure 2.84 Sketch and cross-section of huller/polisher with cleats and knife.
From: R. Wilbaux adapted by CIRAD-CP.

2.7.3.4 Cross-beater Hullers

These machines, of Brazilian origin, can hull parchment or cherry coffee without overheating the beans.

The cross-beater huller, shown in Figs. 2.85 and 2.86, consists of a short horizontal cylinder with perforated sheet screen walls in which a rotor with knives (also called

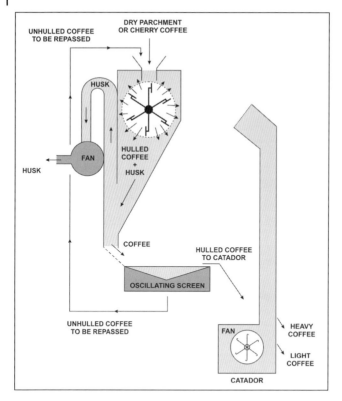

Figure 2.85
Sketch of a cross-beater huller.

Figure 2.86 Hulling screen and rotating knives.

beaters) rotates. Several cylinders may share the same drive shaft in one single machine.

Parchment or cherry coffee is chiefly hulled by the tearing action of the sharp edges of the screen holes which separates the husks from the beans as the coffee is pushed out of the cylinder by the rotating knives. As the pressure only has secondary action, the beans are not overheated. Husks are sucked out by a fan.

Most cross-beater hullers, like the one shown in Fig. 2.87, are equipped with an oscillating screen that separates unhulled beans. These are then taken back to the hulling cylinder for re-passing. The re-passing procedure avoids practically all physical damage to beans and ensures that coffee leaving the machine is 100 % hulled. Most cross-beater hullers also have a built-in catador that separates light beans and performs final cleaning before the coffee leaves the machine.

Different hulling screens may be used for cherry and parchment hullers. Bar screens (Fig. 2.88) were used in the past for parchment hulling and are still used by some manufacturers for either parch-

Figure 2.87 Cross-beater huller with oscillating screen, catador and elevators. From: Pinhalense.

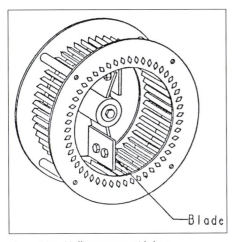

Figure 2.88 Hulling screen with bars. From: CIRAD-CP.

ment or cherry in machines designed for small growers who hull their own coffee.

Cross-beater hullers for small growers may not have the oscillating screen separation but only the hulling cylinder and a fan to remove the husks. Coffee to be re-passed

has to be separated by hand or, as is often the case, all the coffee is fed back into the machine until it is fully hulled. The practice of re-feeding green (hulled) coffee back into the machine causes physical damage that can easily be avoided by hand sorting and re-passing unhulled beans only.

As cross-beater machines are low-pressure cold hullers, there is no weight loss in the process. Quality is likely to be fully preserved as there is no risk of overheating. Physical damage to the beans is lower than in other types of hullers. As a result, the overall coffee yield is found to be higher with this hulling system.

Cross-beater hullers do not polish the coffee beans. Where polishing is required, it should be carried out by a low-pressure polisher installed after the cross-beater huller. The practice of hulling coffee in a cold huller and then polishing the hulled beanscauses much less heating than hulling and polishing coffee in the same machine.

Cross-beater hullers are offered in many sizes, from 200 kg/h for the versions without oscillating screens to 4–5 tons of green coffee/h for full models with oscillating screens and a catador. As Arabica cherry is harder to hull than Robusta cherry which, in turn, is harder to hull than parchment, the hulling capacities are the largest for parchment, intermediate for Robusta cherry and the smallest for Arabica cherry. Capacities may be 25–30 % larger for parchment when compared with Arabica cherry. As a result the power requirement for these hullers varies widely from 15 to 20 hp per ton of green coffee/h when oscillating screens and catadors are included. Where this is not the case, the power requirement is between 12 and 15 hp per ton of green coffee/h, which is the lowest of any hulling system.

2.7.3.5 Combined Cleaning, Hulling and Separation Units

Cross-beater hullers may be combined with pre-cleaners and destoners to make a single machine capable of performing five functions: pre-cleaning, destoning, hulling, re-passing and density separation. These machines are known as combined units or combined hulling units.

Combined units typically include a vibrating conveyor with two screens that pre-clean coffee, a positive-pressure destoner, a cross-beater huller with its oscillating screen, and a catador. The vibrating screen pre-cleaner may be installed above or below ground, under a silo or feed hopper. One or two bucket elevators are used to feed the destoner and the hulling cylinder.

Combined units with a cross-beater huller are offered with capacities of 0.3–2.0 tons of green coffee/h. Capacities and power requirements depend on the type of coffee being hulled. Usually 20 hp is required to perform all five functions for a capacity of 1 ton of green coffee/h.

Combined machines with other types of hullers as well as combined machines that have a cross-beater huller and a polisher have become available recently.

Combined units may be assembled on a truck deck or four-wheeled cart which can be driven from one farm to the other. The mobile machine (Fig. 2.89) offers several advantages: maximum use of its capacity throughout the season; coffee husks are left on the farm for use as fuel for dryers, fertilizer, etc.; the grower learns the actual characteristics of his green coffee before selling it, etc. Mobile machines are usually powered by diesel engines although they may be driven by electric motors that are connected to the power mains at each site.

Capacities and power requirements for mobile and stationary units are the same.

Figure 2.89 Mobile combined hulling unit.

2.7.4
Problems and Impact on Quality

Overheating, physical damage and the discharge of coffee with husks are the main sources of quality and weight losses in hulling and polishing.

The adverse quality effects of overheating are becoming better known year after year and, like drying, hulling/polishing is also perceived as a cause of overheating. Although overheating is easy to control during the drying process, this is not necessarily the case for hulling/polishing, especially if a hot friction huller is used.

Hot hulling should be closely monitored to avoid overheating which causes a loss in quality and weight. Silverskin removal and polishing should be kept at minimum levels of market acceptance so that lower pressures can be used. Excessive pressure at outlet gates, worn-out friction parts and screens, as well as blunt knives are common sources of overheating.

Physical damage to coffee beans is intimately associated with the hulling system and, most important, the moisture content of the coffee being hulled. When the moisture content falls below 12%, physical damage increases dramatically, in the shape of chipped beans, an increased percentage of "ears" and "triangles", broken beans, etc. It is not an overstatement to say that below a moisture content of 9–10 % most hulling systems actually become "coffee grinders", chipping and breaking an inordinate proportion of the beans.

Assuming that incoming coffee is at 12 % moisture, physical damage may still occur as a result of excessive pressure in hot hullers or reduced clearance between knives and screens in cold hullers. High pressure tends to break light beans and ears and even whole beans in extreme cases. Abnormally high pressure is known to flatten beans or to make grooves on their surfaces or both.

Undue pressure sometimes causes damages that are not perceptible because damaged beans may have been pulverized and aspirated by the husk removal system. This takes us to another possible cause of weight losses: poor adjustment of husk aspiration systems.

No huller can hull 100 % of the beans unless excessive pressure is used. The attempt to avoid re-passing by increasing pressure is another cause of physical damage to coffee in hot hullers. The right procedure is to separate unhulled beans and to re-pass them in the same or in a separate machine.

Hullers that can only hull parchment coffee often have problems with cherries that are mixed with parchment. Cherries are damaged or leave the machine unhulled and mixed with the hulled coffee. The correct procedure is to separate the cherries from parchment before hulling. Failing that, they should be separated after hulling and re-passed in a huller compatible with cherries. Unfortunately, many millers use the same parchment huller to re-pass the unhulled cherries at the end of the processing day. In order to hull cherries in the parchment huller, pressure must be unduly increased and the re-

sults are poor with a lot of damaged beans and severe overheating. This procedure should be avoided and replaced by the use of a specific cherry huller.

The choice of a hulling system involves complex decisions that go beyond the technical aspects of each machine and involve both the coffee marketing system and the organization of the coffee business in a given area or country. If, for example a country dries coffee to a moisture content of 14 % expecting a moisture drop to 12 % at hot hulling, cold hullers cannot be used unless drying practices are changed. If hulling is mostly carried out in central mills, there is no reason to use mobile hullers going from farm to farm.

In choosing a huller one must always keep in mind the system currently in use and assess the changes that would be introduced by an alternative system. Is polishing required? To what extent? When changing to the cold hulling of Arabica parchment one must always consider whether a polisher will be required. From a quality standpoint the shift may be justified even if a polisher is required. All the more so if silverskin removal can be avoided as unpolished coffee tends to weigh 0.5–0.7 % more than polished coffee! However, the color and/or overall appearance of the final product may be different when one shifts from hot hulling to cold hulling or to cold hulling followed by polishing. The color and/or overall appearance of the coffee beans may also change when different hot hulling-polishing systems are used.

Robusta coffee cannot be polished at a moisture level of 12 %. Attempts to remove silverskin of Robusta coffee at 12 % or below can only damage the beans. In order to polish hulled Robusta beans their moisture must be increased to well above 12 %. Otherwise, Robusta coffee must be hulled *and* polished at a higher moisture level and then dried back to 12%. If the beans are already at 12% special equipment is required to dampen the coffee, polish it and then dry it again to 12%. If not carried out correctly, this process invariably leads to very unevenly dried coffee and beans with wide color disparities within the same batch.

2.7.5
Comments

In the 1980s and early 1990s a lot of development work was put into machines that hulled parchment coffee by impact, i.e. by throwing the parchment beans against a hard or semi-flexible surface in order to crack the parchment skin and release the beans through a process that does not heat the beans. Most machines consisted of a rotor that threw the beans against the walls of a revolving cylinder. Several types of impact hullers were offered on the market and a few units were sold. However, this machine was not successful because of inefficient hulling and a high percentage of damaged beans.

Research on impact hullers was a reaction to the increasing evidence that overheating damaged coffee. Even though cold impact hullers are yet to be proven viable machines, they had the merit of calling attention to cold hulling and the cross-beater cold huller which was well known in Brazil, but little known elsewhere. Cross-beater hullers are now being used in most areas of the world.

The friction and pressure required to remove the silverskin depends on how firmly the silverskin adheres to the bean. The intensity of silverskin adhesion depends on the coffee variety, the altitude of the plantation, the type of processing (i.e. wet, semi-dry or dry), the coffee moisture content and other less important factors.

2.8
Size Grading

2.8.1
Objective

Coffee is graded by size to enable improved density and color separation, to allow more uniform roasting and, most importantly, to meet client requirements.

Whereas in the past the driving force behind size grading was the fact that roasters had to roast different sizes separately, advances in roasting technology are making it possible to ensure an even roast with coffee of different sizes. Today size grading is required to improve the efficiency of density and color sorting as the market becomes less tolerant of defects. Both density and color sorting are faster and more precise when the beans to be processed are of uniform size.

On the one hand, the market indirectly demands size grading to improve defect removal and quality; on the other hand, it directly demands specific bean sizes as the consumption of specialty coffees increases on a parallel with awareness of qualities, tastes, sizes, shapes, origins, etc.

In fact what is commonly known as coffee grading goes beyond size to include shape. The same process that separates coffee beans by size also separates them into "flat" and "round" beans.

Flat beans, with one flat and one concave surface are the norm. They are the two seeds found in coffee cherries. Round beans, or "peaberries" are found in coffee cherries that contain only one seed instead of two. Their percentage varies from 5% to a maximum of 20% in a few areas. On average, peaberries represent approximately 10% of the crop.

2.8.2
Principle

Green (hulled) coffee is separated by size by sifting the mixed beans in flat screens that vibrate or in round screens that rotate.

Flat beans are separated by screens with round holes. Peaberries are separated by screens with slotted holes.

Screen sizes are denominated in sixty-fourths of one inch (1/64) or in millimeters. When referring to slotted holes, reference is made to their width only. This identifies the corresponding screen size. When sixty-fourths of one inch is used, reference is made only to the number of 64ths, e.g. screen 17 means a screen with round holes with diameter 17/64, screen PB 11 means a screen with slotted holes with width 11/64.

Many countries have their own description systems for their coffees and their grades, like Kenya's AA, Colombia's Supremo, Guatemala's SHB, etc. A grade may be directly associated with a screen size and restricted to it irrespective of any other feature (e.g. Kenya's AA), or it may imply a size and other characteristics (e.g. Colombia's Supremo), or it may not indicate a size at all. However, most countries do consider sizes and shapes in their descriptions. In this case, a grade will be associated with one or several screen sizes. Some countries export ungraded coffees, meaning, among other things that these coffees are not separated by size prior to sale. However, "ungraded" coffee is sometimes said to be above or below a given screen size, meaning that it is a mixture of sizes greater or smaller than that screen. Ethiopia, for instance, exports "ungraded" coffee that is free from oversized and undersized beans, previously separated by size grading.

2.8.3
Techniques and Equipment

2.8.3.1 Flat-screen Graders
Flat-screen graders comprise a set of screens (perforated metal sheets) placed one on top of the other inside a box that vibrates to create the sifting action to separate the product by size.

The round-hole screens are organized according to the size of their holes, with the largest hole at the top and the smallest hole at the bottom of the grading box. If peaberries need to be separated, the slotted-hole screens are inserted between the round-hole screens according to criteria devised by manufacturers.

The sequence of the grading screens has to take into consideration specific country conditions like marketing systems as well as the type of coffee (Arabica or Robusta). Manufacturers often customize these machines as per regional and national requirements.

Coffee is fed onto the top screen and uniformly distributed along its width. The vibration of the box moves the product along the length of the screen causing beans to be retained by the screen whose holes are smaller than the bean size or to pass through to the screens below. The different sizes and shapes of coffee, each corresponding to a given screen, leave the machine through different spouts located either in the front or on the sides of the machine.

Most modern machines have interchangeable screens in order to meet the grading requirements of different consumer countries and markets. The number of screens in one single machine may vary from three to 18 depending on the supplier, the capacity, etc.

Beans of the same size as the holes can clog the screens if they remain stuck in the holes. This is prevented by cleaning brushes or by bouncing rubber balls held by a frame under each screen.

Although flat screen graders have traditionally had a downward flow, upward-flow machines have been launched recently. Coffee moves more slowly along the screen of upward-flow graders because the beans have to push each other in order to reach the outlets at the upper end of the screens. As a result their accuracy is superior, specially when working below full capacity.

Flat-screen graders are available with capacities that range from 0.3 to 16 tons/h. Their power requirements range from 0.3 to 1.5 hp per ton, depending on the number of screens used.

Figure 2.90 Downward-flow-flat-screen grader with two vibrating boxes and 16 grading screens.

2.8.3.2 **Rotary Graders**

Rotary graders have one or several round screens assembled on a spinning shaft. Coffee is fed into the screen and moved along it by cleats placed in helical patterns. Beans larger than the holes of the screen are held back, whereas beans smaller than the holes pass through and are collected below.

Figure 2.91A Upward-flow flat-screen grade.

The single-screen models require as many machines as there are sizes to be separated. The machines are stacked one above the other. The machine with the largest screen is placed on top. Coffee passing through the holes of the top machine is fed into the screen of the second machine and so on down the line.

Multiple-screen models have the screen with the smallest hole placed at the coffee feeding end and the largest screen placed at the discharging end. As coffee moves along the rotary drum, from small to large holes, the different sizes are collected underneath their respective screens, with the largest size remaining inside the screen, to be discharged at the end of the

Figure 2.91 Detail of rubber ball cleaning system.

drum. Three or four sizes are usually combined in a single drum.

Rotary screens are cleaned by stationary brushes or rollers. Rotary coffee graders are offered with capacities from 0.5 to 2.5 tons/h.

2.8.4
Problems and Impact on Quality

Though rotary graders were once the state of the art technology to be found in the industry, they have now mostly been replaced by flat-screen graders. Today, rotary graders are chiefly found in old mills and small processing facilities.

Flat-screen graders are more accurate, more flexible to operate and easier to integrate with other machines in the process flow. Flat-screen graders can handle a wide range of capacities with equal precision, and require little floor space and power.

Modern flat-screen graders use mostly rubber balls instead of brushes to clean the screens. As the specialty coffee market develops and roasters demand better processed coffees, millers have no other choice than to use size-grading systems with interchangeable screens, i.e. flat-screen graders.

Even if coffee is to be sold ungraded, separation in sizes may be required as an intermediate processing step in order to remove light and off-color defective beans. Both gravity separators and color sorters were conceived and are designed to process different sizes separately. Ungraded coffee causes a drop in their efficiency and capacity.

2.8.5
Comments

Few countries export ungraded coffee today, and even those that do still remove oversized and very small beans in an operation that requires graders. Size grading generates improved operational efficiency and additional revenues.

The development of the speciality market and the growth of espresso coffee have in-creased the demand for graded coffee. Not only bean size matters in these markets but also size separation enables a better removal of defects and an improvement in coffee quality.

2.9
Gravity Separation

2.9.1
Objective

The objective of gravity separation is to separate coffee beans according to density. As many coffee defects are associated with loss of density, gravity separation enables the elimination of defective beans.

There is a positive correlation between high bean density and high coffee quality, so gravity separation enables the sorting of the beans into quality classes.

2.9.2
Principle

The acting principle of gravity separation is product flotation induced by air flow. Gravity alone or gravity combined with vibration complete separation after the air flotation process.

2.9.3
Techniques and Equipment

Coffee flotation is carried out either in columns where coffee falls by gravity against an air current or in screen decks where coffee travels at roughly right angles to the air flow. These two somewhat different techniques gave birth to catadors and gravity tables, respectively.

All gravity separation machines work better with graded coffees. In other words, one size is gravity separated at a time instead of separating ungraded coffee. When all sizes are separated simultaneously smaller, sound beans are often lost with light defective beans.

2.9.3.1 Catadors

A catador is a wide vertical column of limited depth assembled on top of one or more fans (positive pressure) or connected to a suction system (negative pressure).

Coffee fed at a given height of the column meets with an upward air current. Light defective products float and are blown upwards till they reach an area where the column expands and decompression takes place. As a result of decompression the light products fall into a parallel column located next to the main column. The light product exits through an outlet at the base of the second column. Good sound coffee, which is heavier, falls down

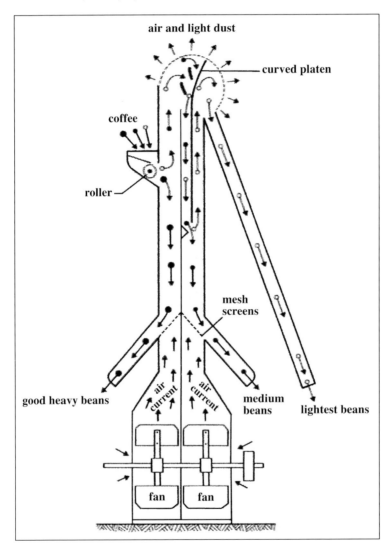

Figure 2.92
Catador.
From:
Wilbeaux.

the main column and exits through an outlet at its base.

Product separation is controlled by means of the strength of the air flow which is adjusted by means of dampers. Adjustment of the dampers is made by visual assessment of the product through an inspection window in the main column and at the product outlets.

The basic two-column catador described above yields two categories of products. However, it is usual to have a third column to re-pass the rejects, i.e. the "lights", from the main column. In this case there will be three outlets, one for the top quality sound product, one for the intermediate fraction and the third for rejects.

Separate catadors may be combined in a single structure to handle different sizes or types of coffee simultaneously. A multiple catador may be used in front of a size grader to separate all grades at the same time. In the 1950s and 1960s, Brazilian machine

Figure 2.93 Multiple catador.

manufacturers built many flat-screen size graders with huge built-in multiple catadors.

Some catadors are offered with a built-in air recycling system to reduce air contamination. Most types of catador blow air out of the machine through a fine screen. This air may carry dust and small particles that are not retained by the screen. Ideally this air should be collected and further filtered through cyclones or other devices.

Catadors range in capacity from a few hundred kilograms to a few tons per hour. Its is difficult to establish a more precise range because catadors are offered in many versions and sizes. Power consumption is between 3 and 5 hp per ton of green coffee.

2.9.3.2 **Gravity separators**

A gravity separator, also called a densimetric table, consists of a vibrating screen deck, inclined along its length and width, assembled below or above one or more powerful fans that create an upward air current which passes through the product to be separated.

Coffee is fed to the upper corner of the inclined screen deck and distributed to cover it completely. As coffee moves down the screen it meets with the upward air current that forces "lighter", mostly defective, beans to float while the "heavier" sound beans remain in contact with the screen. The vibration of the deck is transmitted to the heavier sound beans by the scales, bulbs or cleats of the metallic screen. As the screen vibrates at right angles with the motion of the product that flows along the length of the deck, the "heavy" sound beans are moved sideways against gravity to the upper outlets of the table. The "lighter" beans, which are floating and suffer no influence from the motion

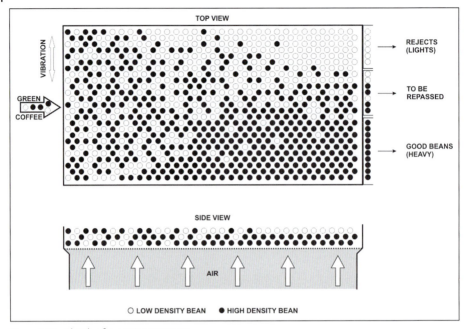

Figure 2.94 Sketch of a gravity separator.

of the deck, flow down by gravity to the lower outlets of the table. There is always an intermediate fraction composed of beans of all qualities that should be repassed.

There are up to six different ways of controlling product separation to obtain the required qualities: feeding rate, longitudinal slope, transversal slope, air pressure, air direction and intensity of vibration. Not all positive-pressure machines have all six controls. Negative-pressure gravity separators have fewer controls available which means they are more difficult to adjust.

Positive-pressure gravity separators (fans under the deck) have either a dust suction column where the coffee enters the machine or a dust suction hood that may cover the full deck. Negative-pressure gravity separators (fan above the deck) must have a hood that both causes flotation and sucks out dust.

Figure 2.95 Gravity separator with dust suction hood. From: Pinhalense.

Gravity separators are offered with capacities from 0.5 to 8.0 ton/h. It is important to note that the actual separating capacity of a gravity separator may be substantially smaller than its nominal capacity depending on the extent of re-passing required.

Figure 2.96 Negative pressure gravity separator. From: Buhler.

The power requirements of gravity separators vary from 1 to 5 hp per ton of green coffee. This wide range depends chiefly on the technology used for product separation as well as on dust control. As a rule, negative-pressure gravity separators tend to be more power intensive.

2.9.4
Problems and Impact on Quality

Gravity separators carry out the same functions as catadors, but their performance is superior and they consume less power. As a result, gravity separators are rapidly replacing catadors in export processing facilities. The last catador stronghold seems to be middle-sized Robusta processing facilities where catadors are still used in front of size graders. However, the conversion is well under way encouraged by the example of large mills that are already using gravity separators.

Some countries still insist on the combination of catadors and gravity separators, one re-passing the other, in either order. This is totally unnecessary if the gravity separators are efficient. The use of the two types of machines entails unnecessarily excessive power consumption that can and must be avoided.

Catadors are, however, likely to be retained in hulling facilities, where they are placed immediately after the hullers to remove husks and light materials that are easy to separate and best discarded at this stage. This is probably the only function where a catador outperforms a gravity separator.

Re-passing of the mixed center fraction of a gravity separator can take place on the same gravity separator or on another one, usually smaller, depending on capacity and quality requirements, type of the incoming product and the machinery layout.

2.9.5
Comments

Gravity separation is one of the areas of coffee processing technology that has evolved the most in recent decades, surpassed only by color sorting.

Advances in gravity separators (not catadors) were related to:

- Deck design, with a view to improving the efficiency of separation.
- Fan design, to reduce noise and power consumption.
- The orientation of the air flow, to increase efficiency and to reduce power consumption and
- Dust control to improve environmental conditions in the mills.

Modern gravity separators can do a fine job of separating defective beans with only small density differences from sound beans.

High-efficiency modern gravity separators should always precede color sorters in every mill. As color sorters are expensive, every effort should be made to remove as many defective beans as possible by mechanical means, i.e. by gravity separators. Many off-color beans also have a lower density than sound beans. This includes over-fermented beans, some black beans and even some stinker beans. It is more economical to remove these beans by means of high efficiency gravity separators.

2.10
Color Sorting

2.10.1
Objective

Color sorting is to remove defective coffee beans that have an undesirable color.

2.10.2
Principle

Coffee beans are color sorted by comparing the wavelengths associated with their color with wavelengths that correspond to acceptable colors. Off-color beans are rejected by a compressed-air ejection system.

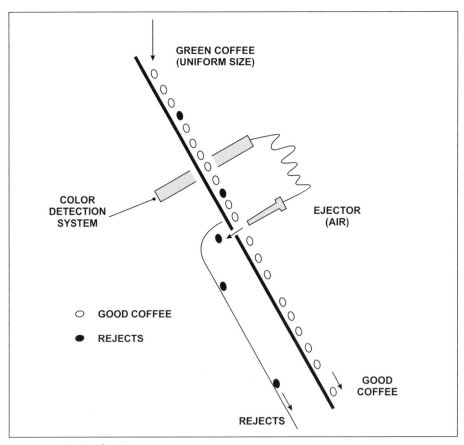

Figure 2.97 Sketch of a color sorter.

The typical defects that may be removed by color sorters are black, dark, pale, white, unripe, waxy and fermented beans. Beans that are chipped, insect-damaged or broken may be removed if their color differs from that of sound beans. Fresh stinker beans may be removed by UV sorters.

Color sorters gain capacity and accuracy when they process beans that have been previously graded by size and density.

2.10.3
Techniques and Equipment

Color sorters have benefited enormously from the developments in electronics, fiber optics and computers that took place in the 1980s, and more importantly in the 1990s. Earlier color sorters remained basically the same for many years. Modern color sorters are being redesigned every few years as new technologies become available. Unfortunately the move from valves to microprocessors has not meant lower prices for the machines as it did in the case of computers. Savings by mass production cannot be achieved because of the small number of machines made. However, machines have become more efficient and accurate.

As there has been so much product development in a relatively few years, added to the introduction of different technologies and the frequent announcements of new things to come, it is difficult to describe the color sorters currently on the market or even to group them as we have done for other machines.

It should be noted, however, that all color sorters basically have the following components:

- Mechanical feeders,
- Gravity or mechanical transport of beans,
- A color detection device,
- A compressed-air ejection system and
- Product outlets ("accept" and "reject").

Beans fed at a constant rate pass through the color-detection device that triggers an air jet. This removes beans that do not conform to the accepted color pattern from the main flow. Accepted and rejected beans leave the machine through different outlets.

With the risk of over-simplifying, we will group the machines currently available according to bean transport technology and sorting technology.

2.10.3.1 Bean Transport Technology
Most color sorters have a vibrating device to feed the machine at adjustable flow rates.

Beans pass through the color detection device after they slide down on an inclined surface or when they are launched by a conveyor belt. Gravity systems include channels, chutes and rollers. Mechanical systems rely on conveyor belts with different widths, speeds and surface designs.

Most earlier color-sorting systems used metallic channels that contained a string of single beans. Each channel had its own color-detection device and its own ejector. This arrangement enables beans to be viewed by the optical device from two or three angles.

In sorters with rollers two inclined stainless steel rollers spinning in opposite directions and almost touching each other receive a string of single beans that may be viewed from two or three angles. The argument in favor of twin rollers is that they force the beans to slide down with their convex side facing the rollers and the flat side facing outwards, which facilitates the viewing of the beans.

Chutes were introduced in the late 1980s and early 1990s. They have a flat inclined

Figure 2.98 Color sorter with channels. From: Elexso.

Figure 2.99 Color sorter with rollers. From: Xeltron.

Figure 2.100 Color sorter with chutes. From: Delta.

surface where beans slide down together in parallel paths that form many strings which are not restricted on their sides as they are in the metallic channels. The set of parallel strings is contained by the two sides of the chute. Beans are viewed as they pass through a battery of closely spaced color-detecting devices. The distance between color-detecting devices corresponds to the distance between the centers of the strings of beans that fall down the chute. Beans may be viewed from one or two angles by lines of color-detecting devices placed either above or below the beans or both.

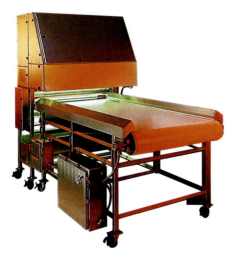

Figure 2.101 Color sorter with belt. From: Allen.

Conveyor belts have been used in color sorters for food products like vegetable and potato chips long before they were introduced in coffee machines. Coffee beans line up in strings along the top of the adjustable speed belt in the same way as they do on an inclined chute. At the end of the belt beans in all strings are simultaneously thrown out and pass through the viewing chamber. Beans may be viewed from one or two angles by one or two batteries of color-detecting devices whose spacing corresponds to the spacing of the strings of beans formed along the belt.

A single channel or roller can handle between 100 and 150 kg/h of green coffee, with larger capacities possible when beans have been previously graded by size and gravity. A single machine may have several chutes or rollers.

Chutes may, in theory, have widely different capacities depending on their width but, in actual fact, most manufacturers tend to limit the width of their chutes to an average capacity of 1 ton/h per individual chute. A single machine may have several chutes.

Belt sorters typically have higher capacities: several tons of green beans per hour. The higher costs of this technology have to be absorbed by processing higher volumes in order to make the investment worthwhile.

2.10.3.2 Color-sorting Technology

Monochromatic sorters measure the intensity of reflected light and can only separate beans that are darker or lighter than average: white, pale, dark or black beans. Monochromatic machines are chiefly used for the sorting of Robusta coffee whose color defects are easier to identify.

Bichromatic sorters measure two wavelengths which makes the machine sensitive to more subtle color differences. They are able to separate pale green, unripe, waxy, yellowish and brownish beans as well as the beans separated by monochromatic sorters (white, pale, dark or black). Bichromatic machines are used more often for Arabica coffees which have a wide range of color defects with smaller color differences.

Several manufacturers currently claim to have sorting technology, including laser, that is even more finely tuned than the bichromatic sorters. These will not be described because they are still specific to different manufacturers.

Ultroviolet (UV) sorters are used mostly to separate moldy or over-fermented beans (e.g. stinkers) which can be identified by their fluorescence provided that sorting is done a short time after harvesting. UV sorting becomes inaccurate after storage because the fluorescence of the beans decreases drastically.

In the past, machines compared the color of the beans with the color of carefully painted patterns. Modern sorters have microprocessors that enable the machine to use predefined digital patterns stored in its memory to define, accept or reject patterns from coffee samples. The microprocessor, which can be linked to a central computer network, can provide statistical data about the sorting operation, e.g. the percentage of rejects.

2.10.4
Problems and Impact on Quality

Color sorters must always be preceded by size graders because they are more precise and faster when they process beans of similar sizes.

Color sorters must also be preceded by efficient gravity separators because, as it has been mentioned before, several color defects are also associated with density losses, e.g. some black beans, some fermented beans, etc. As color sorters are the most expensive coffee-processing machines, they should only sort defects that cannot be sorted by other less expensive machines.

In some countries color sorters are being used to separate stones, twigs and other impurities. This task is easily performed by destoners and pre-cleaners at a much lower cost. On the other hand, it is common to see color sorters operating at unusually low capacity because the coffee had not previously been size graded.

The capacities of all color sorters vary widely depending on the percentage of off-color beans to be sorted. Higher percentages require the speed (capacity) of sorting to be lowered in order to keep the process accurate.

The layout of color sorters may provide for the re-passing of "rejects" or "accepts". Color sorters to be combined in the layout may use the same technology, with different adjustments, or may be different types of machines.

2.10.5
Comments

As color sorters are much more complex than the mechanical machines used elsewhere in coffee processing, it is most important to know what sort of local technical support and services are available before investing in this costly equipment. The servicing necessary for a high-quality mechanical machine can generally be performed by the mill maintenance crew itself, whereas the maintenance of color sorters requires knowledge of electronics, optics, compressed air, computers, etc., which is seldom, if ever, available in-house.

In some countries color defects are still sorted by hand, with workers sitting on the floor, at individual tables or on either side of long tables equipped with mobile belts. When sorting is done by hand, workers may also sort defects other than off-color beans, e.g. broken and hollow beans. There is no technical reason to color sort by hand instead of using machines. Hand sorting is still performed in a few countries in order to maintain employment or because labor is still inexpensive.

Figure 2.102 Sorting defects by hand.

2.11
Blending and Bulking

2.11.1
Objective

Blending means mixing coffees of *different* qualities (sizes, cup qualities, origins, etc.) to obtain a homogeneous coffee lot with the specifications and volume required by the client.

Bulking means mixing coffees of *the same* quality produced by different growers to obtain a homogeneous coffee lot with the volume required by the client.

2.11.2
Principle

The basic principle of both blending and bulking is the same. The component coffees must be brought together or added to each other in given proportions and mixed thoroughly to obtain a final lot that is homogeneous.

2.11.3
Techniques and Equipment

Techniques and equipment for blending or bulking green coffee range from the simple bulking floors common in Kenya to sophisticated blending devices linked to computer systems found in large coffee-roasting facilities. Blending and bulking systems are described below under feeding systems and mixing systems.

2.11.3.1 Feeding Systems

The most commonly used feeding systems are those that involve debagging coffee from stacks that have been pre-arranged using the correct proportions of component coffees. Component coffees should not be debagged separately by type. Instead, they should be debagged simultaneously according to the pre-established proportions and mixed as they are debagged.

When component coffees are in silos the ideal system is to use vibrating or rotary devices at the outlet of the silos so that the right proportion of each component coffee is fed in. This solution is seldom used in green coffee processing facilities in producing countries. It is replaced by opening the valves of the silos gradually, in a trial-and-error process aimed at ensuring that each silo of component coffees is emptied at the same time. Imperfections in proportions are corrected by the mixing system.

A few facilities use automatic scales or load cells in sophisticated arrangements that ensure high precision feeding.

2.11.3.2 Mixing Systems

The simplest way to mix component coffees is with the help of shovels, on a flat smooth floor usually made of wood. The homogeneity of the final product is questionable unless the lots are small.

In the countries of origin, most of the mixing is carried out with elevators or conveyors that feed special silos whose design and inside features help to homogenize the final product. The coffee is frequently fed back into the silos to go through the circuit as often as necessary until homogeneity is considered acceptable.

Revolving drums, similar to concrete mixers, may be used for small lots with good results. Static silos with revolving stirrers of several types have been used with only relative success. The weight of coffee to be revolved makes this solution impractical for middle-sized to large coffee lots like one container load.

2.11.4
Problems and Impact on Quality

The precision required to blend or bulk coffee for export is not high. Homogeneity of the final mix is much more important than the exact percentage of the components. As a result, most blending systems consist of reasonably precise feeding systems coupled with efficient mixing systems composed of transport equipment, mostly elevators, and special silos.

When the products to be mixed are not fed into the machines at roughly the percentages prescribed, it is very difficult for large blending or bulking systems to produce an even mix. As a result, it is highly recommended that feeding and mixing take place simultaneously. Mixing should start when feeding starts and continue well after feeding ends.

Feeding in the products one after the other followed by mixing is only feasible in systems like the revolving drum or else in sophisticated mixers designed for high-volume processing and whose costs are not compatible with the activity of green coffee processing.

Heterogeneous blends or lots that are not properly bulked are frequently a cause of dispute about the quality of the coffee batches exported. When mixing is not homogeneous, the shipping sample will hardly be representative of the entire batch delivered because quality may vary depending on where samples are taken from the lot.

2.11.5
Comments

Whether processing requires blending or bulking, either or both depends on the marketing system adopted by the country as well as on the size of its coffee growers.

Blending or bulking are used to prepare coffee lots in a way that suits both local and overseas end users, i.e. coffee roasters and the soluble industry. Blending is implemented when the client requires coffee mixtures or coffee blends. Bulking is implemented when the client requires lots of coffee of a given size.

The specialty coffee market does not require either blending or bulking when it purchases estate coffees. However, bulking may be necessary when specialty coffees come from a group of small growers in a single region whose coffee quality is the same.

Large roasters and industries are typical clients who benefit from receiving coffee lots that are blended and/or bulked at origin.

2.12
Weighing, Bagging and Bulk Container Loading

2.12.1
Objective

The objective of weighing coffee and placing it in bags or containers is to prepare it for delivery to the end user according to the weight requirements contracted.

2.12.2
Principle

Typically coffee is weighed on scales which have a central post and compare the weight of the coffee against a given standard weight.

The bulk loading of containers can be carried out by elevators, belt conveyors or coffee blowers.

2.12.3
Techniques and Equipment

2.12.3.1   Weighing and Bagging

Platform Scales for Bags and Big-bags
The simplest way to weigh and bag coffee is to place the bag under the spout of a silo or elevator while on top of a platform scale, set to the desired weight of coffee and bag. Coffee falls into the bag after the valve in the spout is opened. The flow is interrupted by closing the valve once the approximate required weight is reached. The final adjustment of weight is performed manually by scooping coffee in or out of the bag. Bags with the exact weight are closed by sewing them up either manually or with a portable sewing machine.

The same operation can be performed with big-bags, whose capacity range from 600 to 1,200 kg each, using larger platform scales. Big-bags transported by fork-lifts are a strong new tendency that saves labor and increases operating efficiency. Big-bags can be used for the storage of raw materials (coffee coming into the mill to be processed), intermediate products and rejects, as well as finished products to be placed into containers or loaded onto trucks.

Automatic Scales
Automatic scales have a weighing bucket that receives coffee from an overhead silo. The weighing bucket is connected by a lever to the standard counterweight with the desired weight, e.g. 60 or 69 kg. The free flow of coffee into the bucket is restricted but not interrupted after coffee reaches a pre-set weight (usually a couple of hundred grams below the desired weight). The smaller, restricted flow then continues for a few seconds, estimated time needed to reach the required weight.

In some automatic scales the bag is fixed by bracings on the bucket itself. In other models the bag is held by an independent structure placed on the floor. If

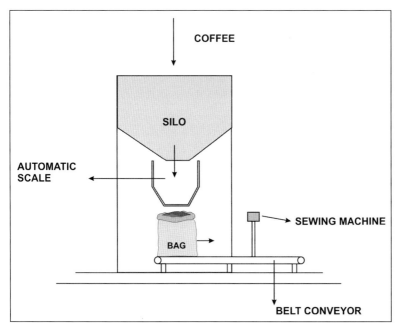

Figure 2.103 Sketch of an automatic scale with sewing machine assembled above a conveyor belt.

the weight of the bag is to be included, the counterweight has to be adjusted accordingly.

Bags with the exact weight are closed by sewing them up either with a portable machine or with a stationary sewing machine assembled above a conveyor belt that forces the top of the bag to slide through the sewing head of the machine.

In order to obtain maximum capacity from automatic scales, they should be coupled to conveyor belts with sewing machines. Manual removal of bags from the scale and their closing with a portable sewing machine create a bottleneck that severely reduces weighing and bagging capacity.

Weighing and bagging stations with conveyor belts and stationary sewing machines pass the top of the bags automatically through the sewing head. This type of arrangement can handle an average of 200–250 bags/h and even more with experienced operators.

Automatic scales are also used to fill big-bags placed directly under them or, more often, with the help of elevators and a special structure that holds the big-bags in place and facilitates the operation of the fork-lifts that transport them.

2.12.3.2 Bulk Container Loading

When coffee started to be was shipped in containers, coffee bags were placed directly in the containers. Later, at the request of roasters, coffee was bulk loaded into containers lined with protective sheets. As a result, more coffee could be shipped in the container, saving on both freight and insurance charges.

Initially, only containers with openings on the roof were bulk loaded and coffee fed in by elevators or conveyors. As the demand for bulk shipments increased some shipping terminals started to tilt standard containers so that their door would face upward to enable feeding by elevators or conveyors. This was mostly done in precarious installations with substantial risks if high-capacity cranes were not available to tilt the containers, to hold them in position while loading and to return the loaded containers to the horizontal position.

Today containers are loaded with bulk coffee chiefly by blowers or high-speed conveyor belts.

The weight of coffee being loaded into a container can be controlled by debagging coffee previously bagged with standard weights of coffee, by using automatic scales in a continuous flow operation (releasing a standard weight of coffee every time it unloads) or by using silos assembled on load cells that determine the weight of coffee in the silo.

Coffee Blowers

The basic component of a coffee blower is a fan that generates a controlled pressure air current that throws coffee into the container without damaging the beans. Elevators, conveyors or silos feed coffee directly in the front of the fan where a chute with adjustable height directs air and coffee into the container.

The most complete coffee blowers include a suction system to remove the dust created by the process of bulk loading and a system of rails or a rotary tower (or both) to move the machine easily from one container to the other.

Efficient coffee blowers load a 20 ft container in 20–25 min. The power required ranges from 20 to 30 hp.

Belt Loaders

Belt loaders are high-speed conveyor belts with a special surface to throw coffee into the container. Where coffee blowers re-

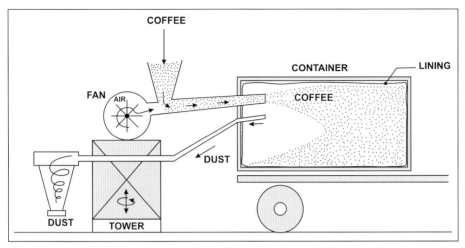

Figure 2.104 Sketch of a blower to load containers in bulk with dust suction system, cyclone and rotary tower.

main mostly outside the container, belt loaders have a structure that allows them to slide into the container to start loading and to back out gradually as loading progresses.

Coffee is fed onto the belt loader by an elevator, another conveyor belt or directly from a silo.

Belt loaders require less power than coffee blowers although they take longer to fill the containers. Some belt loaders cannot fill a container to the same extent as blowers do. As belts are cumbersome to move from one container to the other, trucks have to maneuver to place the containers in a loading position in the front of the belt.

2.12.4
Problems and Impact on Quality

It is amazing that even nowadays many coffee mills around the world still use platform scales instead of automatic weighing and bagging stations at the end of their processing lines. Frequently, processing lines that are otherwise efficient grind to a slow-down because an inefficient bagging operation creates a bottleneck. Added to which, problems related to accuracy of weight and theft control are frequent when coffee is manually scooped in a trial-and-error process.

Some low-cost coffee blowers take a long time to load – 45–60 min or even more. Their option for low-powered fans requires high-speed blowing to obtain desirable coffee flows. High-speed blowing risks damaging the beans. Many of these machines are made in the coffee mills themselves or in small shops according to customers' designs.

In most countries, coffee blowers appear to be more popular than belt loaders. They are easier to move around and can be assembled in a container for transporting from one mill to the other.

A stationary container loader may remain idle for substantial periods while large trucks maneuver to position the container. A mobile loading system (like blowers on rails or on rotary towers) can start loading a new container a few minutes after it finishes the previous one by moving itself to the new container rather than waiting for the truck with the container to maneuver into position.

Many of the early fears related to bean breakage during the bulk-loading process or quality deterioration during the transit period on the sea have now been forgotten. Shipments of coffee in bulk have become a standard practice.

2.12.5
Comments

The trend towards "just-in-time" operations imposed by the industry, with medium-size roasters also installing bulk-receiving facilities, will cause bulk shipments to develop even faster in years to come. However, shipment in bags will remain the preferred option for the specialty coffee sector.

2.13
The Impacts of Processing on Coffee Quality and the Environment

All processing steps described in the previous section are listed in Table 2.2, which also indicates, on its right-hand side, the coffee features that may be affected by processing.

Only three processing steps – pulping, mucilage removal and drying – can in fact change coffee quality. Any other steps may introduce defects, specially if not properly executed, or remove defects either caused by processing itself or developed in the field as the cherries were formed, grew and ripened. Table 2.2 indicates schematically the processing steps that change quality, create defects or remove them.

The way coffee is dried (with or without pulp and/or mucilage), the extent of mucilage removal if any, and if pulping is performed or not will determine if the same

coffee cherries will become natural, semi-washed (pulped natural) or washed coffee, with the respective quality features (body, acidity, aroma, etc.) typical of each system.

Harvesting affects quality as a result of the degree of ripeness of the cherries picked. Processing steps like pulping, hulling and polishing can cause physical damage to coffee beans if not properly performed. Underdrying, overdrying and specially drying that is too fast can affect both physical features (e.g.: aspect, color and uniformity) and the cup quality of coffee as well as the length of time it may remain stored. Adverse storage conditions (e.g.: too dry, too wet, too much light or poor ventilation) can affect physical aspects, cup features and the uniformity of the coffee batch.

Cleaning of fresh cherries, dry parchment, dry cherries, or green coffee removes unwanted materials. Density and color separation remove defective coffee beans,

Table 2.2 Impacts of processing on coffee quality and the environment

Processing steps	Coffee features
▲ Harvesting	
● • Separation	Color
▲ • Pulping	Size
▲ • Demucilaging	Uniformity
▲ • Drying	Physical integrity
▲ • Storage	Absence of defects
● • Precleaning	Cup taste
● • Destoning	– Aroma
▲ • Hulling	– Acidity
▲ • Polishing	– Body
▲ • Size grading	– Other features
● • Density sep.	
● • Color sorting	
• Bulking/blending	
• Weighing/dispatch	

Change quality ▲ Cause defects
●Remove defects ■ Water contamination
■ Air contamination • Noise

e.g., berry-borer affected beans, over-fermented and stinker beans, black beans, etc.

The processing operations performed at a coffee mill affect the working conditions within the mill itself as well as the environment beyond the mill, from neighbors and neighboring properties, that may be more directly affected, to the widespread and far reaching adverse effects that may be caused by water and air contamination. Table 2.2 indicates the processing steps that contaminate water or air and the ones, almost all of them, that produce noise.

2.14
The Organization of Coffee-processing Activities

Each stage of the coffee-processing chain may take place on the farm where coffee is grown or in outside milling centers located in coffee gathering points. Processing of dry parchment or dry cherry coffee may even take place at export harbors. The modern tendency is for processing to take place as much as possible in the production areas themselves.

The place where each processing step occurs varies widely from country to country, depending on coffee marketing systems, regulations controlling the coffee industry and the size of growers. There are cases in Costa Rica, Guatemala and El Salvador where growers deliver their recently harvested cherries to outside mills that perform all processing, from cherry to export, in one single facility outside the farm. Conversely, there are cases in Brazil and Africa where large estates perform all processing stages on the farm itself and deliver export quality coffee to the harbor.

It is nearly impossible to group countries according to the way their coffee processing chain is organized. These organizational systems have evolved over time and taken very specific forms in each country. It is, however, useful to summaries organizational systems used in some producing countries.

In Brazil, most coffee is processed on the farm up to the hulling stage. Only small farmers deliver dry cherry or dry parchment to be stored and hulled in coffee growers cooperatives. Most of the coffee sold by growers and traded within the country is green (hulled) coffee. The major-

Figure 2.105 Reception of cherries on a large coffee estate in Africa.

ity of export processing facilities are now located inland, close to the production areas. Two or three decades ago they were predominantly located in export harbors. A large percentage of coffee is exported in bulk in containers loaded directly in the coffee growing areas.

Irrespective of their size, Colombian coffee growers process their coffee to the dry parchment stage. Even the smallest growers have their manual coffee pulpers or use facilities in cooperatives to deliver dry parchment to the mills located in the towns of coffee-producing regions. All coffee is traded in parchment form, to be hulled and further processed by the central mills immediately before export.

Kenyan farmers also deliver parchment to central mills but most small growers wash and dry their coffee in cooperative societies that have wet-milling lines and drying facilities. Coffee growers are allowed to hull dry cherry coffee and trade the resulting green coffee. Dry parchment can only be milled in government authorized central facilities.

Most countries in Central America have a dual system: small growers deliver fresh cherry to outside millers and large farmers may process their cherries to the dry-parchment stage before delivery to the mill. Some large farmers may even process their coffee further. In some Central American countries, only initial drying is performed on the farm, to be completed in mills located several hours away. The transport of partially wet parchment causes substantial damage to coffee quality because it exposes the beans to undesirable fermentation.

Robusta coffee growers in Indonesia and Uganda deliver green (hulled) coffee to the mills. The moisture of green coffee usually exceeds 12 % and requires additional dry-

Figure 2.106 Transporting coffee to a processing center in Colombia.

Figure 2.107 Green Robusta coffee waiting for processing in Asia.

ing in the mill itself. It is not unusual for coffee to arrive with a moisture content of 15–20 %. This means that the mills need to have large capacity dryers.

Small Robusta growers in Vietnam deliver their dry cherries to micro or small traders who have hullers. These small traders deliver green coffee to collectors/millers and exporters. Large farms have their own hulling facilities and some perform further processing, specially the state-owned farms.

In recent years the coffee business has undergone a process of liberalization in countries like Uganda, India and the Ivory Coast. The organization of the coffee processing chain is reacting accordingly, and changing to adapt to new regulations and marketing systems. As regulation becomes less strict there is a tendency for several organizational arrangements to coexist. This is already the case in countries that liberalized their coffee business earlier.

Changes in import markets also impact the way the processing chain is organized.

The growth of the specialty coffee sector has generated new coffee-processing requirements. The export of estate coffees, the demand for single-origin coffee and the need to have traceability from farm to cup require new processing arrangements

like small capacity milling lines that are flexible enough to process small volumes to very exacting standards while ensuring the identity of the product throughout the process. The processing of organic coffees has requirements similar to specialty coffee and additional ones, like the certification of the processing facilities.

The "just-in-time" practices favored by large roasters have forced their suppliers in producing countries to streamline their mills in order to gain efficiency. Large coffee processors are progressively using big-bags transported by fork-lifts and highly mechanized bulk operations that eliminate conventional bags and reduce labor requirements.

Sustainability concerns in import markets affect processing in several ways. Environmental protection requires improved wet milling that consumes less water and treats the little water that is consumed. Dust and noise control have become basic requirements of modern export processing facilities. The welfare of workers on farms and mills has also become a matter of interest to importers. Working conditions and even the wages paid to workers affect the design of coffee processing facilities and the choice of equipment to be used.

The principles and techniques of coffee processing discussed earlier remain valid whatever the size of the processing facility and the organization of the processing chain. However, equipment may be different depending on the capacity of the processing facility and the amount of manual labor needed to operate it.

Unfortunately, when coffee-processing equipment is reduced in size, the cost of labor required to produce it remains more or less the same. As a result, the price of smaller machines does not fall proportionally to their reduction in processing capacity. This means that small growers pay proportionally more for their coffee processing machinery than large growers.

The relatively high price of processing machinery and the availability of manual labor cause small growers to use processing solutions that are strikingly different from those used by large-scale coffee producing estates. A small coffee grower may pre-select fresh cherries by hand, use a small hand pulper and sun-dry parchment, whereas a neighboring large farm may have a fully mechanized wet mill and coffee dryers. A small grower may have a simple huller that requires the hand sorting of unhulled beans, whereas a large farmer may have a huller with built-in recycling and grading devices.

Frequently, the only way for a small grower to have access to modern wet-milling and drying technology is to share processing facilities with other growers. The so-called central wet mills have a strong economic and coffee quality appeal. It is much less expensive to build and operate a few common facilities than a large number of small separate units, which seldom share the same processing procedures and therefore often produce different qualities that are difficulte to blend into uniform batches. The problem is not so critical at the hulling and sorting stage where it is possible to lower costs for small processors by designing facilities that use basically the same machines, but rely on manual feeding and, to a certain degree, on manual sorting.

2.15
The Shape of Things to Come

There is a generalized misconception that coffee-processing technology has not evolved much in recent decades but, in fact, the way many coffee-processing steps are carried out has undergone substantial changes. New processing systems have been introduced in some countries and there are more changes to come.

A snapshot taken today of a modern wet mill is strikingly different from a snapshot taken only 10 or 15 years ago. Differences may not be as striking in a dry mill, but changes in specific processes, e.g. color sorting and loading of containers with coffee in bulk, may be just as striking.

The shape of things to come in the short term is not very difficult to imagine. The task is harder if one tries to look 10 years ahead.

2.15.1
Harvesting

Lower coffee prices and worldwide competition will accelerate the search for ways to lower harvesting costs. Countries that have traditionally relied on selective picking are being forced to consider some form of stripping as a means to lower production costs. Advances in mechanical harvesting will continue, with greater emphasis on selectivity and product recovery. More and more countries will start using hand-held harvesters sooner than expected.

Fierce competition in the sector of large self-propelled harvesters will tend to bring prices down. A larger market for these machines is likely to have the same effect. However, the impact of these changes will be limited to countries like Brazil, USA (Hawaii) and Australia, as well as a few African countries that are currently considering this technology.

Larger impacts on the way coffee is harvested today will come from the further development of less costly light tractor-driven coffee harvesters. They are better adapted to the topography and planting systems found in a greater number of coffee growing countries. Today these machines are chiefly used in Brazil.

The largest impact on coffee harvesting will, however, come from the improvement of hand-held harvesters that can negotiate any slope and spacing between trees. As hand-held harvesters are developed to pick coffee more selectively, they will appeal to a much wider market and have a great impact on the costs of producing coffee in most countries around the world.

Another area of research and development is equipment to collect coffee off the ground. Most tractor-driven machines and all hand-held harvesters drop coffee cherries to the ground. These cherries must be efficiently collected. Machines to collect coffee off the ground, currently used only in Brazil, are being improved substantially from year to year. Their present level of efficiency indicates that a lot of additional development work may well be possible soon.

2.15.2
Wet Milling

Traditional wet mills occupy a lot of space and consume large volumes of water. Products are usually conveyed with the help of water in open channels which are also used to grade coffee at the end of the line. Fermentation tanks occupy a large area, consume a lot of water and require much labor.

Modern wet mills are much more compact, using a single, combined ecological machine in some cases. The transport of products in channels has been replaced by dry feeding. Fermentation tanks are not necessarily required. Water consumption has fallen dramatically, and the little water that is used is recycled and properly treated and/or disposed of.

As most of these changes have occurred rather recently, more research is required to ascertain their actual impacts on coffee quality. Although a lot of research work has gone into demonstrating that fermentation and mechanical removal of mucilage lead to the same quality results, some experts are still skeptical. The focus of research should be changed from trying to prove that the two systems produce identical results to determining the actual impacts on coffee quality of natural fermentation (wet and dry), the mechanical removal of mucilage and soaking. With these results in hand the processor will be able to decide which process or combination of processes suits his needs and meets the demands of his market.

Not many new developments are to be expected in wet milling in the immediate future as a result of the major changes that have already taken place over the last two decades. However, there is still room for improvement in various areas. Immature cherry separation would be one such area.

The current technology to separate immature cherries is about to be improved to require much smaller volumes of water. But there is a demand for an immature cherry separator that does not require the pulping of the cherries. Color-sorting technology has already been successfully used for this purpose, but the cost-benefit of the machines has not yet been proven acceptable.

Pulping technology itself has not changed in half a century, except for the introduction of the screen pulper (immature cherry separator). The vertical drum pulper is only a rearrangement of the same principles found in horizontal drum pulpers. Pulping without water in a way that does not damage the beans may be another area for further research.

Increased interest in the washing of Robusta coffee may call for new developments in the pulping of these cherries that have less mucilage and are therefore harder to pulp, especially in the absence of water. Is it necessary to remove the mucilage of Robusta parchment or is the semi-washed (pulped natural) system all that this type of coffee requires? Although the Indian experience shows that there is value to be added by fermenting and soaking Robusta coffee, new evidence from Uganda and other countries indicates that the semi-washed system may suffice in their cases. Is this the case for lower grown Robusta elsewhere?

2.15.3
Drying

Modern rotary dryers have positioned themselves as the best machines to dry quality parchment or cherry coffee. The move from vertical to rotary driers has greatly accelerated in the last decade, although rotary driers have not changed much since they were redeveloped in Brazil in the late 1970s and early 1980s. At the time, the aim was to gain efficiency and to be able to dry cherry as well as parchment coffee.

Temperature control in the driers is an area to be improved with the help of microprocessors. Substantial savings in fuel consumption and major gains in product quality are possible if temperature profiles are programmed to suit coffee-drying requirements and the undue oscillations in the heat supply are removed.

With few exceptions, the heat exchangers is another area where development work and even totally new products are required. The heat exchangers on the market today incorporate limited technology. They result chiefly from empirical development rather than scientific research. The combustion of coffee husks must be improved to take advantage of their full calorific value and to control air pollution.

Development work to increase the energy efficiency of dryers is likely to include areas like the solar heating of clean air sucked in by heat exchangers and the collection and reuse of air that leaves a drier without reaching saturation.

2.15.4
Dry Milling

The last decades of the 20th century witnessed many changes in dry milling. Cold cross-beater hullers, whose use was once restricted to Brazil, were introduced into many countries. Rotary graders were replaced by flat-screen graders. Catadors gave way to gravity separators. The capacity of color sorters increased greatly and the machines became much more compact, more efficient and less sensitive to dust. The control of dust in dry mills gained increased attention in more advanced markets.

Incremental changes in processing steps will carry on into the near future as individual machines are improved and redeveloped. The challenge to make small capacity, low cost machines for small processors will soon be taken up.

One important future development may be in the area of small capacity compact units that combine several processing functions in one single machine. Due to savings in elevators and the single structure and drive system, their costs are likely to fall when compared to individual pieces of equipment. New small capacity combined units are being developed to include hulling, polishing, re-passing and grading in one single structure. These new compact units will function as a complete mill for either a small processor or a large farmer.

The single most important change in dry milling in the last decade was the introduction of equipment to load coffee in bulk into containers. This change went beyond the simple introduction of a new machine. It required changes in equipment layout, the addition of new silos and more sophisticated weighing equipment as well as the introduction of better controls.

The trend towards bulk handling of coffee in all stages of dry milling will gain strength in the future. This will create a clear divider, that did not exist before, between small and large mills. Until very recently a small mill used basically the same design approach as a large one. Machines were the same type and the main difference was that the layout of small mills required a higher labor force to handle rejects, to carry bags, etc. These same functions were carried out mechanically in large mills.

The trend towards bulk handling may mean that large and small mills will be substantially different in the future to cater for different markets.

Large mills will be fully integrated, to the extent of receiving coffee in bulk from suppliers and using big-bags and bulk instead of conventional bag storage for raw materials. Intermediate storage will be in silos or big bags. All rejects will be mechanically removed from the machines that produce them and conveyed to silos. Coffee will leave the mill chiefly in containers loaded in bulk.

Small mills will continue to use bags in the reception and in the storage of raw materials in order to preserve the identity and origin of individual coffees. Processing will emphasize quality instead of volume. Size grading and sorting will be very flexible to accommodate specific demands from individual clients. Coffee will leave the mill in bags, often identified by the grower's label.

Today the same coffee mills process large lots to be supplied to middle-size and large roasters and small lots whose final destination is specialty coffee roasters and stores. There is already a trend towards smaller dry processing lines within large mills. Large millers are installing small parallel lines dedicated solely to specialty coffees, large growers are installing dry mills on their estates and small growers' associations are installing small dry mills. The evolution of this trend will create a dual system. On one end of the range, there will be mostly integrated and mechanized, eventually computer controlled, large capacity dry mills oriented to commercial coffees. At the other end, there will be small capacity, highly flexible dry mills that retain the individuality of single origin coffees as required by the specialty coffee market. Between the two, there will be non-specialized middle-size dry mills, very much like the current ones, that cater to both the commercial and specialty markets.

The trend towards traceability in the food industry will become strong in coffee in the future. New dry mills for specialty coffees already take this trend into consideration in their design. However, traceability poses major challenges to the large dry mills that rely heavily on coffee bulking and blending. New technology will have to be developed to face this new challenge in the same way that bulk container loaders were developed to respond to the just-in-time demands of roasters.

Another area where new techniques are likely to be developed is the wet polishing of Robusta coffee. The process of vaporization that is being successfully performed by coffee importers in Europe is likely to develop simpler, lower cost versions to be implemented in producing countries. This will impact the conception and the layout of dry mills in the future.

On a more general note, as Robusta production increases, the design of dry mills is likely to become ever more different for Arabica and Robusta coffee. Pieces of equipment that are specific to Robusta or Arabica, as well as different grading and sorting requirements, will cause dry mills for Arabica and Robusta to abandon the single conceptual approach that prevailed in the past. New Robusta mills built in the last 15 years point clearly to this tendency.

New Robusta mills, specially those in Vietnam, are relying on a fixed-flow layout whereby coffee is divided into a few grades that are processed in a parallel manner, with different sets of gravity separators and color sorters dedicated to each grade. Logistics prevail over quality. New Arabica mills are favoring a more flexible layout, with coffee separated into a larger number of sizes that are processed to different standards in the same set of gravity separators and color sorters in order to yield different product qualities that are supplied to specialty markets, large roasters and the domestic markets in producing countries.

The improvement of working conditions in dry mills will continue. Dust control will gain even greater importance and reach markets where it is not yet a major concern. Noise abatement, which has not yet received much attention, will become as important as dust control.

Finally, computerized controls, which are only marginally used today, will become important in dry processing in the future. This will require self-cleaning equipment, sensors before and after each processing step, automatic samplers and sophisticated weighing equipment that will become ubiquitous in dry mills.

2.16
Acknowledgments

This chapter benefited substantially from an early draft prepared by Mr. Michel Jacquet. His draft contained an excellent review of the coffee processing literature and proposed the framework of analysis used in the chapter. The figures attributed to CIRAD-CP, the institution where Mr. Jacquet then worked, are part of his original draft.

Mr. Jean Wintgens suggested several items that were missing and urged me to improve sections that required additional work. He also contributed many of the photographs in the chapter.

I learned many of the concepts and ideas developed in the text during my 12 years with coffee machinery maker Pinhalense, whose managers and technicians, specially Messrs. Adelcio Piagentini and Lourenço Del Guerra, introduced me to the art and science of coffee processing.

Finally, I wish to thank coffee growers, processors, researchers, cuppers and coffee enthusiasts around the world who shared their knowledge with me during my great coffee adventure that has now lasted for 25 years.

Bibliography

Borém, F. M. *Pós-Colheita do Café*. Editora UFLA, Lavras, 2008.

Ferrão, R. G., Fonseca, A. F. A., Bragança, S. M., Ferrão, M. A. G. and De Muner, L. H. *Café Conilon*. Incaper, Vitória, 2007.

Illy, A. and Viani, R., *Espresso Coffee*, Academic Press, London, 2005.

[1] ANACAFE. *Manual de Beneficiado Húmedo del Café*. ANACAFE, Guatemala, 2000.

[2] Brando, C. H. J. Cereja Descascado, desmucilado, fermentado, despolpado ou lavado? In: *Anais 25th Congresso Brasileiro de Pesquisas Cafeeiras*, Franca, 1999, 342–346.

[3] Cléves, R. *Tecnología en Beneficiado de Café*, 2nd edn. Tecnicafé Internacional, San José, Costa Rica, 1998.

[4] Coffee Board and UPASI. *The Connoisseur's Book of Indian Coffee*. Macmillan India, Bangalore, 2002.

[5] Comite Departamental de Cafeteros de Antioquia. *Construcción de Pequeños Beneficiaderos de Cafe*. Comite Departamental de Cafeteros de Antioquia, Colombia, 1992.

[6] Illy, A. and Viani, R. *Espresso Coffee*, Academic Press, London, 1995.

[7] Instituto Mexicano del Café. *El Cultivo del Cafeto en México*. Instituto Mexicano del Café, Mexico, 1990.

[8] Jacquet, M. *Harvesting and Processing*. CIRAD-CP, Montpellier, 1996.

[9] Matiello, J. B., Santinato, R. and Camargo, A. P. de. *A Moderna Cafeicultura nos Cerrados*. Instituto Brasileiro do Café, Rio de Janeiro, 1987.

[10] Pascoal, L. N. *Aroma of Coffee, A Practical Guide for Coffee Lovers*. Foundation Educar DPaschoal, São Paulo, 2002.

[11] Pinhalense S. A. Máquinas Agrícolas. *Café Cereja Descascado: Manual Técnico*. E. S. Pinhal, Brasil, 2001.

[12] Roa, M. G., Oliveros, T. C. E., Álvarez, G. J., Ramírez, G. C. A., Sanz, U. J. R., Álvarez, H. J. R., Dávila, A. M. T., Zambrano, F. D. A., Puerta, Q. G. I. and Rodríguez, V. N. *Beneficio Ecológico del Café*. Cenicafé, Chinchiná, 1999.

[13] Rothfos, B. *Coffee Production*. Gordian-Max-Rieck, Hamburg, 1990.

[14] Sivetz, M. and Desrosier, N. W. *Coffee Technology*. AVI, Westport, CT, 1979.

[15] Teixeira, A. A., Hashizume, H., Nobre, G.W., Cortez, J. G. and Fazuoli, L.C. Efeito da Temperatura de secagem na caracterização dos efeitos provenientes de frutos colhidos verdes. In: *Proc 10th ASIC Coll.*, 1982, 73–80.

[16] Vasquez, M. R. and Montero, H. M. Pérdida de sólidos del endospermo del café durante el beneficio. In: *Simposio sobre Caficultora Latinoamericana 14*. Ciudad de Panamá, Tegucigalpa, IICA-PROMECAFÉ, 1994, 543–547.

[17] Vásquez-Morera, R. and Hidalgo-Ugalde, G. Influencia del desmucilaginado mecánico del café y de diferentes períodos de espera al secado sobre la calidad. *Revista del Instituto del Café de Costa Rica* **1991**, VI (64), 3–5.

[18] Vincent, J. C. In: *Coffee. Vol. 2: Technology*, eds Clarke, R. J. and Macrae, R. Elsevier, London, 1987.

[19] Wilbaux, R. *Coffee Processing*. FAO, Rome, 1963.

[20] Wintgens, J. N. *Factors Influencing the Quality of Green Coffee*. Nestlé, Vevey, Switzerland, 1993.

[21] Wrigley, G. *Coffee*. Longman, London, 1988.

Technical bulletins, coffee processing manuals and other short publications by research agencies in the following countries: Australia, Brazil, Colombia, Costa Rica, Guatemala, India, Kenya, Mexico, Uganda, USA (Hawaii), Vietnam and Zimbabwe.

3
Ecological Processing of Coffee and Use of Byproducts

R. Cleves S.

3.1
Introduction

High competition in the world economy requires constant improvements in terms of production, processing, packaging, distribution and sales. Coffee growers and processors must update day by day and "competitiveness" is the motto. Added to this, a moral and ecological commitment towards the preservation of water, soil and vegetation is a prevailing consideration.

Coffee processing generates byproducts (erroneously called waste materials) which, due to a lack of technology, eventually end up polluting water and the local environment. A vivid description quoted by the Ministry of Water Development of Kenya in 1975 indicates the magnitude of the problem: "The contamination generated when processing 1 ton of coffee is equivalent to that caused by 2000 people for 1 day".

According to current statistics [9], 1 kg of coffee (fruit) generates 40 g of chemical oxygen demand (COD). Other sources estimate this quantity as high as 60 g of COD/kg. These amounts correspond to soluble solids and those in suspension but do not include coffee pulp. As a whole, it is now accepted that wet-processed coffee generates a contamination of 330 g COD/kg of green beans (coffee pulp not included).

The primary processing of coffee is a delicate issue since large volumes of fruit have to be processed during the dry season, a short spell of only a few months per year. This means that a large quantity of coffee byproducts is produced when the rivers rate their lowest flow. Last, but not least, the constant increase in coffee production and urban development around processing plants are two further factors that considerably aggravate the problem.

3.2
Coffee Components

3.2.1
Balance of Components

Before starting an environmental contamination analysis originated by coffee processing, we will describe the composition of the coffee fruit.

Table 3.1 considers a ratio of 22 m^3 of water to prepare 5.5 tons of coffee cherries (20 l water/kg of green coffee). This ratio has been the norm until recently.

Coffee: Growing, Processing, Sustainable Production, Second Edition. Edited by J. N. Wintgens.
© 2012 WILEY-VCH Verlag GmbH & Co. KGaA. Published 2012 by Wiley-VCH Verlag GmbH & Co. KGaA.

Table 3.1 Mass balance

Coffee fruit (22 % MC)	5.5 ton	Green coffee (11 % MC)	1.00 ton	18.2 %
Water for processing	22.0 ton	Husk (10 % MC)	0.25 ton	4.5 %
		Fresh pulp (82 % MC)	2.25 ton	41.0 %
		Evaporated water[a]	1.10 ton	20.0 %
		Mucilage (65–75 % MC)	0.90 ton	16.3 %
		Residual water[b]	22.00 ton	
Total	27.5 ton	Total	27.50 ton	100.0 %

[a]This table corresponds to the coffee's own water content in humid parchment.
[b]When coffee is dry depulped, or the water recycled, this quantity is even less. From: Cleves 1992.

Our conclusions are:

a) Mature coffee fruit rates 65 % average moisture content, fresh pulp rates 82 %; during the rainy season, the moisture content of the pulp can reach 90 % due to hydration. In a "traditional" pulping process, water is added to facilitate the detachment of the pulp and is also used during the whole process for conveying the coffee.

b) During this process both pulp and mucilage absorb water and part of their solids migrate into the depulping water.

The more prolonged and close the pulp–water contact, the greater the increase in the transfer of solids towards liquids; this results in a higher organic load for the depulping water and a deterioration of pulp quality.

Later on, reference will be made to alterations and/or deterioration of the pulp in a traditional pulping process. For the moment, we will focus on reducing water consumption during the different stages of the process, on the time pulp should remain in contact with the water and, finally, on the processing equipment to be recommended.

3.2.2
The Composition of Coffee Pulp

Coffee pulp represents 40–42 % of the weight of the coffee fruit. It is a voluminous and perishable material with a high sugar and moisture content that produces rapid fermentation. This generates offensive odors and encourages the proliferation of flies. Some processors dispose of the pulp in rivers. This bad habit is lethal for all forms of life and noxious to the environment. It is also a regrettable economic loss as this form of "waste material" can be used as organic fertilizer for the coffee trees, stored in silos in an anaerobic medium or dehydrated to be used as animal feed. In the latter case, coffee pulp is converted into a high-value animal protein.

The chemical composition of dry coffee pulp shown in Table 3.2 calls for the following comments:

- Ash percentage (minerals) of coffee pulp is very important
- Coffee pulp can be considered as a high-value animal food; nevertheless, its consumption is limited due to the presence of caffeine (1 % approximately) and poly-

Table 3.2 Chemical composition of dry coffee pulp (%)

Humidity	12.6	Ethereal extract	2.5
Dry material	87.4	Crude fiber	21.0
	100.0	Nitrogen (N₂)	1.8
		Protein	11.2
		Ashes	8.3
		N₂ free extract	44.4

Source: *Turrialba* 1972, *22*, no. 3.

phenol contents, although this latter can be removed by treatment with hot water (ROTOCEL process).

R.Vásquez considers it is a good option to mix partially dehydrated pulp (20–30 % MC) in equal proportions with dry coffee husks and use it as burning fuel. This option, however, can be discarded since the energy required to dry the pulp is much higher than the energy generated as burning fuel. The only place where this option has turned out to be profitable was in El Salvador, where open-air drying areas are available together with abundant, low-cost labor.

A final observation about "mass balance" refers to the high mucilage content of the coffee fruit (16.3 %). This material, due to its colloidal condition, hydrates and dehydrates very easily according to the rate of rainfall.

3.3
Dry Processing versus Wet Processing of Coffee

There are notable differences between the dry process for Arabica used in Brazil and in a few other countries, and the wet process used in most Arabica-producing coun-

tries. The main differences are found in harvesting methods and in processing systems, which as a result, generate two different types of coffee.

3.3.1
Harvesting

In a dry process, coffee beans are left on the trees until they are over-ripe and partially dry. The objective is to harvest in one pass, picking green, ripe, semi-ripe and dry cherries all together. In certain countries (Brazil, Australia, Hawaii) mechanical harvesters are used and the operation is completed by hand. These methods are possible where the land is flat and are particularly suitable where labor is costly or difficult to recruit.

In countries where coffee is processed by the wet method, picking is selective and only mature cherries are picked. This entails making two to three passes. Immature cherries are sorted either by hand or mechanically and then sold as green bean quality product.

3.3.2
Drying and Processing

Dry processing means that coffee beans are dried with their husk. As a result, the volume processed is larger because of increased humidity. This rudimentary method of drying contrasts with the sophisticated and complex drying techniques that have been developed within traditional wet processing.

In dry processing, the "silverskin" of green Arabica coffee beans is more adherent. When roasted the beans are clear and yellowish with a rather bitter flavor. Over-fermented coffee beans rate a sour and hard cup without acidity. Immature/green or over-ripe beans definitely affect cup quality. Immature cherries are hard and even dirty, over-ripe beans produce a winey brew, sometimes sour and even Rioy when contaminated by fallen fruit collected from the ground.

Dry processing has another disadvantage because by peeling the whole dry fruit, only a single byproduct is obtained which is a combination of pulp, mucilage and parchment. This eliminates any possibility of valorizing the pectin, caffeine, tannins and pulp contents. The husks obtained can only be used as burning fuel or fertilizer.

All the various drawbacks linked to dry processing justify it being no longer considered in this chapter from now on.

3.4
Ecological Coffee-processing Methods

The evolution of coffee-processing technology has led to ecological processing methods which are now being implemented. These methods include rational and ecological handling of pulp, mucilage and effluents.

The concept of ecological processing involves radical changes in the following stages:

- Harvesting and reception of the coffee
- Pulping and green bean separation
- Handling and disposal of the coffee pulp
- Classification
- Treatment of wastewater
- Final treatment and use of effluents and byproducts

The final objective is to implement a series of environmentally friendly methods.

3.4.1
Harvesting Norms

The requirements previously mentioned for selective harvesting are also valid for ecological processing. The maximum acceptance of green fruit on reception at the processing plant is 7%.

The timespan between harvesting and pulping should not exceed 20 h. Cherry hauling should be done under strict control. The process begins with the reception in the "siphon" in order to obtain a first classification by density, and to separate the light and heavy elements.

In recent years, dry cherry reception (almost without water) has become popular. The discharge is made by means of pressurized water. The cherries are then sent to a washer called a "Salvadorian siphon", which efficiently separates low-density cherries (floaters) from the bulk of the fruits.

The use of water for conveying coffee is not relevant because, in the first place, cherries do not pollute water and, secondly, economical draining methods like vertical elevators can be used. Existing structures, if available and rationally used, represent a great economy which should be taken into consideration.

3.4.2
Pulping

Pulping cherries after their reception consists of removing the pulp to which strips of mucilage still adhere.

Cylinder and disk pulpers are the two basic types of pulpers, both of which have been in use for more than 100 years. The cylinder pulper consists of a rotating cylinder mounted on a frame, covered with either a copper or a stainless steel plate depending on the type of coffee to be treated. When rotating at 120 rpm the cylinder pushes the cherries against a concave plate or "breast", and subjects them to pressure and friction. The removed pulp is separated from the cherry by means of adjustable blades. Recently the idea of placing a rubber coating on the breast has improved the system by increasing its capacity and cushioning the contact with the cherries to avoid undue damage.

In disk pulpers, the cylinder is substituted by one to four metallic disks covered with copper or iron tiles, with rounded projections. If the disks are made of cast metal, they must be of top quality in order to keep their shape. Disk pulpers generally last longer and are more reliable. They also offer better resistance to damage caused by metallic objects or stones.

The disk projections or buttons move the cherries to the depulping bars or crests which remove the pulp. A fixed-blade system prevents the passage of the coffee beans and lets the pulp through.

In the traditional process, the cherries are conveyed to the pulpers by means of water, gravity or pumping. If the beans are not uniformly ripe, water helps the pulping process and prevents clogging. Water and gravity then transport the pulp to disposal points from which it is regularly removed. Polluted water seeps from the disposal point to the nearest water stream, particularly if it is out in the open. Bad smells and the proliferation of flies are inevitable. The ensuing degradation of the pulp leads to the loss of its value as foodstuff or fertilizer.

3.4.3
The Separation of Immature Cherries

In order to solve the problem of the combination of green and mature cherries, a Brazilian manufacturer has invented the green cherry separator. The basic component is an iron cylinder with plates welded lengthwise. The cylinder rotates at 350 rpm inside a wire cage. An inner spiral forces the coffee towards the outlet of the separator.

When pressed against the mesh, mature soft beans are pulped and pass through together with the removed pulps. On the other hand, green and dry cherries resist the pressure, and leave through the terminal outlet. This machine is robust, and works smoothly and efficiently. It has, therefore, been widely accepted. The initial idea was to use the separator only at the beginning or at the end of the harvest when small quantities of cherries with a lot of green fruit are received, mainly from Caturra. The green cherry separator combines with a conventional depulper to separate the already removed pulp.

It is now more and more frequent for coffee processors, operating in areas where maturation is not uniform and labor is lacking, to modify their separators and increase their capacity in order to use them solely as pulpers. This entails the inconveniences of high energy consumption and important water contamination. Hence, we are of the opinion that green coffee separators are definitively not compatible with the philosophy of ecological coffee processing.

3.4.4
Pulping without Water

Starting from the investigations that demonstrated that the contact of pulp with water leads to a transfer of solids to the water used as well as a degradation of the pulp, we have drawn the following conclusions:

• Dry pulping using mechanical conveyance is the best option to reduce the polluting effects of coffee processing.
• The recycling and appropriate handling of water is important as well as related measures to which we will revert later on.

- Pulp generated by dry pulping breaks down more rapidly than pulp generated by traditional wet processing and does not produce noxious odors.
- Wet parchment produced by dry pulping needs only 6–10 h to ferment, which means there is less risk of quality deterioration.

Technological progress has created new techniques in pulping to minimize the amount of water needed which reduces the volumes by up to 90 % when compared to traditional processing. The basic idea was to re-engineer the process, to reconvert traditional infrastructure and equipment into technology that minimizes environmental contamination.

As a result, the following measures have been implemented at the processing level:

1) Incorporation of semi-dry or fully dry reception of the cherries.
2) Reduction of the siphons to a quarter of their original capacity.
3) Design and implementation of pulpers that operate without water (pulping in dry or dry pulping).
4) Mechanical feeding of ripe cherries to the pulpers by means of helical servers (feeding in dry).
5) Mechanical conveyance of the pulp by means of conveyor belts and and/or helical screws – this technique avoids undue water absorption which leads to faster decomposition, facilitates transportation and minimizes noxious odors.

3.4.5
Modification of the Pulpers

Most Central American and Caribbean countries have implemented the pulping in dry method based on traditional pulpers, but the following disadvantages can be noted:

- Poor adaptability to different bean sizes.
- High maintenance cost due to damage caused to cylinders by metallic objects, stones, sticks, etc.

Investment costs for the purchase of new pulpers or the retrofitting of the plant are already high, and the need to renew machinery, tools and equipment because of damage during normal use is an added expense which is difficult to accept.

In Colombia, an investigation program was launched to develop the efficiency of dry pulpers. Tests carried out at great expense by the Antioquia Coffee Growers Committee, in association with the National University and a private company, have brought about a number of modifications and led to the release of the ETERNA disk depulper (Pict. 3.1) which offers the following improvements:

- Modification of the hopper by increasing the inclined angles to allow dry alimentation (without water) and removal of the feeding regulator.
- Change in the surface of the punched disk for which it was necessary to cast several pieces with different angles and knob shapes.
- Modification of the tilt of the bars that support the breast and the pulp separators in order to project the coffee with more force towards the openings at the bottom of the pulping channels.
- An increase of the angle of discharge for the pulped beans.
- An acceleration of the operating speed to 220 rpm.

The capacity of a disk pulper varies from 800 to 1400 kg of fresh cherries/h (see Tab. 3.3).

Table 3.3 Quality appreciation of disk pulping

Defect	%
Beans in pulp	0.00
Pulp in beans	3.60
Threshed beans	0.16
Bitten-off beans	0.10
Non-depulped cherries	0.33

From: Eterna Compact Processing.

and fitted with six helical breast plates. The channel has a progressively smaller depth starting from the entrance of the coffee down to the discharge point. This characteristic design allows the big cherries to be pulped first, then regular-sized ones and, finally, the small ones. This eliminates the problem caused by a lack of uniformity in cherry sizes as well as their degree of maturation and is a most outstanding contribution to the technological development of pulping.

As a complement, Penagos also manufactures the upward vertical ascending mucilage remover which operates by friction and consumes only 1 liter of water/kg of dry parchment.

Picture 3.1 ETERNA compact fixed unit for dry pulping and mucilage removal. It includes a disk pulper, a traditional sieve to remove the green cherries and a vertical mucilage remover.

3.4.6
Pulping and Mechanical Mucilage Removal

Another valuable Colombian contribution to dry pulping was carried out by the firm Penagos Hnos and Co. (Pict. 3.2). The vertical conical pulper, with a cylinder of only 10 cm in length and 24 cm in diameter is protected by a stainless steel cover

Picture 3.2 Penagos compact fix unit for dry pulping and mucilage removal. It includes a conical pulper, a post-pulping sieve and a vertical mucilage remover.

Picture 3.3 Rear view of the mobile disk pulper SAESA for dry pulping. The parchment conveyor is cream-colored and the pulp outlet is red-colored.

Picture 3.4 Lateral view of the mobile disk pulper SAESA for dry pulping showing the trailer and the engine with transmission.

Penagos has also come up with a worthwhile statement which should be quoted here: "Ecological processing is not based on decontaminating the sources of water but rather in avoiding to pollute them".

Very recently SAESA Corp. (Costa Rica) designed a novel depulper for dry depulping, equipped with a coated metallic disk (Pict. 3.3 and 3.4).

Two versions are available: either fixed, which is designed to be installed in a pro-cessing plant, or mobile, which can be fixed on a trailer and moved from one production area to another. In this case the pulp is left on site to be used as fertilizing material.

Parchment with practically 50% less weight and volume is far cheaper to transport to processing plants in dryer areas which are closer to exporting points.

Among other innovations, the SAESA pulper is fitted with simple adjustments to compensate for the size and the hardness of the coffee beans.

3.4.7
Natural Fermentation versus Mechanical Mucilage Removing

After pulping, part of the mucilage still adheres to the parchment beans. The mucilage is composed of hyaline tissues rich in sugars and pectins, but free of caffeine and tannins. Picado in Costa Rica determined the composition shown in Tab. 3.4.

The pH of the Arabica mucilage ranges from 5.6 to 5.7 according to Wilbaux. Its moisture content varies (colloidal condition) depending on the climatic conditions that prevail during the harvest. In our investigation with CICAFE, we have observed moisture content extremes ranging from 42.0 to 86.4%. Mucilage represents 15.5–22.0% of the fresh fruit weight and is probably the most important element in the outturn of green coffee to coffee cherries.

Table 3.4 Chemical composition of the mucilage

Total pectins	33%
Reducing sugars	30%
Non reducing sugars	20%
Cellulose, ashes, etc.	17%
Total	100%

The importance of fresh mucilage is that it represents a reliable high-quality source of pectin. Calle states that coffee is the largest potential source of pectin in the world. Pectin can be used as a homogenizer, an emulsifier, a laxative, a hemostatic agent, a plasma culture substrate, a cosmetic and in the pharmaceutical industry in general. This, in itself, is a sound argument in favor of removing mucilage mechanically and avoiding fermentation, which degrades it into contaminant agents.

For more than 100 years, mucilage removal has been carried out by fermentation in tanks. This is done for periods of 6–48 h, depending on the outside temperature, the quality of the water, the maturity of the coffee, microorganisms, water re-circulation, the use of enzymes, etc. The purpose has always been to convert an insoluble hydrogel into a hydrosol.

In coffee fermentation, an abundant microbial flora is highly active and this flora varies from one processing plant to another. Bacteria, fungus and yeasts are present on the tank walls, in the water and even in the air. Fritz [10] and others affirm that the pectic compounds of the mucilage are also hydrolyzed by the action of a pectinase that is present in the coffee fruit itself.

3.4.8
The Disadvantages of Natural Fermentation

Natural fermentation has disadvantages when compared to mechanical mucilage removal. It is a slow and empirical method that has no control of the microorganisms that activate in the fermentation. Over-fermentation can seriously deteriorate coffee quality and generate "stinkers" due to butyric fermentation. In addition, undesirable types of fermentation may occur. The main disadvantages are detailed below:

- The process evolves in batches and is not continuous; as a result, more time, more water and more labor are required
- The metabolism of the beans leads to a lower outturn; soluble solids are also lost by diffusion into the liquid medium
- The degradation of the mucilage pectic materials impedes the extraction of pectin which is a valuable byproduct
- The design of processing plants are more complicated; construction and maintenance rise in price due to the large dimensions and expensive fermentation tank structures – considering the large volumes of coffee that are processed nowadays, this issue is far from negligible

On the other hand, mechanical mucilage removal using an Aquapulpa ELMUS or the latest South American option is recommended since it offers a continuous and rapid system that avoids pollution and enables the recuperation of fresh mucilage.

In Tab. 3.5, we present one of the many versions of the coffee fermentation process. The fermentation is basically alcoholic and lactic. Most investigators think that it is above all lactic and that the lactic acid protects the coffee mass from other undesirable types of fermentation which lead to the degradation of the mucilage (acetic, propionic and butyric fermentations).

Table 3.5 Theory of fermentation in coffee processing

Protopectin	$\rightarrow$	pectin ⟶
Pectin	$\rightarrow$	pectic acid + CH_3OH ⟶ galacturonic acid + CH_3OH
Pectic acid + calcium	$\rightarrow$	calcium pectane ⟶ soluble gel

Source: Technical Resources International Inc.

3.5
Treatment of Residual Water

A few years ago, when studies were initiated on residual water, different sources provided us with information on water issued from pulping and washing. Considerable differences were noted both for values and parameters, which can be attributed to the following factors:

- The volume of water used in the process,
- The recycling and handling of the water,
- Fermentation and/or mechanical mucilage removal.

After years of investigation in different countries, our conclusions and recommendations are the following:

- Reduce the water volume to a minimum and recycle it in a rational way.

- Avoid pulp–water contact and convey the pulp by mechanical means like helical screws or conveyor belts.
- Use sieves in all phases of the process to eliminate organic matter in suspension; this technique is far better than sedimentation or anaerobic digestion.

See also Tab. 3.6.

3.6
Treatment Systems for Residual Water

The previous points have provided an overview of the improvements and modifications that should be implemented in order to reduce to a minimum the sources of pollution generated by coffee processing. There is, however, no "single" solution for the treatment of coffee-processing efflu-

Table 3.6   Characterization of residual water from processing

Parameter	Morales [18] Costa Rica	Ward [24] El Salvador	Brandom [3] Africa	Horton [12] El Salvador	Rolz [20] El Salvador
Depulping water (mg/l and pH)					
oxygen, biochemical or chemical demand	450–11710 (COD)	2360 (COD)	7750 (COD)	3280–15000 (COD)	13900–28020 (COD)
pH	5.9	–	–	4.4	–
solids in suspension	70–850	848	–	625–1055	–
total solids	2687	4960	–	10090–12340	13150–16700
Washing water (mg/l and pH)					
oxygen, biochemical or chemical demand	284–3828	1725	6040	295–3600	2900–10500
pH	5.6	–	–	4.5	–
solids in suspension	100–380	2060	–	235–2385	–
total solids	532	4260	–	885–3140	5060–7280

Source: Cleves [4b].

ents. Each plant presents its own particularities according to its location, climate, processing system and type of coffee. For this reason, only general recommendations can be given.

This topic pertains to the specific area of "Sanitary Engineering". However, as this is a very specialized and complex issue, professionals in this discipline must work in close collaboration with the processing technicians.

Generally speaking, there is a consensus that a primary mechanical treatment, which consists of the mechanical removal of solids in suspension by filtration, centrifuging or by sedimentation, should be completed by a secondary chemical process.

3.6.1
Treatment of the Residual Water in Anaerobic Ponds

Work executed by the Technological Institute of Costa Rica (CEQUIATEC), under the direction of Alma Deloya and the financial support of CAPRE-GTZ.

Primary Treatment by Sedimentation
Sedimentation tanks work in two stages. The first stage is a standard rectangular sedimentation tank with a mud zone accumulator and its respective drains. The second stage consists of a tank for accelerated sedimentation with inclined plates designed to speed-up sedimentation and drainage. The drain of the sedimentation tank is evacuated towards the sludge pond. See Pict. 3.5.

Secondary Treatment in Anaerobic Ponds
The pond receives two types of effluent water: the liquid discharge of the sludge pond and the outflow or overflow of the sedimentation tank.

Picture 3.5 Cylindrical sedimentation tank for residual waters. Capacity: 210 m^3. Equipped with lime injector for pH correction and aeration with air bubbles. A low-molecular-weight flocculating agent is also used. The solids in suspension deposits in the conical bottom.

A good removal averages 94 % COD, and about 93 % BOD is the result obtained in an efficient sludge pond with a two-stage sedimentation system and, finally, a long retention time in the anaerobic pond (28 days).

Another important factor which contributes to the efficiency of the system is the rectangular shape of the anaerobic pond. Rectangular ponds have a higher flow and a more efficient buffering capacity to regulate the pH variations of the water flowing into the tank.

3.6.2
Recent Technology in Anaerobic Biodigestors

This chapter does not seek to describe all the available technology in the anaerobic biodigestion process, but it can be summarized as follows (Pict. 3.6):

• The system has been tested with many devices from simple ponds (2 m depth) to complex structures, with pH and temperature control, methane burner, etc.

Picture 3.6 Five cascaded oxidation lagoons with serial overflow; the walls and the bottom are covered with a 1-mm thick plastic coating to avoid subsoil contamination and infiltration. The pH correction is made with lime. Note the sprinkling device for accelerating the biologic degradation of the contaminants. The lagoons are 3.5 m deep with a capacity of 16 000 m³.

- In the *Memorias del XVIII Latin American Symposio de Caficultura*, Rick Wasser contributed an article on parameter determination of low-cost anaerobic reactors. We give credit to the author, IICA/Promecafe and to the Editorial Coordinator of ICAFE. The reactor was financed by the Dutch Embassy and manufactured by BTG, a Dutch Company. The efficiency of the treatment when completed was approximately 85 % (COD).

3.6.3
Disadvantages of the Anaerobic Systems

With reference to the "continually impelled" up-flow anaerobic sludge blanket (UASB) equivalent to RAFA where, without adding air, it is possible to transform the organic matter in methane, the following drawbacks need to be mentioned:

- The temperature of the residual water has to be raised from 33 to 35 °C
- The pH of the whole system has to be precisely controlled and maintained

- An enormous infrastructure is required (gas deposits, safety valves, sludge separators)
- The system is sensitive to biochemical balance and hydraulic strokes
- Costs are excessively high

3.7
Options for the Treatment of Residual Water

Figure 3.1 shows the options for the treatment of residual water and sheds some light on this issue.

3.8
Integral use of the Byproducts of Coffee

Some years ago the firm Subproductos del Café SA started to develop an ambitious medium-term program, with the participation of Stanford University, INCIENSA from Costa Rica and the firm Gibbs & Hill.

Figure 3.2 presents a summary as covered in this chapter. The only alternative that has not been discussed in detail is the use of parchment as fuel directly on site. It offers a very acceptable thermal value of 4200 kcal/kg and no transport is required.

3.9
Key Words

The author has been particularly cautious when using words and terms which could be difficult to understand for a reader who is not familiar with the topic of ecological processing of coffee.

There are, however, certain terms for which it is necessary to provide the correct interpretation, these are:

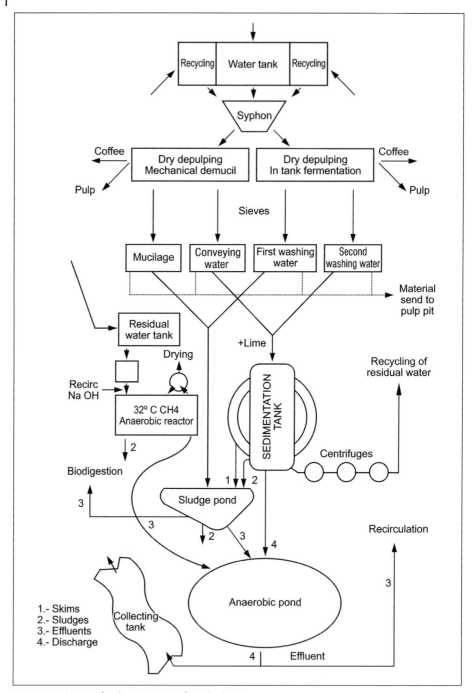

Figure 3.1 Options for the treatment of residual water.

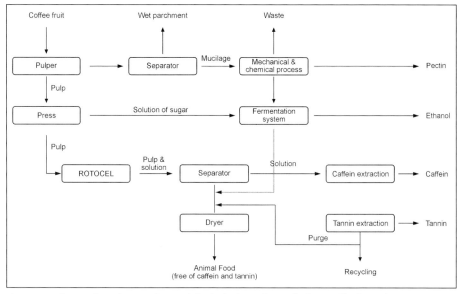

Figure 3.2 Chemical products and animal feed which can be made from coffee byproducts [11].

- *Chemical oxygen demand (COD).* The value of the COD covers everything that may require an input of oxygen, particularly oxidizable mineral salt and most of the organic compounds, whether they are biodegradable or not.
- *Biochemical oxygen demand (BOD).* This is the quantity of oxygen expressed in mg/l and consumed to ensure the biological oxidation of biodegradable organic matter present in water. Biological oxidation takes 21–28 days to be completed – BOD 21 is then obtained.

- When obtaining BOD21 takes too long, it is conventionally replaced by BOD5, which is the quantity of oxygen consumed after 5 days of incubation.
- If all the organic matter present in the water is biodegradable there is an equivalence of the COD which is equal to BOD21.

From: *Water Manual*, Degremont, France.

Bibliography

[1] Álvarez, G. J. *Despulpado de Café sin Agua*. Cenicafé, Colombia.

[2] Braham, J. E. and Bressani, R. *Pulpa de Café*. INCAP, Guatemala, 1978.

[3] Brandom, T. W. Coffee waste waters, treatment and disposal. In: *Sewage Industrial Waters*. Publisher, 1950.

[4] a) Cleves, S. R. *Justificación de un Proyecto para Investigar la Obtención de Pectina a Partir de Mucílago del Café*. Oficina del Café, San José, Costa Rica, 1975; b) Cleves, R. *Technologia en Beneficiado de Café*. TECNICAFE, San José, Costa Rica, 1995.

[5] a) Clifford, M. N. and Willson, K. C. *Coffee – Botany, Biochemistry and Production of Beans and Beverage*. Croom Helm, London, 1985; b) Clifford, M. N. Physical properties of the coffee bean. *Tea and Coffee Trade J.*, USA.

[6] Coto, J. M. *Fundamentos del Proceso de Biodigestión Anaeróbica*. Universidad Nacional de Costa Rica, Costa Rica.

[7] De Giulluly, N. Factors affecting the inherent quality of green coffee. *Advances in Coffee Production*. New York, 1959.

[8] Denton, E. *Coffee*, 1985.

[9] *Encyclcopedia Americana*. Sanitary engineering. American Corp., New York, 1964.

[10] Fritz, A. *Preparación del Café por la vía Húmeda. Estudio de la Fermentación*. AGR, Colon, 1935.

[11] Gibbs & Hill. *Productos Cuímicos y Alimento Animal de los Desechos del Café*.

[12] Horton *et al. Study of Treatment of the Wastes of the Preparation of Coffee*, 1946.

[13] ICAITI. *La Utilización Industrial del Grano de Café y de sus Subproductos*. Otras Publicaciones de ICAITI, Guatemala, 1996

[14] Lingle, R. T. *The Coffee Cuppers Handbook*, 3rd edn. Specialty Coffee Association of America, Long Baech, CA, 1986.

[15] Lopez, M. E. *Aspectos Ambientales de la Biodigestión Anaeróbica*. Costa Rica.

[16] Meier, E. *Estudio de la Depuración de Aguas Residuales de los Beneficios en Costa Rica*. Bisikon, Zurich, 1991.

[17] Menchu, J. F. *La Determinación de la Calidad del Café*. ANACAFE, Guatemala, 1966.

[18] Morales, A. I. *Caracterización de las Aguas Residuales del Beneficiado*. CICAFE. Costa Rica, 1979.

[19] Ramírez, M. H. and Restrepo, G. C. *Evaluación de una Despulpadora de Disco para Café*. Universidad Nacional de Colombia, Colombia, 1993.

[20] Rolz, C., *et al. Producción de Fungi Imperfecti a Partir de las Aguas de Desecho de los Beneficios Húmedos de Café*. Otras Publicaciones de ICAITI, Guatemala.

[21] Sivetz, M. and Foote, H. E. *Coffee Processing Technology*. AVI, Westport, CT, 1963.

[22] Tangarife, L. E. *Evaluación de un Sespulpador de Disco para Café*, 1993.

[23] Thomas, A. A. *Sanitary Engineering*. MIT Press, Cambridge, USA.

[24] Ward, P. C. Industrial coffee wastes in El Salvador. *Sewage Works J.*, 1945.

[25] Wilbaux, R. *Coffee Processing*. FAO, Rome, 1963.

[26] Wintgens, J. N. *Influencia del Ceneficiado sobre la Calidad del Café*. Nestlé, Vevey, Switzerland, 1994.

[27] Zuluaga, V. J., Bonilla, V. and Quijano, R. M. *Contribución al Estudio y Utilización de la Pulpa del Café*. Federación Nacional de Cafeteros de Colombia, Colombia.

Part IV
Storage, Shipment, Quality

1
Green Coffee Storage

J. Rojas

1.1
Introduction

Storage is one of the most important and critical stages in the processing of any agricultural commodity. In the case of coffee storage, the goal to achieve is to maintain its commercial value as long as possible by preserving the integrity of the bean with all its characteristics.

Coffee is the second most important commodity in the world; as a result, it is important to develop storage strategies to preserve its value. Losses caused by inadequate storage could seriously impact many sectors of the world economy, not to mention national economies which depend on coffee for a considerable part of their GNP.

Adequate coffee storage needs to take the following points into consideration:

1) Coffee beans are living entities with their own physiological activity.
2) When used as seeds, their viability depends to a great extent on storage conditions.
3) Food safety has now become an extremely important issue since the effects of toxic substances, which could develop during storage, can cause significant harm to human health.

4) Although coffee does not have a great nutritional value, its price is based on its sensorial value. This is a delicate aspect which can easily be affected if storage is not adequate. Coffee aroma and taste are highly sensitive to contamination. As a result, storage in the proximity of fragrant spices or chemicals with a pervading odor should be avoided at all costs.
5) Due to the inherent imbalance between supply and demand in the coffee market, it is sometimes necessary to store coffee for long periods of time. The length of storage impacts the quality of coffee. This fact is fully acknowledged by the discounts negotiated in contracts for the sale of coffee which has been stored for certain length of time.

Reasons why long-term storage is necessary:

- Coffee production is seasonal, whereas consumption takes place year round. This means that producers, exporters, importers and traders alike tend to keep coffee in the expectation of obtaining better prices
- Due to the systems which govern coffee commodity markeets, certified coffee stocks often have to be stored for

Coffee: Growing, Processing, Sustainable Production, Second Edition. Edited by J. N. Wintgens.
© 2012 WILEY-VCH Verlag GmbH & Co. KGaA. Published 2012 by Wiley-VCH Verlag GmbH & Co. KGaA.

months or years before they can be delivered to the holders of the commodity contracts.

In addition to the above, as in the case of any grain, there are environmental factors, such as relative humidity (RH), moisture content, temperature and gas composition, which directly affect the stability and quality of the coffee beans. As the control of these essential factors improves and develops, storage conditions, in general and storage conditions for coffee, in particular, will improve accordingly.

Specific technologies and procedures according to the available resources in each area and destination have been developed to guarantee that minimum storage conditions, are met. These conditions are not necessarily the same for an importing, as for an exporting country, since each of them has its particular requirements. Consequently, the type of storage will depend on environmental factors, market requirements and the type of coffee, as well as the infrastructure and available financing.

The purpose of the present chapter is to emphasize the relevance of storage within the coffee chain, and to review the technical, physiological and control factors needed to preserve the quality and value of the coffee for the longest possible time while avoiding undue degradation.

1.2
Bean Physiology and Environmental Influences

1.2.1
Intrinsic Physiology

Coffee beans or seeds, just like leaves, stems and roots, are vegetative. They retain all the characteristics and activities of a living being, including respiration and transpiration, among other elements, regardless of whether they are stored as dry cherry, parchment or green coffee.

Respiration is a process by which the oxygen available from the environment is used and the materials available within the bean (starches, carbohydrates, fats and proteins) are consumed by an enzymatic unfolding to produce CO_2 and water in an exothermic reaction. Under field conditions, this activity provides the necessary energy for the bean to germinate. Once the coffee has been picked and processed, however, with the exception of the seeds for sowing, the beans need to remain in their dormant state for as long as possible so as to maintain their commercial value.

The impact of respiration on the bean deterioration can be highlighted by the fact that every 24 h, an average of 4.4 mg of CO_2 are produced by 100 g of coffee beans and, according to Sivetz and Desrosier [19], the 96 cal of heat produced by 44 mg of CO_2 will raise the temperature by 0.25 °C. Consequently, during storage, the temperature will increase sequentially. As shown in Fig. 1.1, the deterioration caused by this effect is incrementally cyclic, which means that, with the combination of moisture and temperature increase, the rate of respiration will accelerate and affect the beans. A high respiration rate, combined with the generation of heat, causes a loss of weight and dry material in the bean as well as the decomposition of components, like fats, which play an important role in the aroma.

The environmental factors which have a direct influence on this process are temperature, RH, moisture content and air composition.

Progressive deterioration process of stored beans

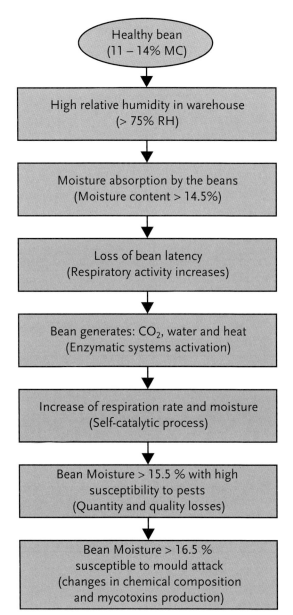

Figure 1.1 Progressive deterioration process of stored beans. From: Othón Serna [14].

1.2.2
Viability

A bean that can germinate is known as a viable bean. Viability depends on the condition of the bean itself as well as on storage conditions. In general, viability remains stable during a certain period of time after which it diminishes at an accelerated speed until none of the beans germinate. This aspect is highly relevant in the case of beans stored for use as seeds for propagation as it has been noted that plants produced from seeds with a low germination rate often present abnormalities, deformations or growth delays.

Physical damage caused during harvesting or processing has a considerable effect on the viability of the bean. A high metabolic or respiratory activity may also significantly affect seed viability. Studies conducted by Duffus and Slaughter [7] have revealed that there is a relation between the O_2 absorption by the seed and its germination capability, measured in the form of the coefficient of CO_2 produced/volume of O_2 absorbed, which reaches high values with bean deterioration.

Robusta coffee, in particular, presents a more accelerated viability loss than Arabica. The germination capacity is better maintained if the seed is stored as parchment or cherry, with a total moisture content of 15–18%. Under adequate conditions, viability is maintained for 5 months, with temperatures of 5–15 °C and a RH of 35–55%.

There are different methods for determining seed viability. The most common being the traditional method, which consists of placing the seeds in germination chambers and counting the number of germinated grains, in order to obtain a percentage with respect to the total. This method is very slow, but it is relatively sure. Tests using indicators like tetrazolium (which indicates the hydrogenase level of activity) or carboxylase are not widely used since they are subject to failure.

1.2.3
Moisture Content and RH

Humidity is the factor which has the highest impact on the speed at which coffee beans deteriorate. Even if beans have been stored with a low moisture content the humidity factor is still very active because they are hygroscopic and tend to balance their moisture content with their immediate surroundings. This phenomenon, generally known as "moisture balance" has been widely studied in cereals and also in coffee by many scientists. It has been ascertained that it enables us to assess the impact of RH, the moisture content and the levels that should be sought to optimize storage conditions. Other more specific factors like the direct influence of temperature, the type of coffee, its origin and the form under which it is stored (parchment, cherry or green bean) should also be taken into account, but their impact on the speed of deterioration of the coffee bean is relatively minor (Fig. 1.2).

It is generally recognized that the ideal coffee moisture content for the preservation of coffee is 12% for Arabica and 13% for Robusta. Beans with a moisture content lower than 9% may be irreversibly damaged in color, as well as in their cup taste and consistency, which means that it is not worth reducing the moisture content to such a low level when drying.

A RH level of 75% corresponds to a moisture content in the bean of 15–16%. According to the Henderson balance, this is the critical level for fungi formation. As a result, the RH level should be kept below 60% because one of the most ob-

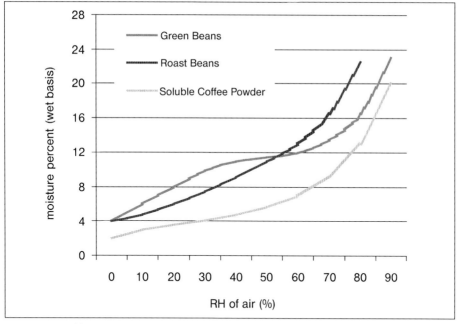

Figure 1.2 Equilibrium moisture content of coffee versus RH of air. From: Sivetz and Desrosier [19].

vious effects of a high RH level, in combination with temperature variations, is the condensation of water which, in turn, contributes to the proliferation of fungi and insects.

According to Narasimhan *et al.* [12], Arabica coffee absorbs moisture faster than Robusta.

There is a genuine concern on the part of carriers, exporters and importers with respect to the loss of moisture and weight, since the loss of humidity during storage or transportation also results in a loss of weight of the coffee and consequently in the profit margin, whether it is due to the commercial weight franchise negotiation, which represents 0.5–1.0%, or to the storage losses, which are reflected on the total manufacturing losses of any roaster or solubilizer. Under ideal conditions, the storage losses, in general, should not exceed 1% on an annual basis.

1.2.4
Temperature

Temperature is the second most important element which affects coffee bean quality. The higher the temperature, the higher the metabolic activity of the seed.

It has been shown that even coffees with a moisture content as low as 11% lose their quality after 6 months under a temperature of 35 °C. On the other hand, a coffee with a moisture content above 15% will maintain its quality at temperatures as low as 10 °C.

Coffee needs to be maintained at low temperatures to reduce its metabolism and respiration. This not only refers to the environmental temperature, which we usually take into account, but also to the intrinsic temperature of the coffee. Calculations made by Sivetz and Desrosier [19], reveal that 8000 stored bags of coffee generate a heat of 210 400 BTU. This highlights the obvious need for ventilation in coffee warehouses and storage premises.

This factor is particularly important in the case of producing countries where average temperatures fluctuate between 20 and 35 °C, and where coffee is normally stored for long periods. A general recommendation is to keep the temperatures below 20 °C, in order to preserve the quality of coffee.

1.2.5
Atmospheric Composition

Although this is an element that normally remains unchanged during storage, we cannot underrate its importance. A high content of molecular O_2 influences the metabolic activity of the seed by accelerating its rate of respiration. The O_2 level varies depending on the altitudes, it is higher at sea level and diminishes at higher altitude. As a result, systems for decreasing O_2 and increasing CO_2 have been developed in order to reduces the rate of respiration and, consequently, increase the shelf-life of the coffee beans.

1.2.6
Altitude

The altitude factor in storage is related to the combination of the most important factors mentioned above. Generally speaking, storage life will be shorter at lower altitudes, i.e. approximately 3 months at 600 m. Whereas at altitudes above 1400 m natural shelf-life can be of 8 months. Inevitably, this means that exporters and industrialists prefer high altitude locations for coffee storage.

1.2.7
Duration

The longer the storage time, the less the preservation of the product characteristics. One year is the generally accepted time for green coffee storage under normal conditions.

In some countries, coffee is stored for more than 1 year but this affects its quality. Countries like Brazil have been known to store coffee lots for more than 4 years. Coffee commodity markets, such as the CSCE of New York and LIFFE of London, store coffee for years but they grant a discount when selling coffee that has been in storage for a long time.

In some cases, however, storage can be beneficial. Robusta, for example, should be stored for a period of 6 months in order to diminish the harsh and woody flavors that are inherent to this species.

Technologies developed to improve the negative impacts of storage time have been developed but, like most sophisticated methods, they generate costs which, in turn, reduce profit margins.

1.2.8
Other Factors

Species of coffee react differently to storage conditions and processing also influences their reactions. Arabica coffee, for example, is more sensitive to adverse conditions, whereas Robusta is more resistant. Similarly, wet-processed coffee may be more easily affected than a dry-processed (natural) coffee.

Coffee may be stored at different stages and under different forms:

1) *Fresh cherry coffee.* Due to its high moisture content, cherry coffee cannot be stored for more than 48 h because of

fermentation as well as the production of mycotoxins.

2) *Dry parchment coffee.* Washed coffee can be preserved in this form for a longer period of time because the parchment protects it from the environment. On the other hand, it requires more space. To store 540 quintals of washed coffee, a net space of 60 m³ or 9 quintals/m³ is required. It can be stored in silos, boxes or bags but it is essential that its moisture content remains homogeneous. Coffee in parchment should be stored 10 days before processing it to homogenize its MC.

3) *Wet parchment.* Even though this is not the best way to store coffee, it is still a common practice in some countries where parchment with 30–40% moisture is stored and traded, respresenting an inefficient and risky procedure, specially due to the generation of mycotoxins and taste deterioration.

4) *Dry or ball cherry coffee.* This is the best form of storage for coffee for long periods since the pericarp fully protects the bean. Obviously, it only applies to natural coffee types. The major drawback of this form of storage is the fact that it is more likely to suffer damage by insects. Under this form storage has a low volumetric efficiency. It is usually made in silos or in jute or propylene bags.

5) *Green or clean coffee.* This is the most usual form for storing commercial coffee. It is also the most vulnerable to environmental factors, since it is ready to be roasted.

Various studies have also been made to identify if there are other factors, like variety, which affect the storage. To date, no differences have been detected between coffee varieties.

Analyses have been made to establish markers or indicators for the measurement of bean deterioration. Studies have been conducted by Bucheli *et al.* [4]. Polyphenol oxydase, for instance, has been widely investigated but there are contradictory reports, particularly because, under industrial storage conditions, its levels are relatively low. Other indicators, such as the formation of coffeeic acid and glucose, are clearer and more specific.

1.3
Main Storage Problems

Coffee is mainly susceptible to attacks by pests and fungi. The damage caused to coffee by these two parasites can be extremely serious both from a financial point of view and with regard to the incidence on consumer health. This does not only involve the pesticide residue level in the bean, but also the level of toxins which may affect human beings. To diminish this risk, adequate controls must be set up during the stages prior to storage because coffee beans have natural physical barriers which can protect them from damage.

As a result, many companies engaged in coffee processing, storage and transportation have set up process control programs,

Table 1.1 Coffee apparent densities

	kg/m³	lb/ft³	Qq/m³
Depulped coffee	840.91	52.43	18.50
Washed parchment	663.64	41.38	14.60
Dry parchment	369.09	23.01	8.12
Green coffee	681.82	42.51	15.00

Source: ANACAFE.

like ISO, in an effort to guarantee the quality and the safety of the bean as a raw material. Similar control systems should be set up to guarantee quality through traceability, good storage practices, identification of critical control points and quality monitoring systems.

1.3.1
Pests

Insects are one of the most important problems in grain storage and coffee is no exception. Damage caused by insects can be lethal to the point that it may destroy the total value of a stored lot. A 3-year study has revealed that in 71% of the cases, damage appeared in imports in the United States. Some of the insects that affect coffee during its storage are described below (for more details on storage pests, see Section II.1).

1.3.1.1   Coffee Berry Borer
The Coffee Berry Borer *Hypothnemus hampei* Ferr. is not only the most important pest in coffee production but it is also a relevant pest in its storage because its biological cycle enables it to continue feeding on the beans for months. Ultimately, it can even cause total loss of the infested beans, which leads to the drop of their commercial value. Losses are also enormous during the roasting, since the beans often become carbonized.

The Coffee Berry Borer prospers when the moisture content is high (above 13%), generally when the bean is stored as dry cherry or in parchment form and when it has been sun-dried. The excrements of the insect can also result in a higher moisture content.

The damage is usually larger in natural Robusta, due to its flowering cycle which generates various maturing conditions

and stages in pest development; added to which, this coffee is generally sun-dried. When coffee is washed and artificially dried, the presence of the pest diminishes.

To a large extent, this pest has been able to spread worldwide, chiefly because of its resistance and its high survival capacity during storage and transportation. Quarantines measures applied in many countries have not totally solved the problem.

1.3.1.2   Khapra Beetle
This pest, *Trogoderma granarium*, has been reported as opportunist. It is not, strictly speaking, a coffee pest, but is important for other products and has generated quarantine measures in many countries because it can affect staple foods like corn and rice.

1.3.1.3   Coffee Bean Weevil
This coleopterous insect, *Araecerus fasciculatus*, can be present in coffee, cocoa and many other crops. Financially speaking, it is one of the most harmful as it even attacks dry coffee cherries. Its larvae develop in environments with a high RH, 80%, and temperatures of 25 °C or more, conditions which generally prevail in tropical and subtropical areas. Investigation has revealed that Arabica coffee is susceptible to attacks by this pest only 3 months after storage, whereas Robusta has resisted for 9 months without being affected.

1.3.1.4   Others
There are other storage pests, mainly from the Lepidoptera order, like the coffee moth *Auximobasis coffeaella*, the almond moth *Corcyra cephalonica*, the Tobacco-moth *Ephestia elutella*, the lepidopterous *Oryzaephillus surinameus* L. and *Laemophloeus ferruginesus* Steph. which affect all beans

whether they are whole, damaged or split. There have also been reports of *Lophocateres pusillus* in Africa and *Sitodrepa panicea* in Colombia.

1.3.2
Fungi

Fungi are another very important concern for storage. Fungi which chiefly attack coffee are *Aspergillus* spp. and *Penicillium verrucosum*. The development of these fungi is favored when the moisture content of the beans is higher than 15 % and the RH above 75 %.

In addition to their impact on the appearance, the aroma and the flavor of coffee, fungi also produce toxic substances which can be harmful when consumed. The most important of these substances are mycotoxins, of which there are various types like aflatoxins and ochratoxins, for which maximum tolerance levels have been ascertained.

Recently, ochratoxin A, which has a carcinogenic effect, has become an issue of the utmost importance, since it is not eliminated by processing and can be detected in green, decaffeinated, roasted and soluble coffee. To date, studies made by private companies have revealed that this toxin is generated by inadequate conditions during the sun-drying of the cherries, particularly when they are damaged or over-ripe. The storage of the cherries during the drying process is critical and placing them in containers or bags without ventilation should be avoided. The drying process should be as short as possible.

Unfortunately, no efficient methods have been developed to eliminate the damage caused by fungi. As a result, control of any kind must be preventive.

The following preventive measures should be considered:

1) The use of mechanical drying methods whenever possible.
2) When sun-drying, coffee should be spread in thin layers and stirred frequently.
3) Lots with different moisture content should not be mixed.
4) Whenever closed containers are used, they should have natural or forced ventilation.
5) Incoming coffee should always be monitored for fungus damage and during storage, it should be checked every 2 weeks, at least.
6) The level of moisture content should be maintained below 15 % and RH should be less than 60 %.

1.3.3
Bacteria

No significant damage to coffee by bacteria is known, since they require a RH higher than 90 % to survive and a medium with a neutral or basic pH. Damage is only incurred in cases where the coffee beans are already considerably deteriorated.

1.3.4
Rats, Mice and Birds

Even though rats, mice and birds are not considered important coffee pests, they often consume the beans if they have nothing else to eat and, through their feces, they can spread diseases which affect the safety and quality of coffee.

1.4
Damage Assessment

Various methods have been designed to assess the damages caused and their commercial evaluation. The systems set up by the CSCE, LIFFE and BM & F commodity markets, as well as those of the ICO and GCA, take the beans damaged by insects and fungi as defective and the imperfections are accounted for in the total result.

The FAO has also developed a simplified system in a 100 gram sample, for weight loss which is expressed as follows:

$$\frac{U_a N - (U+D)\ 100}{U_a N}$$

U = weight of undamaged fraction in the sample, N = total number of grains in the sample, U_a = average weight of one undamaged bean and D = weight of damaged fraction in the sample.

Regardless of the assessment method that is used, it is important that the damage from fungi and insects be permanently evaluated in order to prevent a decrease in the commercial value of coffee.

1.5
Pest Control

Simple and practical recommendations to diminish the damages caused by these pests are described below.

1) Mixing of lots with high infestation levels should be avoided and the damage per unit should be assessed. Whenever it is presumed that the coffee comes from lots picked up from the ground or from cultural control activities like hand picking, repelling or second picking, each lot should be processed separately.
2) Mechanical drying methods should be used and the moisture content of the beans kept below 13 %.
3) New bags should be used for each lot. In cases where bags are reused they have to be fumigated with aluminum phosphide.
4) Rodent traps and low frequency sonic equipment might be used to eliminate feeding sources and any cracks or holes in which rodents may hide should be securely closed.
5) Floors and walls of storage areas must be regularly cleaned to prevent the access of insects. Again, cracks or holes should be filled.
6) Regular fumigation with aluminum phosphide should be carried out for green, parchment and dry cherry coffee alike. Fumigation should be conducted under air-tight conditions, for a period of at least 72 h. Studies have revealed that Robusta coffee absorbs this product better than Arabica, due to its waxy composition. When fumigating with aluminum phosphide, any contact with moisture must be avoided at all costs because of fire risks. Plastic coverings should be used to protect small lots and to ensure an adequate exposure. Silos should be used to store large lots in order to prevent leakage which may be hazardous to staff. Any coffee that has been fumigated must be clearly marked. Applications should be repeated, since they do not eliminate eggs or pupae. In spite of its great efficiency, methyl bromide will no longer be allowed after 2010.
7) Prevent contamination by fumigating with low toxicity products like pyrethroids.
8) The moisture content of the beans in the lot and the percentage of infestation should be monitored every 2 weeks at least.

1.6
Quality Impact

1.6.1
Cup

Although stored coffee may already present flavor damage, the incidence of the already existing damage may become more serious during storage. New damage that did not exist before can also occur.

Potential new damage caused during storage which affects cup flavor can be described as follows:

- *Baggy.* This defect is a flavor absorbed from the bag itself during storage. It is frequently the result of a lengthy storage period where there is a high moisture content (above 13%). It can also be caused by re-using bags or treating them with hydrocarbon oils.
- *Mouldy.* This is a typical fungi flavor which is caused by the development of fungi on the beans. It is one of the worst off-flavors.
- *Earthy.* This results from the combination of drying, handling and storage operations. The earthy flavor and aroma are generated when the coffee comes into direct contact with dusts and earth under half-dried storage conditions.
- *Onion.* This defect arises when coffee is over-fermented, as a result of storage under humid conditions. Propionic acid is generated which gives the product the characteristic "onion" flavor.
- *Old crop.* Even under adequate or optimal storage conditions, coffee beans deteriorate with age. This phenomenon is accelerated when the environment is hot and/or humid and the beans takes on a paper or rancid flavor due to the oxydation of its own fats. When the temperature of the storage area is kept below 20 °C and the

RH is below 60%, this defect is insignificant. The same applies when the coffee is stored in parchment or ball form.
- *Contaminated.* This flavor develops when coffee is stored with other products like sulfur, cardamom, oils, soaps, etc., which have volatile components that are absorbed by the coffee and come out in the cup.

1.6.2
Color

The impact of storage on color is very important. It depends to a great extent on the methods adopted for production and processing which can deteriorate the product during its storage. Indirectly, color reflects storage conditions. A general problem is that the coffee is overdried at temperatures above 80 °C. This diminishes the natural green color giving the beans a grayish hue and when they reabsorb moisture, they bleach starting from the edges. The beans that are dried at adequate temperatures, but which keep a moisture content higher than 12%, may have "wet spots" which bleach in the course of time. There is also the defect called "cardenillo" which is caused by the action of microorganisms that destroy the superficial parts of the bean and leave a dust with a yellowish-reddish color.

1.6.3
Defects or Imperfections

In addition to the impact on green coffee color, the following list describes the main defects due to bad storage.

1) *Mouldy beans.* These beans show the development of mould with white, grey or greenish colors. This will show up in coffee flavor and aroma.

2) *Infested beans.* Coffee which has been badly damaged by insects during storage. Depending on the level of the damage this is considered to be a defect.

3) *Bleached beans.* High temperatures and high levels of moisture will impact bean color, particularly when the beans are stored for a lengthy period.

For more details on storage defects, see Section IV.3.

Figure 1.3 Storing coffee in bags.

1.7
Green Coffee Storage

Based on the organization scheme, storage can be classified under the following systems:

- Home (subsistence)
- Private (direct ownership)
- Community (commercial)
- Centralized (regional or national).

The following factors determine the choice of operational storage systems:

1) The scale or size of the operation
2) Technological and logistical availability
3) Available capital and labor
4) Operating costs
5) Atmospheric and climatic conditions
6) Potential problems with pests and fungi
7) The ratio of value to cost.

The most common storage methods are described below.

1.7.1
Bags

Bags are the most commonly used means for coffee storage around the world, even for the London and New York commodity markets. Most of the warehouses in producing and importing countries use bags

Figure 1.4 Piling pallets with coffee bags.
From: J. N. Wintgens.

because they are adapted to the basic requirements throughout the supply chain (Fig. 1.3).

The bag material is usually jute/sisal and capacity of the bags is 60 or 69 kg. The weight of an empty bag varies from 1 to 2.2 lb. The bags can be positioned with or without pallets. Wooden pallets are the most suitable as they prevent sliding. Between 20 and 30 bags are placed in three to five layers (Fig. 1.4).

The high cost of labour, as well as the need to improve warehousing practices, has promoted the development of technologies such as the automatic unloading, sampling and palletising equipment developed by one big warehouse company in Antwerp. This system has the capacity to unload bags from containers in a totally computerised system that has more than 60 different ways to accomodate bags on the pallet. Benefits in reduction of labour needs, time for unloading, pallet stability and warehousing efficicency make the investment worthwhile.

It is important to avoid direct contact between the bags and the ground and to pile the bags too high for two reasons: (1) to prevent the piles of bags from toppling, and (2) as heat rises and collects in a layer at the top of the storage area, to avoid storing the bags in overheated conditions. Pallets should be cross-piled for better stability.

The density of coffee can vary according to the moisture content. Hence, the use of polypropylene bags is not recommended for green coffee because their ventilation is limited. It should also be noted that they are more slippery and easily cause the piles to slide.

Storage in bags means that the entire storage area cannot be filled. The smaller the lots, the lower the storage efficiency. Generally speaking, it is recommended that only 70 % of the available space is used. The remaining space can be used for corridors and spaces between the lines of piled up bags.

Pallets are generally 20–30 cm in height. Plastic pallets tend to be slippery and may cause the piles to slide. Bags should not lean against the wall because this prevents correct ventilation and the weight could bend the wall structure.

From Table 1.2 it can be seen that the major disadvantages of using bags are related to handling efficiency, labor requirements and the choice of the bag material. On the other hand, as little financing is required, special attention can be given to these factors.

Table 1.2 Advantages and disadvantages of using bags for storage

Advantages	Disadvantages
The same container can be used	High labor and handling requirements
Investment is low	Non-hydrocarbon free bags can cause contamination
Very little specialized handling equipment is required	
The storage possibilities are unlimited	Polypropylene bags do not allow for correct ventilation
It is easier to maintain traceability	When old bags are used there can be losses through holes made by handling hooks
It is possible to separate damaged products	
Suitable bags allow for correct and thorough ventilation	Handling efficiency is low when compared with bulk storage systems
Bags are well adapted to all types of coffee (green, parchment and dry cherry)	In some countries, the weights lifted by workers are limited

1.7.1.1 General Recommendations and a Checklist for Warehouses

- It is better to design specific facilities than to adapt existing ones.
- Air ventilation is indispensable – it can be achieved either by the design and the orientation of the storage facility or by mechanical equipment.
- Birds, rodents, insects and dust entry should be prevented by physical means.
- It is absolutely necessary to level the ground to ensure safe handling and pallet stability.
- Natural light will always be cheaper than artificial light, so translucid material for roofing is preferable.
- Heat traps should be installed on the roof wherever possible.
- All safety and identification signs should be placed inside.
- Permanent, regular cleaning and maintenance is necessary.
- Pallets should not be piled higher than a maximum of four pallets.
- Warehousing must be physically separated from milling facilities.
- Contamination with foreign odors and aromas should be avoided.

Table 1.3 Advantages and disadvantages of using big bags

Advantages	Disadvantages
Reduced labor requirements	Winches are necessary
Increased handling efficiency	Bag washing is necessary
Flexibility	The height of piles on pallets is reduced
Simplified quality control	Sampling is more difficult and less representative
Low cost of storage per kg of coffee	Traceability loss

- The surfaces of walls and ceilings should be smooth in order to prevent dust accumulation and the proliferation of pests.
- If possible, the facilities should be adapted for winches to reduce the in/out loading time.
- No water drainage inside the warehouse should be allowed; permanent inspections are required.
- Lots must not be placed against the walls as this represents a risk for structure stability and labor safety.

1.7.2 Big Bags

This system combines the bulk system and bags. Polypropylene bags of 0.5–2 MT are used to store the product. Reinforced big-bags are more efficient as they are more resistant, they are more space efficient and can also be used for transport in containers, up to 20 MT can be stored in one container. A pallet is used for each bag, with a maximum pile-up capacity ot two levels due to the risk of accidents, unless racks are used. This system is widely used with other products, such as sugar, cocoa, etc. (Fig. 1.5, 1.6).

1.7.3 Silos

Silos are non-ventilated structures built with different materials which allow the bulk storage of green coffee beans. This system is connected with automated ventilation and transporting lines. In this system, air ventilation, initial moisture content, RH and temperature are more relevant than in bag storage. Flexibility can be achieved if the number and capacity of silos are sufficient. Silo changes are required in order to diminish the temperature convection and moisture migration, particularly when the coffee

Figure 1.5 Storing coffee in big bags.

Figure 1.6 Storing coffee in double big bags.

conditions are not optimal. Monitoring equipment for sampling, RH and temperature measuring at different stages and points is absolutely necessary.

Rounded structures with conical tops are the most common. Their diameters range from 5–15 m. Most silos are built of con-

crete or metal structures but concrete is not recommended because it absorbs moisture and its alkaline particles could contaminate the coffee.

Silos are not common in producing countries; they are more frequent in consuming countries. They are justified

Table 1.4 Advantages and disadvantages of storage in silos

Advantages	Disadvantages
Low labor requirements	High initial investment and special equipment required
Single unit handling	
High volume and speed when handling	Pre-blending is necessary
Low running costs if efficient lines are used	Little flexibility
Adapted for blending and vendor inventory systems	Sedimentation of small beans
	Difficult to maintain traceability
	Damage to quality can spread easily if not correctly monitored
	Hot spots, bean damage and compaction are frequent when the capacity has not been correctly calculated

Figure 1.7 Silos for coffee storage.

where large quantities of coffee are stored. In order to avoid contamination, the materials used for traceability purposes should be suited to human consumption needs (e.g. stainless steel).

In silos there is a risk of explosion which is generated by particles suspended in the air. These particles come from the ground and from insect feces which are 91 µm in size with a density of 1.49 g/cm³. An explosion can only occur when there is material in suspension, an oxidating atmosphere and a heat source to start a chain reaction. Initially, a spark causes a dust explosion and a pressure change is created, which causes a re-suspension of the dust for a higher intensity explosion. A source of heat, e.g. fungal activity on high moisture

Table 1.5 Advantages and disadvantages of using containers

Advantages	Disadvantages
Fumigation is possible	Limited ventilation
They are safe and versatile	Great exposure to heat in ports and yards
Traceability is maintained	Bleaching is frequent
	Dependent on the conditions of shippers

beans, generates pyrolic acids which are highly reactive when exposed to any ignition source. Consequently, adequate ventilation, dust removal and static electricity control are recommended (Fig. 1.7).

1.7.4
Containers

Containers are not, strictly speaking, storage facilities but they should be mentioned as large quantities of coffee are transported and stored in containers for periods longer than 15 days.

Containers are usually 20 or 40 feet long and they may or may not be refrigerated. Beans are transported in bags with 250–320 bags per 20 ft container or in bulk with 375 big bags. Transportation should not last longer than 30 days because deterioration may be substantially accelerated if the container is exposed to sunlight and not ventilated. This favors moisture evaporation and condensation due to heat convection inside the container, particularly if the containers are made of steel, which heats up easily (Fig. 1.8).

For more details consult section IV.2.

1.7.5
Controlled-Modified Atmosphere Storage

This system is only used when the value of the coffee makes it worthwhile and when fine quality coffees are to be preserved for a roasted or soluble line.

It is suitable for both bags and bulk and is based on the principle of reduction of the respiration metabolic activity by altering the concentration of air gases (78.08 % N_2, 20.95 % O_2, 0.03 % CO_2), the temperature and the RH. The concentration of O_2 is reduced by combustion or by injection of CO_2 which diminishes the bean's respiratory activity. Refrigeration, ventilation and

Table 1.6 Controlled Atmosphere Storage

Advantages	*Disadvantages*
It greatly increases the length of coffee preservation with minimum deterioration	High initial investment and maintenance costs
	Long-term preservation is costly
It facilitates insect and fungi control	Need for trained staff and specialized technology

1.8 Conclusions

Correct and well-adapted coffee storage depends on the initial quality and sanitary conditions of the bean, on environmental and biological factors and on the infrastructure chosen.

For a long time coffee storage was just considered to be an additional phase in coffee processing. As a result, little research was conducted on the subject during the last century. At present, however, coffee storage has become highly relevant and the center of considerable attention. Significant improvements are now being made in marker biochemistry, material decomposition, controlled atmospheres, etc.

These improvements will undoubtedly lead to the set up of more sophisticated methods for monitoring the deterioration of the various components and the generation of harmful substances, as well as to increasingly technological storage handling. Modern methods like controlled or modified atmospheres will surely be implemented at sites where their higher costs are justified. However, their application will depend, to a large extent, on the value and complexity of the final product, as handling will mostly be carried out in producing countries, where the main limitations are

Figure 1.8 Loading container holding coffee bags.

humidity traps are also used. One variation of this process is the vacuum packing system, where coffee is packed in 25 kg aluminium plasticised bags, then air is extracted and the bags are sealed. Coffee can then be stored for one year without impact on the quality.

Figure 1.9 Vaccum packing.

the prevailing social, technical and financial conditions.

Undoubtedly, the validity of the basic storage principles that are currently applied will be maintained in the future because it has been proved that they are efficient and applicable under a variety of environmental conditions.

Bibliography

[1] ANACAFE. *La Determinación de la Calidad del Café.* ANACAFE, Guatemala, 1971.

[2] ANACAFE. *Manual de Beneficiado del Café.* ANACAFE, Guatemala, 1985.

[3] ANACAFE. *Catación y Clasificación del Café de Guatemala.* ANACAFE, Guatemala, 1986.

[4] Bucheli, P., Meyer, I., Pasquier, M. and Locher, R. Determination of soluble sugars by high performance anion exchange chromatography (HPAE) and pulsed electrochemical detection (PED) in coffee beans upon accelerated storage. Poster presented at the *10th FESPP Meeting,* Florence, Italy, 1996.

[5] Clarke, R. J. and Macrae, R. *Coffee. Vol. 1: Chemistry,* 1985. *Vol. 2: Technology,* 1987. Elsevier, London.

[6] Cleves, R. *Tecnología en Beneficiado de Café.* Tecnica Internacional, San Jose, Costa Rica, 1995.

[7] Duffus, C. and Slaugther, C. *Las Semillas y sus Usos.* AGT Editor SA, México, 1985.

[8] Gecan, J. S., *et al.* Microanalytical quality of imported green coffee beans. *J. Food Protect.* **1988**, *51,* USA.

[9] Goulart, J. *Producao de Café Cereja Descascarado Recomendacões Gerais (Boletim Tecnico).* Guaxupe-Cooxupe, Brasil, 2000.

[10] Guevara, R. *Manual de Plagas y Enfermedades del Café.* Instituto Hondureño del Café, Honduras, 1990.

[11] Kader, A. *Postharvest Technology for Horticultural Crops.* University of California, 1992.

[12] Narasimhan, K. S., Majumder, S. K. and Natajaran, C. P. *Studies on the Storage of Coffee Beans in the Interior Parts of South India.* Indian Coffee, India, 1972.

[13] Nosti, J. *Cacao, Café y Té.* Salvat Editores, Barcelona, 1953.

[14] Othón Serna, S. *Química, Almacenamiento e Industrialización de los Cereales.* AGT Editor, Mexico, 1996.

[15] *Prevention of Post-harvest Losses.* FAO, Rome.

[16] Ramaiah, P. K. *Coffee Guide.* Central Coffee Research Institute, Karnataka, 1985.

[17] Rodriguez-Amaya, D. *Research on Mycotoxins in Latin America.* Faculdade de Engenharia de Alimentos, Universidade Estadual de Campinas, Brasil.

[18] Soares, L. M. V. and Rodriguez-Amaya, D. B. Survey of aflatoxins, ochratoxin A, zearalenone, and sterigmatocystin in some Brazilian foods, utilizing a multi-toxin thin layer chromatographic method. *J. Ass. Official Anal. Chem.* **1989**, *72,* 22–26.

[19] Sivetz, M. and Desrosier, N. W. *Coffee Technology.* AVI, Westport, CT, 1979.

[20] Villaseñor, A. *Caficultura Moderna en México.* Agrocomunicación Sáenz Colín y Asociados, Chapingo, 1987.

[21] Wilbaux, R. *Coffee Processing.* FAO, Rome, 1963.

[22] Wintgens, J. N. *Factors Influencing the Quality of Green Coffee.* Nestlé, Vevey, Switzerland, 1993.

[23] Wrigley, G. *Coffee.* Longman, London, 1988.

[24] Meira, F. *Pos-Colheita do Café.* Edit. UFLA, Lavras, Brazil. 2008.

2
Shipment of Green Coffee

E. Blank

2.1
Introduction

Since coffee drinking became popular in countries other than those producing coffee, the beans had to be shipped by sea and formed an ideal return for the first shipping lines which sailed to these countries.

2.2
Particularities

During transport, high ambient humidity and excessive residual moisture content in the beans can affect their quality.

High ambient temperatures generate "sweating" of the beans, a phenomenon which is emphasized if the remaining moisture in the beans is at the upper limit or above. The resulting humidity has to be evacuated through ventilation or be absorbed by appropriate technical aids. Furthermore, loading during rainfalls, crossing different climatic zones or shipment from tropical areas to countries with a cooler climate can lead to condensation, thus increasing the exposure of the coffee to humidity.

Green coffee may have to travel for several weeks from the countries of origin and mould can develop very quickly under these conditions.

Infestation by insects can also occur and for this reason coffee is generally fumigated prior to shipment to the consumer country. However, legislation can differ from one importing country to another and, therefore, a second fumigation at the port of destination may often be required.

2.3
Bagging

Green coffee is put into jute or sisal bags containing 60, 69 or 70 kg depending on the country of origin. Bags are used for delivering the crop to the exporter's warehouse where sorting, cleaning, etc., takes place and export consignments are prepared.

This has meant that the development of the coffee trade has helped to create a significant jute/sisal industry in the coffee-exporting and other developing countries.

With a view to reducing the handling of numerous small units, big bags containing up to 1200 kg of coffee beans have been used for shipments in containers. How-

Coffee: Growing, Processing, Sustainable Production, Second Edition. Edited by J. N. Wintgens.
© 2012 WILEY-VCH Verlag GmbH & Co. KGaA. Published 2012 by WILEY-VCH Verlag GmbH & Co. KGaA.

ever, units of this size require appropriate mechanized means for handling and their implementation has been limited to re-packing after an intermediary production process such as decaffeination.

2.4
Conventional Shipping Practices

The traditional means of shipping green coffee, prior to the worldwide introduction of containers for sea transport, was to load the bags into the holds of vessels with slings or nets. This very labor-intensive and time-consuming method is still used in a number of countries for important consignments or from ports not yet equipped to handle containers (Fig. 2.2).

2.5
Containers

The introduction of containers for freight shipping in the 1960s also had a consider-able impact on the shipment of green cof-fee.

The basic objective of containers is to re-duce harbor time for seagoing vessels by having the stowage done in the port prior to the ship docking and to hasten loading operations by handling bigger units. This led to considerable investments in vessels and port equipment and generated addi-tional technical constraints, such as the "sweating" of the beans, which had to be addressed and resolved by means of nu-merous trials (Fig. 2.1).

2.5.1
Ventilated Containers

Ventilated containers were the first type of container with the possibility of overcom-ing the problems related to humidity and dampness. Ventilated containers have pro-tected openings at the base and top of the sidewalls, which, through the naturally ris-ing warm air inside the container, create a ventilation effect and allow the evacuation of hot and humid air.

Two major problems are connected with the use of ventilated containers. First, the cost of the special construction is higher than that of standard dry boxes for equip-ment which, due to the openings and the risk of water penetration, are not accepta-ble to exporters of other cargo for the re-turn trip. Therefore, additional cost is in-curred for empty returns. Second, the lim-ited availability of ventilated containers often leads to shortages, particularly during peak shipping periods.

In recent years, shipping companies have been running significant cost-reduc-tion programs in order to successfully com-pete in an industry facing consolidation. Therefore, many carriers have stopped in-vestments in ventilated containers and a possible surcharge may have to be paid in the future for this special equipment.

Other solutions are necessary to satisfy the requirements of all the coffee shippers.

2.5.2
Standard, Dry Containers

Standard dry boxes are usually air and watertight. As a result, they retain the hu-midity of the beans and the air within the container, which can promote the develop-ment of mould.

Adequate precautions can, however, be taken:

• The quality of the container should be ensured by thorough inspection.

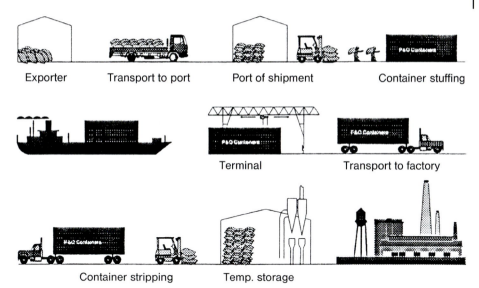

Figure 2.1 Green coffee handling with bags.

Figure 2.2 Loading bags with slings.

- Maximum moisture contents of the beans should be specified in the purchase order (maximum 13 %).
- Cardboard lining of the container floor, walls and door
- Cardboard cover
- Moisture absorbent material (e.g. silicate, dry-bags, etc.) placed in the container
- Stowage "below deck" to attenuate temperature changes (Fig. 2.3).

Numerous trials have proven that green coffee can be shipped in standard, dry containers without excessive risk. This technique enables the use of equipment on a round-trip basis and problems linked to container availability are largely solved.

The loading of the containers is generally carried out in port under the supervision of the shipping line and, if possible, a competent surveyor.

Figure 2.3 Standard dry container with bags.

2.6
Bulk Shipment

Increasing handling costs of the individual bags as well as the increase in surface costs for storage of the bags (due to limited storage height) have led to further research in streamlining and rationalizing the shipping of green coffee.

Bulk shipments, i.e. shipments of loose beans, have been attempted through the following methods (Fig. 2.5). Bulk shipments however are not common today and the majority of green coffee is today shipped by conventional dry containers.

2.6.1
Loose Coffee in Bulk Vessels

These trials have not been very successful due to heavy breakage and humidity prob-lems. In addition, only very large shippers could apply this method as minimum lots of 2500–3000 tons per consignment are required.

2.6.2
Genuine Bulk Containers

These are specially built "tank" containers with manholes for filling at the top and emptying slots at the door. Good results have been achieved with bulk containers of this type.

However, investment costs for such bulk containers are approximately 90 % above the costs for standard dry boxes. Empty returns cannot be avoided and problems linked to shortages during main shipping periods could not be avoided. Furthermore, silos for efficient loading are required in the country of origin (Fig. 2.4).

Figure 2.4 Genuine bulk container.

2.6.3
Standard Dry Containers with Liner Bags

The dry container is lined with a heavy-duty polythene liner bag (for one-way shipment or re-usable) to protect the beans. This is fixed inside on the existing rings at the top of the container walls. The filled bag is secured by steel bars or a wooden construction at the back door (Fig. 2.6, 2.7, 2.8, 2.9, 2.10).

This method has achieved remarkable results without impacting on the quality of the beans and is used increasingly in many producing countries.

The loading of the two bulk container types is done at the exporter's site or in port. For the dry containers with liner bags, very little investment is required in either blowing devices or similar equipment (Fig. 2.11).

Unloading at the destination is generally done by placing the container on a special

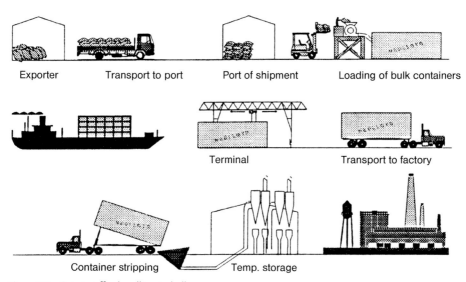

Exporter   Transport to port   Port of shipment   Loading of bulk containers

Terminal   Transport to factory

Container stripping   Temp. storage

Figure 2.5 Green coffee handling in bulk.

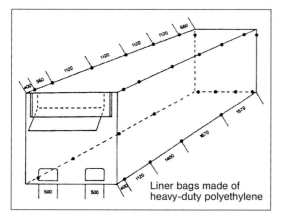

Liner bags made of heavy-duty polyethylene

Figure 2.6 Dry container with liner bags.

Figure 2.7 View of a liner bag.

Figure 2.8 Fixing the liner bag inside the container.

Figure 2.9 Fixed liner bag inside the container.

chassis, which allows the container to tilt and tip the coffee beans out by gravity. At the factory, the beans are stored in silos (Fig. 2.12).

The main advantages of shipping green coffee in dry boxes with liner bags are the following:

- Reduced handling and handling costs, mainly at the destination, providing the beans are delivered in bags by the producers to the exporters; manual unloading of bags takes 1 h for three men, whereas a tilted container can be emptied by one man in 10 min.
- Reduced sea and inland freight due to increased payloads. A bulk container holds 21 tons, whereas a normal container with bags holds approximately 17.5 tons (i.e. +20 %).
- The cost reduction for unused bags is partly offset by the cost for the liner bag which, if used one way only, amounts to approximately 50 % of jute/sisal bags.
- Reduced floor space required for green coffee storage, i.e. less need to use outside warehouse facilities.

Certain disadvantages should also be considered:

- Significant capital investment required in order to install silo reception facilities.
- Difficulties if the quality of the coffee supplied is not up to the required standard.

The fact that the big trading companies as well as important roasters in Europe and the USA are increasingly changing their

Figure 2.10 Loading device for bulk container.

Figure 2.12 Close view of coffee beans emptying out of the liner bag.

shipments from bags to bulk indicates that bulk shipments provide good results in terms of both quality and cost. This trend will therefore become increasingly important for the major coffee producers. Furthermore important traders have started to build and operate silo facilities for bulk coffee in port areas with the objec-

tive of supplying green coffee with more added value (blends). This is mainly geared to the needs of smaller roasters.

The concentration of manufacturing by Nestlé USA has already led to the construction of silos in one factory which delivers ready blends to the other factories with silo lorries on a "just-in-time" base.

Figure 2.11 Tilt chassis in unloading position (sketch and picture).

3
Green Coffee Defects

J. N. Wintgens

3.1
Introduction

Despite improvements made by international organizations and private companies over the past 20 years, the correct identification of physical defects in green coffee remains difficult. The purpose of this chapter is to facilitate the identification of defective beans in green coffee and to point out their effect on cup quality. This should be a great help to agronomists, buyers and quality assurance specialists.

The International Standard ISO 10470 (1993) 'Green Coffee– Defect Reference Chart' was used as a starting point for the preparation of this chapter and it should be noted that the revision of ISO 10740

Figure 3.2 Assessing grade, color and odor of green beans.

Figure 3.1 Green coffee sampling.

Figure 3.3 Cup tasting of roast and ground beans.

Coffee: Growing, Processing, Sustainable Production, Second Edition. Edited by J. N. Wintgens.
© 2012 WILEY-VCH Verlag GmbH & Co. KGaA. Published 2012 by Wiley-VCH Verlag GmbH & Co. KGaA.

which is currently underway is, in turn, using elements included in this chapter. (Personal communication from M. Blanc-Nestlé).

The chapter contains a comprehensive set of true color photographs showing examples of the various defects which may be present in Arabica coffee. To avoid lengthy repetition, similar photographs of defects which may be present in Robusta coffee have not been included as both the aspect of the defects and their influence on cup quality differ only slightly from those of Arabica coffee.

The color photographs are accompanied by the name of the defect, its description, its most probable cause, and its effect on roasting and cup quality. A few microphotographs have also been included to illustrate the damage which may be caused within the bean cells.

Over 200 samples, each consisting entirely of a single type of defect, have been roasted and tasted by a Nestlé expert panel in order to ascertain the effect of each specific defect in the cup. It should be noted, however, that some defects are not apparent in the coffee bean but are detected in the aroma and the cup flavor. This applies, in particular, to foul, dirty, earthy, woody, phenolic or jute-bag-like flavors.

This work is merely one further step towards perfecting the evaluation of green coffee but it appears evident that these efforts still need to be continued, developed and improved.

3.2
Terminology

F   Field damaged beans
Defects which originated in the field and which are related to:

- The coffee tree (genetic)
- The environment (soil, climate)
- Attacks by pests and diseases
- Crop management (water and/or nutritional stress, frost, weed competition, etc.)

FP   Field or process damaged beans
Defects caused by stress due to climatic conditions, water or nutrient deficiencies, inadequate cultivation or harvesting practices, unsatisfactory primary processing.

P   Process damaged beans
Defects caused by inadequate primary processing operations like pulping, washing, drying, hulling, cleaning, etc.

PS   Process or storage damaged beans
Defects caused by inadequate primary processing and/or badly managed storage conditions.

S   Storage damaged beans
Defects caused by imperfect storage conditions and by storage pests.

DP   Dried parts of the coffee fruit
Defects due to poor cleaning operations following dehusking and dehulling.

Scale of negative effect on cup quality
   None
   Low
   Low to medium
   Medium to high
   High
   Very high

3.3
Green Coffee Bean Defects on Arabica

3.3.1
F Field Damaged Beans

F 1: Bean slightly damaged by *Hypothene-mus hampei* (Coffee Berry Borer)

F 2: Bean heavily damaged by *Hypothene-mus hampei* (Coffee Berry Borer)

French	Fève légèrement endommagée par le scolyte du grain
Spanish	Grano levemente dañado por la broca
Portuguese	Grão levemente danificado pela broca

French	Fève fortement endommagée par le scolyte du grain
Spanish	Grano severamente dañado por la broca
Portuguese	Grão fortemente danificado pela broca

Description	Three or less small holes or tunnels inside the bean. The holes are circular and clean cut. Their diameter is 0.3–1.5 mm.
Cause	The cherries are attacked by the pest.
Roasting	When roasted the beans are slightly darker in color than normal beans.
Cup profile	Minor presence of off-flavors. Slightly diminished aroma, flavor and acidity.
Negative effect on cup quality	Low to medium.

Description	More than three small holes or tunnels inside the bean. The holes are circular and clean cut. Their diameter is 0.3–1.5 mm. Sometimes the bean tissue itself has been eaten away by the pest. This gives the bean a ragged aspect.
Cause	The cherries are attacked by the pest.
Roasting	When roasted, the beans are distinctly darker in color than normal beans.
Cup profile	Significant incidence of off-flavors with a predominantly bitter and tarry flavor. Total loss of aroma, flavor and acidity.
Negative effect on cup quality	High to very high.

F 3: Dark brown beans

F 4: Amber bean

French	Fève brun-foncée
Spanish	Grano pardo oscuro
Portuguese	Grão pardo obscuro

Description	Brown to black marked bean, shriveled, crinkled and ragged. Scarred by the Antestia attack or blight whilst immature. Similar attacks on Robusta coffee are very limited.
Cause	Cherries attacked by the Antestia bug which pierce and suck the juice from immature fruit. Also caused by over-ripe cherries.
Roasting	–
Cup profile	Loss of aroma, flavor and acidity with a fruity, sometimes harsh, taste. Occasionally potato flavored.
Negative effect on cup quality	Medium to very high according to the intensity of the damage.

French	Fève ambre
Spanish	Grano ámbar/mantequillo
Portuguese	Grão âmbar

Description	Smooth, yellowish coffee bean, usually semi-translucid.
Cause	Supposed iron deficiency in the soil and/or a high soil pH.
Roasting	–
Cup profile	Slightly diminished aroma, flavor and acidity with a grassy or woody character.
Negative effect on cup quality	Medium.

F 5: Elephant bean
(bullhead, monster bean)

F 6: Triangular bean
(three-cornered)

French	Fève éléphant
Spanish	Grano elefante
Portuguese	Grão elefante

Description	Unusually large bean, spherical in shape. More frequent in Robusta beans.
Cause	Genetic cause consisting of an assembly of two or more beans resulting from false polyembryony.
Roasting	Elephant beans roast unevenly in the presence of normal beans.
Cup profile	No significant effect on cup quality.
Negative effect on cup quality	None.

French	Fève triangulaire
Spanish	Grano triángular
Portuguese	Grão triangular

Description	Triangular in transversal section.
Cause	Genetic cause resulting from the development of three beans per cherry.
Roasting	–
Cup profile	May bestow a slightly immature flavor.
Negative effect on cup quality	None to low.

F 7: Peabean (caracol, caracoli)

F 8: Flaky bean (chips)

French   Fève caracoli
Spanish   Grano caracol
Portuguese   Grão caracol

Description	Coffee bean practically ovaloid in form.
Cause	Genetic cause resulting from the development of a single seed within the fruit. The second bean having aborted.
Roasting	Peabeans roast better when non-mixed with flat beans because they roll more easily during roasting.
Cup profile	Very slight positive effect on cup quality due to more even roasting.
Negative effect on cup quality	None.

French   Fève floconneuse, mince
Spanish   Grano plano/vadijoso/vano
Portuguese   Grão chato/flocoso

Description	Usually very thin beans, light ragged and flaky in appearance and small in size.
Cause	Growth defect arising from nutritional deficiency.
Roasting	Difficult to roast, significant loss of weight. Some beans may be carbonized.
Cup profile	Slight incidence of off-flavors. Diminished aroma, flavor and acidity with a noted woody character.
Negative effect on cup quality	Medium.

F 9: Shell of elephant bean

F 10: Body of elephant bean, core, ear

French	Coquille, oreille de fève d'éléphant
Spanish	Concha de grano elefante
Portuguese	Concha de grão elefante

Description	External part of the elephant bean.
Cause	Splitting of the elephant bean during hulling or husking.
Roasting	May break on roasting and char at the edges.
Cup profile	Diminished aroma and flavor, but otherwise no significant incidence on cup quality when roasting is appropriate.
Negative effect on cup quality	None or low.

French	Corps de fève éléphant, conque
Spanish	Cuerpo de grano elefante (macho)
Portuguese	Corpo de grão elefante

Description	Internal part of the elephant bean.
Cause	Splitting of the elephant bean during hulling or husking.
Roasting	Uneven roasting.
Cup profile	No or little effect on cup quality when the roasting is appropriate.
Negative effect on cup quality	None to low.

F 11: Frost damaged bean

F 12: Immature bean (quaker bean)

French	Fève endommagée par le gel
Spanish	Grano dañado por helada
Portuguese	Grão danificado pela geada

Description	Coffee bean with brown to black color outside and inside. Spotted up to plain brown–black glossy color according to the intensity of frost damage. On dark beans the silverskin may glitter. Strongly adherent silverskin. Robusta beans are very seldom affected because this variety is mostly grown in lowland areas.
Cause	Frost.
Roasting	–
Cup profile	Whether the damage is slight or heavy, it causes a loss of aroma, flavor and acidity with a decrease in body. Heavy frost damage results in intense off-flavors.
Negative effect on cup quality	Medium to very high according to the intensity of the frost damage.

French	Fève immature
Spanish	Grano immaduro/verde
Portuguese	Grão imaturo/verde

Description	Small, "boat-shaped" bean often with a wrinkled surface. Final color of beans ranges from a metallic green to dark green or almost black with a glossy silverskin, depending on the drying conditions. The bean has a very adherent silverskin. Cell walls and internal structure are not fully developed. The beans are smaller than mature beans.
Cause	Growing problems, (drought, stress, fertilization, pests and diseases). Beans from cherries picked prior to ripening. Lower occurrence for dry-processed coffees because immature cherries are partially removed by flotation, but higher with mechanically harvested Arabica.
Roasting	Slow and irregular roast. Beans remain pale in the roast.
Cup profile	Increased bitterness. Diminished aroma, flavor and acidity. Greenish character which can be perceived as a chemical off-flavor. They sometimes have a fermented taste.
Negative effect on cup quality	Medium to high.

F 13: Withered bean (shriveled, ragged)

French	Fève ridée, sèche
Spanish	Grano averanado/arrugado
Portuguese	Grão enrugado

Description	Beans are light in weight and wrinkled.
Cause	Stressed trees. Underdeveloped fruit due to drought.
Roasting	–
Cup profile	Slightly diminished aroma, flavor and acidity. Sometimes immature in flavor.
Negative effect on cup quality	Low to medium.

3.3.2
FP Field or Process Damaged Beans

FP 1: Black Bean

French	Fève noire
Spanish	Grano negro
Portuguese	Grão preto

Description	• Coffee bean of which more than 50 % of the external surface and interior is black. • Coffee bean of which more than 50 % of the external surface is black. Coal-like aspect and dull color with a granulous external surface and often small-sized beans. Adherent silverskin, undesirable appearance, enlarged center-cut, slightly shrunken and often "boat-shaped" (i.e. thin with somewhat pointed ends).
Cause	Six causes: 1) Attacks by pest diseases. 2) Carbohydrate deficiency in the beans due to poor cultural practices and insufficient water during ripening. 3) Over-ripe cherries picked up off the ground. 4) Immature beans affected by faulty drying (i.e. high temperatures). 5) Beans/cherries subjected to over-fermentation by moulds/yeasts and subsequent drying. 6) Poor drying or re-wetting. Higher incidence with dry-processed coffees.
Roasting	Slow to roast. The beans turn dull and yellowish. They rarely reach the second cracking sound during roasting. Lower weight loss.
Cup profile	Total loss of aroma, flavor acidity and body. Significant incidence of off-flavors.
Negative effect on cup quality	High.

FP 2: Partly black bean or semi-black bean

FP 3: Brown bean

French	Fève partiellement noire
Spanish	Grano parcialmente negro
Portuguese	Grão meio preto

Description	• Coffee bean of which 50 % or less of the external surface and interior is black. • Coffee bean of which 50 % or less of the external surface is black.
Cause	The same as for "Black bean".
Roasting	Slow to roast.
Cup profile	Medium incidence of off-flavors. Diminished aroma, flavor and acidity.
Negative effect on cup quality	Medium to high.

French	Fève brune
Spanish	Grano pardo
Portuguese	Grão amarronzado

Description	Beans with brown to dark-brown color.
Cause	Prolonged slow drying, frost damage or die-back on trees.
Roasting	–
Cup profile	Very high incidence of off-flavors, rather similar to stinker beans. Slight loss of body. Total loss of aroma, flavor and acidity.
Negative effect on cup quality	Very high.

FP 4: Waxy bean

FP 5: Foxy bean

French	Fève cireuse
Spanish	Grano ceroso
Portuguese	Grão brunido/cirroso

French	Fève à pelllicule rousse
Spanish	Grano con cuticula rojiza
Portuguese	Grão com cuticula ammarronzada

Description	Coffee bean with translucent, waxy appearance. Range of colors from yellowish green to dark reddish brown, which is the most typical. Very little or no silverskin. Unpleasant odor when shattered. Sometimes a small cavity is visible at the base of the bean and the surface has a decayed fibrous appearance.
Cause	Beans from cherries picked when over-ripe, partly dried. Fermentive effect of bacteria on surface and interior affecting the bean tissue in wet-processed coffees. Prolonged slow drying of over-ripe cherries resulting in fermentation in dry-processed coffees.
Roasting	Tendency to roast quickly, increased roasting loss and darker color.
Cup profile	Ranging from a loss of acidity with low, medium intensity of greenish character to a full loss of aroma and flavor with a high intensity of fermented off-flavor.
Negative effect on cup quality	High.

Description	Beans with foxy (oxidized reddish to light-brown) silverskin The reddish visual color is particularly noticeable in the center-cut. Regular surface structure, normal size and shape. Mainly silverskin is affected.
Cause	Over-ripe. Delay in pulping. Over-fermentation in presence of free skins. Frost damage, improper washing. Collection of effluents in the fermenting mass. Faulty mechanical drying (over-heating, not enough movement).
Roasting	–
Cup profile	Slightly diminished aroma, flavor and acidity. Somewhat greenish in taste.
Negative effect on cup quality	Low.

3.3.3
P Process Damaged Beans

FP 6: Coated bean

P 1: Pulper-nipped bean/pulper-cut bean

French	Fève cuivrée, argentés, bronzée
Spanish	Grano con pelicula plateada/cuticulado
Portuguese	Grão com pelicula plateada

Description	Bean with a strongly adherent silverskin.
Cause	Drought affected beans. Immature beans of normal size. Insufficient fermentation time. Unsuitable drying conditions (too slow). More common in dry-processed coffees.
Roasting	–
Cup profile	Increased bitterness and astringency. Slightly diminished aroma, flavor and acidity. Low to medium incidence of greenish character.
Negative effect on cup quality	Medium.

French	Fève meurtrie, emdommagée au cours du dépulpage
Spanish	Grano mordido por el depspulpador
Portuguese	Grão danificado durante o despolpamento

Description	Wet-processed bean cut or bruised during pulping, often with brown or blackish marks resulting from secondary microbe attacks. Not expected in dry-processed coffees since no pulping is involved.
Cause	Faulty adjustment of pulping machine or feeding with under-ripe or unequal sized cherries.
Roasting	–
Cup profile	Low to medium loss of aroma, flavor and acidity as per the degree of damage. Chemical and/or fermented off-flavors are slight.
Negative effect on cup quality	Medium.

P 2: Bruised bean (crushed)

P 3: Partly depulped cherry

French	Fève meurtrie, écrasée
Spanish	Grano machacado, aplastado
Portuguese	Grão esmagado

French	Cerise partiellement dépupée
Spanish	Cereza media cara
Portuguese	Cereja parcialmente despulpada

Description	Bruised beans, often partly split and faded with center-cut largely open.
Cause	Treading on the beans during drying. Hulling of soft, under-dried beans. Heavy pounding of under-dried beans (manual hulling with a mortar).
Roasting	Uneven roast.
Cup profile	Slight loss of acidity but diminished flavor and aroma. Fermented off-flavors are medium to high.
Negative effect on cup quality	Medium to high.

Description	Only part of the pulp has been removed during depulping.
Cause	Depulping of immature, over-ripe or partially dried cherries. Not expected in dry-processed coffees.
Roasting	–
Cup profile	Low incidence of off flavors. Slightly diminished in aroma, flavor and acidity.
Negative effect on cup quality	Low to medium.

P4: Stinker bean (foul)

P 5: Under-dried soft bean

French	Fève puante
Spanish	Grano hediondo, fétido
Portuguese	Grão fétido

Description	Coffee bean which gives off a very unpleasant odor when crushed or ground. Light-brown, brownish or grayish but always dull in color. Occasionally, it has a waxy appearance which can cause confusion with a waxy bean. The distinction is sometimes difficult but can be ascertained under UV light. Generally a small cavity visible at the base of the bean indicates a decayed embryo.
Cause	Over or repeated fermentation. Beans retained too long or exposed to polluted water. Delay in pulping or faulty drying. Not frequent in dry-processed coffees.
Roasting	–
Cup profile	Total loss of aroma and flavor. High incidence of fermented off-flavor and a foul odor with a rotten fish flavor.
Negative effect on cup quality	Very high.

French	Fève molle, pas assez séchée
Spanish	Grano suave, insuficiente-mente seco, flojo
Portuguese	Grão mole não bastante sêco

Description	Dark bluish–green colored rubbery bean. Easy to cut through with a knife. Turns white during storage.
Cause	Under-drying.
Roasting	Increase in roasting loss.
Cup profile	Slightly diminished aroma and flavor.
Negative effect on cup quality	Low.

P 6: Bean in parchment

French	Fève en parche
Spanish	Grano en pergamino
Portuguese	Grão em pergaminho, marinheiro

Description	Coffee bean entirely or partially enclosed in its parchment (endocarp).
Cause	In wet-processed coffee, the cause is faulty hulling and separation of the parchment. In dry-processed coffee, the cause is generally an accidental depulping by crushing before drying.
Roasting	Risk of fire during roasting.
Cup profile	Diminished flavor, acidity and aroma with a distinctly woody character and perceptible chemical off-flavors.
Negative effect on cup quality	Medium to high.

P 7: Dried cherry, pod

French	Coque, cerise sèche
Spanish	Cereza seca, grano en coco
Portuguese	Cereja sêca, grão em côco

Description	Dried fruit with outside envelopes containing the beans.
Cause	Not expected in dry-processed coffees as pulp is pre-removed but may occur accidentally due to the presence of small cherries and faulty sorting. Incorrect husking allows small pods through and they are not always removed.
Roasting	Risk of fire during roasting.
Cup profile	Slight loss of acidity, aroma and flavor with an occasional incidence of off-flavors.
Negative effect on cup quality	Low to medium.

P 8: Discolored over-dried bean

French	Fève décolorée, trop séchée
Spanish	Grano descolorado, sobreseco
Portuguese	Grão descolorido, misto sêco

Description	Dull, amber or slightly yellowish colored bean which is brittle and shatters under pressure.
Cause	Over-drying (moisture content of 9 % or less).
Roasting	Tendency to char or burn during roasting.
Cup profile	Slightly diminished flavor, acidity and aroma with a greenish character. Occasionally a cereal or woody flavor may develop.
Negative effect on cup quality	Medium.

P 9: Crystallized bean (jade bean)

French	Fève cristallisée, vitreuse
Spanish	Grano cristalizado
Portuguese	Grão cristalizado

Description	Grayish-blue colored bean, brittle and easily broken.
Cause	Drying temperature too high (over 50 °C). Less frequent in Arabica except in the case of mechanical drying.
Roasting	Beans break during roasting.
Cup profile	Slightly diminished aroma, acidity body, and flavor with a greenish taste.
Negative effect on cup quality	Low to medium.

P 10: Bean fragment

P 11: Broken bean

French	Brisure de fève
Spanish	Fragmento de grano
Portuguese	Fragmento de grão

Description	Fragments of coffee beans with volumes of less than half a whole bean.
Cause	General handling. Mostly occurs during dehulling or dehusking operations.
Roasting	Risk of fire and tendency to carbonize during roasting.
Cup profile	Slightly diminished body, which may affect flavor, acidity and aroma.
Negative effect on cup quality	Low to medium.

French	Fève brisée
Spanish	Grano partido, quebrado
Portuguese	Grão partido, quebrado

Description	Fragments of beans with volumes equal to or greater than half a whole bean.
Cause	General handling. Occurs mostly during dehulling or dehusking operations.
Roasting	Uneven roast.
Cup profile	May slightly affect body, acidity, aroma and flavor.
Negative effect on cup quality	Low.

3.3.4
PS Process or Storage Damaged Beans

P 12: Toasted bean

PS 1: "Cardenillo" bean

French	Fève roussie
Spanish	Grano tostado
Portuguese	Grão tostado

Description	Light brown bean.
Cause	Prolonged contact with hot metal surface of dryer elements (perforated plate). Not expected in dry-processed coffee.
Roasting	–
Cup profile	Slightly diminished flavor, acidity and aroma with a woody character. Increased astringency and bitterness.
Negative effect on cup quality	Medium.

French	Fève "Cardenillo"
Spanish	Grano "Cardenillo"
Portuguese	Grão "Cardenillo"

Description	Beans infested with microorganisms and covered with a yellowish or reddish powder (spores).
Cause	Over-fermentation. Prolonged interruption in the drying process. Storing with a too high moisture content. Lower occurrence with dry-processed coffees.
Roasting	–
Cup profile	Diminished aroma, flavor and acidity with distinctly noticeable off-flavors.
Negative effect on cup quality	High.

PS 2: Sour bean (vinegar)

French	Fève sùre, acre, aigre
Spanish	Grano agrio, vinagre, acre
Portuguese	Grão ardido

Description	Beans are yellow to light-brown or reddish to dark-brown both on the external surface and inside. Center-cut free of teguments. Silver-skin tends to become red. Smells of vinegar. Small cavity occasionally visible at the base of the bean indicating a decayed embryo. May also have a waxy appearance.
Cause	Excessive time between harvesting and pulping. Over-fermentation. Dirty fermentation tanks. Use of contaminated water. Storing with a too high moisture content. Fermentation of over-ripe cherries in a slow drying process caused by deep layers which result in the development of excess internal heat and the destruction of embryos. In Brazil, cherries infected by xerophilic moulds (ardido bean) have been observed as a cause for sour beans.
Roasting	–
Cup profile	Significantly diminished aroma and flavor. High incidence of sour, winey or acetic flavors.
Negative effect on cup quality	Very high.

PS 3: Blotchy bean (spotted, mottled)

French	Fève marbrée, bigarrée
Spanish	Grano veteado
Portuguese	Grão pintado

Description	Coffee bean showing irregular greenish, whitish or sometimes yellow patches.
Cause	Faulty drying or re-wetting after drying, often due to broken parchment. Less frequent in dry-processed coffees.
Roasting	–
Cup profile	Slightly diminished aroma, flavor and, especially, acidity. Slightly noticeable mouldy off-flavor.
Negative effect on cup quality	Low to medium.

3.3.5
S Storage Damaged Beans

S 1: Spongy bean, faded, whitish bean

French	Fève spongieuse
Spanish	Grano corcho/esponjoso
Portuguese	Grão esponjoso

Description	Whitish-colored bean with a cork-like consistency. Can be indented with a fingernail.
Cause	Undue moisture absorption during storage or transport leading to deterioration by enzymatic activity.
Roasting	Beans roast rapidly and tend to carbonize.
Cup profile	Slightly diminished aroma, flavor and acidity with a woody or cereal character.
Negative effect on cup quality	Low to medium.

S 2: White floater bean

French	Fève blanche légère
Spanish	Grano blanco, de baja densidad, flotador
Portuguese	Grão chocho, marinheiro

Description	Bean that floats in water because of its very low density when compared to a healthy bean. White in color and more bulky than a normal bean.
Cause	Cause not yet fully identified, but supposed to occur during storage.
Roasting	High roasting loss, the beans present a brown, sooty color.
Cup profile	Diminished flavor, acidity and aroma with a woody character.
Negative effect on cup quality	Medium.

S 3: Stale bean

French	Fève rassise
Spanish	Grano reposado, envejecido
Portuguese	Grão descansado, repousado

Description	Dull in appearance with a surface mottled by lighter spots. Odor of old coffee.
Cause	Prolonged storage, storage under adverse conditions.
Roasting	–
Cup profile	Diminished aroma, flavor and acidity with a distinctly woody character.
Negative effect on cup quality	Medium to high.

S 4: White bean

French	Fève blanche
Spanish	Grano blanco
Portuguese	Grão branco

Description	White-colored bean surface ranging from pale green to light ivory, sometimes with a variegated pattern. Normal density and internal structure. More common in dry-processed coffee.
Cause	Discoloration of the surface due to bacteria of the *Coccus* genus during storage or transport. Generally associated with coffee from old crops. Also caused by re-wetting after drying.
Roasting	–
Cup profile	Diminished flavor and aroma. Stale taste for every kind of coffee.
Negative effect on cup quality	Low to medium.

S 5: Mouldy bean, musty

French	Fève moisie
Spanish	Grano mohoso
Portuguese	Grão mofado

Description	Coffee beans which show mould growth which is visible to the naked eye. Releases typical mouldy odor.
Cause	Faulty temperature and humidity conditions during storage and transportation.
Roasting	–
Cup profile	Total loss of aroma, flavor and acidity with a very high mouldy off-flavor.
Negative effect on cup quality	Very high.

S 6: Bean slightly damaged by storage pests

French	Fève légèrement endomma-gée par les insectes du café au cours de l'entreposage
Spanish	Grano ligeramente dañado por plagas del café almace-nado
Portuguese	Grão levemente danificado pelos pragas do café armaze-nado

Description	Beans with three or less small holes or tunnels. Diameter of the holes larger than 1.5 mm. They differ from Coffee Berry Borer damaged beans because the holes are not clean-cut and there is less secondary infection. No insect presence.
Cause	Beans attacked by storage pests, generally the coffee bean weevil *Araecerus fasciculatus*.
Roasting	–
Cup profile	Diminished flavor, acidity and aroma and slight loss in body. Distinctly noticeable off-flavors.
Negative effect on cup quality	Low to medium.

S 7: Bean heavily damaged by storage pests   **S 8:** Bean infested by storage pests

French	Fève fortement endommagée par les insectes du café au cours de l'entreposage	French	Fève infestée par les insectes du café au cours de l'entreposage
Spanish	Grano fuertemente dañado por plagas del café aimacenado	Spanish	Grano infestado por plagas del café almacenado
Portuguese	Grão fortemente danificado pelos pragas do café armazenado	Portuguese	Grão infestado pelos pragas do café armazenado

Description	Beans with more than three holes or tunnels. Part of the bean tissue has been destroyed. Diameter of the holes larger than 1.5 mm. They differ from Coffee Berry Borer damaged beans because the holes are not clean-cut and there is less secondary infection. No insect presence.	**Description**	Coffee beans that harbor one or more live or dead insects at any stage of development as well as excrements (frass) and insect fragments.
Cause	Attacks by storage pests (SP).	**Cause**	Infestation by storage pests.
Roasting	–	**Roasting**	–
Cup profile	Loss of aroma, flavor, body and acidity with a high incidence of off-flavors.	**Cup profile**	Loss of aroma, flavor, body and acidity with an incidence of off-flavors.
Negative effect on cup quality	Medium to high.	**Negative effect on cup quality**	High.

3.3.6
DP Dried Parts of the Coffee Fruit

S 9: Spotted bean

French	Fève mouchetée
Spanish	Grano moteado
Portuguese	Grão mosqueado

Description	Coffee beans with black specks (small patches of black silverskin remaining).
Cause	Beans stored with high moisture content.
Roasting	–
Cup profile	Loss of aroma, flavor, body and acidity with a high intensity of fermented and chemical off-flavors.
Negative effect on cup quality	Very high.

DP 1: Piece of parchment

French	Petite peau, fragment de parche
Spanish	Cascara de pergamino
Portuguese	Parte de pergaminho

Description	Fragment of dried endocarp (parchment). Parchment is often spotted by pulp pigment in dry-processed coffees.
Cause	Inadequate separation after hulling or husking. More frequent in wet-processed coffees.
Roasting	Risk of fire during roasting.
Cup profile	Slightly diminished aroma and flavor with a slight greenish taste.
Negative effect on cup quality	None to low.

DP 2: Piece of husk

French	Grosse peau, fragment de coque
Spanish	Cascarilla
Portuguese	Fragmenta de casca

Description	Fragment of the dried external envelope (pericarp). Husk and parchment often knit together.
Cause	Inadequate separation after de-husking. More frequent in dry-processed coffees.
Roasting	Risk of fire during roasting.
Cup profile	Diminished aroma with a slight chemical off-flavor.
Negative effect on cup quality	Low to medium.

3.4
Reference Beans

3.4.1
Arabica

A(W)   Arabica Washed
       Wet processed
       Fair quality

A(N)   Arabica Natural
       Dry processed
       Fair quality

A(W)   Arabica Washed
       Wet processed
       Old beans

3.4.2
Robusta

R(W)   Robusta Washed
       Wet processed
       Fair quality

R(N)   Robusta Natural
       Dry processed
       Fair quality

R(N)   Robusta Natural
       Dry processed
       Old beans

3.5
Microphotographs of Coffee Beans

3.5.1
Arabica Washed – Fair Quality Beans:
Ref. A(W)

Surface of the bean – the epiderm is relatively even.

Surface of the bean – presence of only a few patches of adherent silverskin.

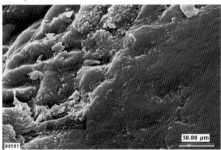

Transversal cut through the bean – tissue and cells have a normal aspect.

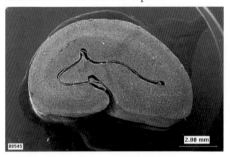

Transversal cut through the bean – cells are well-filled with parenchyma (internal tissue).

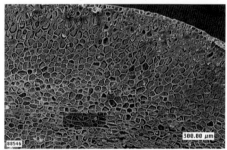

Transversal cut through the cell – cell walls and parenchyma have a normal structure.

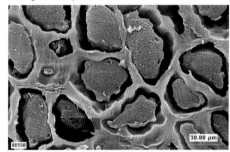

Transversal cut through the cells – slight retraction of the parenchyma due to laboratory preparation.

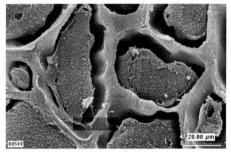

3.5.2
Immature Arabica Bean, ref. F 12

The surface of the bean is cracked.

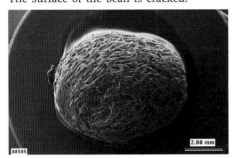

Surface of the bean – shredded silverskin is visible on the surface of the bean.

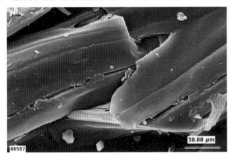

Transversal cut through the bean – patches of silverskin are visible on the surface of the bean.

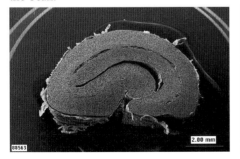

Transversal cut through the bean – cells are almost empty of parenchyma (internal tissue) because the maturation of the bean is incomplete.

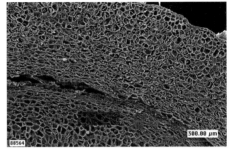

Transversal cut through the cells – cells are almost empty of parenchyma because the maturation of the bean is incomplete.

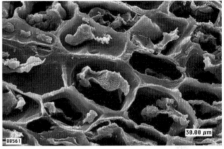

Transversal cut through the cells – cells are almost empty of parenchyma. These destroyed and empty cells explain why immature beans can spoil a coffee brew.

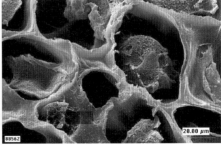

3.5.3
Arabica "Cardenillo" Bean: Ref. PS 1

The surface of the bean is cracked.

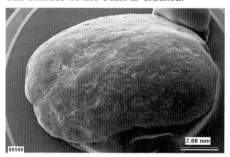

Transversal cut through the bean – cells are almost empty as a result of microbiological attack.

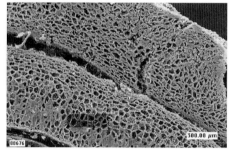

Surface of the bean – a few small patches of silverskin remain. Presence of microorganisms.

Transversal cut through the cells – cell are almost empty as a result of microbiological attack.

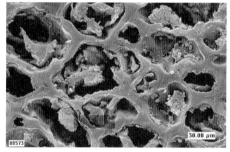

Transversal cut through the bean – tissue of the bean is partly destroyed by microorganisms.

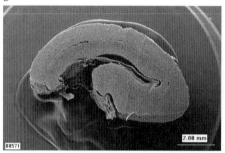

Transversal cut through the cells – cells are almost empty of parenchyma. These destroyed and empty cells explain why "Cardenillo" beans can spoil a coffee brew.

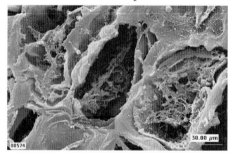

3.5.4
Arabica Black Beans: Ref. FP 1

Transversal cut through the bean. Cells are almost empty as a result of microbiological attack.

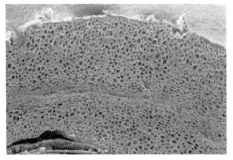

Transversal cut through the cells. These destroyed and empty cells explain why black beans can spoil a coffee brew.

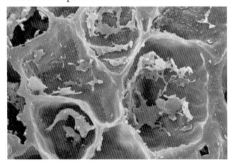

**3.6
Microorganisms which Attack Coffee Beans**

Bacteria

Yeast

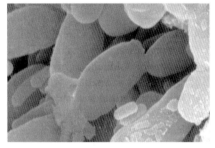

Molds of *Fusarium* spp.

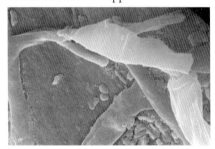

Molds of *Aspergillus flavus* and *A. niger*

Bibliography

[1] ISO 3509. *Coffee and its Products: Vocabulary*, 1989.

[2] ISO 4072. *Green Coffee in Bags: Sampling*, 1982.

[3] ISO 4149. *Green Coffee: Olfactory and Visual Examination and Determination of Foreign Matter and Defects*, 1980.

[4] ISO 4150. *Green Coffee: Size Analysis, Manual Sieving*, 1991.

[5] ISO 5492. *Sensory Analysis: Vocabulary*, 1992.

[6] ISO 9116. *Green Coffee: Guidance of Methods of Specification*, 1992.

[7] ISO 10470. *Green Coffee: Defect Reference Chart*, 1993.

[8] ANACAFE. *Catacion y Clasificacion del Café en Guatemala*. Associacion Nacional del Café, Guatemala, 1986.

[9] Barel, M. and Jacquet, M. *La Qualité du Café*. CIRAD-CP, France, 1994.

[10] Clarke, R. J. and Macrae, R. *Coffee. Vol. 2: Technology*. Elsevier, London, 1987.

[11] Coffee Growers Association. *Coffee Handbook*. Harare, Zimbabwe, 1987.

[12] Comité Departamental de Cafeteros de Antioquia. *El Beneficio del Café*. omité Departamental de Cafeteros de Antioquia, Colombia, 1991.

[13] Coste, R. and Cambrony, H. *Caféiers et Cafés*. Maisonneuve and Larose, Paris, 1989.

[14] Illy, E. and Ruzzier, L. *Propostion d'un Nouveau Système d'évaluation Gravimétrique des Défauts du Café Vert*. Illycaffé, Trieste, 1968.

[15] International Tea and Coffee Brokers. *Coffee Tasting Terminology*.

[16] International Trade Center. *Coffee – An Exporter's Guide*. UNCTAD/GATT, Geneva, 1992.

[17] Jobin, P. *Les Cafés Produits dans le Monde*. Jobin, Le Harve, 1992.

[18] New York Coffee and Sugar Exchange, Inc. *Schedule of Full Imperfections for Coffee under C Contracts*. New York Coffee and Sugar Exchange.

[19] Rothfos, B. *Rohkaffee und Röstkaffee Prüfung*. Gordian-Max-Rieck, Hamburg, 1963.

[20] Shell Agriculture. *A Guide to Arabica Coffee Quality*. Harare, Zimbabwe.

[21] Wintgens, J. N. *Factors Influencing the Quality of Green Coffee*. Nestlé, Vevey, Switzerland, 1993.

4
Factors Influencing the Quality of Green Coffee

J. N. Wintgens

4.1
The Interactions between Market Situation and Quality

For cost-efficient purchasing, the manufacturers and purchasers need to understand the main quality factors and how to assess them. The continual splitting of large coffee estates, the excessive expansion of new plantations and the growing number of intermediaries in the marketing chain has resulted in a deterioration of coffee quality.

An analysis of the coffee market and consumer needs has shown that this situation can be improved if raw material with the required quality is produced, in line with the required quantity. This can probably be achieved through the demand/supply mechanism. The manufacturer has a great impact on this mechanism since he selects the raw material to be processed.

The main factors that influence the quality of green coffee are the genotype, environment, field management, preparation and storage. The manufacturer has to understand these factors and be able to assess the quality. This cannot be done at a glance – it is necessary to take representative samples and test them visually as well as by organoleptic means.

4.2
Influence of the Genotype

The genotype is a key factor, since it determines to a great extent important characteristics such as the size and shape of the beans as well as their color, chemical composition and flavor.

4.2.1
Size and Shape of the Beans

Among the different types of coffee, there are considerable variations of size, shape and density. For instance, Arabica beans are larger and denser than Robusta beans: 100 Arabica beans weigh 18–22 g, 100 Robusta beans only 12–15 g. The SL28 and the Timor hybrid produce large beans; the Caturras and the Rume Sudan produce smaller beans. Table 4.1 shows the characteristics of other varieties.

The shape and structure of the beans (triangular, elephant, peabean and empty beans) are the result of both genotype and environmental factors, e.g. Rume Sudan beans have a large proportion of peabeans (Tab. 4.1). These characteristics, which are shown in Fig. 4.1, are not considered to be true defects as far as flavor is concerned.

Coffee: Growing, Processing, Sustainable Production, Second Edition. Edited by J. N. Wintgens.
© 2012 WILEY-VCH Verlag GmbH & Co. KGaA. Published 2012 by Wiley-VCH Verlag GmbH & Co. KGaA.

Table 4.1 There is a marked difference between bean size and organoleptic quality of different varieties originated from simple or multiple crosses and backcrosses

Variety of cross	Bean size (%)			Cup quality			
	PB	AA	AB	Acidity[a]	Body[a]	Flavor[a]	Standard[b]
SL28	14	46	18	1.0	1.0	2.0	2.0
Caturra	11	15	25	2.0	1.5	3.7	3.8
Rume Sudan (RS)	29	3	27	2.5	2.0	3.3	3.3
Timor Hybrid (TH)	23	39	21	2.0	1.6	4.1	4.0
SL28 × Rume Sudan	35	15	27	1.5	1.0	2.5	3.0
Caturra × Timor Hybrid	31	29	19	3.0	2.5	4.5	3.8
Catimor ex-Colombia F$_3$ prog.2	23	47	15	1.5	2.0	4.0	3.5

PB: peabeans (two beans in one cherry, out of which one has not developed); AA: fraction of beans retained by a no. 18 screen (7.15 mm); AB: fraction of beans retained by a no. 15 screen (5.95 mm).
[a]Marks for acidity, body and flavor: 0–4; 0 = very good, 4 = very low
[b]Standard results: 0–7; 0 = very good, 7 = very low.
All samples were prepared by the wet method, sun-dried, 11 % moisture content.
The liquor was assessed by the same group of tasters.
Source: Van der Vossen and Walyaro [19].

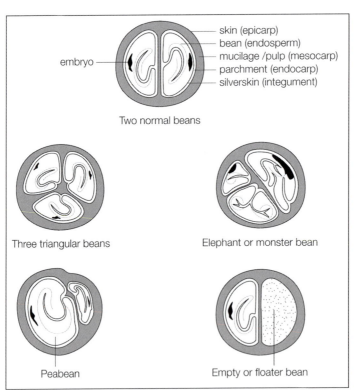

Figure 4.1 Transversal sections of the coffee cherries showing normal and abnormal beans. The shape, size and structure of the cherries are the result of both genotype and environment.

Table 4.2 Percentage of abnormal beans in coffee varieties planted in Colombia

Type of defect	Typcia	Bourbon	Caturra
Empty beans	3.5	3.6	3.6
Peabeans	8.0	7.6	8.7
Triangular beans	1.6	6.4	4.9
Elephant beans	1.1	0.7	1.1

From: Castillo and Moreno [3].

A complete description of defective beans is given in Section IV.3 "Green Coffee Defects".

4.2.2
Color

As shown in Pict. 4.1, some types of coffee beans have a typical color. Robusta, for instance, produces browner beans than the Arabica, which are bluish-green. The color differences between Arabica varieties are very minor.

4.2.3
Chemical Composition

There is a marked difference between Robusta and Arabica coffee. Robusta contains 1.6–2.4 % caffeine, Arabica only 0.9–1.2 %. The selection criteria should be a maximum of 2.5 % of caffeine for Robusta and 1.3 % for Arabica. The Maragogipe variety contains even less caffeine, about 0.6 %.

4.2.4
Flavor

There are also important differences in flavor between the different types of coffee. Robusta produces beans with a bitter, full-bodied taste, but low acidity. Arabica is more aromatic, with more perceptible acidity, but less body. The different varieties can be associated with specific flavor profiles: SL28 produces a milder brew than Kent; Blue Mountain and Bourbon are finer coffees than Catimor; Liberica is bitter and without finesse (Table 4.3).

Several Arabicas like Mocha, SL-28, Blue Mountain and Pluma Hidalgo are famous worldwide because of their very fine quality, but it seems that this is more the result of environmental factors than the genotype. Most of the coffee tasters agree now that there is very little or no difference at all between the Arabica pure breeds cultivated under similar agro-climatic conditions. This is not the case for hybrids like Catimor, Arabusta and Icatu, whose flavors have been noted to be rather inferior compared to Arabica pure breeds.

The association of a genotype with special environmental conditions can produce

Table 4.3 Flavor characteristics of different species and varieties of coffee tested in Turrialba, Costa Rica

Species	Variety	Flavor and characteristics
C. arabica	Bourbon	high acidity, little body
	Red Caturra	light flavor, good body, acid
	Blue Mountain	good body, mild flavor
	Bourbon (Salvador)	good body, flavor and acidity
C. canephora	Robusta	harsh, bitter and earthy when compared to Arabica

From: Gialluly [10].

Picture 4.1 Coffee Species: 10, Arabica "Nouméa" – New Caledonia; 6, Arabica "Maragogipe" – Colombia; 15, Liberica "Prima" – Ivory Coast; 16, Dewerii var. excelsa "Extra Prima" – Central African Republic; 23, Robusta EK1 – Indonesia; 24, Kouillou "Supérieur" – Magadascar (called Conillon in Brazil). From: Coste [9]

outstanding coffee like Blue Mountain in Jamaica, but these very varieties planted under other environmental conditions will not more produce the same quality of coffee. Unfortunately, this relation between genotype and environment has been little investigated up to now.

4.3
Influence of Environmental Factors

4.3.1
Climate

The suitability of the climate for the cultivation of coffee depends on the latitude and the height above sea level (Fig. 4.2). Every 100 m of altitude corresponds to a decrease in temperature of 0.6 °C.

The relative humidity of the air and the rainfall influence the growth of the coffee trees and the development of pests. Hail and frost can damage the plants and cherries and affect the final quality of green coffee.

Periods of prolonged drought, disease and insect attack (such as by the leaf miner and mites) may also result in lower quality beans.

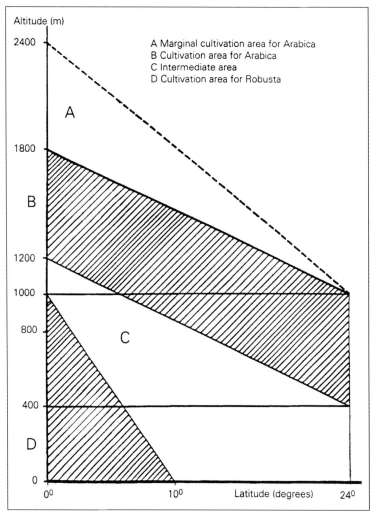

Figure 4.2
The suitability of the climate for the cultivation of coffee. Expressed by the latitude and the height above sea level.
From: Pochet [13A].

4.3.2
Altitude

The positive effect of altitude on coffee quality is well known. According to Wilbaux and De Gialluly [23, 10], the total quality, and particularly the acidity, of Arabica coffee increases with the altitude. It seems that in addition to the lower temperature, this effect is due to the more intense UV radiation.

Beans produced at a higher altitude are denser and harder, and, therefore, more appreciated. As with many other crops, an accelerated maturation in a hot and humid environment has a negative effect on the flavor and the structure of the fruits. On the other hand, it has been found that beans produced at too high an altitude have a silvery greenish skin and low acidity. This phenomenon is often accompanied by a distortion and discoloration of the terminal leaves ("hot and cold effect"). (See also Section II.4.)

In Guatemala, Costa Rica, El Salvador, Honduras and Mexico, countries producing washed mild coffees, the classification is based on the altitude. As a general rule, coffees cultivated at a higher altitude develop more acidity, aroma and flavor though beans produced at very high altitudes are often greenly-blue in color and smaller in size. The bean density is higher and the center-cut after roasting is narrower (see Tables 4.4. and 4.5).

4.3.3
Water Availability

Coffee requires sufficient and well-distributed rainfalls. During the dry periods the physiologic activity decreases. Water shortage during the critical period from week 6 to 16 after the fecundation may cause huge losses because of the formation of empty beans. Furthermore, the remaining beans are smaller due to die-back. This phenomenon reduces the market value of the beans [1].

4.3.4
Soils

Little is known about the influence of the soil on coffee quality, but observation shows that a coffee tree in good growing conditions produces larger beans with a better flavor.

Many coffee varieties known for their high quality and good yields in a particular area do not turn out as well when planted in another zone. For instance, Blue Mountain, when planted in Uganda, and Kenya does not have the same high quality as in Jamaica. This can be explained by the different environmental conditions. Therefore, new planting material should always be tested locally before its extensive propagation.

Table 4.4 Effect of the altitude on the cup quality of two coffee varieties in Costa Rica

Altitude (m)	Body		Acidity	
	Typica	*Bourbon*	*Typica*	*Bourbon*
1400	2.65	2.80	2.20	2.43
1000	1.65	2.35	1.55	1.50
460	1.65	1.80	1.00	1.00

Scale from 0–3: 0 = very bad, 3 = very good.
From: Gialluly [10].

Table 4.5 Classification of mild coffee according to altitude of production site

Altitude (m)						
1600	Estrictamente altura	Antigua* Atitlán*	Estrictamente altura	Tarrasú		
		Estrictamente grano duro	Altura	Estrictamente grano duro		
1400						
1200	Altura	Grano duro		Grano duro	Grano medio duro (Pacifico)	Atlántico de altura
1000						
		Prima lavado				Atlántico de altitud mediana
800	Prima lavado					
600			Normal central			Atlántico de baja altitud
400	Buen lavado	Buen lavado		Lavado		
200						
0						
	Mexicoª	Guatemalaᵇ	Honduras/ El Salvadorᶜ	◄——————— Costa Rica ———————►		

* Origin with special quotation
From: ªINMECAFE (1986), ᵇANACAFE (1991), ᶜUNCTAD/GATT (1992).

4.3.5
Frost and Hail

Hail damage can be reduced if the damaged cherries are treated within a short period of time.

Frost damage on beans can vary according to the severity of the frost. Severe frost affects the whole tissue of the beans; they turn black and loose their commercial value.

4.4
Influence of Cultivation Practices

4.4.1
Fertilization

Northmore [13] found that bean quality is linked to the average bean weight. For this reason, the weight is often used as a quality criterion in experimental works. In general, fertile soils produce larger beans, resulting in a more appreciated final product.

Excessive use of nitrogen can increase production but reduces bean density and quality. According to Mendoza [12], an excess of nitrogen increases the caffeine content, resulting in a more bitter taste of the brew. The same author also states that the caffeine and chlorogenic acid contents of the beans are not affected by the levels of phosphorus, calcium, potassium and magnesium in the soil. A lack of zinc will lead to the production of small light grey-colored beans, which will produce a poor liquor.

A high concentration of calcium and potassium in the beans is associated with a bitter and "hard" taste. On the other hand, there is no correlation between the phosphorus content and the physical and organoleptic quality of the bean [13].

Magnesium deficiency adversely affects coffee quality. In Kenya, repeated applications of elephant grass or livestock manure resulted in an increased percentage of undesirable brown-colored beans and, thus, poor roasting characteristics. This effect was associated with a magnesium deficiency induced by the high potassium content of elephant grass as well as high concentrations of potassium and calcium in manure.

Iron deficiency in soils with a high pH produces an amber or soft bean with reduced quality [15].

Cup quality is affected negatively when the calcium content of the beans exceed 0.11%. In the case of potassium the cup, quality is affected negatively when the bean concentration in this element exceeds 1.75%. According to Wallis [20], it is probable that the potassium is not directly responsible but that the negative effect is caused by a deficiency of magnesium associated to the cationic imbalance induced by the excess potassium. See Tab. 4.6.

4.4.2
Shade

There are two different types of plantations: intensive, mainly large commercial estates where an optimal yield quality is achieved by using all available technological means, and extensive, usually owned by small farmers. In extensive plantations, shading is used to compensate for the lack of technology and resources.

The positive effect of shade, as shown in Tab. 4.7, can be explained by the lower yield. Coffee, unlike many other plants, does not have a regulatory mechanism to

Table 4.6 Influence of potassium and calcium content of the bean on the cup quality

Concentration of Ca and K	Cup quality (%) of analyzed samples			
	Good	Medium	Poor	Total
Calcium				
less than 0.11%	63.2	14.3	22.5	100
more than 0.11%	16.7	50.0	33.3	100
Potassium				
less than 1.75%	52.3	29.5	18.2	100
more than 1.75%	17.5	56.1	26.3	100

Adapted from Northmore [13].

Table 4.7 Shade increases the weight and size of the coffee cherries and improves cup quality

Characteristics recorded	With shade[a]	Without shade[b]
Weight of 1 liter of commercial coffee (g)	716.00	703.00
Average weight of 100 commercial coffee beans (g)	14.56	13.55
Classification by size		
no passage through an 8-mm diameter mesh screen (%)	1.50	0.60
no passage through a 7-mm diameter mesh screen (%)	22.55	9.55
no passage through a 6-mm diameter mesh screen (%)	62.60	75.45
no passage through a 5-mm diameter mesh screen (%)	13.10	13.25
passage through a 5-mm diameter mesh screen (%)	0.25	1.15

[a]Good acidity, body, good brew.
[b]No acidity, poor body, poor brew.
Source: Wilbaux [22].

discard over-ripe fruits. Over-bearing results in nutritional deficiencies and more parasite attacks during the development of the fruit, and, hence, in immature beans.

4.4.3
Crop Management

Good growth conditions (weed control, presence of cover crop, appropriate planting density and pruning) usually have a positive effect on bean size and flavor. The relationship between crop management and total coffee quality, however, has not yet been investigated in detail.

4.4.4
Use of Ripening Hormones

A variety of hormones such as Etephon and giberellic acid are used to synchronize maturation in order to facilitate mechanical harvesting. It has been noted, however, that coffee treated with Etephon matures more rapidly when not shaded. Etephon has a greater effect on the maturation of the pulp than of the bean; this may result in immature beans.

4.4.5
Pest and Diseases

Pests and disease attacks can affect the cherries directly or cause them to deteriorate by debilitating the plants, which will then produce immature or damaged fruits.

The fly *Ceratitis capitata* feeds on the mucilage and the cherry becomes infected with microorganisms; the secondary bacterial infection causes a distinct potato flavor.

The Coffee Berry Borer *Hypothenemus hampii* feeds and reproduces inside the coffee beans and causes their quality to deteriorate.

The Coffee Storage Weevil *Araecerus fasciculatus* feeds on the bean, leaving it completely unusable because it burns during the roasting process.

The Antestia sting bug, as a vector of microorganisms, damages the bean and causes a bitter flavor.

In many cases, the insects can be controlled with pesticides. Certain pesticides, however, can impart an unpleasant flavor and odor to the coffee brew. HCH treatment, for instance, causes a strong musty flavor.

4.4.6
Physiological Damage

Over-bearing is a typical feature of coffee trees cultivated in open sunlight. When accompanied by foliage reduction, over-bearing results in withered branches and immature beans. Water deficiency causes empty and dry cherries to form (die-back).

4.4.7
Harvest

During the final ripening stage, changes occur inside the beans such as:

- Chlorophyll degradation
- Synthesis of pigments (carotenoids, anthocyanins, etc.)
- Reduction of phenolic compounds, i.e. decrease in astringency
- Increase of volatile compounds such as esters, aldehydes, ketones, alcohols etc.,

responsible for the aroma of mature fruits.

Only fully mature berries develop an optimal quality. Immature, green berries have a lower weight and fat content, i.e. less yield and flavor. They produce discolored beans with a harsh and bitter taste (Tab. 4.8). On the other hand, green cherries contain more caffeine than ripe ones [18]. Black cherries, i.e. cherries dried on the tree, produce a brew with a woody flavor.

Mixing of green, partly ripe, red and black cherries produces coffee of inferior quality and should be avoided. At the beginning of the harvest, it is recommended that only the ripe, i.e. red or yellow (Yellow Catuai variety), cherries are picked. As maturation progresses, the partly ripe cherries can be picked. Green berries should be picked only at the end of the harvest.

Harvesting requires a lot of manpower over a short period of time. This problem can be overcome with mechanization, provided that the topography, plantation size, planting density and cost of the equipment are appropriate.

Table 4.8 Influence of fruit maturity on the brew (State of Sao Paulo): in 1959 the cherries were riper than in 1958, which resulted in a better quality brew

	Harvest 1958		Harvest 1959	
	Average points	Quality of the brew	Average points	Quality of the brew
Mild brew model	7.1	apparently mild	8.7	mild
Depulped cherries	8.2	mild/acid	7.3	apparently mild
Non-depulped cherries	6.0	apparently mild	7.3	apparently mild
Fruit dried on the tree	3.4	very hard	4.7	hard
Fruit harvested green	3.0	very hard, immature	4.4	very hard
Rioy brew model	1.6	Rioy	1.1	Rioy
Fruit picked off the ground	0.3	Rioy	0.5	Rioy

Source: Cavallo and Chalfoun (1958, 1959)

Complete maturation is followed by the degradation of the fruits. The intensity and rapidity of this alteration depends on the climate. Therefore, harvested cherries should be processed as rapidly as possible to preserve their quality. The waiting period should never exceed 20 h (see Tab. 4.8).

4.5
Influence of Post-harvest Treatment

There are three procedures to process the cherries: the dry, the semi-dry and the wet method (Fig. 4.3).

When applying the wet method, the cherries must be sorted. Only ripe cherries are depulped, demucilaged and washed. This process washes them thoroughly and reduces bitter compounds.

At the end of the harvesting period, there is usually a large proportion of green, dry and other cherries of inferior quality, which can cause a bitter taste. In this case, the dry method is used, i.e. unsorted cherries are dried without further treatment. This method causes the uncontrolled development of a micro flora that produces unpleasant flavors. Furthermore, the cherries are exposed to rains, dew and droppings from chickens, goats, pigs, etc.

The dry method is also used when there is not enough water, which is the case in certain areas of Brazil, Ethiopia and Yemen.

The semi-dry method takes advantage of a new technique to separate the different bean qualities with minimal use of water. This separation upgrades the market value of the coffee issued from ripe cherries by 20–40 %. In Brazil this method is becoming more and more popular because it improves the quality and the marketing value of coffee.

4.5.1
Pulping

Pulping is the mechanical removal of the pulp from the coffee beans. It has to be carried out as rapidly as possible; a maximum of 20 h after the cherries are picked, depending on the local temperatures.

During this operation, inappropriate use or maladjustment of the pulper can result in nipped and bruised beans. Such beans undergo biological and chemical reactions which cause their quality to deteriorate. This damage can be reduced by means of the vertical Penagos pulper, which does not need to be adjusted due to a special design. Today, it is also possible to separate the green cherries mechanically, which improves the subsequent pulping of the ripe cherries.

Insufficient mucilage in the cherry prevents a complete removal of the pulp, giving rise to the effect known as "media cara" bean. Certain coffees, such as Robusta and especially Liberica, contain less mucilage and require a special adjustment of the pulper.

4.5.2
Mucilage Removal

The mucilage that remains attached to the parchment can be removed by fermentation. Its duration depends on the temperatures and the type of fermentation process (dry or wet). If fermentation lasts too long, a microbial infection of the mass causes the formation of compounds such as propionic and butyric acids, which give the "onion flavor" defect.

The stinker bean is typical of a prolonged or repeated fermentation. It is characterized by the tobacco color of the bean accompagned by frequent expulsion of the embryo, indicated by a small cavity at the base of the bean and the production of

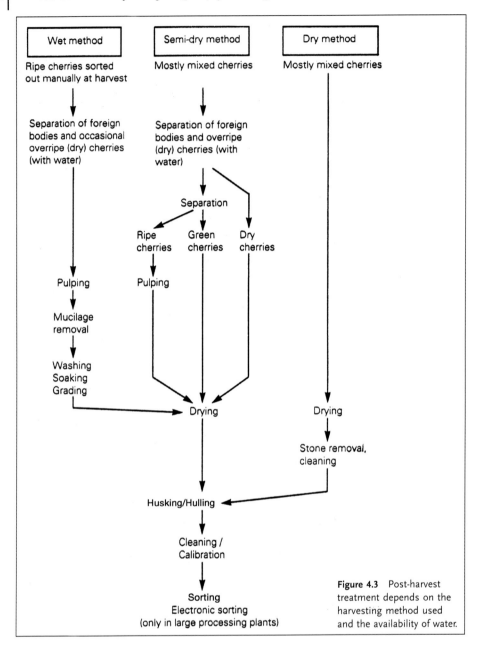

Figure 4.3 Post-harvest treatment depends on the harvesting method used and the availability of water.

foul-smelling substances which escape when the bean is crushed. The stinker bean is a serious defect since it can damage large quantities of beans in the mass. Apart from over-fermentation, the stinker bean can also be a result of delayed pulping, the use of dirty water, lack of cleanliness, etc.

The color of the center-cut of the roasted beans indicates how the fermentation has been carried out:

- White cut: appropriate fermentation conditions
- Dark-yellow cut: insufficient fermentation
- No apparent cut: over-fermented beans or the use of mucilage removing methods other than biological fermentation as well as dry processing.

Apart from fermentation, there are other means for mucilage removal, e.g. mechanically (mucilage removers) or chemically (with sodium hydroxide), etc. The final quality of coffee is not influenced markedly by the method used for mucilage removal, except by the wet fermentation (underwater) that slightly improves the flavor by partially extracting bitter compounds from the beans by ex-osmosis.

4.5.3
Washing, Grading and Soaking

After fermentation, the beans have to be washed with clean water to remove the adhering particles resulting from the degradation of the mucilage.

If the fermentation is incomplete, residues remain in the split of the parchment where they decay and can lead to the "fermented" defect. These residues can also serve as a substrate for the development of microorganisms during storage. This results in a stained parchment, a musty flavor and a "dirty coffee" odor.

Washing coffee with dirty water causes a number of defects, depending on the source of contamination. Water containing residues from coffee processing can produced fermented or stinking coffee; water containing earth produces an earthy flavor.

Density separation of beans in parchment and soaking further improve the quality of the beans. Density separation is separating the beans in canals with circulating water, or with special machines, such as Aargard and Bendig, etc. Generally speaking, there are three categories of parchment, the heaviest having the best quality. Soaking consists of submerging the parchment (i.e. the beans with the adhering parchment) for 10–24 h in clean water to partly extract bitter components such as polyphenols and diterpenes by osmosis. Therefore, the wet method is preferred, especially for Arabica and where possible also Robusta.

The semi-dry process is simple and requires little water. Dry cherries, stones and other foreign bodies are removed first. Prior to pulping, a separator is used to sort the ripe cherries from the green ones. The pulped parchment is dried directly without removal of the mucilage, while the green and over-mature cherries are generally treated separately by the dry method.

4.5.4
Drying

Drying reduces the moisture content of the coffee from 65 to about 12 %. This is necessary to prevent microorganisms from growing during storage.

Coffee cherries and beans can be dried in the sun or in mechanical dryers. Sun drying can take place on the ground (e.g. on earth, covered with plastic film), on concrete or on trays that are elevated to improve air circulation. When drying coffee on the ground, the layer thickness should be no more than 5 cm for fresh cherries and 2.5 cm for wet parchment.

Mechanical driers should be equipped with a heat exchanger to prevent contact of the coffee with the combustion gases, which may cause a smoky flavor except when gas is used as heat source. Mechanical drying is rarely used for the "dry method", since it is more expensive than sun drying and is reserved for better quality coffee.

In some countries of eastern Africa, especially in Kenya, where one of the best coffees is produced, the beans are pre-dried on trays for at least 50 h until the moisture content reaches 20%. Drying can be completed in mechanical dryers. The advantage of pre-drying in the sun is that the beans are exposed for some time to UV radiation, which probably enhances the quality and generates the desired greenish-blue color. To avoid quality deterioration and fissures in the protective parchment, the beans should, however, be covered during the hottest hours of the day by a roof or a plastic film.

During the drying process, the coffee should also be protected from rain and dew, since the parchment and the dried cherries are very hygroscopic. Other damage can be caused by excessive temperatures which includes changes in the chemical composition of the beans or by insufficient drying. The defects that have their origin in the drying process are:

- Crystallized beans characterized by a blue-grey color and a fragile brittle consistency, caused by high temperatures
- Discolored, whitish beans, caused by inappropriate drying, e.g. if the beans are re-wetted by rain, dew or damp
- Blotchy beans (with white blotches on the surface), caused by re-wetting during or after drying
- Burnt beans having a light, foxy yellowish color, caused by excessively high temperatures
- Dark grey bean of a soft consistency, caused by insufficient drying; microorganisms can develop in these beans, causing further deterioration.

For Arabica, the drying temperature should not exceed 45 °C, since higher temperatures cause the quality to deteriorate, mainly through enzymatic and chemical reactions inside the beans. Robusta, on the contrary, requires a high initial air temperature (85–95 °C) and a final temperature of 50–60 °C (Fig. 4.4). This results in darker beans and a milder brew.

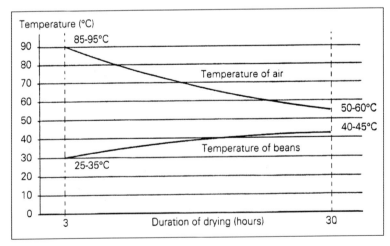

Figure 4.4 Graph of temperatures for Robusta drying.

4.5.5
Hulling, Husking and Sorting

Hulling is the removal of parchment and, occasionally, the silverskins from washed beans, and husking is the removal of all the covers from dry cherries. There are two methods: hulling by friction and hulling by impact. In both cases, the prior removal of stones is recommended.

Hulling can cause different defects. The beans can break, because they are over-dried or because the huller has not been adjusted correctly. Bruising occurs when the beans are soft due to a high moisture content. Elephant or monster beans usually burst during hulling and give rise to the formation of so-called shells and ears. An excessive temperature in the friction hullers can produce whitish or discolored beans. For this reason, hullers operating by impact are preferred.

The beans are usually calibrated on a densimetric table, after having been treated in a pneumatic cleaner (catador).

After calibration, the defective beans have to be removed, either manually or by using electronic (colorimetric) sorters. These machines can separate beans with the following defects: black (unclean, bitter taste), brown (bitter taste), yellowish (bitter-sweet, fruity taste), immature green (harsh, hard, bitter taste), blackish (harsh, unclean taste), discolored, whitish (bland, woody sometimes bitter-sweet and grassy taste) and beans perforated by weevils when the damage is significant.

Sorters that use UV rays can eliminate beans attacked by moulds, even if the damage is not easily detectable on sight, which is the case with "stinkers" and "fouls".

The coffee beans, especially natural Robustas, may be polished in order to rub off the silverskin. This operation is carried out in special machines, sometimes using a little water. It leads to a slight loss of weight, but to a better appearance and to a reduced bitterness as required for certain markets like, for instance, the Japanese market.

4.6
Storage

During storage, an equilibrium must be maintained between the water inside the bean and the humidity of ambient air. As shown in the equilibrium curve (Fig. 4.5), coffee with 11 % humidity at 30 °C is maintained in equilibrium as long as the relative humidity of the air is below 65 %. With a higher relative humidity of the air, the coffee beans will absorb water; if the air is too dry, they will dehydrate.

In hot and humid climates, bulk coffee beans should have less than 11 % moisture and be stored in closed silos, designed for air circulation and for pest control by gassing if necessary.

Beans with a moisture content above 11 % are subject to the action of microorganisms that cause quality to deteriorate. The most important defect caused during the storage of coffee is the "cardenillo" bean; this is the result of moist storage under the action of microorganisms that destroy the superficial parts of the bean and leave a yellow or yellowish-red powder.

Stale coffee is another common defect. The beans have color tones ranging from whitish, cream and yellowish to caramel. The main cause of this discoloration is an excessive relative humidity of the air. The development of stale beans takes a few days to several months, depending on the storage conditions but at a relative humidity of above 80 % the rate of development is accelerated. This defect also causes quality

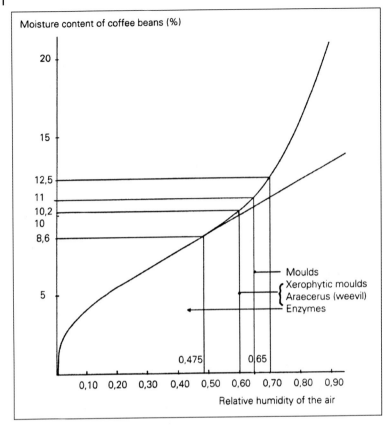

Figure 4.5
Sorption curve for Robusta coffee at 30 °C in hot climates. From: Wilbaux and Halm IFCC (1966).

deterioration due to rupturing of the cell walls.

Perforated beans are caused by insects, especially the weevil *Araecerus fasciculatus*, which feeds on the bean and leaves a yellowish powder as a residue.

The coffee quality can also be impaired by the proximity of materials with an unpleasant odor.

4.7
Summary of Factors affecting the Quality of Green Coffee

4.7.1
Genotype

Key features	Expression	Effect on quality and value	Remedies
Shape, size and uniformity of beans	large	+	
	small	−	change planting material
	non-homogenous	−	change planting material
	peabeans	−	change planting material
	empty beans	−	change planting material
	abnormal beans	−	change planting material
Bean color	blueish-grey	+	
	light-brown	−	change planting material
Chemical composition of beans	high caffeine	−	change planting material
	low caffeine	+	
Flavor	good brew	+	
	poor brew	−	change planting material

4.7.2
Environment

Determining factors	Expression	Effect on quality	Remedies
Altitude			
high	hard beans	+	
low	poor quality beans	−	shade, change area
Heat and dryness	wilting, die-back	−	irrigation, shade
Hail and frost	damaged fruit	−	change area
Parasites	physical damage	−	prevention, control
Soil		−	drainage, change of area
water-logged	poor quality beans	−	drainage, change of area
poor fertility	poor quality beans	−	fertilization

4.7.3
Cultivation Practices

Determining factors	Expression	Effect on quality	Remedies
Fertilization			
adequate	healthy beans	+	
excessive N	low bean density	–	adjust nutrient rates
excessive Ca + K	hard and bitter taste	–	adjust nutrient rates
iron deficiency	amber or soft beans	–	adjust nutrient rates
Mg deficiency	brown beans	–	adjust nutrient rates
Irrigation	healthy beans	+	
Appropriate crop management: pruning, weeding, etc.	healthy beans	+	
Balanced shade	healthy, large beans	+	
Ripening hormones			appropriate planning of hormone application
on underdeveloped fruit	immature beans	–	
on developed fruit	does not affect quality	0	
Insects and diseases	damage to fruit	–	pest and disease control
Physiological stress	damage to fruit	–	assess and correct
Harvest			correct sorting of cherries
green cherries	bitter brew	–	appropriate harvesting
over-ripe cherries	fruity/yeasty taste	–	id

4.7.4
Post-harvest treatment

Operation	Expression	Effect on quality	Remedies
Pulping	green cherries	–	separate green cherries
	fermented cherries	–	avoid lengthy delays before pulping
	nipped, bruised beans, etc.	–	calibrate berries, adjust the pulper or use a special pulper
Mucilage removal	onion taste	–	use clean water
	bitter, fermented taste	–	ensure appropriate removal of mucilage
	stinker beans, etc.	–	control cleanliness, wet fermentation
Washing	stinker	–	use appropriate equipment, control cleanliness
	fermented	–	use clean water
	dirty beans	–	practice densimetric separation
	earthy beans	–	avoid contact with the earth when drying, use clean water
Drying	discolored beans	–	avoid drying on the ground
	burnt, soft beans	–	avoid overheating
	other defects	–	avoid thick layers, adapt drying temperatures; protect against moisture and strong sun rays
Hulling	broken, soft, discolored beans, etc.	–	adjust the huller correctly, control the heating up of the huller, use the impact method

4.7.5
Storage

Determining factors	Expression	Effect on quality	Remedies
Moisture content of the beans > 11%	"cardenillo" beans	–	sufficient drying prior to storage; appropriate storage conditions
Excessive moisture content of the air	stale beans	–	reduce air humidity; improve storage conditions
Weevil damage	perforated beans	–	pest control; avoid re-wetting; sufficient drying prior to storage
Proximity of odorous materials	beans with an unpleasant odor	–	avoid storing coffee in the same premises as odorous products

4.8
Conclusions

The difficulty of finding green coffee of a suitable and constant quality, as well as the complexity of the factors that influence quality means that, nowadays, knowledge of these factors needs to be far greater than in the past.

This situation also calls for more appropriate and uniform quality assessment methods as well as a common terminology.

The main defects that should be known are to be found in Section IV.3 "Green Coffee Defects".

Bibliography

[1] Cannel, M. G. R. Physiology of the coffee crop. In: *Coffee – Botany, Biochemistry and Production of Beans and Beverage*, eds Clifford, M. N. and Willson, K. C., Croom Helm, London, 1985, 109–134.

[2] Carvajal, J. F. *Cultivo y Fertilizacion*, 2nd edn. International Potash Institute, Basel, 1984.

[3] Castillo, Z. and Moreno, R. G. *La Variedad Colombia*. CENICAFE, Colombia, 1988.

[4] Clarke, R. J. and Macrae, R. *Coffee. Vol. 4: Agronomy*. Elsevier, London, 1988.

[5] Clarke, R. J. and Macrae, R. *Coffee. Vol. 2: Technology*. Elsevier, London, 1987.

[6] Cleves, R. *Technologia en Beneficiado de Café*. TECNICAFE, San José, Costa Rica, 1995.

[7] Clifford, M. N. and Willson, K. C. *Coffee – Botany, Biochemistry and Production of Beans and Beverage*. Croom Helm, London, 1985.

[8] Coffee Growers Association. *Coffee Handbook*. Harare, Zimbabwe, 1987.

[9] Coste, R. *Les Caféiers et les Cafés dans le monde*. Ed. Larose, Paris, 1959.

[10] Gialluly, M. Factors affecting the inherent quality of green coffee. *Journal?*, Nov, 12–132. 1958.

[11] Haarer, A. E. *Modern Coffee Production*. Leonard Hill, London, 1958.

[12] Mendoza, L. O. Problematica en la calidad del café verde: del campo a la taza. In: *Memoria del III Simposio Internacional del Café*. CMPC, Xicotepec de Juárez, Puebla, Mexico, 1995, 85–98.

[13] Northmore, J. M. *Some Factors Affecting the Quality of Kenya Coffee 15*. Turrialba, Costa Rica, 184–193, 1965.

[13A] Pochet P. *La qualité du café, de la plantule à la tasse* – AGCD – Bruxelles – 1990.

[14] Robinson, J. B. D. Amber bean. *Kenya Coffee* 1960, 25, 91–93.

[15] Rothfos, B. *Rohkaffee und Röstkaffee Prüfung*. Gordian-Max-Rieck, Hamburg, 1963.

[16] Rothfos, B. *Coffee Production*. Gordian-Max-Rieck, Hamburg, 1990.

[17] Sivetz, M. and Desrosier, N. W. *Coffee Technology*. AVI, Westport, CT, 1979.

[18] Vincent, J. C. *Green Coffee Processing*. Elsevier, Amsterdam, 1987.

[19] Van der Vossen, H. A. M. and Walyaro, D. J. The coffee breeding program in Kenya – review of the progress made since 1977 and plan of action for the coming years. *Kenya Coffee* 1981, 46, 113–130.

[20] Wallis, J. A. N. *La Calidad del Café Arábica en Kenya y Tanzania*. CAFE-IICA. Enero-junio, Lima, Peru, 1967.

[21] Wanyonyi, J. M. *An Outline on Coffee Processing*. Ministry of Agriculture, Nairobi, Kenya, 1972.

[22] Wilbaux, R. *Les Caféiers au Congo Belge. Technologie du café Arabica et Robusta*. La Direction de l'Agriculture des Forêts et de l'Elevage, Bruxelles, 1956.

[23] Wilbaux, R. *Coffee Processing*. FAO, Rome, 1963.

[24] Wintgens, J. N. *Factors Influencing the Quality of Green Coffee*. Nestlé, Vevey, Switzerland, 1993.

[25] Wintgens, J. N. Influencia del beneficiado sobre la calidad del café. In: *Memoria del Primer Seminario Internacional sobre la Reconverción del Beneficiado Húmedo del Café*. Xalapa, México, 1994, 153–170.

[26] Wrigley, G. *Coffee*. Longman, London, 1988.

[27] Zuluaga Vasco, J. *Los Factores que Determinan la Calidad del Café Verde* . Comité Departamental de Cafeteros de Antioquia, Colombia, 1990.

[28] Zuluaga Vasco, J. *El Beneficio del Café*. Comité Departamental de Cafeteros de Antioquia, Colombia, 1991.

5
Coffee Bean Quality Assessment

J. N. Wintgens

5.1
Introduction

Green coffee quality is the result of an interaction among different variables, such as varieties, soil, climate, husbandry, latitude, altitude, luminosity, harvesting, processing, storing, etc. The interaction of all these variables will determine the final quality of the green coffee. Top-quality coffees are produced at high altitudes where ultraviolet light is stronger and growth is gradual. Beans produced at higher altitudes are harder and to illustrate this fact we like to say that "the best coffee is harvested in the sky".

Since many of the aspects affecting quality have already been mentioned in this book, this chapter will concentrate mainly on the assessment of green bean quality.

Coffee quality must be focused on the final utilization and on consumer preferences. For instance, for roasted coffee sold as whole beans, bean size, color and shape play a key role because they influence the general appearance of the coffee. On the other hand, for roast and ground coffee, bean size cannot be appreciated by the consumer and therefore this aspect will be less important.

Quality determination of green coffee can be separated in three main categories: green coffee grading, sensory evaluation and chemical/analytical measurements. In grading, the main objective is to determine the size distribution of the coffee along with the assessment of the amount of defective beans and their color. Sensory evaluation focuses on determining the flavor profile. Finally, the analytical and chemical tools will determine moisture content and some of the chemical components, such as caffeine and possible pesticide residues.

5.2
Green Coffee Grading

Bean size plays an important role for roasted whole coffee beans because many consumers associate bean size to quality; however, larger beans do not necessarily taste better than smaller. It has to be pointed out that for roasting, the more uniform the bean size, the better the heat transfer and consequently the roast. It is also preferable not to roast beans issued from different species, together e.g. Arabica with Robusta.

Coffee: Growing, Processing, Sustainable Production, Second Edition. Edited by J. N. Wintgens.
© 2012 WILEY-VCH Verlag GmbH & Co. KGaA. Published 2012 by Wiley-VCH Verlag GmbH & Co. KGaA.

Bean size distribution is carried out by means of perforated plates commonly called screens. Depending on the shape of the holes, screens can be grouped in two categories: rounded and slotted. Usually screens with rounded holes measure the bean width and slotted screens separate peaberry beans (from now on called peabeans).

The size of the screen hole is usually specified in 1/64 in. For instance, a screen 15 refers to a rounded screen with holes having a diameter of 15/64 in, which is equivalent to 5.95 mm (0.396875 mm × 15). Usually, most green coffee beans will be retained between screens 12 and 19 (Tab. 5.1).

In the case of slotted screens, there are two widths, also specified in 1/64 in, and the length of the slot usually measures 3/4 in (19 mm). The most common slotted screens used for green coffee are in the range 8–11.

The bean size distribution can be shown in two different ways: cumulative or percentage between screens. It is worthwhile mentioning that the moisture content of the green coffee bean has an effect on

bean size: the higher the moisture, the larger the bean.

5.3
Determination of Defective Beans

The number of defective beans has always been associated with quality. It is true that larger amounts of imperfections will increase the probability of finding off-flavors and lesser homogeneity in the cup; however, low amounts of visible defects do not necessarily correlate with higher cup quality.

The assessment of the defective beans count is done by hand picking all the defects from a specified amount of coffee (weight or volume), then grouping similar defects, counting, weighing them and finishing up with one number representing the total amount of defects.

The approach to come up with only one number representing the total amount of defects (defects equivalent) is very simple and is based on defining the equivalency from each defect to a black bean. It implies that not all the defects are equally harmful for the final quality. For instance, five shells are equivalent to one full defect for the Cocoa, Sugar and Coffee Exchange of New York (CSCE), so one shell bean is equivalent to one-fifth of a full defect. Table 5.2 shows some of the most common equivalencies currently used for coffee.

Another approach to determine the amount of defective beans is to weigh them after grouping and then get a percentage (weight basis) of a particular defect. This approach is more frequently used when dealing with coffees having higher amounts of defects, and it is more useful for millers and coffee growers to help them to decide the most appropriate way of processing the beans to meet quality standards.

Table 5.1 Screen size

Screen no.	Screen diameter (mm)	ISO norm	English description Bean size
20	7.94	8.00	very large
19	7.54	7.50	extra large
18	7.14	7.10	large
17	6.75	6.70	bold
16	6.35	6.3	good
15	5.95	6.0	medium
14	5.55	5.6	small
13	5.16	5.0	
12	4.76	4.75	

Table 5.2 Different equivalencies most commonly used in the coffee trade

Defect	LIFFE Robusta	CSCE/NYBOT	SCAA	Brazil
Dried coffee cherry	1	1	1	1
Black	1	1	1	1
Partial black	1/2	1/2 – 1/5	1/3	–
Sour	1/2	1	1	1/2
Partial sour			1/3	
Fungus			1/5	
Severe insect damaged		–	1/5	–
Insect damaged	1/2–1/5	–	1/10	1/2–1/5
Immature	1/5	–	1/5	1/5
Floater	–	1/5	1/5	–
Parchment	–	–	1/5	1/2
Broken (more than half)	1/5	1/5	1/5	1/5
Broken (less than half)	1/5	1/5	1/5	1/5
Shells	1/5	1/5	1/5	1/3
Malformed			1/5	
Withered			1/5	
Foreign matter				
Large husk fragments	1/2	–		1
Medium husk fragments	–	1/3		1/2
Small husk fragments	1/5	–		1/3
Large parchment fragment	1/2	–		–
Medium parchment fragment	–	1/3		–
Small parchment fragment	–	–		–
Large sticks	5	2–3		5
Medium sticks	2	1		2
Small sticks	1/2	1/3		1
Large stone	5	2–3		5
Medium stone	2	1		2
Small stone	1/2	1/3		1

Figures in columns give defect values, i.e. 1/2 means 2 defects equivalent to 1 black bean as the reference.

In terms of defective beans for roasted whole beans, the count is also done by hand-picking quaker (pale beans), black, burnt, insect damaged, broken beans etc.

5.4
Green Bean Color

Green bean color gives a good indication of freshness, moisture and homogeneity. In washed Arabica the green–bluish color is perceived as the highest quality.

In the past, color was determined by looking at one sample representing an acceptable color and using the sample for comparison. Nowadays, color determination is becoming more accurate and reproducible due to the developments in electronics and color meters that measure different color spectrums (red, yellow, green and blue).

5.5
Cup Tasting

Coffee is a beverage where flavor plays the most important role in quality. It could be the nicest looking green coffee, without any defective beans, but in spite of that, flavor could be very marginal or even have off notes. Ultimately, the flavor of coffee is the most important quality parameter independent of the particular utilization (roast and ground, soluble, liquid canned coffee, liquid extract, etc.).

The main goal in cup tasting is to evaluate the coffee in an objective and reproducible way. It is not a matter of whether or not you like the coffee, the objective is to describe the flavor profile by means of words (attributes) and values related to the intensity of each attribute (seek objectivity).

Coffees are assessed by trained tasters that judge the coffee by its flavor, mouth feel and aftertaste. This kind of assessment, based above all on personal experience and memory is in some way, inevitably subjective.

The basic attributes evaluated in cup tasting are: aroma, flavor, body and acidity.

The most used terms to describe the characteristics of coffee include:

- *Aroma* – the fragrance or odor perceived by the nose. There is a clear distinction between aroma at two different stages: aroma of the freshly ground coffee and the "in cup aroma" which is produced when water has been in contact with the ground coffee for 3–4 min.
- *Taste* – which is perceived by the tongue.
- *Flavor* – which is the combination of aroma and taste. The flavor which contributes to the quality of the coffee is described in terms of winey, spicy and fragrant. Off-flavors such as grassy, onion, musty, earthy, etc., reduce coffee quality.
- *Body* is a feeling of the heaviness or richness on the tongue.
- *Acidity* is a sharp and pleasing taste. It can range from sweet to fruity/citrus and is considered as a favorable attribute. Green coffees stored for a longer period described as "aged" may suffer a loss of their acidity. The India Arabica Mysore (plantation A) as well as most Brazilian Arabica coffees are low in acidity. Acidity is affected by roasting; the more intense the roasting, the more acidity is affected.

The International Coffee Organization (ICO) has developed a vocabulary to establish coffee flavor profiles (Tab. 5.3).

There are four basic flavors commonly used in sensory evaluation: acidity, sweetness, saltiness and bitterness. Sweetness and its interaction with acidity provides a broad spectrum of flavors for Arabica coffee. Bitterness and saltiness are associated with Robusta and low-quality, dry-processed Arabica.

Some authors give an overall score for the coffee which might be the sum of the individual scores of intensity for each attribute or a weighed average.

Table 5.3 ICO vocabulary to establish coffee flavor profiles

Aroma	Taste	Mouth feel
Animal-like	Acidity	Astringency
Ashy	Bitterness	Body
Burnt/smoky	Sweetness	
Chemical/medicinal	Saltiness	
Chocolate-like		
Caramel/malty/ toast-like		
Earthy		
Floral		
Fruity/citrus		
Grassy/green/herbal		
Nutty		
Rancid/rotten		
Rubber-like		
Spicy		
Tobacco		
Winey		
Woody		

The most traditional method to prepare green coffee for cup tasting is by roasting, grinding and then pouring hot water into the ground coffee, this method is known as infusion.

In cup tasting, it is very important to know why or what the objective of the testing is:

• Make sure that coffee does not have any off-flavors?
• Make sure that the batch is homogeneous?
• Evaluate how different/similar it is from previously produced batches?
• Is it to fully assess the flavor profile of the coffee?

The answer to these questions will determine the best method of assessing the coffee.

In the case of tasting to look for defects and homogeneity, the best approach is to taste several cups from the same roasting batch. If there are some defective beans, it is very probable that they will show up in one or several cups. For instance, the CSCE "C" coffee graders taste six cups per sample.

In the case of tasting to evaluate if the test sample matches the flavor profile of another sample with a known flavor profile, the approach would be to taste them side by side and score the attributes of the test sample in relation (plus or minus) to the sample of known flavor profile.

In the case of tasting to fully profile a sample, a group of well-trained tasters would be required, in order to score the intensity of each attribute.

5.6
Analytical Techniques

We will talk about moisture determination as an analytical method. Basically moisture refers to the amount of water still remaining in the green coffee bean. A freshly harvested cherry coffee has a moisture content higher than 60% which translates in high water activity and consequently will deteriorate very quickly if not handled properly. In green coffee beans, the final moisture content ranges from 9.5 to 12%. Higher moistures will translate in higher weight losses and faster ageing during storage along with higher roasting losses in manufacturing. On the other hand, over-dried coffee will affect its aromatic profile and some of the components will be lost.

The most accurate way to determine moisture is by means of analytical convection ovens. However, one can also use moisture meters that use the principle of conductance, which correlates rather closely with the oven results.

To overcome the subjectivity factor which inevitably influences the accuracy of the tasting results, modern analytical techniques have been introduced. They identify part of the compounds which are responsible for the desirable flavor of coffee with the expectation to relate the presence of these chemical components of coffee to their profiles. Unfortunately, the very large number of chemicals present in the roasted coffee (about 700) makes this operation still extremely difficult.

5.7
Profiles of Some Coffees (Fig. 5.1)

In washed Arabica coffees, the flavor profile will range from low acidity to high citrus acidity with an aftertaste that could vary from winey to flowery. In the case of dry-pro-

cessed Arabica, the highest quality coffees can be described as sweet with good body. High-grown, lightly roasted Arabicas produce the most acid liquors. Blends containing a large proportion of Arabica are more appreciated in the northern countries of Europe, as opposed to blends rich in Robusta that are preferred in southern Europe, especially for espresso coffee preparation, since Robusta coffees have much more body.

Robusta coffee has a completely different flavor profile to Arabica. It is described as producing a coarse liquor with low or no acidity at all, with harsh and cereal notes and a thick body. Robusta is seldom used alone, but more as a filler in blends.

5.7.1
Brazil

Due to the extension of the production area, Brazilian Arabica coffees present differences in flavour which stem from their origins. Most of these coffees are dry-processed, mostly due to lack of water in the

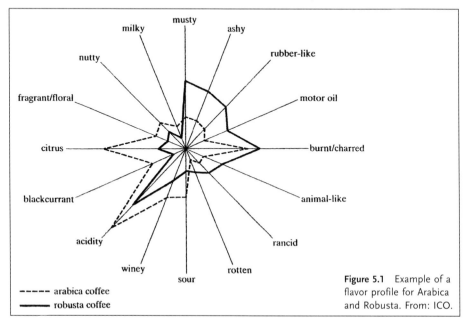

musty
milky   ashy
nutty   rubber-like
fragrant/floral   motor oil
citrus   burnt/charred
blackcurrant   animal-like
acidity   rancid
winey   rotten
sour

- - - - - arabica coffee
———— robusta coffee

Figure 5.1 Example of a flavor profile for Arabica and Robusta. From: ICO.

country, and consequently, they rate a different quality when compared to higher grade washed Arabicas. Currently, however, a number of special, high grade Brazilian Arabicas are appearing on the market as a result of washed and semi-washed processing. These are classified as "Brazil" to distinguish them from the "milds" produced in Kenya, Colombia and Tanzania.

Due to their distinct flavours, Brazils are qualified as 'soft', 'hard' or 'Rio'.

A Rio coffee has a harsh, acid iodine flavour. This characteristic has long been associated with coffees shipped from Rio and Victoria but, in fact, Rio flavour coffees can come from any part of the Brazilian coffee-producing area, not only from the sectors exposed to ocean winds.

The "Rio" flavour can also occur in other countries as well. Surprisingly, these "coffees with a special taste" are greatly appreciated in North Africa, the Middle East and Eastern Europe.

The term "soft" characterizes most of the coffees grown in Sao Paulo and southern Minas Gerais. These coffees are mild with medium body and a fine cup. They are held to be the best Brazilian coffees. This is particularly true for the Yellow Bourbon Santos which is light in acidity, smooth in body and has a pleasantly subtle aroma. Due to their lower costs, "soft" coffees are often used in blends as a filler.

A hard coffee has an astringent mouth feel – Paranà coffees are of this type and are free from Rio flavours. Washed Bahia is of very good quality.

Brazilian coffees stored for a long period catch a better price in certain markets.

5.7.2
Colombia

The higher grade Colombian Arabica coffees compare favourably with high-grown coffees from Central America. They strike a good balance between flavor, acidity and body accompanied by a slight winey flavor.

A significant part of the Colombian coffees are sold as MAM, which means they originate from Mendellin, Armenia and Manizales in the central cordillera of Colombia. Colombian "milds" belong to the group of coffees which also include the high-grown coffees from Kenya and Tanzania.

According to Juan Valdez, Colombian Coffees are the richest coffees in the world. One may wonder whether they truly deserve such praise... In some cases they do, without a doubt, but it is unwise to generalize as a lot of Colombian coffees are not all that outstanding.

5.7.3
Costa Rica

Costa Rica's Arabica coffees are classified according to the altitude at which they are produced. The best quality coffees come from the higher altitudes (1200–1650 m) and is know as "Strictly Hard Bean" (SHB). It is a good coffee, rich in acidity and quite full-bodied with a smooth clean cup. The "Low-Grown Atlantic" (LGA), cultivated on the Atlantic side of the country, produces a grassy cup with less aroma.

5.7.4
Guatemala

In general, Arabica coffees cultivated in Guatemala have a fine flavor – smoky and spicy with a high acidity. Their "SHB" coffees are among the best in the world – complete, medium-bodied with a high acidity and a fragrant cup. The famous growing regions of Antigua reputedly yield some of the world's best coffees but the famous Maragogipe are no longer the very best, special areas in other countries are producing better quality [2].

5.7.5
Dominican Republic and Haiti

The Barahona (Dominican Republic) is a very attractive and good Arabica coffee and the high grown washed Arabicas produced in Northern Haiti have remarkable quality. Haitian coffee is, however, somewhat different because of its slightly sweeter flavor. Both are popular in France and Italy because they are specially suitable for dark roasts.

5.7.6
Mexico

The finest Arabica coffees are produced in Pluma Hidalgo and in the higher parts of southern Mexico. They offer a delicate flavor with medium body and a pleasant, dry, acidy snap. Though not necessarily ranked among the world's most famous, these coffees should certainly not be overlooked. The Mexican Maragogipe have become one the most popular in this category. Other Mexican Arabica coffees are full-bodied and of medium acidity.

Mexico is also a pioneer in the production of organic coffee.

5.7.7
Jamaica

The famous high-grown Blue Mountain. It is grown at high altitudes in a legally defined area. The liquor is full-bodied with moderate acidity, mellow and has an excellent aroma devoid of any off-flavors.

Jamaican Blue Mountain coffee ranks amongst the highest priced coffees in the world (up to 20 times the price of regular coffees). It is supplied in wooden barrels as opposed to bags and most of these barrels go straight to Japan where this type of coffee fetches a very high premium. The high prices are justified by scarcity in supply, high demand and most desirable quality. This coffee is a "must-have" for gourmet shops.

A lot of the "Blue Mountain" imported into the UK is produced in Kenya from Blue Mountain stock but its quality is not as good as its Jamaican counterpart.

5.7.8
Panama

Panamanian coffee from the forest mountains of Boquete is an excellent wet-processed coffee. It is well-balanced, full-bodied and subtly bright with overtones of cocoa and fruit.

The world-famous coffee estate of Hacienda La Esmeralda recently set a new record when its Esmeralda Especial Coffee sold for a stunning USD 130 / lb during an online auction. Panamanian coffee producer, Price Peterson, was rewarded for producing the world's best coffee at the SCAA in 2007.

5.7.9
Kenya

Kenyan coffees are also very highly rated worldwide. These Arabica coffees are acid, smooth and full-bodied with a dry winey aftertaste. The Kenya AA class can be compared to the Blue Mountain and is greatly appreciated for its acidic taste. They are powerful, bright coffees which run the gambit from lemony to peppery, from a fruity blackberry taste to winey richness. They belong to the Colombian mild group.

5.7.10
Ethiopia

Despite their preparation as "naturals", the high-rated Ethiopian Arabica coffees are of excellent quality. The "Harrar" is sometimes sold as Ethiopian Mocha. The brew is soft, mild and floral with a strong winey flavor but it is more acid and has less body than the true Mocha from Yemen. The Djimmah coffee is less acid but still full-bodied and has a singular taste. The washed Arabica coffees produced in the area of Sidamo, Limu, etc., are remarkable: soft, fragrant and well-balanced with a low acidity.

5.7.11
Tanzania

These Arabica coffees belong to the Colombian mild group. The good "Kilimanjaro" (clean cup) can be associated with the finer Keyan coffees. They are medium in body, acid, winey, rich and mellow. On the other hand, the dry-processed Arabica and Robusta are of lower quality.

5.7.12
Kivu, Rwanda and Burundi

These countries produce good washed Arabicas characterized by fair acidity and full body. The liquor is rich with a strong flavor. Sometimes Burundi coffee batches are affected by an unwanted potato flavor" due to an infection developed on injuries caused by the *Ceratitis capitata* fly on the cherries.

5.7.13
India

The Indian Arabica coffees are dark and full-bodied with low acidity. The coffee which comes from the Nilgiris area is of better quality. India also produces the famous monsooned Arabica and Robusta coffees. Monsooned coffee is produced by exposing the coffee to the flow of monsoon winds. After 6 to 7 weeks, the coffee changes color and becomes a golden yellow. At this stage, it is considered to have been fully "monsooned". This thick-bodied, mild and low acidic coffee is particularly appreciated in the Scandinavian countries.

Indian washed Robusta produced at higher altitudes has an excellent quality. Arabica dry-processed coffee is called "cherry" in India whereas wet-processed Arabica is known as "plantation" whether it actually comes from a plantation or not. On the other hand, wet-processed Robusta is called "parchment Robusta".

5.7.14
Indonesia

Arabica coffees are mainly produced in Java, Sumatra and Sulawesi. Good Java coffees are medium-bodied and have a spicy aroma. As a general rule, coffees produced in Sumatra are better than Java coffees as they are richer and mellower and have deep-toned acidity. Sulawesi coffees have the same lovely characteristics to be found in Sumatra coffees, including the full-bodied smoothness, but they feature a more vibrant acidity. All in all, they are truly beautiful coffees.

5.7.15
Papua New Guinea (PNG)

PNG coffee is mellow with a complex combination of aromas, particularly an apple-wineyness, which sets it apart from the earthier Sumatrans and Sulawesians. It is a medium-bodied coffee with moderate acidity and a broad flavor. It is definitely one of the world's finest coffees.

5.7.16
Hawaii

The famous Hawaiian Kona Arabica coffees are grown on juvenile volcanic soils cooled by gently breezes. These ideal conditions produce a coffee that is perfect in many aspects. It is full-bodied and has a high-toned acidity with fine aroma. All in all, an exceptional coffee in terms of appearance, size and density added to which it roasts nicely and very well.

5.7.17
Yemen

The aspect of the Yemen coffee is rather small and irregular, nevertheless, the liquor is unique. It is a coffee which shares medium to full-body, strength, fragrance, aroma and a touch of wildness. The best known is the famous "Mocha" or "Mokha" which is highly priced and chiefly accessible as a gourmet coffee. Yemeni coffees are dry-processed wild coffees and my personal favorites. Mokha itself is the port city where Yemeni coffee beans were loaded into sailing ships which came from Java with the result that first blended coffee ever produced was the Mokha-Java.

5.8
Storage deterioration

Coffee stored in an open atmosphere under normal temperature and humidity behaves in the following ways:

- *Green beans.* Green coffee beans can be stored for up to 3 years without appreciable loss of quality, though Arabica is more sensitive to deterioration than Robusta.
- *Roasted beans.* Roasted beans stored may start deteriorating after 3 weeks.
- *Roast and ground coffee.* A roast and ground coffee deteriorates much faster than roasted beans. After 3 days it starts to express a bitter taste as a consequence of the oxidation of its fatty acids (rancidness).

According to the preceding observations we created the "rule of 3" which can be used as a rough rule of thumb for coffee deterioration. Coffee is more vulnerable to high humidity and oxygen from the air than to high temperatures. Therefore, the safest way to store coffee is in a sealed container placed in a cool environment.

5.9
Conclusions

It is true that some progress has been made in the field of green coffee assessment; nevertheless, the methodology used leaves a significant part to human appreciation, which remains inevitably subjective. It is hoped that in the future and in the interest of a more precise and reproducible assessment, electro-chemical means will become more prominent in the system.

Another weak point in coffee quality assessment is bean defect evaluation and its interpretation, which is far from being unanimous. The ISO, aware of these problems, is tackling this issue, and it is desirable that an amended and unanimous methodology be implemented.

It is obvious that any improvement of the reliability of the methods of assessment of green coffee will contribute indirectly to improve the image of coffee, its quality and, hence, its consumption.

Bibliography

[1] Coste, R. and Cambrony, H. *Caféiers et Cafés*. Maisonneuve and Larose, Paris, 1989.

[2] Jobin, P. *Les Cafés Produits dans le Monde*. Jobin, Le Harve, 1992.

[3] Wintgens, J. N. *Green Coffee Defects Guide for Identification and Evaluation*. Nestlé, Vevey, Switzerland, 1997.

[4] Wrigley, G. *Coffee*. Longman, London, 1988.

Part V
Economics

1
Economic Aspects of Coffee Production

B. Rodriguez P. and M. Vasquez M.

1.1
Introduction

This chapter summarizes a methodology for cost estimation in coffee production. It includes the costs of setting up a plantation and the costs of developing plantations for three types of coffee producers, concluding with recommendations on the most appropriate actions to be adopted by each of them. These recommendations are based on the interaction between financial resources and the international price of coffee.

In the last half of the 20th century, producing countries have seen the way coffee prices, like those of other agricultural products, have undergone strong fluctuations. As a result, millions of producers and consumers throughout the world have radically altered their attitudes and practices.

The major factors that have caused drastic changes in the coffee business have been the suspension of the economic clauses of the International Coffee Organization (ICO) agreement in 1989, the occurrence of frosts in the production areas of Brazil in 1994 and plummeting stocks in 1997. The above variations have brought about fluctuations in the ongoing market price of coffee which, in turn, have caused coffee production to rise or fall as per the changes in the income of the producers. In each of the producing countries, the effects have been different, depending on the type of producer, the production systems used, production costs and the competitiveness of their marketing schemes.

The evolution of the coffee market has even generated a change in the geography of the production areas. Vietnam, for instance, has seen a spectacular growth whereas Ivory Coast has suffered a rapid decrease in coffee production.

The price level of Arabica coffee is generally higher than that of Robusta coffee; however, as a result of supply and demand there are different quotations for the coffee subtypes of each species according to their quality.

Consequently, the economic analysis of coffee growing is based on the ratio between the ongoing market price of coffee and production costs. Therefore, a cost-accounting procedure is an indispensable tool to assist the planter in decision-making.

1.2
Economic Concepts

Unlike the comments that can be made on coffee prices in their different commercial

Coffee: Growing, Processing, Sustainable Production, Second Edition. Edited by J. N. Wintgens.
© 2012 WILEY-VCH Verlag GmbH & Co. KGaA. Published 2012 by Wiley-VCH Verlag GmbH & Co. KGaA.

presentations (cherry, parchment, green, etc.), it is not so easy to set up a standard for production costs and profitability, since any variation, however minor, in the combination of machinery and equipment and/or production factors leads to differences in the resources required. Production costs and profitability vary according to local conditions, technology, size of the estate, etc.

On the other hand, the commercialization and the liberalization of the coffee trade has caused producers to convert their activities in order to increase their competitiveness and lower their production costs.

The indicators used for profitability assessment are: the financial resources needed to set up and develop the estate, the gross profit per cycle, and the plantation profitability rate.

According to the "Agriculture and Cattle Producers' Outlook", the term "production costs" indicates the expenditure made to acquire the machinery, equipment and various materials used in the production process, as well as the cost of labor.

Fixed costs (FC) are those that cannot be reduced once the production process is started (tree felling, field outlining, staking, hole digging, plant purchasing, etc.); these costs do not vary whatever the yield or the quality of the final product.

Variable costs (VC) are those that depend on the level of production (harvesting, hauling, harvesters' transportation, etc.).

Thus, the *total costs (TC)* are the sum of the total fixed costs and the total variable costs ($TC = FC + VC$).

The *annual gross profit (AGP)* is the difference between the annual income and total costs. Gross profit does not include "structural" costs like land taxes, income tax and social security. When these are taken into account and deducted, we are left with the *net profit (NP)*.

The *break-even point (BEP)* is the production level at which the income from sales is exactly the same as the total costs. This is not a procedure to assess the profitability of an investment, but it is an important reference for assessing the risk level, which needs to be taken into account in investment decision making [2]. The break-even point can be calculated by referring the quantities produced, to the expected production level and to the percentage of income expected. The formula used is the following: $BEP = FC/(TI - TVC)$, where TI = total income and TVC = total variable costs.

The break-even point thus calculated determines the percentage of the projected income that must be assigned to cover the fixed costs and the variable costs. It is considered that break-even points between 50 and 80% are acceptable.

1.3
Production Systems

Estates which are totally dedicated to coffee growing exist in countries like Brazil, Colombia, Costa Rica, Kenya, Indonesia, etc.; however, there are also a considerable number of small and medium-sized family farms with diversified agricultural activities which include coffee growing. Hence, the size of coffee estates varies greatly from one region to another, ranging from estates of 10–100 ha to small family holdings of 2–3 ha.

Atmospheric conditions also significantly influence cultivation practices. In areas where rainfall is low, protective covering and complementary irrigation are necessary, whereas in areas where rainfall is abundant, weed control becomes the most important practice. In actual fact, diversified production systems exist in most of the coffee growing areas of the world because coffee growers need subsistence crops and other sources of income.

Regardless of the size of the exploitation or the cultivation design, production systems throughout the world have traditionally been classified according to their technological level. Hence, there are intensive or extensive systems and, according to their size – low-, medium- or high-producing farms.

In Mexico, production systems are classified according to the plantation design, and the first parameter taken into consideration is the presence or absence of shade trees and their density when associated with the plantation. Cultivating systems have been defined as "a typical way of producing coffee that has its own, and immediately recognizable, characteristics" [3].

In general, there is a correlation between the size of a production unit and the production systems practiced. Thus, typically, there are "small producers", "medium-sized producers" and "large producers" (Tab. 1.1).

Small producers are generally family holdings which practice poly-cultivation in order to produce staple crops for food and fuel as well as coffee. In this case, coffee is regarded a complementary income. Labor is mostly provided by members of the family, although minimal extra labor may be hired for the harvest.

Medium-sized producers tend to be more specialized in coffee culture and to plant for commercial purposes. They generally have a more entrepreneurial outlook. Their income is essentially based on the sale of their coffee harvest complemented by the sale of other products like citrus

Table 1.1 Summary of the characteristics of the Arabica coffee types of producers

Item	Small	Medium-sized	Large
Hectares of coffee planted	less than 3	from 3 to 20	over 20
Technology	• traditional cultivation practices • limited use of agrochemicals • replanting carried out with plants from the plantation	• sporadic pest and disease control • traditional fertilization methods • replanting carried out either with plants from the plantation or those bought in local nurseries	• use of specialized technology • intensive pest and disease control • chemical fertilization • plants produced in plantation nurseries
Weed control	manual with machete or hoe	manual and chemical control	predominately chemical control
Yield of green coffee (kg/ha)	552	874	1196
Planting density (plants/ha)	1100–1600	2000–3300	over 3300
Other characteristics	cultivation with varied types of shading	cultivation mostly with specialized, selected types of shade-trees	cultivation with selected types of shade trees or fully sun-exposed

fruits, bananas, fodder, etc. As a result, shade trees are grown for profit, not only for fuel or subsistence purposes. Most of the labor is hired.

Large producers are those that have specialized in coffee growing. They are strictly commercial plantations where, due to their extension, all labor has to be hired. They expect coffee production to be a profitable activity and, consequently, they try to optimize production costs. Their coffee sales are made on an individual basis, either to the domestic market or to specialized markets for high-consuming countries. As their aim is to reduce production costs, they tend to favor the use of agrochemicals and mechanization as opposed to hired labor.

1.4
Set-up and Development Costs

The items taken into account when determining costs differ from one producing area to another within the same country, and, obviously, from one country to another. Where the weather, agro-ecological, social, etc., conditions are different, concepts and definitions have to be adapted to each case.

Basically, to be as accurate as possible, one should never underestimate costs which may appear insignificant because, in the long term, they may well add up to substantial expenses that will reflect on the global result of the operation. This rule applies to factors like the transportation of inputs, machinery or materials within or around the property, the cost of family labor or indirect expenses, like fuel for vehicles, administrative costs, security, etc.

On the other hand, the information required must be compiled systematically as it enables the chronological analysis of

each operation, from the initial land clearance to the first harvest. A questionnaire in which the entire process is broken down, is recommended.

To obtain the information in the field, the questionnaire should be designed so that the person who makes the survey, or the producer himself, can insert the information chronologically, taking into account mainly the labor performed and the equipment, machinery and materials used, from the preparation of the land until the cherry coffee is delivered to the collecting centre. The survey includes four parts: general information, coffee plantation set up, maintenance of the coffee plantation, and harvest data. Once the questionnaire is completed, the information is summarized in tables so that the results can be clearly seen, as shown in Tab. 1.2.

Obviously, the plantation set-up costs are the investments required to start cultivation. They are conducted only once in the lifespan of the plantation and include land clearance, staking out, designing plots, planting, shade set-up, etc., as well as the purchase of equipment, machinery and inputs.

Maintenance expenses incurred during in the first year must be added to the set-up costs. They include activities such as

Table 1.2 Establishing and maintenance costs

Item	Small	Medium-sized	Large
Set-up costs (US$/ha)	1040.02	1971.88	2854.20
Maintenance in the first year (US$/ha)	163.06	254.86	310.27
Total expenditure for the first year (US$/ha)	1203.09	2226.74	3164.47

pest and disease control, weed control, fertilizing, coffee replanting, and further purchases of equipment, machinery and inputs. The resulting total represents the initial outlay for starting up the plantation. The first income from yield can be expected from year 3.

1.5
Annual Production Costs

Expenses incurred for plantation development are different for each type of producer. They are generally lower than those incurred during the first year, because the activities conducted are limited to pest and disease control, weeding, fertilization, pruning, replanting, etc. (Tab. 1.3).

As from the third year, expenditure increases gradually and peaks during the seventh year after the initial set up.

Table 1.4 shows average production costs by calculating average yields per hectare and the annual production costs for each type of producer. It is important to note that these costs are calculated on the basis of delivery at the processing unit, which means that treatment and other expenses involving transport to the export centers inside the country are not included.

The build-up of production costs is important because it illustrates the significance of each type of expenditure on the total production costs. Hence, when it is necessary to highlight a specific type of expenditure, one can calculate it as a percentage of the total expenses of one stable year.

Costs should be grouped into harvesting expenses, labor, and the purchase of further machinery, equipment, materials and other items. Table 1.5 shows that the highest percentages of expenditure for small and medium-sized plantations is the cost of labor whereas for large producers the highest expenditure is for the purchase of machinery, equipment and materials.

Table 1.4 *Coffea arabica* production costs by type of producer

	Small	Medium-sized	Large
Total costs (US$/ha)	848.83	1361.68	1845.34
Yield of green coffee (kg/ha)	552	874	1196
Production costs (US$/kg of green coffee)	1.53	1.55	1.54

Table 1.3 Total disbursements by type of producer

Year	Cost per producer (US$)		
	Small	Medium-sized	Large
1	1203.09	2226.74	3164.47
2	374.92	579.52	607.71
3	497.29	767.92	607.71
4	493.34	944.04	1407.13
5	779.04	1318.06	1636.05
6	752.87	1318.06	1788.66
7	848.83	1361.68	1692.73
8	848.83	1361.68	1845.34

Table 1.5 Production costs (%) by type of producer

Grouped production costs	Small	Medium-sized	Large
Various or other	10	10	12
Inputs (machinery, equipment, materials)	14	21	35
Harvesting	31	30	32
Labor	45	39	21

For the three types of producers, expenses for harvesting are the next in importance.

In view of the importance of Robusta coffee in the international market, production costs for medium-sized producers of this species in Mexico have been calculated as follows:

Various or other	16 %
Inputs	9 %
(machinery, equipment, materials)	
Harvesting	49 %
Labor	26 %

Total costs for Robusta coffee are calculated at US$ 902.34/ha.

Income has been calculated taking into account the different yields for the three types of producers. Sales begin 3 years after the plantation is set up and they increase gradually to reach their highest levels in year 7 or 8.

Income is calculated by multiplying the production by an average selling price of US$ 1.77/kg of green coffee. This is the basis for the annual income figures given in Table 1.6.

1.6
Economic Indicators

It is imperative to be able to assess how much money the production costs and how much money was made in a year or in a cultivation cycle. Annual profit, as shown in Tab. 1.7, for the three types of producers starts from year 4 after the plantation set up, at low levels and increases gradually to peak at an average of US$ 132.63/ha for small producers, US$ 192.30/ha for medium sized producers and US$ 510/ha for large producers.

The break-even point, calculated by taking into account the variable costs, the fixed costs and the total income, for small producers is 79.73 %, for medium-sized producers 81.44 % and for large producers 75.65 % of the total income expected. This indicates that the medium-sized producers have the highest risk level and the large producers have the lowest risk level, inasmuch as the cost of family labor is included in the calculation. If family labor costs are not included, the lowest risk level would, obviously, be incurred by the small family holdings.

Table 1.6 Annual income for the different types of producers (US$/ha)

Year	Small	Medium-sized	Large
3	196.29	392.58	752.55
4	556.16	916.03	1570.34
5	981.46	1553.98	1897.49
6	899.67	1553.98	2355.51
7	981.46	1553.98	1897.49
8	981.46	1553.98	2355.51

Table 1.7 Annual profits by types of producers (US$/ha)

Year	Small	Medium-sized	Large
1	(163.06)	(254.86)	(310.27)
2	(374.92)	(579.52)	(607.71)
3	(300.98)	(375.54)	(363.49)
4	62.83	(28.01)	163.21
5	202.43	235.92	261.44
6	146.84	235.92	566.84
7	132.63	192.30	204.76
8	132.63	192.30	510.16

1.7
Sensitivity Analysis

The above analysis is based on an initial price to the producer of US$ 81.79 for 100 lb of green coffee. When prices fall by 10, 20 or 30 % this has a significant impact on the break-even point for all producers.

The results revealed in Table 1.8 with reference to the break-even point show that the level of risk for each producer increases at an alarming ratio as prices drop. It comes to a point where production is no longer generating an income and may well be carried out at a loss.

It should be remembered that acceptable break-even points should be between 50 and 80 %. This means that even a 10 % drop in selling prices jeopardizes all producers and could well lead to considerable disruption in the coffee growing process.

Table 1.8 Break-even point calculated according to drops in the price offered for cherry coffee

	Price at 100 %	Price at 90 %	Price at 80 %	Price at 70 %
Small producers	79.73	93.80	113.90	144.96
Medium-sized producers	81.44	95.81	116.34	148.06
Large producers	75.65	89.02	108.09	137.56

The Tables of Appendices 1 to 4 present additional quantitative informations.

1.8
Recommendations

As coffee producers have practically no means of influencing the fluctuations of market prices for coffee, they need to limit their investments in order to have enough leeway to be able face the risks incurred by sudden price drops. It is impossible to predict that any given coffee-growing cycle will generate high profits even when yields are considerable and quality is excellent. As a result, it is advisable to manage coffee plantations on a regular, systematic basis for every cycle and to span investments over a long period so that they can be more easily absorbed.

Specific knowledge of the various factors that increase the costs of coffee production will help to recognize sectors where costs can be reduced. Different strategies can be recommended for each type of producer as follows:

Small producers can:

- Reduce weed control from three to two operations per year,
- Diminish the frequency of shade regulation,
- Plant species for commercial purposes,
- Pay special attention to replanting, pruning and desuckering.

Medium-sized producers can:

- Reduce the use of fertilizers,
- Associate their coffee plants with other species of commercial interest, in order to diversify and increase their sources of income.

Large producers can:

- Take advantage of all economies that can possibly be applied to coffee plantations.
- Mechanize labor on the plantation as far as possible.
- Rationalize the use of agrochemicals.
- Conduct the marketing strategies as directly as possible to avoid the added expense of intermediaries.

All producers should implement production systems that allow the generation of products for specialty markets and investigate every opportunity to increase efficiency and reduce costs.

The systematic charting of annual expenses can help to pinpoint areas where production costs can be cut back, and the ongoing research for new technologies for improving production at lower costs should be studied and implemented wherever possible in each individual plantation or estate.

Bibliography

[1] Baca Urbina, G. *Evaluación de Proyectos, Análisis y Administración del Riesgo*, 2nd edn. McGraw-Hill, Mexico, 1994.

[2] Rodríguez, P. B. S., Díaz, C. E., Escamilla, P. J. R., Pérez, P. Y. M. and Suazo, M. *Proyecto de Factibilidad para el Establecimiento de Café Asociado con Cultivos Alternativos en el Municipio de Zentla.* Universidad Autónoma Chapingo, Centro Regional Universitario Oriente, Huatusco, Mexico, 1993.

[3] Escamilla, P. E. *El Café Cereza en México, Tecnología de Producción 44.* Universidad Autónoma Chapingo, CIESTAAM, Mexico, 1993.

[4] Servín, J. R. *Análisis de Costos en Tres Sistemas de Policultivo Comercial con Café en la Zona de Córdoba-Huatusco, Veracruz.* Colegio de Posgraduados, Montecillos, Mexico, 1997.

Appendix 1

Set-up and maintenance costs of 1 ha for a small producer (US$/ha)

Item / activity	Measuring unit	Unit price	Year 1 amount	Year 1 cost	Year 2 amount	Year 2 cost	Year 3 amount	Year 3 cost	Year 4 amount	Year 4 cost	Year 5 amount	Year 5 cost	Year 6 amount	Year 6 cost	Year 7 amount	Year 7 cost	Year 8 amount	Year 8 cost
Labor																		
Land clearance	daily wage	3.68	40.00	147.37														
Marking out and staking	daily wage	3.82	8.00	30.53														
Digging holes	daily wage	3.82	46.00	175.57														
Infilling and fertilizing	daily wage	3.82	40.00	152.67														
Transplanting	daily wage	3.82	16.00	61.07														
Shade set-up	daily wage	3.82	5.00	19.08														
Pest + disease control	daily wage	3.82	3.00	11.45	3.00	11.45	3.00	11.45										
Coffee tree pruning	daily wage	4.36									9.00	39.26	9.00	39.26	12.00	52.34	12.00	52.34
Shade tree pruning	daily wage	3.82									15.00	65.43	15.00	65.43	25.00	109.05	25.00	109.05
Desuckering	daily wage	4.36									9.00	39.26	9.00	39.26	12.00	52.34	12.00	52.34
Weed control	daily wage	3.82	20.00	76.34	50.00	190.84	50.00	190.84	28.00	106.87	28.00	106.87	28.00	106.87	28.00	106.87	28.00	106.87
Fertilizing	daily wage	3.82	2.00	7.63	4.00	15.27	7.00	26.72	7.00	26.72	7.00	29.72	7.00	29.72	7.00	26.72	7.00	26.72
Refilling	daily wage	3.82	1.00	3.82	4.00	15.27	6.00	22.90	9.00	34.35	9.00	34.35	9.00	34.35	9.00	34.35	9.00	34.35
Coffee cherry harvest	Kg	0.09					600	52.34	1'700	148.29	3'000	261.69	2'750	239.88	3'000	261.79	3'000	261.79
Sub-total				685.53		232.82		304.25		316.23		573.58		551.77		643.37		643.37

Numbers in blue = investments; Numbers in red = maintenance costs; Numbers in green = variable costs, starting from the third year (coffee harvesting and transportation).

Appendix 1

Set-up and maintenance costs of 1 ha for a small producer (US$/ha) (continued)

Item / activity	Measuring unit	Unit price	Year 1 amount	Year 1 cost	Year 2 amount	Year 2 cost	Year 3 amount	Year 3 cost	Year 4 amount	Year 4 cost	Year 5 amount	Year 5 cost	Year 6 amount	Year 6 cost	Year 7 amount	Year 7 cost	Year 8 amount	Year 8 cost
Machinery, equipment and inputs																		
Coffee plants	Unit	0.27	1'600	436.27	80	21.81	120	32.72	150	40.90	150	40.90	150	40.90	150	40.90	150	40.90
Urea	MT	239.91	0.20	47.98	0.40	95.97	0.40	95.97										
.18-12-06	MT	163.58							0.50	81.79	0.50	81.79	0.50	81.79	0.50	81.79	0.50	81.79
Organic fertilizers	MT	32.72																
Foliar fertilizers	Kg	5.02	0.30	1.50	1.00	5.02	4.00	20.07										
Copper oxychloride	Kg	3.82	0.50	1.91	1.00	3.82	4.00	15.27										
Sub-total				487.67		126.61		164.02		122.69		122.69		122.69		122.69		122.69
Other expenditures																		
Coffee transportation	MT	21.81		21.81			600	13.09	1'700	37.08	3'000	65.43	2'800	61.07	3'000	65.43	3'000	65.43
Plant transportation	Unit	0.01	1'600	17.45	80	0.87	120	1.31	150	1.64	150	1.64	150	1.64	150	1.64	150	1.64
Fertilizer transportation	MT	10.91	0.20	2.18	0.40	4.36	0.40	4.36	0.50	5.45	0.50	5.45	0.50	5.45	0.50	5.45	0.50	5.45
Tools and equipment				10.25		10.25		10.25		10.25		10.25		10.25		10.25		10.25
Sub-total				29.88		15.49		29.01		82.77		82.77		78.41		82.77		82.77
Total costs				1'203.09		374.92		497.27		493.34		779.04		752.87		848.83		848.83

Numbers in blue = investments; Numbers in red = maintenance costs; Numbers in green = variable costs, starting from the third year (coffee harvesting and transportation).

Appendix 2

Projected income versus expenditure for a small producer (US$/ha) calculated on the basis of US$ 0.327/kg of fresh cherry

	Year														
	1	2	3	4	5	6	7	8	9	10	11	12	13	14	15
Production of fresh cherry (kg/ha)	–	–	600	1700	3000	2750	3000	3000	3000	3000	3000	3000	3000	3000	3000
Total income (TI)	–	–	196.29	556.16	981.46	899.67	981.46	981.46	981.46	981.46	981.46	981.46	981.46	981.46	981.46
Operating costs	163.06	374.92	497.27	493.34	779.04	752.87	848.83	848.83	848.83	848.83	848.83	848.83	848.83	848.83	848.83
Fixed costs (FC)	16.06	374.92	431.85	307.97	451.91	451.91	521.71	521.71	521.71	521.71	521.71	521.71	521.71	521.71	521.71
Total variable Costs (TVC)	–	–	65.42	185.37	327.12	300.95	327.12	327.12	327.12	327.12	327.12	327.12	327.12	327.12	327.12
Operating profits (loss)	(163.06)	(374.92)	(300.98)	62.83	202.43	142.81	132.63	132.63	132.63	132.63	132.63	132.63	132.63	132.63	132.63
Break-even point (BEP)				83.06[a]	69.06	75.48	79.73	79.73	79.73	79.73	79.73	79.73	79.73	79.73	79.73

[a]Example of calculation of break-even point in Year 4: BEP = 307.97 (FC) / [556.16 (TI) − 185.37 (TVC)] = 83.06.

Appendix 3
Structure of production costs for a small producer

Percentage of costs		
45%	US$ 381.68	Labor
31%	US$ 261.69	Harvesting
14%	US$ 122.69	Machinery, equipment, inputs
10%	US$ 82.77	Other costs
Total Cost	US$ 848.83	

Appendix 4
Projected operating profits for a small producer over a period of 10 years (US$/ha)

Year	Total income (TI)	Operating costs	Operating profits
1	–	163.06	(163.06)
2	–	374.92	(374.92)
3	196.29	497.27	(300.92)
4	556.16	493.34	62.83
5	981.46	779.04	202.43
6	899.67	752.87	146.81
7	981.46	848.83	132.63
8	981.46	848.83	132.63
9	981.46	848.83	132.63
10	981.46	848.83	132.63

2
Technology Transfer

F. Martinez, J. Rojas, and G. Castillo F.

"A post-modern system of coffee culture: instead of convincing the producers to change their cultivation techniques, we should provide them with the tools to improve their decision making."

Charles Staver

2.1
Introduction

Throughout the history of agriculture, the transfer of knowledge from one generation to the next has resulted in the gradual evolution of improved farming systems and crop-husbandry techniques, ultimately leading to increased crop yield levels. Until the second half of the 19th century, improvements in coffee cultivation were dependent on the information transferred from father to son or from neighbor to neighbor. However, this gradual process of coffee improvement was obviously ineffective in raising production levels to meet rapidly increasing consumer demands. During the late 19th century and the first half of the 20th century, new agricultural research methods were increasingly applied to coffee in the major coffee-producing areas of the world. The research led to the selection of disease-resis-

tant, high-quality varieties and to major advances in coffee agronomy which greatly increased coffee yield potential. The application of the results of the research programs by coffee farmers raised coffee production levels significantly.

In the second half of the 20th century, the investment in research accelerated as consumer demands for coffee escalated due to population growth, new marketing strategies and ever more diversified consumers tastes. During this period farmers have become increasingly reliant on research and the new technologies being developed in order to be able to compete on world markets and to meet consumer demands for quality coffee. Countries and farmers who have failed to adopt new technologies have become marginalized. Coffee prices have experienced a great volatility in recent years whilst at the same time the costs of essential inputs as well as labour have been rising and consequently profit margins have been variable.

Thus, in order to maintaining profit margins and continuous improvement of the coffee production, the technology transfer from research stations to farmers will be of critical importance in the maintenance of a viable coffee industry in many countries, particularly in countries with a high proportion of smaller coffee producers. The approach to the issues involved in technology transfer and, in a broader context, in agricultural extension will vary depending on the within-country institutional framework and structure of farming. For example, the role of agricultural institutions in technology transfer in the socialist state of Vietnam, where most of the coffee farms are of only a few hectares, is vastly different to the situation in the state of Brazil, where coffee farms are at least 10 times larger than in Vietnam, and the private sector is the provider of most of the technology transfer and agricultural extension.

In the majority of coffee-producing countries, coffee growers are mainly smallholders with an average farm size of less than 5.0 ha. In these countries, government and private agri-services, research institutions and producer cooperatives play an important role in providing producers, in a simple and easily adaptable manner, with the technologies to ensure that the coffee production practices which are adopted by small farmers are sustainable and have no adverse effects on the local environment. Also, when dealing with large numbers of small coffee growers, the introduction of new technologies must take account of the socio-economic circumstances of the smallholders and of the requirements of the markets – involving coffee buyers, exporters, roasters and other participants in the commercial chain.

Coffee production has been developed in numerous countries throughout the tropics and subtropics, and the crop is grown in a wide range of agroclimatic conditions. Coffee growers also represent a wide spectrum of people with very different backgrounds not only in terms of their economic, cultural and social characteristics, but also in terms of their technological achievements. The huge differences in technological achievement is exemplified by comparing Brazil, which is at the forefront for introducing new technology, with many African and Central American countries where production methods have remained largely unchanged for the last 50 years. Thus, technology transfer cannot be dealt with in a generalized way, but each situation must be handled independently.

Technology transfer in Brazil and Hawaii involves all of the most recently developed technologies that take advantage of economies of scale and make full use of sophisticated systems of financing through a dynamic banking sector. The technologies which are being made available to farmers of those countries, largely through private sector organizations, include the provision of improved varieties, the use of fully controlled irrigation systems, the application of mechanized harvesting, as well as the introduction of improved methods of postharvest processing and handling. Some of these new technologies are not appropriate for countries with a high proportion of very small coffee farmers and, for those technologies which are appropriate, prevailing conditions will not allow their introduction.

In the future, the transfer of technology to the small coffee farmers will be essential for their survival. Without the fast adoption of new technologies within a sustainable approach, these small farmers will be unable to obtain the financial support that is crucial to the restoration of the profitability of the crop to something approaching the levels that were attained in the 1970s. It will be necessary to introduce flexible schemes to support the small-scale producers so that they can assimilate and adopt the improvements according to their specific situation. Without such support, many thousands of small coffee growers will be forced out of business in the near future.

2.2
A Brief History of the Development of Coffee Growing

2.2.1
The Origins of Coffee Growing

In the 17th and 18th centuries, Arab traders visiting Africa were introduced to coffee by the indigenous Africans who utilized the seed of the naturally occurring *Coffea* sp. to flavor water and produce a drink. The Arabs took a liking to the drink and some of these traders eventually transported seeds from Ethiopia to Yemen where they established coffee plantations in their native country. The technology used for their exploitation was adapted from that applied to other crops. It was the Yemeni coffee which was widely disseminated, first in the Middle East and later in Europe, that brought the drink international recognition and began the insatiable demand for coffee by the people of the world that has continued to this day.

2.2.2
The Expansion of Coffee during the Colonial Period

It was during the 19th century that coffee production expanded across the tropical and subtropical areas of the world. The European colonial powers were mainly responsible for this expansion. The colonial powers recognized the commercial potential of coffee, and funded large-scale development of plantations in the continents of America, Africa and Asia. Many countries were extensively colonized by Europeans who were responsible for the management of coffee plantations. The managers of coffee estates during the 19th century had only limited knowledge and understanding of coffee husbandry, and had to learn through "trial and error" experiences. During this time, attempts were made to follow models that were not adapted to local conditions. In Guatemala, the Antillean open system developed by the French was introduced, but it had to be changed and adapted to the local conditions of Central America [3]. During the colonial era it was recognized that the production techniques in use were not highly efficient and that the intensive use of the land could reduce soil fertility, but the availability of cheap labor compensated for many of the deficiencies in crop management. In addition, consumer demand in Europe and, later on, in the United States increased continuously through the 19th century and this increasing demand was accompanied by a steady escalation in coffee prices.

2.2.3
The New Order

During the later colonial period in the first half of the 20th century, the indigenous peoples in the colonies became more involved with coffee production. The type and size of the producers' holdings was transformed as smallholders were encouraged to plant coffee. Many nations saw coffee growing as a source of income that could bring about much needed social and economic changes. Producer associations were created to introduce new technologies and to coordinate marketing. The initial technologies to be introduced were concerned with the use of organic (guano) and inorganic fertilizers, the inter-planting of leguminous cover crops for weed control and the production of additional nitrogen. Progress was also made during this period in the field of plant breeding, particularly varietal selection. Technological development and its transfer was mostly carried out by European technicians and the main beneficiaries were the owners of large farms who were able to pay for the services provided.

2.2.4
The Institutional Era

In the second half of the 20th century, after World War II, coffee cultivation underwent an important change of course in response to the rapid growth in consumer demand. As the wealth of nations increased in the post-war period, increasing numbers of people in the developed world were able to purchase coffee which previously had been the preserve of the affluent minorities. The coffee-producing countries identified the commercial opportunities by supplying the coffee to meet that ever-increasing demand and provided considerable financial and technical support to coffee growers. Governments created official agencies and institutions to develop research and new technologies and to transfer the innovations and technologies that were proved effective to the coffee farmers. The government-sponsored organizations generated a substantial increase in production. In addition, in many countries there were large groups of consultants who, with great assiduity, transmitted technical know-how directly to the producers on site. However, not all schemes were effective, e.g. the scheme promoted by the World Bank which was based on a "Training and Visit (T & V)" system [11] failed to produce positive results and the financial benefits barely covered the costs of implementation.

2.2.5
The Globalization Era

Since the late 1980s a series of changes have taken place which radically changed the role of coffee in the economies of many developing nations. Breaches of the economic clauses of the ICO agreement resulted in the collapse of quota restrictions and the subsequent uncontrolled expansion in the areas of coffee in the world and generated large production increases. At the same time there appeared a highly volatile period, where markets, in less than 5 years, reached historical highest and lowest prices. The economies of many coffee-producing countries have been severely affected by the fall in coffee prices, by the related reduction in government revenues and also by corruption in official institutions. Consequently, government expenditures have been cut which has led to the weakening or disappearance of institutional coffee organizations in many countries. Because of the ensuing lack of support, the transfer of new knowledge and technologies is increasingly limited and the coffee growers find themselves trapped in a vicious downward cycle.

The present precarious economic situation, with seemingly ever-declining farm-gate coffee prices, has set the trend for the immediate future. The decline in coffee prices implies that there has to be a compensating rise in productivity if coffee farming is to remain as a viable enterprise on small mixed farms and if large, specialized coffee plantations are to survive. Raising productivity involves the introduction of new technological innovations as well as improvements in management practices, financial controls and plantation administration. During the globalization era two countries have met the challenge of increasing coffee areas and production levels whilst coffee prices have been declining, i.e. Brazil and Vietnam. Both these countries have significantly increased their share of the coffee sold on world markets and this has been achieved by the adoption of new technologies combined with the expansion of coffee areas on newly developed lands. On the other hand, the technologies implemented in those two countries do not necessarily represent sustainable models for the long term.

In the foreseeable future, the volatility of international prices and the need for profit will put pressure on coffee producers to employ technologies to minimize production costs whilst maximizing yields and maintaining quality standards. That is what will ensure plantation sustainability in the long term. To attain these goals, the transfer of necessary technological, financial and administrative knowledge is indispensable. However, producers will no longer be able to rely on a controlling government to extend this knowledge, they will have to become much more self-reliant. In the future, the role of governments is likely to become less directional and more supportive, advisory and promotional. Once producers have adopted new technologies and developed more efficient systems of production they will require support from governments, cooperatives and other institutions to facilititate access to other markets as well as to adapt continuously to the new demands of the consumers and to obtain better payment for their coffee.

2.3
Existing Schemes for Transmitting New Technologies and Developments

2.3.1
Introduction

In different countries and at different times, schemes for the transfer of new technologies and innovations from the applied research stations to coffee growers have been established. Some of these schemes have used models which have been developed for general agricultural extension purposes, whilst others have been organized specifically for coffee extension. All the schemes aim to provide coffee growers with the tools to improve the operational efficiency of their holdings and to raise net farm incomes.

2.3.2
Technological Transfer Groups (TTG)

This model scheme for the transfer of technologies to farmers was developed in Chile for general agricultural extension purposes. Although Chile is not a coffee producer, this model has characteristics that could be applied in coffee-growing countries. The Chilean model is based on the formation of producer groups that meet to share their experiences in dealing with agronomic, financial and marketing problems in the presence of an extension coordinator. The latter helps with advice on how to solve these problems.

A producer group consists of 12–15 producers. Each group has a defined structure with a president, a secretary, members and an external coordinator. Meetings are scheduled each month, generally rotating from place to place. Producers identify problems and, if these problems cannot be dealt with by the group themselves or with the help of the extension coordinator, specialized technicians may be brought in to provide solutions.

Producer groups may be used to assist in undertaking on-farm research and extension activities. Under the supervision of the extension coordinator or research worker they may be responsible for demonstration plots, applied research observation trials and for the testing of experimental management techniques in the field. Although the major objective of the producer groups is to improve management practices and raise farm incomes, in the majority of cases there will be a favorable spin-off in terms of the social impact that closer communal contacts can bring.

2.3.3
T & V System

This system was developed by the World Bank and has been used for decades by governments in a number of countries, covering a wide range of farming systems and many different types of crops. Under this system, an extension worker is responsible for a defined geographical area of land (one or more villages) and including a specified number of farmers. The extension worker meets with his farmers on a regular basis, i.e. a strictly controlled timetable is adhered to. The meetings are usually with groups of farmers. In countries with a high density of farmers and small holding sizes, the extension worker may deal only with selected farmer leaders who are then responsible for conveying information to individual farmers.

The extension messages or new technologies that are transmitted under the T & V system are clearly established and delivered routinely. The messages, which are obviously related to the crops and farming systems within the target area, are developed by the government extension service with any necessary assistance being provided by the government research services. The extension workers are supported by a number of specialist officers (covering such fields as entomology, plant pathology, agronomy, mechanization, vermin control, etc.), who are referred to as subject matter specialists (SMS). The SMSs are much fewer in number and cover whole districts, and they respond to specific requests for assistance rather than providing routine advice to all farmers.

The objective of the T & V system is to provide extension advice in a formal and disciplined manner. The routine work of the extension workers can be easily checked and supervised. The contents of the extension programme are clearly presented and a reliable assessment of the effectiveness of the programme may be readily carried out.

2.3.4
Community Scheme to Support Technology Transfer (MOCATT)

This model scheme was developed in Mexico by INIFAP specifically for coffee growers. The objectives of the scheme were to make technologies available to coffee growers and to accelerate the rate at which these technologies were transferred from applied research stations to the coffee growers. A reliable means of assessment of the effectiveness, and rate, of technology transfer was developed. It is reported that results have been beneficial to all concerned.

In the community scheme to support technology transfer in Mexico (MOCATT), a group of 10–15 producers within an area of 20–30 ha was identified as being representative of an area of up to 500 ha.

The action of the group is based on four pillars:

- Regional characterization (zonification)
- The organization of groups and strategic planning
- Adaptable training and research
- Initial, intermediate and final assessment combined with the communication of information where both the internal and outside experiences of the group are shared and compared; assessment includes agro-ecological, financial and social aspects.

2.3.5
Farmers Participatory Method (FPM)

This scheme or method of agricultural extension was developed and sponsored by the FAO to promote Integrated Pest Management (IPM) in a wide range of situations and involving a number of different crops. The basic objective was to encourage the participation of farmers in IPM programs, and to provide them with an understanding of pests and pesticides that would enable them to make their own decisions with regard to crop-husbandry practices and pesticide use. These decisions have to be based on the agro-ecological conditions on the farm and in surrounding areas. It has been demonstrated that this approach, when applied to the coffee grower, helps him to acquire a self-confidence in decision making which is reflected in an improvement in all aspects of crop management.

Under the FPM, Farmers Field Schools (FFSs) were established to teach the new technologies to local farmers using practical instruction techniques, on-site demonstration plots and other visual extension methods. Some of the training given to farmers whilst attending the FFSs was actually carried out on their own farms. The teachers at the FFSs were initially trained extension workers, but later some of the more enthusiastic farmers were employed, replacing the career extension workers. These farmers had an excellent knowledge of local conditions and were readily listened to by fellow farmers in their area.

2.3.6
Nestlé Integrated Model (NIM)

This extension model, which was developed by Nestlé's agricultural services, incorporates various aspects of the schemes outlined previously. The main objective of the NIM was to provide an efficient scheme (mechanism) for the transfer of new technology to coffee farmers covering all aspects of coffee husbandry from plantation establishment through the crop cycles to tree removal. The NIM also includes the post-harvesting components of quality control and coffee processing, and, finally, coffee marketing. The NIM is a flexible scheme with versions adapted to the needs and conditions of many countries. This model includes:

• New technologies to deal with a range of specific problems with the transfer of technology to farmers effected through a coordinator or extension worker operating with either groups of producers, or individual farmers (Pict. 2.1–2.4).

Picture 2.1 Nestlé extension worker with an individual planter in his Robusta plantation.

Picture 2.2 Nestlé extension worker with a group of coffee growers in Thailand.

Picture 2.3   Nestlé extension worker on a coffee plantation in China.

Picture 2.4   Group of coffee planters visiting a coffee processing plant in Mexico.
From: J. N. Wintgens.

- The establishment of on-farm demonstration plots to introduce new technologies and management practices to farmers. These plots are used for farmer visits with associated conducted explanatory tours.
- Large germplasm banks of Robusta and Arabica coffee cultivars and selections, gathered from throughout the world and representing a huge diversity of genetic material that is vital for future breeding programmes to accommodate new agronomic requirements, new technologies and changing quality standards (Pict. 2.6).
- Research centers where new developments are made available to producers. An example of this is somatic embryogenesis, a technique by which outstanding materials are being made available to producers in different countries.
- Training on sensorial and quality evaluation of coffee, in order to promote continuous improvement.
- Direct purchasing of coffee from individual coffee growers and planter groups.

Picture 2.5   A Nestlé ED farm.

Picture 2.6   Variety trial in a Nestlé ED farm in the Philippines.

- Research and demonstration farms which are used to develop and test new technologies for transfer to local coffee producers farming under similar local conditions (Pict. 2.5).

This system has been successfully promoted in several coffee-producing countries. Direct purchasing provides better prices to the coffee grower and generally contributes to the improvement of coffee

Picture 2.7 Direct coffee procurement in a Nestlé purchasing centre in Thailand.
From: J. N. Wintgens.

quality because prices are also based on quality requirements. Coffee prices are fixed on the basis of close control of bean quality by the purchaser. Bean quality is determined by moisture content, the percentage of defective beans and cup quality. The price premium for good quality coffee and the on-site advice encourages farmers to make the effort and take the necessary steps to improve coffee quality (Pict. 2.7).

2.4
Main Aspects of Technology Transfer to Farmers

2.4.1
Financial and Entrepreneurial Aspects

Producers, extension workers and consultants alike must be prepared to consider the financial aspects of the introduction of new technology on coffee farmers. Coffee farming should be managed on the same basis as any other commercial enterprise. Most, if not all, changes in technology will require some additional investment and it must be clear that the investment will generate a better financial situa-

tion for the grower either in the short or long term. This entails considering new technology costs in relation to potential incremental crop returns and the effect on net income levels.

Once the financial viability of investment in a new technology has been established, the coffee grower will need to have access to the necessary capital needed for the investment. In the vast majority of cases farmers will need to borrow money to finance any extra expenditure on-farm. Thus, the transfer of technology will not be possible unless the farmers have access to credit. In many countries sources of credit are scarce and additional credit may have to be supplied to support any technology transfer programmes. The supply of credit may be organized either through the commercial banking sector or through government agencies.

2.4.2
Producer Needs

The technologies being introduced should take account of the producers needs, so that the results achieved provide an adequate solution to the specific problems of the target farmers. Technological solutions requiring high management skills and high investment levels will not be appropriate on small farms. For example, solutions for irrigated, high-yield varieties of coffee, grown without shade and with high inputs of fertilizers and other agrochemicals, will not be suitable for growing a coffee variety with low yield potential under heavy shade. Thus, technologies which are emerging from government institutions or research centers all too often do not satisfy producers' and consumers' needs.

Many small farmers continue to grow their coffee in low-cost systems of cultivation. In these systems the coffee is grown under intensive shade, with low levels of fertilizer (often restricted to organic mulches), wide spacing, often inter-planted with other crops and minimal use of pesticides. The new technologies that are appropriate for the many small growers should not concentrate on producing high yields, instead the emphasis will be on quality coffee and on meeting the special demands of consumers, e.g. for organic coffee. Technologies will also be required for improving traditional methods of coffee production, e.g. combining coffee with other crops, making the best use of available local materials and diversifying growers' incomes.

2.4.3
Sustainable Production Systems

This concept is related to the long-term outlook for sustaining production levels in the foreseeable future with low impact on the environment. The intensive cultivation practices (which include the absence of shade, weed control by the application of herbicides, the use of irrigation, the application of high rates of fertilizers, the routine use of pesticides, pruning practices to maximize yields, etc.) have produced massive increases in potential yield levels. However, in many situations the high yields have not been maintained over a large number of years despite the continued use of high levels of inputs.

The declining yields are thought to be due in part to deteriorating soil conditions. Soil erosion results in the removal of the important surface layers of soil and losses of organic matter. The exposed surface soil then has poor structure and infiltration rates are reduced. Soil nutrient balances are disturbed by the high yields and are not maintained by the fertilizers being applied. Consequently, trace element deficiencies may be partly responsible for lowering coffee yields. Another factor responsible for declining yields in intensively managed coffee is the physiological deterioration in coffee trees. The premature ageing of coffee trees may result from management practices, which aim to maintain year after year heavy bearing trees, and to over-ride naturally occurring periods of moisture stress as well as tendencies towards biennial bearing.

The recommendations for costly high-input technologies to be adopted for maximizing yields must take account of the local soil and environmental conditions, and of the likely sustainability of high production levels under these conditions. Most low-cost technologies are by their very nature not exploitative and the low production levels are sustainable in the long term.

2.4.4
Influence of the Local Situation on Choice of Technologies

Even within the same region, cultural, educational, social and religious differences exist. Consequently, any scheme, method or strategy set up with a view to improving farming practices and/or the standard of living of rural populations, needs to respect individual and collective traditions, creeds and social criteria (Fig. 2.1).

2.4.5
Practical Learning
(Producers and Consultants)

In many cases, it is very difficult to learn and retain information by means of traditional teaching methods which sometimes even turn out to be anti-pedagogic. It is, therefore, very important to adjust training methods to the producer's conditions in

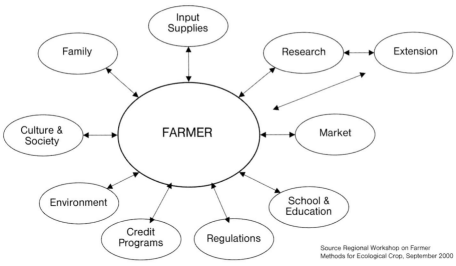

Figure 2.1 Factors which influence farmers in decision making.

Source Regional Workshop on Farmer
Methods for Ecological Crop, September 2000

2.4.6
Responses to Price Fluctuations

The unpredictable behavior of the international market has forced producers and consultants alike to adopt flexible strategies. A certain technology may have been chosen, it may even be perfectly valid, but when coffee prices are low, all efforts may prove useless and profit margins no longer provide enough to live on. When this happens, is necessary to be flexible in order to adapt the system to the circumstances. One way of ensuring other sources of income is to diversify production and to avoid "placing all the eggs in the same basket". Farm management should be guided by the famous quote "only change is permanent".

Similarly, the management schemes developed should adapt to the local ecology as well as to the socio-economic, educational and financial situation of each planter or group of coffee growers.

People generally retain:	10%	20%	30%	50%	80%	90%
of what they read	✳✳✳✳					
of what they hear		✳✳✳✳				
of what they see			✳✳✳✳			
of what they hear and see				✳✳✳✳		
of what they say themselves					✳✳✳✳	
of what they do themselves						✳✳✳✳

Source: Kairo, Moses et.al.

Figure 2.2 Practical learning.

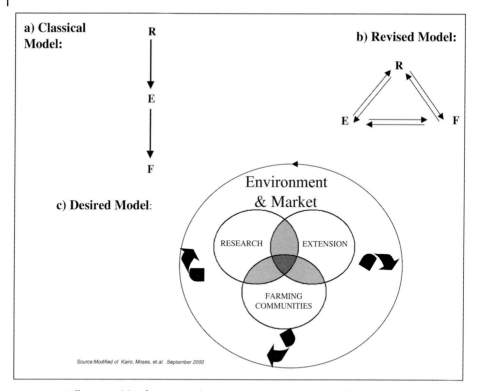

a) Classical Model:

R

E

F

b) Revised Model:

R

E   F

c) Desired Model:

Environment & Market

RESEARCH   EXTENSION

FARMING COMMUNITIES

Source:Modified of Kairo, Moses, et.al. September 2000

Figure 2.3 Different models of interaction between research, extension and farming communities.

2.4.7
Interaction and Dynamics

Traditional methods should always be re-adjusted so that they are geared to a dynamic interaction between all the participants. In this way, technological developments and the transfer of knowledge are mutually beneficial to all the parties involved (Fig. 2.3).

2.4.8
Initiative and Performance

Differences in initiative and availability are inevitable even within the same producer groups, particularly when it comes to implementing new technologies and know-how, as shown in Fig. 2.4.

2.5
Conclusions

Any scheme for the transfer of technology should encourage the active participation of each and every individual coffee grower. A scheme can be considered successful when it has reached the point where the majority of growers are not only aware of their problems, but are also able to find suitable solutions to ensure the future viability of their holdings.

In making the decisions on which new technologies are to be transferred to a specific group of producers, the following aspects should be considered:

• The social and financial status of those producers

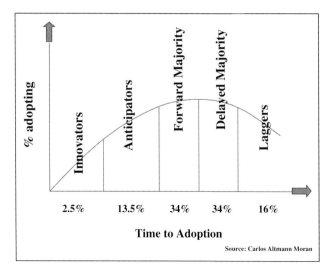

Figure 2.4 Grading of farmers into categories based on their adaptability to new technologies.

- The environmental conditions prevailing in the area under consideration and the crop management requirements in those conditions.
- Present and projected coffee prices, and cost–benefit analysis for each new technology.
- Sources of finance for the farmers to pay for the new technologies.
- The logistics of introducing each new technology.
- Administrative requirements.

The role of the external advisor, extension worker or consultant is to provide the producer with the necessary information and guidance to help him make the right decisions. In return, the producers will provide the feedback to determine the future technological development programmes of institutes and research centers.

Not all farmers have the same sense of initiative or the same capacity to adopt new technologies. The methods of transfer of technology that are employed will need to take account of the differences in the rate of assimilation between individual producers. Advisors, extension workers, agronomists and coordinators have an important role to play in enlisting the participation of the majority of growers in technology transfer. Thus, all interventions at the field level must be made with sensitivity, care and understanding. In this way, the real needs of the growers will be detected, and it will be possible to provide them with the necessary know-how to achieve stability and generate a sustainable income from their coffee.

Technology transfer in the modern era should not only aim at the traditional goal of increasing profits in coffee growing, it should also aim at improving all the other elements which will enable the producer to achieve a long-term competitive and sustainable development.

Bibliography

[1] ANACAFE. *Manual de Caficultura*. ANA-CAFE, Guatemala, 1998.

[2] Altmann, C. *Technological Transfer Groups*. Instituto de Investigaciones Agropecuarias & Sociedad Nacional de Agricultura, Berkeley, Santiago, 1998.

[3] Bertrand, B. and Rapidel, B. *Desafíos de la Caficultura en Centroamérica*. IICA, PROMECAFE, CIRAD, IRD, CCCR, San Jose, Costa Rica, 1999.

[4] Castillo, G. *Modelo Participativo de Transferencia Tecnológica e Investigación para el Mejoramiento de la Cadena Productiva del Café*. Inifap, Xalapa, 2000.

[5] Kairo, M., *et al. Regional Training Workshop on Farmer Participatory Methods for Ecological Crop Management*, Trinidad & Tobago. CAB International, CTA, CARDI Ministry of Agriculture Land and Marine Resources, PROCICARIBE, 11–15 September 2000. Available: http://www.cta.int/pubs/ecm/part1.pdf http://www.cta.int/pubs/ecm/part1.pdf

[6] Lightfoot, C. *Promoting Low External Input Agriculture (LEIA): from LEIA to Intensive Integrated Farming Systems*. International Support Group, Amersfoort, Netherlands, 1997.

[7] Pater, J. *Coffee in the Philippines and Nestlé's Agricultural Involvement*. Nestlé Philippines, 1998.

[8] Proyecto CATIE/INTA-MIP. *Como Implementar MIP en Café con Productores y Técnicos*. Publicaciones Diversas, CATIE, Managua, 1995.

[9] Trujillo, L. E. *Ecología Política de la Caficultura*. Primer Diplomado En Cafeticultura en México, CRUO-UACH, Hustusco, 1999.

[10] Trujillo, L. E. and Guadarrama, C. *Bases Ecológicas de la Agricultura Sostenible*. Primer Diplomado En Cafeticultura en México, CRUO-UACH, Huatusco, 1999.

[11] Williamson, S. Challenges for farmer participation in coffee research and extension. Paper presented at the *Mexican Biocontrol Congress Symposium*, 28 October 1999.

Part VI
Data & Information

1
Units and Conversion Tables

J. N. Wintgens and H. Waldburger

1.1
Introduction

The many units of measurements used in this book – not to mention among our different regions and cultures – reflect the fascinating and beneficial diversity of our planet. Our daily work requires a wide variety of information, often available only in different languages, and it all demands data in different units and in different orders of magnitude. The work often becomes cumbersome because of the very richness of the means we use for measuring. While it is the purpose of Part VI, to provide the reader with the data and information to render the quantitative part of their work easier, this chapter is concerned specifically with the units of measurement.

1.1.1
Presentation of Units

The decimal sign between digits in a number should be a point on the line, not a line-centered dot.

The comma for dividing figures into groups of three for numbers larger than four digits for better readability should be avoided, e.g. 252,004,700 is better written 252 004 700 or 2.52×10^8.

In any number where the decimal sign is placed before the first digit of the number, a zero should always precede it, e.g. 0.251 and not .251.

The combination of a prefix and a symbol for a unit is regarded as a single symbol and the characters should be written with no space between them, e.g. cm and not c m.

Note that the expression μm for the micron (one-millionth of a meter) replaces μ, the symbol often used previously.

When writing the symbol for a derived unit formed from several basic units, the individual symbols should be separated by a solidus (/), a space or a line-centered dot (·), e.g. the unit for velocity—meters per second—is written m/s, m s^{-1}, m·s^{-1} or mps, but not ms^{-1} (as ms would represent a millisecond).

When a unit is raised to a power, the power refers only to the units and not to any number preceding them, e.g. 2.3 cm^3 is the equivalent of 2.3×1 cm^3.

One, but not more than one, solidus (/) may be used instead of a negative power, e.g. g dm^{-3} may be written g/dm^3.

In printed text, both numbers and SI unit symbols are printed in upright type (SI is the international system of weights and measures; from the French, **S**ystème International des Unités). Algebraic symbols

Coffee: Growing, Processing, Sustainable Production, Second Edition. Edited by J. N. Wintgens.
© 2012 WILEY-VCH Verlag GmbH & Co. KGaA. Published 2012 by Wiley-VCH Verlag GmbH & Co. KGaA.

should be printed in *italic type* so they can be distinguished from SI symbols [5]. For further information on the use of SI units please refer to Section 1.6 (Bibliography).

1.1.2
Symbols and Abbreviations

Abbreviation	Name	Used for
2	square	second power
3	cube, cubic	third power
/	per or solidus	
μ	micro (10^{-6})	of units
Å	Ångstrom	length
o	degree	
°C	degrees Celsius	temperature
°F	degree Fahrenheit	temperature
K	Kelvin	temperature
°R	degree Rankine	temperature
°Re	degree Réaumur	temperature
a	are	area
ac	acre	area
arp	arpent	area
atm	atmosphere	pressure
avdp	avoirdupois	weight
bbl	barrel	capacity
Btu	British thermal unit	energy
bu	bushel	volume
cal	calorie	energy
cc	cubic centimeter	volume
cd	candela	luminous intensity
cm	centimeter	length
ch	chain	length
cu	cubic	
CV	metric horsepower	power
cwt	hundredweight	weight
dm	decimeter	length
dr	dram	weight

Abbreviation	Name	Used for
dwt	pennyweight	weight
EC	electrical conductivity	
Ewt	equivalent weight	
fa	fathom	length
F	furlong	length
fl oz	fluid ounce	capacity
ft *or* '	foot	length
g	gram	weight
gal	gallon	capacity
gpm	gallon per minute	flow
gr	grain	weight
h	hour	time
ha	hectare	area
hp	horsepower	power
hph	horsepower-hour	energy
in *or* ''	inch	length
J	Joule	energy
kg	kilogram	weight
km	kilometer	length
kn	knot	speed
kW	kilowatt	power
l	liter	capacity
lb	pound	weight
m	meter	length
MC	moisture content	
meq	milliequivalent	chemical analysis
mi	mile	length
min	minute	time
ml	milliliter	capacity
mS	millisiemens or milliho	conductivity
Mz	manzana	area
oz	ounce	weight
P	poise	viscosity
pH	potential of Hydrogen	chemical analysis
ppb	parts per billion (10^{-9})	density
ppm	parts per million (10^{-6})	density

Abbreviation	Name	Used for
psi	pounds per square inch	pressure
pt	pint	capacity
Q	quintal	weight
qt	quart	capacity
RH (%)	relative humidity (percent)	
rd	rod	length
s	second	time
sq	square	
st	stere	volume
t	ton	weight
tn	short ton	weight
ton	long ton	weight
tr	troy	weight
W	Watt	power
yd	yard	length

1.2
Metric Units of Measures

1.2.1
Prefixes, Symbols and Powers of 10

Prefix	Symbol	Meaning in the US	In other countries	Multiplication factor
exa	E	one quintillion times	trillion	$1\,000\,000\,000\,000\,000\,000 = 10^{18}$
peta	P	one quadrillion times	thousand billions	$1\,000\,000\,000\,000\,000 = 10^{15}$
tera	T	one trillion times	billion	$1\,000\,000\,000\,000 = 10^{12}$
giga	G	one billion times	milliard	$1\,000\,000\,000 = 10^{9}$
mega	M	one million times		$1\,000\,000 = 10^{6}$
kilo	k	one thousand times		$1000 = 10^{3}$
hecto	h	one hundred times		$100 = 10^{2}$
deca	da	ten times		$10 = 10^{1}$
deci	d	one tenth of		$0.1 = 10^{-1}$
centi	c	one hundredth of		$0.01 = 10^{-2}$
milli	m	one thousandth of		$0.001 = 10^{-3}$

Prefixes, Symbols and Powers of 10 (continued)

Prefix	Symbol	Meaning in the US	In other countries	Multiplication factor
micro	μ	one millionth of		$0.000\,001 = 10^{-6}$
nano	n	one billionth of	millardth	$0.000\,000\,001 = 10^{-9}$
pico	p	one trillionth of	billionth	$0.000\,000\,000\,001 = 10^{-12}$
femto	f	one quadrillionth of	thousand billionth	$0.000\,000\,000\,000\,001 = 10^{-15}$
atto	a	one quintillionth of	trillionth	$0.0000\,000\,000\,000\,000\,001 = 10^{-18}$

1.2.2
Length

Unit	Symbol	Equivalent
kilometer	km	1000 m
meter	m	1000 mm
decimeter	dm	100 mm
centimeter	cm	10 mm
millimeter	mm	10^{-3} m
micrometer	μm	10^{-6} m
Ångstrom	Å	10^{-10} m

1.2.3
Area

Unit	Symbol	Equivalent
square kilometer	km^2	10^6 m^2
hectare	ha	100 ac
are	a	100 m^2
square meter	m^2	100 dm^2
square decimeter	dm^2	100 cm^2
square centimeter	cm^2	100 mm^2
square millimeter	mm^2	10^6 μm^2
square micrometer	μm^2	10^{-6} mm^2

1.2.4
Volume and Capacity

Unit	Symbol	Equivalent
Volume		
cubic meter	m^3	$1000\ dm^3$
cubic decimeter	$dm^3\ (= l)$	$1000\ cm^3$
cubic centimeter	cm^3	$1000\ mm^3$
cubic millimeter	mm^3	$10^9\ \mu m^3$
cubic micrometer	μm^3	$10^{-9}\ mm^3$
Capacity		
hectoliter	hl	$100\ l$
liter	$l\ (= dm^3)$	10 dl
deciliter	dl	$0.1\ l$
milliliter	$ml\ (= cm^3)$	$0.001\ l$
microliter	$\mu l\ (= mm^3)$	$10^{-6}\ l$

1.2.5
Weight

Unit	Symbol	Equivalent
metric ton	t	1000 kg
kilogram	kg	1000 g
gram	g	1000 mg
milligram	mg	$1000\ \mu g$
microgram	μg	1000 ng
nanogram	ng	$10^{-3}\ \mu g$

Note: 1 l water weighs 1 kg; 1 ml weighs 1 g.

1.3
UK and US Units of Measures

1.3.1
Length

Unit	Symbol	Equivalent
inch	in	exactly 2.54 cm
foot	ft	12 in
yard	yd	3 ft
rod (or pole or perch)	rd	5.5 yd
chain (Gunter's: 66 ft)	ch (G)	4 rd
chain (Ramden's: 100 ft)	ch (R)	100 ft
furlong	F	10 ch (G)
mile	mi	8 F 80 ch 320 rd 1760 yd 5280 ft 1.609 344 km
fathom	fa	6 ft
nautical mile (1 knot)		6 076 ft 1.853 km

1.3.2
Area

Unit	Symbol	Equivalent
square inch	in^2	6.4516 cm^2
square foot	ft^2	144 in^2
square yard	yd^2	9 ft^2
square rod	rd^2	30.25 yd^2
square chain	ch^2	16 rd^2
acre	ac	10 ch^2 160 rd^2 4840 yd^2 43 560 ft^2
square mile	mi^2	640 ac 2.59 km^2

1.3.3
Volume and Capacity

Unit	Symbol	Equivalent
cubic inch	in³	16.39 cm³
cubic foot	ft³	1728 in³
cubic yard	yd³	27 ft³
UK fluid ounce	fl oz	28.4122 cm³
UK gill	gi	5 fl oz
UK pint	pt	4 gi
UK quart	qt	2 pt
UK gallon	gal	4 qt 160 fl oz 1.201 US gal
UK bushel	bu	8 gal
UK bulk barrel	bbl	36 gal
US fluid ounce	fl oz	29.5727 cm³
US gill	gi	4 fl oz
US pint	pt	4 gi
US quart	qt	2 pt
US gallon	gal	4 qt 128 fl oz 0.833 UK gal
US bushel	bu	9.308 gal
US barrel	bbl	31.5 gal
US dry pint	pt (dry)	550.5 cm³
US dry quart	qt (dry)	2 pt
US peck	pk	8 qt
US bushel	bu	4 pecks

1.3.4
Weight (Avoirdupois)

Unit	Symbol	Equivalent
grain	gr	0.064 807 7 g
dram	dr	27.34 gr
ounce	oz	16 dr
pound	lb	16 oz 7000 gr exactly 0.453 592 37 kg

Unit	Symbol	Equivalent
UK stone	st	14 lb
hundredweight short	cwt short	100 lb
short ton	tn	2000 lb
hundredweight long	cwt long	112 lb
long ton	ton	20 cwt long

1.4
Units of Measures from around the World

The following four lists always represent a unit of the particular measure converted to its metric equivalent; occasionally other equivalents are appended.

1.4.1
Length

Unit	Origin	Metric equivalent	Other equivalents
ana, aun, aune		1.19 m	
Ångström (Å)		10^{-10} m	
archin	Turkey	1.0 m	
archine	Russia	0.71 m	
arpent (length)	Canada	58.52 m	180 French feet ($=$ 12.8 in)
azumbre		5.5 m	
bolt		38.41 m	
bow		1.83 m	
braça	Brazil	2.2 m	2 varas
brasse		1.83 m	approx. 6 ft
braza		1.67 m	
cable	UK	182.88 m	600 ft
cable	US	219.456 m	720 ft
cana		1.56 m	
centimeter (cm)		0.01 m	
ch'ek, ch'ih		0.36 m	
chain (R)		30.48 m	100 ft
chain (G)		20.117 m	66 ft
chi	China	0.33 m	1.0936 ft
cho	Japan	109.09 m	60 ken

Unit	Origin	Metric equivalent	Other equivalents
codo		0.55 m	
coil		204.83 m	
cuadra		125.39 m	
cuarta		0.25 m	
cuerda	Puerto Rico	0.39 ha	
danda		1.83 m	
decameter		10.0 m	
decimeter (dm)		0.1 m	
destre		4.21 m	
dhara		0.48 m	
dhra		0.75 m	
digit		2.31 cm	
diraa		0.75 m	
dvim	Indonesia	2.61 cm	
elle		1.09 m	
estadio		201.168 m	
farsach	Iran	6.72 km	
fathom	nautical	1.8288 m	6 ft
foot	Commonwealth	30.479 cm	
foot	South Africa	31.480 cm	1.033 ft
foot	Mauritius	32.484 cm	1.066 ft
foot (ft) current	UK, US, South Africa, Australia, New Zealand	30.480 cm	12 in
foot, fot	Sweden	29.69 cm	11.689 in
fot	Norway	31.37 cm	12.35 in
furlong	UK	201.168 m	10 chains
gudze, gueza, el, ell, ella		1.0 m	
guz	India	0.9144 m	1 yd
hand		10.16 cm	
hath		0.457 m	
hectometer		100.0 m	
hindaza		0.67 m	
hiro		1.515 m	
inch	Turkey	3.138 cm	1.24 in
inch	Swaziland	2.624 cm	1.033 in

Unit	Origin	Metric equivalent	Other equivalents
inch	El Salvador	2.32 cm	0.913 in
inch (in) current	UK, US	2.54 cm	
jareb		20.12 m	
jo		3.03 m	
karam		1.67 m	1 braza
ken	Japan	1.8 m	6 shahv
keup	Thailand	0.247 m	
kilometer (km)		1 000.0 m	
knot, noeud, nudo (naut.)		1.852 km	1 mi
koss		1.839 km	
koss	India	2.011 km	
league	international nautical	5.556 km	3 nautical mi
league statute		4.828 km	3 mi
legna	Argentina	5.200 km	
legna	Brazil	6.600 km	3000 braças
legna	Bolivia	5.196 km	
legna	Chile	4.514 km	
legna	Colombia, Ecuador	5.000 km	
legna	Cuba	4.240 km	
legna	Guatemala	4.444 km	or 5.572 km
legna	Honduras	4.174 km	
legna	Mexico	4.190 km	
legna	Paraguay	4.330 km	
legna	Peru	5.556 km	
legna	Uruguay	5.154 km	
legna current	Spain, Central and South America	5.572 km	
lei		0.68 km	
li	China	0.50 km	
lieve	Quebec	4.827 km	
lieve	Mauritius	4.000 km	
lieve	Paraguay	4.330 km	
lieve (terr.)		4.828 km	
line		0.21 cm	
link (engineers)		30.479 cm	

Unit	Origin	Metric equivalent	Other equivalents
link (Gunter's)		20.116 cm	
megameter		1 000 km	
micron, micrometer (µm)		0.001 mm	
mile, mille, nautical		1.8523 km	
mile statute, mille (mi)		1.6093 km	
millimeter (mm)		0.001 m	
millimicron		0.001 µm	
miriameter		10.0 km	
mlha	Brazil	2.2 km	1000 braças
niev	Thailand	2.08 cm	
paal		1.51–1.85 km	
pace	(milit.)	0.76 m	
palm		7.6–8.9 cm	
palmo	Brazil	0.22 m	8 polegadas
pé	Brazil	0.33 m	1.5 palmas
perch	UK	5.029 m	
pic, pik		0.686 m	
pie	Egypt	0. 75 m	
pié	Spain	27.9 cm	
pied	France	32.5 cm	
pied, pié		30.479 cm	
point		0.35 mm	
pole	UK	5.029 m	16.5 ft
polegada	Brazil	2.75 cm	
pouce		2.539 cm	
pulgada		2.539 cm	
qasaba	Egypt	3.55 m	
quarter		22.86 cm	0.25 yd
ri	Japan	3.927 km	36 cho
rod (rd)		5.029 m	
rope	UK	6.096 m	20 ft
sauk, sawk	Thailand	0.5 m	
sen	Thailand	40.0 m	
shakukane	Japan	0.301 m	

Unit	Origin	Metric equivalent	Other equivalents
span		0.23–0.28 m	
t'sun		3.58 cm	
thira	East Africa	45.72 cm	18 wanda
tjenkal		3.66 m	
toise		1.94 m	
ugul		1.9 cm	
vaj, wah	Thailand	2.0 m	
vara	Argentina	0.866 m	
vara	Brazil	1.100 m	
vara	Central America	0.861 m	
vara	Chile, Colombia	0.800 m	
vara	Costa Rica	0.836 m	
vara	Cuba, USA	0.848 m	
vara	Ecuador	0.850 m	
vara	Honduras	0.835 m	
vara	Mexico	0.838 m	
vara	Paraguay	0.800 m	
vara current	Spain, Central and South America	0.84 m	
vara-tarea	Domin. Republic	2.75 m	
ver		0.68–0.99 m	
verst	Russia	1.067 km	
voet	Indonesia	31.4 cm	12 diums
wanda	East Africa	2.54 cm	1 in
wari	East Africa	91.44 cm	2 thira
yard (yd)	UK, US	91.44 cm	3 ft

1.4.2
Area

Unit	Origin	Metric equivalent	Other equivalents
acre (ac)		0.404 686 9 ha	160 rd^2 = 4840 yd^2
alqueire mineiro	Brazil	cuadra = 4.84 ha	100 braças × 100 braças
alqueire paulista	Brazil	2.42 ha	100 braças × 50 braças
aranzada		0.37–0.48 ha	
are, area (a)		100 m^2 = 0.01 ha	

Unit	Origin	Metric equivalent	Other equivalents
arpent	France	0.422 ha	
bahu, bahoe, bouw	China, Indonesia	0.71 ha	
baropny		0.16 ha	
bigha	India, Sri Lanka, Bangladesh	0.0809 ha	
bovate		5.67 ha	
braças quadrada	Brazil	4.84 ha	
bu		3.31 m^2	
bunder		1.0 ha	
caballería	Cuba	13.42 ha	
caballería	Mexico	42.8 ha	
caballería	Guatemala, Honduras, El Salvador	45.0 ha	
caballería	Costa Rica	45.2 ha	
caballería	Puerto Rico	78.6 ha	
Cana real		0.61 ha	
cavadura		0.04 ha	
cawny		0.53 ha	
cent	India, Sri Lanka	40.4686 m^2	1/100 ha
centare		1.0 m^2	1/100 a
chattak		4.18 m^2	
cho	Japan	0.992 ha	10 tan
concesión		33.0 ha	
cord		2.74 m^2	
cottah		66.89 m^2	
cuadra	Peru	0.7244 ha	
cuadra	San Paolo	0.08 ha	
cuadra	Uruguay	0.7378 ha	
cuadra	Chile	125.4 m^2	
cuartal		238.4 m^2	
cuarterada		0.71 ha	
cuerda	Puerto Rico	0.39 ha	
cuerda	Guatemala	43.7 m^2	
cuerda	Venezuela	7.14 m^2	
cuerda	Mexico	6.25 m^2	

Unit	Origin	Metric equivalent	Other equivalents
dalag wha	Thailand	$4.0\ m^2$	
deciare		$10.0\ m^2$	
deciatrine	Russia	1.09 ha	
dekare		0.1 ha	
destre		$17.76\ m^2$	
deunam, donum, dunam	Iraq, Lebanon	0.25 ha	
día-buey		0.13 ha	
djerib	Iran	1.0 ha	
djo-eng.	West Java	1.41 ha	
émina		0.06–0.09 ha	
estadal		$11.18\ m^2$	
fanega	Colombia	$64.00\ m^2$	
fanega	Venezuela	0.6987 ha	
fanega española	Peru	2.8978 ha	
fanegada	Central America	0.64 ha	
feddan	Egypt, Sudan	0.418 ha	
feddan massi		0.4201 ha	
ferrado		$437–639.6\ m^2$	
hectare (ha)		$100\ a = 10\,000\ m^2$	2.471 ac
hectárea	El Salvador	0.1 ha	
hide		40.5 ha	
hold		0.575 ha	
huebra		0.22 ha	
jemba		$13.38\ m^2$	
jerib		0.1 ha	
jornal		0.44–0.48 ha	
kanal	India, Sri Lanka	$505.85\ m^2$	
khajjan	North Africa	$1000\ m^2$	
koh	Japan	0.99 ha	
kunte	India	0.16 ha	
legna cuadrada	Ecuador	$31.05\ km^2$	
legna cuadrada	Uruguay	$26.57\ km^2$	
legna cuadrada	Brazil	$43.56\ km^2$	
lima	Zambia	$2\,500\ m^2$	
manzana	Mexico	0.672 ha	

Unit	Origin	Metric equivalent	Other equivalents
manzana	Costa Rica, Guatemala, Honduras, El Salvador	0.699 ha	
manzana	Argentina	1.0 ha	
manzana	Nicaragua	0.750 ha	
marco		0.34 ha	
marla		25.32 m^2	
mishara		0.25 ha	
mojada		0.49 ha	
morabba		10.12 ha	
morgen	South Africa	0.857 ha	
mou, mow	China	0.0674 ha	
mu	China	666.66 m^2	
ngan	Thailand	400 m^2	
obrada		0.39–0.54 ha	
paal		227.09 ha	
pantjar	East Java	2.82 ha	
partido		1.21 ha	
peonada		0.34–0.39 ha	
plaza	Colombia	0.64 ha	
qu'ser	Iran	1.1 m^2	
quadra de Sesmaria	Brazil	87.12 ha	60 braças × 1 legna
quadra quadrada	Brazil	1.742 ha	1.32 m × 1.32 m
rai	Thailand	0.16 ha	
relong	Malaysia	0.28 ha	
robada		0.09 ha	
rood	UK	1011.7 m^2	0.25 ac
sarsani		2.8 m^2	
se	Japan	99.2 m^2	30 tsubo
square foot (ft^2)	UK, US, South Africa, Australia	929 cm^2 (0.0929 m^2)	144 in^2
square inch (in^2) (current)	UK, US, South Africa, Australia	6.4516 cm^2	
square kilometer (km^2)		1 000 000 m^2 (100 ha)	0.3861 mi^2
square mile (mi^2) or section		258.998 ha (2.59 km^2)	640 ac

Unit	Origin	Metric equivalent	Other equivalents
square rod (rd^2)		25.293 m^2	1 perch^2
square yard (yd^2)	UK, US, South Africa, Australia	0.8361 m^2	9 ft^2
stremma		0.1 ha	
tahulla		0.11 ha	
tan	Japan	992.0 m^2	10 se
tanab	Iran	0.44–0.45 ha	
tarca	Dominican Republic	0.625 ha	
tarefa	Brazil	0.4356 ha	50 braças × 30 braças
topo	Peru	0.3494 ha	
topo		0.35 ha	
township		93.233 km^2	36 sections = 23.04 ac
tsubo	Japan	3.31 m^2	
vara (conuquera)		0.64 m^2	
vara, tarca	Mexico	312.5 m^2	
vasana		0.22 ha	
woh^2	Thailand	4 m^2	
yard (yd^2)		12.15 ha	
yoke		0.575 ha	

1.4.3
Volume and Capacity

Unit	Origin	Metric equivalent	Other equivalents
Almud	Peru	16.96 l	
1 second foot	US (vary according to State)		40–50 miners inches
acre × foot			43.56 ft^3 323.136 gal of water
acre × inch			113 t of water (approx.)
Almud	Spain	4.6–5.6 l	
anker	South Africa	34.1 l	
ardeb		197.99 l	
ármina		34.66 l	
arroba	Chile	40 l	

Unit	Origin	Metric equivalent	Other equivalents
arroba	South and Central America	11.5 l	
artabal		65.92 l	
aum	South Africa	40.92 l	
balde		18 l	
ban	Thailand	10.0 hl	
ban	Thailand	1000.0 l	50 sat
barrel	Belize	125 l	
barrel	Haiti	1.10 m^3	
barrel	Argentina	76 l	
barrel	Paraguay	96.928 l	
barrique	Mauritius	227.3 l	
barrique	Haiti	225.0 l	
basket	Myanmar	40.6 l	
bota	South and Central America	484.1 l	
botella		0.65–0.9 l	
botella	Guatemala	0.75 l	
botella	Colombia	0.7 l	
bottle	UK, US	0.758 l	
bouteille		0.8 l	
cahiz	South and Central America	666.0 l	
cañado		32.7–37.0 l	
cántara	South and Central America	16.33 l	
cargueiro		80 l	
cavan		75.0 l	
celemín		4.63 l	
centiliter		0.01 l	
chaldron (dry)		1309.32 l	36 UK bu
chopin		0.85 l	
chopinne		0.4 l	
chupa		0.375 l	
chupak		1.14 l	
coomb	South Africa	145.47 l	4 UK bu

Unit	Origin	Metric equivalent	Other equivalents
copo		25 cm^3	
cord (cd) of wood		3.6246 m^3	8 ft $\times$ 4 ft $\times$ 4 ft = 128 ft^3
cuarta		0.78 l	
cuartán		4.15–18.08 l	
cuartera		69.5–70.8 l	
cuarterola		100.0 l	
cuartillo		0.47–2.46 l	
cuarto		3.2–3.46 l	
cuarto		0.96 l	
cubic centimeter (cm^3)		0.001 l	0.06102338 in^3
cubic foot (ft^3)	UK, South Africa, US, Australia, New Zealand	28.371 l = 0.0283 m^3	1.728 in^3
cubic inch (in^3)	UK, South Africa, US, Australia, New Zealand	16.387 cm^3	
cubic yard (yd^3)	UK, South Africa, US, Australia New Zealand	764.6 l = 0.7645 m^3	27 ft^3
cubic meter (m^3)		1000 l	1.308 yd^3
cup		0.24 l	approx. 8 fl oz
decaliter		0.1 hl = 10 l	
deciliter (dl)		0.1 l	
decimo		40.0 l	
decistère		100.0 l	
din		36.37 l	
dram (fl)		3.696 ml	
émina		18.11 l	
fanega	Spain	57.4 l	
fanega		22.5–137.0 l	
firkin		40.91 l	
gallon	Haiti	3.75 l	
gallon	Honduras	3.456 l	
Gallon Imperial	UK	4.546 l	4 UK qt
gallon	US	3.785 l	4 US qt
galón	Argentina	3.819 l	
galón	Bolivia, Costa Rica, Ecuador	3.364 l	
galón	Peru	4.546 l	1 UK gal

Unit	Origin	Metric equivalent	Other equivalents
galón	Barbados, Cuba, El Salvador	3.78 l	
ganta		3.0 l	
gantang		4.5–8.6 l	
gantang	East Java	1.1 l	
gantang	Malaysia	4.546 l	1 UK gal
gantang	Central Java	8.6 l	
garrafa		5 l	
gill (gi)	UK	0.142 l	5 fl oz
gill (gi)	US	0.118 l	4 fl oz
go		0.18 l	
gwe		18.18 l	
half aum	South Africa	70.463 l	
hectoliter		$0.1 \text{ m}^3 = 100 \text{ l}$	
hogshead		239.48 l	
kadah		2.06 l	
keila		16.5 l	
koku	Japan	180.38 l	100 sho
kwien	Thailand	2000 l	
kwien	Thailand	2000 l	2 ban
last	UK	2907.8 l	
last	The Netherlands	3000 l	
last	South Africa	2909 l	10 UK pt
leaguer	South Africa	581.9 l	128 UK gal
liter (l)		$1 \text{ dm}^3 = 1000.0 \text{ ml}$	
meter cube (m^3)		1000 l	36.31 ft^3
mallal		15.48 l	
microliter		0.001 ml	
milliliter		$1 \text{ cm}^3 = 0.001 \text{ l}$	
mminim		0.062 ml	
moio		12.64 l	
moyo	South and Central America	258.63 l	
mud, muid		$8.0 \text{ l} = 109.0 \text{ l}$	
muid	South Africa	109.140 l	

Unit	Origin	Metric equivalent	Other equivalents
naggin		0.14 l	
nozibn	Myanmar	0.3 l	
octavo		125 cm^3	
ounce fluid (fl oz)	UK	28.41 ml	
ounce fluid (fl oz)	US	29.57 cm^3	
ounce, once, onza,		29.57 cm^3	
pau	0.28 l		
peck (pk)	UK	9.092 l	
peck	US	8.809 l	
pint (pt)	UK (16 fl oz)	0.568 l	20 fl oz
pint	US	0.473 l	16 fl oz
pint dry	US	0.551 l	
pipa	Spain	4.47 hl	
pipa	South and Central America	4.357 l	
pipe		4.16–5.46 hl	
pottle		2.27 l	
pyi	Myanmar	2.5 l	
quart dry (qt)	US	1.101 l	
quart (qt)	US	0.946 l	2 US pt
quart	UK	1.136 l	2 UK pt
quart dry	UK dry	1.136 l	1.2009 cuartus
quarter	UK	2.908 hl	
register ton	UK, US, South Africa, Australia, New Zealand	2.83 m^3	100 ft^3
rob		8.25 l	
robo		28.13 l	
sale		0.57 l	
sat	Thailand	20.0 l	
sayut		4.55 l	
second		UK gal: 34.1 l/s; US gal: 1.7 m^3/min	7.5 gal/s; 450 gal/min
second foot		28.32 l/s	1 ft^3/s flow past a given point
seil		9.09 l	
ser		1.0 l	

Unit	Origin	Metric equivalent	Other equivalents
shaku		0.018 l	
sheng	China	1.0 l	
sho	Japan	1.804 l	
sinquena		20.65 l	
stere		1000.0 l	
strike	South Africa	72.74 l	2 UK bu
tablespoon		15 ml	3 teaspoons
2 tablespoons		30 ml	approx. 1 fl oz
16 tablespoons		0.24 l	approx. 1 cup
tambolha	Mexico	200 l/ha	
tan		113.65 l	
tanan	Thailand	1.0 l	
tanan	Thailand	1.0 l	
tang	Thailand	10.0 l	
tasse, taza		0.237 l	
tate tsubo	Japan	6.01 m^3	
teaspoon		4.93 ml	
thang	Thailand	10.0 l	
to		18.05 l	
tonel		500.0 l	
winchester	US	35.23 l	1.244 ft^3

1.4.4
Weight

Unit	Origin	Metric equivalent	Other equivalents
acheintaya		165.65 kg	
acre × foot of soil		1800 t	
acre × foot of water		1234 m^3	
acre × inch		102.8 m^3	
ardeb groundnut	Sudan	75 kg	
ardeb groundnut	Egypt	74.88 kg	
ardeb maize	Egypt	140 kg	
ardeb maize	Sudan	198 kg	

Unit	Origin	Metric equivalent	Other equivalents
arratel ólibra	Spain	0.46 kg	
arroba	Spain, Colombia	11.5 kg	
arroba	Brazil	15.0 kg	
arroba	Guatemala, Peru, Bolivia, Mexico	11.5 kg	dry green coffee beans
arroba	Honduras	11.34 kg	
bag		45–76.2 kg	
bahar, behar		204.12 kg	
bahra	Malaysia	181.44 kg	3 pikul
baht, bat		8–16 g	
barrel	US	88.9 kg	
bath	Thailand	15 g	
batman	Iran	2.95 g	
beittha		1.66 kg	
box		33.4–40.85 kg	
bunch		14.5–25.4 kg	
bundle	South Africa	3.175 kg	7 lb
bundle	RSA	3.17 kg	
cade		254.01 kg	
caja	Mexico	66 kg	fresh coffee cherries
cajou	Chile	2.941 kg	
canasta	Mexico	10 kg	
candareen		0.38 g	
candy		226–355.62 kg	
candy	India	254.2 kg	
candy	Sri Lanka	254.2 kg	560 lb
cantar	Arabia	44.93 kg	
cantar	Greece, Turkey	56.52 kg	
cantar	Libya, Syria	255.83 kg	
cantar coffee	Sudan	198 kg	
cantaro	Spain, Cuba	46.0 kg	
Cape ton	RSA	907.2 kg	2000 lb
carat		0.19 g	
carga	Colombia	125 kg	2 bags of green coffee
carga	current	91 kg	

Unit	Origin	Metric equivalent	Other equivalents
carga	Bolivia	69.01 kg	
carga algodon	Peru	167.44 kg	
case		18.4–40.82 kg	
cask	US	304.8 kg	
cental	UK, US, South Africa	45.369 kg	100 lb
cental	RSA	50.8 kg	112 lb
centner	metric	100.0 kg	
centner, zentner		50.0 kg	
chang	Thailand	1.2 kg	20 tamlung
chetwert		139.25 kg	
chiquihuite		10 kg	
Chittak	Bangladesh	58.3 kg	5 toflas
clove		3.63 kg	
coyan		1.2 mt	
cran		170.5 kg	
crate		31.75 kg	
dang	China	50 kg	100 jin
decagram		10 g	
decigram		0.1 g	
dirhem	Egypt	3.12 g	
drachme	apoth.	3.888 g	
dram	Avard.	1.772 g	
dram	Greece	0.115 g	
fan		0.38 g	
fanega		51–129.0 kg	approx. 400 l of coffee
fanega	Costa Rica	255 kg	green coffee beans
faunt		1.77 kg	
firkin		25.4–45.4 kg	
firlot		25.8–33.1 kg	
frasilia	East Africa	16.329 kg	36 lb
gisla	East Africa	163 kg	360 lb
gong-jin	China	1 kg	
gong-liang	China	100 g	
grain	Mauritius	53.1 mg	

Unit	Origin	Metric equivalent	Other equivalents
grain (gr), grano	UK, India Pakistan	64.799 mg	current
gram (g)		1000 mg	
haab	Thailand	60 kg	50 chang
heml		249.6 kg	
hiaku-mé		0.375 kg	
hobbet		76.2 kg	
hogga		1.27 kg	
hundredweight long	UK	50.802 kg	112 lb
hundredweight short		45.36 kg	100 lb
kantar, cantaro		44.93–255.83 kg	
kati	China, Thailand, Malaysia	0.604 kg	
kati, kin		0.6–1.2 kg	
kela		3.22 kg	
kemple		199.6 kg	
khavar, kharwar	Iran	294.84 kg	
kilogram (kg)		1000 g	2.205 lb
kin	Japan	0.6 kg	
kintar		288.45 kg	
kintar-shami		256.4 kg	
kip		453.59 kg	
koyan	Malaysia	2419.2 kg	40 pikul
kwan, kan	Japan	3.75 kg	
kwet		16.57 kg	
kyat		8.0–16.0 g	
lata		11.34 kg	
lata	Mexico	18 kg	wet parchment coffee
leung, lliang		37.8 g	
liang	China	50 g	
libra	Spain	350–460.009 g	
libra	Colombia	500 g	
libra	Mexico, Peru, Bolivia, Cuba	460 g	
livre	apoth.	373 g	Troy weight pound
livre	France, Mauritius	460 g	pound

Unit	Origin	Metric equivalent	Other equivalents
livre, libra	avdp	453.592 g	avdp lb
load	East Africa	22.679 kg	50 lb
load	Nairobi	27.215 kg	60 lb
load	Kisuvu	36.287 kg	80 lb
mace		3.77 g	
man	Iran	2 kg	
maund	India, Sri Lanka, Bangladesh	37.324 kg	40 seers
maund	Arabia	25.42 kg	
meskal	Iran	4.54 g	
metermasza		100.0 kg	
microgram (o)		0.001 mg	
milligram (o)		0.001 g	
minim	apoth.	0.065 g	
momme, me	Japan	3.74 g	
muid		68–90.7 kg	
nijo		15.0 g	
oke, okka	Egypt	1.248 kg	
okiya		0.21 kg	
ounce, once, onza (oz)	current	28.3495 g	1 avdp oz
ounce, once, onza	Troy weight	31.104 g	1 Troy or apoth. oz
ounce	Brazil	28.69 g	
	India	28.69 g	
	Damascus	213.5 g	
	Syria	320.5 g	
	Mauritius	30.594 g	
	Zanzibar	28 g	
	Netherlands	100 g	
	Cuba	28.75	
	Philippines	28.75 g	
	Spain	28.75 g	
	Argentina	28.71 g	
	Colombia	31.25 g	
	Dominican Rep.	28.35 g	
	Honduras	28.8 g	
	Mexico	28.765 g	
	Guatemala	28.75 g	
	Peru	28.75 g	
	Paraguay	28.69 g	
	Bulgaria	30 g	
	Pakistan	28.35 g	

Unit	Origin	Metric equivalent	Other equivalents
pack		108.9 kg	
pennyweight		1.555 g	
Pfund	Germany	0.5 kg	
pickul, picul	Indonesia	61.76 kg	
pikul	Malaysia	60.48 kg	
poud, pood, pud	Russia	14.62 kg	
pound (lb)	current	453.592 g	16 oz, 7000 g
pound	The Netherlands, RSA	494.42 g	
pound (lb)	Troy weight	373.241 g	apoth.
pyi		2.13 kg	
quantar	Egypt	45 kg	
quarter	US	12.701 kg	2 stones
quilote	Brazil	0.2 gr	
quintal	Mexico	40 kg 57.5 kg 80 kg 245 kg	dry coffee beans dry parchment coffee dry coffee cherries fresh cherries
quintal	Indonesia	100 kg	
quintal	Spain	46.01 kg	
quintal	Brazil	60 kg	4 arrobas
quintal	Colombia	50 kg	
quintal	Central America	45.36 kg	100 lb
quintal largo	Portugal	58.76 kg	
quintal metric		100 kg	
quintal short		45.36 kg	100 lb
rafa		254.01 kg	
ratel		0.45–3.0 kg	
roba	some Arabian countries	25.4 kg	
rotl, rottolo	Egypt	0.495 kg	
ruba		9.34 kg	
sack		60–127.0 kg	
saco (café)		62.5 kg	
scruple		1.296 g	
seer	India, Sri Lanka, Bangladesh	0.9331 kg	
shih		71.61 kg	

Unit	Origin	Metric equivalent	Other equivalents
sieve		21.8–25.4 kg	
stone	UK, US, Brazil	6.35 kg	14 lb
taam, tam	China	60.45 kg	
tael, tahil		37.8 g	
tamlung	Thailand	60 g	4 baths
tan	China	60.56 kg	
tenate		10 kg	
tercio	Honduras	45.36 kg	100 lb
tical		8.0–16.0 kg	
tierce		138.0–170.1 kg	
tola	India, Bangladesh	11.66 kg	
ton, tonne, tonelada	short	907.185 kg	2000 lb
ton, tonne, tonelada	metric	1000.0 kg	
ton, tonne, tonelada	long	1016.047 kg	2240 lb
tonelada	Portugal	792.88 kg	
tonelada	Argentina	918.8 kg	
tonelada	Chile, Guatemala, Mexico	919.9 kg	
tonelada	Spain	1150 kg	
tonelada	Colombia, Uruguay	1000 kg	
truss		16.3–27.2 kg	
ts'in		3.77 g	
tub		38.1 kg	
Tughar		2.032 mt	
Visham		1.36 kg	
Viss		1.66 kg	
wazna	Iraq	101.6 kg	
wazna	other Arabian countries	1.59 kg	
windle		99.79 kg	

1.5
Conversions

1.5.1
Conversions of UK and US Units to Metric Units

1.5.1.1 Decimal and Metric Equivalents of Fractions of 1 in

1/2	1/4	1/8	1/16	1/32	1/64	inch	mm
					1	0.015 625	0.397
				1	2	0.031 25	0.794
					3	0.046 875	1.191
			1	2	4	0.0625	1.588
					5	0.078 125	1.984
				3	6	0.093 75	2.381
					7	0.109 375	2.778
		1	2	4	8	0.1250	3.175
					9	0.140 625	3.572
				5	10	0.156 25	3.959
					11	0.171 875	4.366
			3	6	12	0.1875	4.762
					13	0.203 125	5.159
				7	14	0.128 75	5.556
					15	0.234 375	5.953
	1	2	4	8	16	0.2500	6.350
					17	0.265 625	6.747
				9	18	0.281 25	7.144
					19	0.296 875	7.541
			5	10	20	0.3125	7.938
					21	0.328 125	8.334
				11	22	0.343 75	8.731
					23	0.359 375	9.128
		3	6	12	24	0.3750	9.525
					25	0.390 625	9.922
				13	26	0.406 25	10.319
					27	0.421 875	10.716
			7	14	28	0.4375	11.112
					29	0.453 125	11.509
				15	30	0.468 75	11.906
					31	0.484 375	12.303

1/2	1/4	1/8	1/16	1/32	1/64	inch	mm
1	2	4	8	16	32	0.5000	12.700
					33	0.515 625	13.097
				17	34	0.531 25	13.494
					35	0.546 875	13.891
			9	18	36	0.5625	14.288
					37	0.578 125	14.684
				19	38	0.593 75	15.081
					39	0.609 375	15.478
		5	10	20	40	0.6250	15.875
					41	0640 625	16.272
				21	42	0.656 25	16.669
					43	0.671 875	17.066
			11	22	44	0.6875	17.462
					45	0.703 125	17.859
				23	46	0.718 75	18.256
					47	0.734 375	18.653
	3	6	12	24	48	0.7500	19.050
					49	0.765 625	19.447
				25	50	0.781 25	19.844
					51	0.796 875	20.241
			13	26	52	0.8125	20.638
					53	0.828 125	21.034
				27	54	0.843 75	21.431
					55	0.859 375	21.828
		7	14	28	56	0.8750	22.225
					57	0.809 625	22.622
				29	58	0.906 25	23.019
					59	0.921 875	23.416
			15	30	60	0.9375	23.812
					61	0.953 125	24.209
				31	62	0.968 75	24.606
					63	0.984 375	25.003
2	4	8	16	32	64	1.000	25.400

1.5.1.2 Length

Unit	Symbol	Metric equivalent
inch	in	exactly 2.54 cm
foot	ft	30.48 cm
yard	yd	91.44 cm
rod (or pole or perch)	rd	5.029 m
chain (Gunter's: 66 ft)	ch (G)	20.12 m
chain (Ramden's: 100 ft)	ch (R)	30.48 m
furlong	F	201.168 m
mile	mi	1.609 344 km
fathom	fa	1.8288 m
UK cable (608 ft)		185.3184 m
US cable (720 ft)		219.456 m
UK nautical mile (6080 ft)		1.853 184 km
mile intern. nautical		1.852 km

1.5.1.3 Area

Unit	Symbol	Metric equivalent
square inch	in^2	$6.4516\ cm^2$
square foot	ft^2	$0.0929\ m^2$
square yard	yd^2	$0.8361\ m^2$
square rod	rd^2	$25.293\ m^2$
square chain	ch^2	4.047 ac
acre	ac	0.4047 ha
square mile	mi^2	$2.590\ km^2$

1.5.1.4 Volume and Capacity

Unit	Symbol	Metric equivalent
cubic inch	in^3	$16.39\ cm^3$
cubic foot	ft^3	$28.317\ dm^3$
cubic yard	yd^3	$0.765\ m^3$
UK fluid ounce	fl oz	28.4122 ml
UK gill	gi	0.142 l
UK pint	pt	0.568 l
UK quart	qt	1.136 l
UK gallon	gal	4.546 l
UK bushel	bu	36.368 l
UK bulk barrel	bbl	163.656 l
US fluid ounce	fl oz	29.5727 ml
US gill	gi	0.118 l
US pint	pt	0.473 l
US quart	qt	0.946 l
US gallon	gal	3.785 l
US bushel	bu	35.241 l
US barrel	bbl	119.228 l
US dry pint	pt (dry)	$550.6\ cm^3$
US dry quart	qt (dry)	$1.101\ dm^3$
US peck	pk	$8.80965\ dm^3$
US bushel	bu	$35.283\ dm^3$

1.5.1.5 Weight (Avoirdupois)

Unit	Symbol	Equivalent
grain	gr	64.799 mg
dram	dr	1.772 g
ounce	oz	28.349 g
pound	lb	exactly 0.453 592 37- kg
UK stone	st	6.350 kg
hundredweight short	cwt short	45.359 kg
hundredweight long	cwt long	50.802 kg
short ton	tn	0.907 t
long ton	ton	1.016 t

1.5.1.6   **Household Measures**

Unit	Equivalent metric	Equivalents
teaspoon (tsp)	5 ml	60–80 drops
dessert spoon	8 ml	
tablespoon (tbsp)	15 ml	0.5 fl oz
glass	60–120 ml	2–4 fl oz
1 teacup	1.2 dl	4 fl oz
1 cup	2.4 dl	8 fl oz

1.5.2
Volume per Area

Unit	ml/ha	l/ha	US gal/ac	US pt/ac	US fl oz/ac	UK gal/ac	UK pt/ac	UK fl oz/ac
ml/ha	1	0.001			0.013			0.014
l/ha	1000	1	0.107	0.855	13.68	0.089	0.712	14.24
US gal/ac		9.353	1	8 000	128.0	0.833	6.661	133.2
US pt/ac		1.169	0.125	1	16.00	0.104	0.833	16.65
US fl oz/ac		0.073	0.008	0.062	1	0.006	0.052	1.041
UK gal/ac		11.23	1.201	9.608	153.7	1	8 000	160.0
UK pt/ac		1.404	0.150	1.201	19.22	0.125	1	20.00
UK fl oz/ac		0.070	0.008	0.060	0.961	0.006	0.050	1

1.5.3
Weight per Area

1.5.3.1   **Application Rate and Yield**

Unit	Equivalents		
1 kg/ha	0.892 lb/ac	14.275 oz/ac	
1 lb/ac	1.121 kg/ha	16.000 oz/ac	
1 oz/ac	0.070 kg/ha	0.062 lb/ac	
100 kg/ha	0.796 cwt long/ac		
1 bu US/ac	87.075 l/ha		
1 bu UK/ac	89.89 l/ha		
1 quintal (100 lb)/ha	45.35 kg/ha		
1 quintal (100 kg)/ha	100 kg/ha		

Unit	Equivalents		
1 quintal (100 lb)/acre	112 kg/ha		
1 quintal (100 lb)/manzana	65 kg/ha		
1 μg/cm^2	100 g/ha		
1 g/ha	0.01 μg/cm^2		
1 lb/100 ft^2	488.172 kg/ha		
1 lb/yd^2	0.543 kg/m^2		
1 kg/m^2	0.205 lb/ft^2		
1 kg/m^2	1.842 lb/yd^2		
1 kg/feddam	2.38 kg/ha	2.123 lb/ac	33.875 oz/ha
1 lb/feddam	1.079 kg/ha	0.962 lb/ac	13.731 oz/ha

1.5.3.2 Double Conversion

kg/ha	kg or lb	lb/ac
1.121	1	0.892
2.242	2	1.784
3.363	3	2.677
4.483	4	3.569
5.604	5	4.461
6.725	6	5.353
7.846	7	6.245
8.967	8	7.137
10.088	9	8.030
11.209	10	8.922
22.417	20	17.844
33.626	30	26.765
44.834	40	35.687
56.043	50	44.609
67.251	60	53.531
78.460	70	62.453
89.668	80	71.374
100.877	90	80.296
112.085	100	89.218

The central figure refers to either the left or the right-hand column (i.e. **3** kg/ha = 2.677 lb/ac or **3** lb/ac = 3.363 kg/ha).

1.5.4
Volume per Volume
(Concentration or Dilution)

Metric units	UK units
623 ml/100 l	1 fl oz/gal
78 ml/100 l	1 fl oz/bu
125 ml/100 l	1 pt/100 gal
290 ml/100 l	1 pt/bbl
99.6 g/100 l	1 lb/gal
1.248 kg/100 l	1 lb/bu

Metric units	US units
782 ml/100 l	1 fl oz/gal
82.5 ml/100 l	1 fl oz/bu
125 ml/100 l	1 pt/100 gal
396 ml/100 l	1 pt/bbl
120 g/100 l	1 lb/gal
1.288 kg/100 l	1 lb/bu
17.14 g/m^3	1 gr/gal

1.5.5
Weight per Volume

UK and US units	Metric units
1 long ton/yd^3	1.329 t/m^3
1 short ton/yd^3	1.186 t/m^3
1 lb/yd^3	0.593 kg/m^3
1 lb/in^3	0.027 kg/cm^3
1 oz/in^3	1.73 g/cm^3
1 lb/Imperial gal	0.0998 kg/dm^3
1 lb/US gal	0.1198 kg/dm^3
1 lb/pt	0.9584 kg/dm^3

1.5.6
Concentrations per Parts

Parts	Symbol	Weight/volume	Weight/weight
per cent	%	10 g/l = 1 kg/100 l	10 g/kg = 1kg/100 kg
per mille	‰	1 g/l = 1 kg/1000 l	1 g/kg = 1 kg/1000 kg
part per million	ppm	1 µg/ml = 1 mg/l	1 µg/g = 1 mg/kg
part per billion (US)	ppb	1 ng/ml = 1 µg/ml	1 ng/g = 1 µg/kg

1.5.7
Temperature

1.5.7.1 **Temperature conversion formulas**

degree Celsius (°C)	= (°F − 32) × 5/9
	= (°Re × 5/4))
	= (°K − 273.15)
degree Fahrenheit (°F)	= (°C × 9/5) + 32
	= (°Re × 9/4) + 32
	= (1.8 × K) − 459.67
Kelvin (K)	= (°C + 273.15)
	= (°F + 459.67) ÷ 1.8
	= (°R ÷ 1.8)
degree Rankine (°R)	= (°F + 459.67)
	= (K × 1.8)
degree Réaumur (°Re)	= (°C × 4/5)
	= (°F − 32) × 4/9

Note: the word "degree" is not used with Kelvin; the abbreviation is "K".

1.5.7.2 Examples of Temperature Conversion

100 degrees Celsius	$= (100 \times 1.8) + 32°F$	$= 212°F$
212 degrees Fahrenheit	$= (212 - 32) \div 1.8$	$= 100\,°C$
212 degrees Fahrenheit	$= (212 + 459.67) \div 1.8$	$= 373.15\ K$
80 degrees Réaumur	$= (80 \times 1.25)$	$= 100\,°C$

1.5.7.3 Scale Comparison

	Kelvin	Celsius	Fahrenheit
boiling point of water	373.15	100	212
freezing point of water	273.15	0	32
absolute zero	0	−273.15	−459.67

1.5.7.4 Boiling Temperatures of Water and Barometric Pressures at Various Altitudes

Altitude		Temperature		Pressure		
ft	m	F	C	mm Hg	in Hg	psi
0	0	212.0	100.0	760	29.92	14.70
3000	914	206.4	96.9	677	26.65	13.09
5000	1524	202.6	94.8	629	24.76	12.16
7000	2134	198.8	92.7	582	22.91	11.25

1.5.8
Pressure

Unit	mm Hg	in H_2O	ft H_2O	in Hg	lb/in^2	kg/cm^2	atm
mm of mercury	1	0.2353	0.04460	0.03937	0.01934	0.00136	0.00132
inches of water	1.868	1	0.08333	0.07355	0.03613	0.00254	0.00246
feet of water	22.42	12	1	0.8826	0.4335	0.03048	0.02950
inches of mercury	25.40	13.60	1.133	1	0.4912	0.03453	0.03342
lb/in^2	51.71	27.67	2.307	2.04	1	0.07031	0.06805
kg/cm^2	735.6	393.7	32.81	28.96	14.22	1	0.9678
Atmospheres	760	406.8	33.90	29.92	14.70	1.033	1

Note: lb/in^2 = psi

1.5.9
Energy

1.5.9.1 **Units of Work**

Units	J	kWh	kcal	Btu	hph
Joule	1	2.778×10^{-7}	2.388×10^{-4}	9.478×10^{-4}	3.725×10^{-7}
kilowatt-hour	3.600×10^{6}	1	859.8	3412	1.341
kilocalorie	4187	1.163×10^{-3}	1	3.968	1.560×10^{-3}
British thermal unit	1055	2.931×10^{-4}	0.252	1	3.930×10^{-4}
horsepower-hour	2.684	0.746	641.2	2544	1

1.5.9.2 **Units of Power**

Units	Watt	kcal/h	hp	Btu/h
Watt	1	0.860	1.341×10^{-3}	3.412
kilocalorie/hour	1.163	1	1.560×10^{-3}	3.968
horsepower	745.7	641.2	1	2544
Btu/hour	0.293	0.251	3.930×10^{-5}	1

1.5.10
Wind (Beaufort Scale)

Wind strength	Description	Visible signs	Approximate air speed	
			mph	km/h
force 0	calm	smoke goes up vertically	0–0.5	0–0.7
force 1	light air	direction shown by smoke drift	up to 1	up to 1.5
force 2	light breeze	leaves rustle, wind felt on face	2–4	3–6.5
force 3	gentle breeze	leaves and small twigs in constant motion	4–6	6.5–10
force 4	moderate	small branches move, raising dust and loose paper	6–9	10–15

1.5.11
Speed

Units	km/h	m/min	m/s	mi/h	ft/min	kn
kilometer/hour	1	16.667	0.278	0.621	54.682	0.540
meter/minute	0.060	1	0.017	0.037	3.208	0.032
meter/second	3.600	60.000	1	2.237	196.850	1.944
miles/hour	1.609	26.817	0.447	1	88.000	0.869
feet/minute	0.018	0.305	~0.005	0.011	1	~0.010
knot	1.852	30.866	0.514	1.151	101.269	1
miles/minute	96.56	1609.34	26.822	60	5280	52.138

1.5.12
Fuel Consumption

Units	Equivalent
l/km	1.609 l/mi
l/km	1.852 l/kn
l/km	0.425 US gal/mi
l/km	0.489 US gal/kn
l/km	0.354 UK gal/mi
l/km	0.407 UK gal/kn
US gal/mi	2.352 l/km
US gal/kn	2.044 l/km
UK gal/mi	2.825 l/km
UK gal/kn	2.455 l/km

Note: 1 knot = 1 nautical mile, but also used as unit of speed for nautical miles per hour.

1.5.13
Precipitation and Irrigation

1 inch of rain or irrigation per acre	= 22 613 UK gal/ac = 27 158 US gal/ac = 102 798 l/ha
1 mm of rain or irrigation per ha	= 10 000 l/ha or 1 l/m^2 = 1070 US gal/ac = 890 UK gal/ac

1.5.14
Flow Rates

Units	Equivalent US gpm (1 US gpm = 0.227 m³/h)	Equivalent UK gpm (1 UK gpm = 0.333 m³/h)
ft³/s	448.8	373.8
l/s	15.88	13.23
l/h	0.0044	0.0037
m³/min	264.2	220.5
m³/h	4.4	3.67
ac × in/h	452.6	377.0
ac × ft/day	226.3	188.6
1000 UK gal/day	0.834	0.694

1.5.15
Flow Rate Calculation
Flow rates can be calculated by using the following formula:
Flow rate = forward speed × run width × application rate
= (length per time) × length × (volume per area) = volume per time
while using conversion factors according to the different units of measurements involved:

Calculation examples

$$l/min = \frac{m/min \times width\ in\ m}{10\,000} \times l/ha$$

$$l/min = \frac{km/min \times width\ in\ m}{600} \times l/ha$$

$$gal/min = \frac{mph \times width\ in\ ft}{495} \times gal/ac$$

$$l/min = \frac{2.7 \times mph \times width\ in\ m}{1000} \times l/ha$$

$$l/min = \frac{3.1 \times kts/min \times width\ in\ m}{1000} \times l/ha$$

$$gal/min = \frac{2.0 \times mph/min \times width\ in\ ft}{1000} \times gal/ac$$

1.5.16
Water Analysis Equivalents of (CaCO₃)

Units	ppm	g/gal UK	g/gal US
ppm or mg/l	1	0.07	0.058
grains per UK gallon or Clark degrees	14.3	1	0.83
grains per US gallon or German degrees (practically)	17.1	1.2	1

1.5.17
Soil Chemistry: Common Soil-related Conversions

		Data to convert	Multiply by	Result
1	Soluble ions concentrations	meq/l	saturation % × 0.001	meq/100 g
2	Conductivity	EC (mS/cm)	10.0	meq/l (covers range 0.1–5.0 mS/cm)
		EC mS/cm)	0.36	atm osmotic pressure (covers range 3–30 mS/cm)
		EC (mS/cm)	640 (approx)	total soluble salts in ppm or mg/l (covers range 0.1–5 mS/in water)
3	Concentrations	mg/100g	10	ppm
		meq/100 g	10 × Ewt	ppm
		meq/l	Ewt	ppm
4	Chemical concentrations by weight			
	Calcium	CaO	0.715	Ca
		Ca	1.399	CaO
		meq/100 g Ca	200.400	ppm Ca
	Magnesium	MgO	0.603	Mg
		Mg	1.658	MgO
		meq/100 g Mg	121.600	ppm Mg
	Nitrogen	N	6.067	$NaNO_3$
		N	7.218	KNO_3
		N	4.717	$(NH_4)_2SO_4$
		$NaNONaNO_3$	0.165	N
		KNO_3	0.138	N
		$(NH_4)_2SO_4$	0.212	N
	Phosphorous	P_2O_5	0.436	P
		P	2.292	P_2O_5
	Potassium	K_2O	0.830	K
		K	1.205	K_2O
		meq/100 g K	391.000	ppm K
	Sodium	Na_2O	0.742	Na
		Na	1.348	Na_2O
		meq/100 g NA	230.000	ppm Na
	Sulfur	S	2.497	SO_3
		SO_3	0.400	S
5	Gypsum concentrations	meq/100 g	0.086	% by weight

1.5.18
Dry Soil

Unit dry soil	weight
1 ft³ muck	25–30 lb
1 ft³ clay and silt	68–80 lb
1 ft³ sand	100–110 lb
1 ft³ loam	80–95 lb
1 ft³ of average soil	80–90 lb

1 acre × feet³ = 43.560 ft³ or 2000 tons = 3 500 000–4 000 000 lb.
A surface plow depth of 6.66 inches is usually calculated at 2 million lb or 1000 tons per acre.
The volume of compact soil increases by approximately 20 % when it is excavated or tilled.

1.5.19
Slope Data

1.5.19.1 **Degrees to Percent and Percent to Degrees**

Degrees and half degrees to percent			Percent to degrees	
degrees (°)	percent % of (°)	% of (° + 30')	percent %	degrees and minutes (°)(')
0	0	0.87	1	0°34'
1	1.75	2.62	2	1°09'
2	3.49	4.37	3	1°43'
3	5.24	6.12	4	2°18'
4	6.99	7.87	5	2°52'
5	8.75	9.63	6	3°26'
6	10.51	11.39	7	4°00'
7	12.28	13.17	8	4°34'
8	14.05	14.95	9	5°09'
9	15.84	16.73	10	5°43'
10	17.63	18.53	11	6°17'
11	19.44	20.35	12	6°51'
12	21.16	22.17	13	7°24'
13	2309	24.01	14	7°58'
14	24.93	25.86	15	8°32'
15	26.80	27.73	16	9°05'
16	28.68	29.62	17	9°39'

Degrees and half degrees to percent			Percent to degrees	
degrees (°)	percent % of (°)	% of (° + 30')	percent %	degrees and minutes (°)(')
17	30.57	31.53	18	10°12'
18	32.49	33.46	19	10°45'
19	34.43	35.41	20	11°19'
20	36.40	37.39	25	14°02'
21	38.39	39.39	30	16°42'
22	40.40	41.42	35	19°45'
23	42.45	43.48	40	21°48'
24	44.52	44.57	45	24°14'
25	46.63	47.70	50	26°34'
26	48.77	49.86	55	28°49'
27	50.95	52.06	60	30°38'
28	53.17	54.30	65	33°01'
29	55.43	56.58	70	35°00'
30	57.74	58.91	75	36°52'
31	60.09	61.28	80	38°40'
32	62.49	63.71	85	40°2'
33	64.94	66.19	90	42°00'
34	67.45	68.73	95	43°32'
35	70.02	71.33	100	45°00'
36	72.65	74.00		
37	75.36	76.33		
38	78.13	79.54		
39	80.98	82.43		
40	83.91	85.41		
41	86.93	88.47		
42	90.04	91.63		
43	93.25	94.90		
44	96.57	98.27		
45	100			

Note: Percentage slope = natural tangent $\times$ 100, e.g. 1 : 2 = 50%.

1.5.19.2 Natural Tangent to Degrees

Ratio (tan)	Degrees (rounded)
1:0.25	76
1:0.5	63
1:1	45
1:1.5	34
1:.2	27
1:2.5	22
1:3	18
1:3.5	16
1:4	14

Note: the first figure refers to the height, the second figure refers to the horizontal distance

In the following example with a 1:2 smooth slope, there is a climb (or a descent) of 5 m for every 10 m traversed horizontally along a line parallel to the maximum gradient.

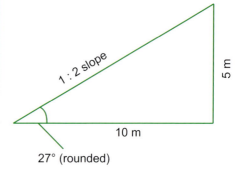

27° (rounded)

1.5.20
Maps

Scale	Distance in m or km represented by 1 cm		Distance in cm representing 1 km	Area in ha km² represented by 1 cm²		Area in cm² representing 1000 ha
1:1000	10		100	0.01		10^5
1:2500	25		40	0.0625		1.6×10^4
1:5000	50		20	0.25		4×10^3
1:7500	75		13	0.56		1786
1:10000	100		10	1.00		10^3
1:15000	150		6.7	2.25		444
1:20000	200		5.0	4.00		250
1:25000	250		4.0	6.25		160
1:30000	300		3.3	9.00		111
1:40000	400		2.5	16.0		62.5
1:50000	500		2.0	25.0		40
1:100000	1000	1 km	1.0	100	1.0 km²	10
1:250000		2.5 km	0.4		6.25 km²	1.6
1:500000		5 km	0.2		25 km²	0.4
1:750000		7.5 km	0.1		56.25 km²	0.18
1:1000000		10 km	0.1		100 km²	0.1

1 cm² representing 1 ha is, effectively, the minimum mappable area for planning work.

1.5.21
Coffee Classification

1.5.21.1 Bean Sizing

Screen number	Diameter of screen in mm[†]	AFNOR and ISO norm	English description for Brazilian coffees
20	7.94	8.00	Very large beans
19.5	7.74	7.75	
19	7.54	7.50	Extra large beans
18.5	7.34	7.25	
18	7 14	7.10	Large beans
17	6.75	6.70	Bold beans
16	6.35	6.30	Good beans
15	5.95	6.00	Medium beans
14	5.55	5.60	Small beans
13	5.16	5.00	
12	4.76	4.75	
11	4.36	4.50	
10	3.97	4.00	
9	3.57	3.50	
8	3.17	3.00	

[†] number of screen $\times$ 1/64 inch, i.e. = number of screen $\times$ 0.396875 mm.

1.5.21.2 Flat Bean Separation

Roasting of coffee beans in their natural range of sizes is not practical, because small beans roast faster than large beans. Hence roasted bean colors – and particularly flavor development – would be non-uniform. Therefore the flat beans are separated by sieving through perforated screens with round holes or equivalent meshes. Thereby they are categorized by size into groups.

1.5.21.3 Peabean Separation

Peabeans, which have no flat faces, but are spherically oval beans, are first separated from the flat beans by letting them roll off an inclined belt. Then they are separated and categorized by sieving through screens with slotted holes. These slots, normally 1/2 in long, have different widths such as 14S/5.5; 13S/5.0; 12S/4.5; 11S/4.0 and 10S/3.5. The first number indicating the slot width in 1/64 in, the second number in millimeters.

Each bean size range is roasted separately. As broken and small beans fall through or jam the 12/64 inch roaster cylinder perforations, they are roasted in a non-perforated cylinder. After roasting the beans are reblended before grinding. Dust and chaff are removed at all stages.

1.5.22
Net Weight per Bag used in Producing Countries/Regions

Country/region	kg	packing	UK (lb)	Tare (kg)
Angola	60	J	132	1.00
Brazil	60	J	132	0.50
Cameroon	60	J	132	0.43–0.90
Colombia	70	S	154	0.70–0.90
Congo (DRC)	60	J	132	0.90–1.00
Costa Rica	69	S, K	152	0.69
Cuba	60	J	132	0.69
Dominican Republic	75	S	165	1.00
Ecuador	60	S	132	0.50
El Salvdor	69	J	152	0.46
Ethiopia	60	J	132	0.80
Guatemala	69	J, K	152	0.69
Haiti	60	S	132	0.90
Hawaii	45,3	J	100	0.35
Honduras	69	J	152	1.00
India, Malabar	50	J	132	0.80
Indonesia	60	J	132	1.00–1.20
Ivory Coast	60	J, S	132	0.80–0.90
Jamaica	60	J	132	1.00
Kenya	60	S	132	1.10
Madagascar	90	J, P, S	132	0.85
Mexico	69	S	152	1.00
Nicaragua	69	J, K	152	0.60 (Maragogipe 1.00 kg)
Peru	69	J	152	1.00
Tanzania	60	S	132	1.10
Togo	60	S	132	1.00
Uganda	60	J	132	1.00
Venezuela	60	S	132	0.70
West Africa region	60	J, S	132	variable
Yemen	60	J	132	1.00

J = Jute, K = Kenaf, S = Sisal, P = Paka

Bibliography

[1] BROCKHAUS ENZYKLOPÄDIE. F. A. Brockhaus, Wiesbaden, 1966.

[2] Ciba-Geigy Agricultural Division. *Conversion Tables and Definitions in Application Techniques.* Basel, Switzerland.

[3] De Sola, R. *Abbreviations Dictionary.* Elsevier, Amsterdam, 1978.

[4] Ellis, G. *Units – Symbols and Abbreviations.* Royal Society of Medicine, London, 1971.

[5] ENCYCLOPAEDIA BRITANNICA, William Benton, Chicago, 1971.

[6] EUROCONSULT, ed. *Agricultural Compendium.* Elsevier, Amsterdam, 1989.

[7] Gaboury, J. *SI Units and Conversion Factors including World Weights and Measures.* J. A. M. Gaboury, Montreal, 1980.

[8] Hesse, P. R. *SI Units and Nomenclature in Soil Science.* FAO, Rome, 1975.

[9] Incoll, L. D., Long, S. P. and Ashmore, M. R. SI units in plant science. *Curr. Adv. Plant Sci.* **1977**, *9*, 331–343.

[10] Jobin, P. *[The Coffee Produced Throughout the World].* P. Jobin & Cie, Le Harve, 1992.

[11] Landau, S. I., ed. *Webster Concise Dictionary,* Trident Press International, Naples, FL, 1997.

[12] Landon, J. R., ed. *Booker Tropical Soil Manual: A Handbook for Soil Survey and Agricultural Land Evaluation in the Tropics and Subtropics.* Booker Agriculture, London, 1984.

[13] Lorenz, A. O. and Maynard, D. N. *Handbook for Vegetable Growers.* Wiley, New York, 1988.

[14] McGlashan, M. L. *Physicochemical Quantities and Units*, 2nd edn. Royal Institute of Chemistry, London, 1971.

[15] Peterson, M. S. Recommendations for use of SI units in hydraulic engineering. *Proc. Am. Soc. Civ. Eng.* **1980**, *106*, HY12, 1981–1994

[16] Rothfos, B. *Rohkaffee & Röstkaffee Prüfung.* Gordian Verlag, Hamburg, 1963.

2
Information Sources

C. Fardeau

2.1
List of Coffee Machinery Suppliers

Name	Address	Telecom	Activity
Agromática	Via Nazionale 31, 33040 Pradamano, UD, Italy	Tel: +39 0432 671-413; Fax: +39 0432 671-415; E-mail: spi-dy@agromatica.com	Coffee harvesting equipment
Bendig Indústrias SA	PO Box 290, 2400 Desamparados, San José, Costa Rica	Tel: +506 259 7379; Fax: +506 259 0110	General coffee processing, pulper, dryers, coolers, sorting, grading equipment
Consortium Français de Constructeurs pour l'agro-industrie (CFAI)	Rue Parmentier BP 36, 60290 Rantigny, France	Tel: +33 44 730822; Fax: +33 44 735990	Post-harvest coffee engineering
D'Andrea SA	Avenida Souza Queiróz, 267 Caixa Postal 455, 13480 Limeira, SP, Brazil	Tel: +55 19 441 3026; Fax: +55 19 441 3026	Coffee processing, sorting, grading equipment, materials handling
Delta Technology Corp.	1602 Townhurst, Houston, TX 77043, USA	Tel: +1 713 464 7407; Fax: +1 713 461 6753; Website: www.deltatechnology.com	Sorting machines (color sorting)
Denlab International	Hill Crest House 4, Market Hill, Maldon, Essex CM9 7PZ, UK	Tel: +44 1621 858944; Fax: +44 1621 857733	Postharvest machinery driers, heat generators
Elexso Sortiertechnik AG	Hans-Duncker-Strasse 1, PO Box 8005 47, Hamburg 21005, Germany	Tel: +49 40 7342 4196; Fax: +49 40 7342 4132; E-mail: sales@elexso.com	Sorting machines (color sorting)

Name	Address	Telecom	Activity
Gauthier SA	Parc Scientifique Agropolis, 34397 Montpellier Cedex 5, France	Tel: +33 67611156; Fax: +33 67547390	Pulpers, "Bars-Cadge" hullers, size sorting, destoners
Gerico France	2 rue du Colonel Driant, 75001, Paris, France	Tel: +33 1 42 36 1922; Fax: +33 1 42 36 1816	On-farm processing equipment
Jacto Maquinas Agrícolas	Rua Dr Luis Miranda, 1650, CP 35, Pompéia, SP 17580-000, Brazil	Tel: +55 14 452 1811; Fax: +55 14 452 2744; E-mail: jacto@jacto.com.br	Coffee harvesting equipment
Key Technology Inc.	150 Avery Street, Walla Walla, WA 99362, USA	Tel: +1 509 529-2161; Fax: +1 509 527-1331; E-mail: productinfo @keyww.com	Sorting machines (color sorting)
Korvan Industries Inc.	270 Birch Bay Lynden Road, Lynden, WA 98264, USA	Tel: +1 360 354-1500; Fax: +1 360 354-1300; E-mail: info@korvan.com	Coffee harvesting equipment
Mc Kinnon India Pvt Ltd	25 Mettupalayman Road, Narasimhansikenpalay-man, Coimbatore 641 031, India	Tel: +91 4222 14981; Fax: +91 4222 12246	Hulling equipment, pulpers
Mercator Agrícola, SA de CV	Plaza de la Republica, 49 Oficina 06030, Mexico, DF 06030	Tel: +52 5 546 5165; Fax: +52 5 546 9196	Coffee equipment and machinery, clean-ers (green coffee)
Oliver Manufacturing Co. Inc.	PO Box 512, Rocky Ford, CO 81067, USA	Tel: +1 719 254-7814; Fax: +1 719 254-6371; E-mail: oliver@ria.net	Cleaners (green coffee), destoners, graders
Paul Kaack and Co.	Lerchenberg 69A, 22359, Hamburg, Germany	Tel: +49 40 644 19 150; Fax: +49 40 511 7013; E-mail: info@kaack-maschinen.com meyer@kaack-maschinen.com	Cleaners (green coffee), coffee plan-tation machinery
Penagos	Calle 28, No. 20–80, PO Box 689, Bucaramanga, Colombia	Tel: +57 7 630 1600; Fax: +57 7 630 2795; E-mail: penagos @b-manga.multi.net	Pulpers, de-mucila-ging equipment
Pinhalense SA Máquinas Agrícolas	c/o P and A International Marketing, Praça Rio Branco, 13 CP 83, E. S. Pinhal, SP 13990-000, Brazil	Tel: +55 19 3651 3233; Fax: +55 19 3651 2887; Website: www.pinhalense.com.br	Coffee machinery for plantation, drying and export (pulpers, mucilage removers, dryers, cleaners, hullers, graders)
Sanmak	Rua Francisco Vahldieck 377, Blumenau, SC 89058-000, Brazil	Tel: +55 47 339 2700; Fax: +55 47 339 2700	Electronic color sorters

Name	Address	Telecom	Activity
Satake USA Inc.	9800 Townpark Drive, Houston, TX 77036, USA	Tel: +1 713 772-8400; Fax: +1 713-981 8704; E-mail: vision@satake-usa.com	Sorting machines (color sorting)
Siraga	BP 1436500, Buzançais, France	Tel: +33 54840930; Fax: +33 54840513	Coffee hullers
Sortex Ltd	Pudding Mill Lane, London E15 2PJ, UK	Tel: +44 181 519 0525; Fax: +44 181 519 5614; E-mail: sales@sortex.com	Sorting machines (color sorting)
Talleres Industriales	San Carlos Carretera A Ciudad Vieya el Panorama Cote 9 C, AP 413, Antigua Guatemala	Tel: +502 9323159; Fax: +502 9320953	Post-harvest machinery reduction of water consumption, coffee engineering
Tec Machine	Rue Benoît Fourneyron ZI Sud, 42166 Andrezieux Bouthéon, France	Tel: +33 77555222; Fax: +33 77555200	"Bars-Cadge" hullers
Xeltron	PO Box 6953-1000, San José, Costa Rica	Tel: +506 279 5777; Fax: +506 279 5799; E-mail: ventasx@xeltron.com	Sorting machines (color sorting)

2.2
Shade, Forest and Cover Plant Seed Sources

Name	Address	Telecom
ARGENTINIA		0054
CARDI Forage Seed Project	PO Box 766 Friars Hill, St John Antigua	Tel: +462 0661/462 1666
AUSTRALIA		0061
Australian Seed Co.	Albert Park, 119 Ashworth Street, Melbourne, Vic 3206	
Australia Tree Seed Center, CSIRO Division of Foresty and Forest Products	PO Box 4008, Queen Victoria Terrace, Canbera, ACT 2600	Tel: +062 818203
Australian Tropical Plant Supplies	PO Box 5, Mount Molloy, Qld 4880 or Pinnacle Road, Julaten, Qld 4880	
Conservation Commission of the Northern Territory, Forestry Unit	PO Box 1046, Alice Springs, NT 5750	
Conservation Commission of the Northen Territory, Forestry Unit	PO Box 38496, Winnelie, Darwin, NT 5789	Tel: +220211

Name	Address	Telecom
Conservation and Land Management Department	50 Hayman Road Como, WA 6152	Tel: +09 3676333
Dendros Seed Supplies	PO Box 144, Weston Creek, ACT 2611	Tel: +062 881 490
Department of Conservation, Forests and Lands	GPO 4018, Melbourne, Vic 3001	Tel: +03 6179222
Division of Forestry Research CSIRO	Yarralumia, 2601 ACT	
Division of Tropical Crops and Pastures, Cunningham Laboratory	306 Carmody Road, Sta Lucia, Qld 4067	
D. Orriell – Seed Exporter	47 Frape Avenue, Mt Yokim, Perth, WA 6060	Tel: +09 344 2290
Ellison Horticultural Pty Ltd	PO Box 365, Norwa, NSW 2541	Tel: +044 214255
Flamingo Seeds Pty Ltd	PO Box 1046, East Norwa, 2541 NSW	
Forest Department, Western Australia	PO 104, Como, WA 6152	
Forestry Commission of New South Wales	PO J19, Coffs Harbour Jetty, NSW 2451	Tel: +066 528900
Forestry Commission	PO 207B, Hobart, Tas 7001	
Forestry Commission	PO 4018, Melbourne, Vic 3001	
Fruit Spirit Research Nursery and Gardens	Dorroughby, NSW 2480	Tel: +066 895 192
H. Grant – Australian Seeds	2, Sandra Place, Dubbo, NSW 2830	Tel: +068 82 4003
H. G. Kershaw Pty Ltd	PO Box 84, Terry Hills, NSW 2084	Tel: +02 450 2444
J. H. Williams and Sons Pty Ltd	PO Box 102, Murwillumbah 2484	
Kimberly Seeds	51 King Edward Road, Osborne Park, WA 6017	Tel: +09 4464377
Kingston Rural Management Pty	PO Box 47, Maryborough, Qld 4650	
Kylisa Native Seed Supplier	1 Gould Court, Traralgon, Vic 3844	
M. L. Farrar Pty Ltd	PO Box 1046, Bomaderry NSW 254	Tel: +044 217692
Nindethana Seed Service	Narrikup, WA 6326	
North Australian Native Seed	PO Box 40003 Casaurina, NT 5792	
Phoenix Seeds Association	PO Box 9, Stanley, 7331 Tasmania	Tel: +004 581105
Queensland Department of Forestry	PO Box 944, Brisbane, Qld 4001	Tel: +07 224 8335
Royston Petrie Seeds Pty LtdL	PO Box 77, Dural, NSW 2158	Tel: +2 6512658/ 6541186
Seeds Australia Pty Ltd	PO Box 30, Cloverdale WA 6105	

Name	Address	Telecom
Seeds for Preservation	Badgerup Road, Wanneroo WA 6065	
Silvan Improvements Pty Ltd	PO Box 42, Watson, ACT 2602	
Southern Seed Sales	35 Wilhelmina Ave., Lauceston, Tas 7250	
Tasmanian Forest Seeds	Summerleas Farm, Kingston, Tas 7150	
Vaughans Wildflower Seeds	PO Box 66, Canberra, ACT 2600	
Western Wildlife Supply	Terrara, PO Box 90, Gilgandra, NSW 2827	
Yates Agricultural Seeds	PO Box 616, Toowoomba, Qld 4350	Tel: +076 34 2088
BANGLADESH		
Forest Research Institute	PO Box 273, Chittagong	
BELIZE		
Ministry of Agriculture	E. Block, Belmopan	
BOLIVIA		
Centro Desarollo Forestal	Casilla 3548, La Paz	
BRAZIL		0055
Ministerio de AA CC y Agropecuarios	CP 09, 13.400 Piracicaba, SP	
Dept de Silvicultura, ESALQ, University of Sao Paulo	13.400 Piracicaba, SP	
EMBRAPA/IAPBS	Km 47, 23460 Seropedica, RJ	
Empresa Brasileira de Pesquisa Agropecuaria (EMBRAPA)	Estrada Da Ribeira, Km 111, Caixa Postal 3319, 80000 Curitiba, PR	Tel: +041256 2233
EPAMIG	Caiza Postal 295, CEP35 700, Sete Lagaos, Minas Gerais	
Reserva Florestal da CVRD	Caixa Postal 91 CEP, 29900 Linhares, ES	
CANADA		001
Canadian Forestry Service, Petawawa National Forestry Institute	Chalk River, Ontario	Tel: +613 589-2880
Ontario Ministry of Natural Resources, Ontario Tree Improvement and Forest Biomass Institute	Maple, Ontario L0J 1E0	

Name	Address	Telecom
CHILE		**0056**
Intituto Forestal (INFOR)	Huerfanos 554, Casilla 3085 Santiago	Tel: +396189
Centro de Semillas Forestales, Corporacion Nacional Forestal	Castilla 5, Chillan	Tel: +23888
COLOMBIA		**0057**
Banco de Semillas Forestales INDERINA Estacion Forestales "La Florida"	Apartado Aereo 57596, Bogota DE	
Centro Internacional Agricultura Tropical	Apartado Aereo 6713, Cali	
INDERENA, Banco de Semillas	Apartado Aereo 13458, Bogota, DE	Tel: +245 61 48
COSTA RICA		**00506**
Banco de Semillas Forestales, Direccion General Forestal	Apartado 10094, San Jose 1000	
Banco Latinoamericano de Semillas Forestales (BLSF)	CATIE Turrialba	
Centro Agroeconomico Tropical de Investigaciones y Ensenanzas (CATIE)	Turrialba	
Latin American Forest Tree Seeds Bank CATIE	Turrialba	Tel: +56 6021
CUBA		
Centro de Investigaciones Forestales	Calle 174, No. 1723 E/17, B y 17C Siboney, Playa, Cuidad, Havana	
CYPRUS		**00357**
Department of Forestry, Director of the Department of Forests, Ministry of Agriculture and Natural Resources	Nicosia	Tel: +402261
DENMARK		**0045**
DANIDA Forest Seed Center	Krogerupvej 3A, 3050 Humlebaek	Tel: +45 2190500
ECUADOR		**00593**
The Director, Dept de Capacitacion y Experimentacion, Direccion de Desarollo Forestal, Ministerio de Agricultura y Ganaderia	Quito	

Name	Address	Telecom
EL SALVADOR		00503
Banco de Semilas Forestales, Dept. Conser. de Res. Nat., Ministerio de Agr. y Ganaderia	Sazopango, San	
Centro de Recursos Naturales (CENREN)	Apartado Postal 2265, Canton el Matasano, Soyapango	Tel: +27 0622
ETHIOPIA		00251
Forestry Research Centre	PO Box 1034, Addis Ababa	Tel: +185444/185445
International Livestock Centre for Africa (ILCA)	PO Box 5689, Addis Ababa	Tel: +183215
FRANCE		0033
Centre Technique Forestier Tropical	45 bis, Ave. de la Belle Gabrielle, 94130 Nogent sur Marne	
VERSEPUY	43000 Le Puy	Tel: +71093213
GUATEMALA		00502
Banco de Semillas Forestales	7, Ave. 7-00, Zona 13, Guatemala City	Tel: +314754
Banco de Semillas Forestales (BANSEFOR), Instituto Nacional Forestal (INAFOR)	Edificio Galerias de España, 7a, Ave. 11-63, Zona 9, Guatemala City	
Exportacion de Semillas – Seed Export	Apartado Postal 543, Guatemala City	Tel: +536491/515247
HONDURAS		00504
Banco de Semilas – ESNACIFOR	Apartado Postal 2, Siguatepeque, Dept De Comayagua	Tel: +73 20 18 ext. 35
INDIA		0091
Aggarwal Nursery and Seed Stores	Panditwari, PO Prem Nagar, Dehra Dun 248007, UP	
Agronomist and Principal Investigator, AICRP on Agroforestry, Agricultural College	Dharwad 5	Tel: +80191
Bharatiya Agro-Industr. Fund	Kamdhenu, Senapati Bapat Marg, Pune 411016, Maharashtra	
Central Arid Zone Research Institute	Jadhpur, Rajasthan 342003	
Forest Research Institute and Colleges	PO New Forest, Dehra Dun 248006, UP	
India Nursery and Seeds Sales Corp.	PO Box 4314, 36/962 DDA Flats Kalkaji, New Delhi 19	

Name	Address	Telecom
Indian Grassland and Fodder Research	Jhansi 284003, UP	
Kumar International	Ajitmal 206121 Etawah, UP	
Nand-Prakash 7 Co.	PO Nex Forest, Dehra Dun 248006, UP	
National Botanical Research Institute RANA Pratab Marg	Lucknow 226001, UP	
Nimbcar Seeds Pvt Ltd	Phaltan Maharashtra 415523	
PAL Seed Traders	PO Kaulagarh, Dehra Dun, UP	
Pratap Nursery and Seed Stores	PO Box 91 Panditwari PO, Premnagar, Dehra Dun 248007, UP	
Regional Research Laboratory	Canal Road, Jammu Tawi	
Shivalik Seeds Corp.	Panditwari, PO Prem Nagar, Dehra Dun 248007, UP	
Tamil Nadu Forest Department, Forest Genetic Division	Bharathi Park Road, Coimbatore 43, Madras	
The Bharatiya Agro Industries Foundation "Khamdenu"	Senapati Bapat Marg, Pune 411016	Tel: +0212 52621
Titaghur Paper Mills Co. Ltd	95, Park Street, Calcutta 700016	
Tosha Trading Co.	161, Indira Nagar Colony, PO Indira Nagar Colony, Dehra Dun 248001, UP	Tel: +3430
Vijay Seed Stores	PO Ranjhawala (Raipur), Dehra Dun 248008, UP	
UAS/KDDC Fodder Project, University of Agricultural Sciences	Hebbal, Bangalore 560024	
INDONESIA		**0062**
Agricultural Polytechnic, University of Nusa Cendana, Jl.	Adisucipto, Penfui, PO Box 153, Kupang 85001	
Biotrop, SEAMEO Research Center for Tropical Biology	PO Box 17, Bogor	
Firma Pangkalredjo	JIMH Thamrin 57, Semarang	
Forest Research Institute	PO Box 66, Bogor	
Hortimary Utama PT	Jl. Taman Tanah Abang III 31, Tanau, Abang, Jakarta	
INTI ARGOPURA	Jl. Kebon Jeruk III 154 Sawah Besar, Jakarta	
PT Pradjika Patria	Jl. Imam, Bongol 8, Sala	
PARTA TANI	Jl. Senopati 43 A, Jakarta 12190	Tel: +583038/583048

Name	Address	Telecom
Proyek Kalikonto	Jl. Guntur 27, Malang 65112	
PT Bibit Baru	PO Box 632, Medan 20001, North Sumatra	
State Forest Corp.	Jalan JenderalGatot, Subroto 17–18, Jakarta	
Toko Pertama	Jl. Baluwerti 32, Surabaya	
Yayasan Tananua	Tromolpos 3, Waingapu Sumba Timur NTT	
ISRAEL		**00972**
Forestry Division, Agricultural Research Organization	"Ilanot", Doar Na Lev, Hasharon 42805	
Arava Seeds Suppliers	PO Box, Haifa 45109	
Institute for Applied Research, Ben Gurion University of the Negev	PO Box 1025, Beer Sheva 84110	Tel: +057 78382
Israflora, Forest Seeds of Arid Zones	PO Box 502, Kiryat-Blalik 2700	Tel: +04 737155
ITALY		**0039**
Florasilva ansaloni	Casella Postale 2100, 40100 Bologna	Tel: +051 455 218
JAMAICA		**001876**
Department of Forestry and Soil Conservation	173 Constant Spring Road, Kingston 8	Tel: +092 42667
KENYA		**00254**
Baobab Farm Ltd	PO Box 90202, Mombasa	Tel: +485 729
East Pokot Agricultural Project, Kositei	PO Marigat Via Nakuru	
Environment and Silvicultural Division, Agricultural Research Institute	PO Box 74, Kikuyu	
Kenia Forestry Seed Center	PO Box 74, Nairobi	Tel: +0154 32540/1
Tree Seed Program, Ministry of Energy and Regional Development	PO Box 21552, Nairobi	Tel: +565232
UNESCO IPAL	PO Box 30592, Nairobi	
KOREA		**0082**
Forest Research Institute, Chungyangni-Dong	Dongdaemun-Ku, Seoul	Tel: +966 8961 5
MALAWI		**00265**
Forest Research Institute of Malawi	PO Box 270, Zomba	Tel: +5228666
National Seed Co. of Malawi Ltd	PO Box 30050, Lilongwe 3	

Name	Address	Telecom
MALAYSIA		**0060**
Forest Research Center	PO Box 1407, Sandakan, Sabah	Tel: +214179
Forest Research Institute, Kepong, Selangor Malaysia Pointomuo Agri-Forest Service Station	WDT 8, Penamang Sabah	
Senior Research Officer	Pusat Penyelidi-kan Hutan, PO Box 1407, Sandakam	
Sabah Softwoods Sdn. Bhd	PO Box 137, Tawau, Sabah	
MEXICO		
Banco de Germoplasmo, Instituto Nacional de Investigacion Sobre Recursos Bioticos (INIREB)	AP 63, 91000 Xalapa Veracruz	
Banco De Mexico	Calle 58 # 389 Merida, Yucatan	
Chiapas/INIA/SARH, Dept Plantaciones y Mejoramiento Genetico	Apdo 98, Tapachula Chiapas	
Instituto Nacional de Investigacion Forestal	Sur No. 694 10 Piso, Col del Valle, 03100 DF	
Instituto Nal. de Investigaciones, Dept de Plantaciones	Av. Progreso No. 5, Viveros de Coyoacan, 04110 DF	
Centro de Investigacion Esenanza y Extension en Granderia Tropical	Vera Cruz	
NEPAL		**0097**
Tree Seed Unit, Hattisar, Naxal	Kathmandou	Tel: +412004
NETHERLANDS		**0031**
Medigran	PO Box 731, 1180 AS Amstelveen	
SETROPA Ltd	PO Box 203, 1400 AE Bussum	Tel: +21 5258754
Timmers and Leyer	PO Box 17, 2100 AA Heemstede	Tel: +23 284340
NEW ZEALAND		**0064**
Peter B. Down and Co., Dowseeds	PO Box 696, Gisborne	Tel: +83 408
NICARAGUA		**00505**
Banco de Semillas Forestales	IRENA AP 5123, Managua	Tel: +315948
PAKISTAN		**0097**
Elite Nurseries Inc.,	3 Street 67, Sector f7/3, Islamabad	Tel: +051 826523
Pakistan Forest Institute	Gatwala, Faisalabad	

Name	Address	Telecom
PANAMA		
Banco de Semillas Forestales Renare – Paraiso	Corregimiento Ancon, AP 2016	
PAPUA NEW GUINEA		
Office of Forests	PO Box 5055, Boroko	
PHILIPPINES		**0063**
Agtalon-Agro-Technical Assistance and Livelihood Opportunities in the North	2nd St Gracia Village, Urdaneta 2428, Pangasinan	
Abat Ramos, Barangay Quirino	Bacnotan La Union	
Bureau of Plant Industry (BPI)	San Andres Street, Malate, Manila	
Dave Deppner, New Forest Project	241 Caballero Street, Pozorrublo, Pangasinan	
Department of Environmental and Natural Resourses (DENR)	Region X, Cagayan de oro City	
Department of Horticulture	UPLB College, Laguna 4031	
Development Alternatives Inc.	Co/DENR Visayas Avenue, Quezon City	
Ecosystem Research and Development Bureau (ERDB) Seed Bank, College of Forestry UPLB	College, Laguna 4031	
Ernesto Cadaweng, DENR	Visayas Avenue, Quezon City	
Evergreen Gardens Inc.	Osmena Village, Maquikay, Mandaue City	
Export Import Trading Corporation or Greenworld Agri-Farm International Center	809 San Andres Street, Malate, Manila	Tel: +59 10 98
Institute of Plant Breeding, UPLB College	Laguna 4031	
International Institute of Rural Reconstruction (IIRR)	Silang Cavite	
Keystone Seedhouse Corp.	807 San Andres Street, Malate, Manila or PO Box SM6	Tel: +58 56 81/59 75 43
Magabaul Foundation Inc.	Bismartz, Masipag Don Carlos, Bukidnon	
Magsaysay Foundation	1640 Roxas Boulevard, Manila	
Makinnabang Entreprise	470 Makibang Baliwag, Bulacan	Tel: +23 27 85/90 40 11
Manila Seedling Bank Foundation Inc.	Cor. E. de los Santos Quezon City	Tel: +99 50 51/99 50 52

Name	Address	Telecom
Matalini Reforestation Project	Negros Occidental	
Mindanao Baptist Rural Life Center (MBRLC)	PO Box 94, Bansalan, Davao del Sur	
Nasipit LUMBER Company (NASIPIT)	Agusan	
Nestle Phil Inc.	Bukidnon	
Paper Industries Corporation of the Phillipines (PICOP) Forest Research Department	Tabon Bislig, Surigao del Sur 8311	
PAT Marketing	8 Forestry Street, Proj. 6, Quezon City	
Rayos Marketing	78 Molave Street, Proj. 3, Quezon City	
Visayan Seedling Center	PO Box 879, Rm 219, LK Bldg, Subangdaku, Mandaue City, Cebu	Tel: +8 15 26/8 17 09
Wood Development Corp.	Kitting Building, Busa, Cagayan de Oro City	
World Neighbours MAG-UUGMAD Foundation Inc.	524 CP del Rosario Ext., Cebu City 6000	Tel: +94003
PUERTO RICO		**001787**
Southern Forest Experiment Station, Institute of Tropical Forestry	PO Box AQ, Rio Piedras, PR 00928	Tel: +809 753 4335
RWANDA		
ISAR. Département de Foresterie Centrale de Graines Forestières	BP 617, Butare	
SENEGAL		**00221**
Centre National de Recherches Forestières	BP 312, Dakar	
Institut Sénégalais de Recherches Agricoles, Centre National de Recherches, Laboratoires de Graines	Parc Forestier de Hann, BP 2312, Dakar	Tel: +213219
SIERRA LEONE		**00232**
Forestry Division MANR	Youyl Building, Brookfields, Freetown	Tel: +23445
SINGAPORE		**0065**
The Inland and Foreign Trading Corp. Ltd	PO Box 2098, Maxwell Road Post Office Singapore 9040	Tel: +2722711/2782193
SOLOMON ISLANDS		
Forestry Division	PO Box 79, Munda	
National Agricultural Training Institute	PO Box 18, Auki	

Name	Address	Telecom
SOUTH AFRICA		0027
Directorate of Forestry Seed Section	PO Box 727, Pretoria 0001	Tel: +012 287120
SUDAN		
Forestry Research Center	PO Box 658, Soba Khartoum	
TAIWAN		
Yilin Co.	24 FU-jen, 2nd St Huallen	
TANZANIA		00255
Department of Forest Biology, Faculty of Forestry, Sokoine University of Agriculture	Box 3009, Morogoro	Tel: +2511
Tanzania Forestry Research Institute, Silvicultural Research Centre	PO Box 95, Lushoto	
THAILAND		
ASEAN–CANADA Forest Tree Seed Center (ACFTSC)	Muak-Lek, Saraburi 18180	
Benjavan Rerkasem Multiple Cropping Project	Chiang Mai University, PO Box 29, Chiang Mai	
Rupert Nelson	PO Box 29, Chiang Mai 50000	
UK		0044
Kins Plants Ltd	Woodcote Grove, Ashley Road, Epsom, Surrey kT18 5BM	
Commonwealth Forestry Institute	South Park Road, Oxford OX4 1EP	Tel: +865 511431
Department of Plant Sciences, University of Oxford	South Park Road, Oxford OXI 3RB	Tel: +865 275000
Henry Doubleday Research Association (HDRA)	Ryton-on-Dunsmore, Coventry CV8 3LG	
Royal Botanic Garden	Kew Richmond, Surrey TW9 3AB	
Unit of Tropical Silviculture, Department of Forestry, Commonwealth Forestry Institute	South Park Road, Oxford OX1 3RB	
USA		001
Carter Seed Co.	475, Mar Vista Drive Vista, CA 92083	Tel: +714 724-5931
Hudson Seedsman	Redwood City, CA	
Hurov's Tropical Seeds	PO Box 1596, Chula Vista, CA 92012	Tel: +619 690-0496
ECHO	17430 Durrance Road, North Fort Mayers, FL 33917	

Name	Address	Telecom
USDA – ATS, Regional Plant Introduction Station, Georgia Experiment Station	GA 30212	Tel: +404 228-7255
National Tree Seed Laboratory Seed Bank	RT 1, Box 182B, Dry Branch, GA 31020	
Amorient Inc.	PO Box 131, Kahuku, HI 96731	
Department of Agronomy and Soil Science, University of Hawaii at Manoa	3190 Maile Way, Honolulu, HI 96822	
Niftal Project University of Hawaii	PO Box "0", Paia, HI 96779	Tel: +808 579-9568
Nitrogen Fixing Tree Association (NFTA)	PO Box 680, Wainmanalo, HI 96795	
Tree Seed International	2402 Esther Court, Silver Springs, Madison, WI 20910	
Plant Introduction Office	USDA-ARS Building 001, Rm 322, BARC-West Beltsville, MD 20705	
Department of Forestry, Michigan State University	East Landing, MI 48824	
University of Minnessota	Alderman Hall, 1970 Fowell Avenue, St Paul, MN 55108	
Caesar Kleburg Wildlife Research Centre	Campus Box 218, Texas A & I University, Kingville, TX 78363	
Douglass W. King Co.	PO Box 200320, 4627 Emil Road, San Antonio, TX 78220	Tel: +512 661-4191
Native Plants	360, Wakara Way, Salt Lake City, UT 48108	
WINROCK International, F/FRED Project	1611 N. Kent, Arlington, VA 22209	
Tree Seed International	1015, 18th Street, NW, Suite 802, Washington, DC 20036	Tel: +202 822-8817
URUGUAY		
Coordinator Depto Forestal, Facultad de Agronomia	Avda Garzon 780, Montevideo	
VENEZUELA		
Estacion Experimental de Semillas Forestales	MARNR, El Limon, Maracay Estadoo de Aragua	
VIETNAM		
Faculty of Forestry, University of Agriculture and Forestry	Thu'Duic, Ho Chi Minh City	

Name	Address	Telecom
ZAMBIA		
Forest Geneticist, Division of Forest Research	PO Box 22099, Kitwe	
ZIMBABWE		
Divisional Manager	PO Box HG 595, Highlands, Harare	
National Herbarium and Botanic Garden	PO Box 8100, Causeway, Harare	

2.3
Coffee Books, Manuals and Reports

Ahrens, L. and Vandenput, R. *La Lutte Contre les Ennemis des Principales Cultures Pérennes de la Cuvette Centrale Congolaise.* Publication de la Direction de l'Agriculture, des Forêts de l'Elevage et de la Colonisation, Bruxelles, 1952.

AMAE-IFOAM-UACH. *Conferencia International sobre Café Organico (Memorias).* AMAE-IFOAM-UACH, 1995.

ANACAFE. *Manual de Cafeticultura Organica.* Associacion Nacional del Café, Guatemala, 1998, 1999.

Associacao Brasileira para Pecquisa da Potasa e do Fosfato. *Cultura do Cafeeiro,* 1986.

Australian Coffee Research and Development Team. *Coffee Growing in Australia.* DPI, Queensland, 1995.

Bertrand, B. and Rapidel, B. *Desafios do la Cafoticultura en Centroamerica.* CIRAD IRD, IICA-PROMECAFE CCCR, Francia San José, Costa Rica, 1999.

Bouharmont, P. L'Utilisation des plantes de couverture et du paillage dans la culture du caféier arabica au Cameroun. *Café, Cacao, Thé* **1979**, *XXIII* (2), 75–102.

Boyd, J. *Tools for Agriculture – A Buyers Guide to Low-cost Agricultural Implements,* 2nd edn. Intermediate Technology Publications, London, 1976.

Brian, H. and Robinson, D. *Coffee in Yemen – A Practical Guide. Rural Development Project in Al-Mahwit Province.* Klaus Schwarz, Berlin, 1993.

Cambrony, H. R. *Le Caféier Le Technicien d'Agriculture Tropicale.* Maisonneuve and Larose, Paris, 1989.

Camilo, D. A. *Manual de la Caficultura Dominicana.* Subsecretaria de Estado de Agricultura Dep. De Café, 1987.

Carvajal, F. *Cafeto – Cultivo y Fertilización,* 2nd edn. International Potash Institute, Basel, 1984.

Clarke, R. J. and Macrae, R. *Coffee. Volume 1: Chemistry,* 985. *Volume 2: Technology,* 1987. *Volume 3: Physiology,* 1988. *Volume 4: Agronomy,* 1988. *Volume 5: Related Beverages,* 1987. *Volume 6: Commercial and Technico-Legal Aspects,* 1988. Elsevier, London.

Clevez, R. *Technologia en Beneficiado de Café.* Impresora Tica SA, Costa Rica, 1995.

Clifford, M. N. and Willson, K. C. *Coffee: Botany, Biochemistry and Production of Beans and Beverage.* Croom Helm, London, 1985.

Coffee grower's Handbook, revised edn. Coffee Grower's Association, Harare, 1987.

Coffee Research Foundation. *An Atlas of Pests and Diseases.* Coffee Research Foundation, Ruiru, Kenya, 1989.

Coolhaas, C. and de Fluiter, H. J. *Kaffee.* Ferdinand Enke, Stuttgart, 1960.

Coste, R. and Cambrony, H. *Caféiers et Cafés.* Maisonneuve and Larose, Paris, 1989.

Coste R., *Les caféiers et les Cafés dans le monde.* Tome 1, les caféiers – 1995 – 382 p. Tome 2, Les Cafés Vol. & – 372 p. 1959, Ed. Larose Paris Vol. 2 397 p. 1961.

CR. Kenya. *An Atlas of Pests and Diseases* – 1989 – Publ. by Coffee Research Foundation Ruiru – Kenya.

Cultura de Café do Brasil. *Pequeno manual de recomendações.* Editada pela Seção de Programação/DEPET/DIPRO/IBC, 1986.

Daviron, F. and Lerin, B. *Le Café.* Economica, Paris, 1990.

De Graaf, J. *The Economics of Coffee.* Pudoc, Wageningen, The Netherlands, 1986.

De Smet, J. *Manuel de Gestion des Stations de Lavage Café.* République du Burundi/Ministère de l'Agriculture et de l'Élevage. Projet filière Café, 1993.

El Beneficio del Café. Publicacíon del Comité Departamental de Cafeteros de Antioquia. Edinalco, Medellín, Colombia, 1991.

El Cultivo del Cafeto en Mexico. Instituto Mexicano del Café/Nestlé, Lafuente, 1990.

Escamilla, E. and Diaz, S. *Sistemas de Cultivo de Café en Mexico.* VACH, Mexico, 2002.

Federacion Nacional de Cafeteros de Colombia. *Cartilla Cafetera.* Café Comité Departamental de Cafeteros del Risaralda, Division Tecnica, Servicio de Extension, Colombia, 1999.

Finney, A. Le système de taille en tiges multiples écimées pour le caféier Arabica *Café, Cacao, Thé* **1988**, *XXXII* (1), 33–43.

Fischersworring Hömberg, B. and Rosskamp Ripken, R. *Guia para la Caficultura Ecológica.* GTZ, Eschborn, 2001.

Forestier, J. Relations entre l'alimentation du caféier Robusta et les caractéristiques analytiques des sols. *Café, Cacao, Thé* **1964**, *VIII* (2), 89–112.

Fuchs, A. *Nutzpflanzen der Tropen und Subtropen.* Institut für tropische Landwirtschaft, S. Hirzel, Leipzig, 1980.

Gaie, W. and Flémal, J. *La Culture du Caféier d'Arabie au Burundi.* Institut des Sciences Agronomiques du Burundi, Isabu. Pub. du Service Agricole no. 14. AGCD, Bruxelles, 1988.

Guharay, F., Monterrey, J. Momnterroso, D. and Staver, Ch. *Manejo Integrado de Plagas en el Cultivo del Café.* CATIE, Managua, Nicaragua, 2000.

Guizol, P. *L'Erosion des Sols au Burundi. Résultats après trois années de mesures sur les parcelles expérimentales de Rushubi.* ISABU, Burundi, 1983.

Haarer, A. E. *Modern Coffee Production.* Leonard Hill, London, 1962.

IBC. *Cultura de Café do Brasil.* IBC, Rio de Janerio, 1986.

INIFAP Mexico Tecnologia para la producción de café en Mexico. *Foll. Tecnico* **1997**, 8, 90.

Instituto del Café de Costa Rica. *Manual de Recomendaciónes para el Cultivo del Café.* San José, Costa Rica, 1998.

Jariya, V. Ecological study of coffee stem borer *Xylotrechus quadripes* Chevrolet (Coleoptera, Cerambycidae) in Northern Thailand II: mating and production. *J. Agric.* **1992**, *I* (2), 211–225.

Jariya, V. Personal comment. Department of Entomology. Faculty of Agriculture, Chian Mai University, 1993.

Jobin, P. *Les Cafés Produits dans le Monde.* Jobin, Le Harve, 1992.

Kenya Coffee. Coffee establishment in Kenya. Technical Circular no. 42. *Kenya Coffee* **1988**, 53 (615), 247–253.

Krug, C. A. and de Poerck, R. A. *World Coffee Survey.* FAO, Rome, 1968.

L'Administration Générale de la Coopération au Développement. *Les Principales Cultures en Afrique Centrale*. R. Vandenput Imprimé par LESAFFRE, Bruxelles, 1981.

Le-Pelley, R. H. *The Pest of Coffee*. Longmans, London, 1968.

Malavolta, E. Nutrition, fertilisation et chaulage pour le caféier. In *Cultura do Cafeeiro*. Associação Brasiliera para Pesquisa de Potasa e do Fosfato, Piracicaba, SP, 1986.

Malavolta, E. *Seja O Doutor do Seu Cafezal*. Arqu. do Agronomo 3, Informaç. Agronomicas 64, 1993.

Mansingh, A. Limitations of insecticides in the management of the coffee berry borer *Hypothenemus hampei* (Ferr.) *J. Coffee Res.* **1991**, *21* (2), 67–98.

Manual del Cafetero Colombiano, 4th edn. Sección de Divulgación Científica de CENICAFE, Colombia, 1979.

Matériels pour l'Agriculture, 4th éd. 1500 Références pour l'Equipement des Petites et Moyennes Exploitations. Groupe de Recherche et d'Echanges Technologiques (GRET) en association avec Intermediate Technology Development Group (ITDG), 1992.

Matthews, G. *Pesticide Application Methods*, 2nd edn. Blackwell, Oxford, 1982.

Matthews, G. and Hislop, E. C. *Application Technology for Crop Protection*. CAB International, Wallingford, 1993.

Mitchell, H. W. Cultivation and harvesting of the arabica coffee tree. In *Coffee, Vol 4: Agronomy*. Clarke, R. J. and Macrae, R. (eds). Elsevier, London, 1987, 43–90.

Morallo-Rejesus, B. and Baldos, E. The biology of coffee berry borer *Hypothenemus hampei* (Ferr.) (Scolytidae, Coleoptera) and its incidence in the Southern Tagalog Provinces, Philippines. *Entomologist* **1980**, *4* (4), 303–316.

Mwangi, C. N. *Coffee Production Recommendation (Handbook)*. Coffee Research Foundation, Ruiru, Kenya 1987.

Mwangi, C. N. *Coffee Grower's Handbook*. Ruira, Kenya, 1987.

National Academy Press. *Vetiver Grass: A Thin Green Line against Erosion*. National Research Council, Washington, DC, 1993.

Nestlé Ltd. (Wintgens J. N.), Green Coffee Defects Guide for identification and evaluation, 1997.

Nova Scotia Federation of Agriculture. *Pesticide Safety Manual*. NSFA, 1991.

Pochet, P. *Le Bouturage du Caféier Robusta [Robusta Propagation by Cuttings]*. Publications Agricoles, 1987.

Pochet, P. *La Trachéomycose du Caféier Robusta [Tracheomycosis in Robusta Coffee Bushes]*. Publications Agricoles, 1988.

Pochet, P. *How to Prune Your Coffee Bushes. Pluriannual pruning cycle for Robusta*, 2nd edn. Agricultural publications no. 2, 1989.

Pochet, P. *The Quality of Coffee from Plantlet to Cup*. Agricultural publications no. 21. Published with support of Department of Overseas Aid AGCD, Brussels, 1990.

Pohlan, J. *Mexico y la Caficultor Chiapaneca*. Shaker Verlag, Aachen, 2002.

PROCAFE. *Manual del Caficultar Salvadoréño*. PROCAFE, San Salvador, 1997.

Rehm, S. X. and Epsig, G. *Die Kulturpflanzen der Tropen und Subtropen*. Verlag E. Ulmer, Stuttgart, 1976.

Rena, A. B., Malavolta, E., Rocha, M. and Yamada, T. *Cultura do Cafeeiro*. Associação Brasiliera para Pesquisa de Potassa e do Fosfato, Piracicaba, SP, 1986.

Rohkaffee und Röstkaffee Prüfung. 1963. Gordian Verlag, Hamburg – 366 p.

Rothfos, B. *Coffee Production*. Gordien-MaxRieck, Hamburg, 1980.

Sivetz, M. and Desrosier, N. W. *Coffee Technology*. AVI, Westport, CT, 1979.

Snoeck, D., Bitoga, J. P. and Barantwarir-ije, C. Avantages et inconvénients des divers modes de couverture dans les caféiers au Burundi. 1994. *Café, Cacao, Thé* **1994**, *XXXVIII* (1), 41–48.

Snoeck, D. and Snoeck, J. Programme Informatisé pour la Détermination des Besoins en Engrais Minéraux pour les Caféiers Arabica et Robusta à Partir des Analyses du Sol. 1988. *Café, Cacao, Thé* **1988**, *XXXII* (3), 201–218.

Snoeck, J. Le caféier d'Arabie à Rubona. *Bull. d'Inf. de l'INEAC* **1959**, *VIII* (2), 69–99.

Snoeck, J. Cultivation and harvesting of the Robusta coffee tree. In *Coffee, Vol 4: Agronomy.* Clarke, R. J. and Macrae, R. (eds). Elsevier, London, 1987, pp. 91–127.

Snoeck, J. *Proc. 8th ASIC Coll.* 1977, 463–487.

Snoeck, J. *Rapport de Mission auprès de l'Academie des Cultures Tropicales de la Chine du Sud (SCATC).* IRCC-CIRAD, Montpellier, 1987.

Snoeck, J. and Jadin, P. Mode de calcul pour l'étude de la fertilisation minérale des caféiers basée sur l'analyse du sol. *Café, Cacao, Thé* **1990** *XXXIV* (1), 3–21.

UNICAFE. *Manual de Caficultura de Nicaragua.* Union Nicaraguense de Cafetaleros, Managua, 1996.

US National Academy of Science and Philippine Council for Agriculture and Resources Research. *International Consultation on IPIL.* IPIL Research, 1981.

Vallaeys, G. *La Pratique de la Taille du Caféier Robusta.* Publication de la Direction de l'Agriculture des Forêts et de l'Elevage, Bruxelles, 1959.

Van Hilten, H. J., Fischer, P. J., Wheeler, M. A., Scholer, M., Coffee: An exporter's guide, ITC. Centre, Geneva, 2002.

Wanyonyi, J. J. M. *An Outline of Coffee Processing.* Ministry of Agriculture, Kenya, 1972.

Waterhouse, D. F. and Norris, K. R. *Hypothenemus hampei* (Ferrari). In *Biological Control Pacific Prospects Supplement I.* ACIAR, Canberra, 1989, 57–76.

Wellman, L. *Coffee, Botany, Cultivation and Utilization.* Leonard Hill/Interscience, New York, 1961.

Wilbaux, R. *Coffee Processing.* Agricultural Engineering Branch, Land and Water Development Division, FAO, Rome, 1963.

Wilbaux, R. *Technologie du Café Arabica et Robusta. Les Caféiers au Congo Belge.* Publication de la Direction de l'Agriculture des Forêts et d'Elevage, Bruxelles, 1956.

Willson, K. C. *Coffee, Cocoa and Tea.* CAB International, Wallingford, 1999.

Wintgens, J. N. *Factors Influencing the Quality of Green Coffee.* Nestlé, Vevey, Switzerland, 1993.

Wintgens. J. N. *Green Coffee Defects Guide for Identification and Evaluation.* Nestlé, Vevey, Switzerland, 1997.

Wrigley, G. *Coffee.* Longman/Wiley, New York, 1988.

Zamarripa, A. and Escamilla, E. *Variedades de Café en Mexico.* VACH, Mexico, 2002.

Zapata, C. and Ruiz, M. *La Variedad Colombia: Selección de un Cultivar Compuesto Resistente a la Roya del Cafeto.* CENICAFE, Colombia, 1988.

2.4
Coffee Periodicals

Agricultura de la Americas, Keller International, Great-Neck, NY. Tel: +1 516 829-9210

Annual Reports, Coffee Research Foundation, Nairobi, Kenya

CAB International, Wallingford, UK.
  Tel: +44 1491 832111
Cenicafé, Federación Nacional de Cafeteros
  de Colombia, Caldas, Colombia
Coffee and Cocoa International (bimonthly),
  UK
Coffee International Directory (annual), Ding
  World Media, Redhill, UK
Coffee, Cocoa and Tea
Directory and Buyers' Guide (annual),
  Tea and Coffee Trade Journal
FAO Production Yearbook, Natural
  Resources Institute, FAO, Rome, Italy.
  Tel: +39 06 57971
Far Eastern Agriculture, Alai Charles, Hong
  Kong. Tel: +852 5846212
F. O. Licht's International Coffee Yearbook,
  F.O. Licht, Ratzeburg, Germany
G. Gordon Paton – Coffee Annual
ICO International Coffee Organization:
  Publications
Journal of Coffee Research, Central Coffee
  Research Institute, Balehonnur, India
Kaffee and Tee Markt, Max Rieck, Hamburg,
  Germany
Kenya Coffee, Coffee Board of Kenya,
  Nairobi, Kenya
Natural Resources Institute:
  Publications, Chatham Maritime, UK.
  Tel: +44 6348 80088
OECD: Publications, Paris, France.
  Tel: +33 1 45248200
Plantations, Recherche, Développement,
  CIRAD, Paris, France.
  Tel: +33 1 53702260
Tea and Coffee Trade Journal, Whitestone,
  NY
Technical Circular, Coffee Research
  Foundation, Ruiru, Kenya
Uckers's International Tea and Coffee
World Coffee and Tea, McKeand,
  West Haven, CT

2.5
Coffee Associations, Organizations and Teaching Centers

ANACAFE – Asociación Nacional del Café,
  Guatemala
CATIE – Centro Agronomico Tropical de
  Investigación y Enseñanza, Costa Rica
CENICAFE – Centro Nacional de
  Investigación de Café, Colombia
Coffee Board of Kenya, Kenya
Coffee Research Foundation – Kenya
FEDECAFE – Federación Nacional de
  Cafeteros de Colombia, Colombia
Fundacão Instituto Brasileiro de Geografica
  e Estatistica, Brazil
ICAFE – Instituto del Café de Costa Rica,
  Costa Rica
IHCAFE – Instituto Hondureño del Café,
  Honduras
IICA – Inter-American Institute for
  Cooperation on Agriculture, Costa Rica
OAMCAF – Organisation Africaine
  et Malgache du Café, France
OIAC – Organisation Interafricaine
  du Café, Abidjan, Ivory Coast
PROCAFE – Fundación Salvadoreña para
  Investigaciones del Café, El Salvador
Tropical Agriculture Teaching Centers/
  Universities/Colleges in the UK,
  Belgium, France, Germany, Holland,
  Portugal, USA, Mexico, Costa Rica,
  Columbia, Brazil, etc.

2.6
Coffee Events

AGRICULTURAL EXHIBITIONS, FAIRS, etc.
England (Royal show), Frankfurt, Paris,
  Israel (Agritech), China, etc.
ASIC – Association Scientifique Interna-
  tionale du Café, France (biennal confer-
  ence – about 500 participants)

CIRAD – Centre de Coopération International en Recherche Agronomique pour le Développement, France

Coffee and Cocoa International UK – organizes conferences on coffee

Federacion Cafetalera Centro America–Mexico–El Caribe

ICO (International Coffee Organization), London

SPECIAL MEETINGS

Pan-American Coffee Conference

PROMECAFE – Programa Coperativo Regional para el Desarollo Tecnologico y Modernización de la Caficultura (symposia and conferences in Latin America)

SCAA – Specialty Coffee Association of America USA (annual conference and exhibition – about 8000 participants)

Tea and Coffee Trade Journal (regular conventions – about 2000 participants).

Vereinigung für Angewandte Botanik: "International Symposium" (Tropische Pflanzen), Universität Hamburg

World Coffee Congress

World Speciality Coffee Conferences and Exhibitions

2.7
Important Coffee Research Organizations

Name	*Address*	*Telecom*
BRAZIL		**0055**
Fundação Instituto Brasileiro de Geografica e Estadistica	Rua General Canabarro 666, CEP 20271, Rio de Janeiro	Tel: 2844597 Fax: 2846189
Instituto Agronomico Secao de Genetica – IAC	Avenida Barao de Itapura 1481, CEP 13100, Campinhas, SP	
COLOMBIA		**0057**
Federación Nacional de Cafeteros de Colombia	Calle 73, No. 8-13, Bogota DE	Tel: +1 2170600; Fax: +1 2171021
COSTA RICA		**00506**
Centro Agronómico Tropical de Investigación y Enseñanza	c/o Instituto del Café Costa Rica (ICAFE), Calle 1, Ave. 18-20, Aptdo. 37, 1000 San Jose	Tel: +506 226411; Fax: +506 222838
Inter-American Institute for Cooperation on Agriculture (IICA)	PO Box 55, 2200 Coronado, San Jose	Tel: +506 290222; Fax: +506 294741
EL SALVADOR		
PROCAFE – Fundacion Salvadoreña para investigaciones del Café		

Name	Address	Telecom
FRANCE		**0033**
(CIRAD) Centre de coopération international en recherche agronomique pour le développement	42 rue Scheffer, 75116 Paris	Tel: +1 47043215; Fax: +1 47271947
IRD (ex. ORTSOM) – Institut de Recherche et Développemnt	213 rue Lafayette, 75480 Paris Cedex 10	Tel: +1 480377
INRA Institut National de la Recherche Agronomique	147 rue de l'Université, 75341 Paris Cedex 07	Tel: +1 42759000
GUATEMALA		
ANACAFE – Associacion Nacional del café	5a calle 0-50, zona 14, Edificio Las Americas, Guatemala City	Tel: +5022370133
INDIA		
Central Coffee Research Institute	Balehonnur, Chikmagalur District, Mysore State	
INDONESIA		
Perbebunan/Pusat Ct Ps		
MEXICO		
INIFAP – Rosario Izapa		
TAPACHULA CHIAPAS		
PORTUGAL		
Centro de Investigacão des Ferrugens do Cafeeiro	Quinta do Marques, Oireas 2780	

3
Data on Coffee

J. N. Wintgens

3.1
Environment for Growing Coffee

Temperatures

Arabica	Optimal annual average	18–22 °C
	Minimal annual average	16 °C
	Maximal annual average	26 °C
	Extremes	7–30 °C
Robusta	Optimal annual average	20–25 °C
	Extremes	14–30 °C

Rainfall

Arabica	1500–1800 mm with 2–3 months drought (max. 6 months drought); extreme 4000 mm
Robusta	1500–3000 mm with 1–3 months drought (max. 4 months drought)

Hours of sunshine (Arabica × Robusta)
1900–2200 h sunshine per year

Relative Humidity

Arabica	60–75 %
Robusta	80–90 %

Coffee: Growing, Processing, Sustainable Production, Second Edition. Edited by J. N. Wintgens.
© 2012 WILEY-VCH Verlag GmbH & Co. KGaA. Published 2012 by Wiley-VCH Verlag GmbH & Co. KGaA.

3.2
Characteristics of Coffee by Variety

Characteristics	Arabica	Robusta	Liberica
Origin	East Africa Ethiopia high altitude 1000–2400 m	Central and West Africa groups congensis guineensis low altitude	West Africa dryer areas low to middle altitude
Date of earliest plantations	6th century	1900–1910	1864–1881
Free growing height (m) Cultivated height (m)	5 2.5–3.0	3–8 2.5–4.0	15–18 2.5–9.0
Fecundation	autogamy (self-pollination) with occasionnal insect pollination	allogamy (cross-fertilization) wind and insect pollination	allogamy (cross-fertilization) wind and insect pollination
Propagation	seeds vegetative for hybrids	vegetative recommended	vegetative recommended
Chromosome number	tetraploid ($2n = 44$)	diploid ($2n = 22$)	diploid ($2n = 22$)
Fruit length (mm) width (mm) thickness (mm) shape (mm)	 12–20 8–14 7–10 elliptical	 9–16 7–13 5–12 elliptical	 15–28 12–20 ovoid
Root distribution	deep – no distinct tap root	shallow rooted in top layer – distinct tap root	deep
Planting density (plants/ha)	1000–5000	800–2000	500–760
Time from flower to fruit (months)	8–9	10–11	12
Bearing age (years)	3	2–3	4–5
Yield (kg/ha) low medium high very high	 300 600 1200 over 2000	 300 600–1200 1200–2000 over 2300	 300–500 500–900 900–1400 over 1400
Altitude for optimal production in tropical areas (m)	900–1800	200–900	0–1200
Care needed for production	significant	medium	least
Drought resistance	fair	medium	high
Wilting during drought	little	most rapid	medium
Ripe cherry drop	drops cherries	holds cherries	holds cherries

Continued ▶

Characteristics	Arabica	Robusta	Liberica
Plan of flowering	generally new wood	generally new wood	both old and new wood
Resistance to diseases			
coffee leaf rust	susceptible	tolerant to resistant	immune
koleroga	susceptible	tolerant	tolerant
nematodes	susceptible	tolerant to resistant	resistant
tracheomycosis	resistant	susceptible	
blister spot virus	susceptible	tolerant	
coffee berry disease	susceptible	resistant	
pH of (GB)	5.26–6.11	5.27–6.13	
Free acidity of GB (ml n-NaOH per 100 g)	28.5–39.5	36.5–43.5	
Mineral cont. of GB (%)	3.5–4.5	3.9–4.5	
Fat cont. of GB (%)	13–17	7.2–11	
Caffeine cont. of GB (%)	0.7–2.2 (aver. 1.4)	1.5–2.8 (aver. 2.2)	1.2–1.6 (aver. 1.4)
Chlorogenic acid cont. of GB (%)	4.80–6.14	5.34–6.41	
N substance cont. of GB (%)	aver. 13		

GB = green beans.

3.3
Common Characteristics of the Coffee Plant

Plantlet
- Appearance of first branches (cross) after 6–8 months
- Age for transplanting 3–18 months after sowing
- The most common is to use plants at 8–10 months with six or seven pairs of leaves, one or two pairs of primaries and a height of 30–60 cm

Flowers

Flower life	few hours
Blossom duration	2–3 days
Pollen viability	30 h

Fruit

Germination time under appropriate conditions

fresh seeds	1 month
dry seeds	2 months

Seed viability under appropriate storage conditions
after 4 months   90%
after 5 months   80%
after 5–12 months viability decreases drastically

Apparent density
fresh cherries   620 kg/m^3
dry cherries   420 kg/m^3
dry green beans   650–780 kg/m^3
dry parchment: 385 kg/m^3

Moisture content
ripe cherry   60–65%
washed undried parchment 52–55% (once the surface water is drained off)
dry green coffee   11–12%
fresh coffee pulp   76%

Production Life Cycle

The potential productive life of a coffee tree is about 30 to 50 years under favorable growing conditions and good management. Nevertheless, productive trees over 100 years old have been reported.

3.4
Labor Requirements for Coffee Operations

Harvesting (equivalent green coffee)	1.25 man-hours/kg green coffee
Weed control	60 trees/man-day
Fungicide spraying	120 trees/man-day
Pruning	40 trees/man-day
Handling (desuckering)	65 trees/man-day
Fertilizer application	450 trees/man-day
Manure application	40 trees/man-day

Where there is little or no mechanization for maintenance, one permanent worker per hectare is required.

3.5
Coffee Post-harvest Treatment

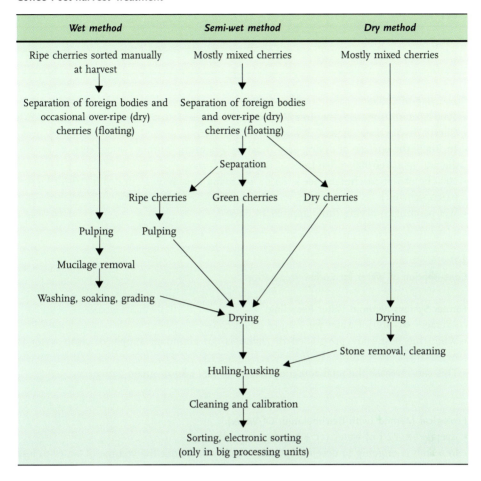

Wet method	Semi-wet method	Dry method
Ripe cherries sorted manually at harvest	Mostly mixed cherries	Mostly mixed cherries

Separation of foreign bodies and occasional over-ripe (dry) cherries (floating)

Separation of foreign bodies and over-ripe (dry) cherries (floating)

Separation

Ripe cherries   Green cherries   Dry cherries

Pulping   Pulping

Mucilage removal

Washing, soaking, grading

Drying   Drying

Stone removal, cleaning

Hulling-husking

Cleaning and calibration

Sorting, electronic sorting (only in big processing units)

3.6
Pulping and Drying

Pulping
- A single disk of a pulper can process approximately 1000 kg of cherries per hour
- An Argaard pre-grader can grade 3000 kg of cherries per hour
- A three-disk pulper needs:
  16 fermentation tanks
  20 skin drying tables, 15 m × 1.8 m
  80 drying tables, 23 m × 1.8 m

This corresponds to 100 t of cherries per week

Performance of Conventional Depulpers

	Drum/Bentall	Vertical drum/Penagos	Disc/McKinnon/J.Gordon
Capacity (kg cherry/h)	600	1000–2500	1000–2500 per disk
Water requirement (l/h)	500	750–1500	1500

Sun-drying of Coffee
- 1000 kg of fresh cherries requires 20 m^2 of drying area at 5–6 cm depth (40 kg cherries/m^2)
- Cherry drying time: approx. 3 weeks
- In Brazil the surface of drying area must be 5 % of the number of coffee trees in production (e.g. 100 000 coffee trees $\times$ 0.05 = 5000 m^2)
- 1000 kg of wet parchment needs 80 m^2 of drying area at 3–4 cm depth (10–12 kg wet parchment/m^2)
- Wet parchment drying time: approx. 14 days

3.7
Consumption of Water for Coffee Preparation

Former System (without Water Recycling)
- Average: 40–80 l of water for 1 kg of green coffee
- When the beans are pre-graded in channels, an abundant supply of clean water is necessary
- This can increase the total consumption up to 200 l/kg of green coffee

Ecological Method (with Re-circulation Of Water)
- Average: 8–22 l of water for 1 kg of green coffee
- Research is ongoing to develop systems which will reduce the volume of water to less than 5 l/kg of green coffee

3.8
Outturn of Coffee

Maximum and Minimum Outturn of 100 kg of Arabica red-ripe cherry (Wilbaux)
100 kg of Arabica red-ripe cherry contains:
    45–65 % fresh skin and mucilage
    35–55 % fresh parchment coffee

1000 kg of fresh cherries produce	0.5 m^3 of wet parchment
1000 kg of dry parchment produces	800–850 kg dry green coffee
1000 kg of dry green coffee produces	840 kg roasted coffee
1000 kg of dry green coffee produces	388 kg of soluble powder coffee (ratio 2.6 : 1)

Balance of coffee constituents during preparation (Arabica by wet process)

Coffee cherry (65 % MC)	5.5 t	Green coffee (11 % MC)	1.0 t (18.2 % of 5.5 t)
Water needed for processing these cherries	22.5 t	Husks (10 % MC)	0.25 t (4.5 % of 5.5 t)
		Fresh pulp (82 % MC)	2.25 t (41 % of 5.5 t)
		Water content in wet parchment[a]	1.1 t (20 % of 5.5 t)
		Mucilage (65–75 % MC)	0.9 t (16.3 % of 5.5 t)
		Residual water[b]	22.0 t
Total	27.5 t	Total	27.5 t

[a]Corresponds to the composition water of the wet parchment evaporated during the drying process.
[b]This corresponds to 22 l/kg of green coffee (22 : 1). Outturn green coffee/cherries – 18.2 %.

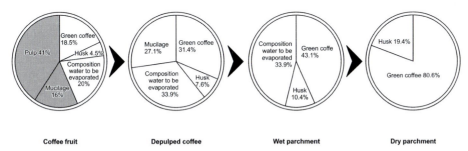

Figure 3.1 Composition of coffee. The percentages of pulp and mucilage are based on fresh weight. The percentages of green coffee and husk are based on dry product with 11 % MC. Adapted from: R. Cléves (1995).

Outturn of 100 kg Robusta Red-ripe cherry (Knaus)

	kg	
Dry cherry	40–45	
Pulped beans	74	
Washed parchment	52	
Washed and centrifuged parchment	47–49	
Drained parchment	49	
Washed and pre-dried parchment	44	
Dried parchment coffee	26	
Clean dry coffee	22	(22 % outturn)

Byproducts	kg	
Dried husks (dry method)	20	
Fresh pulp (after pulping)	50–60	
Dried pulp (wet method)	12–15	
Parchment (wet method)	3–5	

Composition and outturn of Liberica cherry in Spanish Guinea (Nosti Nava):
100 cherries weight 500 g
Pulp removed by pulper: 312.5 g (62.5 %) with 77.55 % moisture
Pulped beans: 187.5 g (37.5 %) produces:
                            89.9 g of mucilage which contains 72 % moisture
                            97.6 g of washed beans yields 58 g of clean dry coffee
                            (11.6 % outturn)

3.9
Storage

- One bag of green/clean coffee weighs 60 kg net, 60.5–61 kg gross
- Ten bags of dry parchment require a storage area of 1 m^2 floor surface stacked at 8 bags (maximum authorized)

3.10
Composition of the Coffee Fruit

Dry Coffee Pulp Composition (%)

Moisture	12.6	Etheric extract	2.5
Dry Matter	87.4	Crude fiber	21.0
Total	100.0	Nitrogen (N_2)	1.8
		Protein	11.2
		Ash	8.3
		Free N_2 extract	44.4

From: *Turrialba*, Vol. 22 – No. 3 – Jul.–Sept. 1972.

Chemical Composition of Green Arabica and Robusta Coffee

Composition	Variations of Content of Dry Matter (%)	
	Arabica	Robusta
Kahweol	0.7–1.1	NA
Caffeine	0.6–1.5	2.2–2.7
Chlorogenic acids	6.2–7.9	7.4–11.2
Sucrose and reducing sugars	5.3–9.3	3.7–7.1
Total free amino acids	0.4–2.4	0.8–0.9
Strecker-active	0.1–0.5	0.2–0.3
Araban	9.0–13.0	6.0–8.0
Reserve Mannane	25.0–30.0	19–22
Reserve Galactan	4.0–6.0	10.0–14.0
Other polysaccharides	8.0–10.0	8.0–10.0
Tryglycerides	10.0–14.0	8.0–10.0
Proteins	12.0	12.0
Trigonelline	1.0	1.0
Other lipids	2.0	2.0
Other acids	2.0	2.0
Ash	4.0	4.0
Totals*	90.0–114.0	86.0–107.0

*These totals of the lower and upper volues reflect the scope of the variations of the 100% of dry matter in particular coffees.
From: Cliffort and Willson (1985), p. 356.

3.11
Weight of Coffee Beans (Average Weight (g) of 1000 Cured Flat Beans)

C. arabica	Yemen – Hodeida	127
	Ethiopia – Harrar short	140
	Guatemala B	163
	Guatemala – Maracaibo A	176
	Santos – retained on screen 17.5	179
	Mexico – Tapachula A	180
	Java – Kalisat	189
	Colombia – Excelso	193
	Costa Rica A	203
	Colombia Supremo	213
	Guatemala – Maragogipe	291
C. robusta	Angola	161
	Java WIB	167
	Java – graded	196 (Maitland)
	Uganda ordinary	157 (Maitland)

From: Rothfos (1980).

3.12
Coffee Extraction Rate

Soluble Coffee
- In 1972 the EEC agreed for customs purposes a ratio of 1:2.5 (40%)
- According to ICO, the ratio is now 1 kg of soluble coffee from 2.6 kg of green coffee, i.e. a 38.46% yield
- It is usual for coffee makers to extract about 45–52% for Robusta and 46% for Arabica

Coffee Brew
For a cup of coffee, the average extraction rate is 20% of soluble solids. Extraction over 20% increases the bitterness of the brew and a lower extraction produces a light cup.

> 10 g of R & G coffee are used for one cup of 180 ml
> at 20% extraction rate these 10 g of coffee give 2 g of soluble solids
> 2 g of soluble solids for 180 g, correspond to 1.11% brew concentration

3.13
Caffeine Content in Different Brews

Brew (150 ml)	Range (mg)	Average (mg)
R & G coffee: drip method	60–180	115
R & G coffee: percolator	40–170	80
Instant coffee	30–120	65
Tea	25–110	60
Cola drink (12 fluid oz)	30–46	–

From: US FDA.

3.14
Coffee by Quality Group

Colombian milds

> Wet processed Arabicas – acid, fine flavor
> Origin: Colombia, Kenya, Tanzania

Other milds

> Wet processed Arabicas – less acidity
> Origin: Central America, Mexico, Bolivia, Burundi, Cuba, Dominican Republic, Ecuador, El Salvador, Haiti, India, Jamaica, Malawi, Papua New Guinea, Paraguay, Peru, Rwanda, Venezuela, Zambia, Zimbabwe, Hawaii, DRC, etc.

Brazilian and other Arabicas

> Dry processed Arabicas
> Origin: Brazil, Ethiopia, Yemen (Mocha), etc.

Robustas

> Usually dry processed – body and strength
> Origin: Cameroon, Uganda, DRC, Indonesia, Ivory Coast, Philippines, Thailand, Vietnam, Madagascar, etc.

3.15
Basic criteria for Coffee Quality Evaluation

- Moisture content
- Bean color and odor
- Bean size (screen)
- Defects (black, damaged, etc.)
- Cup taste (determining test)

3.16
Comment

The author would like to point out that the main objective of this chapter is to provide the reader with a general frame of coffee information. The data cannot be taken as mathematically precise, since it is based on observations and controls made on objects that differ considerably in their geographical location, their type, genetics and husbandry.

Bibliography

[1] Cléves, R. *Technologia en Beneficio de Café*. Impresora Tica, San Jose, Costa Rica, 1995.

[2] Cliffort, M. N. and Willson, K. C. *Coffee, Botany, Biochemistry and Production of Beans and Beverage*. Croom Helm, London, 1987.

[3] Coolhaas, C., de Fluiter H. J. and Koenig H. P. *Kaffee*. Ferdinand Enke, Stuttgart, 1960.

[4] Coste, R. *Les Caféiers*. Larose, Paris, 1955.

[5] CRF. *Coffee Production Recommendations (Handbook)*. Coffee Research Foundation, Ruiru, Kenya, 1987.

[6] Rothfos, B. *Coffee Production*. Gordian-Max-Rieck, Hamburg, 1980.

[7] Wellman, F. L. *Coffee*. Leonard Hill Interscience, London, 1961.

[8] Wilbeaux, R. *Coffee Processing*. FAO, Rome, 1963.

[9] Wrigley, G. *Coffee*. Longman, London, 1988.

4
Acronyms and Terms used in Coffee Production

S. Bonnet-Evans

4.1
Terms used in Coffee Production

Active ingredient (a.i.)	The biologically active portion of a pesticide formulation.
Adsorption	The attraction of ions or compounds to the surface of solid soil colloids (clay, humus). Adsorption complexes adsorb a high proportion of primary plant nutrients.
Agrochemicals	Chemicals or chemical compounds used in agriculture for crop fertilization, for the control of pests and diseases or for weed control.
Agroecology	The application of ecological concepts to the design and management of cultural practices.
Agroforestry	Land use systems in which woody perennials are grown in mixed stands with annual crops.
Airveying	Movement of solid material (e.g. coffee beans) through pipes using air flow.
Allele	One of two, or more, loci on a particular chromosome influencing identical factors.
Allelopathic	Populations or species which are mutually exclusive.
Alley cropping	An agroforestry system where annual food crops are grown in alleys between hedgerows of trees or shrubs which recycle nutrients and fix nitrogen.
Allogamous/allogamy	Refers to the need for cross-pollination to produce viable seed as opposed to autogamous.
Alluvial soils	Soils made up of deposits of particles transported from another place by water.
Alternate host	A plant (or animal) that is host to pests or diseases which affect cultivated crops (or farm livestock).
Anaerobic microorganisms	Microorganisms that are found in airless, oxygen-free environments.

Coffee: Growing, Processing, Sustainable Production, Second Edition. Edited by J. N. Wintgens.

Anion-exchange capacity	The sum total of exchangeable anions that the soil can adsorb.
Annual plant	Plant which completes its life cycle in less than 1 year.
Anthesis	Period of expansion in flowers.
Apex	Pointed end of a branch.
Aphids	Sucking insects of the Hemiptera Family that feed on the sap of young tissues of a wide variety of different plant species.
Apical dominance	Dominance exerted by an apical bud which prevents the development of lower buds.
Aquifer	Geological formations of pervious rocks and materials that hold groundwater.
Arable	Adjective referring to land suitable for cropping as opposed to pasture or woodland.
Attendant ants	Ants which nest in the soil and feed on the secretions of aphids scale insects and mealy bugs (honeydew). When present, ants protect these parasites and aggravate the infestation.
Auger (soil)	A soil auger is a device used in soil surveys to sample the soil.
Autogamy	Self-fertilization or fecundation of the ovule of a flower by its own pollen.
Autotrophy	Refers to self-nourishing plants able to build their own nutritive substances from simple chemical compounds, mostly by photosynthesis or chemosynthesis.
Auxins	Growth hormones that promote the elongation of shoots and roots in all plants.
Available water	Amount of water that plants can extract from the soil as long as moisture tension remains below the permanent wilting point (about 4.2 = 15 atm.). Coffee trees are adversely affected by moisture stress at a lower moisture tension than 15 atm.
Axil	Angle between the leaf petiole or leaf and the supporting stem. Term also used for the angle between branch, pedicel and its main stem.
Back-crossing	The technique which entails crossing an individual with one of its parents in order to introduce a desirable gene into a cultivated variety.
Base saturation %	Exchangeable cations (Ca, Mg, K and Na) as a percentage of total cation-exchange capacity (CEC).
Bedrock	Solid rock underlying soils as parent rock or below surface deposits.
Beneficial insects	Insects that parasitize insect pests (mainly caterpillars).
Berry	Common name for the coffee fruit. The correct botanical name is a cherry or a drupe.
Biocontrol/biological control	The control of crop pests by the introduction of natural predators, parasites or other living organisms without harmful side-effects on the crop plants.
Biodiversity	Refers to the numbers of species, variety and variability of living populations of organisms within specific environments.

Biomass	The total quantity of organisms present in a given area. Also in common usage to describe organic materials/waste used as fuel or organic fertilizer.
Biotic	Related to life or to living things.
Biotope	Particular habitat where a community of different kinds of animals and plants live together and interact.
Biotype	Group of organisms with the same hereditary characteristics (genotype).
Blade or lamina	The expanded part of the leaf fed by its central and lateral veins.
Boundary planting	The practice of establishing trees, hedges or shrubs and live fences as boundary markers.
Bracts	Small leaf-like part of the plant or scale located at the base of the flower or inflorescence, usually below the calyx.
Broadleaf forests	A type of forest where broadleaf species predominate (as opposed to coniferous species). Broadleaf trees are often referred to as "hardwoods".
Bud	Terminal or axilliary rudiment of leaf cluster foliage or flower on a stem; leaf or flower not fully open.
Bud initiation	The beginning of the formation of a bud.
Callus	Mass of parenchymatous cells that form on a wounded surface from which new plantlets can be developed.
Canopy	A more or less continuous cover provided by the leaves of tall-growing vegetation, such as the upper branches and crowns of trees.
Capillary water	The water retained in the capillary spaces of a soil.
Carbon fixation	The conversion of atmospheric carbon dioxide into organic compounds through photosynthesis within the plant tissues.
Carbon:nitrogen ratio or C:N	The ratio of the weight of organic carbon (C) to the weight of total nitrogen (N) in a soil or in organic material.
Cash crop	Crops produced for sale – as opposed to crops produced for food, fodder or seed.
Catch crop	Annual crops intersown in young coffee planting or stumped plots at a wide spacing as a temporary measure to produce additional revenue.
Caterpillar	The larvae of a butterfly or a moth.
Cation exchange	Exchange between a cation in solution and another cation on the surface of a colloidal complex, as clay or organic material.
Cation-exchange capacity (CEC)	The sum total of exchangeable cations that a soil can absorb. Also called "total exchange capacity" or "cation adsorption capacity". Like the anion-exchange capacity it is expressed in milliequivalents per 100 g of soil.
Cell (plant)	Smallest living unit of biological structure and function. It is comprised of a cell wall, a nucleus and cytoplasmic material.
Chemical control	Reducing the population of pests, fungi, bacteria, nematodes or weeds by means of treatment with chemical compounds.
Chlorosis	Yellowing of the foliage of a plant.

Chromosome	Rod-like or thread-like structure occurring in pairs within the nucleus of animal and plant cells which contain genes.
Clipping	Cutting off or removing branches or twigs from a plant with shears.
Clone	Population of genetically identical individuals obtained by asexual propagation.
Closed forest	Forest where the crowns of the trees practically touch each other.
Collar region	Region of a plant located at ground level where roots change to stem.
Colluvium	A deposit of soil material accumulated at the base of steep slopes as a result of gravitational action.
Compaction (of the soil)	The compression of soil particles due to the transit of farm equipment or the impact of rainfall and pedestrian traffic. Compaction reduces the soil porosity.
Contact pesticide	Pesticide which is only effective if it is in direct contact with the body of the pest.
Contour cropping	Crops planted in rows which run parallel to the contours of the land to divert the flow of run-off down the slope and prevent soil erosion.
Coppice	A thicket, a grove or a growth of small trees or bushes grown for periodic cutting.
Cortex (bark)	Tissue located between the epidermis and the vascular tissue in a root or a stem.
Cotyledon	The first leaf, or leaves of the embryo in seed plants.
Cova	Small group of coffee trees planted together (frequent in Brazil).
Cover crop or living mulch	Crop grown as ground cover to protect the soil and reduce erosion, provide additional soil nitrogen, and improve soil structure by adding organic matter.
Crusting	The formation of a hard, brittle surface layer on soils which restricts necessary air exchange, infiltration of rainfall and hampers seedling emergence.
Cultivar/cultivated variety	A variety of plant produced by selective breeding (c.v.).
Cultural control	The use of cultural practices (mulching, pruning, tilling, etc.) to provide conditions which are unfavorable to insect pests and which improve crop resistance to disease.
Cutworms	The larvae of certain noctuid moths which live in the soil and feed on young plants during the night, and cut them off at or near ground level.
Cytokinins	Collective term for growth substances which chiefly stimulate cell division and regulate development.
Cytoplasm	Protoplasmic content of the cell other than the nucleus.
Decussate	A term used to describe a leaf formation which has a pair of opposite leaves with succeeding pairs at right-angles.
Deficiency	A lack or a shortage of nutrients for the plant.
Deflation	Action where the fine soil particles are removed by wind erosion.

Deforestation	The removal of forest growth or woodland.
Denitrification	Loss of nitrogen which takes place in reduced oxygen conditions within the nitrates reduced to nitrites and nitrogen compounds.
Die back	Describes symptom which begins with the appearance of necrotic tissues in the young shoots or branches of a plant, spreading to older tissues, ultimately affecting the main stem and finally causing the death of the plant.
Diploid	An organism with two sets of chromosomes.
Diptera	Family of insects – two-winged flies.
Diurnal	Describes events which takes place during daytime. Not nocturnal.
Dormancy	A period during which the plant or a seed is inactive, generally due to physiological and ambiental reasons.
Dry or rain fed farming	Method of raising crops without the use of irrigation in areas of low rainfall.
Dust	Pesticide mixed with other powders applied dry and undiluted.
Ecology	Deals with the interrelation between living organisms and their natural environment.
Ecosystem	Community of organisms which interact with one another and with the environment in which they live.
Ectoparasite	Parasite that attacks plants and animals externally.
Embryo culture	Growth of isolated plant embryos in a suitable medium (*in vitro*).
Endemic	Restricted or peculiar to a site or region.
Endocarp	The outer covering of the coffee bean, otherwise known as "parchment".
Endogenous	Growing or proceeding from within.
Endoparasite	Animal parasite that lives and feeds inside another cell or organism.
Endosperm	Nutritive starchy tissue of most seeds.
Environmental	Situation or circumstances relating to elements which directly surround and interact with living organisms.
Enzyme	A protein which acts as a catalyst in biochemical reactions.
Epidermis	Outer layer of cells which form the integument of the seeds of plants and ferns.
Epiphytes	Aerial plants that derive their moisture and nutrients from air and rain, usually attached to another plant which provides support.
Estate	The land on which the coffee is grown and all the infrastructures which surround the coffee growing and processing business. Ownership can be individual, corporate or cooperative. "*Plantation*" in French, "*finca*" in Spanish and "*fazenda*" in Portuguese.
Evapotranspiration	Refers to total loss of water from a cropped area including transpiration loss through the stomata of the leaves of the plants and evaporation loss from the soil.

Exocarp	Outer covering of the coffee cherry. It can be red or yellow when ripe (skin).
Explantation	Removal of parts of living organisms for culture in a suitable nutrient medium (*in vitro*) – hence explants.
Exponential	Term describing a continually increasing or decreasing growth rate, but which changes geometrically rather than arithmetically.
Extension services	Services to farmers that involve technical advice, the introduction of new technologies and the organization of the supply of agricultural inputs.
Extensive agriculture	A low cost system of farming which uses large areas and minimal inputs to raise livestock or crops.
F_1 generation	First filial progeny from a cross between two homogenous individuals. This generation is necessarily heterozygous.
F_2 generation	Second filial generation produced by selfing the first generation or allowing the individuals to interbreed.
Fallow	Land which is left uncultivated for a period in order to allow it to recover its fertility naturally.
Field capacity	The amount of water the soil retains after free drainage from a saturated soil.
First crop	Refers to the first coffee picking of the season.
Fixed phosphorous	Phosphorous which has been changed to an immobile form as a result of reaction – usually in combinations with free iron and aluminum.
Floral atrophy	Withering or distortion of the flower.
Foliar application	Spraying leaves and the canopy with agrochemicals to control pests or diseases, or to feed plants.
Frass	Solid insect excreta.
Fuelwood	Wood used as fuel for cooking, heating or generating power.
Fungicide	A natural or chemical substance which has the property of killing fungi.
Gallery forest	A forest that grows along a watercourse in a region otherwise devoid of trees.
Gamete, germ cell	Reproductive cell, normally haploid, capable of fusing with a cell of similar origin but of a different sex to generate a zygote.
Gamopetalous	A flower with united petals forming a tube, at least at the base, e.g. the coffee flower, as opposed to polypetalous, e.g. the buttercup.
Genome	The complete set of genes on the chromosomes carried by an organism or virus.
Genotype	The genetic constitution of an organism.
Germplasm	Cells from which a new plant or animal can be regenerated as in collections of plants in banks.
Gibberellins	A plant growth regulator, first discovered in the fungus *Gibberella fujikuroi*.

Girdling	The bark and the cambium cut in a ring around the tree trunk to interrupt the movement of sap and to kill the tree.
Growth regulators	A substance produced by a living organism to regulate growth.
Grub	A static type of larvae with very small legs or no legs at all.
Guard cell	One of a pair of cells which causes the stoma to open or to close.
Gully erosion	The erosion process whereby rainfall run-off is concentrated in narrow channels, soil is removed by the water and the channels deepen to form gullies.
Habit (habitus)	The external appearance of a tree.
Handling	When applied to coffee cultivation practices, this term means removing green suckers by hand (desuckering).
Haploid	Cell with a single set of unpaired chromosomes; organism having this in its somatic cells.
Head	A vertical stem, including its lateral branches.
Hedgerow	Row of shrubs or trees grown to separate fields or mark boundaries.
Hemiptera	An order of insects that usually have two pairs of membranous wings like bees or wasps, for example.
Herbaceous	Vegetation that has little or no woody tissue.
Herbicide	A chemical compound for eliminating undesirable plants which can either be applied to the plants or to the soil.
Heterosis or hybrid vigor	Tendency of cross-bred individuals to perform better than both parents.
Heterozygous	Product of the fusion of two unlike gametes.
High grown	Arabica coffees grown at high altitudes, over 1000 m (approximately). These coffees are usually superior to those grown at low altitudes. The term "high grown" also figures in many grade descriptions.
Homozygous	Product of the fusion of two like gametes.
Honeydew	A sweet, sticky fluid produced by aphids, scale insects and mealy bugs.
Host plant	A plant on which a particular insect, disease or fungus feeds, develops and reproduces during part of its life.
Humid tropics	Regions where the mean annual biotemperature in the lowlands is over 24 °C and where the annual rainfall exceeds or equals the potential evaporative return of water to the atmosphere. They include all lowland areas where the annual rainfall is over 1500 mm.
Hydroponics	System of producing crops under irrigation in which most plants nutrients are provided in the irrigation water and an artificial rooting medium is used.
Immature	Unripe.
In situ	In the original location.
Indigenous	Native to a specific area, as opposed to exotic or introduced.
Infection	The penetration of an undesirable organism into a host plant.

Infiltration rate	The rate at which water enters the soil under controlled conditions.
Inflorescence	Flowering head in flowering plants.
Inputs	Items purchased to carry out farming operations. Inputs include fertilizers, pesticides, herbicides, seeds, fuel, etc.
Insecticide	A natural or synthetic chemical substance used for killing insects.
Insecticide formulation	An insecticide mixed with other powders or liquids, according to a formula, to make an efficient mixture for spraying or dusting.
Insolation	Exposure to suns rays.
Instar	A stage in the growth of a insect larva or a nymph – the end of each instar is marked by a molt.
Integrated pest management	An integrated approach to the control of pests or diseases in which chemical, cultural, biological and any other methods of control are combined.
Intensive cultivation practices	Involves the use of high levels of inputs of agrochemicals, irrigation, mechanization and other modern technologies. Costs are high so profitability depends on high yields.
Intercropping	The growing of different crops in mixed stands (as opposed to mono-culture).
Internode	Portion of a plant stem or branch between two successive nodes.
Interspecific hybrid	The progeny of two parents of a different species.
Interveinal	Refers to spaces between the veins of leaves.
Introgressive hybridization	Diffusion of genes of one species into another related species.
Irrigation	The artificial application of water to the soil.
Isoptera	An order of social insects otherwise known as termites.
Lanceolate	Leaf shaped like a lance, i.e. a blade of grass.
Larva	Young growing stage of various insects which does not resemble the adult insect. Different types of larvae are more commonly known as caterpillars, grubs, etc.
Latosol	Tropical soil typically of humid regions with high kaolinitic clay content. Usually reddish in color though yellow forms occur in some areas. Relatively high in aluminum, manganese and iron. The intensity of the red color depends on the iron content.
Leaching	The removal of soluble constituents of the soil, including plant nutrients, by water moving downwards through the soil by percolation.
Leaf area index	The total leaf area of the crop divided by the surface area of the ground covered by the crop.
Legume	Any plant which is a member of the Leguminosae Family.
Leguminous	Refers to species which belong to the Leguminosae Family, including herbs, shrubs and trees, all of which bear nitrogen fixing nodules on their roots.

Lepidoptera	An order of insects comprised of butterflies and moths.
Lixivation	Leaching of alkaline salts from the topsoil.
Loam	The textural class name for soil containing 7–27 % clay, 28–50 % silt and less than 52 % of sand.
Locus	The chromosomal position of a gene on a chromosome as determined by its place in relation to the other genes on the same chromosome.
Loess	Fine soil particles transported and deposited by wind.
Looper	A caterpillar that moves by forming loops lifting sections of the body.
Lopping	Practice of cutting branches from a tree or shrub for animal feed, mulching, etc.
Lung branch	Mature coffee stem that remains on the tree when all the other stems have been stumped, also called "breather".
Marginal land	Relatively infertile land which will only be productive with significant levels of inputs such as fertilizers and irrigation. Farming this land may not be profitable.
Mealy bugs	Sucking insects with sedentary habit body usually covered with a mealy white wax.
Meiosis	Type of nuclear division which results in daughter nuclei each containing half the number of chromosomes of the parent.
Meristem	Zone of undifferentiated cells in plants where cell division takes place.
Microclimate	The climate in a very small area or in a particular habitat.
Microorganism	Organism which is only visible with the aid of a microscope.
Milpas	Small field, cleared from the tropical forest which is cropped for one or two seasons and then abandoned.
Miscible liquid (ML)	A liquid containing pesticides mixed with oils, wetters, spreaders, etc., which is miscible to water.
Mitosis	Cell division producing daughter nuclei which contain identical chromosome numbers and are genetically identical to each other and the parent nucleus from which they stem.
Monocropping	Cultivation practice which consists of growing a single crop in a specific area.
Mottling (soil)	Spots or blotches of different colors or shades interspersed with the dominant soil color, which are symptomatic of waterlogged conditions.
Mulching	The practice of spreading fresh or decayed plant material on the surface of the soil with a view to decreasing weed growth, reducing evaporation losses from the soil surface and ultimately increasing soil organic matter levels.
Multiple stem	Coffee tree with more than one stem developed through selective pruning.
Mycelium	Vegetative part of a fungus, consisting of microscopic thread-like hyphae.

Mychorriza	Vegetative mass formed by the symbiotic association of fungus with the roots of a plant.
Nematode or eelworm	Non-segmented round worm. Plant parasitic nematodes are microscopic in size.
Net assimilation rate	Rate of growth in a plant as measured by dry weight per unit of leaf area.
Nitrification	The biochemical oxidation of ammonium to nitrate.
Nitrogen fixation	The conversion of atmospheric nitrogen gas to ammonia, nitrates and other compounds which contain nitrogen by means of nitrogen-fixing bacteria, photosynthetic bacteria and algae.
Node	Point on stem of a plant from which leaf buds and flowers emerge.
Nucleus	Central part of a cell containing the genetic material. Nuclei are present in practically every cell of plants and animals.
Nutrient depletion	The process of decreasing levels of essential plant nutrients from the soil.
Nutrient recycling	Nutrients are recycled in the natural world – a living plant takes nutrients from the soil, dies and decays, and nutrients are recycled through the next generation of plants.
Nymph	Stage in the development of various insects which generally precedes the emergence of the adult insect.
Organic matter	Carbohydrate materials originally produced as a result of photosynthesis and the combination of gaseous carbon dioxide and oxygen.
Orthoptera	Order of insects which includes grasshoppers and crickets.
Orthotropic growth	Vertical growth from ground level, either upwards as in the case of stems and trunks, or downwards, as in the case of roots.
Ovule	Female gamete which develops into the seed after fertilization.
Oxidation	A chemical reaction which increases the oxygen content of a compound.
Oxisols	Laterite soils with an oxic horizon within 2 m of surface. Deep, well-drained, mainly red soils. Oxisols are generally acidic and of low fertility.
Parasite	An animal or plant which is totally dependent on another living organism for its survival. Endoparasites live within the host whilst exoparasites are attached to the outside of the host.
Parasitoid	Insects which introduce their eggs into an other animal in which they grow and develop in a slow manner using the host resources without killing it immediately (Ex. Parasitoid wasp *cotesia congregata*).
Parenchyma	The fundamental (soft) cellular tissue of plants as in the softer parts of leaves, the pulp of fruit, the pith of stems, etc.
Parent rock	The rock from which residual soils are formed by weathering.
Pathogen	An undesirable organism which penetrates a host plant and causes a disease.
Pectic enzyme	A chemical of vegetal origin used for accelerating the fermentation process.

Pedicel or peduncle	Stalk which attaches individual flowers to the main axis of the inflorescence.
Pedigree breeding	Long-term breeding technique used to create new improved varieties which incorporate the most desirable qualities of selected existing varieties.
Perennial	A plant that lives for more than 2 years.
Pest	Any organism which attacks a plant or an animal. This includes fungi, viruses, bacteria, insects, nematodes, ants, mammals, and all other micro flora and fauna.
pH	Indicates the degree of acidity or alkalinity. A neutral pH = 7. Below 7, it is acid and above 7, it is alkaline.
Phenotype	The outward appearance of an organism.
Phloem	Soft tissues of the plant through which nutrients and sap are conducted.
Photosynthesis	Synthesis by plants and some bacteria of organic compounds from the atmospheric gases, carbon dioxide and hydrogen. Synthesis takes place in cells which contain chlorophyll when they are exposed to sunlight.
Phytotoxic	Toxic to plants.
Phytotron	Integrated group of facilities which provide a controlled environment for the study of environmental effects on the development and reproduction of plants.
Pioneer species	The first plant or animal species to become established after the localized destruction of an environment, e.g. after the clearance of forest.
Placenta	Tissue by which ovules are attached to the maternal tissue.
Plagiotropic growth	Tendency to grow horizontally or obliquely from the trunk or the tap root (branches or lateral roots).
Pollarding	The cultivation practice which consists of cutting a tree back to the trunk in order to encourage the growth of a dense head of foliage.
Polymorphism	The existence of different forms in a given population.
Population	A local community of organisms which share a common gene pool.
Predator	Insect which feeds on other insects. Unlike parasites, its movements are autonomous so predators do not depend solely on an individual host.
Primary branch	A lateral branch arising directly from a vertical stem.
Primary forest	Forest that has been untouched by human activity.
Primary industry	Term applied to all types of agriculture, forestry, fishing, mining and quarrying industries that provide food, fuel and raw materials.
Profile	A vertical section of the soil through all its horizons which extends, where applicable, to the parent material.
Propagule	Part of the vegetative structure of the plant from which new individuals are propagated (spores, sclerotia etc.).
Protoplast	The actively metabolizing part of the cell which includes the cytoplasm, a nucleus, cell membranes and organelles, all fully organized.

Protoplast fusion	Fusion of wall-less plant cells under culture conditions.
Pruning	Cultivation practice which consists of removing wood from a plant in order to increase the fruit-bearing branches or to shape it.
Pseudococcidae	Insect Family which includes the species referred to as mealy bugs.
Pupa	Form taken by an insect in its torpid stage of passive development between larva and imago where there is no feeding.
Reforestation	The replacement of trees in cleared forest land.
Relative humidity (RH)	The ratio (%) of the water vapor in the atmosphere to the amount required to saturate it at the same temperature.
Resistance	The power of an organism to overcome the effects of a pathogen or any other specific damaging factor.
Retrogradation	Phosphate fertilizers immobilized in the soil which cannot be assimilated by the plants.
Rhizobium	A genus of bacteria found in root nodules which is able to fix nitrogen in return for carbon from the host plant.
Rhizomorph	Root-like mass growth of fungal hyphae spread through the soil.
Rhizosphere	The immediate environment of the roots.
Rill erosion	An erosion process where numerous small channels are formed. It occurs mainly on recently cultivated soils.
Root mealy bugs	Mealy bugs which feed on the roots of the plant.
Root-stock	A seedling tree providing a base for grafting another variety.
Rubiaceae	Botanical Family which includes coffee (*Coffea* sp.).
Run-off	Occurs when the rate of precipitation exceeds the infiltration rate. Surface run-off is generally removed by the surface water drainage system and does not contribute either to stored soil moisture or to the aquifer.
Salinization	Salinization is an accumulation of salt in the soil due to the evaporation of saline water.
Scale insects	A group of sedentary bugs which feed on plants, otherwise known as cochineal insects.
Scleroti	Hard propagating organs of certain fungi.
Screen	Frame with a perforated metal plate or wire mesh with openings of different sizes, used to grade coffee beans by size.
Secondary branch	A branch arising directly from a primary branch.
Secondary forest	Natural forest regrowth after the destruction of the primary forest.
Secondary industry	Manufacturing and construction industries which convert the products of primary industries.
Seed at stake	Planting by placing seeds directly in the field instead of transplanting from seed beds.
Seedling	Plant issued from a seed.
Selective harvesting	Consists of hand-picking of ripe cherries only.

Self-incompatibility/self-sterility	The failure of gametes of the same plant to generate a viable embryo.
Serial buds	A number of buds arranged in series.
Shedding	The loss of unripe cherries which fall to the ground.
Sheet erosion	The removal of a relatively uniform layer of soil by run-off water.
Shifting cultivation	Land is cultivated and cropped for several years and then returned to fallow for a similar period. It is synonymous with "slash and burn" or "swidden" agriculture.
Shoot	Young orthopedic stem formation.
Single stem	Coffee tree maintained on one stem only.
Soil aggregate	Clay and fine silt particles held in a single mass.
Soil amendment	Substances such as lime, sulfur, gypsum, etc., used to improve the soil condition.
Soil hardpan	An indurate, impermeable horizon in the soil profile often cemented by chemical compounds such as aluminum silicate, silica, $CaCo_3$, iron sesquioxides, gypsum, etc.
Soil reaction	Refers to the pH of the soil – acid, neutral or alkaline condition.
Soil texture	Defined by the proportion of particle sizes – sand, silt and clay – in the soil.
Soluble powder (SP)	A powder which contains a fixed percentage of an insecticide or a fungicide or any other ingredients in powder form which will dissolve easily in water to produce a solution for spraying.
Somatic cell	Cells which take part in the formation of the body and which become differentiated into the various tissues, organs, etc.
Sooty mould	A thin black crust of fungus found on the upper side of leaves which develops on the honeydew dripped down from scale infestation.
Species	A group of interbreeding individuals having some common characteristics not normally able to interbreed with other such groups. Species are subdivided in subspecies, races and varieties.
Spodosols	Soils with an eluvial horizon (spodic) in which active organic matter and amorphous oxides of aluminium and iron have precipitated.
Spore	The reproductive unit of a fungus which is equivalent to a seed in flowering plants.
Staple food	A commodity that forms the majority of the diet of a population in a given area.
Stele	The central core of the stems and roots of vascular plants.
Stem	The upright branch of a tree which bears leaves and flowers.
Stipule	One of a pair of leaf-like appendages located at the base of a leaf petiole occurring in many plants.

Stoma	Minute orifice in the epidermis of plant tissues, mainly located underneath the leaves, through which gaseous interchange takes place, particularly water vapor and carbon dioxide. Plural = stomata.
Stripping	Also known as "milking", consists of removing all the coffee cherries present on the branch irrespective of their degree of ripeness.
Stumping	Cutting a tree back, leaving only the stump.
Subsistence agriculture	Agricultural production which fails to attain a surplus beyond what is required for consumption by the producer and his family.
Subsistence farming	Farming that provides all or almost all the goods required by a farming family, usually without any significant surplus for sale.
Subsoil	Less fertile soil below the top layers (topsoil).
Sucker	Orthopedic shoot arising from the stem, the root, the axil or from a branch.
Susceptibility	Vulnerable to a given disease (Sensitivity).
Suspension culture	Culture consisting of cells or cell aggregates in a suitable medium, agitated by bubble aeration and stirring so that the cells do not settle.
Sustainable agriculture	Agricultural production system which enables the farmer to maintain productivity at levels that are economically viable, ecologically sound and culturally acceptable in the long term. Under such a system, resources are managed efficiently with minimal damage to the environment and human health.
Symbiosis	Beneficial partnership of two mutually dependent organisms, e.g. the nitrogen-fixing bacteria on the roots of leguminous plants.
Symptom	A characteristic sign of infection or deficiency.
Systemic pesticide	Pesticide which enters the plant tissues before being fatally ingested by the pest.
Tap root	Main root which grows downwards from the radicle and generates lateral roots.
Temperate zone	Area between the tropic of Cancer and the Arctic Circle and between the Tropic of Capricorn and the Antarctic Circle.
Tensiometer	A device for measuring the suction force required to extract water from the soil.
Terracing	Land forming practice to reduce the rate of run-off and erosion. Consists of steps of level land on the contour, on which crops are planted.
Tetraploid	An organism with four sets of chromosomes in its cells.
Thrips	Very small insects with feather-like wings fringed with hairs.
Tillage	Cultivation practice which entails loosening up the soil and preparing it for planting. Tillage can be carried out manually or mechanically.
Tipping	Pruning the ends of primary branches or stems.
Tomentose (adj.)	Refers to dense covering of short hairs (e.g. leaf surface, stems, etc.).

Topping, pollarding, capping	Cutting off the terminal end of an upright stem.
Topsoil	The layer of soil cultivated in agriculture.
Tracheomycosis	Disease which attacks the vascular tissues of the roots and stems of the coffee tree.
Trimming	To reduce a tree or a plant to a neat, orderly shape by pruning or clipping.
Tripetidae	A family of insects also known as fruit flies.
Tropical zone	Area located on either side of the equator between the Tropic of Cancer and the Tropic of Capricorn. These areas originally supported broadleaf evergreen forests. They are frost-free and generally have only 2 "dry" months per year where the rainfall is less than 100 mm.
Trunk	The upright part of the tree from which the branches stem.
Twig	Small branch.
Understory	Vegetation which grows under the shade of taller plants.
Uredospore	An orange–red spore of the rust fungi.
Vacuole	Small spherical space within a cell cytoplasm surrounded by a membrane which contains liquid and/or solid matter.
Veins (in leaves)	Ribs of leaves.
Virescence or chloranthy	Failing to become green due to insufficient chlorophyll.
Wasp	A member of the Order Hymenoptera. The females have a tube-like extension of the abdomen which may be used for egg laying and/or a sting. Some species are important for biological control of the larval stage of important crop pests.
Watershed	Area of land from which the rainfall run-off drains into a river or into a larger body of water (lake). Sometimes referred to as a catchment.
Web blight	Plant disease where the affected tissues are covered with web-like mycelial strands.
Weeding	Weed control by physical or chemical means. There are several types of weeding. (1) Clean weeding: to remove all vegetation, apart from the crop, from the soil surface. (2) Strip weeding: to remove weeds in lines or furloughs. (3) Ring weeding: to remove weeds in rings around the base of the coffee trees. (4) Selective weeding: to remove harmful weeds and leave useful weeds *in situ*.
Weeds	Plants that grow where humans do not want them to grow because they compete with crop plants for space, nutrients and water.
Weevil	Type of small beetle with a beak-like head and L-shaped antennae.
Wettable powder (WP)	Finely ground powder containing a fixed percentage of a pesticide which mixes easily with water to form a suspension suitable for spraying.
Wild species	Species of faunae or flora which has not been manipulated by mankind.
Wilting point	Point at which plants are unable to absorb any more soil moisture.

Windrowing	System used for clearing forests which consists of stacking the parts of the felled trees and shrubs (trunks, branches, etc.) in rows.
Xylem	The lignified part of the vascular tissues of plants.
Yield	The weight or volume of the harvest which has an economic value.
Zygote	Product of the fusion of two gametes which will form the embryo.

4.2
Terms related to Green and Roasted coffee

A	"A" is used to describe the bean size. "A" differs from country to country. In Peru, for example, the largest bean is "AAA"; in Kenya, Tanzania and New Guinea, it is "AA"; in India it is "A".
Aged coffee	Coffee from previous crops stored in special warehouses for long periods. Ageing reduces acidity and increases body.
Arabica coffee	Coffee of the botanical species *Coffea arabica*.
Arabusta coffee	Hybrid interspecific of *Coffea arabica* and *Coffea canephora*.
Bean	Endosperm or seed of the coffee fruit.
Bean in parchment	Coffee bean enclosed in its parchment.
Bean sizes: flat beans	Screened through round holes of different sizes ranging from size 13 to 20. Very large (size 20), extra large (size 19), large (size 18), bold (size 17), good (size 16), medium (size 15), small (size 14–13).
Black Jack	Coffee which has turned black, either during the curing process, in the hold of a vessel during transport or due to a blight disease.
Blend	A mixture of two or more coffee varieties producing a recognizable and reproducible quality of coffee liquor.
Blue Mountain (1)	A district in Jamaica which produces the finest coffee grown in the island for which there is a very high demand on the international market, especially the Japanese market.
Blue Mountain (2)	A coffee grown in Kenya, mostly for export to the UK, which does not have the same quality characteristics as the Blue Mountain produced in Jamaica.
Bold	A large to very large well formed and even coffee bean.
Caffeine	An alkaloidal compound present in the coffee tissues but more concentrated in the bean (average 1.5%) When isolated, it forms silky white needles.
Caffeol, coffee oil	The volatile, oily substance developed in the coffee bean during roasting.
Catimor coffee	Hybrid of *C. arabica* and *C. canephora* (Timor hybrid) with *C. arabica* var. *caturra*.
Centre cut	Cleft or groove on the flat side of the bean.
Clean coffee	A well graded coffee, free of defects.
Cleaning	Removal of foreign matter, fragments of coffee and defective beans from green coffee.

Coffee bean	Commercial term used for the dried seed of the coffee tree.
Coffee cherry	The fresh fruit of the coffee tree.
Color	From blue–green to yellow–green and brown depending on origin, species, age, method of processing, fruit maturity, harvesting and conditions of storage and transport.
Conillon	*Coffea canephora* var. *kouillou* is mostly cultivated in the Espirito Santo region of Brazil. The word "Conillon" is a corruption of the word "*kouillou*".
Decaffeinated coffee	Coffee from which the caffeine content has been removed.
Defect	Any impairment of the coffee bean that could cause a deterioration in quality. Further details can be found in section 4.3 "Green Coffee Defects".
Dried coffee cherry	Also known as coffee in pod or husk coffee.
Dry fermenting	After pulping, the coffee is fermented without water.
Drying of parchment coffee	Drying parchment coffee to reduce its moisture content in order to condition it for storage and further hulling.
Dry-processing	Treatment of coffee cherries whereby they are dried under the sun or mechanically. This process produces husk coffee. Drying is usually followed by the mechanical removal of the husk. The result is "natural" or "unwashed" green coffee.
Ethiopian Mocha	False Mocha produced in Harrar. Lighter in body and more acid than the true Mocha from Yemen.
Excelsa coffee	Coffee of the botanical species *Coffea dewertii*.
Fair	Term which refers to quality: practically no defects.
Fancy	Special quality of coffee.
Fermentation	Biological treatment which consists of degrading the mucilage which still adheres to pulped coffee.
Flat bean	Coffee bean with one visibly flat side.
Foreign matter	Any mineral, animal or vegetable matter which does not come from the coffee cherry (stones, sticks, clods, metallic residues, etc.).
Grinding	Mechanical operation which fragments roasted coffee beans and produces ground coffee.
Ground coffee	Roasted coffee which has undergone grinding.
High-grown Atlantic (HGA)	Costa Rica Arabica coffee produced at approximately 900 m altitude. Good quality coffee but not up to the standard of SHB.
Hull	Dried parchment of the coffee fruit.
Hulling or dehulling	The mechanical removal of the dried endocarp from parchment coffee to produce green coffee.
Husk	Assembled external envelopes of the dried coffee fruit.
Husking or dehusking	The mechanical removal of the husks from dry coffee cherries.

Kenya AA	This quality is regarded as equal to the Kenya Blue Mountain.
Kona	A district on the west side of the island of Hawaii. Produces the best Hawaiian coffees. It has a large blue flinty bean. In the cup it is mildly acid with striking character.
Liberica coffee	Coffee of the botanical species *Coffea liberica*.
Low grown Atlantic (LGA)	Costa Rica Arabica coffee grown on the Atlantic side of the country at about 150–600 m. Described as having an undesirable grassy cup quality with little body and aroma.
Luwak coffee	Considered to be a "gourmet" coffee. Production is very little and its price is 5–6 times higher than normal coffees. It is often used to improve special blends. Luwak coffee comes from Indonesia where the *Viverro musanga* or coffee rat consumes only the sweetest ripe cherries and digests the mucilage. The parchment beans remain intact and are collected from the luwak's droppings which are carefully washed. The unique quality of this coffee is due to the optimum maturity of the cherries selected by the luwak and the special action of its gastric juices on the parchment coffee.
Malabar	Name given to the best of the Indian coffees. It has a small, blue–green bean. In the cup it has a strong flavor and deep color.
MAM	Combines coffee from three regions in Colombia – Medellin, Armenia and Manizales.
Mandheling	Coffee produced in the Mandheling district of east Sumatra. One of the finest and highest priced coffees in the world.
Maragogipe	A mutant of *C. arabica* discovered near the town of Maragogipe in Brazil. It is a low bearer but produces very large, flat beans highly appreciated on the coffee market.
M'buni or Buni	A Swahili term used in East Africa to describe coffee prepared by dry processing.
Mild coffee (1)	A quality trade term used for high-quality beans.
Mild coffee (2)	Washed Arabica coffee which comes from countries other than Brazil.
Mocha coffee quality	Coffee exported in the past through the port of Mocha in Yemen before the port was closed in by a sand bar. Yemen is the only source of true Mocha coffee which has become recognized throughout the world as one of the best coffees.
Mocha coffee (description)	The beans are small and irregular. Their color is olive green to pale yellow. The roast is poor and irregular, but in the cup, the finest qualities have a unique acid character, heavy body and a smooth, delicious flavor.
Mocha coffee (production)	Production is low. To meet the continued quality demand, Mocha Ethiopian coffee, usually from Harrar, has been used as a substitute. It is similar but cannot really compete in quality. Only shipment via Hodeidah and Aden guarantees the authenticity of Mocha today.
Mokka coffee (plant)	Mutant of *C. arabica*. The time and place of origin are unknown though it is believed to have transited through Reunion (Bourbon Island) where it had been introduced from Mokka.

Monsooned coffee	Indian coffee which swells and turs yellow after storage in damp conditions. It is mostly appreciated in Scandinavian countries.
OIB	Oost Indische Bereiding = dry-processing.
Open	Beans where the centre cut tends to widen during roasting. Indicates beans with soft tissues, grown at lower altitudes and also light beans.
Pales	A term used to describe discolored beans. They can come from old stocks, from immature beans or from coffee afflicted by drought. Amber beans often cause pales in the roast.
Peaberry coffees	Beans are graded through oval shaped screens, sizes 9–13.
Polished coffee	Green coffee where the silverskin is removed by mechanical means to improve the appearance and to provide an extra gloss.
Polishing	The mechanical removal of the silverskin from green coffee.
Pulp	Part of the coffee cherry which is eliminated during pulping and fermentation. It is composed of the skin and part of the mucilage.
Pulping or depulping	Operation which consists of removing the pulp and part of the mucilage by mechanical means. Part of the mucilage generally remains adhering to the parchment.
Roasting	The use of heat to generate fundamental chemical and physical modifications in the structure and composition of green coffee in order to darken the beans and develop its characteristic flavor.
Robusta coffee	Coffee of the botanical species *Coffea Canephora* var. Robusta.
Santos	Santos is the main shipping port for coffee in Brazil. It gives its name to the coffee which is shipped via this port. Santos coffees are considered to be the best in Brazil.
Santos (processing)	The bulk of the Santos coffees are dry-processed, though the production of washed and semi-washed Santos is on the increase.
Selection	Operation which consists of sorting coffee cherries according to size, density and degree of maturity, and, at the same time, eliminating foreign matter.
Sidamo	Coffee from South Central Ethiopia. It is inferior to plantation-grown Harrar coffees.
Silverskin	Dried seed coat of the coffee bean. It is usually silver or copper colored.
Sound coffee	A good marketable coffee.
Straight coffee	An unblended coffee, either from a single country, a single region or a single crop.
Strictly hard bean (SHB)	The best quality Arabica coffee produced in high altitude plantations (1200–1650 m) on the Pacific side of Costa Rica. Noted for its acidity, body and aroma.
The drying of cherry coffee	Drying coffee cherries to reduce their moisture content in order to remove their husks and to condition them for storage.
Usual good quality (UGQ)	Term which refers to a new crop assessed by certified graders and arbitrators.

Vintage Colombian	Colombian coffee that has been stored up to 8 years, during which time the acid characteristics are replaced by a sweet richness of flavor.
Washed-and-cleaned coffee	Dry-processed green coffee from which the silverskin has been removed by mechanical means and using water.
Washing	The use of water to remove the degraded mucilage from the parchment.
Wet-processed coffee	Green coffee that is wet-processed is known as washed or semi-washed coffee. Washed coffee is green coffee from which the mucilage has been totally removed and semi-washed coffee is green coffee where most of the mucilage still adheres to the parchment.

4.3
Terms related to Coffee Liquor

Term	*Description*	*Cause/particularity*
Acidity	smooth, rich with verve, snap, life	does not refer to a great amount of actual acid but to a pleasing taste
Acidy	nippy, sharp taste	acidy coffees command a higher price
Acrid	sharp, bitter, sometimes irritating burnt flavour	high level of sour acids typical of unwashed Rio coffee
American roast	also known as medium or regular roast, medium brown in color	coffee roasted to American tastes
Aroma	fragrance of freshly brewed coffee	Pleasant-smelling substances with the nice odour of coffee
Baggy	undesirable taint remaining, the smell of an old jute bag	frequent in coffees that have been unsuitably stored for a long time in jute bags
Baked	a tainted odor that gives the coffee an insipid taste and a flat bouquet	underdevelopment of the bean during roasting because of low heating
Balanced	a good combination of acidity and body with a subtle verve	top-grade, well-blended coffees for connoisseurs
Bitterness	moderate bitterness can give added verve to the flavor. When strong it gives out an unpleasant sharp taste like quinine	a taste associated with darker roasts more frequently found in Robustas
Body	full and rich flavour with feeling of a certain viscosity	usually found in heavy, mature coffee
Bouquet	the total aromatic profile of a coffee brew	from the sensation produced by the gases and vapors on the smell to the fragrance, aroma, nose and aftertaste of the brew
Burnt	coffee with a tarry or smoky odor and flavor	too much heat for too long which burns the bean fiber during roasting
Carbony	aroma, reminiscent of creosol similar to a burnt substance	

Term	Description	Cause/particularity
Carmelized	slightly burnt flavor, like carmelized sugar. Not undesirable taste	can complement the taste of some coffees. Associated with spray dried coffees
Caustic	harsh, burning, sour sensation on the side of the tongue at first sip	displeasing and sour as the brew cools
Coarse	a raspy harsh flavor lacking in finesse	
Common	poor liquor, lacking acidity but with full body	usually associated with coated raw beans, softs and pales in roast
Continental roast	term used to describe a degree of roasting of green coffee which generates a dark brown roast when the fat (oils) of the beans are brought to the surface	French roasts are dark-colored and Italian roasts are even darker, bordering on burnt or black
Creosty	bitter taste, scratching at the back of the tongue	typical of dark roast coffees
Cup	refers to quality in liquor	
Cup testing	judging the merit of a coffee by roasting and brewing it to determine its profile, aspect, odor and taste	the brew is sipped and held in the mouth only long enough to get the full strength of the flavor; tasters should then spit it out
Dirty	similar to "earthy"	unclean preparation conditions
Dull	coffee which has lost its original lustre	associated with processing and age
Earthy	odor and taste of soil dirt like after-taste sensation	fats in the coffee bean have absorbed organic material from the earth during the drying process
Fermented	highly unpleasant sensation on the tongue, reminiscent of fermented pulp	due to unclean fermentation conditions and washing
Fine	distinguished gourmet quality	excellent body, acidity and aroma
Flat	lack of character, and consistency in the brew	lack of aroma gases and vapors in the fragrance, the aroma, the nose and the aftertaste
Foul	liquor often similar in taste to rotten coffee pulp, sometimes present in the more advanced stages of fruity and sour coffees	pulper-nipped beans can give foul cup to an otherwise good liquor. Bad preparation or the use of polluted water
Fresh	an aromatic highlight	not stale, sour or decayed
Fruity	over-ripe flavor, first stage of sourness	coffees left too long in the cherry or fermentation in the presence of too many skins
Full	full, balanced taste	all natural gases and vapor present in the fragrance, the aroma, the nose and the aftertaste

Term	Description	Cause/particularity
Grassy	greenish or green flavor, reminiscent of grass	common in early picking
Hard	firm, individual characteristics indicative of good roasting qualities	generally found in mild coffees grown in higher altitudes with a long period of maturation on the tree
Harsh	bitter, astringent, harsh taste, Liberica is a typical harsh coffee	coffee originated from over-bearing or drought-stricken trees producing mottled cherries
Herby	distinct aroma and flavor of herbs ranging from onion to cabbage	
Light	a fluid coffee, not too dark in color	a low level of fine particles, bean fiber and insoluble proteins
Light roast	coffee roasted to a lighter degree than the North American norm	
Mellow	rounded, smooth taste lacking acidity	not too acid, not too bitter, but dense
Mouldy	odor and flavor of mold	caused by storing coffee in premises with a high degree of moisture
Muddy	dull, thickish flavor	contains sediments often caused by stirring the grounds
Neutral	indistinct, no dominant flavor smooth, lacking coffee flavour	good or blending
Nippy	sweet nipping taste at first sip, sweet sensation as the brew cools	
Nose	the aroma sensation of the vapors released from brewed coffee	
Old	flat taste with a slight flavor of hay	lack of freshness, stale coffee
Onion flavor	self-explanatory, often bordering on foul	associated with lengthy fermentation which generates propionic and butyric acid or with the use of polluted water
Potato flavor	self-explanatory	caused by a secondary infection which occurs after the cherry has been attacked by fruit flies (*Ceratitis capitata*)
Pungent	prickly, stinging or piercing sensation, not necessarily unpleasant	sensation of overall bitterness of the brew
Pyrolysis	the chemical breakdown, during roasting of fats and carbohydrates, into the delicate compounds which provide all of the coffee aroma and most of its flavor	
Rancid	disagreeable taste of spoiled oil	poor storage conditions
Rich	an overall full-bodied flavor	complete set of gases and vapors in the fragrance, aroma, nose and aftertaste

Term	Description	Cause/particularity
Rioy	Iodine/phenolic flavor, mostly found in Brazil, but sometimes present elsewhere in fancy mild coffees	delayed cherry picking, with some cherries partially dried on the tree
Rubbery	term often used to describe a taste of rubber	usualy present in fresh Robusta
Smoky	a slightly caustic flavor of smoke	defective drying operations
Smooth	a full bodied coffee	low in acidity
Soft	well-rounded flavor	lacking harshness or acidity
Sour, sourish	unpleasant acidic, biting flavor	
Strong	unbalanced liquor where body predominates to the point of being tainted	said to be peculiar to soil climatic conditions or method of growth
Sweet	a clean, soft coffee free from harshness	well-conditioned, undamaged coffee
Taint	coffee with an undesirable taste and aroma, foreign to a clean liquor	Poorly-conditioned, damaged beans which generate flavors that are not easily recognizable and are, therefore, difficult to define
Thin	flat, lifeless coffee, lacking body	underbrewing is a possible cause
Turkish coffee	also known as Middle Eastern coffee	coffee ground to a fine powder, sweetened, brought to a boil and served complete with coffee grounds
Winey	rich, rounded, smooth full-bodied coffee, fruit taste similar to red wine	prevalent in Colombian coffees
Woody	hard, wood-like flavor	green beans stored for too long or at high temperatures and humidity

4.4
Acronyms

ABEC	Association Burundaise des Exportateurs de Café
ACP	African–Caribbean–Pacific countries (CEE)
ACTA	Association de Coordination Technique Agricole
AEKI	Association of Indonesian Coffee Exporters
AFNIC	Association Française du Négoce International du Café
AID (IDA)	Association Internationale de Développement (BIRD)
AIDS	Agro-Industrial Development Scheme
AIS	Agronomy and Information Services (KIT)
ANACAFE	Asociacion Nacional del Café (Guatemala)
ANACAFEH	Asociacion Nacional de Caficultores de Honduras

ANAGENTI	Associazione degli Agenti e Rappresentanti di Caffé Droghe e Coloniali (Rome)
ANCAFE	Agrupación Española del Café (Madrid)
ASEAN	Association for Southeast Asian Nations
ASIC	Association Scientifique Internationale du Café
ATA	Abstracts on Tropical Agriculture (KIT)
AVC	Asociacion Venezolana de Cafeteros
BADC	Belgian Agency for Development Co-operation
BCF	Belgian Coffee Federation
CAB	Commonwealth Agricultural Bureaux
CAP	Common Agricultural Policy (EEC)
CATIE	Centro Agronómico Tropical de Investigación y Enseñanza (Turrialba, Costa Rica)
CDC	Commonwealth Development Corporation
CEA	Commission Economique pour l'Afrique
CEAP	Coffee Exporters Association of the Philippines
CECA	Committee of European Coffee Associations
CEEMAT	Centre d'Etudes et d'Expérimentations du Machinisme Agricole Tropical (CIRAD)
CENICAFE	Centro Nacional de Investigaciones de Café (Colombia)
CFTC	The Commodity Future Trading Commission
CGIAR	Consultative Group on International Agricultural Research (USA)
CIAE	Central Institute of Agricultural Engineering (India)
CIBC	CAB International Institute of Biological Control (UK)
CIDARC	Centre d'Information et de Documentation en Agronomie des Régions Chaudes (Montpellier, France)
CIDAT	Centre d'Informatique appliquée au Développement et à l'Agriculture Tropicale (Belgium)
CIFC	Centro de Investigação das Ferrugens do Cafeeiero (Portugal)
CIRAD	Centre de Cooperation Internationale en Recherche Agronomique pour le Développement
CMB	Coffee Marketing Board
CNC	Comisión Nacional de Café (Mexico City)
CNRS	Centre National de Recherche Scientifique (CIRAD)
COBOLCA	Comité Boliviano del Café
COLUMNA	Comité Français de Lutte contre les Mauvaises Herbes
Commonwealth	The community of 34 countries of the former British Empire
CSCSA	Compania Salvadoreña de Café
CTA	Centre Technique de Cooperation Agricole et Rurale (Convention de Lomé)
CTA	Coffee Trade Association (UK)

CTFT	Centre Technique Forestier Tropical (CIRAD)
DIMACC	Développement Intégré dans les Marges de l'aire Caféière (Xalapa-Coatepec/Mexico)
ECF	European Coffee Federation
EDF	European Development Fund
EEC	European Economic Community
ENCAFE	Empresa Nicaraguense del Café
EU	European Union
EUCRA	European Federation of Coffee Roasters Associations
FAC	Fonds d'Aide et de Coopération (Paris)
FAD	Fonds Africain de Développement
FAO	Food and Agriculture Organization
FEBEC	Brazilian Federation of Coffee Exporters
FEECAME	Federacion Cafetalera de America
FIPA	Fédération Internationale des Producteurs Agricoles
FMO	Financieringsmaatschappij voor Ontwikkelingslanden
FNCC	Federación Nacional de Cafeteros de Colombia
FNCCV	Fédération Nationale du Commerce des Cafés Verts
FUSAGRI	Fundación Servicio para el Agricultor (Venezuelan Farmers Service Foundation)
GATE	German Appropriate Technology Exchange
GATT	General Agreement on Tarifs and Trade
GEMS	Global Environment Monitoring System (UN)
GENAGRO	Gestion Environnement Agronome
GERDAT	Département de Gestion, Recherche, Documentation et Appui Technique (CIRAD)
GRID	Global Resource Information Data Base (operated by UNEP)
GRU	Genetic Resources Unit
GTZ	German Agency for Technical Cooperation
HACCP	Hazard Analysis and Critical Control Points
HCTA	Hard Coffee Trade Association
IAC	International Advisory Committee of Research in Natural Sciences (UNESCO)
IACO	Interafrican Coffee Organization
IBC	Instituto Brasileiro do Café (Rio de Janeiro)
IBPGR	International Board for Plant Genetic Resources
ICAFE	Instituto del Café de Costa Rica
ICIPE	International Centre for Insect Physiology and Ecology (Kenya)

ICO	International Coffee Organization
ICRA	International Course on Research for Agricultural Development
ICRAF	International Council for Research in Agroforestry (Kenya)
ICTA (CIAT)	International Centre for Tropical Agriculture
IDA	International Development Agency
IDEFOR	Insitutut des Forêts, Département Café, Cacao et Autres Plantes Stimulantes
IDRC	International Development Research Centre (Canada)
IHCAFE	Instituto Hondureño del Café
IICA	Inter-American Institute for Cooperation in Agriculture
IIMI	International Institute for the Management of Irrigation
IITA	International Institute for Tropical Agriculture (Nigeria)
ILO	International Labor Organization
IMF	International Monetary Fund
INCA	Instituto Nacional do Café de Angola
INCAFE	Instituto Nacional del Cafe (El Salvador)
INIAP	Instituto Nacional de Investigaciones Agropecuarias (Ecuador)
INIFAP	Instituto Nacional de Investigaciones Forestales y Agropecuarias (Mexico)
INRA	Institut National de la Recherche Agronomique (France)
IPPC	International Plant Protection Centre
ISABU	Institut des Sciences Agronomiques du Burundi
ISF	Institute for Soil Fertility
ISIC	Instituto Salvadoreno de Investigaciones del Cafe
ISO	International Standards Organization
ITC	International Institute for Aerial Survey and Earth Sciences
KIT	Koninklijk Insitute voor de Tropen (Royal Tropical Institute, The Netherlands)
LCE	London Commodity Exchange
MAMS	New York Exchange contract for Colombian Coffees (Medellin, Antioquia, Manizales)
MCTA	Mild Coffee Trade Association of East Africa
NCA	National Coffee Association of USA
NGO	Non-Governmental Organization
NRI	Natural Resources Institute
OAMCAF	Organization Africaine et Malgache du Café
OCDE	Oficina Central de Café (Guatemala)
OCIBU	Office des Cafés du Burundi
OCIRU	Office des Cafés Indigènes du Rwanda–Burundi

OECD	Organization for Economic Cooperation and Development
ONCC	Office National du Café et du Cacao (Cameroun)
ORSTOM	Institut Français de Recherche Scientifique pour le Développement en Coopération
PACB	Pan-American Coffee Bureau
PROMECAFE	Programa de Protección y Modernización de la Caficultura en Centro-America
SKV	Schweizerische Kaffeehändler Vereinigung (Switzerland)
SOCA2	Société de Réalisation et de Gestion des Projets Café et Cacao (Paris)
SOCAGAB	Société de Caféiculture et de Cacao culture du Gabon
SRCC	Société Nationale pour la Rénovation et le Développement de la Cacaoyère et de la Caféière Togloaises (Gabon)
STABEX	Stabilization of Export Earnings (EEC)
STRI	Smithsonian Tropical Research Institute (USA)
TCMB	Tanzania Coffee Marketing Board
TCTA	Tanganika Co-operative Trading Agency
TPI	Tropical Products Institute
UN	United Nations
UNCA	Uganda National Coffee Association
UNCSTD	United Nations Conference on Science and Technology for Development
UNCTAD	United Nations Conference on Trade and Development
UNDP	United Nations Development Program
UNEP	United Nations Environment Program
UNESCO	United Nations Educational, Scientific and Cultural Organization
UNIDO	United Nations Industrial Development Organization
UPCAIC	Union Professionnelle de Commerce Anversois d'Importaion du Café
USAID	United States Agency for International Development
USCIC	Union Syndicale du Commerce International du Café
USDA	United States Department of Agriculture
WB	World Bank
WHO	World Health Organization
WMO	World Meteorological Organization
SCAA	Specialty Coffee Association of America
ITA	International Trade Center
TPI	Tropical Product Institute
NFTA	Nitrogen Fixing Trees Association (Hawaii)
WTO	World Trade Organization

4.5
Abbreviations

AFLP	amplified fragment length polymorphism DNA
aq	aqueous solution
as is	composition based on total weight of sample, no correction for water content
db	on dry basis
GMT	Greenwich Mean Time
GNP	gross national product
meq%	milliequivalent per 100 g
MIP	minimum import price
RAPD	random amplified fragment length polymorphism
SD	standard deviation
vpm	volume per million
v/v	volume per volume (only if active ingredient is a liquid)
w/v	weight per volume (only if active ingredient is a mass)
w/w	weight per weight

Index

Coffee: Growing, Processing, Sustainable Production, Second Edition. Edited by J. N. Wintgens.
© 2012 WILEY-VCH Verlag GmbH & Co. KGaA. Published 2012 by Wiley-VCH Verlag GmbH & Co. KGaA.